Important Aqueous Species (alphabetical by key element symbol)

Ag^+, silver ion (argentous ion)
Ag^{2+}, silver (II) ion (argentic ion)
Al^{3+}, aluminum ion
H_3AsO_3, arsenous acid
H_3AsO_4, arsenic acid
Au^+, aurous ion
Au^{3+}, auric ion
H_3BO_3, boric acid
Ba^{2+}, barium ion
Br^-, bromide ion
Br_2, bromine
BrO_3^-, bromate ion
$HCOOH$, formic acid
$H_2C_2O_4$, oxalic acid
CO_3^{2-}, carbonate ion
CH_3COOH, acetic acid
Ca^{2+}, calcium ion
Cd^{2+}, cadmium ion
Ce^{3+}, cerous ion
Ce^{4+}, ceric ion
Cl^-, chloride ion
Cl_2, chlorine
$HOCl$, hypochlorous acid
ClO_4^-, perchlorate ion
Cr^{2+}, chromous ion
Cr^{3+}, chromic ion
CrO_4^{2-}, chromate ion
Cu^+, cuprous ion
Cu^{2+}, cupric ion
F^-, fluoride ion
Fe^{2+}, ferrous ion
Fe^{3+}, ferric ion
H_2, hydrogen

H_3O^+ (abbrev. H^+) hydronium (or hydrogen) ion
Hg_2^{2+}, mercurous ion
Hg^{2+}, mercuric ion
I^-, iodide ion
I_2, iodine
I_3^-, triiodide ion
IO_3^-, iodate ion
K^+, potassium ion
Li^+, lithium ion
Mg^{2+}, magnesium ion
Mn^{2+}, manganous ion
MnO_4^-, permanganate ion
NH_3, ammonia
N_2H_4, hydrazine
NH_2OH, hydroxylamine
$HONO$, nitrous acid
NO_3^-, nitrate ion
Na^+, sodium ion
Ni^{2+}, nickel ion
OH^-, hydroxide ion
H_2O_2, hydrogen peroxide
O_2, oxygen
H_3PO_3, phosphorous acid
H_3PO_4, phosphoric acid
Pb^{2+}, lead ion
H_2S, hydrogen sulfide
H_2SO_3, sulfurous acid
SO_4^{2-}, sulfate ion
$S_2O_3^{2-}$, thiosulfate ion
$S_4O_6^{2-}$, tetrathionate ion
Sn^{2+}, stannous ion
Sn^{4+}, stannic ion
Zn^{2+}, zinc ion

CHEMICAL EQUILIBRIUM AND ANALYSIS

Richard W. Ramette
Carleton College

▲ ADDISON-WESLEY PUBLISHING COMPANY
Reading, Massachusetts • Menlo Park, California
London • Amsterdam • Don Mills, Ontario • Sydney

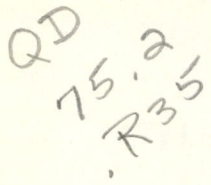

This book is in the ADDISON-WESLEY SERIES IN CHEMISTRY

Library of Congress Cataloging in Publication Data

Ramette, Richard W.   1927–
  Chemical equilibrium and analysis.

  Includes index.
  1. Chemistry, Analytic.   2. Chemical equilibrium
I. Title.
QD75.2.R35   543   80-11937
ISBN 0-201-06107-4

Reprinted with corrections, October 1981

ISBN 0-201-06107-4
ABCDEFGHIJ-DO-8987654321

This book is gratefully dedicated to

David Leader, the biology teacher at Hartford Public High School who demonstrated the Fehling's analytical test for glucose one day and instantly transformed me into a boy chemist;

Alfred R. A. Brooks, HPHS science club adviser who encouraged us to fool around in the lab after school to learn how things work;

Joel and Grace Ramette, my patient parents who let me make those awful smells and bangs in the attic lab, and who sacrificed much in sending me to college;

Gil Burford and George Matsuyama, Wesleyan University professors who helped me through the initial bewilderments of analytical and physical chemistry;

Ernest B. Sandell, my thesis adviser at the University of Minnesota, who demonstrated so clearly in his research that chemical equilibrium and analysis are a single subject;

Isaac M. Kolthoff, who inspired all of us at Minnesota with a steady flow of principles and creative ideas including his favorite maxim: "Theory guides, experiment decides";

Laurence M. Gould, extraordinary President of Carleton College, who hired me on some sort of faith in 1954 and constantly encouraged the faculty to work toward excellence in college teaching in "Science and the Other Humanities";

My students, who have educated me while I have tried to teach them, and whose questions, anxieties, suggestions, and successes continue to provide a stimulating workshop;

My wife, Lee, whose understanding support and encouragement have helped tremendously to make this book-writing project possible.

PREFACE

This is a textbook for an introductory course in analytical chemistry that enrolls premedical, biology, and geology students as well as chemistry majors. I believe that such courses should treat the subject as a fundamental science, with a central focus on the healthy symbiosis between applied chemical analysis and theoretical models of aqueous equilibrium systems. Traditional analytical texts have made good use of tabulated values of equilibrium constants and electrode potentials, but students have a tendency to think of these tables as exact and unchanging. Most textbooks have not encouraged students to think about the origin of equilibrium data, which therefore take on the character of carved-stone tablets handed down by some forgotten scientific ancestor. Over the years I have tried to find many ways to convince my students that such "constants" are still subject to revision, that even the identities of solution species may be questioned, and that studies of new equilibrium systems are important in many research areas. Progress in all fields of chemical science, including oceanography, geochemistry, and environmental science, depends strongly on current and past research on chemical equilibrium and analysis, including studies of speciation.

I have chosen the important classical methods of analysis: titration, gravimetry, spectrophotometry, and potentiometry, and have discussed them in conjunction with the chief types of aqueous equilibria: acid–base, oxidation–reduction, metal–ligand complexes, precipitation–solubility, and distribution. I believe we should teach chemical equilibrium and analysis as complementary facets of a single subject. The treatment in this text is a blend of theoretical principles with practical applications, of descriptive chemistry with precision measurements, and of algebraic derivations with numerical calculations. In some sections hypothetical chemical models are translated into their mathematical counterparts, while elsewhere the anatomy of analytical methods is revealed by discussion of underlying equilibria.

One unique feature of this book is the liberal use of Case Studies based on data taken from the research literature. This approach not only places ideas in a realistic context but minimizes our natural tendency to oversimplify. For example, the effect of ionic strength on equilibria becomes obvious when we deal with real data, and we learn quickly that the assumption "activity coefficients can be neglected for our purposes" is not always valid. Very soon the students become uncomfortable about ignoring such important effects and, much to their credit, they would rather not have to find out later that their instruction was inadequate. As Alfred North Whitehead said: "They should feel that they are studying something and are not merely executing intellectual minuets."

The first chapter describes the main concepts of chemical analysis and defines common terms and methods. In Chapter 2, I have provided a practical approach to aqueous solution stoichiometry that will serve as a strong bridge from the basic ideas of the freshman course to the precise calculations of analysis. In Chapter 3, I have kept simple the statistical treatment of experimental data. Chapter 4 on chemical equilibrium and reaction tendency is especially important in showing how to handle the nonideal chemical systems that are so common in experimental work. I have demonstrated the need for activity coefficients and introduced a useful distinction between the equilibrium constant K and its molarity counterpart Q. The topics of spectrophotometry and galvanic cells are placed early in the book, so they may be used to advantage in the equilibrium discussions that follow. Acid–base systems are treated before other equilibria because they are relatively simple and because pH effects are so important in systems involving complexes and solubility. Considerable space is given to metal–ligand complexes because of their great significance in many areas of chemistry in addition to their value in chemical analysis. It is logical to delay the discussion of solubility equilibria because it depends strongly on both acid–base and metal–ligand reactions. I believe that the traditional overdependence on the simple solubility-product model has hampered chemists' understanding of solubility. The metal–ligand complexes and the solubility chapters offer rich opportunities to integrate all of the topics treated earlier in the book.

The concluding chapter on distribution equilibria and solvent extraction is optional in an introductory course. Further, when time is short, an instructor may choose to omit various parts of Chapters 7, 11, 13, 15, and 18.

The Appendixes are novel and instructive in the way they summarize data on equilibrium constants and reduction potentials. By grouping information about each element we gain an overview of important properties. A great deal of descriptive chemistry is implicit in these data summaries.

Another unusual feature of this book is its Laboratory Program. It includes some experiments of a more traditional type together with several experiments that support the main theme of the book. The laboratory work may readily be scheduled to correlate closely with the lectures. In comparison with the common "quant unknowns," some experiments will be more difficult for the instructor to grade. For example, it may not be the student's fault that, due to tarnish, the response of the silver-wire electrode departs from the Nernst equation. However, these experiments are worthwhile because of the gain in understanding achieved when students interpret their own laboratory data in terms of the same fundamental principles they are struggling with in homework.

The first version of the manuscript took shape in 1975–1976, mainly during a sabbatical leave in Gainesville, Florida, and I thank Carleton College, the University of Florida, and Roger G. Bates for encouraging support. Much of the later writing was done during summers in Madison, Wisconsin, and it is a pleasure to acknowledge the excellent professional environment at the University of Wisconsin and the creative help given by Robert Lavine. I also appreciate the secretarial talents of Wendy Zimmerman and the duplicating skills of Marion Leidner and Loretta Springer, while the manuscript evolved through ten revisions in five years. Among the many reviewers who offered valuable suggestions I am grateful to Wilmer Stratton (Earlham College), Dennis Evans (University of Wisconsin/Madison), William Guenther (University of the South), Thomas Dunne (Reed College), Arthur Hubbard (University of California/Santa Barbara), Dennis Johnson (Iowa State University), and my colleagues at Carleton, Delores Bowers and David Smith, who actually taught from the manuscript. The excellent editorial contributions by Rima Zolina at Addison-Wesley have been remarkably valuable in the final metamorphosis. Finally, I am indebted to the many students in Northfield and Madison who studied from the preliminary editions. Their tolerance and helpful suggestions during the prolonged process of testing and revision have stimulated many improvements in the book.

R. W. Ramette,
Laurence McKinley Gould
Professor of Chemistry

Northfield, Minnesota
September, 1980

CONTENTS

CONCEPTS OF CHEMICAL ANALYSIS

1

I often say that when you can measure what you are speaking about, and express it in numbers, you know something about it; but when you cannot express it in numbers, your knowledge is of a meagre and unsatisfactory kind; it may be the beginning of knowledge, but you have scarcely, in your thoughts, advanced to the stage of *Science*, whatever the matter may be.

Lord Kelvin

I like analytical chemistry. It not only uses and nourishes other fields of natural science and mathematics, but is a constantly growing blend of elegant theory, precise technique, and practical relevance. Analytical chemistry is so varied in content and so broad in scope that no one can master it fully or ignore it completely.

Although this textbook concentrates on the classical approaches of gravimetric, titrimetric, spectrophotometric, and potentiometric determinations, the underlying concepts apply to all of the modern and sophisticated procedures as well. Such concepts will be identified in this chapter and then used repeatedly in examples throughout the text. As I.M. Kolthoff said*:

> ... The aims and objectives of analytical chemistry are to determine the composition of any simple or complex compound or mixture of compounds ... The progress and advance of analytical chemistry depend to a great extent upon an intelligent application of the fundamentals of physical chemistry and the close relation between physical and analytical chemistry.

Perhaps the first famous chemical analysis was performed by Archimedes in ancient Greece. The King had given an artisan some gold for making a beautiful crown but suspected (without proof) that the goldsmith had cheated by replacing part of the gold with silver. Archimedes was the head scientist and the King wanted an answer. After considerable fretting without the slightest idea of how to test the crown, Archimedes decided to take a bath and, as he lowered his body into the tub, the water flowed over the edge. The answer came in an intuitive flash. "Eureka!" (I have found it!), cried Archimedes as he leaped from the tub and ran naked into the street.

Archimedes went to the King and got some pure gold that weighed as much as the crown and also some pure silver that weighed the same. When carefully lowered into a vessel full of water, the pure gold caused less overflow than did the pure silver (gold is almost twice as dense as silver). Then came the key step—he submerged the crown and noticed that the water overflow was intermediate between that caused by the gold and by the silver. This proved to the King's satisfaction that the artisan had made an alloy which still looked like gold but contained a substantial amount of silver. Archimedes went on to further discoveries and fame, but there is no record of what happened to the goldsmith. We still make regular use of Archimedes' Principle in the modern form: a body immersed in a fluid is buoyed up with a force equal to the weight of the displaced fluid.

---

*I. M. Kolthoff, "Analytical Chemistry as a Technique and as a Science," *Chemical and Engineering News, 28,* p. 2882, August 21, 1950.

1.1   MAIN STEPS IN CHEMICAL ANALYSIS

The story of Archimedes illustrates the main steps in analytical procedure:

1. Statement of the problem.
2. Obtaining the sample for analysis.
3. Preparation of materials, including reference standards.
4. Sample treatment, including separations if necessary.
5. The determinative step.
6. Interpretation and conclusions.
7. Actions.

Each of these steps deserves some discussion to establish concepts and to define important terms:

1. Proper Statement of the Problem

Suppose you run an analytical service and have two customers who want you to determine the percentage of mercury in some material. The statements of their problems might be very different.

Customer A wants to bid on a boatload of high-grade mercury-containing ore, can provide you with as large a sample as you wish, states that the mercury is a major constituent at about the 5% level, and wants a representative figure of high accuracy for the mercury content of the load.

By contrast, customer B has one specimen of an ancient copper coin, suspects that the mercury content is at the part-per-million level, and tells you that the coin is not to be damaged, but that the determination need be accurate only to one significant digit.

Even though the overall problem is identical (determining the amount of mercury), the analytical approaches will have great differences. The ore will require statistical sampling techniques, and has enough mercury to permit the accurate methods of gravimetric analysis or titration to be used. The coin might be subjected to neutron bombardment followed by gamma-ray spectroscopy.

Proper statement of the problem is helpful as a guide to the analytical chemist who must choose from an arsenal of methods. It is also important to the person seeking information, so that the right sort of results are obtained. Further discussion at this stage may rule against the determination being carried out: for example, money may be no barrier to the mercury ore buyer, but the archaeologist with the rare coin must ask, "Is it really so vital for me to know the trace mercury content of this coin, given that the neutron activation analysis will cost about $500?"

As another example, suppose a physical chemist wants to know the concentration of chlorine in a mixture of gases used in a kinetics study. It makes a great difference if the determination must be carried out at 10-millisecond intervals following compression by a shock wave compared to the situation where the gas sample may be removed and analyzed without haste.

The variety of needs that can be met by chemical analysis seems limitless. In many cases these needs, though important, are no longer novel and require only routine and well-established procedures. An example is the regular checking of commercial fish samples for their pesticide content by using gas chromatography. However, new problems are constantly put before analytical chemists and these require close communication to answer questions about the nature of the sample, the content level of the constituents, the available technology for the analysis, the accuracy required and, *not* least of all, the wisdom of bothering with the determination in the first place. For example, there is little point in determining the oxygen content of well water unless the sample has been scrupulously protected from any contact with the atmosphere from the moment it was obtained.

2. Obtaining the Sample for Analysis

When we say the **sample** we typically refer to a small portion selected for examination out of some much larger amount of material. Only rarely would the sample comprise the entire supply of material to be analyzed. Distributed within the sample are the **constituents**—the substances we may seek to *determine* by analytical procedures. For example, a sample of clean air contains **major** constituents (nitrogen and oxygen), **minor** constituents (water vapor, carbon dioxide, and argon), and **trace** constituents (such as neon, helium, methane, krypton, hydrogen, nitrous oxide, and xenon). In addition there are countless **ultratrace** constituents which are mostly, but not entirely, due to industrial pollution.

The quantity of a constituent available for a chosen analytical procedure depends both on **constituent level** and **sample size.** When a **macroscopic** sample is examined for a major constituent (for example, the determination of chloride in seawater), there may be a gram or so of constituent. As sample size decreases, through **semimicro-,** to **micro-** and **submicroscopic** quantities, the quantities of the constituent become smaller and smaller, making it necessary to use more sophisticated techniques for analysis.

In any case there must be serious concern that the portion of a total body of material taken for the analysis should be *representative* of the composition of the body as a whole. Carelessness in this concern can be disastrous. For example, Digoxin, a drug used to stimulate a weak heart, must be administered in tablets containing the precise amount intended by the physician, because toxic effects are produced by bloodstream concentrations only slightly greater than the therapeutic level. For this reason, the FDA National Center for Drug Analysis checks a minimum of 60 tablets or capsules chosen randomly from each tested drug batch.

Again, suppose you are a geologist for a mining company. They take you to a rugged mountainous area in a helicopter, unload your camping gear and portable analytical lab and say, "Tell us if we should spend a lot of money in

exploration for commercial mining of kryptonite in these hills." Your sampling procedure would have to be quite complex to lead to useful recommendations.

Once a representative sample is obtained (or, more likely, several samples), the effect of time sets in and sample **stability** becomes a major concern.
For example, a specimen of urine should be analyzed by the clinician before it
is exposed to air and the natural biochemical decomposition has a chance to
change the levels or even identities of the dissolved constituents.

The foregoing was not an attempt to teach procedures for sampling,
though this is a topic of great importance; rather, the purpose has been to
mention the conceptual basis for this first step, with which the ultimate
reliability and usefulness of an analytical procedure begins.

3. Chemical Preparation of Analytical Reagents

It could be argued that everything consists of chemicals, but the term
reagent refers to a substance or chemical mixture used for an intended
purpose in a **procedure**. The development of a good analytical procedure is an
interesting story in itself, but assuming the analyst has a well-tested method to
follow, it is important to make a judgment about the **purity** of the chemicals to
be used in making the reagent solutions.

Perhaps the purest substance made routinely in modern technology is the
silicon grown in crystal form for use in electronic integrated circuits. But a
sample of this material that is $99.999999 \pm 0.000001\%$ pure ("a purity of eight
nines") contains about 200 trillion impurity atoms per gram of silicon.

In the old days it was necessary for chemists to resort to tedious
procedures (distillation, recrystallization, etc.) to remove unwanted impurities
from their reagents. This is still necessary in many cases but now, more often
than not, the commercial manufacturers of reagent chemicals offer highly
purified materials (admittedly at highly increased prices) which may be used
with confidence and convenience. For example, if you wish to carry out some
experiments with large quantities of sucrose, it might be practical and wisely
economical to visit the local supermarket where sugar is sold at about $0.20
per pound. This commercial form of sucrose is actually quite pure for most
purposes. However, if you need the sugar for some experiments that require
low levels of impurities, you will probably be willing to pay about $1.80 per
pound for a product that is certified to meet American Chemical Society
specifications. The almost tenfold price increase results from the costs of both
purification and testing of the product.

To go still further, if you undertake some highly critical biological-
metabolism studies requiring sucrose with extremely low levels of metal
impurities and if you have the funds in your research grant, you might even
buy J. T. Baker Co. Ultrex® quality sucrose at the price of $227.00 per
pound! This price includes the costs of certification by testing with sensitive
and sophisticated procedures.

The normal procedure in most laboratories is to use Analytical Reagent Grade chemicals prepared by manufacturers according to the specifications of the American Chemical Society as given in the periodically revised monograph Reagent Chemicals (5th edition, 1974). For example, the specifications for sucrose are as follows:

Specific rotation	between $+66.3$ and $66.8°$
Insoluble matter	not more than 0.005%
Loss on drying at 105°C	not more than 0.03%
Residue after ignition	not more than 0.01%
Acidity (as CH_3COOH)	not more than 0.005%
Chloride	not more than 0.005%
Sulfate and sulfite	not more than 0.005%
Heavy metals (as Pb)	not more than 0.0005%
Iron	not more than 0.0005%
Invert sugar	not more than 0.05%

The book then gives detailed procedures for testing a sugar sample for each of these criteria.

The foregoing discussion has emphasized the significance of **impurities** in reagents because in many applications a small amount of impurity can have serious effects on the experiment. However it is often more important to worry about the percentage of the **active ingredient**, i.e., the **assay** value. For example, when a sample of copper metal foil is to be weighed and dissolved in nitric acid to prepare a solution with a precise copper content, it is vital to know that the metal contains, say 99.97% copper, but it may be of no interest that the other 0.03% consists of iron, or zinc, or lead, or whatever if these impurities have no effect on the intended application.

Other factors to be considered in the preparation of an analytical reagent solution are its **accuracy** and its **stability**. For many applications it is not necessary to make a solution with particular accuracy, while in titrations, for example, the solution should have an uncertainty of no more than 0.1% in its content of active ingredient. As for stability, some reagent solutions decompose in a matter of hours and must be prepared fresh, just before use, while others can remain in storage virtually unchanged for years.

REFERENCE STANDARDS

Before attempting a chemical analysis you must have confidence in the technical procedure, i.e., a good reason to believe that the method will give the correct result. The only way to gain this confidence is by trying out the entire procedure on samples that (a) are similar in composition to the unknown sample and (b) have a precisely known content of the constituent to be determined.

Especially when an analytical chemist is trying to develop and test a new procedure for analysis it is vital to pay close attention to the use of standards and to the effects of other constituents that are likely to be in the sample and may act as chemical **interferences** causing incorrect results. A major program of the U. S. National Bureau of Standards is the rigorous testing, certification, and sale of highly reliable Reference Standard Materials of many types (metal alloys, geological and biological samples, glasses, pure substances, etc.).

It is important that the standards be treated in the same way as the unknown. For example, suppose a sample of city drinking water is to be examined for its fluoride content, by using a fluoride ion-selective electrode. It would be wrong to calibrate the electrode response with simple solutions of sodium fluoride in distilled water, because the drinking water contains calcium, magnesium, iron, etc., which would interfere with the ability of the electrode to sense the total fluoride content. What is done in this case is simple: both the sodium fluoride standards and the drinking water samples are treated with a concentrated buffer solution that fixes the pH and complexes the interferences. Then the electrode is working in the same chemical environment at all times.

For well-tested methods in gravimetric and titrimetric analysis we sometimes assume that the testing of standards may be omitted. However standards are always included in spectrophotometric, potentiometric, polarographic, fluorescence, etc., methods that require the interpretation of data from an instrument readout device.

A special standard is the **blank**—a simulated unknown that contains none of the constituent sought in the determination. The particular importance of the blank, which should always be run along with the regular standards, is that it gives a check on the reagents used in the analytical procedure. Ideally, the result from the blank should be "zero," but if one or more of the reagents includes some of the sought constituent as an impurity there will be a response in the final measurement. It may also be that, even though the reagents are quite pure, they may themselves have a slight color or some other property that can contribute to a false indication in the final measurement step.

The magnitude of the blank response also indicates the lowest level of the constituent that can be determined by the method. This is called the **limit of detection** of the method and is an important concept in trace analysis.

Suppose that trace amounts of bromide are to be determined in a sample of limestone, which must be dissolved in hydrochloric acid first. Ordinary reagent-grade hydrochloric acid may contain up to 50 parts per million of bromide as impurity, or 500 micrograms in a 10-milliliter portion. If the bromide level in the limestone is, say, 200 micrograms, then the blank is larger than the constituent, an undesirable situation. But if the sample is dissolved with Baker "Ultrex" hydrochloric acid, certified to contain less than 5 parts per million of bromide, the determination will be much more reliable.

4. Sample Treatment

Often the sample contains, in addition to the constituent to be determined, one or more constituents that would interfere in some way and thereby spoil the accuracy of the procedure. Sometimes these interferences can be rendered harmless, even if they are not removed by adjusting the pH, or by adding a soluble complexing agent, or by changing the oxidation state. In many cases, however, either the interfering substances must be removed from the solution before the measurement step, or else the desired constituent must be removed from the interferences and placed in a new solution.

The theory and practice of such separations is an integral part of analytical chemistry and often involves some ingenious procedures. Some of the main categories of separation methods are as follows:

Precipitation The constituent is made to react with a reagent forming an insoluble compound that is removed by filtration. If an interference is thus separated, it may be discarded, but if it is the desired constituent that is precipitated, either it must be filtered off and redissolved or, in some cases, the mass of the precipitate is determined after drying and weighing (gravimetric determination).

Distillation If the constituent is volatile it may be distilled and collected in a cold trap. Volatile interferences may simply be allowed to escape into the hood exhaust.

Extraction By using an immiscible solvent, such as chloroform, it is possible to extract some molecular constituents quantitatively out of the aqueous phase, leaving all the rest unchanged.

Chromatography Separation is achieved by passing the mixture through a packed column, with individual constituents moving through at different rates because of their inherently different absorption behavior.

5. The Determinative Step

The "moment of truth" arrives once the sample and standards have been put into the right chemical environment. It is time for the crucial chemical or instrumental measurement that yields the precise quantity of the constituent sought in the unknown sample. The choice is made from a wide variety of possible techniques according to the factors listed in Section 1.2.

The main categories of analytical determinations, with some of their subcategories, are as follows:

a) **Gravimetric determination**—the constituent is converted to a form that can be precisely weighed: electroplating, precipitation, loss by volatilization.

b) **Titrimetric determination**—the constituent is made to react in a known and precise stoichiometric ratio with a standardized reagent: acid–base, precipitation, chelometric, oxidation–reduction.

c) **Spectroscopic determination**—the constituent is made to absorb or to emit electromagnetic radiation: visible and ultraviolet absorption (solutions), infrared absorption, visible and ultraviolet atomic absorption, fluorescence, flame, arc, and spark emission, Raman spectroscopy.

d) **Mass spectroscopy**—sample constituents are examined in mass spectra after travel through electric and magnetic fields.

e) **Electrochemical determination**—the constituent affects the redox equilibrium at a suitable electrode: potentiometry, coulometry, polarography, anodic stripping.

f) **Chromatographic determination**—the constituent is separated by passage through a selectively absorbing column and then sensed by a detector: gas chromatography, liquid chromatography.

g) **Radiochemical determination**—the constituent is determined by measurement of beta-particle or gamma-radiation emission from unstable isotopes: naturally radioactive substances, isotopic dilution technique, isotopic labeling, neutron activation analysis.

6. Interpretation and Conclusions

A good observer is one who sees what is there even when he isn't looking for it and doesn't see what isn't there even when he is looking for it. This definition can be modified to fit the interpretation of scientific data, including results of chemical analysis, which too often are accepted without concern for their reliability.

We should distinguish two aspects of analytical-data interpretation. The first concerns the reliability of the measurements themselves, with due care for errors in sampling, interferences, variations in chemical procedure, accuracy of instruments, and statistical measures of precision. In other words, "How well-done was the entire laboratory project?"

The second aspect (once reliable data are at hand) is their application to solving the problem, which was the original purpose for the work. It is here that the hard work of intelligent judgment is essential. For example, "Do these results really prove that the steel mills of Gary, Indiana, are causing a significant concentration of sulfuric acid in the rains?" Another example, "Can we deduce reliable values from these data for the equilibrium constant of the reaction under study?"

Without interpretation and conclusions as a follow-up, the chemical analysis should never have been made for it would be mere data-gathering and a waste of good time. This consideration should be applied at the beginning of an analytical project in the form of the question, "Is it reasonable to expect these analyses to give results that will be helpful in solving the problem?" Unfortunately, this question is not asked (or realistically answered) as often as you might think. Just because Lord Kelvin once said, "To measure is to know," it does not follow that discrimination in measurement isn't essential and wise.

7. Actions

This step is included (without discussion) as a reminder that the laboratory efforts and the thoughtful conclusions of the entire analytical process should lead to something. The results might lead to an improvement in health care, or manufacturing quality, or a criminal conviction, or the authentication of an artistic work, or to a deeper understanding of scientific fundamentals, which in turn would lead to new ideas and experimental projects.

1.2 CHARACTERISTICS OF AN ANALYTICAL PROCEDURE

In the discussion of the proper statement of the problem we have used the example of two customers, each wanting a mercury analysis. But the sample types and customer needs were totally different and required different laboratory approaches. The question is: "How does an analytical chemist decide on the best choice of method?"

In practice the decision will most likely not be systematic or optimal, except for rather standard situations. Most chemists feel more competent and confident in certain areas (e.g., electrochemistry) and have an understandable tendency to solve problems that way. Also, even within a given area a chemist cannot have thorough awareness of all published methods and procedure variations, and will choose according to what comes to mind or what can be found readily in available references. Finally, the "best" procedure may have to be ruled out on the basis of cost, time requirement, lack of instruments or reagents, or competence (it takes training and experience to obtain a differential pulse polarogram, for example).

However, it is possible to list a number of characteristics of analytical procedures and these will be important in choosing the method for solving a given problem:

a) Is the method empirical or is it based on sound theoretical principles? Unless the analyst has a fundamental knowledge of how and why the procedure works, it may not be possible to make the appropriate adjustments and judgments required for a nonroutine sample. This must include the understanding of the effects of interfering substances and of how to deal with them. It is specifically important to understand the chemical-reaction stoichiometry, the equilibrium systems involved, the kinetic behavior, the reagent stability, the critical parameters (such as temperature and pH) and so on. Much of the fundamental research in analytical chemistry is directed at providing this kind of sound basis for laboratory analysis.

b) Does the method require *destruction of the sample*? In most cases the sample must be sacrificed, but some methods can provide information without destruction (electron microprobe, neutron activation, x-ray fluorescence, microscopic examination) and others require so little sample that a valuable specimen (e.g., an archaeological artifact) can be sampled without visible damage.

c) Is the method *easy in practice* or will the procedure require hours of tedious attention, and specialized and exacting techniques?

d) Can the analysis be done with *equipment* normally found in a well-equipped laboratory or is it necessary to have unusual and expensive instruments?

e) What will the analysis *cost*? We must consider not only the personal time of the analyst, but the cost of "exotic" reagents.

f) Is the procedure *selective*? This refers chiefly to the elimination of interferences, so that only the sought constituent will control the result. Very few methods are *specific*, i.e., responsive to one and only one constituent, but by controlling the experimental conditions it is often possible to approach this ideal.

g) Is the procedure *rugged*? This term refers to "fussiness." If it is necessary to control the temperature to $45 \pm 1°C$, the pH to 5.6 ± 0.1, the rate of addition of reagent, the precise time for certain steps, the precise volumes, etc., then there are too many chances for error. A "rugged" procedure is not so dependent upon the operator's skill, except of course in the determinative step where there can be no substitute for competence.

h) What is the *sensitivity* of the method? This question is the most important one for trace analysis or microanalysis. Gravimetric determinations are the least sensitive, simply because there must be enough precipitate for accurate weighing. If the balance can provide a weight determination to the nearest 0.0001 g, then the precipitate must weigh at least 0.01 g (for an uncertainty of about one percent). If about half the precipitate is the desired constituent, then the sensitivity (the smallest quantity that can be determined with reasonable accuracy) is about 0.005 g, or 5 mg. At the other end of the sensitivity scale are methods (gas chromatography, neutron activation, spark-source mass spectrometry, anodic stripping analysis, etc.) that boast sensitivities of a few nanograms ($1 \text{ ng} = 10^{-9} \text{ g}$) or even picograms ($1 \text{ pg} = 10^{-12} \text{ g}$). We are beginning to talk in terms of femtograms ($1 \text{ fg} = 10^{-15} \text{ g}$). However, it should be remembered that as sensitivity improves, accuracy often decreases on account of specific difficulties in dealing with trace quantities.

i) Last but not least, and perhaps usually the first question asked, is "How *reliable* is the procedure?" Here there are two aspects, *precision* and *accuracy*. If successive results from repeated analysis agree very closely with *each other*, we speak of high precision. The method then gives very nearly the same result each time. If, in addition to having high precision, the results are believed to be very close to the *true value*, then the method is said to be accurate. Note that it would be possible to run a series of analyses with high precision but low accuracy if there were a systematic error such as an incorrectly standardized titrating solution or a set of reference standards prepared from an impure material.

1.3 CASE STUDY: DETERMINATION OF MERCURY IN URINE*

In the late 1960's a graduate student made the shattering discovery that various edible fish species contained enough mercury to suggest a threat to human health. The FDA and others mounted crash programs for analytical survey. New, sensitive procedures for mercury had to be devised. The most successful

---

*After J. Toffaletti, J. Savory (University of North Carolina), *Analytical Chemistry*, **47**, No. 13, p. 2091, November 1975.

of these is the **flameless atomic absorption technique,** one version of which is discussed below.

Fish with oily flesh, especially those at the top of the food chain (e.g., swordfish) may contain several parts per million of mercury (several milligrams of mercury per kilogram of flesh). The toxicology of low mercury levels in the human body is not yet clear but there are examples of chronic illness and even death when the major part of a diet is based on contaminated fish. Industrial workers who use much mercury may also suffer from exposure; the expression "mad as a hatter" derives from the mental illness caused by use of mercury compounds in processing felt.

The FDA decided that an appropriate regulation would be to prohibit the sale of fish with mercury content in excess of 0.5 part per million. Of course, it is not possible to inspect the entire output of the fisheries, but a continuous screening of random samples provides a measure of safety. This explains the saddening disappearance of a favorite seafood, swordfish, from most restaurant menus.

No doubt *some* of the mercury in fish, mammals, and plants is due to environmental pollution by industry. Chlorine gas is produced by using mercury electrodes in brine, and the vast paper industry uses mercury-containing fungicides and slimicides. But Mother Nature has been doing a steady job of mercury pollution all along, and samples of fish preserved in museums since before the Industrial Revolution have also been found to contain mercury.

Anyway, the new methods for trace mercury analysis have made it possible for medical scientists to study the low levels of mercury in the human body. One survey showed a geographical dependence for the level of mercury in human hair. Workers in plants that use mercury are routinely checked for exposure, which still exists in spite of tightened safety procedures. A convenient sampling technique is to analyze urine collected over a 24-hour period, on the assumption that its mercury content will be somewhat proportional to the total mercury load present in the body.

Chemical Principles of the Method

The mercury present in body fluids is in the form of compounds such as mercuric chloride ($HgCl_2$), methylmercury chloride (CH_3HgCl), and phenylmercury acetate ($C_6H_5HgOCOCH_3$). Such mercury, in the $+2$ oxidation state, can be reduced to the zero oxidation state by a number of reducing agents. Toffaletti and Savory studied the use of sodium borohydride for the reduction:

$$BH_4^- + 4CH_3HgCl + 3H_2O \rightarrow H_3BO_3 + 4Hg(aq) + 4CH_4 + 4Cl^- + 3H^+.$$

When only trace amounts of mercury are reduced in this way, the element is formed in its atomic state dissolved in the solution, and not as droplets of the usual shiny metal. If an inert gas is bubbled through the solution without delay, the atomic mercury is swept out of solution and into the gas stream:

$$Hg(aq) \xrightarrow{\text{argon}} Hg(g),$$

Apparatus and Measuring Principle

Free (chemically uncombined) mercury atoms in the gas phase in their lowest energy state (the "ground state") have an electronic configuration of

$$1s^2 2s^2 2p^6 3s^2 3p^6 3d^{10} 4s^2 4p^6 4d^{10} 4f^{14} 5s^2 5p^6 5d^{10} 6s^2.$$

For one of the $6s$ electrons to be excited to a higher energy state $6p$ it is necessary for the atom to absorb exactly the right quantum of energy, and this can be provided in the laboratory by using ultraviolet radiation of a precise wavelength, namely 253.7 nm. Consider the simplified diagram shown in Fig. 1.1.

The apparatus is used as follows (see below for preparation of reagents): 1.00 mL of urine and 1 mL of the pH = 6.5 buffer are added to the generation cylinder and argon gas bubbling is started to establish a baseline reading from

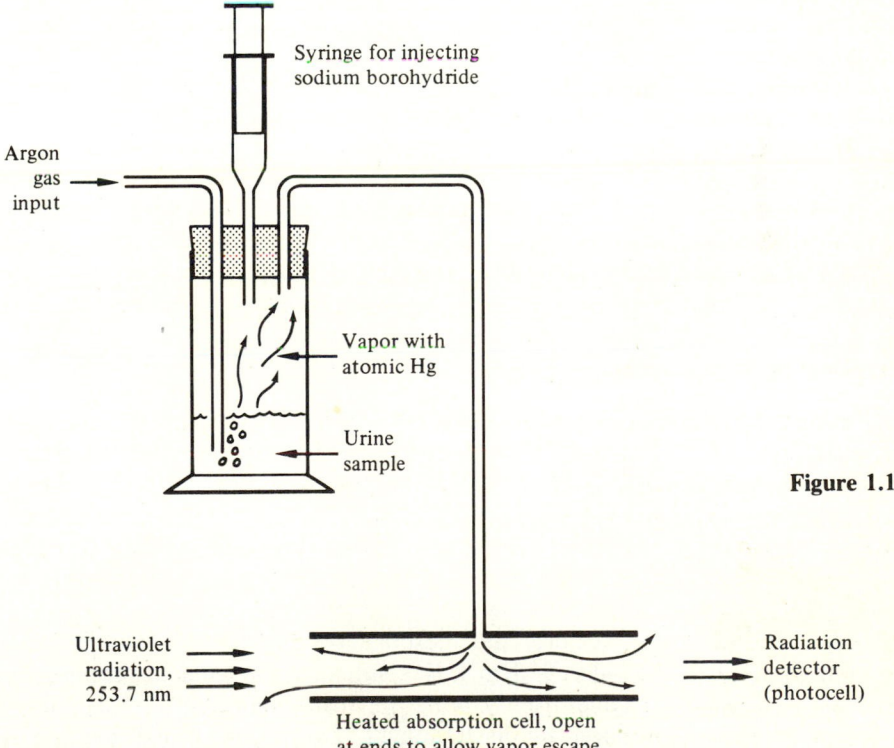

Argon gas input

Syringe for injecting sodium borohydride

Vapor with atomic Hg

Urine sample

Figure 1.1

Ultraviolet radiation, 253.7 nm

Heated absorption cell, open at ends to allow vapor escape

Radiation detector (photocell)

the radiation detector. Then 1.5 mL of the sodium borohydride solution is quickly injected by the syringe, the chemical reduction of mercury proceeds rapidly, and the argon gas sweeps the atomic mercury through the connecting tube and into the absorption cell.

When no mercury atoms were present in the cell (prior to the addition of sodium borohydride), the ultraviolet radiation passed through the cell without appreciable loss before it reached the detector. But once the mercury vapor enters the cell in the gas stream, the mercury atoms interact with the radiation, absorbing photons to produce atoms in an excited state. Thus, the detector is no longer receiving as much radiation as before and the electronic circuits cause the chart recorder to show the change (Fig. 1.2).

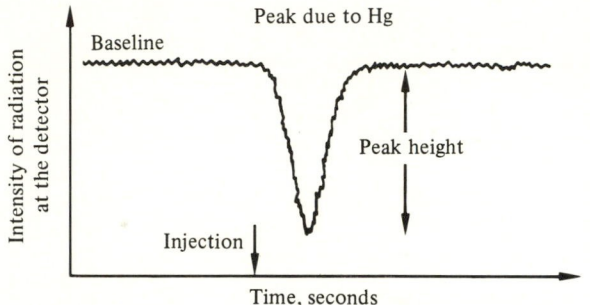

Figure 1.2

Of course, not all the mercury appears in the absorption cell at once, so the signal is spread out over a few seconds of time. For highest accuracy in the estimation of the amount of mercury present in the sample the total *area* of the peak should be determined by electronic integration. However, Toffaletti and Savory showed that simple measurement of peak *height* was reasonably proportional to the mercury content of the sample provided that conditions, especially bubbling rate, were closely reproduced for both standards and sample.

Reagents and Standards

1. Sodium borohydride solution: dissolve 1.5 g of $NaBH_4$ in 30 mL of water.
2. Buffer solution: dissolve 35 g of sodium dihydrogen phosphate (NaH_2PO_4) in 500 mL of water and add 10 molar sodium hydroxide until the pH is 6.5 (about 20 mL required).
3. Catalyst solution: dissolve 250 mg copper sulfate ($CuSO_4 \cdot 5H_2O$) in 100 mL of water. Add 10 μL (microliter) of this to the 1-mL urine sample as a catalyst for the borohydride–mercury reaction.
4. Mercury standards: Because the mercury species in the body is largely methylmercuric chloride CH_3HgCl, this substance is used to prepare standards that are run concurrently with the urine samples. A concentrated stock solution is prepared by dissolving 12.52 mg of CH_3HgCl in water and

diluting to precisely 100 mL in a volumetric flask. Then a dilute stock solution containing 1.00 μg of mercury per milliliter is prepared by diluting the concentrated stock. For the analytical determinations, portions of this dilute stock are taken to provide quantities of mercury in the range from 10 to 200 ng. Only by such successive dilutions can these very dilute solutions be prepared accurately.

Experimental Results

When the peak heights obtained with various quantities of the methylmercuric chloride standard are plotted versus the amount of mercury used, the result is a straight line with an acceptable small scatter of experimental points (Fig. 1.3).

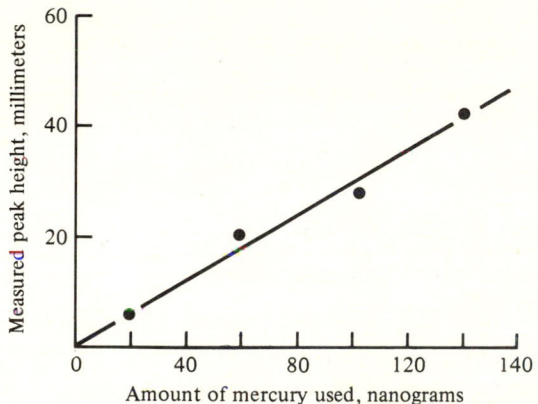

Figure 1.3

Such a graph is called a **standard curve,** or an **analytical curve** (some methods yield curved plots rather than straight lines). It serves to calibrate the entire analytical procedure, including the response of the measuring instrument. Toffaletti and Savory showed that the same response was obtained regardless of what was used: a simple methylmercuric chloride solution in water or samples of mercury-free urine "spiked" with known quantities of mercury.

When an unknown urine sample is run through the procedure, the peak height is simply measured, and by reading the standard curve it is easy to deduce the number of nanograms of mercury present. As a test of precision, the authors used 20 different normal urine samples, each spiked with 100 ng of mercury. The average result was 98 ± 4 ng, which is excellent, considering the low levels and the fact that such a clinical analysis does not require a higher precision of accuracy.

The authors further showed that none of the major constituents of urine interfered with quantitative reduction of mercury, and it seems likely that this method will find regular use in clinical analysis.

QUESTIONS AND PROBLEMS

1. The density of pure gold is 19.32 g/mL, while that of silver is 10.50 g/mL. If the King's crown weighs 0.431 kg and displaces 24.8 mL of water (density 0.997 g/mL), what is the approximate percentage of gold in the crown? Assume that the density of the alloy varies linearly with the percentage of gold.

2. A sample of sucrose weighing 44.851 g is found to weigh 44.838 g after drying at 105°. After strong heating (to decompose it to carbon dioxide and water) the residue weighs 3.8 mg. Analysis shows that the chloride ion content of a 50.0 gram sample is 2.1 mg and that the iron content is 440 μg. Discuss these results in terms of the A.C.S. specifications for reagent-grade sucrose.

3. A sample of copper sulfate pentahydrate $CuSO_4 \cdot 5H_2O$ is assayed by electrolytic deposition and weighing of the metallic copper. If a sample of the salt weighing 2.3381 g yields 0.5932 g of copper, what is its apparent purity? Consult the table of atomic and molecular weights.

4. Make a list of new technical terms used in Chapter 1 and give a brief definition of each.

5. Chemical elements of particular importance in introductory analytical chemistry are as follows:

Ag, Al, As, Au, B, Ba, Br, C, Ca, Cd, Ce, Cl, Cr, Cu, F, Fe,
H, Hg, I, K, Li, Mg, Mn, N, Na, Ni, O, P, Pb, Pt, S, Sn, Zn.

Draw a circle around each of these elements in the Periodic Chart and memorize the name and symbol for each.

6. An electronic calculator that handles logarithms and displays exponents is essential for the problems in this text. Find the following values for x:

   a) $x = 1.23 \cdot 10^{-9} + 2.35 \cdot 10^{-14}/(5.54 \cdot 10^{-5} - 9.36 \cdot 10^{-6})$

   b) $x = -0.361 - \dfrac{0.0592}{5} \log \dfrac{0.1}{0.0335 \cdot (2.26 \cdot 10^{-3})^8}$

   c) $x = 10^{-3.19}/(2 - 10^{0.225})$

   d) $x = (2.32 \cdot 10^{-8}/4)^{1/3}$

   e) Make up a complicated arithmetic expression for x, solve it, and trade with a friend to see if you both get the same answers.

7. Algebraic manipulations and derivations are used frequently in this text. For each of the following, find the expression for x in terms of the variable y and the constants a,b,c,d,e:

   a) $w^2 = a - \dfrac{1}{x}$,   where $w = 8y^3$

   b) $x^2 = w^{1/2}(b - x^2)$,   where $w = 9/y^4$

   c) $ax + b = cw^{1/2}$,   where $w = (9z)^{1/4}$ and $z = 9y^8$

   d) $y = a + b \log \dfrac{1}{x^2}$

   e) $10^{ay} = b \cdot 10^{cx}$

8. If two variables x and y are related so that y is a function of x^2, it is always possible to write the relationship in the standard form:

$$ax^2 + bx + c = y.$$

We frequently need to determine the roots of this type of relationship, i.e., the values of x that correspond to a zero value for y. This is done by using the quadratic formula

$$x = \frac{-b \pm (b^2 - 4ac)^{1/2}}{2a}.$$

When this formula is applied to problems in equilibrium and analysis, there is only one correct root. The other root is impossible in the context of the problem (e.g., a molarity cannot be negative, nor can it be higher than permitted by composition of the solution). Here are a few examples to try out for practice. Find the value for x in each case:

a) $1.2 \cdot 10^{-2} = \dfrac{x^2}{0.1 - x}$

b) $2x^2 + 4x - 3 = 0$

c) $3 \cdot 10^5 = \dfrac{x}{(0.01 - x)(0.02 - x)}$

d) $7 \cdot 10^{-3} = \dfrac{x(0.01 + x)}{0.005 - x}$

e) How must b^2 compare with $4ac$ for the quadratic formula to yield real roots?

f) Rewrite the quadratic formula for the situation of $c = 0$. What are the roots in terms of a and b?

AQUEOUS-SOLUTION STOICHIOMETRY

2

A good analyst can estimate.

George Matsuyama

Stoichiometry* is the study of mole and mass relationships in chemical reactions and therefore is at the heart of quantitative chemical analysis. All chemists make use of balanced equations, which we may discuss in the somewhat general symbolic form

$$a A + b B \rightarrow c C + d D,$$

where the small letters are the stoichiometric coefficients, typically small integers, and the capital letters are the formulas for the reactants and products. A chemical reaction is said to have **clean stoichiometry** if, in the laboratory, substance A reacts with substance B in precisely the a/b ratio. For analytical purposes this is certainly of great importance. However it is not uncommon for more than one reaction to be possible when A is mixed with B. Suppose that the following chemical change can also occur:

$$x A + y B \rightarrow e E + f F.$$

In the reaction mixture, C, D, E, and F will be formed in relative amounts which may depend upon temperature, rate of addition of A to B, acidity, the presence or absence of a catalyst, etc. Thus, the stoichiometry may be variable instead of clean, and the reacting ratio between A and B will be somewhere between a/b and x/y. Such undesirable behavior cannot always be avoided and is generally (though reluctantly) accepted in synthetic organic chemistry, where the important goal is the preparation of, say, substance C. If the competing reaction produces E and F instead of C and D, the percentage yield is lower but the chemist may well be satisfied with even a small quantity of C because the main goal may be, for example, to determine its crystal structure.

By contrast, chemical analysis usually requires clean stoichiometry with essentially 100% yields. Reactions that depart significantly from this ideal are generally avoided. The search for reactions, conditions, catalysts, etc., which enable highly reliable stoichiometry, is an ongoing and interesting aspect of research in analytical chemistry.

2.1 FOUR FUNDAMENTAL RELATIONSHIPS IN STOICHIOMETRY

The calculations of chemical analysis depend very heavily upon combinations of a few basic concepts. Four important relationships are as follows:

1. The relative amounts (moles) of reactants and products that participate in a chemical change are easily seen from the balanced chemical equation for the reaction.

---

*From the Greek *stoikheinon* (element) + -*metry* (measurement).

Using the reaction symbolized by the balanced equation $a\mathrm{A} + b\mathrm{B} \rightarrow c\mathrm{C}$, we may write:

$$n_\mathrm{A} = \frac{a}{b}n_\mathrm{B} = \frac{a}{c}n_\mathrm{C} \quad \text{or} \quad n_\mathrm{C} = \frac{c}{a}n_\mathrm{A} = \frac{c}{b}n_\mathrm{B},$$

where n_A are moles of substance A used up and n_C are moles of substance C produced.

2. The absolute amount (moles) of a substance is directly proportional to the mass of the substance.

The relationship involves the gram formula weight (GFW), which is the mass (in grams) numerically equal to the formula weight of the substance:

$$n_\mathrm{A} = \frac{w_\mathrm{A}}{\mathrm{GFW}_\mathrm{A}} \quad \text{or} \quad w_\mathrm{A} = n_\mathrm{A} \cdot \mathrm{GFW}_\mathrm{A}.$$

3. The absolute amount (moles) of a solute in a solution is directly proportional to the concentration (molarity) of the solution and to its volume.

Molarity is discussed in detail in a later section. The useful relationship is as follows:

$$n_\mathrm{sol} = V(\mathrm{L}) \cdot M(\mathrm{mol/L}).$$

4. One mole of electrons is called a Faraday; it corresponds to an electrical charge of 96,487 coulombs. One coulomb corresponds to an electrical current of one ampere flowing for one second.

The important relationships in electroanalytical chemistry are:

$$Q(\text{coulombs}) = i(\text{amp}) \cdot t(\text{sec}) \qquad \text{(for constant current)},$$
$$n_\mathrm{electrons} = Q/96{,}487.$$

2.2 AQUEOUS SOLUTIONS

A high proportion of analytical procedures involves the use of water solutions of reagents, often having precisely known concentrations. As B. R. Sundheim put it in his article,*

> Chemists tend to concentrate their attention on chemical reactions taking place near room temperature and, by and large, in aqueous solutions. This is doubtless a reflection of the fact that the experimenter, viewed as a chemical system, is also an aqueous solution thermostatted at 37°C and at a constant pressure of one atmosphere.

---

*B. R. Sundheim, in *International Science and Technology*, November 1962.

Concentration of a Solution

In addition to the solvent (water, in most cases in this text), a solution contains one or more solutes. There are many ways to describe the composition of a solution in terms of the quantities of solvent and solute(s) present, but we will consider only the three most common terms: molarity, molality, and weight/volume percent.

MOLARITY

This measure of concentration is by far the most important for our purposes, both for equilibrium and for analysis. It is defined simply as the amount of solute (in moles) per liter of solution (not solvent). The usual symbol for molarity is M, but C is also used. Thus,

$$M_{solute} = \frac{n(\text{moles of solute})}{V(\text{liters of solution})},$$

or, since we usually work with smaller quantities,

$$M = \frac{n(\text{mmol})}{V(\text{mL})}.$$

This expression and its two useful rearrangements ($n = MV$, $V = n/M$) find constant application in the discussions of analytical stoichiometry. For example, if 0.055 mole of sodium chloride is dissolved in water and diluted to 500 mL, then

$$M = \frac{0.055 \text{ mole}}{0.500 \text{ liter}} = 0.11 \text{ mol/L.}$$

If a pipet is used to take 25.0 mL of this solution, the amount of NaCl taken is

$$n = 25.0 \cdot 0.11 = 2.75 \text{ mmol.}$$

If it is desired to measure out enough of the solution to obtain 13.6 mmol of NaCl, the volume required would be

$$V = 13.6/0.11 = 124 \text{ mL.}$$

In addition to the symbol M it is common practice to use square brackets [ ] to indicate molarity of a species in the solution. However, brackets must not be used *except* for actual solution species. For example, for the 0.11M solution of NaCl we may write

$$M_{\text{NaCl}} = 0.11 \qquad \text{(this is correct)}$$

but not

$$[\text{NaCl}] = 0.11 \qquad \text{(this is wrong).}$$

The point is that we believe sodium chloride to be completely dissociated into its ions in solution, so that there are no NaCl "molecules" existing as an actual solution species. It would be correct, however, to write

$$[Na^+] = 0.11 \text{ mol/L} \quad \text{and} \quad [Cl^-] = 0.11 \text{ mol/L},$$
$$[NaCl] = 0 \text{ mol/L}.$$

MOLALITY

The molality of a solute in solution is defined as the amount (moles) of solute per kilogram of solvent (not solution). As with molarity, we may use molality both for the compound placed in solution and for individual species present as a result of dissociation. Thus,

$$m_{solute} = \frac{n(\text{moles of solute})}{K(\text{kilograms of solvent})}.$$

No other symbolism (such as C, or [ ]) is used for molality.

Since molality is defined entirely on a mass basis, the use of convenient volume-measuring equipment (burets, pipets, volumetric flasks) is ruled out. Therefore, most chemical work is based on the easier system of molarity, and molality is employed only in work of the highest accuracy. It is unrealistic to try to use volumetric glassware with an accuracy of better than about 99.9% (or with an error of less than 1 part per 1000). However, in research with quality analytical balances accurate to 0.01 mg, the imprecision of a single weighing approaches one part per 100,000 when samples as small as 1 g are weighed.

Also, molality has the convenient characteristic that it does not depend on the temperature of the solution, since the masses of solute and solvent are invariant. By contrast, solution molarity will decrease about 1 part per 1000 for a 4° temperature increase because of the expansion of solution volume.

The numerical relationship between molarity (C) and molality (m) is easily derived: For a given solution of solute x, the molarity is

$$C_x = \frac{n_x(\text{moles})}{V_{sol}(\text{liters})}.$$

The solution has a density

$$d(\text{kg/L}) = \frac{\text{Mass of solution}}{V_{sol}},$$

so

$$C_x = \frac{n_x \cdot d}{\text{Mass of solution}} = \frac{n_x \cdot d}{\text{Mass of solvent} + \text{Mass of x}}.$$

Now, the mass of x is simply ($n_x \cdot \text{GFW}_x/1000$) kg. If we consider a quantity of this solution such that it contains exactly 1 kilogram of solvent, then by the definition of molality, $n_x = \text{Molality}_x$ or m_x. Therefore,

$$C_x = \frac{m_x \cdot d}{1 + m_x \cdot \text{GFW}_x/1000}.$$

Thus, given the molality and the density of a solution, one can calculate the molarity. The reader should rearrange the equation to give the molality as a function of molarity, density, and GFW.

As the molality of a solution decreases, so, of course, does the molarity. Inspection of the equation shows that the relationship approaches a limit and so at "infinite" dilution we obtain

$$C_x \to m \cdot d.$$

Since the density of water is 0.997 g/mL at 25°, very dilute solutions of solutes in water have approximately the same values of molality and molarity.

WEIGHT/VOLUME PERCENT

Instead of using units of moles (or millimoles) for the quantity of solute present per unit volume of solution, we may use the mass of the solute in grams (or milligrams). Because it is so easy in the laboratory to measure solutions by volume (pipets, graduated cylinders), it is convenient to keep the volume units for the solution. Therefore, on the weight/volume basis we define the percent (parts per hundred) composition of the solution as follows:

$$P\% = \frac{\text{Solute mass (grams)}}{\text{Solution volume (mL)}} \cdot 100.$$

For example, a 7% solution of a solute will contain 7 grams of solute per 100 mL, or 70 grams per liter. To prepare a $P\%$ solution we simply weigh out a quantity of solute having a mass of $PV/100$, where V is the desired final volume in milliliters. The solute is dissolved and then diluted to this final volume.

Two widely used and related units for solution composition are *parts per thousand* (ppt) and *parts per million* (ppm). These are defined as follows:

$$\text{Concentration of solute in ppt} = \frac{\text{Solute mass (grams)}}{\text{Solution volume (liters)}},$$

$$\text{Concentration of solute in ppm} = \frac{\text{Solute mass (milligrams)}}{\text{Solution volume (liters)}}.$$

These units are commonly used in trace analysis involving very dilute solutions, as in environmental studies.

Preparation of Solutions of Desired Molarity

Suppose a solution of calcium chloride is needed with a molarity of 0.30. The experiments to be performed will require no more than 250 mL of this solution.

Step 1.   Calculate the quantity of $CaCl_2$ needed: $(0.25L)(0.3M) = 0.075$ mole.

Step 2.   Find what chemical form is available. Suppose the stockroom has only the dihydrate, $CaCl_2 \cdot 2H_2O$ (GFW = 147.02).

Step 3.   Weigh out the appropriate quantity: $(147.02)(0.075) = 11.0$ grams.

Step 4.   Dissolve this sample in water, and dilute to 250 mL in a volumetric flask. Mix well to ensure homogeneous composition.

MOLARITIES OF COMMERCIAL STOCK SOLUTIONS
OF ACIDS AND BASES

The commonly used acids and bases are usually available as very concentrated solutions with specifications of percentage composition and solution density. It is important to know how to calculate the molarity of these stock solutions, so that more dilute solutions may be prepared from them. For example, perchloric acid $HClO_4$ is available as a 70.5% solution having a density of 1.67 g/mL. The molecular weight of $HClO_4$ is 100.47. The molarity may be calculated by using the following steps:

1. One liter of the stock solution weighs 1670 g and contains

$$1670 \text{ g stock} \cdot \frac{0.705 \text{ g } HClO_4}{\text{g stock}} = 1178 \text{ g } HClO_4.$$

2. The number of moles of $HClO_4$ per liter (i.e., the molarity) is thus

$$1178 \text{ g/L} \cdot \frac{1 \text{ mol}}{100.47 \text{ g}} = 11.7 \text{ molar.}$$

The data for other concentrated reagents are given in Table 2.1.

Table 2.1

Compound	Molecular weight	Density	Percent by mass	Molarity
HCl, hydrochloric acid	36.46	1.19	37.2	12.1
HNO_3, nitric acid	63.01	1.42	70.4	15.9
HF, hydrofluoric acid	20.0	1.18	49.0	28.9
$HClO_4$, perchloric acid	100.47	1.67	70.5	11.7
CH_3COOH, acetic acid	60.05	1.05	99.8	17.4
HCOOH, formic acid	46.03	1.20	90.5	23.6
H_2SO_4, sulfuric acid	98.08	1.84	96.0	18.0
H_3PO_4, phosphoric acid	98.10	1.70	85.5	14.8
NH_3, ammonia	17.03	0.90	28.0	14.5
NaOH, sodium hydroxide	40.00	1.54	50.5	19.4
KOH, potassium hydroxide	56.11	1.46	45.0	11.7

PREPARATION OF A SOLUTION BY DILUTION
OF A MORE CONCENTRATED SOLUTION

Frequently a solution of a reagent is already available (as with the above commercial stock solutions) but is too concentrated for an intended purpose. Simply by adding water, one may obtain a diluted solution of the desired molarity. The following examples will show both the approximate and the precise approach to this common operation:

a) Given a stock solution of concentrated perchloric acid, it is desired to make one liter of about $2M$ acid. This requires

$$1\,L \cdot 2\ \text{mol/L} = 2\ \text{mol of } HClO_4.$$

This quantity may be obtained by measuring (in a graduated cylinder)

$$\frac{2\ \text{mol}}{11.7\ \text{mol/L}} = 0.171\ L$$

of the concentrated acid. The desired solution can be easily prepared by placing about 830 mL of water in a beaker and then adding 170 mL of the concentrated acid (with stirring).

b) Suppose that the $2M$ solution of $HClO_4$ was later standardized by titration and found to have a precise molarity of 1.984. For a research application it is now desired to have 500 mL of a precisely known solution with a molarity close to 0.080. The amount needed is

$$500\ \text{mL} \cdot 0.080\ \text{mmol/mL} = 40\ \text{mmol}.$$

This may be obtained by measuring

$$\frac{40\ \text{mmol}}{1.984\ \text{mmol/mL}} = 20.15\ \text{mL}.$$

One approach would be to pipet a portion with a calibrated 20-mL pipet into a 500-mL volumetric flask. When diluted to the mark and mixed well, the precise molarity would be

$$\frac{20.00\ \text{mL} \cdot 1.984\ \text{mmol/mL}}{500.0\ \text{mL}} = 0.0794\ M,$$

assuming that the pipet and volumetric flask were quite accurate. If one wanted the diluted solution to be precisely $0.0800\,M$, then the necessary 20.15 mL could be measured either with a calibrated buret or by using a 1-mL graduated pipet to add 0.15 mL to the 20.00 delivered by the transfer pipet.

Conclusion   When a solution of molarity M_1 is diluted to a molarity of M_2 by adding water V_w to an original volume V_1 producing a final volume V_2, then:

1. The amount (millimoles) of solute n remains unchanged, so that

$$n = M_1 V_1 = M_2 V_2.$$

2. The final volume V_2 will only approximately be equal to the sum $V_1 + V_w$, so that it is necessary to use a volumetric flask for a precise dilution.

Important Aqueous Species (alphabetical by key element symbol)

Ag^+, silver ion (argentous ion)
Ag^{2+}, silver (II) ion
   (argentic ion)
Al^{3+}, aluminum ion
H_3AsO_3, arsenous acid
H_3AsO_4, arsenic acid
Au^+, aurous ion
Au^{3+}, auric ion
H_3BO_3, boric acid
Ba^{2+}, barium ion
Br^-, bromide ion
Br_2, bromine
BrO_3^-, bromate ion
$HCOOH$, formic acid
$H_2C_2O_4$, oxalic acid
CO_3^{2-}, carbonate ion
CH_3COOH, acetic acid
Ca^{2+}, calcium ion
Cd^{2+}, cadmium ion
Ce^{3+}, cerous ion
Ce^{4+}, ceric ion
Cl^-, chloride ion
Cl_2, chlorine
$HOCl$, hypochlorous acid
ClO_4^-, perchlorate ion
Cr^{2+}, chromous ion
Cr^{3+}, chromic ion
CrO_4^{2-}, chromate ion
Cu^+, cuprous ion
Cu^{2+}, cupric ion
F^-, fluoride ion
Fe^{2+}, ferrous ion
Fe^{3+}, ferric ion
H_2, hydrogen

H_3O^+ (abbrev. H^+) hydronium
   (or hydrogen) ion
Hg_2^{2+}, mercurous ion
Hg^{2+}, mercuric ion
I^-, iodide ion
I_2, iodine
I_3^-, triiodide ion
IO_3^-, iodate ion
K^+, potassium ion
Li^+, lithium ion
Mg^{2+}, magnesium ion
Mn^{2+}, manganous ion
MnO_4^-, permanganate ion
NH_3, ammonia
N_2H_4, hydrazine
NH_2OH, hydroxylamine
$HONO$, nitrous acid
NO_3^-, nitrate ion
Na^+, sodium ion
Ni^{2+}, nickel ion
OH^-, hydroxide ion
H_2O_2, hydrogen peroxide
O_2, oxygen
H_3PO_3, phosphorous acid
H_3PO_4, phosphoric acid
Pb^{2+}, lead ion
H_2S, hydrogen sulfide
H_2SO_3, sulfurous acid
SO_4^{2-}, sulfate ion
$S_2O_3^{2-}$, thiosulfate ion
$S_4O_6^{2-}$, tetrathionate ion
Sn^{2+}, stannous ion
Sn^{4+}, stannic ion
Zn^{2+}, zinc ion

Strong Electrolytes

In working with aqueous equilibrium systems it is vital to take proper account of the dissociation of electrolytes into ions. Many compounds are only partially dissociated, but a few, the so-called **strong electrolytes**, are usually considered to be 100% dissociated into ionic species. The following list shows the most common categories of strong electrolytes and should be memorized:

A. Electrolytes with these anions:
   perchlorate, ClO_4^- bromate, BrO_3^-
   nitrate, NO_3^- iodate, IO_3^-
   bisulfate, HSO_4^- bicarbonate, HCO_3^-
   sulfate, SO_4^{2-} carbonate, CO_3^{2-}

B. Electrolytes with these cations:
   ammonium, NH_4^+ magnesium, Mg^{2+}
   lithium, Li^+ calcium, Ca^{2+}
   sodium, Na^+ strontium, Sr^{2+}
   potassium, K^+ barium, Ba^{2+}
                                       lanthanum, La^{3+}

C. The common strong acids:
   perchloric acid, $HClO_4$ hydrobromic acid, HBr
   nitric acid, HNO_3 hydriodic acid, HI
   hydrochloric acid, HCl   sulfamic acid, NH_2SO_3H
   sulfuric acid, H_2SO_4 (first dissociation step)

D. The common strong bases:
   sodium hydroxide, NaOH   potassium hydroxide, KOH

The Electroneutrality Law

An aqueous solution of several different solutes may contain a large number
of ions of various concentrations and electrostatic charges. But it is a natural
requirement that, overall, the solution must be electrically neutral. This
follows from the stoichiometric concept that dissociation of an electrolyte
into its ions must produce as many positive charges as negative charges. For
example:

$$H_2O \rightleftarrows H^+ + OH^-,$$
$$NaCl \rightarrow Na^+ + Cl^-,$$
$$Mg(NO_3)_2 \rightarrow Mg^{2+} + 2NO_3^-.$$

This requirement (that any aqueous solution must contain equal numbers
of positive and negative charges) may be expressed in terms of the molarities of
the ions carrying those charges. In pure water, the only ions are H^+ and OH^-, and
the electroneutrality law is at its simplest:

$$[H^+] = [OH^-].$$

If some sodium chloride is added to the solution, the law is written as follows:

$$[H^+] + [Na^+] = [OH^-] + [Cl^-].$$

In a solution of magnesium nitrate:

$$[H^+] + 2[Mg^{2+}] = [OH^-] + [NO_3^-].$$

Since each magnesium ion carries two charges, it is necessary to multiply its molarity by two to count properly its contribution to the positive charges present in the solution.

In general, to write the electroneutrality law one first identifies all ionic species present in the solution and then sums the molarities of all cations, each multiplied by the charge on the ion. This sum is set equal to the sum of the anion molarities, again with each multiplied by the charge.

Example Write the electroneutrality expression for a solution of ammonium dihydrogen phosphate $NH_4H_2PO_4$. The ions present are: H^+ and OH^- (present in all aqueous solutions), NH_4^+, $H_2PO_4^-$, and both HPO_4^{2-} and PO_4^{3-} (the latter two due to reaction of $H_2PO_4^-$ with the solvent). The result is as follows:

$$[H^+] + [NH_4^+] = [OH^-] + [H_2PO_4^-] + 2[HPO_4^{2-}] + 3[PO_4^{3-}].$$

2.3 ACID–BASE STOICHIOMETRY

Acid–base reactions are defined as those in which a proton (hydrogen ion H^+) is transferred from one chemical species to another. The acid species (proton donor) is thereby converted to its conjugate base and the base (proton acceptor) becomes protonated to form its conjugate acid. A simple example is:

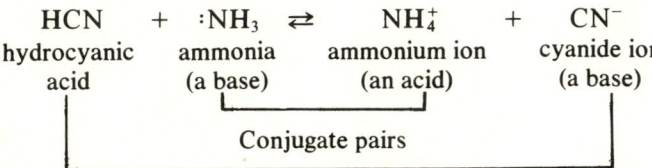

In one sense, acid–base reactions always have a $1:1$ stoichiometry (the ratio of reacting acid to base is unity) because only one proton can be transferred to each basic site (the unshared pair of electrons on the base species). However, it is certainly possible to have a series of proton transfers between species, so that the *net* stoichiometry is not $1:1$. For example, the stoichiometry of the reaction(s) between phosphoric acid (a triprotic acid) with sodium hydroxide depends on the amounts (moles) of the substances that are mixed together:

$1:1$ stoichiometry: $H_3PO_4 + OH^- \rightleftharpoons H_2PO_4^- + H_2O,$

$1:2$ stoichiometry: $H_3PO_4 + 2\,OH^- \rightleftharpoons HPO_4^{2-} + 2\,H_2O,$

$1:3$ stoichiometry: $H_3PO_4 + 3\,OH^- \rightleftharpoons PO_4^{3-} + 3\,H_2O.$

Because acid–base reactions are always reversible, the stoichiometry of such polyprotic systems is not perfectly "clean." For example, when 1.000 mmol of phosphoric acid is mixed with 2.000 mmol of sodium hydroxide (in solution), the result is not precisely 1.000 mmol of HPO_4^{2-}. The solution will contain small quantities of $H_2PO_4^-$ and PO_4^{3-}. If 1.500 mmol of sodium hydroxide had been used, the final solution would contain close to 0.500 mmol each of $H_2PO_4^-$ and HPO_4^{2-}. Nevertheless, such "variable stoichiometry" systems can be quite useful in chemical analysis because acids and bases are generally very stable and their reactions are highly reproducible.

2.4 OXIDATION–REDUCTION STOICHIOMETRY

The first problem is finding a good definition of oxidation–reduction (redox). One of the traditional statements is that a **redox reaction** is one that proceeds by transfer of electrons from a reducing agent to an oxidizing agent. It is true that many redox reactions are believed to occur through the mechanism of electron transfer. For example, the reduction of chlorine by zinc metal

$$Zn + Cl_2 \rightarrow Zn^{2+} + 2\,Cl^-$$

is explained by postulating the transfer of two electrons from the zinc atom to the chlorine molecule, which splits into the more stable singly charged chloride ions.

But it is not always that an electron transfer model is useful or valid. For example, one reaction which would be classified as redox by any chemist is the oxidation of nitrite ion by hypochlorite ion:

$$OCl^- + NO_2^- \rightarrow NO_3^- + Cl^-.$$

But in this case the mechanism is the transfer of an oxygen atom with the formation of the intermediate complex

$$[Cl—O—NO_2]^{2-}$$

which then dissociates into the nitrate ion and the chloride ion, while electron transfer is not involved.

Consider also the oxidation of hydrogen by chlorine:

$$H \colon H + \colon \ddot{C}l \colon \ddot{C}l \colon \rightarrow H \colon \ddot{C}l \colon + H \colon \ddot{C}l \colon$$

Can one say that the hydrogen has lost electrons or that the chlorine has gained electrons? Yes, but not without invoking arguments about electronegativity and the unequal sharing of the electrons in the covalent bonds. Such an argument, however, would be unwise, because it would automatically classify *all* chemical reactions as oxidation–reduction, because they all involve changes in electronic distribution in bonds.

A full and very interesting discussion of the difficulties of defining redox is given by VanderWerf who illustrates his points with some provocative examples.* His conclusion, which appears to be inescapable, is this: *redox reactions are those in which one or more elements undergo changes in oxidation state.* It is then necessary to define oxidation state, also called oxidation number, by a series of arbitrary rules summarized in Table 2.2.

All redox reactions have one requirement in common, and this is the key to oxidation–reduction stoichiometry: *The total increase in oxidation state must equal the total decrease in oxidation state.*

Journal of Chemical Education, **25, 547 (1948).*

Table 2.2   Rules for assigning oxidation numbers

Rule

1. Elements in the free state, uncombined with other elements, have oxidation numbers of zero.

2. In ions containing only one element the oxidation number of the element is equal to the ionic charge divided by the number of atoms.

3. If hydrogen is known to exist as the hydride ion: H^-, in a compound it is given an oxidation number of -1. Otherwise, bound hydrogen always has an oxidation number of $+1$.

4. If oxygen is known to be bonded to another oxygen, in the peroxide linkage —O—O—, it has an oxidation number of -1. Otherwise, oxygen is always assigned a value of -2.

5. If sulfur is known to be bonded to another sulfur, in the disulfide linkage —S—S—, it has an oxidation number of -1. In other compounds (see rules below) sulfur may have different oxidation numbers.

6. In species containing at least three atoms we may distinguish between "outer" and "central" atoms. If the outer atom is a nonmetal, it is assigned its lowest possible oxidation number, which is simply G-8, where G is the periodic-group number. This rule does not apply to hydrogen.

7. Once the outer atoms have been assigned, the inner atom is given an oxidation number such that the sum of all the numbers must equal the ionic charge on the species.

8. It will be found that carbon shows a large variety of oxidation numbers. It is best to assign all other elements first and then calculate the value for carbon according to rule 7.

9. The maximum value for an oxidation number is the periodic group number G, except for copper, silver, and gold, all of which are in group IB.

10. Ligands in metal complexes are considered to have elements in the same oxidation states as when they are not bound to the metal.

Do Oxidation Numbers Have Any Fundamental Meaning?

In the case of simple ions, such as H^+, H^-, Na^+, Al^{3+}, etc., there is a clear correlation between the oxidation numbers and the electrostatic charge on the species. But when multi-element species (with covalent bonds) are examined, there is no satisfactory relationship between the oxidation numbers and some physical reality. In the perchlorate ion, ClO_4^- for example, the arbitrary rules

---

Examples

Na, H_2, Cl_2, P_4, S_8

K^+(ox. no. +1), S^{2-}(ox. no. −2), Mg^{2+}(+2), Fe^{3+}(+3), I_3^-(−1/3), Hg_2^{2+}(+1)

Sodium hydride Na^+H^-(sodium +1; hydrogen −1) H_2O, CH_4, CH_3COOH, NH_3 (hydrogen +1 in all cases)

H_2O_2 (H—O—O—H, hydrogen peroxide: oxygen −1; hydrogen +1), Na_2O, NaOH, H_2O, CO_2, C_2H_5OH (oxygen −2 in all cases)

Cystine [$HO_2CCH(NH_2)CH_2S—$]$_2$

Thiocyanate ion SCN^-: the sulfur is assigned $(6-8) = -2$, and the nitrogen, also an outer atom, is assigned $(5-8) = -3$. Thionyl chloride O=SCl_2: oxygen is −2 and Cl is −1. Chloroform $HCCl_3$: hydrogen is +1 and Cl is −1.

SCN^-: the carbon must be +4 (see above for the S and N). O=SCl_2: the sulfur must be +4. $HCCl_3$: the carbon must be +2.

CO(+2), CO_2(+4), $C_{12}H_{22}O_{11}$(zero), CH_4(−4), CH_3Cl(−2). Cyanide ion CN^- has N(−3) and therefore C(+2)

Al^{3+} (group III, 3+), SO_4^{2-} (S is +6), ClO_4^- (Cl is +7), HNO_3 (N is +5), $S_2O_8^{2-}$ (S apparently +7, but actually is +6 because the species contains a peroxide linkage [O_3S—O—O—SO_3]$^{2-}$ and so there are two kinds of oxygen: −1 and −2). But Cu^+, Ag^+, and Au^+ are all +1, while we also know of Ag^{2+}, Cu^{2+}, and Au^{3+}.

$Cu(NH_3)_4^{2+}$ is considered as Cu^{2+}(+2) and NH_3, of which N is −3 and H is +1.

---

give the chlorine a +7 oxidation number. This certainly does *not* mean that there is a +7 electrostatic charge in the center of the ion and four −2 charges around the edges on the oxygen atoms! As another example, aside from the arbitrary rules there is no way to explain why carbon in methane CH_4 should have an oxidation number different from carbon in butane C_4H_{10}. There is no merit in trying to relate oxidation numbers to some deeper, physical idea.

Are Oxidation Numbers Useful?

YES! In spite of their arbitrary nature, oxidation numbers are an important part of chemical nomenclature.

a)   We may use them to classify reactions as redox or nonredox. If there are changes in oxidation number, then by definition the reaction involves oxidation–reduction, whether or not it involves electron transfer. In the hypochlorite–nitrite example, we find the following oxidation numbers:

	$OCl^- + NO_2^- \rightarrow Cl^- + NO_3^-$		
Chlorine	$+1$	-1	
Oxygen	-2	-2	-2
Nitrogen		$+3$	$+5$

Both the chlorine and the nitrogen have changed oxidation numbers, and so this is properly classified as a redox reaction. The dissolving of arsenic trioxide in sodium hydroxide solution goes according to the following equation:

	$As_2O_3 + 2OH^- \rightarrow 2AsO_2^- + H_2O$		
Oxygen	-2	-2	-2	-2
Hydrogen		$+1$		$+1$
Arsenic	$+3$		$+3$	

None of the elements show changes in oxidation number, and so the reaction is not a case of oxidation–reduction.

b)   Oxidation numbers are widely used in inorganic chemistry as a way of characterizing the chemical condition of an element. The oxidation number immediately shows whether an element has been oxidized ($+$ values) or reduced ($-$ values) in comparison with its elemental (0 value) chemical state. For example, chromium(VI) signifies that the chromium is in a high-oxidation state of $+6$ and therefore may be expected to serve as an oxidizing agent in a reaction. Quite often the actual species of an element in a solution may not be known, although the oxidation state can be deduced. In such a case the simple name, such as uranium(III), may be the only way to name the substance.

c)   It is an absolute rule that, in a redox reaction, the total increase and the total decrease in oxidation numbers must be equal. This is essential for experimental deductions of what products are formed (see below) and it is also useful in balancing equations with simplified symbols for the species involved. For example, permanganate ion MnO_4^- will oxidize arsenious acid H_3AsO_3 to form manganous ion Mn^{2+} and arsenic acid H_3AsO_4. The equation may be balanced in terms of these actual species, of course. But for purposes of stoichiometry alone, in a titration for example, it may be quite adequate to use the simplified symbols with oxidation numbers:

$$Mn(VII) + As(III) \rightarrow Mn(II) + As(V).$$

In oxidation number, the manganese decreases 5 units per atom and the arsenic increases 2 units per atom. The total increase must equal the total decrease. The smallest common multiple is $5 \cdot 2 = 10$, and so the balanced simplified equation must be

$$2Mn(VII) + 5As(III) \rightarrow 2Mn(II) + 5As(V).$$

Balancing the Complete Redox Equation

The best procedure is summarized in Table 2.3 and uses the idea of half-reactions based on the model of electron transfer. It is interesting that this technique works perfectly well even for redox reactions that do not actually involve electron transfer. It is also interesting that no assignment of oxidation numbers is necessary for any step in the procedure.

Deduction of Oxidation State from Experimental Stoichiometry

If the identities of all reacting species but one are known and if the experimental measurements reveal the reaction stoichiometry (reacting mole ratios), then it is possible to calculate the oxidation number of the element in the unknown species. From this information it may be possible to make a good guess of the chemical identity of that species. One example will serve for illustration: It has been found that vanadium(III) sulfate will reduce perchlorate ion to chloride ion if the mixture is boiled in $7M$ sulfuric acid in the presence of OsO_4 catalyst. The unbalanced equation is:

$$V_2(SO_4)_3 + ClO_4^- \rightarrow Cl^- + V(?),$$

where $V(?)$ indicates an unidentified reaction product species, containing vanadium in an unspecified oxidation state. If experimental measurements show that it takes 3.2 mmoles of vanadium(III) sulfate to react with 0.800 mmole of perchlorate ion, then what must be the oxidation state of the vanadium in the product?

Solution   The reacting ratio is $3.2/0.8 = 4:1$. Assuming that the product contains only one V atom per ion (or molecule), we may write

$$4V_2(SO_4)_3(s) + ClO_4^- \rightarrow Cl^- + 12SO_4^{2-} + 8V(?).$$

The oxidation number for the chlorine has changed from $+7$ in ClO_4^- to -1 in Cl^-, a total *decrease* of 8 units. Therefore the total change in vanadium's oxidation number must be an *increase* of 8 units. We note that in the reactant, vanadium(III) sulfate, the oxidation number for V is $+3$. There are 8 vanadium atoms which must share the total increase of 8 units, and so each must increase 1 unit, i.e., from $+3$ to $+4$. The oxidation number of V in the product is thereby deduced to be $+4$.

Table 2.3   Procedure for balancing equations for redox reactions

I. *Acidic-solution conditions.* For example, dichromate ion oxidizes iodide ion, forming triiodide ion and chromic ion:

$$Cr_2O_7^{2-} + I^- \rightarrow Cr^{3+} + I_3^-.$$

Step	Chromium half-reaction	Iodide half-reaction
1. Separate the whole reaction into two half-reactions:	$Cr_2O_7^{2-} \rightarrow Cr^{3+}$	$I^- \rightarrow I_3^-$
2. Balance atoms other than H and O by using coefficients:	$Cr_2O_7^{2-} \rightarrow 2Cr^{3+}$	$3I^- \rightarrow I_3^-$
3. Balance O by adding the necessary number of H_2O:	$Cr_2O_7^{2-} \rightarrow 2Cr^{3+} + 7H_2O$	not needed
4. Balance H by adding H^+:	$Cr_2O_7^{2-} + 14H^+ \rightarrow 2Cr^{3+} + 7H_2O$	not needed
5. Balance charge by adding electrons as needed:	$Cr_2O_7^{2-} + 14H^+ + 6e \rightarrow 2Cr^{3+} + 7H_2O$	$3I^- \rightarrow I_3^- + 2e$
At this point, each half-reaction is properly balanced.		
6. For each half-reaction, find the smallest multiple so that each has the same number of e:	$Cr_2O_7^{2-} + 14H^+ + 6e \rightarrow 2Cr^{3+} + 7H_2O$	$3[3I^- \rightarrow I_3^- + 2e]$
7. Add these multiples to find the desired whole reaction and cancel electrons. If both half-reactions have H_2O and H^+, consolidate as needed:		

$$Cr_2O_7^{2-} + 14H^+ + 9I^- \rightarrow 2Cr^{3+} + 3I_3^- + 7H_2O$$

II. *Basic-solution conditions.* For example, hydrogen peroxide oxidizes chromium(III), forming chromate ion and water:

$$H_2O_2 + Cr(OH)_3 \rightarrow H_2O + CrO_4^{2-}$$

Step	Peroxide half-reaction	Chromium half-reaction
1. Write half-reactions:	$H_2O_2 \rightarrow H_2O$	$Cr(OH)_3 \rightarrow CrO_4^{2-}$
2. Balance atoms other than H and O:	Not needed	Not needed
3. Balance O by adding H_2O:	$H_2O_2 \rightarrow 2H_2O$	$Cr(OH)_3 + H_2O \rightarrow CrO_4^{2-}$
4. Balance H by adding H^+:	$H_2O_2 + 2H^+ \rightarrow 2H_2O$	$Cr(OH)_3 + H_2O \rightarrow CrO_4^{2-} + 5H^+$
5. Balance charge with e:	$H_2O_2 + 2H^+ + 2e \rightarrow 2H_2O$	$Cr(OH)_3 + H_2O \rightarrow CrO_4^{2-} + 5H^+ + 3e$
At this point the reactions are balanced as if they were taking place in acidic solution.		
6. For each H^+, add OH^- to each side of the half-reaction:	$H_2O_2 + 2H^+ + 2OH^- + 2e$ $\rightarrow 2H_2O + 2OH^-$	$Cr(OH)_3 + H_2O + 5OH^-$ $\rightarrow CrO_4^{2-} + 5H^+ + 5OH^- + 3e$
7. Let the H^+ "react" with the OH^- to form H_2O, and consolidate water from both sides:	$H_2O_2 + 2e \rightarrow 2OH^-$	$Cr(OH)^3 + 5OH^- \rightarrow CrO_4^{2-} + 4H_2O + 3e$
8. Find the smallest multiples for equal numbers of e:	$3[H_2O_2 + 2e \rightarrow 2OH^-]$	$2[Cr(OH)_3 + 5OH^- \rightarrow CrO_4^{2-} + 4H_2O + 3e]$
9. Add and consolidate H_2O and OH^-, to find desired whole reaction:	$3H_2O_2 + 2Cr(OH)_3 + 4OH^- \rightarrow 2CrO_4^{2-} + 8H_2O$	

In looking at the solution chemistry of vanadium we find these species are known to exist: V^{2+}, V^{3+}, VO^{2+}, and VO_2^+. Only VO^{2+} has the required oxidation state and therefore is the product of the reaction.

2.5 STOICHIOMETRY OF TITRATIONS

In performing a titration as the determinative step in an analysis, one makes systematic and careful additions of a solution containing the **titrant** to another solution containing the sample. The goal is to add enough and only *just* enough titrant to react with the desired constituent in the sample. Again we may refer to the general balanced equation:

$$a\mathrm{A} + b\mathrm{B} \rightarrow \mathrm{Products},$$

where A is the constituent being titrated and B is the titrant added. In contrast to gravimetric determinations, where one deliberately adds an excess of precipitant, in titrations it is necessary to strive for addition of precisely b moles of titrant B for each a moles of constituent A present.

We refer to the **stoichiometric point** in a titration as the condition for this exact stoichiometry. The **endpoint** in a given titration is our experimental estimate of the stoichiometric point; it corresponds to the buret reading when we think the correct amount of titrant has been added.

The calculations related to a titration may be summarized in terms of the fundamental relationships given at the beginning of this chapter. At the stoichiometric point (also called the **equivalence point**) it must be true that

$$n_\mathrm{A} = \frac{a}{b} \cdot n_\mathrm{B}.$$

Since the titrant B is added in the form of an aqueous solution of certain molarity M_B, we have

$$n_\mathrm{B} = M_\mathrm{B} \cdot V_\mathrm{ep},$$

where V_ep is the actual buret reading at the endpoint. If it is desired to calculate the mass of the constituent, then

$$w_\mathrm{A} = n_\mathrm{A} \cdot \mathrm{GFW}_\mathrm{A}.$$

A few typical uses of the titration technique and examples of the calculations are as follows:

Application 1   *Standardizing a titrant solution by use of a known sample*

A sample of pure sulfamic acid NH_2SO_3H (MW = 97.09) weighing 1.0483 g is dissolved in 50 mL of water and titrated with a solution of sodium hydroxide of unknown molarity. The reaction is of 1:1 stoichiometry:

$$NH_2SO_3H + NaOH \rightarrow NH_2SO_3Na + H_2O.$$

It required 35.64 mL of the NaOH solution to reach the endpoint. What is the NaOH molarity?

a)   The amount of sulfamic acid taken is $1.0483/0.09709 = 10.80$ mmol.

b)   Since the reaction is $1:1$, this is also the amount of NaOH used.

c)   Therefore, the NaOH molarity is $M = 10.80/35.64 = 0.3030$.

Application 2   *Determination of the stoichiometric ratio in a reaction*

A 50.02-mL portion of a 0.0700 M oxalic acid $H_2C_2O_4$ solution is titrated with a solution of potassium permanganate $KMnO_4$ having a molarity of 0.0328. It requires 42.56 mL of the $KMnO_4$ solution to react with the oxalic acid. What coefficients should appear in the balanced equation for the reaction?

a)   The amount of oxalic acid used is $50.02 \cdot 0.0700 = 3.50$ mmol.

b)   The amount of potassium permanganate used is $42.56 \cdot 0.0328 = 1.40$ mmol.

c)   The ratio $H_2C_2O_4/MnO_4^-$ is therefore $3.50/1.40 = 2.50$.

d)   Avoiding fractional coefficients, we see that 5 mol of oxalic acid must react with 2 mol of permanganate. No further conclusion may be reached from the above data, but in fact the actual reaction is

$$5H_2C_2O_4 + 2MnO_4^- + 6H^+ \rightarrow 10CO_2 + 2Mn^{2+} + 8H_2O.$$

Application 3   *Determination of molecular weight*

An organic chemist has synthesized a new compound that is known to have *two* sulfonic acid groups and no other acidic or basic functional groups. When a sample weighing 0.364 g is dissolved in water and titrated with a standard solution of NaOH of molarity 0.0943, it requires 28.80 mL to reach the endpoint. Calculate the molecular weight of the compound.

a)   The amount of NaOH used is $0.0943 \cdot 28.80 = 2.72$ mmol.

b)   Since the compound has two acid groups, the amount present must have been half the amount of NaOH, or 1.36 mmol $= 1.36 \cdot 10^{-3}$ mol.

c)   Hence, the gram formula weight is: GFW $= 0.364/1.36 \cdot 10^{-3} = 268$ g/mol.

Application 4   *Determination of purity*

We define purity as the percentage of a certain substance in a sample that is chiefly composed of that substance. For example, magnesium chloride is available as the hexahydrate $MgCl_2 \cdot 6H_2O$, but commercial preparations are not 100% pure because they usually contain some excess water. Suppose that a sample taken from a bottle in the stockroom is found to weigh 0.1282 g and that it is titrated with $0.03443M$ silver nitrate solution according to the reaction:

$$MgCl_2 + 2AgNO_3 \rightarrow Mg(NO_3)_2 \text{ (in solution)} + 2AgCl \text{ (precipitate)}.$$

If the equivalence point is reached when 36.29 mL of the silver nitrate solution have been added, what is the calculated purity of the sample?

a) The amount of silver nitrate used was $0.03443 \cdot 36.29 = 1.249$ mmol.

b) With the 1:2 stoichiometry, there must have been half this amount of $MgCl_2$ present in the sample, or 0.6245 mmol.

c) The molecular weight of the hexahydrate is 203.3, and therefore the sample contained $203.3 \cdot 0.6245 = 127.0$ mg of $MgCl_2 \cdot 6H_2O$.

d) The purity of the sample is therefore: $(127.0/128.2) \cdot 100 = 99.1\%$.

[Note the importance of keeping track of units. The reader should go through this example and add the proper units to each numerical quantity.]

Application 5 *Standardization of a solution by titration with some other already standardized solution*

An approximate solution of phosphoric acid H_3PO_4 is prepared by diluting 5 mL of the concentrated stock (85%) to one liter with distilled water. (What is the approximate molarity?) When a 20.03-mL portion of the diluted solution is titrated with 0.0920 M potassium hydroxide, it requires 32.35 mL of the KOH to reach the second equivalence point. That is, the titration is based on the reaction

$$H_3PO_4 + 2KOH \rightarrow K_2HPO_4 + 2H_2O.$$

What is the precise molarity of the phosphoric acid solution?

a) The amount of KOH used was $32.35 \cdot 0.0920 = 2.976$ mmol.

b) The 1:2 stoichiometry indicates that half this amount, 1.488 mmol of phosphoric acid was present.

c) Therefore, the molarity of the acid is $1.488/20.03 = 0.0743$ mol/L.

2.6 IMPORTANT CRITERIA FOR TITRATION REACTIONS

It's not easy for a reaction to qualify for use in a titration. The following requirements must be met or else the titration technique will be inaccurate or even impossible. First, it is essential that the stoichiometry of the reaction be clean, meaning that the titrant reacts with the titrand in a definite b/a ratio without any loss by side reactions, decomposition, volatility, etc. For example, titration of copper ion with ammonia gives a mixture of products:

$$Cu^{2+} + ?NH_3 \rightarrow CuNH_3^{2+} + \cdots + Cu(NH_3)_4^{2+}.$$

Second, the titration reaction must be rapid and must reach equilibrium in a short time (i.e., less than a second after addition of titrant). Otherwise the titration would be impractically slow.

Third, there must be some means of knowing precisely when the stoichiometric quantity of titrant has been added, i.e., when the equivalence point has been reached. We make a useful distinction between the **true stoichiometric (or equivalence) point** in a titration and the titration **endpoint**. One always makes a judgment that the titration is complete and reads the buret to find the **endpoint volume**, i.e., the volume of titrant added to that point. This will always differ somewhat from the true equivalence-point volume, and the resulting error in the determination is:

$$\text{Titration error, } \% = \frac{V_{\text{endpoint}} - V_{\text{equiv. point}}}{V_{\text{equiv. point}}} \cdot 100.$$

Titration error may be either positive or negative. The most common source of positive errors is the addition (when the equivalence point is very near) of a couple of drops of titrant instead of the necessary fraction of one drop. The use of an unstandardized buret or an inappropriate indicator can lead to either positive or negative errors.

Finally, high precision in endpoint detection requires not only that a suitable instrument or internal indicator be applicable, but also that the titration reaction proceed nearly to completion. Otherwise the titration will not have what is called a "sharp endpoint," meaning that there is a clear signal when and only when the added volume of titrant is very close to the equivalence-point value. As an approximate rule, the reaction should be at least 99.9% complete at the equivalence point, that is, only 0.1% of the reactants remain.

Endpoint Detection

No matter how cleanly, rapidly, or extensively a titration reaction may proceed, it will not be practical unless we can devise a warning signal that will first tell when the endpoint is near and then give us the means of finding a precise value for the endpoint volume of titrant.

A simple example is the titration of triiodide ion with sodium thiosulfate solution. The reaction

$$I_3^- + 2S_2O_3^{2-} \rightarrow 3I^- + S_4O_6^{2-}$$
$$\text{yellow}$$

goes quickly, with a precise 1:2 stoichiometry and has a large equilibrium constant. The triiodide ion is the only colored species present and therefore can act as its own endpoint detection device: as the thiosulfate is added and stirred, the yellow color steadily decreases as the I_3^- is used up. The titrant may be added quickly until the color is nearly gone, and then very small increments can be added until the last trace of yellow has just disappeared. Visual sensitivity can be enhanced by adding a small amount of starch as an *indicator*, because it forms a deep blue complex with unreacted triiodide ions.

Another example is the titration of iron(II) with potassium permanganate:

$$5Fe^{2+} + MnO_4^- + 8H^+ \rightarrow 5Fe^{3+} + Mn^{2+} + 4H_2O.$$
$$\underset{\substack{\text{intense} \\ \text{purple}}}{}$$

At the early stage of titration the deep color of the permanganate disappears almost as quickly as the solution can be stirred, but one knows the endpoint is near when it takes a second or two for a small squirt to swirl around and react. The endpoint is taken as the first permanent appearance of a very pale purple-pink color, indicating that a very small excess of permanganate has been added. Of course, this implies a positive titration error, but a small one.

In both of these examples the human eye is capable of detecting when the endpoint has been reached. More dilute solutions can be titrated by means of a spectrophotometer which can measure the color of the solution. In a typical setup, the solution in the titration vessel is simply circulated through an absorption cell, and color intensity (called **absorbance**) is recorded at intervals in the titration. The endpoint can then be found by using a plot of absorbance versus volume of titrant. For the two examples given, the results are presented in Fig. 2.1.

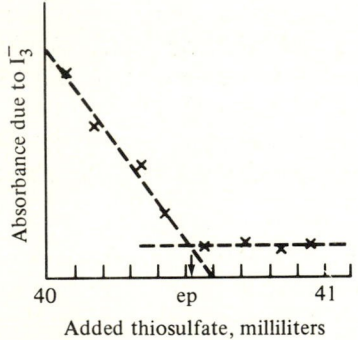 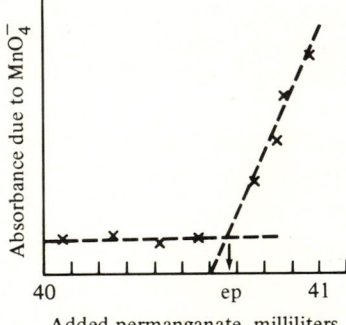

Figure 2.1

When several points are taken in the immediate vicinity of the endpoint (the sketches assume that the endpoint is between 40 and 41 mL), the intersection of the straight lines connecting the points gives a very reliable indication of the precise endpoint. This technique is often called **photometric titration**, but this is something of a misnomer. It is really **photometric endpoint detection** and may be applied to all types of reactions if a suitable wavelength can be found for monitoring.

Other instrumental methods of endpoint detection include measurement of solution resistance (conductimetric titration), electrode potential (potentiometric titration), electrolysis current (amperometric titration), and so on.

Such methods are capable of giving the most precise results possible for a titration, but they require auxiliary equipment and typically are more troublesome and time-consuming than the simple technique of using internal titration indicators.

An indicator is a highly colored substance which is added to the titration solution in such a small quantity that it may be ignored as far as stoichiometry is concerned. The outstanding property of a well-chosen indicator is that it changes color rather abruptly in the immediate vicinity of the equivalence point. As a warning, it shows fleeting color changes when the endpoint is near. The titration is finished when the indicator has changed color permanently.

Indicators are available for titrations involving acids and bases, complex formation, precipitation, and oxidation–reduction. The general principles used to explain indicator action (and hence to guide the researcher to the choice of the best indicator for a given case) are quite similar for all these types and are discussed in later chapters.

2.7   LECTURE DEMONSTRATION:
SOLUTION PREPARATION AND TITRATION

Summary A solution of potassium permanganate $KMnO_4$ is prepared by dissolving a weighed sample in distilled water. This is used to titrate a weighed sample of $Fe(NH_4)_2(SO_4)_2 \cdot 6H_2O$, ferrous ammonium sulfate (**Mohr's salt**), which is first dissolved in $1M$ sulfuric acid. Assuming that the Mohr's salt is pure and that the ferrous/permanganate stoichiometry is precisely $5:1$, we calculate the molarity of the $KMnO_4$ solution. This solution is then used to titrate a sample of a solution of arsenious acid H_3AsO_3. The final oxidation state of the arsenic is deduced from the observed stoichiometry.

1. Prepare a $0.02\ M$ $KMnO_4$ solution. Take 1.6 grams of finely ground salt (GFW = 158.04 g/mol) and demonstrate quantitative transfer from the beaker to a 500-mL volumetric flask using funnel and wash bottle. Discuss the importance of thorough mixing after dilution to the mark and the problem of knowing when the entire amount of it has dissolved since the solution is so dark. Calculate the approximate molarity from the actual weight used, but explain that the usual reagent grade salt is not 100% pure because of inherent tendency to decompose slowly, giving a small amount of MnO_2. Fill a 50-mL buret with the $KMnO_4$ solution.

2. Prepare 100 mL of $1M$ sulfuric acid. Explain the hazard of adding concentrated acid to water. Show that 5–6 mL of $18M$ H_2SO_4 is needed, and demonstrate slow addition, with stirring, to about 100 mL of water (goggles).

3. Have a previously weighed sample of Mohr's salt (GFW = 392.14 g/mol). Explain why this sample should be about 1.5 gram if it is desired to use about 40 mL of the $KMnO_4$ solution for the titration and assuming a $5:1$ stoichiometry. Show the quantitative transfer of this sample to a titration flask using the 100 mL of $1M$ sulfuric acid.

4. Titrate the ferrous ion solution with the permanganate solution. Show the technique of holding a little of the solution in a medicine dropper (until the preliminary endpoint is reached) to minimize the chance of overtitrating. Explain the advantage of permanganate as a titrant, in that it serves as its own indicator, but point out that the titration error is necessarily positive. The error can be estimated by titrating a blank (no ferrous) if desired.

5. From the titraiion endpoint, calculate the precise molarity of the $KMnO_4$ solution assuming the Mohr's salt to be 100% pure. Compare this result with the molarity expected from the weight of the $KMnO_4$ used and thereby find the purity of the $KMnO_4$ solid.

6. Have a previously prepared solution of arsenious acid, about $0.1M$ (accurately known). Demonstrate proper pipet technique and pipet 10 mL into a titration flask. Add 100 mL of water and about 10 mL of concentrated HCl. Show that this makes the HCl about $1M$. Begin to titrate with the $KMnO_4$ solution and observe the slow fading of the color. Add one drop of 0.002% KIO_3 as a catalyst and then continue the titration until the pale-purple endpoint color is observed.

From the titration result, deduce the stoichiometry of the reaction

$$a\ MnO_4^- + b\ H_3AsO_3 \rightarrow a\ Mn^{2+} + b\ \text{(As product)}.$$

From the observed values of a and b, deduce the oxidation state of the oxidized arsenic species and propose a formula for it.

2.8   LECTURE DEMONSTRATION: STOICHIOMETRY OF THE BROMATE–HYDROXYLAMINE REACTION

Summary   In a strongly acidic solution the bromate ion is reduced to bromide ion by hydroxylamine:

$$a\ BrO_3^- + b\ NH_3OH^+ \rightarrow a\ Br^- + b\ \text{(N product)}.$$

The object of this experiment is to determine the a/b ratio and then to deduce the oxidation state of nitrogen in the product and guess its probable identity.

1. To a titration flask add: 100 mL of $2M$ HCl, 5.00 mL of $0.100M$ hydroxylamine hydrochloride, and 5.00 mL of $0.250M$ potassium bromate. Stopper, swirl to mix, and let stand for 5 minutes; this is long enough for all the hydroxylamine to be used up by the bromate, which is in deliberate excess.

2. Now it is necessary to find out how much bromate ion is left over from the reaction. A two-step analysis is performed: first add about 2 g of solid potassium iodide, which causes the reaction

$$BrO_3^- + 6H^+ + 9I^- \rightarrow Br^- + 3I_3^- + 3H_2O.$$

Thus, 3 mmol of triiodide ion are formed for each millimole of bromate ion.

Finally, titrate the triiodide ion with a previously standardized solution of $0.2M$ sodium thiosulfate $Na_2S_2O_3$. The titration reaction is:

$$2S_2O_3^{2-} + I_3^- \rightarrow S_4O_6^{2-} + 3I^-.$$

Thus, it requires 6 mmol of thiosulfate for each millimole of bromate ion left over from the initial reaction with hydroxylamine. The titration endpoint is taken as the volume of sodium thiosulfate solution required to make the last bit of yellow color (caused by I_3^-) disappear.

3. Calculate the amount (millimoles) of bromate used up in the reaction with the 0.500 mmol of hydroxylamine. This gives the a/b ratio. Deduce the oxidation state and the probable identity of the nitrogen oxidation product.

PROBLEMS

Aqueous solutions*

1. What mass of solute is present in each milliliter of the following:

   a) $0.1M$ nitric acid   b) concentrated sulfuric acid
   c) 1 millimolar potassium chloride   d) $19M$ sodium hydroxide

2. What volume of the commercial concentrated reagent (see Table 2.1) should be used to prepare the following diluted solutions?

   a) 250 mL of $1.0M$ perchloric acid   b) 5 L of $20mM$ ammonia
   c) 10 mL of $8M$ hydrochloric acid   d) 0.5 L of $0.3M$ acetic acid

3. What mass of solid solute is required for preparation of the following?

   a) 1 L of $0.015M$ $Pb(NO_3)_2$ b) 2 L of 35 millimolar Na_2SO_4
   c) 100 mL of $0.50M$ $LiClO_4$ d) 250 mL of $0.10M$ KOH

4. What is the molarity, after mixing, of the following?

   a) 5.0 g $Cr(NO_3)_3 \cdot 9H_2O$ diluted to 500 mL
   b) 25 mL of concentrated H_3PO_4 diluted to 1 L
   c) 10 mL of ethanol (density 0.79 g/mL) diluted to 100 mL
   d) 0.10 mol of NH_3 gas dissolved in 350 mL of water

5. What volume of each of the following should be diluted to prepare 2 L of $0.10M$ solution?

   a) concentrated acetic acid   b) $1.5M$ HNO_3
   c) $6.7M$ NaOH   d) $0.25M$ hydrofluoric acid

6. Express the concentration of water in pure water in terms of

   a) molarity   b) molality   c) mole fraction   d) parts per thousand

7. Assuming complete ionic dissociation, what are the ionic molarities in the following?

   a) $CdCl_2$, 100 parts per million   b) $Fe(NH_4)_2(SO_4)_2$, 1 g/100 mL
   c) $La(IO_3)_3$, $1.1 \cdot 10^{-5}M$ d) Na_2CO_3, 12 millimolar

---

*Refer to Appendixes 2 and 3 for atomic and molecular weights.

Names and formulas

8. Complete the table.

Chemical name	Formula	Oxidation states
Fluoride ion		
	CaF_2	
Chloride ion		
	I^-	
Cadmium iodide		
Nitrate ion		
	HNO_3	
Perchloric acid		
Ammonium perchlorate		
Carbonate ion		
	SO_4^{2-}	
Perchlorate ion		
Iodate ion		
	copper(II) iodate	
Acetate ion		
Acetic acid		
	OH^-	
Potassium hydroxide		
Lead carbonate		
	$CrCl_3$	
	H_3PO_4	
Trisodium phosphate		
Bromate ion		
	$Ag_2C_2O_4$	
Chromate ion		
	Ag_2CrO_4	
Aluminum nitrate		
Ferrous oxide		
	$BaBr_2$	
Calcium acetate		
	$LiOH$	
Ammonium oxalate		
	$HSCN$	
Bismuth sulfate		
	$Cu(ClO_4)_2$	
Potassium permanganate		
	Hg_2I_2	

Equation balancing

Write and balance a chemical equation for each of the following reactions. Consult the Appendix for formulas of species.

9. In the presence of chloride ion, ferric ion is quantitatively reduced to ferrous ion by silver metal, the latter forming silver chloride precipitate.

10. Ceric ion will oxidize arsenious acid to arsenic acid. The cerium is reduced to cerous ion.

11. In acidic solution, bromate ion oxidizes iodide ion to form iodine and bromide ion.

12. When a dilute solution of iodide ion is treated with an excess of bromine, the iodide is converted to iodate ion and the bromine is reduced to bromide ion.

13. Oxalic acid may be titrated with a solution of permanganate ion. In acidic solution the products are carbon dioxide and manganous ion.

14. In a basic solution, chromic ion is precipitated as chromic hydroxide. The addition of hydrogen peroxide, followed by boiling, causes oxidation to chromate ion. The peroxide is reduced to water.

15. Copper metal will dissolve upon treatment with concentrated nitric acid. The solution turns blue due to cupric ion and the brown gas that is evolved is nitrogen dioxide.

16. Iron metal will dissolve in an acidic solution to form ferrous ion and hydrogen gas.

17. A useful titration of ferrous ion is based on its oxidation to ferric ion by dichromate ion, which is reduced to chromic ion in acidic solution.

18. One of the most important reactions in titration methods is the reduction of triiodide ion by thiosulfate ion, during which iodide ion and tetrathionate ion are formed.

19. One way to remove dissolved oxygen in a basic solution is to add some sulfite ion, which is oxidized to sulfate ion while the oxygen is reduced to water.

20. In the presence of chloride ion, mercuric ion is reduced to form calomel (mercurous chloride) when stannous ion is added. The tin is oxidized to stannic ion.

21. When an excess of iodide ion is added to an acidic solution of iodate ion, triiodide ion is formed.

22. Manganous ion can be oxidized to permanganate ion in a hot solution of periodate ion, which is reduced to iodate ion.

23. Chromic ion is oxidized to hydrogen chromate ion when heated with bromate ion, the latter forming bromine.

Titration stoichiometry

24. Using a standardized pipet, 1.005 mL of glacial acetic acid (nearly pure acid, called **glacial** because upon freezing it forms ice-like crystals) is pipetted into 50 mL of water in a titration flask. It requires 34.95 mL of $0.5000M$ NaOH to reach the endpoint:

$$NaOH + CH_3COOH \rightarrow Products.$$

What is the molarity of the glacial acid?

25. An unknown base sample weighing 5.3692 g is dissolved and diluted to 250 mL. A 50-mL portion requires 27.66 mL of 0.2136M HCl to reach the titration endpoint:

$$B + HCl \rightarrow Products.$$

What is the molecular weight of the base?

26. What is the percentage purity of a sample of $FeSO_4 \cdot 7H_2O$ (MW = 278.05) if a sample weighing 0.3616 g requires 12.53 mL of 0.02013M $KMnO_4$?

$$5Fe^{2+} + MnO_4^- \rightarrow Products.$$

27. What is the molarity of an EDTA solution if 48.11 mL are required to titrate a solution that is known to contain 0.1922 g of zinc?

$$Zn^{2+} + EDTA \rightarrow 1:1 \text{ complex.}$$

28. A solution of NaOH, approximately 0.15M, is to be standardized by titration of a portion of 0.06082M H_2SO_4:

$$2NaOH + H_2SO_4 \rightarrow Products.$$

What volume of the sulfuric acid solution should be pipetted if the titration endpoint is to be at about 40 mL of NaOH solution?

29. Titration of 1.00 mL of a concentrated HCl stock solution requires 9.5 mL of 1.15M NaOH. How many milliliters of the HCl stock should be used to prepare, by dilution with water, 5 L of 0.075M HCl solution?

30. Suppose an analyst has to determine the molarities of a large number of hydrogen peroxide solutions by titration with potassium permanganate:

$$H_2O_2 + MnO_4^- \rightarrow O_2 + Mn^{2+} \text{ (unbalanced).}$$

To avoid calculations, it would be nice if the buret reading at the endpoint were simply 100 times the peroxide molarity. If the peroxide samples are always 10.00 mL, what molarity of $KMnO_4$ should be used?

Redox stoichiometry and deduction of product oxidation state

31. At room temperature there is no appreciable reaction between iron(II) and nitrate ion. However, when the solution is boiled in the presence of a molybdate catalyst the nitrate ion is reduced in accordance with the following stoichiometry:

$$3Fe(II) + NO_3^- \rightarrow 3Fe(III) + ?$$

Deduce the oxidation state and probable identity of the nitrate reduction product.

32. When precipitated in colloidal form, elementary selenium can be oxidized to selenous acid H_2SeO_3 by titration with a solution of potassium bromate. For each millimole of Se present, $\frac{2}{3}$ mmol of bromate is required. What is the reduction product?

33. The reaction between hypochlorite ion and iodide ion produces chloride ion and some species containing iodine:

$$3OCl^- + I^- \rightarrow 3Cl^- + ?$$

What is the oxidation state and probable identity of the iodine product?

34. An experiment is performed to determine the stoichiometry of the reaction between hydrazine N_2H_4 and iodine. To 20.0 mL of $0.0120M$ hydrazine, 9.00 mL of $0.0800M$ iodine is added. The latter is known to be in excess, so it may be assumed that the hydrazine is completely oxidized:

$$?N_2H_4 + ?I_2(\text{excess}) \rightarrow ?? + ?I^-.$$

After the reaction, the excess iodine is determined by a titration with standard sodium thiosulfate solution according to the titration reaction:

$$I_2 + 2S_2O_3^{2-} \rightarrow 2I^- + S_4O_6^{2-}.$$

If 47.5 mL of $0.0100M$ sodium thiosulfate are required to reach the endpoint, what must be the coefficients in the equation for the hydrazine–iodine reaction? What is the oxidation state of nitrogen in the reaction product and what is the probable identity of that product?

35. When 50 mL of $0.015M$ hydroxylamine and 50 mL of $0.025M$ potassium bromate are mixed and made strongly acidic, a redox reaction occurs:

$$?NH_2OH + ?BrO_3^- \rightarrow ?Br^- + ?$$

There is an excess of bromate ion in this mixture and it may be assumed that the hydroxylamine is completely oxidized. After the reaction has ceased, the amount of excess bromate is determined iodometrically. That is, first an excess of potassium iodide is added to reduce the bromate:

$$BrO_3^- + I^- \rightarrow Br^- + I_3^- \quad (\text{unbalanced}).$$

Then a standard solution of $0.100M$ sodium thiosulfate is used to titrate the triiodide ion according to the reaction:

$$2S_2O_3^{2-} + I_3^- \rightarrow S_4O_6^{2-} + 3I^-.$$

If 30.0 mL of the sodium thiosulfate solution were required to reach the endpoint, what must have been the reacting ratio between hydroxylamine and bromate and what is the probable identity of the nitrogen-containing product?

36. A basic solution of bismuthite ion BiO_2^- will oxidize vanadyl ion VO^{2+} to vanadate ion VO_3^-. If it is found that 20.0 mL of $0.050M$ sodium bismuthite are required to react with 10.0 mL of $0.30M$ vanadyl chloride solution, what is the final state of the bismuth?

37. A solution of vanadium in the +5 oxidation state was prepared by dissolving 8.25 mmol of V_2O_5 in strong acid and diluting to 250 mL. A 25-mL portion of the solution was then treated by bubbling SO_2 gas until the color turned to a clear blue, and the excess SO_2 was removed by boiling. Then the vanadium was oxidized back to the +5 state by titration with $0.0200M$ potassium permanganate of which 16.5 mL was required:

$$?MnO_4^- + ?V(\text{unknown state}) \rightarrow ?Mn^{2+} + ?VO_3^-.$$

What is the oxidation state of the vanadium in the clear blue solution?

LIMITATIONS OF EXPERIMENTAL MEASUREMENTS

3

... in this world nothing is certain but death and taxes

Ben Franklin, 1789

3.1   ACCURACY AND PRECISION

Scientists have a natural desire to incorporate their laboratory data into Franklin's brief list of certainties, but it is even more natural that this should be impossible. Typical laboratory measurements seek to quantify the values for variables which are continuous in nature, and any attempt to measure a continuous quantity with exactitude will be subject to error (or at least doubt) for several reasons: a) there are limits to human capabilities of observation and judgement, b) no instrument can be engineered to give an absolutely correct readout, c) the material under study may be changing with time, even during the experiment, and d) the technical measurement is usually subject to variation due to interferences from substances or phenomena which are not part of the system being studied.

      Therefore it is essential to adopt a critical, even suspicious attitude toward quantitative data, especially one's own data, and particularly when it

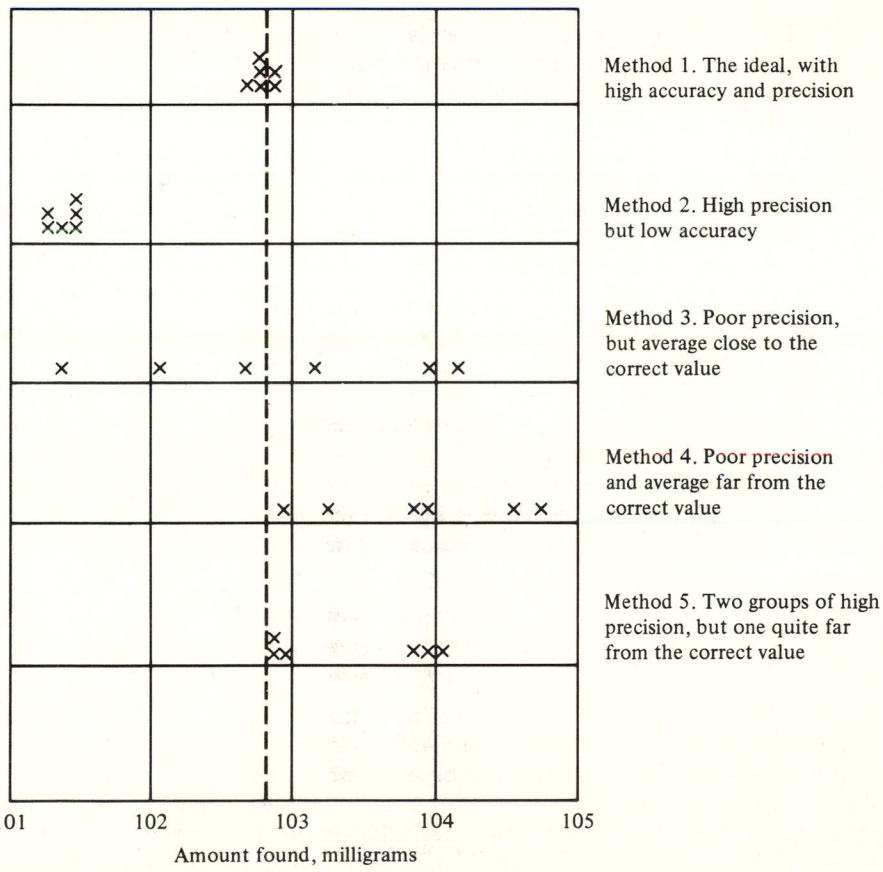

Method 1. The ideal, with high accuracy and precision

Method 2. High precision but low accuracy

Method 3. Poor precision, but average close to the correct value

Method 4. Poor precision and average far from the correct value

Method 5. Two groups of high precision, but one quite far from the correct value

101   102   103   104   105

Amount found, milligrams

Figure 3.1

appears to be superb. The general problem has been reviewed by John Gall in his book *Systemantics: How Systems Work and Especially How They Fail* (Pocket Books, 1978). Beginning with the observation that "things aren't working very well," along with the fundamental theorem (known as Murphy's Law) "If anything can go wrong it will", Mr. Gall offers a humorous but perceptive discussion of how and why complex systems display antics.

In this chapter we cannot treat the important topic of how one should design highly reliable experimental procedures. We will deal only with the evaluation of data already obtained, our aim being to show how one may view results objectively. We will consider accuracy, i.e., the extent to which a result agrees with what is believed to be the true value of the quantity. We will also discuss precision, i.e., the extent to which a group of repeated measurements agree with their average, or mean, value. We will consider ways to assign confidence limits, or uncertainties, to experimental quantities.

Ideally, a procedure should be both precise and accurate, giving results that are consistently very close to the truth. This ideal and deviations from it are considered in the chart of Fig. 3.1. It shows hypothetical results obtained by five different methods of determining the mass of vitamin C in pharmaceutical tablets; the presumed correct value is 102.8 mg and six measurements (indicated by x marks) are obtained with each procedure.

3.2   TWO TYPES OF ERROR

A careless laboratory worker can ruin any procedure, no matter how reliable it may be otherwise. Large errors may occur through failure to mix solutions thoroughly, by spilling or contamination, by using the wrong amount of a reagent, by inverting the order of additions of reagents, by reading scales wrong, and so on. Within the scope of this book we are not able to treat these types of errors systematically. In the following discussion it is assumed that human errors are reduced to a minimum expected from skilled analysts. In addition to this residual human factor, there will be errors due to slight malfunction of instruments and errors inherent in the design of the particular analytical procedure used. It is convenient to discuss two main types of error: systematic errors and random errors.

Systematic Errors

A systematic, or determinate, error is one that consistently causes the result to be wrong in the same direction. For example, part of a precipitate collected in a filter crucible will always be lost since the process of transfer from beaker to crucible involves rinsing with water, causing some loss due to the slight solubility of the precipitate. On the other hand, some precipitates retain absorbed water so tenaciously that, even after hours in the oven, they weigh a little too much.

When systematic errors are essentially constant from sample to sample, the results may be as shown in Fig. 3.1 for method 2. The precision may be high, but there is a constant deviation from the true value.

If the errors in a procedure are subject to variation, perhaps because some factor (such as a waiting time or temperature) is not under control, the results may be as shown for methods 3, 4, or 5. This situation calls for research of the procedure to identify and correct the problems.

The results for method 5 deserve further comment. There is a systematic error in three of the results only. The analyst should look for differences in the procedure. For example, perhaps there were two bottles of a certain reagent, used indiscriminately. If one bottle had become contaminated, or was old, it might be the source of the problem.

Random Errors

Even when the systematic errors are under control, as illustrated by the results for Method 1, there will always be some lack of agreement in the results. There are numerous sources of random, or indeterminate, errors, which sometimes cause overestimated results and other times cause underestimated results. Examples include such unpredictable things as building vibrations (which can affect the operation of sensitive optical equipment), atmospheric conditions such as temperature, pressure, and humidity (which can affect accurate weighing due to buoyancy changes and films of moisture), variation in the power line voltage (which might cause an error in a digital readout), human judgment in reading a liquid meniscus in a buret or a needle position on a meter, slight variations in solution pH due to small differences in the reagents added to a reacting mixture, and many other sources. A skilled analyst has learned tricks of the trade and subtleties in technique to keep these problems to a minimum. Some procedures have been refined to the point where neither systematic nor random errors are large enough to affect the results significantly, given the purpose of the analysis. For example, it is not difficult to standardize a solution of hydrochloric acid to any desired degree of accuracy.

Consider methods 1 and 3 in the above chart. In discussing random errors we are concerned with the slight spread of values around the true value as shown in method 1. Since the average of the values for method 3 is also close to the true value, one might think that the spread of individual values may be attributed to random error. This may be true, but it is more realistic to regard such large variations as a sign of systematic errors which are not under control. Of course, it *is* possible to have large random errors, as with a defective and erratic electronic instrument, but we should expect that this sort of problem would show up during routine checking of the instrument under controlled conditions.

One final comment about the results of method 3. Just because the average of several poor results happens to be close to the correct value, this is no justification for classifying this method as an accurate one. With a small number of determinations it would be quite fortuitous for the average to be close to that obtained with a large number of results. *It is possible to have high precision with low accuracy, but it is not likely that high accuracy will be achieved (for the average value) when the precision is low.*

3.3   STATISTICAL TREATMENT OF DATA

In the following discussion we deal with N attempts to measure a certain quantity, the true value of which is T. Each of the N attempts yields one experimental estimate of T called x. For our statistical interpretations to be reliable, the value of N should be large, perhaps greater than 30. More typically we deal with analytical determinations on only three to five samples, and our statistical inferences are rather shaky.

The Average, or Mean, Value

Given N values of x, we define the average $\bar{x}$ as:

$$\bar{x} = \frac{x_1 + x_2 + x_3 + \cdots + x_N}{N} = \frac{\Sigma x}{N}.$$

Example Five determinations of the percentage of chloride in a sample of pure sodium chloride gave us 60.53, 60.63, 60.64, 60.61, and 60.59. Show that the average value is 60.60%. As N becomes very large approaching infinity, $\bar{x}$ approaches a limiting value μ.

Individual Deviation from the Average

Once the average value for a series of results is calculated, the deviation d of each individual value may be calculated:

$$d = x - \bar{x}.$$

Example For the above data, the individual deviations are: -0.07, $+0.03$, $+0.04$, $+0.01$, and -0.01.

Average Deviation

If the deviations are simply averaged in the usual way, the result would always be zero because of the plus and minus signs. To find the average deviation $\bar{d}$ it is necessary to *ignore* the signs and to use the absolute values:

$$\bar{d} \equiv \frac{\Sigma |d|}{N}.$$

Example The average deviation for the above data is calculated as follows:

$$\bar{d} = \frac{0.07 + 0.03 + 0.04 + 0.01 + 0.01}{5} = 0.03.$$

The correct name used for this quantity is **average absolute deviation**, but it is widespread practice to call it simply **average deviation** or **deviation**. The average deviation is one way to express the precision of a series of measurements; it gives equal weight to each individual deviation and certainly is easy to compute. However, for some statistical purposes it is preferable to use the standard deviation (see later).

The Median

If a series of values is arranged in increasing order, the **median** value is simply the middle value, with an equal number of values above and below it. If the data set has an even number of values, the median is taken as the average of the two middle values. For the above data set, the median value is 60.61%.

The Range

A simple measure of precision is the **range**, which is the difference between the highest and lowest values in a data set. By itself, the range is only a crude indication of the precision of the set of measurements. However, the range is useful in applying the *Q-test* to see whether a suspicious value should be rejected from the set.

Rejection of a Suspicious Value: the *Q*-test

Often an analyst will finish a series of measurements and note that one of the values is an outlier, seemingly far out of agreement with the rest of the values. The only truly sound basis for discarding this value is the knowledge that something went wrong with that particular determination. For example, it may have been noted that a bit of the solution was spilled, or that it was accidentally boiled when the directions called for only gentle warming, etc. It is especially important to check all calculations carefully to see if a numerical mistake has been made. If no such error can be blamed for the discordant result then it is reasonable to apply the *Q*-test.*

One first calculates the value of *Q*, defined as follows:

$$Q = \frac{(\text{Suspicious value}) - (\text{Nearest neighbor value})}{\text{Range}}.$$

If the absolute (ignoring sign) value of *Q* is greater than that shown in Table 3.1 for the appropriate number of data values, then the suspicious value may be rejected with 90% confidence that the decision is correct. If the *Q*-test indicates that the suspicious value should be retained as a valid member of the set, it is recommended that the final result be reported as the median, rather than the average of the values.

---

*R. B. Dean, W. J. Dixon, *Analytical Chemistry*, **23**, 636 (1951).

Once the Q-test has been used to reject value, it is not valid to use it again to see if a second value might be rejected also.

Table 3.1   Minimum Q values for 90% confidence in rejection

N	Q
3	0.94
4	0.76
5	0.64
6	0.56
7	0.51
8	0.47
9	0.44
10	0.41

Example We note that in the set of five values for the percentage of chloride 60.53 has the largest individual deviation from the average. The Q-test is applied:

$$Q = \frac{60.59 - 60.53}{60.64 - 60.53} = 0.55.$$

For $N = 5$, the table shows that Q must be at least 0.64 for the rejection to be justified. Therefore we keep the value of 60.53 and include it with all the calculations of the average, average deviation, etc.

Error and Average Error

The *error* E of an individual measurement is simply its difference from the true value:

$$E = x - T;$$

for N determinations we may calculate the *average error*:

$$\bar{E} = \frac{\Sigma E}{N} = \frac{\Sigma x}{N} - T = \bar{x} - T.$$

Of course, if the quantity we are attempting to measure is really an unknown, we have no way of knowing the true value T. However it is often possible to test a procedure by using a sample or system so well characterized by other studies that the true value of the measured quantity can be estimated with high confidence.

Note that there are no absolute-value symbols in the definition of $\bar{E}$ and so the positive and negative errors will tend to cancel, unless there is some systematic error that consistently throws the observations off in one direction.

If the errors are random, we expect $\bar{E}$ to be small, especially if the data consist of many values. If $\bar{E}$ is significantly different from zero even for large data sets, this is an indication that the procedure has some bias due to a systematic error.

Example It was stated that pure sodium chloride was the sample used for the five determinations described above. The true value for the percentage of chloride may therefore be calculated from atomic weights:

$$\frac{\text{AW chlorine}}{\text{MW sodium chloride}} \cdot 100 = \frac{35.453}{58.443} \cdot 100 = 60.663\%\text{Cl.}$$

The average absolute error is:

$$\bar{E} = \bar{x} - T = 60.60 - 60.663 = -0.06\%.$$

Such a small average error of only one part per thousand indicates work of high quality. It may also indicate that there is a small systematic error in the procedure and/or that the sample of sodium chloride may not be quite as pure as assumed.

Relative Values of Deviation and Error

It is common practice to express the deviation (or error) of an experimental result in comparison with the measured quantity, rather than in absolute terms. For example, if the mass of an object is reported as 8.5327 grams (average of five weighings) with an absolute average deviation of 0.0005 gram, then the **relative average deviation** is calculated as follows:

$$\bar{d}_{rel} = \frac{\bar{d}_{abs}}{\bar{x}} \cdot M = \frac{0.0005}{8.5327} \cdot M = 6 \cdot 10^{-5} \cdot M,$$

where M is a multiplier equal to 100 if the relative deviation is desired in terms of percentage. To express d_{rel} in terms of parts per thousand, use $M = 1000$.

A similar approach is used for expressing relative error, except that the absolute error is compared with the true value for the quantity being measured:

$$E_{rel} = \frac{E_{abs}}{T} \cdot M.$$

Thus, for the five chloride determinations used in preceding examples,

$$\bar{d}_{rel} = \frac{0.03}{60.60} \cdot 1000 = 0.5 \text{ ppt,}$$

$$\bar{E}_{rel} = \frac{-0.06}{60.66} \cdot 1000 = -1 \text{ ppt.}$$

Note that the algebraic sign must be shown for the error calculation.

Standard Deviation

Although the average deviation (see above) is frequently used as an indication of the precision of a set of data and is very easy to compute, it is statistically more appropriate to use the **sample standard deviation** s which is defined by the formula:

$$s = \sqrt{\frac{\Sigma(x - \bar{x})^2}{N - 1}} = \sqrt{\frac{\Sigma d^2}{N - 1}}, \tag{3.1}$$

or, stated in words, the sample standard deviation is the square root of the sum of the squares of the deviations, divided by one less than the number of data values in the set.

For small data sets consisting, for example, of 3 to 5 points, it does not particularly matter whether one uses average deviation or sample standard deviation. For larger data sets there is a convenient relationship between s and $\bar{d}$ that holds quite well:

$$s \approx 1.25 \, \bar{d}. \tag{3.2}$$

It is a little easier to calculate the average deviation and then to multiply it by 1.25, than to calculate the standard deviation directly.

Example Referring to the chloride data again, we have:

$|d|$	d^2
0.07	0.0049
0.03	0.0009
0.04	0.0016
0.01	0.0001
0.01	0.0001

$$\bar{d} = 0.032 \qquad 0.0076 = \Sigma d^2$$

From Eq. (3.1) we get $s = \sqrt{0.0076/(5 - 1)} = 0.044 = 0.04$. By using Eq. (3.2) for comparison, we again get $1.25 \, \bar{d} = 0.04$.

3.4 PATTERNS SHOWN BY RANDOM DEVIATIONS

Let us assume that there exists a 50-mL pipet that delivers 50.015 mL of water if all random errors happen to be zero. Even in the hands of a skilled technician this pipet will actually deliver varying volumes with deviations up to several hundreths of a milliliter on each side of the average value. But it is more likely that a given deviation will be small (less than 0.02 mL) rather than large (greater than 0.06 mL), partly because of the skill of the technician and partly because of the tendency for the various random errors to offset each other. A large deviation could be caused by the random errors being simultaneously at their worst and in the same direction, which is a less likely situation.

Suppose the technician patiently makes repeated determinations of the volume of water delivered in successive trials. The following set of histograms shows what might occur with subtotals of 4, 20, 50, and 200 trials (here one × was used for each result). These "data" were generated by a computer program.

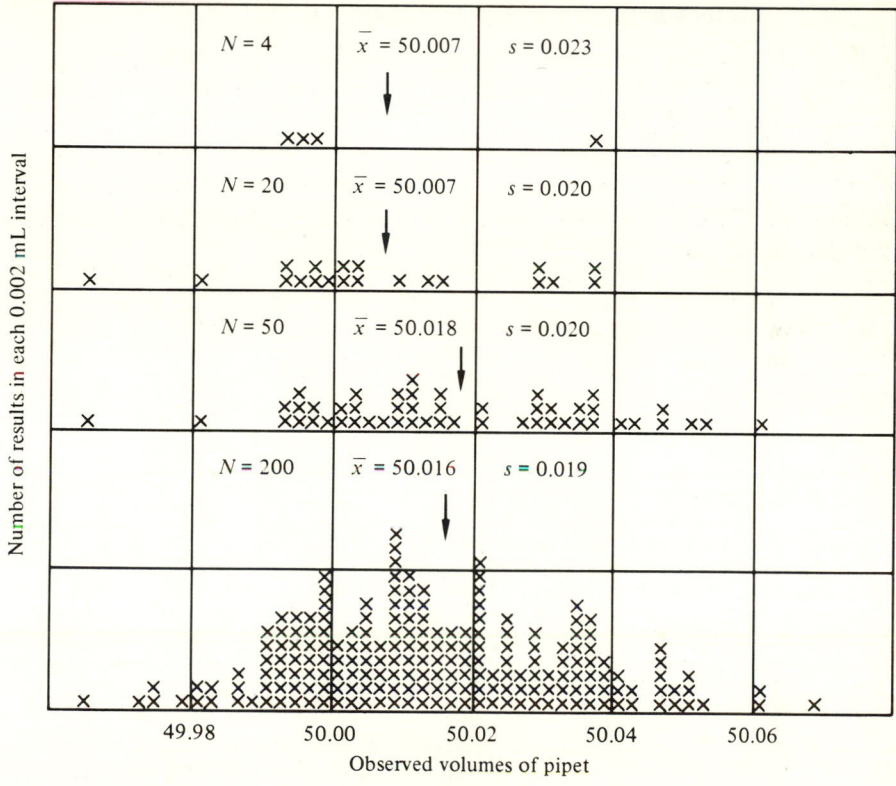

Observed volumes of pipet

Figure 3.2

The numerical quantities shown with each histogram are the average value $\bar{x}$ for the set of volumes and the standard deviation s. The latter is a measure of the precision for the set.

Consider the first four values obtained. Three of them cluster tightly around 49.994 mL, while one apparent outlier is at 50.036. The overall average for the four values is 50.007, but few chemists would hesitate to reject the high value on the assumption that it was due to an unseen error. The folly of such a decision may be shown only by making additional measurements, as in the 20-value set where it appears that it was merely a fluke that the three values agreed so well with each other.

By the time 50 values have accumulated, there are several high values in the set, and we begin to see a pattern that suggests, as assumed earlier, that there is a higher probability for small, rather than large deviations from the average, which now has become 50.018 mL.

By sticking doggedly to the task, the technician piles up 200 values for the histogram shown at the bottom of the figure.

A few rather "bad" values have appeared at the high and low extremes, but most of the time the results are close to the average, and we might even imagine that a somewhat symmetrical distribution of deviations is evident. The average of all these data, 50.016 mL, has become rather stable and quite close to the true value of 50.015 mL. Note that the value for s is also rather stable.

In the interest of science and per order of the lab director the technician eventually turns in a final histogram consisting of no fewer than 1000 determinations, a prodigious (if foolish) accomplishment. When these are tallied in intervals of 0.002 mL, the histogram acquires the form shown in Fig. 3.3.

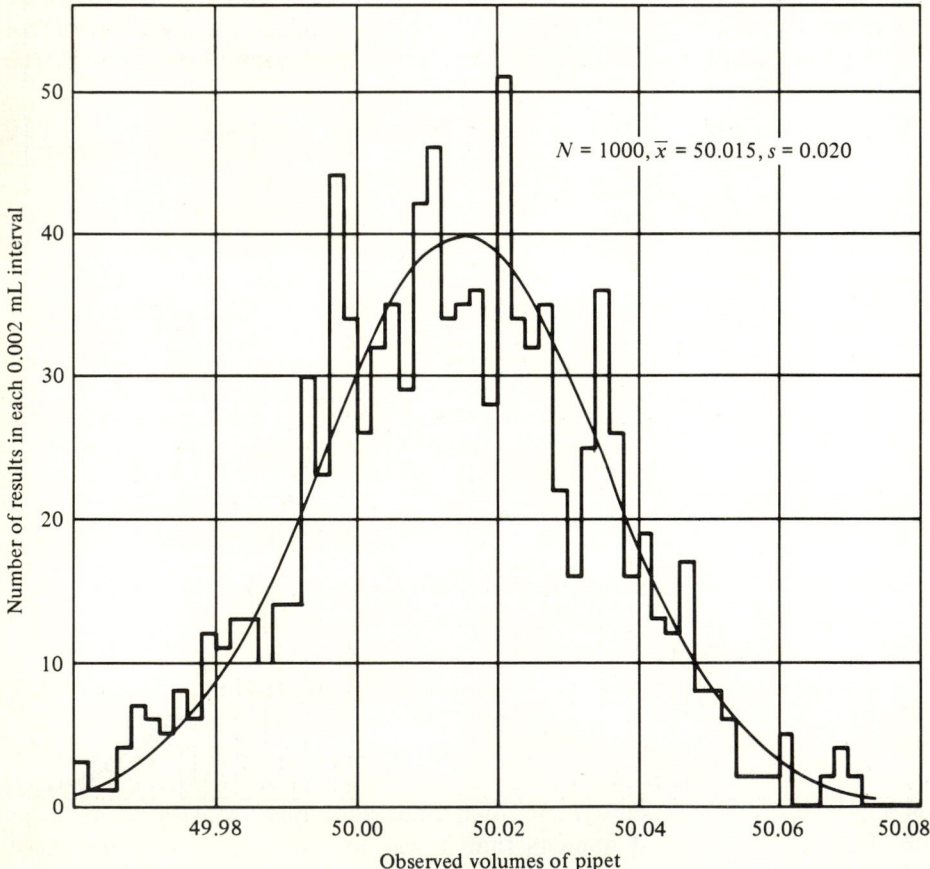

Figure 3.3

Superimposed on the histogram is a smooth bell-shaped curve calculated according to the distribution equation devised by Karl Friedrich Gauss (1777–1855). Called either the **gaussian** distribution, or the **normal distribution**, this mathematical model for the distribution of random errors has been widely adopted by mathematicians, social scientists, biologists, chemists, physicists, etc. Even though the 1000-point histogram is ragged, it certainly does not seem inconsistent with the gaussian model. Of course, there are numerous other curves which could describe the data just as well, *and in most experimental work the true distribution function remains unknown because too few measurements are taken to allow the pattern to smooth out.* Nevertheless, it is commonly assumed that the gaussian model is applicable unless there is a specific reason for choosing an alternative.

The Gaussian Distribution Function

Given a mean value μ for a very large set and a value for the standard deviation s, we can predict the distribution of results as a function of the deviation of values from the mean. One widely used model is provided by the Gauss equation:

$$P = \frac{100}{s\sqrt{2\pi}} e^{-(x-\mu)^2/2s^2} \, dx,$$

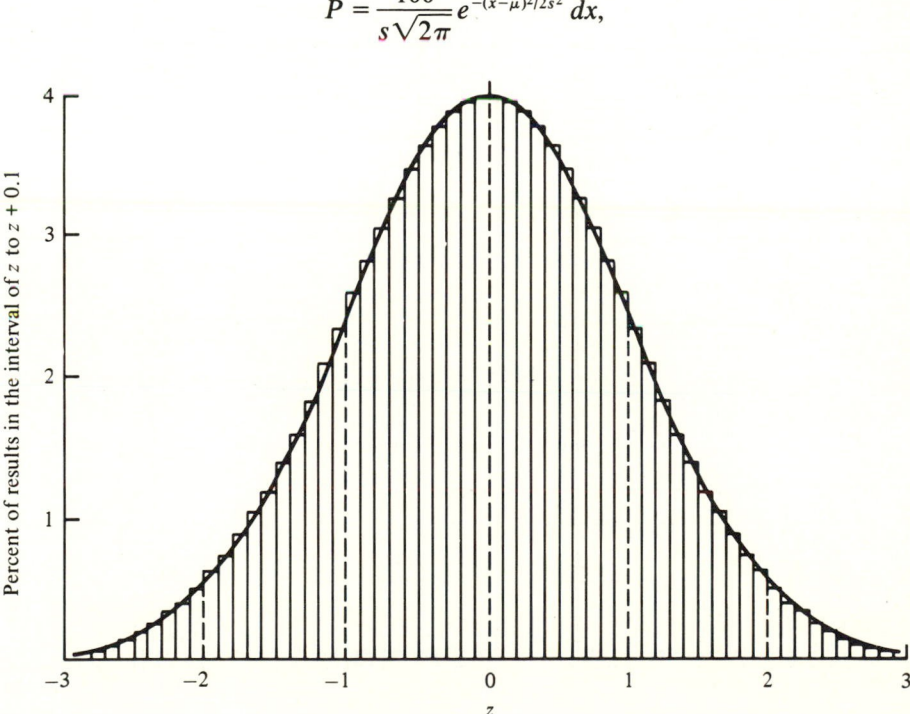

Figure 3.4

where P is the percentage of values that should fall in the interval between x and $x + dx$, and dx is small compared to s.

According to this equation, the horizontal axis for the histogram must represent the value of the quantity x. All histograms based on this model can be normalized by defining the ratio z:

$$z = (x - \mu)/s.$$

We see that z is the ratio of an individual deviation from the mean to the standard deviation. The normalized equation is as follows:

$$P = \frac{100}{\sqrt{2\pi}} e^{-z^2/2} dz. \tag{3.3}$$

An example of a normalized histogram with $dz = 0.1$ is depicted in Fig. 3.4. As the value for dz is made smaller, the sawtooth appearance becomes less pronounced and the histogram approaches the smooth curve. It is useful to consider the percentage of all values that may be expected to fall within certain limits from the mean value. To calculate such results from Eq. (3.3) it is necessary to use the techniques of numerical integration. One may find extensive tables of results in handbooks, and only a few examples are shown in Table 3.2

Table 3.2

Value of z	Percentage of results between $-z$ and $+z$
0.01	0.8
0.02	1.6
0.05	4
0.1	8
0.2	16
0.5	38
0.67	50
1	68
1.65	90
2	95.5
3	99.7

The table provides direct answers to a number of significant questions (see also Fig. 3.5).

a) What percentage of values may be expected to fall within the standard deviation from the mean value? *Answer*: About 68%.

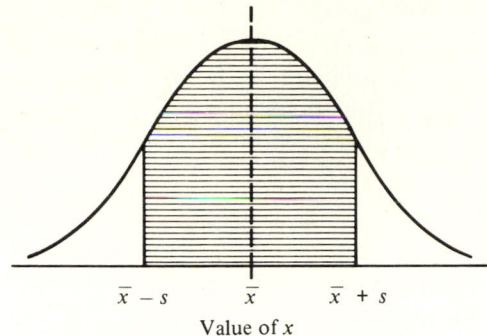

Fig. 3.5 The shaded part contains 68% of results.

b) If a pipet delivers, on the average, 25.04 mL with a standard deviation of 0.03 mL, what is the probability that it will deliver more than 25.10 mL on occasion? *Answer*: The deviation would be greater than +0.06, which is twice the standard deviation. Since 95.5% of the results should be within $\pm 2s$ only $4.5/2 = 2.3\%$ of results should exceed the mean by more than $+2s$, or +0.06 mL.

c) How close to the mean do we expect half the results to be? *Answer*: They should be within about $\pm 0.67s$, regardless of what the value for s happens to be.

The Uncertainty Interval of an Experimental Average Result

It is only when a very large number of results is obtained that the experimental average value $\bar{x}$ can be expected to approach a limiting and constant value called μ. In the case of the 50.015-mL pipet discussed earlier, the limiting value of 50.015 was obtained (at least as a value rounded to three decimal places) when 1000 data points had been accumulated. In general, with a very large number of results we may feel very confident that the mean $\bar{x}$ is a very good estimate of μ, so that

$$\mu = \bar{x}.$$

More often we must deal with small data sets comprising only three to five values. It is not realistic to assume that $\bar{x}$ is then a reliable estimate of μ. Our doubts about $\bar{x}$ can be expressed in terms of an *uncertainty interval* Δ as follows:

$$\mu = \bar{x} \pm \Delta$$

Putting this idea in words, we say that there is a certain probability that the true mean μ (which we would obtain from a very large number of measurements) lies in the interval between $(\bar{x} - \Delta)$ and $(\bar{x} + \Delta)$, called the **confidence limits**.

An approach for calculating values for the uncertainty interval was devised in 1908 by W. S. Gosset, who wrote under the pen name of Student. He proposed the use of a parameter t whose value depends not only on the size of the data set, but also on the degree of confidence desired by the experimenter. The value for the uncertainty interval is calculated by relationship:

$$\Delta = \frac{t \cdot s}{\sqrt{N}},$$

where N is the number of values in the data set, s is the standard deviation calculated for that set, and the value for t is found from Table 3.3.

Table 3.3 Values of t for given confidence

Degrees of freedom	N	99%	95%	90%	50%
1	2	63.7	12.7	6.3	1.0
2	3	9.9	4.3	2.9	0.82
3	4	5.8	3.2	2.4	0.77
4	5	4.6	2.8	2.1	0.74
5	6	4.0	2.6	2.0	0.73
6	7	3.7	2.5	1.9	0.72
7	8	3.5	2.4	1.9	0.71
8	9	3.4	2.3	1.9	0.71
9	10	3.3	2.3	1.8	0.70
	...				
	∞	2.58	1.96	1.65	0.67

Example calculation A pipet is found to deliver 9.992 mL of water as an average of eight trials, the standard deviation being 0.007 mL. Calculate the 95% confidence limits.

From the table of t values we find for $N = 8$ and 95% confidence that $t = 2.4$. Therefore,

$$V_{\text{pipet}} = 9.992 \pm \frac{2.4 \cdot 0.007}{\sqrt{8}} = 9.992 \pm 0.006 \text{ mL}.$$

SIMPLE APPROXIMATIONS FOR Δ

To express 90% confidence limits for small data sets, one may as well use either the average deviation or the standard deviation. For $N = 5$,

$$\Delta = \pm \frac{2.1 \, s}{\sqrt{5}} = 0.94 \, s \approx s.$$

Since it is also true that $s = 1.25 \, \bar{d}$ (at least for large sets), we see that

$$\Delta = 0.94 \cdot 1.25 \cdot \bar{d} = 1.17 \, \bar{d} \approx \bar{d}.$$

Because the s value for small data sets is rather uncertain anyway, these approximations are valid.

Calculations Involving Uncertainties

Virtually all calculations of chemical analysis involve the use of imprecise experimental quantities with a $\pm\Delta$ attached. We still carry out the usual addition, subraction, multiplication, and division of the quantities, but it is important to deal correctly with the uncertainties, so that the final calculated result will have a realistic uncertainty interval. The reader probably has learned to deal with significant digits for calculations, and that approach is a useful, if crude, version of the following.

ADDITION AND SUBTRACTION

Suppose it is desired to find the result R which is the algebraic sum

$$R = A + B - C, \tag{3.4}$$

where A, B, and C are experimentally measured quantities. Actually, we are concerned with the operation

$$R \pm r = A \pm a + B \pm b - C \pm c,$$

where the small letters indicate the uncertainty of each quantity. The uncertainty might be expressed as Δ, or s, or $\bar{d}$.

The desired value for r is found by the formula

$$r = \sqrt{a^2 + b^2 + c^2}. \tag{3.5}$$

Note that the squares of the individual uncertainties are *added*, even though C is subtracted in the calculation of R in Eq. (3.4).

Equation (3.5) can be extended to as many quantities as desired. To illustrate, let us find the sum of the four quantities: (1.06 ± 0.02), (2.20 ± 0.03), (0.55 ± 0.01), and (0.00924 ± 0.00008). From Eq. (3.4), the nominal sum, rounded to two decimal places, is 3.82. From Eq. (3.5) we find the uncertainty:

$$r = \sqrt{0.02^2 + 0.03^2 + 0.01^2 + 0.00008^2} = 0.037.$$

Since it is rarely justified to show more than one significant digit in an uncertainty, our final result is

$$\text{Sum} = 3.82 \pm 0.04.$$

MULTIPLICATION AND DIVISION

Suppose we desire to find the result R of the calculation:

$$R = A \cdot B/C.$$

Again, let each of the quantities have uncertainties r, a, b, c, respectively. Given the values for a, b, and c, we find the value for r by using an equation involving the *relative* uncertainties:

$$r = R \cdot \sqrt{\left(\frac{a}{A}\right)^2 + \left(\frac{b}{B}\right)^2 + \left(\frac{c}{C}\right)^2}. \tag{3.6}$$

To illustrate, suppose a solution of copper sulfate has been accurately standardized, the molarity being 0.1284 ± 0.0001. A 25-mL pipet of class A is used to deliver a portion of the solution into a 250-mL volumetric flask, also of class A. The manufacturer's specifications on the glassware are ± 0.03 mL for the pipet and ± 0.12 mL for the flask. We first do the nominal calculation of the diluted concentration:

$$\text{Concentration} = 0.1284 \cdot 25/250 = 0.01284 \; M.$$

Then the uncertainty is found by using Eq. (3.6):

$$0.01284 \cdot \sqrt{\left(\frac{0.0001}{0.1284}\right)^2 + \left(\frac{0.03}{25}\right)^2 + \left(\frac{0.12}{250}\right)^2} = 1.9 \cdot 10^{-5}.$$

Therefore, the final result should be written as

$$\text{Concentration} = (1.284 \pm 0.002) \cdot 10^{-2} \; M.$$

COMBINATIONS OF ADDITION AND MULTIPLICATION

When the result to be calculated involves successive operations, as in the relationship

$$R = A \cdot B + C(D + E),$$

we go one step at a time. In the present example, first find $(D + E)$ and its uncertainty. Then find the results for the two products. Finally, carry out the addition.

Example calculation Find the value for Y and its uncertainty if

$$Y = (3.16 \pm 0.01)(7.5 \pm 0.2) + (16.9 \pm 0.4).$$

First we deal with the multiplication

$$3.16 \cdot 7.5 = 23.70.$$

Then we find its uncertainty:

$$\text{Uncertainty} = 23.7 \sqrt{\left(\frac{0.01}{3.16}\right)^2 + \left(\frac{0.2}{7.5}\right)^2} = 0.64.$$

Now the problem is only to deal with the addition:

$$Y = (23.70 \pm 0.64) + (16.9 \pm 0.4).$$
$$\text{Nominal value} = 23.70 + 16.9 = 40.6.$$
$$\text{Uncertainty} = \sqrt{0.64^2 + 0.4^2} = 0.75 \approx 0.8 \; (\text{rounded}).$$

Therefore, the final result of the calculation is:

$$Y = 40.6 \pm 0.8.$$

SIMPLE APPROXIMATE APPROACH

The foregoing method of finding the square root of the sum of the squares of the relative uncertainties is often more troublesome than it is worth. When several quantities are combined in multiplication and division, it is usually adequate to use the **weakest-link approach.** That is, the quantity with the largest relative uncertainty is identified, and the same relative uncertainty is assigned to the calculated result. For example, suppose the following calculation is to be done, assuming that each quantity has an uncertainty of one digit in its last place:

$$X = \frac{1013.5 \cdot 0.06624 \cdot 4.55}{0.223 \cdot 3.3} = 4.150846 \cdot 10^2.$$

We note that the relative uncertainty is worst in the quantity 3.3 ± 0.1. This uncertainty is 1 part in 33, or 3 parts per 100. Therefore the overall result is realistically written as $4.2 \cdot 10^2$, since 1 part in 42 is comparable to 1 part in 33.

3.5   INTERPRETATION OF STRAIGHT-LINE DATA

Scientists have an inner compulsion to look for linear relationships between experimental variables. No doubt this is partly because straight lines are easier to draw than curves. Also, they are easier to extrapolate (a risky but useful practice) into other regions of the graph where the conditions differ from those in the laboratory. And the calculations associated with straight lines are easier to manage than for quadratic, cubic, etc., relationships.

The fundamental linear relation is as follows: there is an experimental variable X that is, as a rule, accurately controlled by the investigator. For each value of X there is an experimental (measured) result Y which is subject to both systematic and random errors. The experimental results are often compared with a chosen descriptive model that has the following algebraic counterpart:

$$Y = A + B \cdot X,$$

where A and B are unknown constants, usually of some significant chemical or physical meaning. To find the numerical values for these constants we must interpret the X, Y data. The standard technique is to make a graph: suppose that Y has the values 5, 7, and 9 when the values of X are 0.01, 0.02, and 0.03, respectively. The graph is presented by Fig. 3.6.

The plotted points are connected by the solid straight line, which is extrapolated (dashed line) to its intercept with the Y-axis. The value of this intercept is, of course, the desired value for the constant A. This is simply the value one would expect for Y, were X equal to zero.

The value of the constant B is found by calculating the slope of the straight line:

$$B = \text{Slope} = \frac{\Delta Y}{\Delta X} = \frac{9-5}{0.03-0.01} = 200.$$

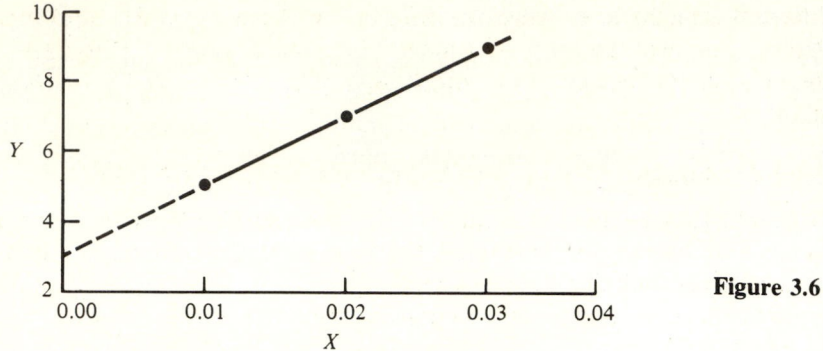

Figure 3.6

Thus the equation for the straight line describing this particular data set is:

$$Y = 3 + 200\, X.$$

Finding the Best Straight Line through Erratic Data

The above example was useful for defining terms, but was unrealistically simple. In experimental work both the X and Y values are always subject to uncertainties and errors. In the following discussion we will make the typical assumption that the X-values are known so accurately that we may ignore their errors in comparison with the larger errors showing up in the Y-values. The graph in Fig. 3.7 is for the following set of (X, Y) data: $(10, 0.0902)$, $(20, 0.1251)$, $(30, 0.1840)$, $(40, 0.2507)$, $(50, 0.2886)$.

If five different persons plotted these data and then drew "by eye" what they considered to be the best straight line, there would be five slightly

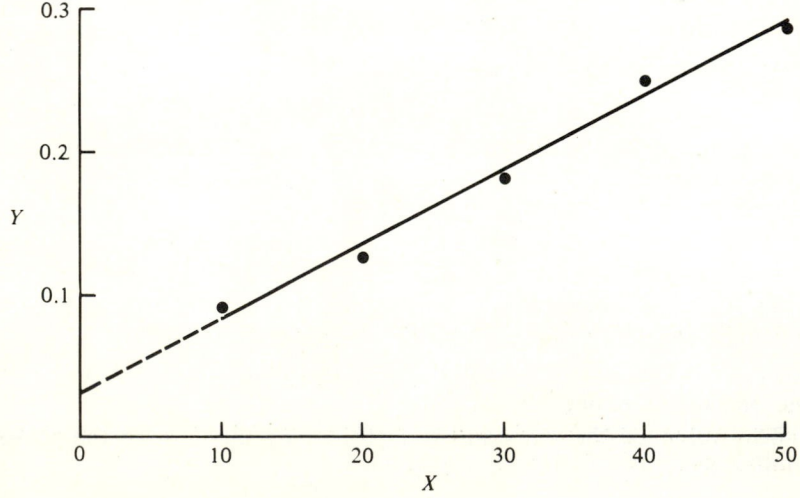

Figure 3.7

different straight lines with five different intercepts and five different slopes. Probably none of them would coincide with the line shown on the graph, which *is* the best line according to the widely accepted criterion of the *method of least squares.*

The Least-Squares Method (also Called Linear-Regression Analysis)

Conceptually, the best straight line through the (X, Y) data is the one that comes as close as possible to all the points. Each Y-value, being subject to random error, will show a certain deviation from the straight line. Of course, some of these deviations will be positive (points above the line) and some will be negative (points below the line). We eliminate the problem of $+$ and $-$ signs in the deviations by using the *squares of the deviations.* Thus, the best straight line is *defined* as the one for which the sum of the squares of the deviations of the Y-values from the line is minimal. The derivation of the equations used for the calculations requires partial differentiation with respect to the values of A and B. The procedure for doing these calculations is as follows:

1. Find the following sums:

   ΣX, the sum of all X-values,
   ΣY, the sum of all Y-values,
   ΣXY, the sum of all $X \cdot Y$ products,
   ΣX^2, the sum of the squares of all X-values,
   ΣY^2, the sum of the squares of all Y-values.

2. Calculate the following quantities (N is the number of data points):

$$D = \Sigma X^2 - (\Sigma X)^2/N,$$
$$E = \Sigma Y^2 - (\Sigma Y)^2/N.$$

3. Calculate the value for the slope B:

$$B = \frac{\Sigma XY - \Sigma X \cdot \Sigma Y/N}{D}.$$

4. Calculate the value for the intercept A:

$$A = \frac{\Sigma Y - B \cdot \Sigma X}{N}.$$

5. Calculate the standard deviation of the Y-values:

$$s = \sqrt{(E - B^2 \cdot D)/(N - 2)}.$$

6. Calculate the value for Student's t for 90% confidence limits or look it up in a table, using $N - 2$ degrees of freedom:

$$t = 1.643 + \frac{1.592}{N - 2} + \frac{0.798}{(N - 2)^2} + \frac{2.277}{(N - 2)^3}$$

7. Calculate the 90% confidence limits for the slope:

$$b = \pm t \cdot s / \sqrt{D}$$

8. Calculate the 90% confidence limits for the intercept:

$$a = \pm t \cdot s \sqrt{\Sigma X^2 / ND}.$$

After all this, the equation for the best straight line is:

$$Y = (A \pm a) + (B \pm b)X.$$

When these calculations are applied to the data used for the above graph the result is:

$$Y = (0.031 \pm 0.025) + (0.00522 \pm 0.00076)X.$$

Note the importance of including the calculation of the 90% confidence limits: without them there might be a temptation to assume an unrealistic accuracy for the A and B values.

It is perhaps obvious that the only sensible way to do the least-squares calculations is by using a computer or a programmed calculator. Many scientific calculators are available with hard-wired linear regression, making it simple to obtain least-squares results.

Caution: Computer least-squares fits may be hazardous to your conclusions.

There are at least three reasons for being cautious about using a computer program to find the best straight line through a set of data. All of these reasons point to the importance of *also* inspecting an actual plot of the data or at least a table of the deviations of each Y-value from the calculated straight line. In the following examples, the dashed line is the linear least-squares fit. The solid line is the more appropriate fit.

1. The data may include points that deviate so far that they should be omitted from the data set before doing the least-squares fit. Inspection of individual deviations is essential. The algebraic signs of the deviations should show alternations (Fig. 3.8).

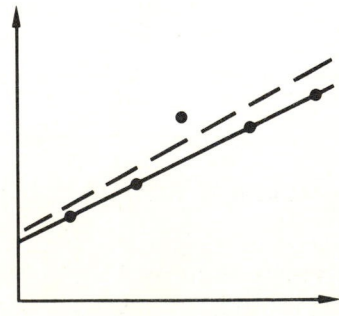

Figure 3.8

2. There may be a good linear relationship over part of the data range, but some other chemical factor may become important at one end, causing a definite trend that should not be included in the linear fit (Fig. 3.9).

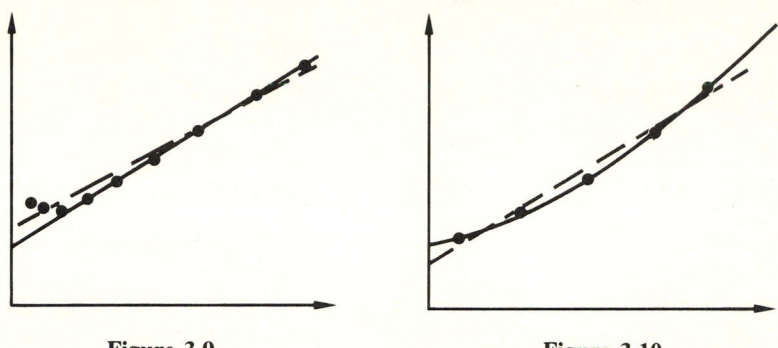

Figure 3.9   Figure 3.10

3. The data may have a definite curvature, showing that the linear model is inadequate. In this case the algebraic signs of the deviations will not alternate, but will show a pattern, such as $+++-----+++++$ (Fig. 3.10).

Working with Nonlinear X, Y Data

When a plot of Y versus X is not linear, it may still be possible to discover the algebraic relationship between the variables. We will consider only a few common situations. The reader is referred to works in numerical analysis for a detailed treatment. In the following examples, the letters A, B, C stand, as before, for numerical constants to be determined. The letters X, Y, W, and Z stand for various experimental quantities available from the data for each measurement. The common ingredient in all the examples is easy to state: the experimental quantities are grouped and operated on to define new variables Y' and X' that bear a linear relationship.

Example 1 The theoretical model for the data suggests that the expected relationship between X and Y is

$$Y = A + B/X.$$

A plot of Y versus X is a hyperbola. It is easily converted to a straight line by defining $X' = 1/X$, so that

$$Y = A + BX'.$$

A plot of Y versus X' is then a straight line, and the least-squares method may be used.

Example 2 The model suggests that $Y = W + A + B \cdot X \cdot Z^2$.

To cast this into linear form, it is necessary to define $Y' = (Y - W)$ and $X' = (X \cdot Z^2)$. With a plot of Y' versus X', the intercept should be A and the slope should be B.

Example 3   A power relationship is expected to occur, such that

$$Y = A X^B.$$

In this case it is useful to take logarithms of each side, obtaining

$$\log Y = \log A + B \log X.$$

If we let $Y' = \log Y$ and $X' = \log X$, the linear plot yields $\log A$ as intercept and B as slope.

3.6   STANDARD CURVES IN CHEMICAL ANALYSIS

First we should distinguish between *absolute* and *relative* (comparative) methods of chemical analysis. The only absolute method of analysis is the determination of the mass of the desired constituent or the mass of a pure compound which is known to contain that constituent in a precise stoichiometric percentage. Such methods are discussed in Section 6.1 on gravimetric determinations. All other methods of chemical analysis involve comparisons of the amount of constituent present in the sample with known amounts of other materials.

In titration methods it is always necessary to have a standard solution to be used as the titrant. We do not measure the constituent being titrated, but rather the amount of titrant required to react with the constituent. Then, by application of stoichiometry, we deduce the amount of constituent present in the sample. Thus, titration methods are relative. To establish the validity of a titration method, it is essential to prove that the correct results are obtained when known (i.e., standard) amounts of the constituent substance are carried through the titration procedure. However, once this has been done and reliable conditions are established, it is not necessary to repeat the proof each time a titration is to be performed. We rely on the reproducibility of the stoichiometric process without rechecking the standards. In this sense, titration methods come closer to absolute methods of analysis.

In this section we are concerned with analytical methods which are not so clearly based on stoichiometry. These methods involve the use of optical and/or electronic instruments for measurement, not of the constituent as such, but of some sort of phenomenon such as color, fluorescence, radioactivity, electrode potential, chromatographic peak, etc., which can be attributed to the constituent. In these methods it is essential that the response of the instrument be repeatedly checked by the use of standard reference materials and that the response due to a constituent in a sample be directly compared with standard responses. This comparison is typically carried out with the help of a *standard curve*.

Establishment of a Standard Curve

This discussion will be general and the specific applications will be treated in later chapters. Let us assume that a sample containing a constituent X with a concentration C_x is placed in an instrument that yields a response R, proportional to the value of C_x. Generally there will be some response, usually small, even when $C_x = 0$, and we will refer to this as the blank response R_b. We may express these ideas in terms of a simple linear equation:

$$R = R_b + kC_x,$$

where k is the proportionality constant. The higher the value of k, the more sensitive is the instrument to the presence of the constituent. If we were to do nothing but place our sample in the instrument, noting the value of R, we would be unable to deduce the value for C_x because the equation contains two unknown parameters: R_b and k. First it is necessary to find the values of these parameters by using known values for the concentration, i.e., by using reference standards.

Therefore, before measuring the sample response, we prepare a series of standards which ideally should be similar in nature to the sample. For instance, if the sample is contained in an alcohol solution, we would prepare the standards in alcohol, not in water. The series of standards will contain substance X in concentrations that cover the range expected for the value of C_x in the sample. We will also include a *blank* that has everything present but substance X.

Having set the controls of the instrument, we then find the response values R for the series of standards. These are plotted in Fig. 3.11.

For the above data we note that the blank response, 70 units, is consistent with the straight line drawn through the set of points. One might use the method of least squares to find this line, of course. The response is 420 units when the concentration of the standard is 0.5 units, and so the proportionality constant, which is the slope of the line, may be calculated:

$$k = \frac{420 - 70}{0.5 - 0} = 700.$$

Therefore, the equation for the standard curve may be written:

$$R = 70 + 700 \cdot C_s.$$

All response values may be corrected for the effect of the blank by subtracting 70 units. Then the corrected response is the one that corresponds to the amount of substance X in the standards:

$$R_x = (R - R_b) = kC_x.$$

Once the instrument is thus calibrated, the determination of C_x in the unknown sample consists of finding the response for the sample and correcting it for the blank (assuming that the correction will be the same as was found for the standards). The value for C_x (sample) is obtained either by using the established equation

$$C_x = R_x/k$$

or by finding the point on the graph that corresponds to the response. The

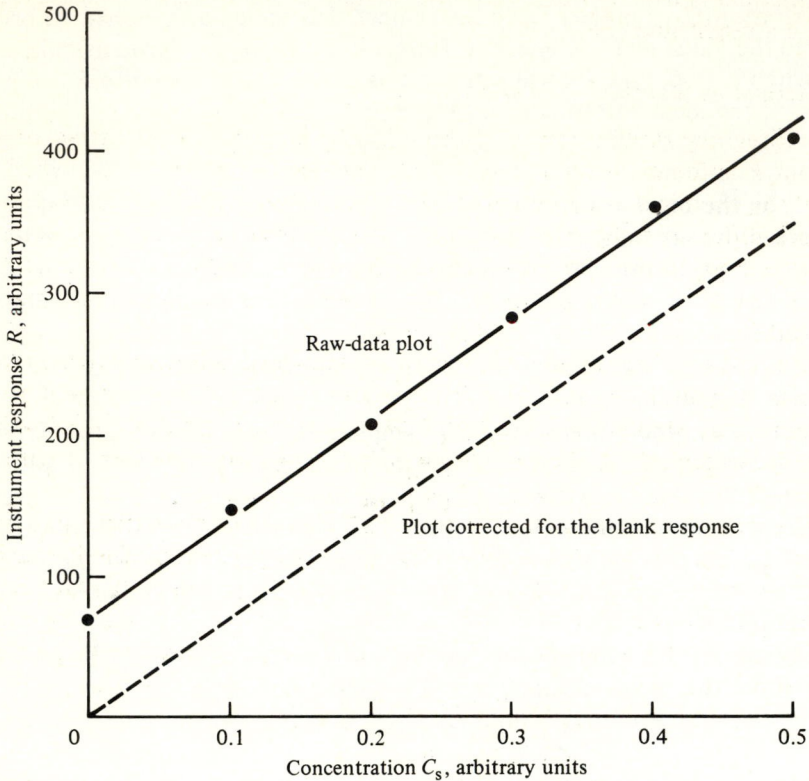

Figure 3.11

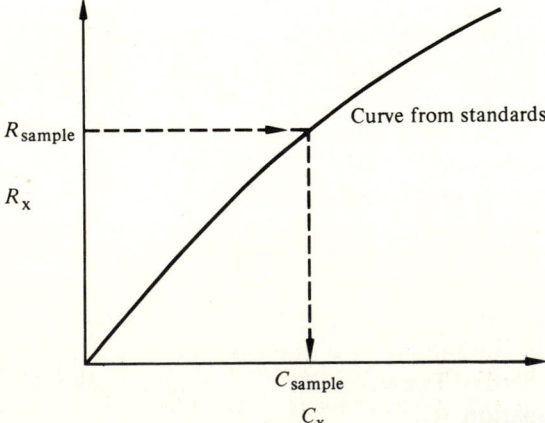

Figure 3.12

latter method is used in those (not uncommon) cases when the standard curve
is not a straight line (Fig. 3.12).

The Method of Standard Additions

In the foregoing standard-curve method it is assumed that the proportionality
constant k is identical for the standards and the unknown. However, it may
be that the chemical environment of the unknown sample (e.g., seawater) may
be quite different from that used for the standards (e.g., distilled water). It
would be best to prepare standards in the same chemical environment, but
what is one to do if the nature of that chemical environment is unknown or
uncertain?

The answer lies in the technique of **standard additions**, whereby the
unknown sample itself *provides* the constant chemical environment for the
standards. The standards (minute volumes of a stock solution of constituent
x) are added directly to the sample and the response R is measured with each
addition.

The solutions will contain substance X from two sources: a constant
amount C_x in the sample and a varying amount C_s from the added standards.

Since the chemical medium remains virtually unchanged by the addition
of small quantities of the standard solution, we may assume that the proportionality constant k is constant. Therefore, assuming the blank response has
been subtracted, we get

$$R = k(C_x + C_s) = kC_x + kC_s.$$

If C_s is deliberately varied to obtain a series of R-values, a plot of R versus C_s
will have the appearance shown in Fig. 3.13 (remember that kC_x is a constant,
even if it is unknown).

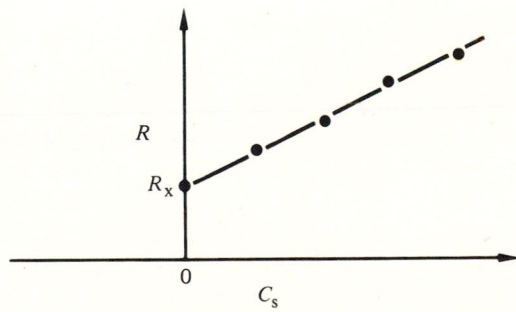

Figure 3.13

To find the value of C_x (which after all is the object of the measurements) one merely extrapolates the straight line to the intersection with the
horizontal axis, where R is zero (Fig. 3.14).

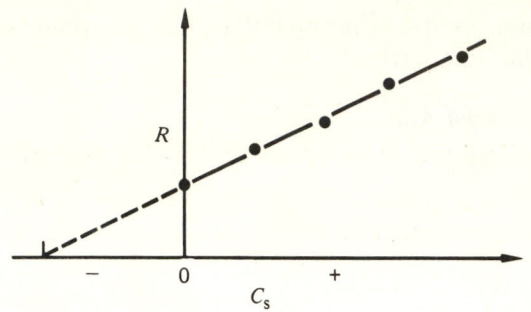

Figure 3.14

It is clear from the above equation that $C_x = -C_s$ when $R = 0$, where $-C_s$ is the hypothetical value corresponding to the *removal* of the constituent until $R = 0$. The technique of standard additions is especially helpful in trace analysis methods that use spectrophotometry, atomic absorption spectroscopy, polarography, and measurements by means of ion selective electrodes.

3.7   LIMITATIONS OF MASS AND VOLUME MEASUREMENTS

A modern analytical balance is precise to ±0.1 milligram, while volumetric glassware is precise to about ±0.01 milliliter. Since 0.01 mL of water has a mass of 10 mg, we see that weighing is 100 times more precise than volume measurements. We now consider ways to get the most out of these two important techniques.

Reliable Mass Determinations

A modern balance contains chromium–nickel steel "weights" which are accurately manufactured and have known masses. Before the object is placed on the balance pan, all these weights are in place and the balance is adjusted to read 000.0000 gram. When the object is placed on the pan, it is necessary to remove some of the weights to restore the balance to the same position. In using a balance, we are actually comparing weights, not masses. However, since the gravitational field is the same for the object and the weights, we are actually comparing their masses also. It is important to understand the distinction between weight and mass, but in practice most chemists use the terms interchangeably, meaning mass in both cases.

A small but often significant factor in precise weighing is the buoyancy effect of the air surrounding the weights and the object. According to Archimedes' principle, a body immersed in a fluid is buoyed up with a force equal to the weight of the displaced fluid (or gas). When the balance is "zeroed," it is the weights in the balance that are buoyed up. But when these weights are replaced by the object being measured, the buoyant force is different, unless the object happens to have the same volume as the replaced weights. Thus, when we have achieved an output reading on the balance we have actually made the *net* weight of the replaced weights equal to the *net* weight of the object being weighed:

$$\text{Net weight of weights} = \text{Net weight of object.}$$

The net weight differs from the true weight by the buoyant force due to air displacement:

$$\text{Net weight} = \text{True weight} - \text{Weight of displaced air.}$$

The buoyant force (weight of displaced air) may be expressed in terms of the volume and density of the air:

$$\text{Weight of air} = \text{Volume of air} \times \text{Density of air.}$$

We may now express the net weights of the steel weights and of the object in terms of these quantities, remembering that the volume of displaced air is the same as the volume of the weights (or object):

$$\text{Net weight of weights} = W_s - V_s \cdot d_a,$$
$$\text{Net weight of object} = W_o - V_o \cdot d_a,$$

where the subscripts "s" and "o" refer to steel weights and the object, respectively, and d_a is the air density at the time of the measurement. The volume of the weights is related to their true weight and their density:

$$V_s = W_s/d_s.$$

Similarly, for the object,

$$V_o = W_o/d_o.$$

Finally, we may substitute these relationships in the foregoing equations, set the net weights equal to each other, and solve to find the true weight of the object in terms of all the other quantities:

$$W_o = W_s \frac{(1 - d_a/d_s)}{(1 - d_a/d_o)} = W_s \cdot B.$$

Thus, we can find the true weight (mass) of the object by multiplying the observed value for W_s (which is the quantity we read from the balance display) by the buoyancy factor B. The value for B is unity (exactly 1.0) only when the object has the same density as the weights in the balance.

HOW SIGNIFICANT IS THE BUOYANCY EFFECT?

The density of the stainless steel weights used in modern analytical balances is specified as 7.7 g/cm^3. The density of laboratory air depends upon the existing temperature and barometric pressure according to the ideal gas law:

$$d_a = 0.001293 \, \frac{273}{273 + t} \cdot \frac{P(\text{mm of Hg})}{760} \, g/cm^3,$$

where t is the Celsius temperature and P is the corrected barometer reading.

The value 0.001293 g/cm^3 is simply the density of dry air at conditions of 0°C and 760 mm of Hg. It is rare (or should be) that a modern laboratory is operated outside the temperature range of 15–30°C or that the atmospheric pressure is outside the range 720–770 mm of Hg. The density of air in this range of conditions is close to 0.0012 g/cm^3. Then, given the density of the steel weights (7.7 g/cm^3), it is only necessary to calculate the values for B corresponding to the weighing of various objects of different densities. The results are presented in Table 3.4.

Table 3.4

d_{object}	B	d_{object}	B
0.7	1.00154	2.0	1.00044
0.8	1.00133	3.0	1.00024
0.9	1.00116	4.0	1.00014
1.0	1.00103	5.0	1.00008
1.1	1.00092	7.0	1.00002
1.2	1.00083	10.0	0.99996
1.5	1.00064	20.0	0.99991

Suppose we want to weigh out precisely 0.1 mole of sodium chloride. Taking the molecular weight to be 58.443, we find the desired mass (if the sample is truly pure) as 5.8443 grams. If the buoyancy effect is applied (which is not common in ordinary work) then the apparent mass to be weighed out is

$$W = 5.8443/1.00039 = 5.8420 \text{ grams.}$$

The buoyancy factor 1.00039 was obtained by using 2.165 g/cm^3 as the density of sodium chloride. Note that the effect is not negligible, but is 2.3 mg, or 23 times the sensitivity of a typical balance. On a *relative* basis, however, the neglect of the buoyancy factor in this case introduces an error of only 39 parts in 100 000, which would be of concern only in the most precise research.

It is clear from the table of calculated B-values that the buoyancy effect becomes increasingly important as the density of the weighed material becomes smaller.

Example   What is the true mass of a benzoic acid sample (density 1.27 g/cm^3) that appears to weigh 137.2294 grams?

The value of B, by interpolation, is 1.00079. The true mass is therefore

$$W = 137.2294 \cdot 1.00079 = 137.3378 \text{ grams.}$$

Because the error due to buoyancy rarely exceeds 1 part per thousand, this effect is commonly ignored. However it is essential to make the buoyancy correction in high-accuracy measurements. The buoyancy effect is also significant when volumetric glassware is calibrated by weighing the water it contains or delivers, since the density of water is only about 1.0.

Reliable Volume Measurements of Solution

The National Bureau of Standards has established tolerance standards for the manufacture of class A volumetric ware used for research and accurate analytical work (Circular C-602). For the four principal types of glass volumetric ware these tolerances (in milliliters) are as follows:

Table 3.5

Capacity in mL	Burets	Measuring pipets	Transfer pipets	Volumetric flasks
1		0.02	0.006	
5		0.04	0.01	
10	0.02	0.06	0.02	0.02
25	0.03		0.03	0.03
50	0.05		0.05	0.05
100	0.10		0.08	0.08
200			0.10	0.10
250				0.12
500				0.20
1000				0.30
2000				0.50
4000				1.00
6000				1.50

For example, a class A transfer pipet with a nominal value of 50 mL should, when used properly, deliver a volume of water between 49.95 and 50.05 mL. That is, the maximum permitted error is only 0.05 mL, or one part per 1000 of the nominal value. For contrast, note that a 1-mL measuring pipet may be in error by 2% and still be within specifications. In routine analytical work, and even for most research work, these tolerances are acceptable. However, if the glassware is to be used in measurements of highest quality, it is necessary to find out for oneself just how well each piece performs. A skilled lab analyst can use a 10-mL pipet with an imprecision of only about 0.002 mL, which is 10 times better than the tolerance.

The glassware manufacturer places calibration marks (etched lines, often colored) on the glassware. When we want to use the glassware with the highest possible precision, we do not attempt to change the position of those marks to make them correspond more closely with the nominal value of the volume; rather, we standardize the glassware, using the original calibration marks.

TESTING (STANDARDIZATION) OF VOLUMETRIC GLASSWARE

To standardize a piece of volumetric ware it is only necessary to weigh the water contained in or delivered by the piece. Volumetric flasks are first dried (not in the oven, for that could change their volumes slightly and permanently) and weighed and then filled with pure water to the calibration mark and weighed again. The content of pipets and burets is delivered into a previously weighed vessel, which is stoppered (to prevent evaporation) and reweighed.

Of course, the first datum is really the apparent weight of the water, and because of its low density it is necessary to correct for buoyancy. The buoyancy factor is 1.00103, as seen from the previous table. The resulting true weight is then divided by the precise density of water (at the temperature it was used) to find the volume of water:

$$V_{H_2O} = W_{H_2O}/d_{H_2O} = (\text{Apparent mass } H_2O) \cdot B/d_{H_2O}.$$

Note that in this case it is convenient to apply the combined density factor B/d, where B is the buoyancy correction. Table 3.6 gives the values for the water density and the combined factor over the appropriate temperature range.

Table 3.6

$t°C$	d_{H_2O}	B/d
18	0.998595	1.00244
19	0.998405	1.00263
20	0.998203	1.00283
21	0.997992	1.00304
22	0.997770	1.00327
23	0.997538	1.00350
24	0.997296	1.00374
25	0.997044	1.00400
26	0.996783	1.00426
27	0.996512	1.00453
28	0.996232	1.00482

For example, suppose that a 50-mL pipet is found to deliver 49.866 grams (apparent mass) of water at 23° (average of five trials). Then the volume of this pipet is

$$V = 49.866 \cdot 1.00350 = 50.041 \text{ or } 50.04 \text{ mL}.$$

EFFECT OF TEMPERATURE ON THE GLASSWARE VOLUME

If a piece of glassware has been standardized at any typical laboratory temperature, it may be used at other temperatures without restandardization. This is because the coefficient of cubical expansion for borosilicate glass is only 25 parts per million per degree. Thus, if a pipet is standardized at 23° and later used at 27°, the relative increase in its volume due to the expansion of the glass would be only

$$4 \cdot 0.000025 = 0.0001,$$

or one part per 10,000. A 50.000-mL pipet calibrated at 23° would deliver 50.005 mL at 27°. For most applications this effect is negligible.

EFFECT OF TEMPERATURE ON SOLUTION VOLUMES

Of greater importance in accurate work is the expansion of aqueous solutions with temperature. As a good approximation we may presume that dilute solutions change volume in the same way water does, and we use the densities of water as a guide. Suppose that a solution of nitric acid contains exactly 1.0000 gram of HNO_3 in a 50-mL portion taken at 22°. The next day the temperature is 26° and the same pipet is used to obtain a portion of the solution. Because of expansion, the nitric acid solution now contains slightly less than 1.0000 gram of HNO_3 in the 50-mL volume. We may use the ratio of the water densities at the two temperatures to find the new concentration:

$$C_{26°} = C_{22°} \cdot \frac{d_{26}}{d_{22}} = 1.0000 \cdot \frac{0.996783}{0.997770} = 0.9990 \text{ gram/50 mL.}$$

A simple approach to this calculation is this: remember that a solution increases in volume about 1 part per 1000 for a 4° temperature increase. Thus, a 1° increase will cause a concentration to decrease by 0.25 part per 1000, etc.

PROBLEMS

Accuracy and precision

1. Refer to table of chemical atomic weights with indications of uncertainties in the values. Calculate the molar mass (and its uncertainty) for each of the following compounds:

   a) methyl iodide CH_3I
   b) osmium tetroxide OsO_4
   c) calcium carbide CaC_2
   d) mercuric selenide $HgSe$

2. Calculate the molarity (and confidence limits) of a solution prepared by dissolving 6.173 (± 0.005) gram of KCl (purity 99.2 $\pm$ 0.2%) in water, followed by dilution to a final volume of 100.00 $\pm$ 0.06 mL. Assume that the molecular weight of KCl is 74.56.

3. A clean dry beaker was found to weigh 34.9621 g; after a sample of potassium nitrate was added, the weight was 36.1224 g. After drying in an oven overnight and cooling in a desiccator, the weight was 36.1196 g. If each weighing has an uncertainty of 0.2 mg, what was the percentage (and its uncertainty) of volatile material in the sample?

4. Five determinations of the density of an organic liquid gave results of 0.9132, 0.9138, 0.9129, 0.9131, and 0.9131 g/cm³. Calculate the average weight and the average deviation; decide whether the value of 0.9138 should be rejected from the set by using the Q-test. Calculate the relative average deviation, in parts per thousand, for the retained values.

5. In testing a new analytical procedure it was found that the percentage of copper in a certain alloy was 23.66%. If the true amount of copper, obtained by other means, is believed to be 23.75%, what is the average error (relative, in parts per thousand), of the new procedure?

6. A skilled analyst is asked to evaluate two different methods for the determination of trace concentrations of lead in glacial acetic acid and is given a large bottle containing (unknown to her) precisely 1.282 mg Pb per liter. Eight determinations by each method were performed, with the following results on Pb concentration (in ppm):

Method A	1.34	1.33	1.32	1.35	1.32	1.34	1.34	1.31
Method B	1.30	1.26	1.30	1.33	1.20	1.24	1.24	1.33

Compare the two methods with regard to precision and accuracy. Can you recommend one method over the other?

7. A 100-mL pipet is used to deliver a portion of $0.1293M$ HCl into a 500-mL volumetric flask and water is added to the calibration mark at 23°C. Previous work had shown that the flask contained 498.8 g of water at 23° and that the pipet delivered 99.78 g of water at 23° (both apparent masses).

a) Are the pieces of glassware accurate to within class A specifications?
b) What is the molarity of the diluted HCl after mixing?

8. A solution of sodium chloride has a molarity of 0.11544 at 20°. What is its molarity at 28°?

9. What is the true mass of a sample of aluminum metal (density 2.7 g/mL) that weighs 25.0119 g (uncorrected for buoyancy)?

Gaussian statistics

10. Define each of the following as used in the text:

$$\mu, d, \bar{d}, \bar{x}, s, t, \bar{E}.$$

11. If 200 repeated measurements are made, how many may be expected to fall within ±1 standard deviation of the mean? How many between +2 and +3 standard deviations?

12. How many measurements must be made before an individual deviation (from the mean) of three standard deviations becomes *expected*? How many measurements must be made before such a deviation becomes *possible*?

13. Nine repeated determinations of the molar solubility of an organic acid in water at 25°C resulted in a mean value 0.00673 mol/L with a standard deviation of 0.00008 mol/L. Find the 90% and also the 99% confidence limits for the mean value.

14. State, in words, what is meant by "90% confidence limits."

15. Given the 200-value set for the pipet volume (see p. 57), find the 95% confidence limits for the pipet volume. What is the (approximate) probability that a single use of this pipet will deliver a volume smaller than 50.00 mL?

16. A student reports 20.33, 20.36, and 20.35 mL for buret readings of a triplicate titration. With the help of a hand calculator he obtained an average of 20.346667 mL. What should the instructor's comment be?

17. Give four examples of everyday phenomena that are subject to gaussian distribution.

18. Suppose that the flameless atomic absorption technique used for the determination of mercury in urine and described on p. 11 gave the following instrument response:

Hg in the standard, nanograms	Peak height, millimeters
0 (the blank)	4.1
40	16.2
80	27.6
120	40.1
160	52.0

a) Plot these data and draw a standard curve after subtracting the value obtained for the blank from each of the peak heights.

b) How many nanograms of Hg were in the sample of a patient's urine if it showed a peak height of 37.0 mm when treated in the same way as the standards?

19. In applying the method of standard additions to an analysis, the following data were obtained:

Added standard, μg	Instrument response
0	0.60 (the sample alone)
0.5	0.84
1.0	1.09
1.5	1.32

a) Assuming that the response was zero for the blank, find the number of micrograms of the constituent in the sample by plotting response versus added standard, as discussed in the text.

b) If the blank gave a response of 0.50, how would you reinterpret the data?

20. Values for an experimental quantity Y were determined by using several controlled values of a variable X. It is expected that a linear relationship exists:

$$Y = A + B \cdot X.$$

The data were as follows:

X	1	2	3	4
Y	0.16	0.21	0.26	0.30

a) Make a careful plot of Y versus X and draw by eye the best straight line through the points. Read the intercept and calculate the slope of your line.
b) Use the method of least-squares to find the linear fit for the data. Compare the results for slope and intercept with your own estimates from (a).
c) Use the procedure outlined in the text to assign confidence limits (90%) to the least-squares values of slope and intercept.
d) What would you predict for the value of Y when $X = 5$? What uncertainty should be attached?

21. At 22°C, a 5-mL pipet is found to deliver the following apparent weights of water: 4.995, 4.993, and 4.992 grams. What is the average volume of water delivered?

22. A 50-mL volumetric flask is standardized in two ways. First, by weighing it empty and then filled to the mark with water at 26° it is found to contain 49.75 g (apparent weight, not corrected for buoyancy). Then, after thorough drying of the flask, a standardized pipet known to deliver 9.984 ± 0.005 mL is used to place five portions of water into the flask. The resulting meniscus is slightly below the mark, and so a small graduated pipet is used to adjust it, with 0.035 mL required. Calculate the flask volume for each approach. Which approach do you favor for accuracy?

CHEMICAL EQUILIBRIUM AND REACTION TENDENCY

4

The lingering value of college:
    It's so that later on in life when you knock on yourself, somebody answers.

                                        J. Arbolino

4.1   REVERSIBLE REACTIONS

When we write the equation for a chemical change we imply, by long-standing agreement, that the species shown at the left are **reactants**, while those on the right are **products**. For example, the reaction of gasoline combustion can be written as

$$2C_8H_{18} + 25O_2 \rightarrow 16CO_2 + 18H_2O,$$

where the single arrow indicates the unidirectional nature of this chemical transformation. There is no likelihood that a mixture of carbon dioxide and water will spontaneously interact to resynthesize gasoline. This reaction, like many others involving extensive molecular rearrangements and complex mechanisms, is totally **irreversible**. When oxygen is in excess, the gasoline is totally consumed.

However, a large fraction of the chemical reactions that play such an important role in analytical chemistry involve very simple mechanistic pathways and can proceed in both directions, to the left and to the right, as implied by the chemical equation. Consider the reduction of ferric ion by silver metal:

$$Ag(s) + Fe^{3+}(aq) \rightleftarrows Ag^+(aq) + Fe^{2+}(aq),$$

where (s) and (aq) mean solid, and dissolved in water, respectively. As written, the equation emphasizes the direction from left to right, with the formation of ferrous ion and silver ion. If silver metal is placed in contact with an acidic solution of ferric ion this change does take place, for the resulting solution can be shown to contain both Ag^+ and Fe^{2+} by simple tests with reagents. However, it is also easy to show that the solution still contains some unreacted Fe^{3+}, indicating that the reaction does not go to completion even when an excess of silver metal is present. Further, it is also observed that a mixture of only Ag^+ and Fe^{2+}, with no silver metal or Fe^{3+} present, will react to form these species. This property of **reversibility** is implied by the use of the double arrow $\rightleftarrows$ in the above equation; the equal sign = can also be used.

Therefore, in the Ag, Fe^{3+}, Fe^{2+}, Ag^+ system it is ambiguous as to which species are reactants and which are products, for the actual net change depends upon what particular mixture of the ingredients is prepared.

The Concept of Chemical Equilibrium

For reversible reactions, i.e., reactions that are able to proceed in both directions, it is inevitable that there will be some "middle ground" in which the chemical composition of the reaction mixture is said to be at **equilibrium**. Let us consider four criteria that must be met for this important condition:

1. At the macroscopic level, we say that a system is in equilibrium if direct observations reveal no changes in the bulk properties with the passage of

time, i.e., the system appears to be static. Of course, this could be misleading in the case of a very slow reaction.

2. At the microscopic (i.e., molecular) level we compare the rates at which the forward and reverse reactions are taking place. For the system to be at equilibrium it is necessary for each forward rate to be equal to its reverse counterpart. From this viewpoint, reactions at equilibrium are far from static, for the mechanistic steps may be proceeding at fast rates. However, if they are going at the same rates in both directions there will be no *net change* in properties with time and the macroscopic criterion will be met.

3. From the viewpoint of energy change we say that a chemical reaction reaches its equilibrium state when the composition corresponds to a minimum in the energy of the reactants and products. For some reactions this may mean a composition that contains mostly products, while for other reactions it may be mostly reactants. It is not so easy to describe the nature of the energy involved here (it is the Gibbs, or "free" energy), but the idea is akin to that of a mountain stream seeking its minimum energy level by flowing into a lake.

4. Our most practical and widely used criterion for equilibrium requires a mathematical expression called the **equilibrium constant**. A reaction is at equilibrium if the concentrations of the species are in accord with the numerical value of this constant, which is different for each chemical reaction. Major portions of our study of chemical equilibrium and analysis will be based on applications of this principle.

4.2 THE STATES OF REACTING SPECIES

The various species participating in a chemical reaction may be present in different physical states: a species may be present as a gas (g), perhaps mixed with other gases, or as a liquid (l), perhaps mixed with other liquids, or as a solid (s), perhaps in solid solution with other solids. In addition, the species may be dissolved in a liquid solvent (e.g., water). When we write equations for the reversible reactions we include information about the physical state of each species:

$$NH_3(g) \rightleftarrows NH_3(aq)$$

A water solution (aq means aqueous) of ammonia in equilibrium with gaseous ammonia in the space above the solution.

$$H_2O(s) \rightleftarrows H_2O(l)$$

Ice in equilibrium with liquid water at 0°C

$$H_2O(l) \rightleftarrows H^+(aq) + OH^-(aq)$$

Dissociation of water into its ions.

$$I_2(CCl_4) \rightleftharpoons I_2(aq)$$ Distribution of iodine between two liquid phases,
one of carbon tetra-
chloride and the other of
water.

$$NaCl(s) \rightleftharpoons Na^+(aq) + Cl^-(aq)$$ Saturated solution of salt
in water

$$Ag(s) + Fe^{3+}(aq) \rightleftharpoons Ag^+(aq) + Fe^{2+}(aq)$$ A reversible redox reaction

If it is clear from the context of a discussion that water solutions are being used, it is common practice to omit the (aq) symbol for **ionic** species, since these do not exist as separated species except in solution. For molecular (uncharged) species it is good practice to indicate the state. For example, the equation

$$I_2 + I^- \rightleftharpoons I_3^-$$

is ambiguous, for it does not tell whether the iodine is present only as a dissolved species, $I_2(aq)$, or as a solid, $I_2(s)$, or even as vapor, $I_2(g)$.

Expressing the Concentrations of Reacting Species

The amounts of the reacting species present in a given mixture can be expressed by a variety of concentration conventions. The most common practice is as follows:

Gases Use the partial pressure P_i of the gas as it exists in the gaseous phase in contact with the reaction system; the pressure is measured in atmospheres. For example, the total air pressure under normal conditions is 1.00 atm, while the partial pressure of O_2 is 0.21 atm and that of N_2 is 0.78 atm.

Liquids Use the mole fraction of the liquid in the liquid phase which is in contact with the reaction system. If this is simply a pure liquid, its mole fraction is unity, 1.000. If the liquid phase consists of a mixture, then each component will have its individual mole fraction

$$X_i = \frac{\text{Moles of substance } i}{\text{Total moles in the liquid phase}}.$$

Solids Use the mole fraction* of the solid in the solid phase which is in contact with the reaction system. For pure solids, $X_i = 1.000$, but for solid solutions X_i is less than unity.

Solutes Use the molarity of the solute species in the solution, $[i]$, in units of moles per liter.

---

*For an enlightening discussion of the usefulness of mole fractions the reader is referred to the article "On Mole Fractions in Equilibrium Constants" by C. Delaney and L. Nash in *Journal of Chemical Education*, **54**, 151 (1977).

4.3 IDEALIZED STATEMENT OF THE EQUILIBRIUM LAW

It is best to begin with an idealized approach, similar to the use of the ideal-gas law: $PV = nRT$. This law is valid only for the hypothetical case of an ideal gas with infinitesimal molecules that show no attractions or repulsions. Nevertheless, we find that this law works fairly well for real gases at low pressures (e.g., below 1 atm). For gases at higher pressures it is necessary to introduce "correction terms" that account for molecular size and interactions. Similar considerations hold for aqueous equilibrium.

The idealized statement of the Equilibrium Law is as follows: given the stoichiometrically balanced equation, for example,

$$a\,A(s) + b\,B(aq) \rightleftarrows d\,D(aq) + e\,E(g),$$

for any reversible reaction, we may write the **ratio function** r:

$$r = \frac{[D]^d P_E^e}{X_A^a [B]^b}.$$

The numerator is the product of the concentration terms for the species on the right side of the equation with each term raised to a power corresponding to the stoichiometric coefficient in the balanced equation. The denominator is written similarly for the species on the left side of the equation.

There is an infinite variety of reaction mixtures of A, B, D, and E that may be created. If one uses only A and B to start with, then at the instant of mixing there is no D or E present, and so the numerical value of r is zero. If the choice is to mix only D and E, the value of r is temporarily infinity. In the first case, the above reaction has no choice but to proceed to the right, while in the second case it must proceed to the left. We may call r the **nonequilibrium** ratio function, and its numerical value will change continuously as the chemical reaction proceeds toward equilibrium. Once chemical equilibrium is attained, whether it is approached from the left or from the right, *the ratio function r will become constant at a value that is characteristic of the particular reaction.* We refer to this important value of r as the **equilibrium quotient** Q:

$$\text{Eventual value of } r = Q = \frac{[D]^d P_E^e}{X_A^a [B]^b}, \tag{4.1}$$

where only equilibrium concentrations are used.

Definition of the Stoichiometric Unit

In writing the expressions for r and for Q it is important to use the proper coefficients in the balanced chemical equation. There must be an exact correspondence between these coefficients and the exponents to which the individual concentrations are raised in the Q-expression.

A balanced equation may be correctly written in many ways, using different multiples of the coefficients. For example, the reaction between ferric ion and iodide ion to form ferrous ion and iodine may be written as

$$Fe^{3+} + I^- \rightleftharpoons Fe^{2+} + \tfrac{1}{2}I_2(aq);$$

in this form it designates a reduction of one mole of ferric ion by one mole of iodide ion. We may say that this particular stoichiometric unit corresponds to these amounts. The equilibrium quotient expression for *this* stoichiometric unit is

$$Q = \frac{[Fe^{2+}][I_2]^{1/2}}{[Fe^{3+}][I^-]}$$

and its numerical value is approximately $Q = 7 \cdot 10^3$.

If we prefer to avoid the fractional coefficient 1/2, the stoichiometric unit can be doubled and the reaction becomes

$$2Fe^{3+} + 2I^- \rightleftharpoons 2Fe^{2+} + I_2(aq).$$

Then the expression for Q corresponding to *this* unit is

$$Q = \frac{[Fe^{2+}]^2[I_2]}{[Fe^{3+}]^2[I^-]^2}.$$

Since this is the square of the previous expression, it follows that the numerical value must also be squared: $Q = (7 \cdot 10^3)^2 = 5 \cdot 10^7$.

Therefore, given *only* the statement "The Q-value for the ferric iodide reaction is $7 \cdot 10^3$," no conclusion may be drawn. It would be vital to ask "To what balanced equation (stoichiometric unit) does this Q-value refer?"

Equilibrium under Ideal Conditions: the Infinitely Dilute Solution

Equation (4.1), the idealized version of the Equilibrium Law, is one of the most important generalizations of theoretical chemistry. In introductory textbooks it is usually derived by using kinetic arguments and expressions for the rates of reactions. The Equilibrium Law also follows from rigorous deductions of chemical thermodynamics. However, the many studies carried out on reversible reactions have shown conclusively that Eq. (4.1) must be regarded as a *limiting law* that is valid only under experimental conditions of extremely low concentrations of the reacting species. At the concentrations normally used in routine chemical work, we find that the numerical values for Q are not truly constant for a given chemical reaction, but depend significantly upon the choice of concentration conditions.

The variation in the Q-value is due to deviations in the reactive behavior of chemical species from what may be defined as ideal behavior. For gaseous species we define **ideal behavior** as that shown by the species under conditions of virtually zero gas pressure. For species that are present as solutes in a solution phase we use a similar definition, the infinitely dilute solution, in which the dissolved species are present at such a low concentration that they are affected only by the properties of the pure solvent and not by each other. In this text we shall deal chiefly with aqueous equilibria, so that the reference condition of the infinitely dilute solution is simply pure water. Particularly for ionic species with strong electrostatic fields, deviations from ideal behavior are important as concentrations become finite.

In pure water itself there is an important ionic equilibrium based on the reversible dissociation of water molecules into hydronium and hydroxide ions:

$$2H_2O \rightleftarrows H_3O^+ + OH^-,$$

$$Q = \frac{[H_3O^+][OH^-]}{X_{H_2O}^2} = [H_3O^+][OH^-] = 1.008 \cdot 10^{-14} \quad \text{(at 25°)}.$$

In this case the mole fraction of water molecules has been taken as unity because of the very low concentrations of the ions, each present at a molarity of $1.004 \cdot 10^{-7}$ at 25°C. Although this is not truly "infinitely dilute," it is close enough to be regarded as ideal. Therefore, the value of $Q = 1.008 \cdot 10^{-14}$ may be taken as the limiting value, which we call the **equilibrium constant** symbolized by K.

4.4 EQUILIBRIUM UNDER NONIDEAL CONDITIONS

To illustrate the nonideal behavior of ions in the presence of electrostatic fields due to other ions in the same solution, we shall consider the effects of adding very small amounts of sodium chloride to pure water. Sodium chloride is a strong electrolyte that does not react in any chemical way with the H_3O^+ or OH^- ions. We presume that its only effect on the water equilibrium is through the electrostatic forces that its ions Na^+ and Cl^- exert on the water ions. The results are summarized in Table 4.1.

Table 4.1

Molarity of added NaCl	Equilibrium $[OH^-] = [H_3O^+]$	$Q = [OH^-][H_3O^+]$
0	$1.004 \cdot 10^{-7}$	$1.008 \cdot 10^{-14}*$
$1 \cdot 10^{-6}$	$1.005 \cdot 10^{-7}$	$1.010 \cdot 10^{-14}$
$1 \cdot 10^{-5}$	$1.008 \cdot 10^{-7}$	$1.016 \cdot 10^{-14}$
$1 \cdot 10^{-4}$	$1.016 \cdot 10^{-7}$	$1.032 \cdot 10^{-14}$
$1 \cdot 10^{-3}$	$1.040 \cdot 10^{-7}$	$1.082 \cdot 10^{-14}$
$1 \cdot 10^{-2}$	$1.113 \cdot 10^{-7}$	$1.239 \cdot 10^{-14}$
$1 \cdot 10^{-1}$	$1.286 \cdot 10^{-7}$	$1.654 \cdot 10^{-14}$

*Limiting value, K

There is a steady departure of the Q value from the limiting value observed for pure water. At a sodium chloride molarity of 0.1, typical of the concentrations we must work with in chemical analysis, the change in Q is over 60% from the ideal value.

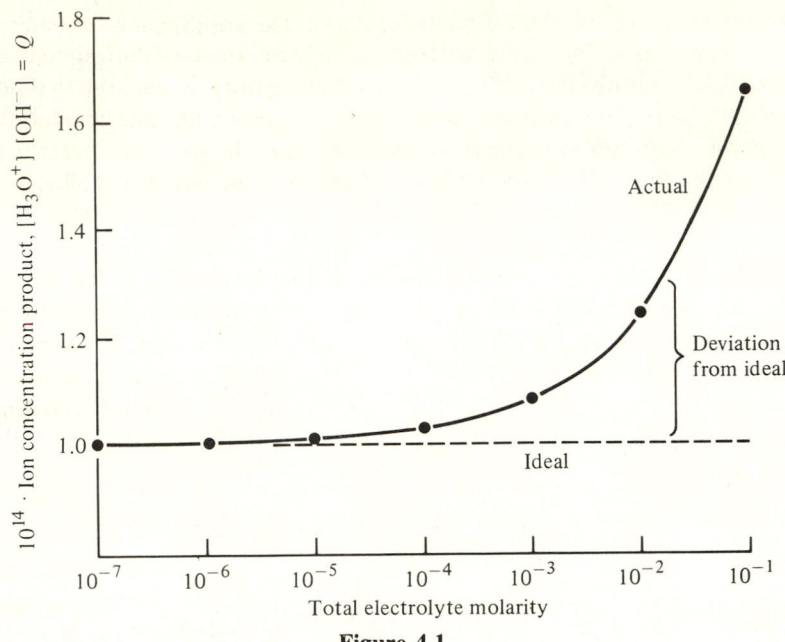

Figure 4.1

The above results of Table 4.1 are shown in Fig. 4.1, with the Q-values plotted versus the total electrolyte molarity of the solution. Because of the wide range of concentration, the molarity axis is plotted on a logarithmic scale.

It is easy to give a qualitative explanation for this behavior, although the theoretical (quantitative) basis is extraordinarily complex. The electrostatic forces due to the added sodium and chloride ions interfere with the kinetic abilities of the H_3O^+ and OH^- ions to recombine. These ions are not un-hindered (as in pure water) because they tend to be surrounded by a loose atmosphere of ions of opposite charge and must drag this atmosphere with them. Therefore, the rate of the recombination reaction (the reverse of dissociation) is lowered by the presence of the inert Na^+ and Cl^- ions, while the rate of the forward reaction is not particularly affected. Therefore the equilibrium is positioned farther to the right in the presence of NaCl or any other ion-producing substance.

Empirical Definition of Activity Coefficient

As with the dissociation of water into its ions, every reversible reaction involving ionic species departs from ideal behavior because of electrostatic effects when the solution contains appreciable concentrations of ionic substances. Thus, for each reaction at equilibrium there is a different value of the equilibrium quotient Q for each particular set of solution conditions.

However, there can be only one value of the equilibrium constant K for each reversible reaction because this is defined as the limiting value of Q when the concentrations of all solute species approach zero in the infinitely dilute solution. The relationship between K, which we call the thermodynamic value of the equilibrium constant, and Q, which we may regard as the practical value for particular conditions, may be expressed as follows:

$$K = Q \cdot F,$$

where F is a factor that expresses the deviation from ideal behavior.

It is reasonable to assign part of the deviation from ideal behavior to each of the chemical species involved in the reversible reaction. This means that the factor F may be subdivided into a cluster of f values called **activity coefficients**. It is simple to express this mathematically: for each concentration term in the expression for Q we introduce an activity coefficient f_i. For the dissociation of water, we write:

$$K = \frac{[H_3O^+]f_H\,[OH^-]f_{OH}}{X_{H_2O}^2\,f_{H_2O}^2} = Q\,\frac{f_H f_{OH}}{f_{H_2O}^2},$$

where the subscripts on the f-values merely indicate the particular species to which that f-value applies. Note that although the mole fraction of water may be essentially unity, its activity coefficient may nevertheless show a deviation from ideal behavior. In the infinitely dilute solution, where ideal behavior is defined to exist, all values of activity coefficients are simply unity and $K = Q$.

We have used three versions of the concentration ratio: r, Q, and K. The relationships between them are important: $r \rightarrow Q$ as time tends to infinity and $Q \rightarrow K$ as concentrations tend to zero.

The Ionic Strength of an Electrolyte Solution

By evaluation of a large amount of accumulated data on the nonideal behavior of electrolytes in the presence of other inert electrolytes, Gilbert Newton Lewis found an important correlation.* It is not only the concentration of the electrolyte that determines deviations from ideality, but also the magnitude of the ionic charges on the ions of the electrolyte. He defined the **ionic strength** as a general measure of the nonideality of the solution environment. The equation proposed by Lewis is still used today and is as follows:

$$\text{Ionic strength } I = \tfrac{1}{2}\sum c_i Z_i^2,$$

where for all *ionic species in the solution* there is a summation of the products of ion molarity c_i and the square of the ionic charge Z_i. This shows that doubly charged ions, such as Mg^{2+}, SO_4^{2-}, etc., have four times the nonideal

---

*University of California, Berkeley, ca. 1918.

effect of a singly charged ion such as Na^+ or Cl^-. In many publications the Greek letter μ is used instead of I to symbolize ionic strength.

For example, in a solution containing only 0.02 mole of calcium chloride per liter,

$$[Ca^{2+}] = 0.02, \qquad Z_{Ca} = 2,$$
$$[Cl^-] = 0.04, \qquad Z_{Cl} = 1,$$

and the ionic strength is

$$I = 0.5(0.02 \cdot 2^2 + 0.04 \cdot 1^2) = 0.06.$$

It is easy to see that the definition of ionic strength leads to simple relationships between the molarity of an electrolyte and its charge type. For a $1:1$ charge type, such as NaCl, KNO_3, HBr, etc., the ionic strength is simply equal to the molarity. As the above calculation shows, for a $2:1$ charge type, such as $CaCl_2$, $Mg(NO_3)_2$, Na_2SO_4, etc., the ionic strength is three times the molarity. The reader should check the values given in Table 4.2.

Table 4.2

Strong-electrolyte charge type	$\dfrac{I}{C}$	Example
$1:1$	1	KI, $HClO_4$
$2:1$ or $1:2$	3	Li_2SO_4, $MgCl_2$
$2:2$	4	$CaSO_4$
$3:2$ or $2:3$	15	$Fe_2(SO_4)_3$
$1:3$ or $3:1$	6	Na_3PO_4, $Al(NO_3)_3$

Ionic strengths are additive. When a solution contains two or more strong electrolytes, the total ionic strength of the solution is the sum of the two or more contributions. Thus, in a mixture of $0.04M$ KOH and $0.02M$ Na_2SO_4, the ionic strength is $0.04 + 3 \cdot 0.02 = 0.10$.

Care must be taken that only *ionic* concentrations are used in the calculation of ionic strength. For example, in a solution of $0.10M$ mercuric chloride $HgCl_2$, the ionic strength is *not* 0.30 as one would calculate for a strong electrolyte of the $2:1$ charge type, because this particular salt is almost completely in the *un*dissociated state in water solution, existing as $HgCl_2$ neutral molecules. The ionic strength is virtually zero. Similarly, in $0.05M$ acetic acid, the ionic strength is only about 0.0005, because only about 1% of this weak acid is dissociated into its ions.

Theoretical Calculation of Ionic Activity Coefficients

THE DEBYE–HÜCKEL EQUATION

In a brilliant theoretical analysis starting from the first principles of electrostatics, P. Debye and E. Hückel (1923) showed that for *very* dilute solutions of strong electrolytes it was possible to predict activity coefficient values in close agreement with those obtained experimentally. In the first stage of their approach, when ions were considered to be merely point charges with infinitesimal diameters and at relatively great distances from each other, a simple equation emerged. (The derivation itself is far from simple, however.) Called the **Debye–Hückel Limiting Law** because it holds only for very low concentrations, the equation is:

$$\log f_z = - AZ^2 I^{1/2},$$

where f_Z is the activity coefficient for an ion carrying a charge equal to Z (that is, $Z = 1$ for sodium ion, 2 for sulfate·ion, 3 for ferric ion). Because the value for Z is squared in the equation, it doesn't matter whether a cation or anion is being considered. The symbol A is a collection of fundamental physical quantities, including temperature and the dielectric constant of the solvent (the theory is not restricted to water solutions). For water solutions at 25° the value for A is 0.512. For other temperatures t it is satisfactory to use the relationship:

$$A = 0.4917 + 6.709 \cdot 10^{-4} t + 3.5213 \cdot 10^{-6} t^2.$$

When Debye and Hückel added the more realistic assumption that ions are of significant size and are usually present in solutions of ionic strength higher than assumed for the derivation of the limiting law, the theoretical relationships became *much* more complex, and many pages of advanced calculus are required to derive the **extended Debye–Hückel equation**:

$$\log f_z = \frac{- AZ^2 I^{1/2}}{1 + aBI^{1/2}}. \tag{4.2}$$

In this equation, A, z, and I have the same meaning as before. The symbol a stands for the size of the ion, i.e., its diameter expressed in units of angstroms ($1 \text{ Å} = 10^{-10}$ m). The symbol B, like A, is a physical constant that depends on the temperature and solvent. For water at 25°, the value of B is 0.328.

CALCULATIONS WITH THE EXTENDED D–H EQUATION

To illustrate the values of activity coefficients obtained from Eq. (4.2) we may use some of the ion sizes proposed by Kielland.* The results are summarized in Table 4.3. The large value for the size of the hydrogen ion might seem surprising, until it is remembered that the bare proton does not exist in solution. Rather, it is strongly hydrated to form the hydronium ion H_3O^+ which, in turn, is hydrogen-bonded to perhaps four other water molecules. Similarly, the large size of the lithium ion is due to hydration.

---

*J. Kielland, *Journal of the American Chemical Society*, *59*, 1675 (1937).

Table 4.3 Activity coefficients from the extended Debye–Hückel equation

Size a. Å	Examples of common ions	Ionic strength			
		0.001	0.01	0.1	1
	Charge = 1				
9	H_3O^+	0.966	0.913	0.825	0.742
6	Li^+	0.966	0.906	0.795	0.672
4	Na^+, IO_3^-, HCO_3^-, $H_2PO_4^-$, CH_3COO^-	0.965	0.901	0.768	0.601
3	K^+, NH_4^+, Ag^+, OH^-, F^-, CNS^-, ClO_4^-, BrO_3^-, Cl^-, Br^-, I^-, CN^-, NO_3^-, $HCOO^-$	0.964	0.898	0.753	0.552
	Charge = 2				
8	Mg^{2+}	0.871	0.688	0.443	0.272
6	Ca^{2+}, Cu^{2+}, Zn^{2+}, Sn^{2+}, Mn^{2+}, Fe^{2+}	0.869	0.674	0.399	0.204
5	Sr^{2+}, Ba^{2+}, Cd^{2+}, Hg^{2+}, Pb^{2+}, CO_3^{2-}	0.868	0.667	0.375	0.168
4	Hg_2^{2+}, SO_4^{2-}, CrO_4^{2-}, HPO_4^{2-}	0.867	0.659	0.349	0.130
	Charge = 3				
9	Al^{3+}, Fe^{3+}, Cr^{3+}, La^{3+}, Ce^{3+}	0.736	0.441	0.176	0.068
4	$Cr(NH_3)_6^{3+}$, PO_4^{3-}, $Fe(CN)_6^{3-}$	0.725	0.391	0.093	0.010
	Charge = 4				
11	Th^{4+}, Ce^{4+}, Sn^{4+}	0.585	0.250	0.062	0.017
5	$Fe(CN)_6^{4-}$	0.567	0.198	0.020	0.001

The above calculations are mainly to illustrate the properties of the extended Debye–Hückel equation. The coefficient values are increasingly unreliable when the ionic strengths used are higher than about 0.01. Of special concern is the inability to predict the observed minimum in experimental values of activity coefficients in the region of 0.5–1.0 ionic strength by means of Eq. (4.2) which indicates only a continuing decrease in f-values with increasing ionic strength. It has not proved possible to extend the purely theoretical treatment of electrostatic forces to ionic strengths higher than about 0.01, and so Hückel proposed that an empirical linear term be added to Eq. (4.2):

$$\log f_z = -\frac{AZ^2 I^{1/2}}{1 + aBI^{1/2}} + cI. \tag{4.3}$$

The parameter c would have to be determined by trial and error, with a different value for each case. Clearly, this is not useful for the researcher who wants to calculate reasonable values of activity coefficients for situations and ions that have not been previously studied.

THE DAVIES EQUATION

In 1938 C. W. Davies reviewed the accumulated data on experimental values of mean activity coefficients. He proposed that an equation similar to Eq. (4.3) could serve rather well for typical values of activity coefficients if only one

"average" value for the parameter c was used. He revised his equation*
slightly in 1962 and settled on the following:

$$\log f_Z = Z^2 \left(-\frac{A \cdot I^{1/2}}{1 + I^{1/2}} + 0.15I \right),$$

where $A = 0.51$ at 25°C.

Note that there is no recognition of differences in ionic sizes. In this
equation it is assumed that the effective average ion size is about $3 \cdot 10^{-10}$ m,
making the aB product equal to 1.0. The value of 0.15 for the parameter c
was chosen by Davies as best representing typical behavior. The equation is
therefore a compromise between theoretical and experimental findings and
cannot be regarded as totally reliable. For problems involving low ionic
strength and ions of known size one might be better off to use Eq. (4.2). Most
of the time this is not possible, however, and the Davies equation is the best
resource for finding the approximate values of activity coefficients.

It's an easy equation to work with: note that there are only two variables,
Z and I, and that they are separated in the equation. The entire term in
parentheses depends only on ionic strength and temperature and may be
called I'. Then,

$$\log f_Z = Z^2 I' \qquad \text{or} \qquad f_Z = 10^{Z^2 I'}.$$

Values of I' have been computed for the entire practical range of ionic
strengths; they are presented in Table 4.4 along with the corresponding
values of activity coefficients for ions of charges 1, 2, 3, 4, and 5. This table
serves as a handy reference: once the ionic strength of a solution is known,
the necessary activity coefficients can easily be determined by using interpolation for ionic strengths between the listed values.

A word of caution: the table was generated all the way to $I = 1.00$ and to
the improbable charge of 5; therefore the figures in the lower and the right
portions of the table are not reliable.

Activity coefficients calculated with the Davies equation are depicted in Figs.
4.2 and 4.3; note the very large effect of ionic charge on the values. The first set of
graphs (Fig. 4.2) is an expanded plot of the lower ionic strength range, 0 to 0.1,
where the Davies equation may be used with fair confidence. The second set of
graphs (Fig. 4.3) is extended to ionic strength 1.0, where the equation yields the
minimum value of the activity coefficients (solid lines). In addition, the
experimental values of mean activity coefficients for some common 1:1 strong
electrolytes are shown for comparison (broken lines). Note that the calculated
values are in rather good agreement at lower ionic strengths, but the experimental values show larger and individualized deviations with increasing ionic
strength.

---

*C. W. Davies, *Ion Association*, Butterworths, London, 1962, p. 41.

Table 4.4 Activity coefficients calculated according to the Davies equation

$$\log f_Z \approx Z^2\left(-\frac{0.51I^{1/2}}{1+I^{1/2}}+0.15I\right) = Z^2I',$$

where Z is ionic charge and I is ionic strength

I	I'	f_1	f_2	f_3	f_4	f_5
0.00001	−0.0016	0.996	0.985	0.967	0.943	0.912
0.00002	−0.0023	0.995	0.979	0.954	0.920	0.878
0.00003	−0.0028	0.994	0.975	0.944	0.903	0.852
0.00004	−0.0032	0.993	0.971	0.936	0.889	0.832
0.00005	−0.0036	0.992	0.968	0.929	0.877	0.814
0.00006	−0.0039	0.991	0.965	0.922	0.866	0.80
0.00007	−0.0042	0.990	0.962	0.916	0.856	0.78
0.00008	−0.0045	0.990	0.959	0.911	0.847	0.77
0.00009	−0.0048	0.989	0.957	0.906	0.839	0.76
0.00010	−0.0050	0.988	0.955	0.901	0.831	0.75
0.00020	−0.0071	0.984	0.937	0.863	0.77	0.67
0.00030	−0.0086	0.980	0.924	0.836	0.73	0.61
0.00040	−0.0099	0.977	0.913	0.814	0.69	0.56
0.00050	−0.0111	0.975	0.903	0.79	0.66	0.53
0.00060	−0.0121	0.973	0.895	0.78	0.64	0.50
0.00070	−0.0130	0.970	0.887	0.76	0.62	0.47
0.00080	−0.0139	0.968	0.880	0.75	0.60	0.45
0.00090	−0.0147	0.967	0.873	0.74	0.58	0.43
0.00100	−0.0155	0.965	0.867	0.73	0.57	0.41
0.00200	−0.0215	0.952	0.820	0.64	0.45	0.29
0.00300	−0.0260	0.942	0.79	0.58	0.38	0.22
0.00400	−0.0297	0.934	0.76	0.54	0.33	0.18
0.00500	−0.0329	0.927	0.74	0.51	0.30	0.15
0.00600	−0.0358	0.921	0.72	0.48	0.27	0.13
0.00700	−0.0383	0.916	0.70	0.45	0.24	0.11
0.00800	−0.0407	0.911	0.69	0.43	0.22	0.10
0.00900	−0.0428	0.906	0.67	0.41	0.21	0.08
0.01000	−0.0449	0.902	0.66	0.39	0.19	0.08
0.02000	−0.0602	0.871	0.57	0.29	0.11	0.03
0.03000	−0.0708	0.850	0.52	0.23	0.07	0.02
0.04000	−0.0790	0.834	0.48	0.19	0.05	0.01
0.05000	−0.0857	0.821	0.45	0.17	0.04	0.007
0.06000	−0.0913	0.810	0.43	0.15	0.03	0.005
0.07000	−0.0962	0.801	0.41	0.14	0.03	0.004
0.08000	−0.1004	0.79	0.40	0.12	0.02	0.003
0.09000	−0.1042	0.79	0.38	0.12	0.02	0.002
0.10000	−0.1075	0.78	0.37	0.11	0.02	0.002
0.20000	−0.1276	0.75	0.31	0.07	0.009	0.0006
0.30000	−0.1355	0.73	0.29	0.06	0.007	0.0004
0.40000	−0.1376	0.73	0.28	0.06	0.006	0.0004
0.50000	−0.1362	0.73	0.29	0.06	0.007	0.0004
0.60000	−0.1326	0.74	0.29	0.06	0.008	0.0005
0.70000	−0.1273	0.75	0.31	0.07	0.009	0.0007
0.80000	−0.1208	0.76	0.33	0.08	0.01	0.001
0.90000	−0.1133	0.77	0.35	0.10	0.02	0.001
1.00000	−0.1050	0.79	0.38	0.11	0.02	0.002

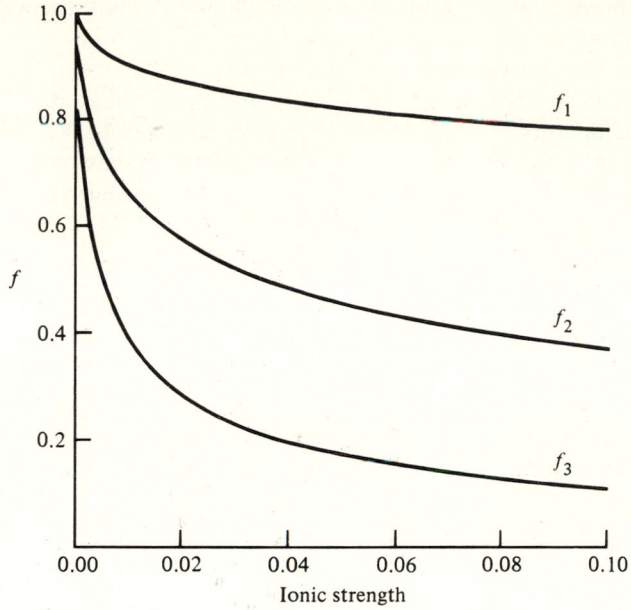

Fig. 4.2 Activity coefficients calculated with the Davies equation.

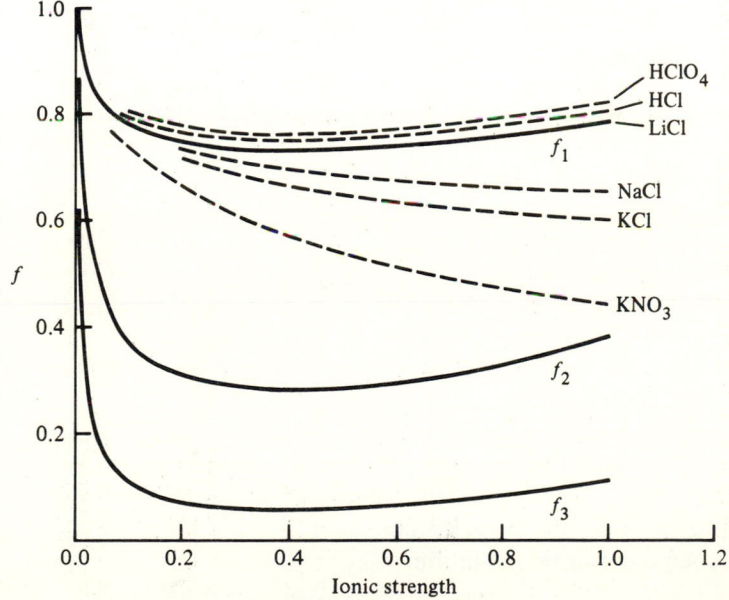

Fig. 4.3 Activity coefficients at higher ionic strengths. *Solid lines*: calculated with the Davies equation; *dashed lines*: experimental values of mean-activity coefficients.

The observation that the mean activity coefficient for potassium nitrate deviates more than the others suggests that there may be some ion-pairing of potassium and nitrate ions in the solutions.

In any case, it is risky to depend on theoretically calculated values for ionic activity coefficients at ionic strengths higher than about 0.1. All that can be said is this: it is better to make an imperfect estimate of the activity coefficient for an ion than to ignore the effect of ionic strength completely.

Molecular Activity Coefficients: Example of Iodine Solubility

In comparison with the strong electrostatic forces acting upon ions in a solution, the effect of ionic strength upon uncharged molecular species is minor. This is fortunate, because there is no useful theory for the quantitative prediction of numerical values for molecular activity coefficients and it is not possible to make experimental determinations of their values over the tremendous range of solution conditions used in practice. To illustrate the problem we will consider molecular iodine I_2. The element iodine is a purple-black crystalline solid that is only slightly soluble in water. The molarity of a saturated solution at 25° is 0.00132 mole/L; this is the value of solubility S.

The law of chemical equilibrium is quite properly applied to the solubility reaction:

$$I_2(s) \rightleftarrows I_2(aq)$$
$$K = [I_2]f_0/1 \qquad \text{(mole fraction of solid } I_2 \text{ is 1)}$$
$$= Sf_0,$$

where f_0 is the activity coefficient of the uncharged molecule and S is the solubility value of iodine, that is, its molarity in the saturated solution in contact with an excess of solid iodine.

Now, in the absence of any other solutes, the ionic strength of the solution of iodine in water is zero (the dissociation of water itself is disregarded), and so it is reasonable to assign f_0 a value of 1.000 in this solution. This means that $K = S = 0.00132$, and we have made a direct determination of the value of K.

When the pure water in contact with solid iodine is replaced by a $1.00M$ solution of sodium perchlorate (a strong electrolyte that gives the solution an ionic strength of 1.00), the solubility of molecular iodine is found by analysis to be 0.00104, which is distinctly smaller than in pure water. This is attributed to electrostatic forces and to the specific individual effects of sodium and perchlorate ions on the activity coefficient of iodine. We can calculate the activity coefficient in this solution:

$$f_0 = K/S = 0.00132/0.00104 = 1.27.$$

Molecular activity coefficients are often greater than unity. Because the sodium perchlorate has caused the solubility to be less, the effect is often called **salting out**.

Even when the effect of a specific electrolyte has been determined, as with sodium perchlorate in the present case, it is not possible to generalize about the effects of other electrolytes. For example, when solid iodine is dissolved in a solution of $1.00M$ perchloric acid, we find experimentally that the solubility is $0.00127M$. Even though this solution has the same ionic strength as the sodium perchlorate solution, the effect on the iodine molecular activity coefficient is different:

$$f_0(\text{in } 1M \text{ HClO}_4) = K/S = 0.00132/0.00127 = 1.04.$$

Observations of this type have been made on numerous molecular species, and it has not proved possible to devise either a theory or an empirical set of relationships that will allow prediction of values for molecular activity coefficients. However unsatisfying it may be, knowing that molecular activity coefficients are not really equal to 1.00 in most cases, we have little choice in making that assumption in typical equilibrium calculations.*

Another New Term: Activity

When we refer to the **activity** of a chemical species, we are speaking of a hypothetical concentration based on the ideal behavior shown by the species in the reference state of the infinitely dilute solution. Rigorous definitions of activity are an important part of a study of chemical thermodynamics, and no attempt to duplicate that will be made here. It is perfectly sufficient for us to regard activity as the product of molarity and the activity coefficient:

$$\text{Activity } A_i = [i]f_i.$$

For example, for a singly charged ion at a molarity of 0.025 in a solution with ionic strength 0.100 we have

$$A = 0.025 \cdot f_1 = 0.025 \cdot 0.78 = 0.0195.$$

In terms of the Equilibrium Law, we may refer to K as the **equilibrium expression in terms of activities**, while Q is the **equilibrium expression in terms of molarities**.

4.5   NONIDEALIZED STATEMENT OF THE EQUILIBRIUM LAW

We have seen that the environmental effects (chiefly electrostatic forces from ions) causing deviations from ideal behavior may be expressed in terms of individual factors called **activity coefficients**. The activity of a species, often

---

*For further discussion of the problem, see F. A. Long, W. F. McDevit, *Chemical Reviews, 119* (1952).

called its **effective concentration**, is simply the product of molarity and the activity coefficient. For a chemical reaction at equilibrium the value of the equilibrium quotient Q varies continuously with changing concentration conditions and only approaches its limiting value in the infinitely dilute solution. This limiting value is called the **equilibrium constant** K and is unique for each reaction.

From the way activity coefficients were defined, it follows that the numerical value for K, as determined for ideal conditions, will also hold true for nonideal conditions. In other words, in the relationship $K = Q \cdot F$, both Q and F will vary but in opposite directions as conditions vary from the ideal, but K will remain constant (at a given temperature).

Conversion of K-Values to Q-Values and Vice-Versa

Experimental work in the laboratory typically deals with concentrations, not with activities of reacting species. Therefore it is very useful to have values for the equilibrium quotients Q. Since most reference tables are set up with the values of equilibrium constants K, we must be able to convert as needed. The relationship is:

$$K = Q \cdot F \qquad \text{or} \qquad Q = \frac{K}{F},$$

where F is the appropriate cluster of individual activity coefficients.

For example, suppose we are concerned with the equilibrium reaction

$$HA^{2-} \rightleftharpoons H^+ + A^{3-},$$

which is known to have an equilibrium constant value of $4.4 \cdot 10^{-5}$:

$$K = 4.4 \cdot 10^{-5} = \frac{[H^+][A^{3-}]}{[HA^{2-}]} \cdot \frac{f_1 f_3}{f_2} = Q \frac{f_1 f_3}{f_2}.$$

Note that the first step is to write the complete equilibrium expression for the reaction, including an activity coefficient for each participating ionic species. Now, the f-values will depend upon the ionic strength of the solution. For illustration, we shall assume that the solution in question has an ionic strength of 0.030, but in general it will be necessary to perform some calculations to find the ionic strength for the case under study.

By consulting the chart of activity coefficients (or by using $I = 0.03$ in the Davies equation), we find:

$$f_1 = 0.85, \qquad f_2 = 0.52, \qquad f_3 = 0.23.$$

Therefore, the value for Q at this ionic strength is:

$$Q = K \frac{f_2}{f_1 f_3} = 4.4 \cdot 10^{-5} \frac{0.52}{0.85 \cdot 0.23} = 1.2 \cdot 10^{-4}.$$

Another example A solution is prepared by dissolving 0.0371 mole of potassium dichromate $K_2Cr_2O_7$ and diluting the solution to one liter. Spectrophotometric measurements show that the equilibrium solution contains

$$[Cr_2O_7^{2-}] = 0.0259 \text{ mole/L},$$
$$[HCrO_4^-] = 0.0224 \text{ mole/L}.$$

We assume that the K^+ ions are completely dissociated from the negative ions and that the system may be described by a single reversible reaction:

$$H_2O + Cr_2O_7^{2-} \rightleftharpoons 2HCrO_4^-$$

with an equilibrium constant

$$K = \frac{[HCrO_4^-]^2}{[Cr_2O_7^{2-}]} \cdot \frac{f_1^2}{f_2}.$$

In this particular solution the ionic strength is found by adding the contributions of the three ions:

$$I = 0.5 \cdot ([K^+] \cdot 1^2 + [HCrO_4^-] \cdot 1^2 + [Cr_2O_7^{2-}] \cdot 2^2)$$
$$= 0.5 \cdot (0.0742 + 0.0224 + 0.0259 \cdot 4) = 0.100.$$

The value for the equilibrium quotient Q is readily calculated:

$$Q = \frac{[HCrO_4^-]^2}{[Cr_2O_7^{2-}]} = \frac{0.0224^2}{0.0259} = 0.0194.$$

To find the value for K it is necessary to look up the values for the activity coefficients. At $I = 0.1$, we have $f_1 = 0.78$ and $f_2 = 0.37$. Therefore,

$$K = Q \cdot \frac{f_1^2}{f_2} = 0.0194 \cdot \frac{0.78^2}{0.37} = 0.032.$$

What is the Q-value for this equilibrium in a solution of ionic strength 0.02? *Answer*: At $I = 0.02$ we find $f_1 = 0.87$ and $f_2 = 0.57$. Therefore,

$$Q(I = 0.02) = 0.032 \cdot \frac{0.57}{0.87^2} = 0.024.$$

4.6 LOGARITHMIC VALUES FOR EQUILIBRIUM CONSTANTS

Numerical values for equilibrium constants vary over a tremendous range: from very small values such as 10^{-25} to very large values such as 10^{25}. It is common practice to use logarithmic values in reference tables and in discussion of relative values for the equilibrium constants. Usually, when K is greater than 1, we use $\log_{10}(K)$, and when K is less than 1, we use $-\log_{10}(K)$, which is given the symbol pK. A few examples will serve:

K	$\log(K)$	pK
$2 \cdot 10^{-14}$		13.7
$1 \cdot 10^{-6}$		6.0
50	1.7	
$4 \cdot 10^{23}$	23.6	

To convert a pK value to the corresponding K value, one simply uses the relationship

$$K = 10^{(-pK)} = 10^{(\log K)}$$

4.7   BRIEF SURVEY OF TYPES OF AQUEOUS EQUILIBRIA IMPORTANT IN ANALYSIS

Even though this anticipates the detailed discussions in later chapters, it seems appropriate to illustrate the application of the law of chemical equilibrium to the four main classes of aqueous reactions that are so important in chemical analysis.

Oxidation–Reduction (Redox) Reactions

Redox reactions are so variable in their stoichiometry that we might as well rely on the old standard general equation

$$a\,A + b\,B + \cdots \rightleftarrows c\,C + d\,D + \cdots$$

to represent them. In contrast to the other groups, there are no "special" definitions for equilibrium constants in this class. It is particularly true with redox reactions, much more than with the other groups, that one must be cautious about thermodynamic conclusions, for very often the redox reaction mechanisms are such that the reactions are very slow in spite of large equilibrium constant values. In Chapter 2 we discussed the half-reaction concepts and stoichiometry of redox reactions. Even though many reactions involving a change in oxidation numbers (and therefore called **redox**) do not proceed via an electron transfer mechanism, we may accept the working hypothesis that electrons are transferred and write two half-reactions, one involving a reduction and the other an oxidation.

As explained in Chapter 8, it is customary to tabulate redox equilibrium constants in the equivalent form of standard potentials, that is, E^0 values, rather than as K- or $\log(K)$-values.

Redox reactions are of enormous value in every aspect of chemical analysis including titrations, gravimetry, spectrophotometry, electroanalytical chemistry, separations, and sample treatment. This is the reason for treating this important topic prior to the other material.

We speak of **oxidizing agents** and **reducing agents** as substances having thermodynamic *and* kinetic tendencies to undergo reduction and oxidation, respectively. In terms of a half-reaction, an oxidizing agent reacts as follows:

$$Ox + hH^+ + ne \rightleftarrows Red + \frac{h}{2}H_2O.$$

Redox couple

In some reactions the H^+ and H_2O are reversed.

A substance such as permanganate ion, MnO_4^-, is referred to as a **strong oxidizing agent** or **strong oxidant**, because it can increase the oxidation state of many other substances.

Acid–Base (Proton-Transfer) Reactions

In comparison with the other categories, acid–base reactions are the least complicated. Reaction rates are fast, stoichiometry is exact, and equilibrium systems are not too involved. Through the measurement of the hydrogen ion activity it is possible to make exceptionally precise determination of equilibrium constant values. Because of the importance of H^+ concentration in controlling the solution conditions, it is necessary to understand how to calculate this factor in various solutions by using ideas of stoichiometry in combination with tables of equilibrium constants.

An **acid** is defined as a substance capable of transferring a proton (hydrogen ion H^+) to a **base**, which is simply a substance capable of accepting a proton. From the bonding concepts it is apparent that any base must have at least one pair of unshared electrons in its molecular structure:

$$H:A \quad + \quad :B \quad \rightleftarrows \quad H:B \quad + \quad :A.$$

| acid | base | conjugate acid of B | conjugate base of HA |

Conjugate pairs

A **strong acid** is one with a strong thermodynamic tendency to transfer its proton. To compare the **strengths** (not to be confused with **concentrations**) of different acids we use a reference base, water, as substance B:

$$HA + H_2O \rightleftarrows H_3O^+ + A.$$

Then the equilibrium constant is called the **acid dissociation constant**:

$$\text{Acid dissociation constant } K_a = \frac{[H_3O^+][A]f_Hf_A}{[HA]f_{HA}}.$$

In these equations the ionic charges have been omitted both for simplicity and to indicate that acids may be either molecules, or cations, or anions.

The most common molecular functional groups involved in proton transfer are: carboxyl —CO(OH), hydroxyl —OH, sulfonic —SO$_2$(OH), amino —NH$_2$, sulfhydryl —SH.

Acid–base equilibria are considered in detail in Chapter 9, while applications to titration are covered in Chapter 10.

Metal–Ligand Complex Formation Reactions

Aqueous solutions of so-called **simple metal ions** actually contain coordination complexes of the metal ion with water molecules. When we casually write, for example, Cu^{2+} or Al^{3+}, we are merely using shorthand for the existing species which are Cu(H$_2$O)$_4^{2+}$ and Al(H$_2$O)$_6^{3+}$. The **aquo-complexes**, as they are called, exist because the unshared electron pairs on the oxygen atom of a water molecule are able to form a coordinate covalent bond using one of the unfilled orbitals in the metal ion.

This means that substances (other than water) that have unshared pairs of electrons may compete with water for these bonding positions. Such substances are called **ligands** (from the Latin *ligare*, to bind or tie up) and may be successful in displacing one or more of the water molecules from the aquo-complex:

$$M(H_2O)_n + :L \rightleftarrows ML(H_2O)_{n-1} + H_2O.$$

In such reactions the equilibrium constant is called the **formation constant**:

$$K_f = \frac{[ML]f_{ML}}{[M][L]f_M f_L}.$$

The H$_2$O molecules are usually omitted for simplicity.

Because this class of reactions is often studied or used under conditions of high ionic strength, it is more realistic to ignore the activity coefficients and to apply Q-expressions instead of K-expressions.

Various metal ions differ greatly from each other in the way they interact with ligands. Also, a given ligand will react very differently with various metal ions. In some cases, for example, Cu^{2+} with Cl$^-$, the complex formation is weak enough to be neglected in a dilute solution. By contrast, the complexation of mercuric ion Hg^{2+} with Cl$^-$ is very strong.

When a ligand is capable of displacing more than one water molecule from the aquo-complex, we speak of **successive steps** in complex formation, with each step having its equilibrium quotient:

Step 1 Cu^{2+} + NH$_2$CH$_2$CH$_2$NH$_2$ $\rightleftarrows$ Cu en^{2+}, $Q_1 = \dfrac{[\text{Cu en}^{2+}]}{[\text{Cu}^{2+}][\text{en}]}$;
                  ethylenediamine (en)

Step 2 Cu en^{2+} + en $\rightleftarrows$ Cu(en)$^{2+}$, $Q_2 = \dfrac{[\text{Cu(en)}^{2+}]}{[\text{Cu en}^{2+}][\text{en}]}$.

As stated above, for a ligand to interact with a metal ion it must have an unshared pair of electrons. But this means that it is also able to act as a base, i.e., to accept a proton from whatever acid might also be in the solution. Therefore the study of metal–ligand complexing is inherently complicated by the competition between metal ion and hydrogen ion for the electron pair on the ligand. If the hydrogen ion concentration is made very small in order to eliminate its competition, this automatically means that the hydroxide ion concentration of the solution is made larger. Another complication arises: the hydroxide ion, acting as a ligand itself, may be able to compete effectively with the ligand for the metal ion.

It may be clear at this point why we should postpone the discussion of metal–ligand complexing to Chapter 14, after a thorough study of the principles of acid–base equilibria.

Precipitation–Solubility Reactions

In principle we may assume that *any* substance has some finite solubility in water solutions, even if it is too small to measure with existing analytical techniques. Therefore, in principle, we should not use the term **insoluble**. However it is a lot easier to say "insoluble" than to meticulously enunciate phrases such as "very slightly soluble," "virtually insoluble," "sparingly soluble," and the like. As a rough guide, "insoluble" refers to substances that can have solution concentrations no greater than, say, $0.01 M$.

Precipitation is the formation of an insoluble substance either by changing the nature of the solvent or by reaction between species in the solution. For example, a substance soluble in water but not in alcohol may be caused to precipitate from its aqueous solution simply by pouring in an equal volume of alcohol. More commonly for our purposes, one of the solutions will contain a metal cation in the form of one of its soluble salts, while the other solution contains an anion in the form of one of its soluble salts. When the two solutions are mixed, the metal ion and anion may form an insoluble complex:

$$M + A \rightleftarrows MA(solid).$$

Thus, precipitation is really just a special case of metal–ligand complex formation, with the complex being insoluble. This has important implications for chemical analysis in terms of separations and for gravimetric determinations of amounts of either the metal ion or the anion.

It certainly would be possible to define equilibrium constants in terms of the precipitation process but it is more of a custom to use the insoluble substance as the "starting point" and to think in terms of *solubility* rather than *precipitation*. There are only two fundamental types of equilibrium constants that apply to solubility as such:

1. Sticking with the simple 1:1 compound MA for convenience, we may first speak of its **intrinsic solubility**. This refers to the passage into the aqueous solution of molecules (or ion pairs) of the substance from the crystal lattice:

$$MA(solid) \rightleftarrows MA(aqueous).$$

The identity and integrity of the species has been preserved and the equilibrium constant for this step may be called K_0, the intrinsic solubility:

$$K_0 = \frac{[MA]f_0}{X_{MA(s)}} = [MA]$$

if $f_0 = 1.0$ and $X_{MA} = 1.0$.

Regardless of what else might happen to the species through interaction with other substances in the solution, this step must be independently satisfied in accordance with the law of chemical equilibrium. That is why it is called the intrinsic (due to its essential nature) solubility.

Admittedly, the intrinsic solubility of many substances is so small that current analytical techniques are not sensitive enough to measure it. In such cases it is necessary to regard K_0 as being virtually zero and unknown.

2. The second type of equilibrium that must be considered is the dissociation of MA(aqueous) into its constituent ions:

$$MA(aq) \rightleftarrows M + A.$$

We speak of the **dissociation constant** for MA(aq):

$$K_d = \frac{[M][A]f_M f_A}{[MA]},$$

again letting $f_0 = 1$.

This process is clearly just the reverse of the metal–ligand complex formation discussed in the previous section. The value of K_d is very small for some substances and virtually infinity for others (for example, NaCl).

There is an important combination of circumstances, not uncommon, which make it useful to define a third type of equilibrium constant pertaining to solubility. When the intrinsic solubility is negligibly low, while the dissociation into ions is essentially complete, the *net* chemical change in the solubility process may be written without any evidence of the molecular species:

$$MA(s) \rightleftarrows M + A.$$

We speak of a substance showing this behavior as a **strong electrolyte**, meaning that for practical purposes it is completely dissociated into its ions. The equilibrium constant for this overall change is given the name **solubility product** and is written

$$K_{sp} = [M][A]f_M f_A.$$

For insoluble salts not having a 1:1 composition we write:

$$M_m A_a(s) \rightleftarrows mM + aA, \qquad K_{sp} = [M]^m [A]^a f_M^m f_A^a.$$

It may be that such a salt will undergo stepwise dissociation until it finally produces the M and A ions. There will be a dissociation constant for each step. The reader may care to prove that the solubility product K_{sp} is equal to the algebraic product $K_0 K_{d1} K_{d2} \ldots K_{dn}$, where n is the number of successive dissociation steps.

Solubility product constants are very useful in planning conditions for analytical precipitations, provided they do not mislead the user into neglecting the possible effects of intrinsic solubility and incomplete dissociation.

In this text, solubility equilibria are studied rather late, in Chapters 17 and 18, because of the important effects that can occur *beyond* the simple and fundamental steps discussed above. The concentration of the metal ion may be strongly affected by the presence of complexing ligands in the solution, while the concentration of the anion may be affected by hydrogen ion, forming HA. In fact, the topic brings together all of the other equilibrium and analysis principles studied previously.

4.8   LECTURE DEMONSTRATIONS: TYPES OF EQUILIBRIA

1. Oxidation–Reduction.

To 400 mL of $0.01M$ ferric nitrate ($0.1M$ nitric acid) add 1 mL of $0.4M$ potassium iodide (or 4 mL of $0.1M$). A yellow color develops due to the reversible redox reaction.

$$Fe^{3+} \quad + \quad I^- \quad \rightleftarrows \quad Fe^{2+} \quad + \quad I_2(aq)$$

nearly	colorless	ferrous ion,	iodine
colorless		colorless	yellow

This reaction may be forced to reverse itself if the equilibrium molarity of iodide ion is strongly decreased. This is accomplished by adding 10 mL of $0.1M$ silver nitrate. The precipitation reaction, which is also a reversible equilibrium, is:

$$I^- + Ag^+ \rightleftarrows AgI(solid).$$

2. Acid–Base, or Proton, Transfer

To 400 mL of distilled water add one drop of $1M$ hydrochloric acid; this produces a weakly acidic solution. Now add, dropwise, bromcresol purple (or bromcresol green) indicator solution until the color is sufficient. The reversible reaction is:

$$H_3O^+ + \quad In^- \quad \rightleftarrows H_2O + \quad HIn$$

	base-form		acid-form
	purple		yellow

Add more acid if necessary to transform the indicator to its acid form. Show the reversibility of the reaction by alternately adding drops of $1M$ sodium hydroxide and hydrochloric acid, using a magnetic stirrer to speed mixing.

Explain how the addition of NaOH controls the equilibrium concentration of H_3O^+, thus in turn controlling the extent of the indicator transformation.

3. Metal–Ligand Complexing

a)　To 400 mL of $0.01M$ ferric nitrate ($0.1M$ nitric acid) add successive drops of $0.1M$ potassium thiocyanate. The reaction is the formation of the complex ion:

$$Fe^{3+} + \quad SCN^- \quad \rightleftarrows FeSCN^{2+}$$
$$\text{thiocyanate ion,} \qquad \text{red}$$
$$\text{colorless}$$

As more thiocyanate is added, the color deepens due to increased formation of the complex. As in the first demonstration, this reaction may be reversed by the addition of enough silver ion to precipitate the thiocyanate:

$$SCN^- + Ag^+ \rightleftarrows AgSCN(\text{solid}).$$

b)　To 400 mL of $0.01M$ copper sulfate add successive portions of concentrated ammonia solution. The deepening of the color is due to the formation of a series of copper–ammonia complexes:

$$Cu^{2+} \quad + NH_3 \rightleftarrows CuNH_3^{2+} \cdots Cu(NH_3)_4^{2+}.$$
$$\text{light blue} \qquad \text{deep blue}$$

The reversibility of these reactions may be demonstrated by adding enough hydrochloric acid to react with the ammonia.

4. Precipitation

To 400 mL of $0.01M$ mercuric nitrate ($0.01M$ nitric acid) add successive 1-mL portions of $1M$ potassium iodide, while stirring vigorously on a magnetic stirrer. In the vortex there is formation of the mercuric iodide precipitate

$$Hg^{2+} + 2I^- \rightleftarrows HgI_2(\text{solid});$$

the reversibility of this reaction is shown by the fact that the red-orange precipitate redissolves as the solution becomes mixed. When enough potassium iodide has been added, the precipitate will not redissolve upon continued stirring.

If more potassium iodide solution is then added, the red-orange precipitate will eventually dissolve due to a metal–ligand equilibrium:

$$HgI_2(s) + 2I^- \rightleftarrows \quad HgI_4^{2-}.$$
$$\text{colorless}$$

4.9   CASE STUDY: EFFECT OF POTASSIUM CHLORIDE ON CALCIUM IODATE SOLUBILITY*

Calcium iodate hexahydrate $Ca(IO_3)_2 \cdot 6H_2O$ is a slightly soluble salt. If we assume that it is completely dissociated into its ions in solution, then the equilibrium system in the presence of the inert electrolyte potassium chloride is:

$$Ca(IO_3)_2 \cdot 6H_2O(\text{solid}) \overset{\text{slight}}{\rightleftarrows} Ca^{2+} + 2IO_3^-,$$

$$KCl \overset{}{\underset{100\%}{\rightarrow}} K^+ + Cl^-.$$

There is no evidence that the potassium or chloride ions interact *chemically* with the calcium or iodate ions. Their only effect is assumed to be the establishment of a varying ionic strength. Therefore, the only equilibrium constant involved in this system is the solubility product constant

$$K = \underbrace{[Ca^{2+}][IO_3^-]^2}_{Q} f_{Ca} f_{IO_3}^2$$

if the mole fraction of the solid is taken to be unity.

Establishing the Equilibrium Mixtures

A sufficient quantity of solid calcium iodate was added to each of a series of solutions of potassium chloride, ranging in concentration from zero (i.e., pure water) to $2.00M$. The stoppered bottles were shaken for several days while submerged in a water bath maintained at precisely $25°$, so that the solubility equilibrium was reached. The excess solid calcium iodate was then removed by filtration, so that the saturated solutions could be analyzed to determine how much had dissolved.

Analytical Principle

Once the solid calcium iodate had been removed, portions of the saturated solutions were treated with hydrochloric acid and potassium iodide. All of the iodate ions that had entered the solution were therefore converted to triiodide ions by the oxidation–reduction reaction

$$IO_3^- + 6H^+ + 8I^- \leftrightarrows 3I_3^- + 3H_2O.$$

The resulting triiodide ion could then be determined by titration with a

---

* After P. Gross, S. Klinghoffer, *Monatshefte für Chemie*, **55**, 338 (1930), as reported in W. Linke *Solubilities of Inorganic and Metal Organic Compounds*, 4th ed., Vol. 1, p. 609, 1958.

standardized solution of sodium thiosulfate $Na_2S_2O_3$ by using the reaction

$$I_3^- + 2S_2O_3^{2-} \rightarrow 3I^- + S_4O_6^{2-}.$$
$$\text{tetrathionate ion}$$

The endpoint of the titration is marked by the disappearance of the last visible yellow color due to the triiodide ion.

Now, each dissolved millimole of calcium iodate dissociates to form 2 mmoles of iodate ion. But each millimole of iodate ion yields 3 mmoles of triiodide ion and each millimole of triiodide ion requires 2 mmoles of thiosulfate ion in the titration. Overall, it is clear that each millimole of calcium iodate ultimately requires 12 mmoles of thiosulfate ion. Here is a sample calculation of solubility:

Suppose that a 10.00-mL sample of the saturated solution of calcium iodate in water (no KCl present), after treatment as described above, required 48.0 mL of $0.02000M$ sodium thiosulfate for titration. This is 0.960 mmole of thiosulfate, which corresponds to $0.960/12 = 0.0800$ mmole of calcium iodate. The molarity of calcium iodate in the sample (and hence the solubility of calcium iodate in pure water) is therefore $0.0800/10 = 0.00800$ mole/L.

Calculation of Q-Values from Solubilities

If the molar solubility of calcium iodate is S and the dissociation gives two iodate ions for each calcium ion, then

$$[Ca^{2+}] = S, \quad [IO_3^-] = 2S, \quad \text{and} \quad Q = (S)(2S)^2 = 4S^3.$$

For the example just given for the solubility in pure water,

$$[Ca^{2+}] = 0.00800, \quad [IO_3^-] = 0.01600,$$
$$Q = (0.00800)(0.01600)^2 = 2.048 \cdot 10^{-6}.$$

Research Results

The values for S and Q found by Gross and Klinghoffer are given in Table 4.5.

Table 4.5

Molarity of KCl	Solubility of $Ca(IO_3)_2$	Ionic strength	Solubility product, Q
0.00	$7.976 \cdot 10^{-3}$	0.02393	$2.030 \cdot 10^{-6}$
0.050	$9.551 \cdot 10^{-3}$	0.07865	$3.485 \cdot 10^{-6}$
0.100	$10.60 \cdot 10^{-3}$	0.1318	$4.764 \cdot 10^{-6}$
0.150	$11.39 \cdot 10^{-3}$	0.1842	$5.911 \cdot 10^{-6}$
0.300	$13.26 \cdot 10^{-3}$	0.3398	$9.326 \cdot 10^{-6}$
0.500	$15.16 \cdot 10^{-3}$	0.5455	$1.394 \cdot 10^{-5}$
0.750	$17.51 \cdot 10^{-3}$	0.8025	$2.147 \cdot 10^{-5}$
1.00	$18.78 \cdot 10^{-3}$	1.056	$2.649 \cdot 10^{-5}$
1.50	$22.75 \cdot 10^{-3}$	1.568	$4.710 \cdot 10^{-5}$
2.00	$25.59 \cdot 10^{-3}$	2.077	$6.703 \cdot 10^{-5}$

The values of ionic strength were calculated with the realization that the calcium iodate is a $2:1$ electrolyte, so that

$$I = M_{KCl} + 3M_{Ca(IO_3)_2} = M_{KCl} + 3S.$$

The effect of increasing ionic strength on the solubility is obviously important: S increases threefold over the range of 0 to $2M$ KCl. Far from being constant, the equilibrium quotient increases by a factor of 33 as the solution deviates further from ideality. To find the thermodynamic value of the equilibrium constant K, it is necessary to extrapolate the Q-values to the hypothetical condition of zero ionic strength.

Extrapolation of Results

We shall consider two approaches for the determination of K by extrapolation of the Q-value. The simple, direct approach is to plot Q versus I, draw a smooth curve connecting the experimental points, and then extend the curve until it cuts the Y-axis at $I = 0$. This plot is shown in Fig. 4.4 (top curve) using only the four lowest points in the data set. The approach appears to be quite satisfactory, with an intercept value of $1.35 \cdot 10^{-6}$. The four points certainly lie on a smooth curve and the extrapolation (dashed part of the curve) is not very

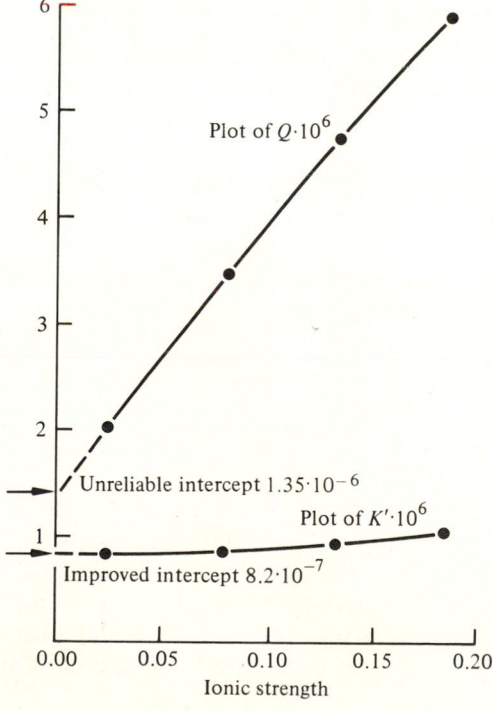

Figure 4.4

long. Unfortunately, there is a hidden problem. Before discussing it let us consider the alternative, more reliable approach to carrying out the extrapolation.

This approach requires that the experimental Q-values be first converted to a set of estimates of K. This is done by using the relationship,

$$K' = Qf_2 \cdot f_1^2,$$

where the activity coefficients f_2 and f_1 are obtained from the Davies equation for the value of ionic strength corresponding to the Q-value. The symbol K' is used (instead of K) as a reminder that the Davies equation is not perfect, so that the result is merely an approximation to K. As an illustration of the calculation, consider the first data point, where $I = 0.02393$ and $Q = 2.030 \cdot 10^{-6}$. At this ionic strength,

$$f_1 = 0.862 \qquad \text{and} \qquad f_2 = 0.551.$$

Therefore,

$$K' = 2.030 \cdot 10^{-6} \cdot 0.551 \cdot 0.862^2 = 8.300 \cdot 10^{-7}.$$

Similar calculations for the next three data points give K' values of $8.765 \cdot 10^{-7}$, $9.585 \cdot 10^{-7}$, and $10.44 \cdot 10^{-7}$. The plot of K' versus I is shown as the lower curve of Fig. 4.4.

Note the large difference between the intercept values of the two plots. The hidden problem in the plot of Q vs. I is the absence of experimental points in the region of very low ionic strength, and it is precisely in this region that the values of Q are changing most rapidly. Inspection of the table of activity coefficients shows that their values change very quickly at first and then more gradually. In other words, we should expect the plot of Q vs. I to turn downward more and more sharply as the ionic strength decreases. It is not possible to anticipate this from the inspection of the data obtained at higher ionic strengths, and therefore we end up with an intercept that is too high.

On the other hand, the conversion of Q-values to K'-values tends to eliminate the effect of activity coefficients. If the Davies equation were perfect, then the K'-values would all be the same and equal to the true K-value. The deviation of K' as the ionic strength increases is merely evidence of the increasing error in the Davies equation. However, and this is a most important point, we may expect the Davies equation to improve as the ionic strength decreases, and therefore *we should expect the plot of K' vs. I to approach a horizontal limiting slope* upon extrapolation to zero ionic strength. The line in the figure has been drawn with this guiding principle. The intercept $8.2 \cdot 10^{-7}$ thus serves as a more reliable determination of the equilibrium constant.

Calculation of Mean-Activity Coefficient from Experimental Data

The mean-activity coefficient $f_\pm$ for a $2:1$ electrolyte such as calcium iodate is defined by the equation,

$$f_\pm = (f_{Ca}\, f_{IO_3}^2)^{1/3}.$$

Once the value of K has been determined by extrapolation, the relationship between K and Q may be used to calculate $f_\pm$ as follows:

$$\frac{K}{Q} = f_{Ca}\, f_{IO_3}^2, \qquad \text{therefore} \qquad f_\pm = \left(\frac{K}{Q}\right)^{1/3}$$

For example, taking the data at ionic strength 2.077 (the last data point), we get

$$f_\pm = \left(\frac{8.2 \cdot 10^{-7}}{6.70 \cdot 10^{-5}}\right)^{1/3} = 0.230.$$

There is no certain way to resolve this value into separate activity coefficients for the individual ions, calcium and iodate.

4.10 REACTION TENDENCY

When we write a balanced equation for a chemical reaction

$$a\,A + b\,B \rightarrow c\,C + d\,D,$$

we are simply making a statement about stoichiometry. This is not as simple as it seems because it is vital to know the identities of reactant and product species and the molar amounts involved. But the equation itself tells us nothing about the tendency for the reaction to take place and, after all, it is the occurrence (or perhaps nonoccurrence) of a chemical change that is of supreme importance to a laboratory chemist who is working in either chemical synthesis or chemical analysis.

There are two aspects to reaction tendency and they must be considered separately: one is reaction rate (kinetic tendency) and the other is the change in Gibbs free energy (thermodynamic tendency). The success of experimental chemistry depends on the chemist's ability to influence each of these factors by controlling the conditions of the experiments.

Kinetic Tendency

The control of reaction rates is of tremendous importance, sometimes because it is advantageous to speed up a desired reaction and at other times because an undesired reaction has to be suppressed. There are three principal approaches to consider:

1. It is well known that the rate of chemical reactions increases with temperature. In some cases this may be due simply to the faster motions of the species and the greater number of collisions per second, but it is frequently true that higher temperatures will change the nature of the reacting species (e.g., through dissociation) and thus make possible a different mechanistic pathway with a smaller activation energy barrier.

For example, the oxidation of chromium(III) by bromate ion, used for the precipitation of insoluble metal chromates from homogeneous solution, is very slow at room temperature but can be completed in a few minutes in boiling solution. There are many examples of analytical titrations that must be carried out in hot solution. In other situations a chemist may deliberately carry out a reaction in an ice-water bath to keep it from "taking off" and destroying the lab.

The effect of light on certain chemical reactions may be similar to that of heat due to the formation of free radicals. For example, the importance of distinguishing between thermodynamic and kinetic tendency is shown in a colorful account written in 1808 by the French chemists Gay-Lussac and Thenard:

> We made new mixtures of hydrogen and chlorine gases and placed them in darkness, awaiting some moments of bright light. Two days after having made the mixtures we were able to expose them to the sun. Scarcely had they been exposed, when they suddenly enflamed with a very loud detonation, and the jars were reduced to splinters and projected to a great distance.

It seems clear that the reaction $H_2 + Cl_2 \rightarrow 2HCl$ has a rather large thermodynamic tendency, judging from the considerable release of energy when the change did occur. But in the absence of sunlight the mixtures sat around for two days without seeming to change at all, thus showing virtually zero kinetic tendency until the ultraviolet light initiated a free-radical mechanism.

2. Many chemical reactions with a strong thermodynamic tendency proceed at negligible or slow rates when the reactant species are mixed together. It is often possible to speed up the rate without changing the temperature, through addition of a **catalyst**. A catalyst is a substance that provides a mechanistic pathway with smaller activation barriers by acting as an intermediary between the reacting species. The catalyst is usually required only in small quantity because it is rapidly recycled.

For example, the oxidation of manganese(II) by cerium(IV) is slow:

$$Mn^{2+} + 5Ce^{4+} \rightarrow MnO_4^- + 5Ce^{3+} \qquad \text{(not fully balanced)}.$$

Yet, due to the deep purple color of permanganate ion MnO_4^-, this reaction can be used in the spectrophotometric determination of manganese by simply adding a trace of silver nitrate to the solution. Evidently the silver facilitates the transfer of electrons by two main steps:

$$Ag^+ \quad + \quad Ce^{4+} \quad \rightarrow \quad Ag^{2+} \quad + \quad Ce^{3+},$$
$$\text{argentic ion}$$

$$5Ag^{2+} + Mn^{2+} \rightarrow MnO_4^- + 5Ag^+.$$

The mechanism involves the formation of intermediate oxidation states of manganese, but the important point is that the silver ions are used over and over again as they are cycled between the +1 and +2 states.

    As another example, a concentrated (30%) solution of hydrogen peroxide is fairly stable at room temperature in spite of a huge thermodynamic tendency to decompose according to the equation:

$$H_2O_2 \rightarrow \tfrac{1}{2}O_2 + H_2O.$$

The results are spectacular when a pinch of solid manganese dioxide MnO_2 is added to a few milliliters of the H_2O_2 solution.

3. The third experimental approach to controlling reaction rates in the laboratory is to choose concentration conditions. Even a very fast reaction can occur only as quickly as the reactants are allowed to come into contact. By adding a solution of one reactant very slowly to a solution of the other reactants, or by using very dilute concentrations in a homogeneous mixture, or by generating one of the reactants slowly *in situ*, a chemical change may be slowed down as desired. This is particularly important in the formation of precipitates for analytical determinations.

    On the other hand, the concentration effect may be used to speed up a reaction. For example, the important reaction between iodate and iodide

$$IO_3^- + 8I^- + 6H^+ \rightleftharpoons 3I_3^- + 3H_2O$$

is slow in weak acidic solutions, but by increasing the concentration of H^+ it can be made to go very quickly.

    The foregoing discussion has emphasized reactions with a strong thermodynamic tendency that are slow unless particular steps are taken to speed them up. As a final comment we should note that the opposite behavior is also possible: a reaction may be kinetically very fast, but thermodynamically unfavorable. The most familiar example of this is the proton transfer between two molecules in pure water:

$$2H_2O \rightleftharpoons H_3O^+ + OH^-.$$

The dissociation of water into ions is fast, i.e., has a strong kinetic tendency. However, the recombination reaction is even faster, and so in pure water the equilibrium is far to the left in the above equation.

    The point to remember is this: there is no way to predict the reaction rate from chemical thermodynamics.

4.11   LECTURE DEMONSTRATIONS: REACTION TENDENCY

1. The Hydrogen–Chlorine Reaction

The classroom performance of this demonstration is described in *Tested Demonstrations in Chemistry* (6th ed., p. 154), but I prefer to do it outside on a sunny day. Equal volumes of hydrogen and chlorine are injected into a small balloon which is then placed in a box to keep the light out. The box is taken outside and the covers are lifted by using long strings to expose the balloon to sunlight. The effect is well worth the trouble.

2. Catalysis in Redox Reactions

a)   Pour about 15 mL of 30% hydrogen peroxide into a large (for example, 3 liter) erlenmeyer flask. There is a strong thermodynamic tendency for decomposition into oxygen and water, but the kinetics are very slow. Now add a tiny scoop of manganese dioxide MnO_2 to catalyze the decomposition. Swirl to mix and see the genie come out of the bottle.

b)   To 250 mL of $1M$ nitric acid add 1 mL of $0.1M$ manganous sulfate solution. Then add 25 mL of $0.1M$ ceric ammonium nitrate (in $1M$ nitric acid). The reaction to oxidize the manganese to the $+7$ state is slow:

$$Mn^{2+} + 5Ce^{4+} \rightarrow MnO_4^- + 5Ce^{3+}.$$

Repeat the experiment, but this time also add 10 mL of $0.1M$ silver nitrate. Compare the rate of reaction, as judged by the formation of the purple color due to the permanganate ion.

3. A Reaction Strongly Dependent upon Acidity

To 250 mL of water add 10 mL of $0.025M$ potassium iodate and 10 mL of $0.2M$ potassium iodide. No visible change occurs. Now add 10 mL of $2M$ hydrochloric acid, which causes the immediate formation of triiodide ion.

This reaction can be reversed by the addition of sodium hydroxide, since it removes the hydrogen ion necessary for the change.

4. A Reaction Requiring Heating

To 250 mL of water add 10 mL of $0.5M$ chromic nitrate. The purplish color is due to the hydrated chromium(III) ion Cr^{3+}. Add 10 mL of $6M$ sodium hydroxide, causing the precipitation of chromic hydroxide:

$$Cr^{3+} + 3OH^- \rightleftarrows Cr(OH)_3(s).$$

Now add 10 mL of 30% hydrogen peroxide. At room temperature there is no rapid change, but when the solution is heated to near boiling the chromium is oxidized to the chromate ion. This is the reaction used in Chapter 2 to illustrate the method of balancing an equation for a reaction in basic solution.

4.12   PREDICTION OF THE REACTION DIRECTION
FOR A NONEQUILIBRIUM MIXTURE

If the numerical value for the equilibrium quotient Q is known for a particular reaction, it is easy to predict the direction of chemical change for any mixture of the species. For example, consider the proton-transfer reaction between cyanide ion and hydrogen sulfide molecules:

$$CN^- + H_2S \rightleftharpoons HCN + HS^-.$$

The value of Q is known to be about 250:

$$Q = \frac{[HCN][HS^-]}{[CN^-][H_2S]} = 250$$

when equilibrium values are used for the molarities.

Let us create a mixture of these four species with the following non-equilibrium concentrations:

$$[CN^-] = 1 \cdot 10^{-3}, \qquad [H_2S] = 1 \cdot 10^{-4}, \qquad [HCN] = 5 \cdot 10^{-2}, \qquad [HS^-] = 1 \cdot 10^{-2}.$$

Will the reaction go to the right producing more HCN and HS$^-$ or will it go to the left to form more CN$^-$ and H$_2$S? It is merely necessary to calculate the nonequilibrium ratio r and to compare it to the equilibrium value Q:

$$r = \frac{(5 \cdot 10^{-2})(1 \cdot 10^{-2})}{(1 \cdot 10^{-3})(1 \cdot 10^{-4})} = 500.$$

Since r is greater than Q, the reaction must proceed to the left, so that the ratio will become smaller, eventually reaching a value of 250.

If the initial mixture consisted of $0.01M$ concentrations of each of the four species, the initial value of r would be 1.0, which is smaller than $Q = 250$. The reaction would go to the right, decreasing the molarities of CN$^-$ and H$_2$S and increasing those for HS$^-$ and HCN, until the concentration ratio became equal to 250.

What would happen if a mixture were made with the following molarities:

$$[HCN] = 2.5 \cdot 10^{-2}, \quad [HS^-] = 1 \cdot 10^{-2}, \quad [CN^-] = 1 \cdot 10^{-4}, \quad [H_2S] = 1 \cdot 10^{-2}?$$

4.13   INTRODUCTORY EQUILIBRIUM CALCULATIONS

A fairly high percentage of reversible reactions of interest in chemical analysis fit the simple model

$$A + B \rightleftharpoons D$$

with its corresponding equilibrium quotient expression

$$Q = \frac{[D]}{[A][B]}. \tag{4.4}$$

Suppose we create a mixture of these three species such that their molarities (before any reaction occurs) are C_A, C_B and C_D. The initial value for the nonequilibrium ratio is

$$r = \frac{C_D}{C_A C_B}$$

and, typically, this will differ from the value for Q. The reaction will go either to the right or to the left, as necessary to establish equilibrium. Suppose it goes to the right and that the molarity of substance D increases by the amount x. We may show the before and after status of the system as follows:

$$
\begin{array}{llll}
\text{Reaction:} & A & + & B & \rightleftarrows & D \\
\text{Initial molarity:} & C_A & & C_B & & C_D \\
& \downarrow & & \downarrow & & \downarrow \\
\text{Final Molarity:} & C_A - x & & C_B - x & & C_D + x
\end{array}
$$

Note that the molarities of A and B have to decrease equally and the change is equal to the change in the molarity of D. Of course, if the chemical equation had unequal stoichiometric coefficients, this would have to be taken into account.

We can place the expressions for the equilibrium molarities into the expression for the equilibrium quotient:

$$Q = \frac{(C_D + x)}{(C_A - x)(C_B - x)}. \tag{4.5}$$

If the reaction had proceeded to the left to establish equilibrium, then the molarity of substance D would have decreased, and that of A and B would have increased. If we again let x be the change in molarity, we would write:

$$Q = \frac{(C_D - x)}{(C_A + x)(C_B + x)}.$$

In either case, if we wished to find the value for x, so that we could then calculate the values for the final (equilibrium) molarities, the equation may be rearranged into the standard quadratic form. For Eq. (4.5) this would be:

$$Qx^2 - [Q(C_A + C_B) + 1]x + (QC_A C_B - C_D) = 0. \tag{4.6}$$

Therefore, if we know the value for Q and the initial values for the concentrations, it is not particularly difficult to find the equilibrium molarities of the species.

Example calculation A solution is prepared consisting of $0.02M$ cupric perchlorate and $0.1M$ hydrochloric acid. The salt is fully dissociated into its ions, but there is a weak tendency to form a complex ion:

$$
\begin{array}{lll}
Cu^{2+} + Cl^- \rightleftarrows CuCl^+, & Q = 1.3. \\
\text{Blue} \qquad\qquad \text{Green}
\end{array}
$$

What are the equilibrium molarities of the species? Using the above model, we get $C_A = C_{Cu} = 0.02$, $C_B = C_{Cl} = 0.1$, and $C_D = C_{CuCl} = 0$. That is, we may describe the initial conditions in terms of the cupric and chloride ions alone. By solving Eq. (4.6) we find:

$$x = [CuCl^+] = 0.00225.$$

The other root is 0.887, which is impossible with these molarities. Then, by difference,

$$[Cu^{2+}] = 0.02 - x = 0.01775,$$
$$[Cl^-] = 0.1 \ - x = 0.09775.$$

As a check on the correctness of the calculation we can use these molarities in Eq. (4.4) to see if the correct Q value is obtained:

$$\frac{0.00225}{0.01775 \cdot 0.09775} = 1.3 \quad \text{(as expected)}.$$

Another example, when Q is very large Whereas cupric ion forms only a weak chloro-complex, citrate ion forms a very strong complex:

$$Cu^{2+} + Cit^{3-} \rightleftarrows CuCit^-, \quad Q = 1.6 \cdot 10^{14}.$$

Let us again specify initial concentrations of 0.02 for the Cu^{2+} ion, and 0.1 for the complexing ligand, citrate ion. Solution of Eq. (4.6) yields

$$[CuCit^-] = 0.100000000 \quad \text{or} \quad 0.020000000.$$

Of course, only the value of 0.02 can be correct, given the initial molarities. The reason for giving the nine decimal places in the above answer was to show the impossibility of calculating the equilibrium molarity of the Cu^{2+} ion:

$$[Cu^{2+}] = \underset{\text{initial}}{0.02} - \underset{\text{reacted}}{0.02} = \text{``0''} \quad \text{(i.e., very small)}.$$

The equilibrium molarity cannot actually be zero for a reversible reaction having a finite (though admittedly fairly large) equilibrium quotient. If we want to calculate the value for $[Cu^{2+}]$ it will be necessary to change our approach. This problem will occur whenever a reaction goes nearly to completion. The way to attack it is to change the definition of x. Knowing that $[Cu^{2+}]$ is very small, we will let it be called x. Then, the before-and-after scheme becomes:

Reaction:	Cu^{2+}	$+$	Cit^{3-}	$\rightleftarrows$	$CuCit^-$
Initial molarity:	0.02		0.1		0
	$\downarrow$		$\downarrow$		$\downarrow$
Final molarity:	x		$0.08 + x$		$0.02 - x$

That is, the citrate ion concentration *would* decrease to 0.08 if the reaction went to completion. But we have defined x as the unreacted Cu^{2+}, and so the equilibrium molarity for citrate ion is not 0.08, but $0.08 + x$. The complex does not quite form with molarity 0.02, but rather with $0.02 - x$ because of the small amount of copper, which doesn't react.

The new algebraic set up is:

$$Q = \frac{0.02 - x}{x(0.08 + x)} = 1.6 \cdot 10^{14}.$$

Rearrangement to the quadratic form and solution of the equation yield

$$x = [Cu^{2+}] = 0.000000000 \qquad (!\,!\,!).$$

It appears that x is so small that even a 10-digit calculator cannot keep track of it. Well, if it's *that* small, then surely we can make the approximation that it is negligible in comparison to 0.02 and 0.08. This will allow us to write the equation as follows:

$$\frac{0.02}{x(0.08)} = 1.6 \cdot 10^{14}, \qquad \text{whence} \qquad x = [Cu^{2+}] = 1.56 \cdot 10^{-15}.$$

General recommendation When doing equilibrium calculations, it is always good practice to see whether the quadratic form of the equation may be simplified by neglecting quantities that are certain to be very small in comparison to the quantities they are being added to or subtracted from. When the equation can thus be simplified to a linear form it is easy to solve.

Final example We mix equal volumes of $0.100M$ ammonia and $0.0400M$ hydrochloric acid. The reaction forms ammonium ion, and the before-and-after set up is as follows:

$$\text{Reaction:} \quad NH_3 \;+\; H^+ \rightleftarrows NH_4^+ \qquad Q = 2 \cdot 10^9.$$

$$\text{Initial molarity:} \quad 0.05 \qquad 0.02 \qquad 0$$
$$\downarrow \qquad \downarrow \qquad \downarrow$$
$$\text{Final molarity:} \quad 0.03 + x \qquad x \qquad 0.02 - x$$

We choose to set the equilibrium molarity of H^+ to be the x because (a) the equilibrium quotient for the reaction is rather large, indicating that the reaction will proceed to the right until one of the reactants is essentially used up, and (b) it is the H^+ that is the limiting reactant, being present at an initial value of 0.02 compared to 0.05 for the ammonia.

Placing the equilibrium molarity values in the Q-expression, we have

$$\frac{0.02 - x}{(0.03 + x)x} = 2 \cdot 10^9.$$

Now, this *could* be solved by using the quadratic approach, but it is much simpler to use our "chemical sense" to realize that x will very likely be negligible compared to 0.02 and 0.03. Let us make that approximation, at least tentatively:

$$\frac{0.02}{0.03x} = 2 \cdot 10^9 \qquad \text{whence} \qquad x = [H^+] = 3.3 \cdot 10^{-10}.$$

From this answer we note that the approximation was quite justified. If it had turned out that x was not negligible compared to 0.02, we would have had to solve the quadratic after all, but it is always worth a try to simplify the equation if the conditions seem favorable.

4.14   EXPERIMENTAL CONTROL OF THE EQUILIBRIUM POSITION

Reversible reactions differ greatly in the magnitudes of their equilibrium constants, which are indicators of the reaction's thermodynamic tendency to proceed to the right. A very small K-value tells us that, left to itself, the reaction will reach equilibrium with a high ratio of reactants to products. For example, silver iodide has little tendency to dissolve in pure water:

$$\text{AgI(s)} \rightleftarrows \text{Ag}^+ + \text{I}^-, \qquad K_{sp} = \frac{[\text{Ag}^+][\text{I}^-]f_i^2}{1} = 8 \cdot 10^{-17}.$$

An example of a reaction with a very large equilibrium constant is the complexation of silver ion by cyanide ion:

$$\text{Ag}^+ + 2\text{CN}^- \rightleftarrows \text{Ag(CN)}_2^-, \qquad K = \frac{[\text{Ag(CN)}_2^-]}{[\text{Ag}^+][\text{CN}^-]^2} \cdot \frac{1}{f_i^2} = 8 \cdot 10^{21}.$$

In experimental work it is frequently desired to "shift the position of equilibrium." For example, we may want to make the silver iodide dissolve in spite of its small solubility product constant, or we may want to prevent the cyanide ion from complexing silver ion in spite of its strong tendency to do so. Inspection of the equilibrium constant expression shows that there are three possible approaches:

1. At a given temperature the value for K is constant. Therefore, if we change the ionic strength of the solution to cause a change in the activity coefficients, it follows that the molarity ratio must adjust accordingly. Although this effect is important in performing precise equilibrium calculations, it is rarely useful in effecting a large shift in the equilibrium position for a reversible reaction.

2. Since the value for K is dependent upon temperature (see Chapter 7), it is sometimes possible to shift an equilibrium position greatly by controlling temperature. A common application of this effect is found in the technique of recrystallization, where a substance is dissolved in hot solution and then deposits its crystals upon cooling in an ice bath. For example, potassium perchlorate $KClO_4$ has only a small solubility at $0°$ (7.5 grams per liter of water), but at $100°$ we can dissolve about 218 grams per liter. This is chiefly due to an increase in the equilibrium constant for the reaction

$$KClO_4(s) \rightleftarrows K^+ + ClO_4^-.$$

3. By far the most important approach in shifting an equilibrium is based on using competing chemical reactions that can strongly influence the equilibrium molarities of one or more of the species involved in the reversible system. We will now illustrate this idea with examples.

Returning to the reactions of silver at the start of this section, we may use the strong complexing ability of cyanide ion to advantage. Given a saturated pure-water solution of silver iodide, virtually none of the solid is dissolved, the equilibrium molarity of silver ion is very small, about $10^{-8}M$, and this is also the molarity of the iodide ion. If we now add some sodium cyanide, we will greatly lower this already small silver-ion molarity by converting the silver to the complex:

$$AgI(s) \rightleftarrows Ag^+ + I^-$$
$$+$$
$$2CN^-$$
$$\updownarrow$$
$$Ag(CN)_2^-$$

The silver-ion molarity will drop, say, to about $10^{-15}M$ and therefore the silver iodide equilibrium will be disrupted. To reestablish the equilibrium, it will be necessary for more silver iodide to dissolve, but the cyanide ion keeps complexing it. Eventually the system *as a whole* will reach equilibrium when most of the cyanide is used up. By that time a considerable amount of silver iodide will have dissolved.

In terms of the solubility product constant $K_{sp} = [Ag^+][I^-]f_1^2$, we make it necessary for the iodide-ion molarity to increase to offset the decrease in silver-ion molarity, for it is necessary that the product of these molarities be constant (ignoring the relatively minor change in the activity coefficients); and the only way for the iodide-ion molarity to increase is for more solid silver iodide to dissolve.

As a final example we may actually *prevent* the cyanide ion from reacting with the silver ion by making use of the competing reaction with hydrogen ion. Hydrocyanic acid HCN is rather weak in its dissociation:

$$HCN \rightleftarrows H^+ + CN^-, \qquad K_a = 4 \cdot 10^{-10}.$$

Therefore, by making a solution strongly acidic, e.g., by adding perchloric acid, we may greatly reduce the equilibrium molarity of cyanide ion by shifting the HCN dissociation reaction to the left. Then it will have little or no capability to complex the silver ion.

The foregoing examples include two somewhat different ideas. First, it is possible to shift the equilibrium position of a reaction by using another, competing reaction to lower the concentration of one of the species. Secondly, it is possible to effect the shift not by using another reaction, but simply by adding more of one of the species from a different source. This second approach is often called the **common-ion effect**. In the example above we used it to repress the dissociation of HCN, and it is especially valuable in the methods of gravimetric analysis (see Chapter 6).

But in both cases the basic idea was the same: by controlling the molarity of one species in the equilibrium reaction we force the molarities of the other species to adjust, so that equilibrium is reestablished under the new conditions. This concept was recognized long ago by the French chemist Le Chatelier and it is often referred to as **Le Chatelier's Principle**: *When stress is applied to a system at equilibrium, the system will adjust in a manner that tends to offset the stress.* This principle finds wide applications, such as in political lobbying, labor contracting, and machine maintenance, where "the squeaky wheel gets the grease."

4.15   OBTAINING NUMERICAL VALUES FOR EQUILIBRIUM CONSTANTS

Before we can understand the aqueous solutions we work with in chemical analysis and through that understanding devise ways to control the reactions to our advantage, it is necessary to have the following information:

a)   the identities of the species in the solution,

b)   the chemical interactions between the species,

c)   some knowledge of the rates at which the possible reactions occur,

d)   the values of the equilibrium constants for the chemical reactions that reach equilibrium within the time frame of our experiments.

Points (a), (b), and (c) are part of the factual knowledge of chemistry that each person gradually accumulates. We are concerned here with point (d), i.e., obtaining the numerical values for equilibrium constants. Four approaches will be briefly mentioned:

1. *Check the literature.* Most of the time we may simply look up a desired K-value in reference tables. The results of previous experimental researches have been compiled in several places, including the Appendix of this book. The familiar handbooks include several tables of approximate values. For more comprehensive and critical tabulations refer to:

   Stability Constants, Special Publication No. 17. Eds. L. G. Sillen, A. E. Martell, The Chemical Society, London, 1964. Supplement: Special Publication No. 25, 1971.

   Handbook of Biochemistry, Chemical Rubber Publishing Co., 1970.

Critical Stability Constants (4 volumes). A. E. Martell, R. M. Smith, Plenum Press, New York, 1974–1977.

2. *Calculate K, using the standard Gibbs free energies of formation.* This method will be discussed in Chapter 7.

3. *Direct estimation of non standard Gibbs free energy.* In Chapter 8 we will consider in detail the use of galvanic cells for this method.

4. *Experimental determination of Q by analysis of an equilibrium mixture.*

Example To illustrate the study of a chemical reaction at equilibrium let us consider the reduction of ferric ion Fe^{3+} by silver metal:

$$Ag(s) + Fe^{3+} \rightleftarrows Ag^+ + Fe^{2+}, \qquad Q = \frac{[Ag^+][Fe^{2+}]}{[Fe^{3+}]}.$$

One could create a mixture of, say, $0.0500M$ ferric perchlorate (in the presence of $0.10M$ perchloric acid to prevent appreciable formation of $FeOH^{2+}$) with an excess of powdered silver metal. After vigorous stirring until the reaction reaches equilibrium, the excess silver metal could be filtered off. The remaining equilibrium solution contains the three ionic species, and a portion may easily be analyzed for the ferrous ion content by titration of a portion with a standard solution of potassium permanganate using the reaction

$$5Fe^{2+} + MnO_4^- + 8H^+ \rightarrow 5Fe^{3+} + Mn^{2+}.$$

Another portion could be analyzed for its silver content by precipitating and weighing the silver chloride. One would expect that the equilibrium concentration of silver ion would be equal to that of ferrous ion, since the reaction requires that they be formed in equal amounts. However, it is always wise to have corroboration.

The equilibrium concentration of ferric ion is also needed for calculating the equilibrium constant. But since this species was introduced in known concentration at the start (good planning), it is not really necessary to analyze for it because it follows from the concept of material balance that

$$\text{Total Fe} = [Fe^{3+}] + [Fe^{2+}] = 0.0500.$$

Therefore, $[Fe^{3+}]$ may be found by calculating the difference, once the ferrous molarity has been determined. Finally,

$$Q = \frac{[Fe^{2+}] \cdot [Ag^+]}{0.05 - [Fe^{2+}]}.$$

where $[Fe^{2+}]$ is found from titration and $[Ag^+]$ is found from gravimetry.

It is important to realize that experimental work of this sort yields a value for the equilibrium quotient Q rather than for K. The conversion of Q-values to K-values may be accomplished only by using estimated values for activity coefficients.

PROBLEMS

Writing reversible reactions and equilibrium constant expressions

1. For each of the following cases write a balanced chemical equation consistent with the verbal statement. Then write the equilibrium constant expression, including activity coefficients and showing when any of the latter cancel.

Example Solid calcium nitrate is in contact with its saturated water solution where it is completely dissociated into ions.

Answer The chemical equation is: $Ca(NO_3)_2(s) \rightleftarrows Ca^{2+} + 2NO_3^-$.
   The equilibrium constant expression is: $K = [Ca^{2+}][NO_3^-]^2 f_2 f_1^2/1$.

a) A concentrated solution of potassium chloride is immersed in an ice bath. Crystallization of solid KCl occurs and the mixture is kept at 0°.

b) A small amount of carbon tetrachloride is dissolved in pure water. The solution is placed in a stoppered bottle and the CCl_4 in solution establishes an equilibrium vapor pressure in the air space above the solution.

c) An excess of mercury metal is shaken with an aqueous solution of ferric ion causing formation of ferrous and mercurous ions.

d) When ammonia gas is dissolved in water it reacts slightly with the water, taking a proton to form ammonium ion and hydroxide ion.

e) When hydrochloric acid is added to a blue solution of cupric ion, the color changes to green because of the association of cupric and chloride ions.

2. For each of the following chemical equations, write a verbal statement describing the system and write the equilibrium constant expression, including activity coefficients.

a) $Hg_2Cl_2(s) \rightleftarrows Hg_2^{2+} + 2Cl^-$

b) $Cu^{2+} + 4NH_3(aq) \rightleftarrows Cu(NH_3)_4^{2+}$

c) $HCOOH + H_2O \rightleftarrows H_3O^+ + HCOO^-$

d) $HgCl_2(s) \rightleftarrows HgCl_2(aq)$

e) $Hg^{2+} + Zn(s) \rightleftarrows Hg(l) + Zn^{2+}$

Calculation of ionic strength

3. For each of the following cases calculate the ionic strength of the solution, allowing for whatever chemical reactions are indicated. Also, use the table of activity coefficients (found from the Davies equation) to determine values for the ions present in the solutions.

a) A solution contains $0.015M$ potassium chloride and $0.020M$ calcium chloride. No chemical reaction occurs.

b) 100 mL of $0.80M$ magnesium perchlorate is added to 900 mL of $0.30M$ perchloric acid. No reaction.

c) A buffer solution is prepared by mixing 50 mL of $0.080M$ potassium dihydrogen phosphate with 50 mL of $0.060M$ potassium monohydrogen phosphate. No reaction occurs.

d) Equal volumes of $0.0600M$ hydrochloric acid and $0.0600M$ sodium hydroxide are mixed. The hydronium ion reacts with the hydroxide ion to form water.

e) 25 mL of $0.0200M$ potassium iodide are mixed with 75 mL of $0.0300M$ silver nitrate. Silver iodide precipitates.

f) Analysis of a solution shows that 1 L contains the following amounts of ions, in millimoles:

Anions	Cations
Bromide: 50 Chloride: 60 Nitrate: 50	Hydronium: 10 Magnesium: 5 Lithium: 40 Potassium: 20 Calcium: 40

Are these amounts consistent with the requirement that an aqueous solution must be electrically neutral? Calculate the ionic strength.

g) Consider the case study on the solubility of calcium iodate discussed in this chapter (p. 109). Is the ionic strength shown in the table of results for the $0.15M$ potassium chloride solution correct?

Calculation of ionic-activity coefficients

4. For an ionic strength of 0.0025, calculate the activity coefficient of zinc ion Zn^{2+}, using each of the three equations discussed (Debye–Hückel limiting law, extended D–H equation, Davies equation). Repeat the calculation for an ionic strength of 0.25 and compare the results.

5. Use the Davies equation to verify the values of f_1 to f_5 as shown in the reference table for $I = 0.01$ and $I = 0.1$.

6. Prove that, for a given ionic strength, $f_2 = f_1^4$. Show that in general $f_z = f_1^{z^2}$.

7. For an ionic strength of 0.1, calculate the mean ionic-activity coefficient for each of the following electrolytes:

$$1:1\ NaCl, \quad 2:1\ Na_2SO_4, \quad 1:2\ MgCl_2, \quad 2:2\ BaSO_4,$$
$$3:1\ K_3PO_4, \quad 1:3\ Al(ClO_4)_3, \quad 2:3\ Al_2(SO_4)_3$$

Calculation of molecular-activity coefficients

8. At 25° the solubility of benzene (a neutral nonpolar liquid) in solutions of perchloric acid and sodium perchlorate varies as follows:

$M\ HClO_4$: $M\ C_6H_6$:	0.235 0.0232	0.478 0.0238	0.782 0.0245	
$M\ NaClO_4$: $M\ C_6H_6$:	0.145 0.0219	0.375 0.0208	0.762 0.0189	0.904 0.0181

Since each of these solutions is saturated with benzene, the following equilibrium must be satisfied:

$$C_6H_6(l) \rightleftarrows C_6H_6(aq), \qquad K = [C_6H_6(aq)]f_0.$$

Therefore the activity of dissolved benzene must be the same in all cases. The solubility of benzene in pure water is 0.02273 mole/L. If we assume that in this case, with no electrolyte present, the behavior of dissolved benzene is ideal, then we may take 0.02273 as the activity for all saturated solutions.

Calculate the value for f_0 in each of the solutions and plot $\log f_0$ versus the electrolyte molarity. Calculate the slopes of the plots.

The purpose of this problem is to illustrate that there are some regular relationships for molecular activity coefficients, but that it is not possible to generalize in terms of ionic strength because of specific electrolyte effects.

Conversion of K-values to Q-values

9. For each of the following reactions, convert the equilibrium constant value K to its Q-counterpart at the indicated ionic strength. Also, find pK and pQ.

   a) $CN^- + H_2O \rightleftarrows HCN + OH^-$, $K = 2.5 \cdot 10^{-5}$, $I = 0.12$

   b) $CH_3COOH + H_2O \rightleftarrows CH_3COO^- + H_3O^+$, $K = 1.75 \cdot 10^{-5}$, $I = 0.25$

   c) $CaF_2(s) \rightleftarrows Ca^{2+} + 2F^-$, $K = 5 \cdot 10^{-11}$, $I = 0.050$

   d) $Ag(s) + Fe^{3+} \rightleftarrows Ag^+ + Fe^{2+}$, $K = 0.13$, $I = 0.50$

   e) $HPO_4^{2-} + H_2O \rightleftarrows PO_4^{3-} + H_3O^+$, $K = 5 \cdot 10^{-13}$, $I = 0.04$

Redox stoichiometry

10. Refer to the case study on solubility of calcium iodate (p. 109). If a solution of $0.0543\,M$ sodium thiosulfate was used for the titration of a 10-mL portion of the saturated calcium iodate solution (in $2M$ KCl) according to the triiodide procedure described, what volume of sodium thiosulfate would have been required, given the solubility reported in the table of results?

Reactions in nonequilibrium mixtures and equilibrium calculations

For each of the following problems, assume that the ionic strength remains constant, so that the given Q value is also constant.

11. 0.050 mmole of solid iodine is added to 50 mL of $0.00500M$ potassium iodide and is dissolved completely. The triiodide ion is formed:

$$I_2(aq) + I^- \rightleftarrows I_3^-, \qquad Q = 7.2 \cdot 10^2.$$

Calculate the equilibrium molarities of the three species.

12. A solution is prepared by mixing 10 mL of $0.95M$ hydrochloric acid, 25 mL of $0.00400M$ sodium sulfate, and 65 mL of $0.00308M$ potassium hydrogen sulfate. Equilibrium is reached according to the proton transfer reaction:

$$HSO_4^- + H_2O \rightleftarrows H_3O^+ + SO_4^{2-}, \qquad Q = 0.027.$$

Compare the value of r (found with the initial nonequilibrium molarities) with Q and predict the reaction direction. Then calculate the equilibrium molarities for the reacting species.

13. Aqueous solutions of chloracetic acid reach proton-transfer equilibrium as follows:

$$ClCH_2COOH + H_2O \rightleftarrows H_3O^+ + ClCH_2COO^-, \quad pQ = 2.7.$$

a) When 0.03 mole of the acid is dissolved in 500 mL of water, the above reaction must go to the right, forming equal concentrations of the ions. What are the equilibrium molarities?

b) Suppose 0.03 mole of the acid is dissolved in 500 mL of a solution of $0.100M$ sodium chloroacetate. What will be the equilibrium molarity of hydronium ion?

c) If 50 mL of $0.05M$ hydrochloric acid is mixed with 50 mL of $0.02M$ sodium chloroacetate, what will be the equilibrium molarities of the three species?

14. Silver bromate is not very soluble. When excess solid $AgBrO_3$ is allowed to reach equilibrium with aqueous solutions, the reaction is:

$$AgBrO_3(s) \rightleftarrows Ag^+ + BrO_3^-, \quad Q = 1 \cdot 10^{-4}.$$

a) What are the molarities of the ions in a saturated solution of silver bromate in pure water?

b) If solid silver bromate is stirred with a solution of $0.04M$ potassium bromate, what will be the equilibrium molarity of silver ion?

c) When 50 mL of $0.005M$ silver nitrate is mixed with 50 mL of $0.10M$ sodium bromate, will a precipitate form? What will be the equilibrium amount of the solid silver bromate, expressed in grams?

15. The equilibrium quotient for the oxidation of iodide ion by ferric ion is fairly large:

$$Fe^{3+} + I^- \rightleftarrows Fe^{2+} + \frac{1}{2}I_2(aq), \quad Q = 400.$$

A trace of potassium iodide is added to 100 mL of a solution that contains both ferric ion $(0.01M)$ and ferrous ion $(0.02M)$. Calculate the equilibrium molarity of $I_2(aq)$ if the equilibrium molarity of I^- is $1.0 \cdot 10^{-4}$.

Including ionic-activity coefficients in the equilibrium calculation

16. Chlorine reacts with water as follows:

$$Cl_2(g) + H_2O(l) \rightleftarrows \quad \underset{\substack{\text{hypochlorous} \\ \text{acid}}}{HOCl} \quad + Cl^- + H^+.$$

The equilibrium constant for the reaction is small: $K = 2.7 \cdot 10^{-5}$.

Suppose chlorine is maintained at a partial pressure of 0.010 atm over a solution containing strong electrolytes $HCl(0.050M)$ and $CaCl_2(0.010M)$. Assuming that the above reaction has only a very minor effect upon solution composition, find the following:

a) Calculate the molarities of H^+, Cl^-, and the ionic strength.

b) Taking the activity coefficient of a singly charged ion to be 0.79 in this solution, calculate the value for the equilibrium quotient Q.

c) Assuming that the chlorine reaction proceeds only to a very slight extent, calculate the equilibrium molarity of hypochlorous acid in the solution.

17. A solution of potassium dichromate was prepared by dissolving 1.00 mmole of pure $K_2Cr_2O_7$ in 1 L of $0.100M$ perchloric acid. The dichromate ion partly dissociates to reach equilibrium with hydrogen chromate ion (the monomer form):

$$Cr_2O_7^{2-} + H_2O \rightleftarrows 2HCrO_4^-, \qquad K = 0.032.$$

Assuming the ionic strength to be simply 0.10, since the chromium(VI) concentration is rather small compared to that of the perchloric acid, calculate the percentage of Cr(VI) remaining in the dichromate ion form at equilibrium.

18. Silver iodate is a slightly soluble salt:

$$AgIO_3(s) \rightleftarrows Ag^+ + IO_3^-, \qquad K = 3.0 \cdot 10^{-8}.$$

a) Calculate the solubility of silver iodate, i.e., the silver ion molarity, when excess $AgIO_3$ is stirred with pure water. Assume that the ionic strength is so low in this solution that activity coefficients may be taken as unity (1.00).

b) Instead of pure water, a solution of $0.10M$ sodium perchlorate is used to equilibrate with solid silver iodate. Taking the ionic strength to be 0.10, calculate the solubility of silver iodate in this solution. What is the percentage increase compared to the water solution?

Determination of equilibrium constants
by chemical analysis of systems at equilibrium

19. An excess of solid silver bromate was stirred with $0.0250M$ perchloric acid at $25°$, so that the following equilibrium was established:

$$AgBrO_3(s) \rightleftarrows Ag^+ + BrO_3^-, \qquad K = [Ag^+][BrO_3^-]f_1^2.$$

The excess solid was removed, and a 100-mL portion of the saturated solution was treated with an excess of potassium iodide. The following reactions occurred:

$$BrO_3^- + I^- \rightleftarrows Br^- + I_3^- \qquad \text{(unbalanced),}$$
$$Ag^+ + I^- \rightleftarrows AgI(s).$$

The precipitate (MW = 234.77) was collected, dried, and found to weigh 0.2060 gram. The triiodide ion in the solution was then determined by titration with $0.1500M$ sodium thiosulfate:

$$I_3^- + S_2O_3^{2-} \rightarrow I^- + S_4O_6^{2-} \qquad \text{(unbalanced).}$$

It required 34.92 mL to reach the endpoint.

a) Calculate the solubility of silver bromate in the solution using the observed weight of the silver iodide precipitate.

b) Balance the redox equations.

c) Calculate the solubility of silver bromate by using the titration data and compare with the result from (a).

d) Calculate the ionic strength of the saturated solution before the addition of potassium iodide.

e) Calculate the value of the equilibrium constant for the solubility reaction taking activity coefficients into account.

SPECTROPHOTOMETRIC DETERMINATIONS

5

Are not gross bodies and light convertible into one another; and may not bodies receive much of their activity from the particles of light which enter into their composition? The changing of bodies into light, and light into bodies, is very comformable to the course of Nature, which seems delighted with transmutations.

[Sir] Isaac Newton
Opticks **1704, Query 30**

5.1 ELECTROMAGNETIC RADIATION

In this chapter we will discuss analytical uses of a very small portion of the electromagnetic spectrum, namely, ultraviolet and visible radiation. The chart in Table 5.1 emphasizes the continuous range of electromagnetic radiation from very high-energy cosmic rays to low-energy radio waves. What makes visible radiation (i.e., "light") special is not that it is different in kind from the other portions of the spectrum, but rather that the human vision system responds to this narrow range of wavelengths.

One of the most perplexing puzzles of early 20th-century physics was the apparent dual nature of radiation. In some experiments light exhibited the properties of waves (Fig. 5.1).

The **wavelength**, symbolized by the Greek letter λ (lambda) is the distance between successive points on the wave that are "in phase." The unit used for wavelength is the meter or, more commonly, the nanometer (1 nm = 10^{-9} m).

The **speed of light** (c) has been measured by ingenious experiments and found to be about $3 \cdot 10^8$ m/sec in a vacuum and lower when passing through matter.

The **wave frequency**, symbolized by the Greek letter ν (nu), is the number of complete oscillations per second. The unit used for wave frequency is $\sec^{-1}$ and the simple relationship between wavelength, speed, and frequency is

$$\lambda = \frac{c}{\nu}.$$

Table 5.1 Spectrum of electromagnetic radiation

Range	Wavelength, nm		Color	Complement
Cosmic rays	10^{-5}			
	10^{-4}			
Gamma rays	10^{-3}	400	violet	yellow–green
	10^{-2}	450		
X rays	10^{-1}		blue	yellow
	1		blue–green	orange to red
	10	500	green	purple (red–violet)
Far ultraviolet	100	550	yellow–green	violet
Ultraviolet			yellow	blue
Visible		600		
Near infrared	1000		orange	blue–green
	10^4	650		
Infrared	10^5		red	blue–green
	10^6	700		
Microwave	10^7			
	10^8			
Radio waves	10^9			

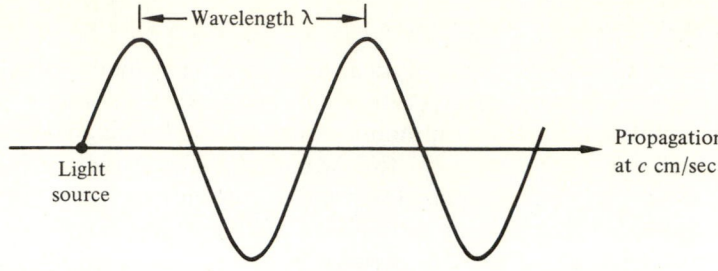

Figure 5.1

Since the frequency is strictly determined by the original source of radiation, it follows that the wavelength will vary according to the speed, which changes when light passes from one medium to another, as from air to water.

The wave theory of light seemed ideal for explaining certain physical observations, but quite inadequate for understanding why radiation is absorbed by some substances and transmitted unchanged by others. Why should a diamond transmit light freely, refracting it into little rainbows in accordance with the properties of waves, while another form of pure carbon, graphite, is opaque because it absorbs all the light?

It was necessary to accept a nonwave concept of radiation as well and to postulate a particle theory. The wave theory and the particle theory are not competitors. Rather, both theories of light are needed to explain all the observed phenomena. In the particle theory we think of electromagnetic radiation as a stream of photons having different energies. **Photons** are very small particles with a zero rest mass; they travel in straight lines until absorbed by matter (Fig. 5.2).

There is a simple relationship between photon energy E and its wave frequency:

$$E = h\nu,$$

where h is Planck's constant ($h = 6.63 \cdot 10^{-34}$ joule sec. Thus, a photon of

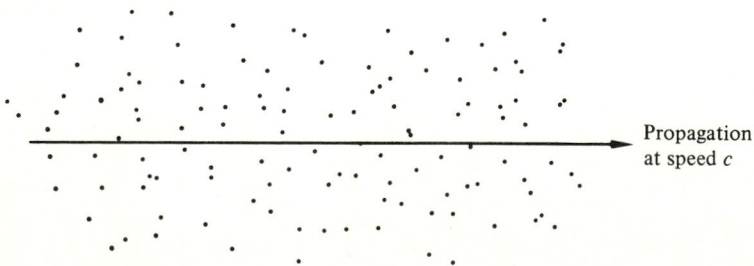

Figure 5.2

green light with a wavelength of 550 nm would have a frequency

$$\nu = \frac{c}{\lambda} = \frac{3 \cdot 10^8}{550 \cdot 10^{-9}} = 5.5 \cdot 10^{14} \text{ sec}^{-1}$$

and an energy

$$E = h\nu = 6.63 \cdot 10^{-34} \cdot 5.5 \cdot 10^{14} = 3.6 \cdot 10^{-19} \text{ joule.}$$

This seems frightfully small, but a mole of such photons would have a total energy

$$E \text{ (per mole)} = 3.6 \cdot 10^{-19} \cdot 6.02 \cdot 10^{23} = 2.2 \cdot 10^5 \text{ joules} = 5.3 \cdot 10^4 \text{ cal,}$$

which is considerable.

5.2   SELECTIVE ABSORPTION OF PHOTONS BY MATTER

Matter, whether in the form of molecules, atoms, or ions, and whether solid, liquid, gaseous, or dissolved in solution, owes its chemical properties and its photon-absorbing ability to the electronic structure. According to the quantum theory, electrons in matter can exist only in certain discrete energy states. Let us consider an individual molecule: the **ground state** for this molecule is the lowest possible energy level. The electronic energy of the molecule can vary as one or more of its electrons become **excited** to higher energy states. Electronic excitation can be caused by absorption of a photon, provided the photon energy exactly matches the resulting change in molecular energy

In the illustration given by Fig. 5.3 the molecule can exist in a particular excited state which is exactly $h\nu$ joules higher than the ground state, and so photon absorption is allowed. If the same molecule were struck by a photon of higher or lower energy, there would have been no absorption unless other suitable excited states were allowed.

Actually the energy requirements are not as rigid as just implied because the molecule has small energies due to vibration and rotation, caused by collisions with other molecules. These energy states are superimposed upon the electronic states, giving a range of suitable energy transitions that may be effected by photons (Fig. 5.4).

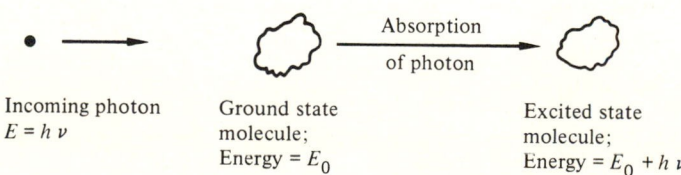

Incoming photon   Ground state   Excited state
$E = h\nu$ molecule;   molecule;
                      Energy $= E_0$ Energy $= E_0 + h\nu$

Absorption of photon

Figure 5.3

Without vibration
or rotation, only
one value of $h\nu$
can excite

Vibration and
rotation make
"fuzzy" levels;
a range of $h\nu$ can excite

Figure 5.4

White Light and Colored Solutions

The chart of the electromagnetic spectrum in Table 5.1 shows an expansion of the small portion we call **visible light**. Ranging from about 400 to 700 nm in wavelength, the color varies from violet to red with intermediate hues of blue, green, yellow, and orange. A beam of white light can be refracted into such a "rainbow" simply by passing it through a glass prism, as Sir Isaac Newton found three centuries ago. Another common way of spreading out the colors is to use a diffraction grating consisting of a large number of closely spaced lines ruled on a reflective surface.

From these observations we realize that white light must consist of a complex mixture of photons of different energies; the intervals between these energy levels are so minute that even upon refraction we find no "gaps" in the spectrum. When we let white light (e.g., sunlight) pass through a colored glass or aqueous solution, it is the phenomenon of selective absorption that gives the medium its apparent color. A solution of potassium permanganate looks purple, meaning that it **transmits** purple light. But what is purple light? It is a mixture of red and blue light, and this means that the solution is **not transmitting** the green light that is in the middle of the visible spectrum. The photons with energies corresponding to the 500–575 nm wavelength range are selectively **absorbed** by the permanganate ions. What passes through a colored solution is the spectrum of light **complementary** to the spectrum of the absorbed light.

When we say that water is colorless, we mean that it does not have significant ability to absorb photons in the visible portion of the spectrum. In fact, water does not absorb photons even in the ultraviolet region. Ordinary glass is colorless and therefore does not absorb light in the visible region. However, glass begins to absorb significantly when the wavelength of the radiation drops below about 320 nm.

The precise makeup of white light is somewhat indefinite, for the "mix" of various photon energies can differ somewhat and still result in what our eyes evaluate as white. The simplest way to obtain white light for a laboratory experiment is to use a regular tungsten filament bulb. The electronic transitions that occur rapidly in the "white hot" filament are so closely spaced that they produce a continuous spectrum of radiation in the visible range. The heat generated by the electric current produces countless excited states in the tungsten, and when they return to lower states, the various photons are emitted.

5.3   QUANTITATIVE VIEW OF PHOTON ABSORPTION

Using an Absorption Cell

For purposes of chemical analysis we will place a colored solution in a' glass or silica **absorption cell**, two common forms of which are shown in Fig. 5.5.

    We refer to **cell breadth** b as the distance that the light travels through the solution contained in the cell. Thus, b is the internal dimension and is usually expressed in centimeters. The sides of the cells through which the light beam passes must be optically polished and parallel, as well as scrupulously clean.

    Suppose we put a dilute $(0.00030M)$ solution of potassium permanganate in a 1.00-cm cell and direct a beam of green light (550 nm) through the absorption cell. With an exaggerated thickness of the glass, the situation is depicted in Fig. 5.6.

    Suppose the incident beam had a power (intensity) corresponding to 10,000 photons (550 nm) per second. The power of the exit beam would be significantly lower than this for several reasons:

    Going from left to right, the first loss in power is caused by reflection at the air/glass interface. This loss is easily calculated by the equation

$$\text{Loss, } \% = \left(\frac{G-A}{G+A}\right)^2 \cdot 100,$$

where G and A are the refractive indices of glass and air, respectively. Taking

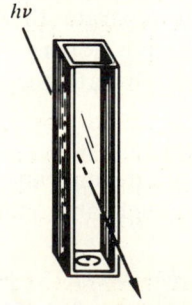

1.00 cm rectangular cell

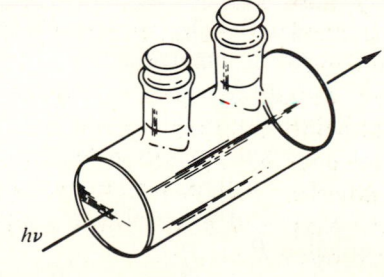

5.00 or 10.00 cm cylindrical cell

Figure 5.5

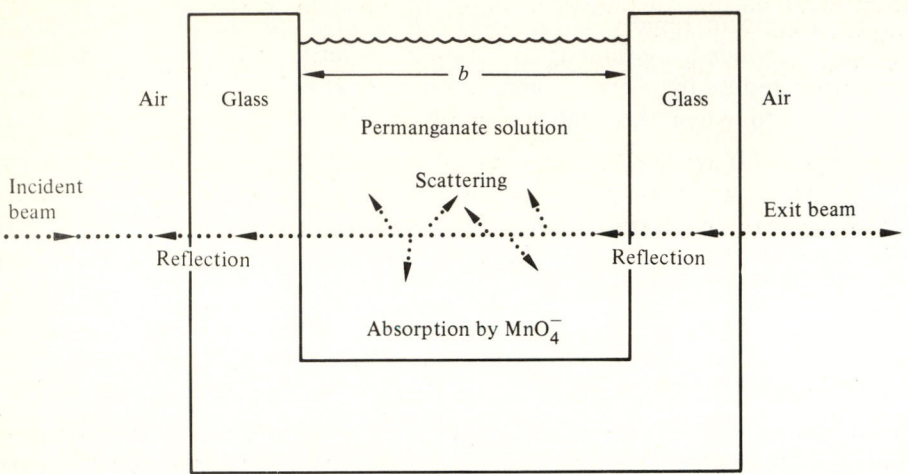

Fig. 5.6 Decrease in light power upon passing through an absorption cell containing water and a colored solute.

$G = 1.5$ and $A = 1.0$, we predict a 4% reflection loss at each of the two air/glass interfaces:

At the glass/solution interface the difference in refractive index is smaller and the reflection loss amounts to about 0.4% at each interface.

As the beam passes through the solution there may be appreciable loss due to tiny particles of dust, etc., which scatter some of the photons in the beam.

Finally, what is of special importance from the analytical viewpoint, there will be a loss of power due to selective absorption of the photons by the permanganate ions dissolved in the solution.

The Need for a Reference Cell

If we were to measure the power of the exit beam using only a single cell (as in the above discussion), we would be unable to separate the effects of reflection and scattering from the loss due to selective absorption by permanganate. Yet for analytical purposes it is vital to isolate the absorption effect. The procedure is rather simple.

First, an absorption cell is filled with distilled water and placed in the light beam. The beam is weakened by reflection, scattering, and slight absorption by glass and water. Let the power of the exit beam emerging from this cell be called P_0.

Then, replace the reference cell with the cell containing the sample (the two cells should be as similar as possible in their physical characteristics). We presume that the light beam undergoes precisely the same reflection, scattering, etc., losses as before, and that the only difference in the power of the exit beam must be due to the presence of the sample in the solution. Let the power of the exit beam emerging from this cell be called P.

The ratio of the two powers is called the solution **transmittance** T:

$$T = P/P_0.$$

More commonly we use the percent transmittance $\%T$, which is simply $100\ T$.

In precise work the analyst will usually fill both cells with distilled water to check for slight differences in manufacture, such as nonparallel sides. Any difference in the exit beams may then be taken into account in later measurements using the same cells.

5.4   BEER'S LAW (THE BOUGEUR—LAMBERT—BEER LAW)

It was shown long ago that the transmittance of light through a solution is an exponential function of the length of the path (Bougeur, 1729, and Lambert, 1760) and of the concentration of the absorbing species (Beer, 1852). The combination of these findings has become known as **Beer's Law**, perhaps because it is the concentration dependence that is most widely applied in chemical analysis. Beer's law may be stated either in exponential form:

$$\frac{P}{P_0} = T = 10^{-abC}$$

or in the logarithmic version:

$$A \equiv \log\left(\frac{1}{T}\right) = \log\left(\frac{100}{\%T}\right) = abC,$$

where a is the **absorptivity**, a proportionality constant that varies with the wavelength of the radiation used (it is a molecular property of the light-absorbing species); b is the cell path, the breadth of the solution expressed in centimeters; C is the concentration of the absorbing species; A is the **absorbance**, a popular quantity because, unlike the transmittance, it is directly proportional to the concentration.

Choice of Units for Absorptivity and Concentration

Clearly, A is a dimensionless quantity because it is the logarithm of a ratio of beam powers, and the units cancel. We have already stated that b is measured in centimeters. The units for absorptivity are thus dependent upon the choice of the units for concentration C. If the concentration of the absorbing species is expressed, say as grams per liter, then the units for a must be L/g · cm, so that the product abC comes out dimensionless, as required for the absorbance.

Because of the importance of molarity as a concentration unit in analytical chemistry we use a special symbol for the absorptivity in this case. The symbol is the Greek letter epsilon ϵ, and is called the **molar absorptivity**. It has the units L/mol · cm. Thus, a commonly used statement of Beer's Law is:

$$A = \epsilon bC.$$

Older Symbols and Terms

The terminology presented above was recommended in 1966 by the Nomenclature Committee of the ACS Division of Analytical Chemistry. However, in older works one encounters a variety of terms and symbols. A few of them are summarized in Table 5.2 as an aid to translation.

Table 5.2

Approved term and symbol		No longer approved	
Absorbance	A	Optical density	D or $O.D.$
		Extinction	E
		Absorbancy	
Absorptivity	a	Absorbancy index	k
Molar absorptivity	ϵ	Extinction coefficient	
		Absorption coefficient	
		Specific extinction	
Transmittance	T	Transmittancy	
		Transmission	
Path length	b		l or d
Nanometer	nm	Millimicron	$m\mu$
		Angstrom (0.1 nm)	Å
Power	P	Intensity	I

5.5   COMPONENTS OF A SIMPLE SPECTROPHOTOMETER: THE B&L SPECTRONIC 20

The Bausch & Lomb Spectronic 20 and other instruments in its relatively low price range are frequently used in applications that do not require high precision. The B&L Model 20 is a single-beam spectrophotometer with a wavelength range of 340–600 nm. This may be extended to 900 nm when a red-sensitive phototube detector is substituted. A simplified diagram is depicted in Fig. 5.7.

A tungsten filament lamp is the source of white light focussed by a lens on an entrance slit and directed to a movable diffraction grating. The grating is a reflective surface covered by precisely ruled lines with a density of 600 lines/mm. The white light is dispersed into a divergent spectrum and directed at the exit slit.

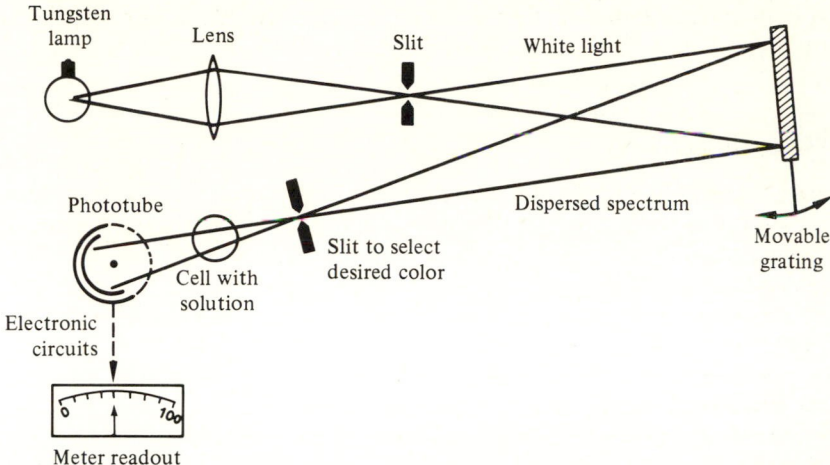

Figure 5.7

The grating can be rotated by means of a front-panel knob, so that light of the desired wavelength is centered on the exit slit. The exit slit passes wavelengths in a band about 20 nm wide. Thus, when the grating scale is set to, say, 500 nm, the exit light beam consists of wavelengths in the 490–510 nm range. This wide band places severe limits on the use of the B&L 20 in some applications.

Not shown in the diagram is a small opaque shutter that automatically falls in front of the exit slit when no absorption cell is present. This blocks off the entire beam and provides a reference point of zero transmittance, as will be discussed later.

The absorption cell normally supplied with the B&L 20 is cylindrical (like a small test tube) with an internal diameter of about 1 cm. A small mark on the side permits the user to orient the cell (also called a cuvette) in a similar way each time it is used, to minimize errors caused by differences in reflection and path length. It is also possible to use a rectangular cell with a special adaptor.

When the exit beam has passed through the cell filled with the sample, it falls on the light sensitive surface of a phototube, generating a small electric current amplified by electronic circuits. The current output of the phototube is proportional to the power of the light beam.

The readout device is a traditional meter with a swinging needle. The meter face has two scales, one for percent transmittance ($100T$) and one for absorbance (Fig. 5.8).

Because the absorbance scale is nonlinear (logarithmic) and runs "backwards," it is preferable to read the linear transmittance scale as accurately as possible and then to convert to absorbance if desired. Later we will discuss the errors arising from the compression of the absorbance scale at higher

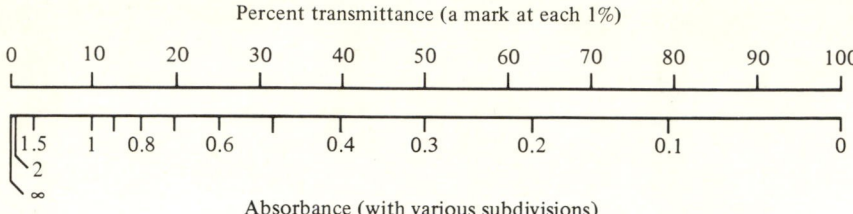

Figure 5.8

values; but for now it should be obvious that the B&L 20 was not intended for use with solutions whose absorbance is greater than about 0.8.

5.6   PROCEDURE FOR ABSORBANCE (OR %T) MEASUREMENTS

Now that we have discussed the photon absorption process, established the necessary nomenclature, and outlined the basic instrumentation, we may list three very simple steps in measurement:

1. With the shutter preventing the light beam from reaching the phototube, turn the "zero control" to make the meter needle read $\%T = 0$ and $A = \infty$. This merely adjusts the base-level output of the electronic amplifier.

2. Set the desired wavelength and place the reference cell containing distilled water in the cell holder. Turn the "100%T" control to make the meter needle read $\%T = 100$ and $A = 0$. This adjustment simply defines $T = 100$ when the exit beam from the cell has the power P_0. Repeat steps (1) and (2) as necessary, since the controls interact somewhat.

3. Without changing the adjustments, replace the reference cell with the cell containing the colored solution. The meter needle will move to an intermediate position somewhere between $\%T = 0$ and $\%T = 100$. Read the value of $\%T$ to the nearest $0.1\%T$ if it is sufficiently steady.

   For increased reliability it is important to repeat these steps two or three times and to take an average of the several readings.

5.7   OBTAINING AN ABSORPTION SPECTRUM

To find how the absorbance of a solution depends on the wavelength of the light, it is merely necessary to carry out the above steps for a series of wavelengths spaced, for example, at intervals of 10 nm. Such a spectrum obtained on the B&L 20 for the acid–base indicator called methyl red in $0.01M$ HCl solution is shown in Fig. 5.9.

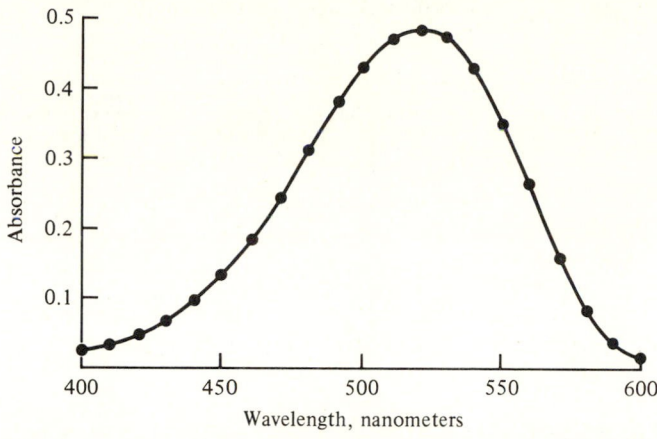

Figure 5.9

5.8   GOING FIRST CLASS:
THE CARY MODEL 118 UV-VISIBLE SPECTROPHOTOMETER

This is an excellent example of a modern high-quality instrument, up-to-date in its electronic components, including digital readout to four places in absorbance, and superbly designed in its optical system. The Cary Model 118 is a "double-beam" spectrophotometer, meaning that the light beam is directed alternately (30 times per second) between the reference cell and the sample cell. This eliminates the need for cell handling after the two cells are in the instrument.

The optical diagram of the Cary 118 is given in Fig. 5.10. The following is a description taken from its operation manual.

> The light originates at either the ultraviolet deuterium lamp or the visible tungsten–halogen lamp, depending on which lamp is selected by the operator. The light is focussed on the entrance slit S_1 by mirrors A and B and lenses C and D.
>
> The light proceeds through entrance slit S_1 to collimating mirror E, which reflects it to prism P_1. The light passes through the front face of the prism, is reflected from its aluminized back face, and leaves the prism through the front face. The light leaving the prism has been dispersed, i.e., spread out into its component colors or wavelengths. The dispersed light returns to mirror E, which now acts as a "telescope" to focus the light on the intermediate slit S_2, after reflection by mirror F.
>
> The prism is rotated (by the wavelength mechanism) so that light of the particular wavelength desired (indicated by the *wavelength counter* on the instrument panel) passes through the center of slit S_2.

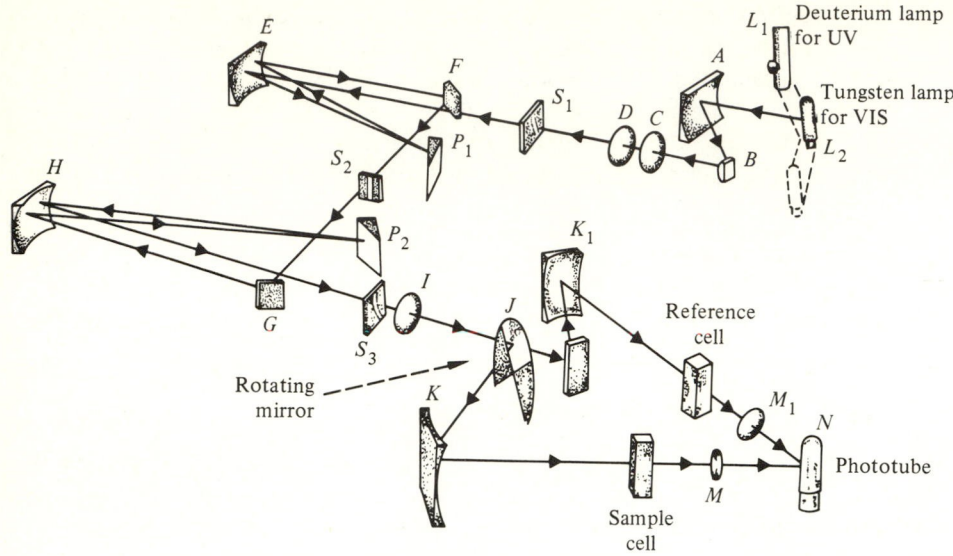

Figure 5.10

Slit S_2 admits the desired light into the second half of the monochromator, while rejecting most of the scattered light of other wavelengths. The light then traverses similar optical elements in reverse order, so that the principal aberrations of the off-axis spherical mirrors E and H are cancelled. However, the prisms are arranged so that the dispersions of the two systems are additive.

A very narrow waveband of light (narrow band of colors) leaves the monochromator by way of exit slit S_3 and passes through lens I. This light of selected wavelength falls on "chopper" J, which consists of a rotating disk divided into four sectors: one sector carries a plane mirror, two sectors are blackened, and the fourth is cut away so as to be completely "transparent." This disk rotates 30 times per second. Depending on the position of the chopper disk at any instant, the beam is either (1) deflected to toroidal mirror K, or (2) passed to the auxiliary plane mirror and then toroidal mirror K_1, or (3) blocked off entirely. The toroids K and K_1 direct the beams to the Sample Compartment, and through lenses M and M_1 to phototube N. The beam from K to M passes through the "sample" side of the compartment; the beam from K_1 to M_1 passes through the "reference" side. Beams arrive at the phototube alternately from sample and reference, producing corresponding electrical currents in the phototube.

Thus the output of the phototube consists of rectangular current pulses on which are superimposed the noise and the steady "dark current" generated in the phototube. It is the function of the electronic equipment to indicate the ratio of the amplitude of the sample current pulses to the amplitude of the reference current pulses. The entire measurement process is repeated at a rate of about eight times per second.

The measured absorbance (or %T if desired) is continuously displayed on a digital readout and simultaneously on a strip chart recorder if desired. Thus, an entire absorption spectrum may be recorded automatically on the chart.

5.9   ABSORPTION SPECTRUM OF BENZENE

The capability of high-quality instruments such as the Cary Model 188 is illustrated by Fig. 5.11. It gives the ultraviolet absorption spectra of benzene C_6H_6 under two conditions. In Fig. 5.11(a) the spectrum of benzene vapor is shown; it was obtained by putting a little liquid benzene in the bottom of the stoppered 1-cm cell letting it evaporate to mix with the air in the cell. The reference cell contained just air.

The second spectrum was obtained by shaking some liquid benzene with distilled water, separating the two phases, and then diluting the aqueous phase fourfold. Since the solubility of benzene in water is about 0.023M, this solution has a molarity of about 0.0058.

The vapor spectrum reveals a much finer structure, showing that the electron energy levels are partially resolved. This is due in part to a relatively small contribution from vibration and rotation in the vapor phase, when the molecules are far apart, and in part due to the excellent resolving power of the instrument with its nearly monochromatic light (spectral bandwidth about 0.1 nm, compared to 20 nm for the B&L 20).

The loss of fine structure in the aqueous spectrum indicates that the benzene molecules, in constant collisions and interactions with the solvent water molecules, have much "fuzzier" electron energy levels.

In fact, even a finer structure can be observed for the vapor spectrum when the Cary 118 is set to use a still smaller slit width and a slower scan rate. Such detailed absorption spectra can be valuable to the theoretician who studies quantum transitions, but pushing the resolution to its limit would be counterproductive for chemical analysis.

These spectra illustrate an important fact: the absorption spectrum of a chemical species is unique for that species. Therefore a spectrum may serve as a "fingerprint," which is valuable in identifying an unknown substance by comparing its spectrum with those obtained previously for known compounds. Catalogs of spectra are available for this purpose.

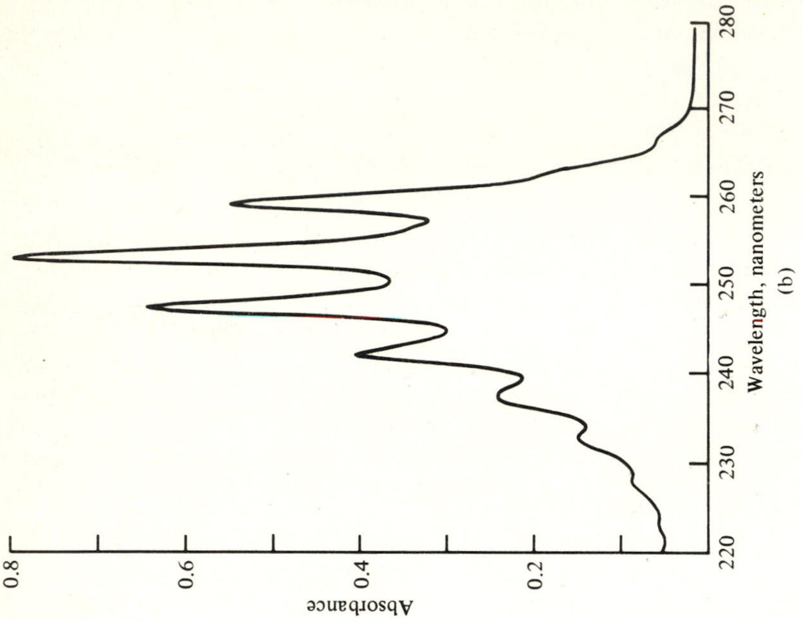

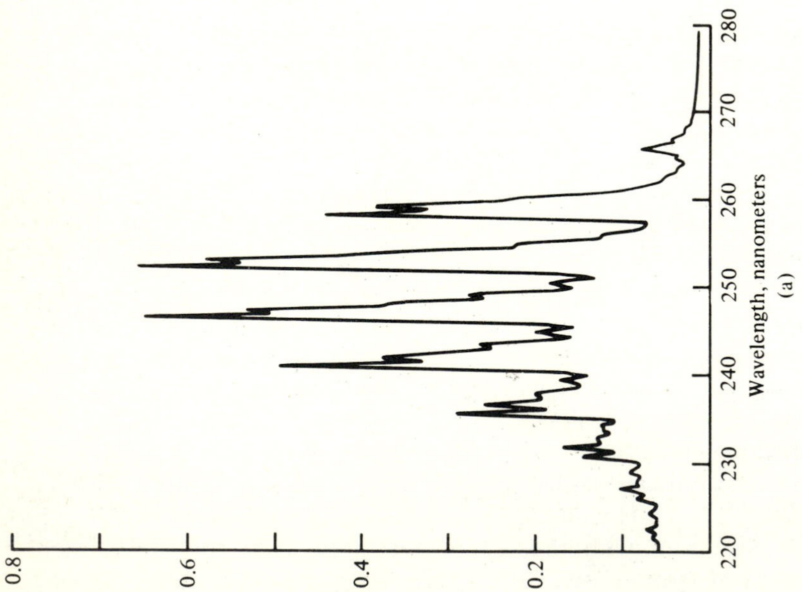

Fig. 5.11 (a) Absorption spectrum of benzene vapor in air above liquid benzene (cell path 1.00 cm, slit width about 0.3 mm, Cary Model 118 spectrophotometer). (b) Absorption spectrum of benzene dissolved in water (molarity 0.0058, cell path 1.00 cm, slit width about 0.3 mm, Cary Model 118).

5.10 DIRECT SPECTROPHOTOMETRIC DETERMINATION OF CONCENTRATION

So that we may have something specific to discuss for illustration, suppose our task is to determine the concentration of manganese (present as manganous ion Mn^{2+}) in a natural water supply. Tolerances for manganese in industrial water supplies are low (less than 0.2 mg/L), particularly in textile dyeing, food processing, distilling and brewing, photography, and paper and plastics production. The best method might be to use an atomic absorption spectrophotometer; but for the sake of discussion let us consider the possibility of using ultraviolet or visible spectrophotometry.

First of all, knowing from experience that manganese in the +7 oxidation state (permanganate ion MnO_4^-) has an intense purple color, we decide to convert the manganese to this species. Several redox reactions are available to us, but we shall use the periodate method: the water sample is made strongly acidic with nitric acid, a little solid potassium periodate is added, and the solution is heated to effect the reaction

$$2Mn^{2+} + 5IO_4^- + 3H_2O \rightarrow 2MnO_4^- + 5IO_3^- + 6H^+.$$

Absorption Spectrum and Choice of Wavelength

The first step in a spectrophotometric determination is to find the absorption characteristics of the species to be measured. Since potassium permanganate is a readily available compound, it is a simple matter to make up a solution of known concentration and to determine its absorption spectrum.

The spectrum obtained by means of 1.00-cm quartz cell with a water reference cell for the ultraviolet and visible regions of a solution of potassium permanganate of molarity 0.000300 is presented in Fig. 5.12. We may note several things: first, the absorbance is higher in the visible region than in the ultraviolet, so we may gain a small sensitivity advantage by working in the visible range. More specifically, we note that the absorption maximum occurs at a wavelength of 523 nm, and therefore this will be the wavelength to use for the determination. It is generally best to use a wavelength corresponding to maximum absorbance, partly because of the higher sensitivity, but also because slight variations in the setting of the wavelength control have little effect compared to the case of trying to set the wavelength accurately on a steeply rising portion of the spectrum, where the problem of finite bandwidth is also serious, often leading to apparent deviations from Beer's Law.

Calculation of Absorptivity

Noting from the spectrum that the maximum absorbance is 0.735, we find the molar absorptivity from Beer's Law:

$$\epsilon = \frac{A}{bC} = \frac{0.735}{1 \cdot 0.00030} = 2450 \text{ L/mol} \cdot \text{cm}.$$

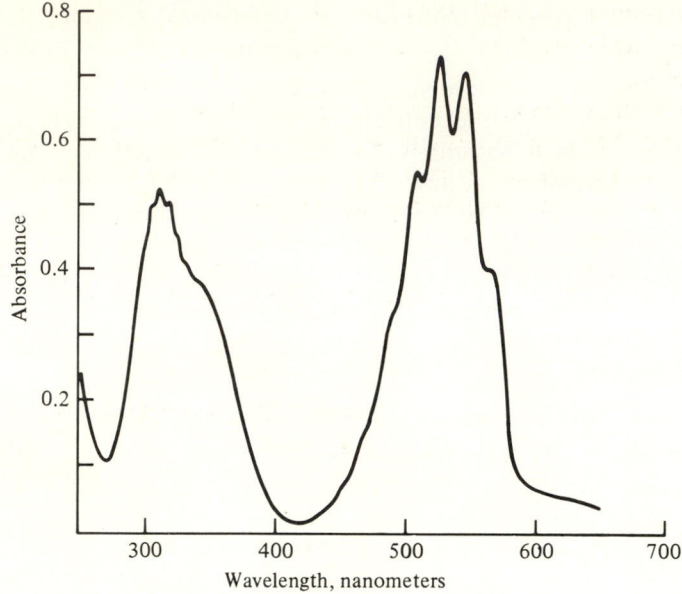

Fig. 5.12 Absorption spectrum of permanganate ion (0.000300M, 1.00-cm cell, Cary Model 118).

However, in the analysis of the natural water we will prefer to use concentration units of mg/L or ppm. The absorptivity corresponding to this is:

$$a = 2450 \frac{L}{mol \cdot cm} \cdot \frac{1 \text{ mol } MnO_4^-}{1 \text{ mol Mn}} \cdot \frac{1 \text{ mol Mn}}{54.94 \text{ g}} \cdot \frac{1 \text{ g}}{1000 \text{ mg}}$$

$$= 0.0446 \text{ L/mg} \cdot cm.$$

Feasibility of the Proposed Method

Before doing any tests on the natural-water samples, it is well to perform a simple calculation to see whether the expected levels of manganese are sufficient to give a measurable absorbance. We are concerned with manganese levels in the 0.1–0.5 ppm range and will be quite content to make the determinations with an uncertainty of about 10%. Let us choose the worst case (0.1 ppm or 0.1 mg/L) to see what absorbance may be expected. Further, we know that the sample treatment (with acid and periodate) will cause an increase in volume and therefore a decrease in manganese concentration. Suppose we double the volume, making the final manganese concentration equal to 0.05 mg/L for the lowest case.

By Beer's Law, if we use a 1-cm cell, the absorbance will be:

$$A = abC = 0.0446 \cdot 1 \cdot 0.05 = 0.0022.$$

This is simply too small, because our absorbance measurements are only precise to within 0.002 units, and therefore the uncertainty would be as large as the signal.

Rather than abandoning the approach, however, we can suggest two possibilities. First of all, there is nothing wrong with concentrating the natural-water sample by evaporation. Suppose we take a 100-mL portion and boil it gently until the volume is only 10 mL. This gives a tenfold increase in concentration. Secondly, we may use a 10-cm cell instead of the 1-cm cell, thus gaining another tenfold improvement. These two steps together should result in an absorbance reading of 0.22 instead of 0.0022, making the method quite feasible indeed.

Our procedure therefore takes the following form:

1. Obtain a 100-mL sample of natural water and evaporate it to about 10 mL.

2. Add 10 mL of $2M$ nitric acid and a small scoop of KIO_4. Boil for a few minutes to fully develop the purple color of the permanganate ion.

3. Dilute the cooled solution to precisely 25 mL in a volumetric flask.

4. Measure the absorbance at 523 nm, using a 10-cm cell.

5. Calculate the concentration of MnO_4^- in the solution using Beer's Law:

$$C = \frac{A}{0.0446 \cdot 10}.$$

6. Calculate the concentration of manganese in the original sample, taking account of dilution and evaporation:

$$C_{initial} = C \cdot \frac{25}{100}.$$

Determining the Standard Curve

There are three shortcomings in the simple approach just outlined:

1. We assumed that Beer's Law holds perfectly and that the absorptivity remains constant for different concentrations of permanganate. These assumptions are often not valid when a spectrophotometer of average quality is used, not because there is anything wrong with Beer's Law itself, but rather because the instrument may not be linear in its response to the power of the light beam. Further, even with a high-quality instrument there are uncertainties connected with each measurement of absorbance. The "two-point" approach suggested above (one measurement of the standard and one of the unknown) does not permit an estimate of the confidence limits for the analytical determination.

2. We assumed that our chemical procedure will work perfectly, that all the manganese in the natural-water sample will be quantitatively converted to

permanganate ion. Such hopes are not always fulfilled in practice. Further, because of such chemical uncertainties, it is best practice to use standards that are put through the same chemical procedure as the unknowns. In the above, however, we used a water solution of potassium permanganate for the standard and a chemical treatment in nitric acid for the unknown.

3. We assumed that the "blank" is zero. Many otherwise good analyses have suffered because of neglect of the blank. We may define the blank as the response (in this case an absorbance reading) that is obtained when a blank sample (such as distilled water) is carried through exactly the same procedure as the standards and the unknowns. This simple step will show whether the chemical reagents themselves can cause an appreciable response, even when the sample is free of the constituent being determined. All analysts should have a neatly lettered card on the lab bench saying *"Don't forget to run the blank!!"*

ILLUSTRATION OF A STANDARD CURVE

Suppose it is desired to determine the molar solubility of a strong electrolyte salt MX in water. The equilibrium is

$$MX(s) = M^+ + X^-, \qquad K_{sp} = [M^+][X^-]f_i^2.$$

The solubility is expected to be in the range of 0.004–0.006 mol/L, and X^- is colored ($\epsilon = 60$ at λ_{max}) while M^+ does not absorb at that wavelength.

To set up a standard curve, we make a few accurate solutions of the soluble salt Na^+X^- and measure their absorbances; the results are presented in Table 5.3.

Table 5.3

Molarity of NaX	Absorbance
0.000	0.001*
0.00200	0.126
0.00400	0.237
0.00600	0.363
0.00800	0.484
0.01000	0.597

*This is the blank when distilled water is used. The reading is not zero perhaps because of a slight mismatch in the two cells.

The analysis of the saturated solution of MX is simple in this case. A portion of the solution is filtered to remove any solid MX, which otherwise would cause light loss by scattering; and the absorbance is found to be 0.300,

0.296, 0.302, 0.304, and 0.301 in successive trials. We decide to take the result as 0.300 ± 0.004 absorbance units.

A plot of the data for the standard curve is shown in Fig. 5.13. There is some scattering of the points, as may always be expected, but the overall linearity is good. This shows that Beer's Law is applicable to this system.

The straight line drawn through the points was calculated by the least-squares method and included calculation of 90% confidence limits (see p. 67). The results are:

$$\text{Slope} = 59.7 \pm 1.0,$$

$$\text{Intercept} = 0.003 \pm 0.006,$$

$$\text{Standard deviation of } A\text{-values} = 0.004.$$

Now we may find the concentration of X^- in the unknown. The most simple and direct approach is to "read" the standard curve: to find what concentration corresponds to the observed absorbance of 0.300. This technique is often recommended but may cause additional error if the plot is not

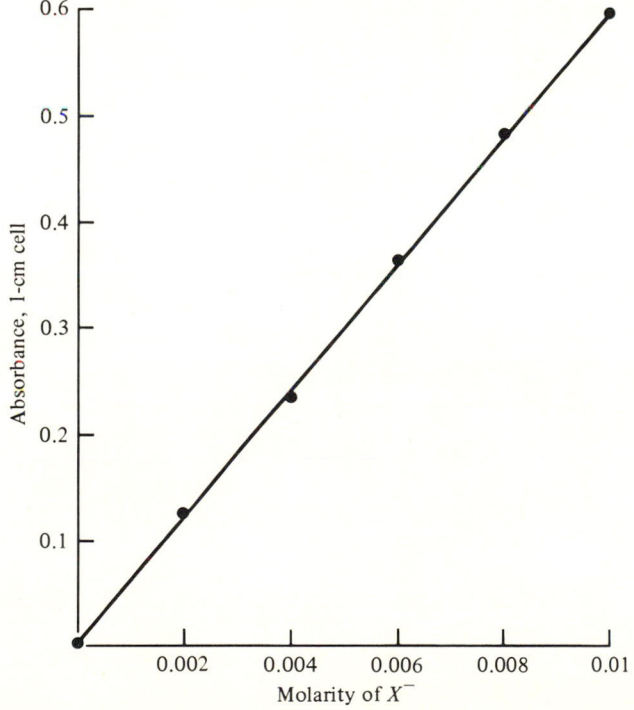

Fig. 5.13 Standard curve for spectrophotometric determination of X^- with solutions of NaX used as reference standard. The least-squares line with 90% confidence limits is

$$A = (0.003 \pm 0.006) + (59.7 \pm 1.0) \cdot [X^-].$$

prepared with care and if the line is "eyeballed" instead of being in accord with the least-squares fit.

But if a least-squares fit *has* been done, then why not use the resulting equation instead of trying to read a graph? If we (temporarily) ignore confidence limits, the result for a saturated solution of MX is

$$[X^-] = \frac{0.300 - 0.003}{59.7} = 0.00497\,M.$$

However, since we have confidence limits on each of the quantities, it makes sense to include them in the calculation. The setup is:

$$[X^-] = \frac{(0.300 \pm 0.004) - (0.003 \pm 0.006)}{59.7 \pm 1.0}.$$

First, the numerator is evaluated by using the absolute values of the confidence limits because a simple subtraction is all that is required:

$$(0.300 - 0.003) \pm \sqrt{0.004^2 + 0.006^2} = 0.297 \pm 0.007.$$

Then the quotient is evaluated by using relative values of the confidence limits:

$$\frac{0.297}{59.7}\left\{1 \pm \sqrt{\left(\frac{0.007}{0.297}\right)^2 + \left(\frac{1.0}{59.7}\right)^2}\right\} = 0.00497 \pm 0.00014.$$

Thus, the 90% confidence limits for the solubility correspond to an uncertainty of about 3% in the solubility value.

The reader should calculate the solubility product constant for MX, taking account of the above confidence limits. *Answer*: $(2.12 \pm 0.08) \cdot 10^{-5}$.

UNWELCOME TYPES OF STANDARD CURVES

The illustration in Fig. 5.13 is typical of a good standard curve, with the points lying close to a straight line and with an intercept close to zero. Some other possibilities are reflected in Figs. 5.14–5.16.

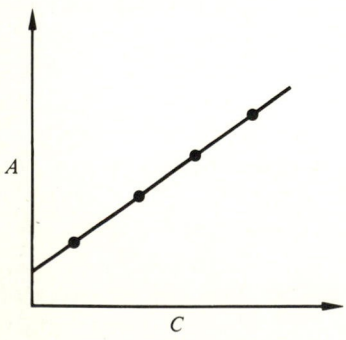

Figure 5.14

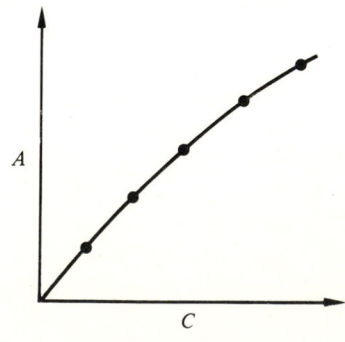

Figure 5.15

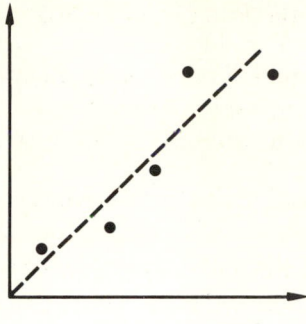

Figure 5.16

In Fig. 5.14 the plot is quite linear but the intercept is far too high. It is likely that this high "blank" is caused by substances in the reagent(s), which may be purified before use. Or the reagent(s) may have a color that might be avoided by working at a different wavelength.

In Fig. 5.15 the curvature of the plot is undesirable though satisfactory results may be obtained by "reading" the plot. The instrument may be nonlinear or perhaps in need of repair. It may be that the chemical system actually changes with concentration, as with dimer formation. There may be too much stray light in the instrument.

The type of disaster reflected in Fig. 5.16 can have two causes in addition to personal incompetence. It may be that the instrument is erratic and in need of repair. Also, it may be that the chemical system is not under control, in which case the effects of variables (such as pH, temperature, waiting time, etc.) have to be investigated.

5.11 DIFFERENTIAL ABSORBANCE MEASUREMENTS FOR HIGHER PRECISION

When we attempt to make absorbance measurements on strongly colored solutions, at least two problems arise to plague us. The first is the difficulty of making accurate readings of the meter (see p. 140), given the logarithmic compression of the absorbance scale. It doesn't help at all to read the linear transmittance scale because at low $\%T$ small errors become more significant upon conversion to absorbance. It can be shown that the optimum region of the absorbance scale, as far as relative accuracy is concerned, is from 0.3 to 0.6. When dealing with a solution of high absorbance such as 0.9, it is better to dilute the solution accurately before measuring the absorbance.

The other problem is in the nonlinearity of the spectrophotometer under conditions of a weak (low-power) light beam coming from the sample cell. Not only can the photocell behavior be nonlinear, but whatever stray light is bouncing around inside the instrument has a proportionately larger effect on the reading when it has to compete only with a low-power beam. The percentage of stray light compared to the full-power beam depends very much on the quality of the instrument. For the B&L Spectronic 20 the specification for stray-light content is 0.5%, and for the Cary Model 118 it is only 0.001%. This is one good reason why the Cary 118 can give reliable absorbance measurements as high as $A = 3$.

Suppose you wish to measure the absorbance of a series of solutions of copper ion Cu^{2+}, knowing that the copper concentrations are likely to be slightly above $0.20 M$. Since the molar absorptivity of Cu^{2+} at 800 nm is about $11 \, L/mol \cdot cm$, one would predict absorbance values of about

$$A = 11 \cdot 1 \cdot 0.2 = 2.2$$

if a 1-cm cell is used. With an average spectrophotometer this is simply too high to permit high accuracy in the results for the copper concentrations.

One possibility is to use a cell of shorter path length. With a 1-mm cell path, the absorbance would be in the satisfactory range of about $A = 0.2$. Another possibility is to dilute the solutions, as mentioned above, so that the absorbances will be about 0.4. However, in both of these approaches the measurement will become what we may call "normal spectrophotometry" and, as shown by the previous example dealing with the solubility of MX, the overall uncertainty in the results is likely to be a few percent.

If we desire to improve the accuracy of the determination, we may use *differential spectrophotometry*. To understand the principle it is only necessary to realize that a spectrophotometer neither knows nor cares what you choose for a reference solution. The output absorbance, whether on a digital readout, a recorder chart, or a meter with needle, is simply the *difference* in absorbance between the sample and the reference solutions. In ordinary spectrophotometry we routinely use distilled water as the reference and by setting the controls we effectively define the absorbance of the water as zero. Then the absorbance of the sample solution is read from the dial.

Now, back to the problem of making accurate determinations of the copper ion solutions with molarities slightly greater than 0.20. Instead of distilled water, suppose we use as a reference a very accurate solution of copper ion with a molarity of precisely 0.2000. Perhaps this solution has been standardized by a gravimetric procedure or made up by weighing and dissolving a sample of very pure copper foil. In any case we will assume that we can make accurate reference standard solutions of copper ion in the 0.20 to $0.25 M$ range.

Using the $0.2000 M$ solution as a reference, we simply turn the knob on the instrument to make the absorbance read zero when this solution is in the cell. Then we measure a few of the other standard copper solutions against this reference. The results are given in Table 5.4.

These absorbance values are not remarkable in themselves and are subject to the same uncertainties as ordinary spectrophotometric data.

Table 5.4

Solution in sample cell, M	Observed absorbance	Concentration difference
0.2000	0.002	0
0.2100	0.021	0.0100
0.2200	0.039	0.0200
0.2300	0.051	0.0300

However, remember that they represent only the difference in concentration (see Fig. 5.17).

The data points show the usual degree of random scattering; the fact that the plot is curved indicates that the spectrophotometer has a nonlinear response. Nevertheless, a smooth curve may be drawn as shown; and although it is possible to use a quadratic or cubic least-squares fit to find the equation for the curve, we may as well simply make a direct reading from the plot.

To illustrate, let the first of the unknown copper solutions give a differential absorbance reading of 0.044. We find that this corresponds to a concentration difference of 0.0235. Therefore, the concentration of the unknown solution is

$$C = 0.2000 + 0.0235 = 0.2235\,M.$$

Even with the uncertainty related to the scatter of points on the standard

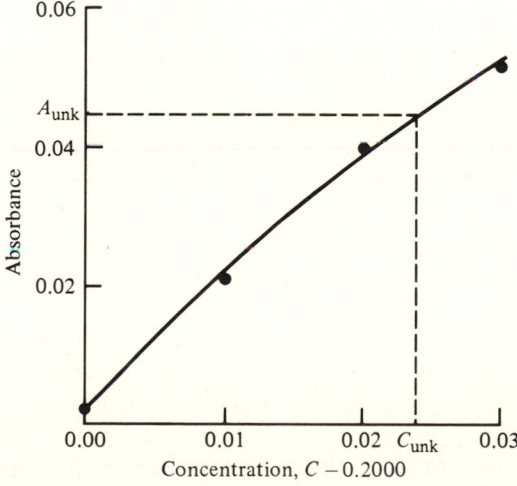

Fig. 5.17 Standard curve for differential measurements on copper-ion solution.

curve, the overall precision of this determination is far better than could be hoped for in ordinary spectrophotometry with a water reference. In favorable cases, differential spectrophotometry rivals gravimetric and titrimetric techniques for accuracy and precision.

5.12   CHEMICAL ASPECTS OF SPECTROPHOTOMETRIC DETERMINATIONS

Before an absorbance measurement can be used for chemical analysis, the constituent being determined must have a suitable absorptivity. Although many substances can be determined with a minimum of preparation (since they show characteristic and useful absorption spectra), countless others must be put through some chemical treatment to convert them to an absorbing species. The literature is replete with clever, well-tested procedures for accomplishing this goal for a huge variety of materials. To devise such procedures and to understand their basis in terms of chemical reactions and equilibria is a challenging and ongoing task of chemical analysts. In an introductory text there is no point in trying to give a comprehensive review of procedures for drugs, food additives, environmental samples, biological fluids, trace metals in geological specimens, etc. All of these and many other types of samples are routinely analyzed with the help of spectrophotometry, but let us consider merely a small selection of illustrative ideas.

Change in the Oxidation State of an Element

We have already discussed one example of this type, the conversion of the nearly colorless manganous ion to the intensely purple permanganate ion.

A second important example is used for the determination of chromium. The Cr^{3+} ion is only weakly absorbing, but when it is heated in basic solution with an excess of hydrogen peroxide it is converted to the yellow chromate ion:

$$Cr^{3+} \xrightarrow{\ OH^-\ } Cr(OH)_3(s) \xrightarrow{\ H_2O_2\ } CrO_4^{2-}.$$

$$\underset{\substack{\text{absorbs strongly} \\ \text{in the near UV}}}{}$$

Small amounts of oxidizing agents are easily determined spectrophotometrically even if they do not absorb light themselves. Advantage is taken of the reaction with iodide ion:

$$\underset{\substack{\text{added in} \\ \text{excess}}}{I^-} + \text{Oxidant} = I_3^- + \text{Reduced species.}$$

The triiodide ion is formed in direct proportion to the amount of originally present oxidant. It has a strong absorption maximum ($\epsilon = 2.5 \cdot 10^4$) at 350 nm and an even larger peak at 285 nm. This is shown in the set of absorption

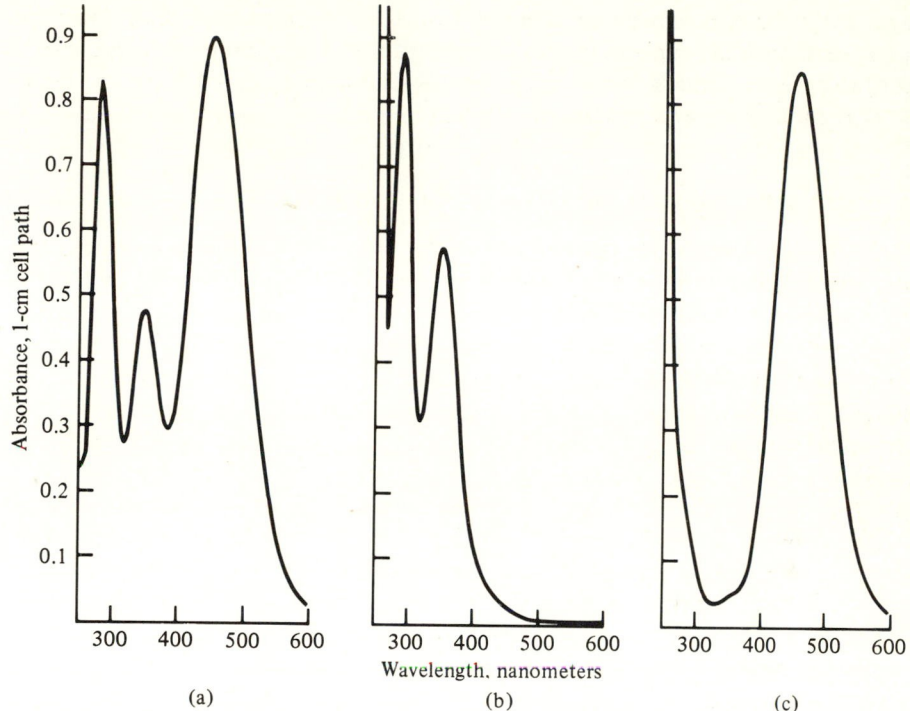

Wavelength, nanometers

(a)   (b)   (c)

Fig. 5.18 (a) Saturated solution of iodine in water. (b) Iodine in $0.1M$ KI, 1/50 dilution of saturated solution. (c) Saturated solution of iodine in the presence of KIO_3 and H_2SO_4.

spectra depicted in Fig. 5.18. The spectrum in Fig. 5.18(a) was obtained by using a filtered sample of distilled water that had been shaken for a few minutes with excess solid iodine. This solution has a brownish color and shows three peaks in the spectrum.

That two of these peaks must be due to the presence of a little triiodide ion is proved by the spectrum in Fig. 5.18(b), because in the presence of $0.1M$ potassium iodide there is nearly complete conversion of iodine to triiodide ion I_3^- (note the 50-fold dilution):

$$I_2(aq) + I^- = I_3^-, \qquad K = \frac{[I_3^-]}{[I_2][I^-]} = 720 \quad (\text{at } 25°C).$$

Why is I_3^- present in the first solution when there is *no* potassium iodide added? There is a small tendency for iodine to react with water by undergoing hydrolysis:

$$3I_2(aq) + 3H_2O \rightleftarrows 5I^- + IO_3^- + 6H^+,$$

slight

I_3^-

The small amount of iodide ion I^- formed by this hydrolysis (or **disproportionation**) reaction is partly converted to triiodide ion, and hence the additional peaks appear in the absorption spectrum.

To demonstrate this further, small amounts of potassium iodate and sulfuric acid are added directly to the absorption cell containing the solution. This prevents an appreciable concentration of I^- by shifting the hydrolysis reaction to the left.

In Fig. 5.18(c) we see that the I_3^- peaks have disappeared except for two very slight shoulders on the curve at 285 and 350 nm; hence this shows the true absorption spectrum of the I_2 species.

Change in Solution pH

For many organic acids and bases, either the acid form or the conjugate base or both show significant absorption in either the UV or visible region. The sample treatment simply adjusts the solution pH to force the substance into the desired form. For example,

$$HD \quad + \quad OH^- \quad = \quad D^- \quad + \quad H_2O.$$

HD			D⁻		
dinitrophenol,			conjugate base,		
colorless			yellow		

This technique, as well as the others mentioned in this section, is sometimes used to transform a colored substance to a colorless one, so as to eliminate it as an absorbing interference in the determination of some other species.

Formation of a Colored Metal–Ligand Complex

This type of sample treatment is perhaps the most important for trace-metal determinations by the spectrophotometric method, although in recent years it has been largely supplanted by the technique of atomic absorption spectroscopy. The outstanding work by E. B. Sandell and H. Onishi should be consulted for both fundamental and applied aspects of the subject.*

The determination of iron in beer or ale will serve as an example: first the beer is degassed by shaking and then two 25-mL samples are pipetted into flasks. To each is added 25 mg of ascorbic acid (vitamin C) to reduce the iron from the +3 state to the +2 state (ferrous ion Fe^{2+}). Now, to one flask is added 2 mL of the color-forming reagent, which is a solution of 2.2′-dipyridyl. To the other flask, which will serve as a blank, 2 mL of water are added. Both flasks are allowed to stand for 30 minutes, while the color of the ferrous–dipyridyl complex forms:

$$Fe^{2+} + 3L = FeL_3^{2+}, \quad \text{where L is dipyridyl:}$$

Colorimetric Metal Analysis, 4th ed., Interscience, 1978.

The absorbance of the colored solution is measured versus the blank and compared with a standard curve prepared by using known amounts of ferrous ammonium sulfate in the same procedure. The complex is stable and the color remains constant for hours.

5.13 PHOTOMETRIC TITRATION

The progress of many titrations may be followed by measuring the solution absorbance as a function of the volume of added titrant. This technique is usually referred to as **photometric titration** but it would perhaps be better to call it **photometric determination of titration endpoints**. For the present it will suffice to set forth the main ideas of the technique for titrations that do not involve added indicators.

Consider the general form of the chemical equation for a titration reaction:

$$a\text{A} \quad + b\text{B} = cC$$
$$\text{sample} \qquad \text{titrant} \quad \text{product}$$
$$\text{constituent}$$

Suppose that the absorption spectra for substances A, B, and C are as shown qualitatively in Fig. 5.19.

When we perform this titration, adding substance B from a buret into a solution containing substance A, we can watch the resulting solution change color from beginning to end. However, if the titration is carried out in a spectrophotometer cell, the changes in measured absorbance will depend greatly on the chosen wavelength. If the spectrophotometer is set to operate at the lowest indicated wavelength λ_1, where only substance A absorbs

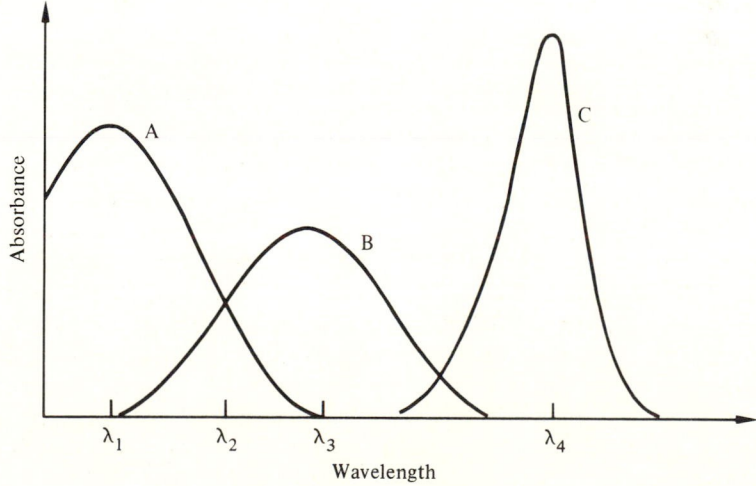

Wavelength

Figure 5.19

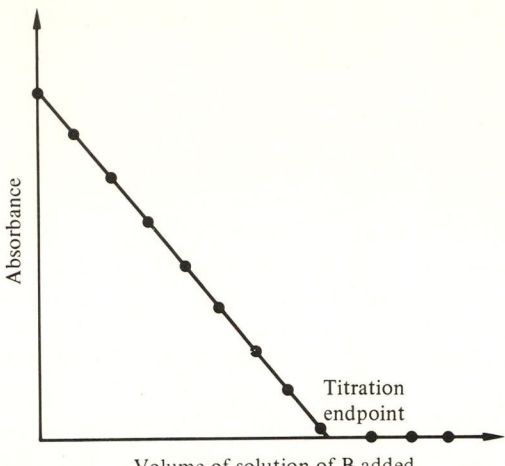

Figure 5.20

radiation, we would obtain the titration curve (a plot of absorbance versus titrant volume) presented in Fig. 5.20.

To obtain the data, we simply add small increments of titrant, measuring the absorbance after each addition. Because of the increase in solution volume with each addition, the concentration of substance A, and therefore its absorbance A, is diminished by dilution as well as by the chemical reaction. To correct the data for the dilution effect, it is only necessary to multiply each absorbance by a dilution factor:

$$A_{cor} = A_{obs} \cdot \frac{V + V_0}{V_0},$$

where V_0 is the original solution volume and V is the volume of added titrant. The several points obtained are fitted by a straight line and extrapolated to the volume axis to determine the titration endpoint. Once the endpoint has been determined, the stoichiometric calculations are quite normal:

$$n_a = \frac{a}{b} \cdot n_B = \frac{a}{b} \cdot V_{ep} \cdot C_B.$$

The sketches in Figs. 5.21–5.23 show the titration curves that would be obtained at the other wavelengths.

In Fig. 5.21, the titration curve for λ_2 prior to the endpoint is qualitatively the same as for λ_1 because substance A is the only absorbing species. It is only after the endpoint has been reached that the solution contains increasing amounts of substance B, which also absorbs at this wavelength.

In Fig. 5.22, there are no absorbing species at λ_3 before the endpoint, but the addition of excess titrant causes the absorbance to rise once the endpoint is passed.

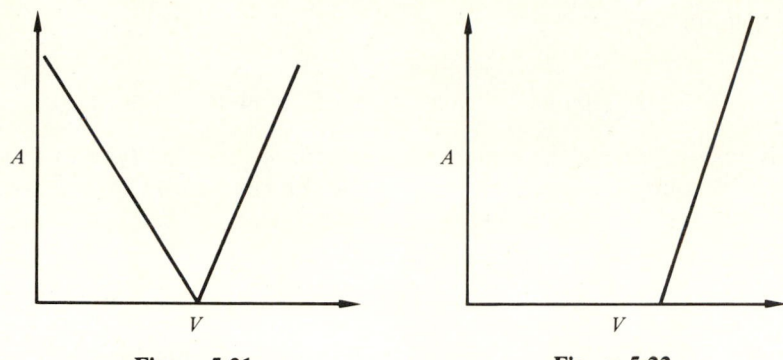

Figure 5.21   Figure 5.22

In Fig. 5.23, the absorbance rises continuously at λ_4 as the absorbing substance C is produced by the titration reaction. Once the endpoint has been reached, no more C can be produced and substance B does not absorb. Therefore the absorbance levels off at a constant value. This wavelength is less suitable than the others because of the dependence on absorbance values, which may become too high to be reliable.

Effect of Incomplete Reaction

The foregoing sketches correspond to the favorable cases of titration reactions with large equilibrium constants. That is, the added titrant is completely used up before the endpoint, and after the endpoint there is no unreacted constituent. There are many practical examples in this category, especially those using oxidation–reduction reactions or very stable chelate-formation reactions (see Chapter 16).

However it is important to consider titrations in which the reaction is not complete in the vicinity of the equivalence point, either because the solution is very dilute or because the equilibrium quotient for the titration reaction is not particularly large. The most important category of such reactions involve

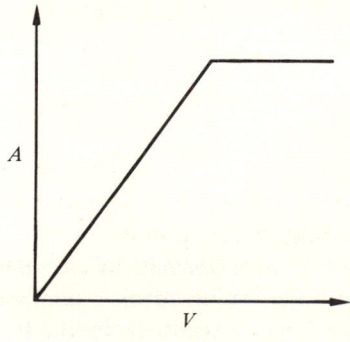

Figure 5.23

a 1:1 stoichiometry:

$$\begin{array}{ccccc} D & + & X & = & DX. \\ \text{titrant} & & \text{constituent} & & \text{product} \end{array}$$

It is easy to derive theoretical titration curves for this type of reaction: let the initial solution contain C_X molar X, which reacts with D in accordance with the equilibrium quotient Q:

$$Q = \frac{[DX]}{[D][X]}.$$

For simplicity we shall ignore any dilution of the solution that occurs upon addition of titrant (imagine that substance D is added in the form of a rather concentrated solution, so that very small volumes are needed). The endpoint of the titration is reached when the added concentration of D, represented by C_D, is equal to the initial concentration C_X of X. At any point in the titration it must be true from considerations of material balance that

$$[X] = C_X - [DX] \quad \text{and} \quad [D] = C_D - [DX].$$

When these relationships are substituted into the equilibrium quotient expression, we find:

$$Q = \frac{[DX]}{(C_D - [DX])(C_X - [DX])}.$$

This is rearranged to the standard quadratic form and may be solved by the quadratic formula to find [DX] for any desired combination of Q, C_X, and C_D. Three titration curves calculated for the titration of $0.001M$ X are depicted in Fig. 5.24 for Q values equal to 10^6, 10^5, and 10^4, respectively.

In these calculations, for each of the Q-values, C_X was held constant at $0.001M$ and C_D was varied in $0.0002M$ increments. Thus, the quadratic formula was solved ten times for each curve to find the value for [DX] that would be present for each point. By assuming arbitrarily that $\epsilon_{DX} = 200$ and that the cell path is equal to 5 cm, the hypothetical absorbance for each point was calculated from Beer's Law. As Q decreases, the deterioration in the usefulness of the plots becomes quite evident. The curvature is not entirely due to the Q-value, however. Titration of a more concentrated solution of X would yield more nearly linear plots. The reader may wish to prove this by calculation.

It is important to know that titration endpoints may be determined with fair accuracy by the photometric technique even for cases of incomplete reaction, provided the data in the early part of the titration (when there is still enough excess X to force the reaction to near-completion) and well after the endpoint (when there is enough excess D to do the same) are reasonably linear. One may then connect the points in the linear sections of the plot and

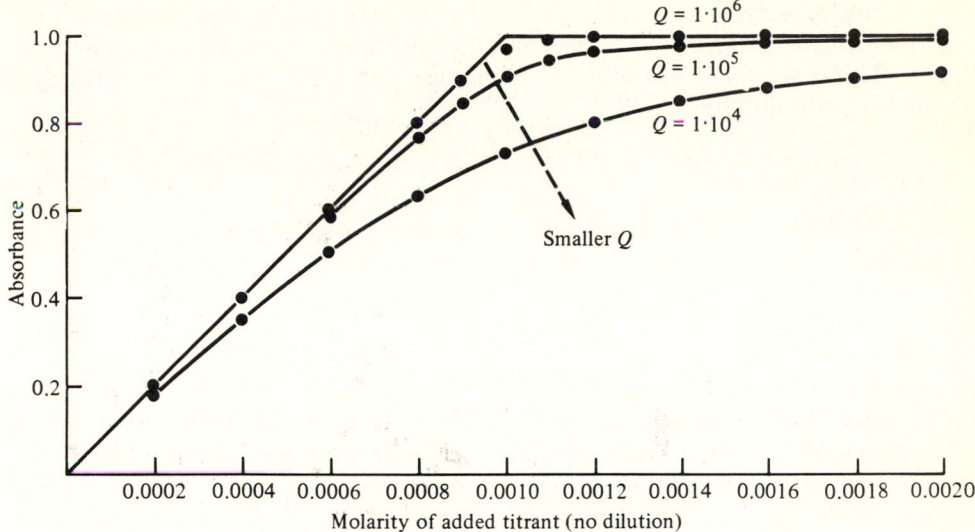

Fig. 5.24 Photometric titrations (theoretical) of $0.001M$ X:

$$D \quad + \quad X \quad = \quad DX.$$
$$\text{titrant} \qquad \text{titrand} \qquad \text{product}$$
$$\text{(colorless)} \quad \text{(colorless)} \quad \text{(colored)}$$

Straight lines show the absorbance when the reaction is complete (large Q).

extrapolate to find the intersection (see Problems 1 and 2 on photometric titration).

Simple Rules for Predicting Titration Curves When All Species Absorb Light

Suppose that D, X, and DX all have significant absorptivities at the wavelength used for titration. Only three absorbances need be calculated, as follows:

1. The initial absorbance is due only to substance X present at a concentration C_X. By Beer's Law,

$$A_{init} = \epsilon_X b C_X.$$

2. The absorbance at the endpoint is due only to the product DX present at a concentration C_X (ignoring dilution). Therefore,

$$A_{ep} = \epsilon_{DX} b C_X.$$

3. The absorbance when the amount of added titrant is twice as high as that required to reach the endpoint is due to the DX already formed plus that due to the excess titrant. We have:

$$A_{200\%} = A_{ep} + \epsilon_D b C_X.$$

One merely plots these three points on graph paper and connects them with straight lines; it is assumed that the reaction goes nearly to completion.

There are three operational techniques that work well for performing a photometric titration:

1. A spectrophotometer may be modified to accommodate a beaker placed in the cell compartment. A magnetic stirrer is built in to keep the solution mixed, and increments of titrant are added with a buret. Care must be taken to keep extraneous light from getting into the cell compartment.

2. The titration vessel and buret may be external to the instrument, and the solution must be continuously circulated through a flow-through cell in the cell compartment. Noncontaminating pump arrangements are available.

3. The titration may be carried out on a small scale, using only a few milliliters of sample in a regular spectrophotometric cell. The titrant is added in the form of a relatively concentrated solution from a microburet capable of measuring very small volumes with high precision. The cell is taken in and out of the instrument for each addition and its contents are mixed (by using a very small magnetic stirring bar) in the cell while holding it over a regular magnetic stirrer.

Photometric titrations often have great advantages over other methods of detecting the endpoint, particularly for dilute solutions or for reactions that do not have a large equilibrium constant, and for cases where no suitable visual indicator is available.

PROBLEMS

Spectrophotometric determinations

1. A sample of pure salicylic acid HOC_6H_4COOH weighing 0.0691 gram was dissolved in water and diluted to 250 mL. A 5-mL portion of the solution was further diluted to 100 mL, using an acetate buffer of pH 5. When this solution was placed in a 2.00-cm quartz absorption cell, the ultraviolet absorption spectrum shown in Table 5.5 was obtained.

Table 5.5

Wavelength, nm	Transmittance	Wavelength, nm	Transmittance
250	0.759	295	0.200
260	0.851	300	0.209
270	0.830	310	0.408
280	0.398	320	0.777
290	0.229	340	0.998

a) Make a plot of absorbance *versus* wavelength, connecting the points with a smooth curve.

b) Calculate the absorptivity at 295 nm in units of mL/mg · cm.

c) Calculate the molar absorptivity for salicylic acid at the wavelength of maximum absorbance.

2. In determining the concentration of salicylic acid in a certain aqueous solution, a spectrophotometer was first checked by measuring the absorbance of some standards. At 295 nm, using a 1.00-cm cell, the results were as follows:

Concentration of standard, mol/L: $3.00 \cdot 10^{-4}$ $1.80 \cdot 10^{-4}$ $1.00 \cdot 10^{-4}$ $0.40 \cdot 10^{-4}$
Absorbance: 1.020   0.628   0.350   0.138

The unknown solution gave an absorbance reading of 0.517.

a) Plot the standard curve of absorbance *versus* concentration. Comment on the applicability of Beer's Law.

b) Find the concentration of salicylic acid in the unknown solution.

3. One procedure for the determination of small quantities of manganese (e.g., in a sample of a steel alloy) is based on the oxidation by periodate ion:

$$Mn^{2+} \quad + IO_4^- = MnO_4^- + IO_3^-.$$
in acid solution

The excess periodate ion and the iodate ion are colorless, while the permanganate ion is intensely purple, with a molar absorptivity of 958 at 535 nm.

a) Balance the redox equation.

b) If a steel sample that weighs 0.1605 gram is treated according to this procedure, including a final dilution to 100 mL, what percentage of manganese does it contain if the absorbance is 0.075 when a 2-cm absorption cell is used?

4. A novel way of determining the approximate molecular weight of simple organic bases is to combine them with picric acid to form amine salts:

$$R—NH_2 + HO—C_6H_2(NO_2)_3 = R—NH_3^+Pic^-.$$
"HPic"

The picrate salt is separated and purified by recrystallizing. In one application involving R—NH$_2$ of unknown structure, 0.0505 gram of the purified picrate salt was dissolved in 100 mL of water, and 10 mL of this solution was further diluted to 250 mL. The absorbance at 380 nm was 0.488 with a 1-cm cell. This is due entirely to the yellow picrate ion, which has a molar absorptivity of 13,440 at this wavelength. The molecular weight of picric acid is 229. Calculate the molecular weight of the amine R—NH$_2$.

5. A certain weak acid HA is colorless but its conjugate base A$^-$ absorbs strongly at 570 nm, having a molar absorptivity of $2.24 \cdot 10^4$. A buffered solution of the acid with a concentration of $1.00 \cdot 10^{-4}\,M$ is found to have an absorbance of 0.741 at 570 nm by means of a 5-cm absorption cell. By use of a pH-meter it is determined that the [H$^+$] in this solution is $1.55 \cdot 10^{-6}$. Find the value for the acid dissociation

quotient Q,

$$Q = \frac{[H^+][A^-]}{[HA]}$$

Photometric titration

6. When a 100-mL sample of a solution of substance X was titrated with a $0.50M$ titrant T using a 5-cm cell path, the following data were obtained:

Volume of titrant, mL	Absorbance	Volume of titrant, mL	Absorbance
0	0.661	10	0.050
2	0.509	12	0.043
4	0.367	14	0.041
6	0.232	16	0.040
8	0.102		

Titration reaction: $2X + T \rightarrow$ Products.

Correct the data for dilution and find the titration endpoint from a plot of absorbance versus titrant volume. Calculate the concentration of X in the sample and estimate its molar absorptivity at the wavelength used. Also estimate the absorptivity of the titrant. Is there any evidence from the plot that the product of the titration reaction also absorbs?

7. A 100-mL portion of a standard solution of substance Y with a molarity of 0.0036 was titrated with a titrant solution of unknown molarity. Calculate the titrant molarity from the data obtained with a 5-cm cell path.

Volume of titrant, mL	Absorbance	Volume of titrant, mL	Absorbance
0	0.234	3	0.070
0.5	0.202	3.5	0.157
1	0.169	4	0.242
1.5	0.132	4.5	0.333
2	0.102	5	0.420
2.5	0.070	5.5	0.508

Titration reaction: $Y + T \rightarrow$ Products.

DETERMINATIONS BASED UPON PRECIPITATION

6

I think it was in a course in quantitative chemical analysis that an appreciation of the scientific method and its rigours began really to take hold of me.... There were no short cuts to beat clear thinking, careful technique and endless patience. Later on, I found that the same unnatural methods are always required in those activities commonly called "research."

R. S. Mulliken

First we shall consider the gravimetric (based on weight) methods and then a few classical titration methods that depend upon formation of precipitates.

6.1   GRAVIMETRIC DETERMINATIONS

The quantitative foundations of chemistry were established in the 19th century by observations of the change of mass when one substance is converted into another. Gravimetric* analysis requires only a precise balance, and procedures have been developed for nearly all elements in a wide variety of materials. Although modern instrumental methods are now preferred for most rapid work, they usually cannot rival the inherent accuracy of a careful gravimetric procedure.

Each particular gravimetric procedure has its own technical problems relating to the chemical transformations used. One must do some research in the literature to find the details of a desired procedure. The present discussion is concerned only with the general principles common to most gravimetric procedures.

First, consider the initial and final states of the substance that is to be determined by a gravimetric procedure:

A *sample* containing an unknown quantity of the *constituent* to be determined.

The sample may be a solid material (which is weighed) or a measured volume of a solution. The constituent must be present in a weighable amount.

Suitable chemical treatment

The treatment consists of a series of steps that are different for each sample type and for each constituent.

A *weighing form*, which is a pure substance of known composition whose mass can be used to find the mass of the constituent in the original sample.

The weighing form will usually contain all of the constituent, but there are exceptions to this.

Electrodeposition of a Metallic Constituent

When the substance to be determined is a metallic species that can be plated out on a cathode in the form of the pure metallic element, the gravimetric approach may be both rapid and accurate. By weighing the cathode before and after electrodeposition, we can determine the mass of the element very simply. The principles of this important technique are discussed in Chapter 8, and for now only the stoichiometry is considered.

Example An inorganic chemist has synthesized a coordination complex of copper and decides to determine the copper content to verify its composition. A sample of the solid complex weighing 2.705 g is dissolved and the copper is

---

* From the Latin *gravis*—heavy.

plated onto a platinum cathode. The mass of copper is 0.7255 g. Does this result support the chemist's hope that the compound is tetramminecopper(II) chloride?

The amount of Cu is calculated using the atomic weight of copper:

$$n_{Cu} = 0.7255/63.54 = 1.142 \cdot 10^{-2} \text{ mol.}$$

If the complex contains one atom of Cu per molecule, then its gram-formula weight may be calculated as

$$GFW = \frac{2.705 \text{ g}}{1.142 \cdot 10^{-2} \text{ mol}} = 236.9 \text{ g/mole.}$$

Since the molecular weight of $Cu(NH_3)_4Cl_2$ is 202.6, the result is discouraging.

Loss of a Volatile Constituent

In some cases the sample contains a constituent that is the only volatile substance present. In this situation it may be possible merely to heat the sample to drive off the constituent, leaving everything else as a weighable residue. The constituent is determined simply by noting the loss in mass.

Example A bottle of recrystallized potassium chloride is checked out of the stockroom and is to be used to prepare a very accurate solution of KCl. A sample weighing 5.6634 g is placed in an oven at 110° for 4 hours and then is reweighed after cooling in a desiccator. The mass is found to be 5.6459 g. Calculate the percentage of water in the original sample:

$$\%H_2O = \frac{5.6634 - 5.6459}{5.6634} \cdot 100 = 0.31\%.$$

Precipitation of the Desired Constituent

This class of procedures is what most chemists think of as gravimetric analysis. The literature contains many hundreds of precipitation procedures, with applications to nearly all chemical elements and to many more complex substances. Most of these procedures have the following steps in common:

1. Given a sample of material (which may be biological, geological, or chemical in origin), it is desired to determine the amount of a certain constituent present. In the original sample, the constituent may be present in an unknown chemical form. The first step is to use a chemical treatment that will transform the constituent to a known chemical form. A geological sample may be digested with concentrated acids, for example, to release metal ions from a silicate matrix.

2. Once the modified sample is in aqueous solution with a desired set of conditions, a precipitating agent (precipitant) is added that reacts selectively with the desired constituent to form a precipitate. An excess of the precipitant is deliberately added, to be sure that virtually all of the constituent is incorporated into the precipitate. It is important in this step that no other insoluble substances be formed.

3. The precipitate is collected by pouring the entire mixture through a filter, typically a porous glass filter crucible. A small amount of water is used to rinse away the last part of the solution that clings to the precipitate.

4. Upon drying in an oven, typically at 110°, the excess water is removed from the precipitate. In many cases it is merely necessary to find the weight of the dry precipitate at this point.

5. In cases where simple oven drying does not produce a dry precipitate of definite composition, it is necessary to resort to higher-temperature drying, called **ignition**, in order to decompose the precipitate to a simpler form that does have a definite composition. It is this substance that is actually used as the weighing form.

The stoichiometric relationships in these transformations may be symbolized in general terms as follows.

$$a\,\text{A} \quad + \quad b\,\text{B} \quad \xrightarrow{\text{heat}} \quad c\,\text{C} \quad \rightarrow \quad d\,\text{D}.$$

| constituent | precipitant | precipitate | weighing form |

precipitate
(often serves as
weighing form)

The overall transformation is simply

$$a\,\text{A} \rightarrow d\,\text{D}.$$

In terms of the amounts (moles) of A and D, we can write

$$n_\text{A}\ (\text{moles of } A \text{ in the sample}) = \frac{a}{d} \cdot n_\text{D}\ (\text{moles of } D \text{ found by weighing}).$$

Since the amount (moles) is proportional to the mass (grams) by the relationship

$$n = w/\text{GFW},$$

we find the mass relationship between A and D:

$$w_\text{A} = \frac{a}{d} \cdot \frac{\text{GFW}_\text{A}}{\text{GFW}_\text{D}} \cdot w_\text{D}.$$

Thus, once the mass of the weighing form has been determined, the mass of the element A in the sample is readily determined by multiplying by a constant. This constant is called the **gravimetric factor** for the particular procedure.

Example To check the composition of the nickel–copper alloy used in making coins, a chemist at a U.S. Mint dissolves a 0.5482-g sample in nitric acid, adjusts conditions appropriately, and precipitates the nickel with the organic reagent dimethylglyoxime. The red precipitate $Ni(C_4H_7O_2N_2)_2$ (with GFW = 288.94) has a mass of 0.6766 g. Calculate the percentage of Ni in the alloy.

The amount of the precipitate is:

$$n_{ppt} = \frac{0.6766}{288.94} = 2.342 \cdot 10^{-3} \text{ mol.}$$

Because the precipitate contains only one atom of Ni per molecule, this is also the amount of nickel present in the sample. Therefore the mass of nickel can be calculated as

$$w_{Ni} = n_{Ni} \cdot GAW_{Ni} = 2.342 \cdot 10^{-3} \cdot 58.70 = 0.1375 \text{ g.}$$

Finally, the percentage of nickel in the alloy is calculated as

$$\%Ni = \frac{w_{Ni}}{w_{sample}} \cdot 100 = \frac{100 \cdot 0.1375}{0.5482} = 25.08\%.$$

Example An aqueous solution containing uranium is analyzed for its UO_2^{2+} content by precipitating the latter as the ammonium salt $(NH_4)_2U_2O_7$, which is separated and then heated strongly to be converted to the pure oxide U_3O_8 (with GFW = 842.09). If a 50.00-mL portion of solution yields 0.2459 g of the oxide, what is the molarity of UO_2^{2+}?

The amount of oxide is:

$$n_{U_3O_8} = 0.2459/842.09 = 2.920 \cdot 10^{-4} \text{ mol.}$$

The overall transformation is:

$$3UO_2^{2+} \rightarrow 1U_3O_8.$$

Therefore,

$$n_{UO_2^{2+}} = 3n_{U_3O_8} = 8.76 \cdot 10^{-4} \text{ mol} = 0.876 \text{ mmol.}$$

The molarity of this species in the solution is, then:

$$M = n/V = 0.876/50 = 0.0175.$$

This is an example of a gravimetric procedure in which the precipitation form serves only to isolate the constituent from the solution and is then converted to a more reliable weighing form for the final measurement.

PREDICTING THE FORMATION OF A PRECIPITATE

When a chemist considers a mixture of solutes in an aqueous solution, how is it possible to say whether or not any of the solute species will combine to form an insoluble substance, i.e., a precipitate?

The first step is to figure out the molarities of all the species under the assumption that they are homogeneously mixed, without yet assuming that a precipitate will form.

The second step is to postulate the various possible precipitates and to look up K_{sp} values for the solubility products for those compounds. If a value for K_{sp} is not available in the table, there can be two explanations:

1. The compound is soluble;
2. The table is incomplete in that the value for K_{sp} has been omitted or perhaps has never been determined.

By looking elsewhere it may be feasible to find either a value for K_{sp} or at least some quantitative information about the solubility of the compound.

The third step is to find whether the existing molarities of the ions that might precipitate are high enough. This is determined by comparing the product of the molarities with the value of the solubility product.

An example will make this clear. Suppose that 50 mL of $0.08 M$ potassium iodide are mixed with 50 mL of $0.05 M$ lead nitrate. If precipitation did *not* occur the ion molarities would be as follows,

$$[K^+] = 0.04, \quad [Pb^{2+}] = 0.025, \quad [NO_3^-] = 0.05, \quad [I^-] = 0.04.$$

Will potassium nitrate precipitate? Will lead iodide precipitate? A check of a handbook shows that the solubility of KNO_3 is quite high. At these low concentrations of K^+ and NO_3^- no precipitate will form. For lead iodide, however, we find that $K_{sp} = 6.3 \cdot 10^{-9}$, which immediately suggests that precipitation is likely. If, for simplicity, we ignore activity coefficients, we get:

$$Q_{sp} = 6.3 \cdot 10^{-9} = [Pb^{2+}][I^-]^2 \quad \text{(equilibrium molarities only)}.$$

This is the relationship that will exist at equilibrium if a precipitate of PbI_2 does form. The existing molarities of Pb^{2+} and I^-, when multiplied in this way, give the following value:

$$\text{(Molarities before precipitation) } [Pb^{2+}][I^-]^2 = (0.025)(0.04)^2 = 4.0 \cdot 10^{-5}.$$

This value $4 \cdot 10^{-5}$ is *larger* than the value for K_{sp} and therefore a precipitate of lead iodide will form when the two solutions are mixed.

SOME REASONS WHY THIS REASONING MAY BE FAULTY

Tacit assumptions have been the undoing of many a logical conclusion. Here are some things that might be wrong with the foregoing example:

1. Perhaps the intrinsic solubility Q_0 for the compound is high enough so that the molecular species will form and stay in solution. For example, mercuric chloride $HgCl_2$ has a small value for $K_{sp} = 4 \cdot 10^{-14}$. Yet when $0.1 M$ solutions of mercuric nitrate and potassium chloride are mixed, there is no precipitate because the intrinsic solubility for $HgCl_2$ is about $0.25 M$.
2. Even if the equilibrium calculation is done correctly and Q_0 is taken into account, we may not predict the kinetics of the precipitation process. The rate of the reaction may be very slow and there are many examples of

supersaturated solutions that seem to be stable indefinitely. This is particularly observed when very dilute solutions are mixed, even though the ion concentrations are high enough to exceed K_{sp} and Q_0.

3. The illustration ignored the importance of ionic-activity coefficients, which typically have the effect of making the substance more soluble at finite ionic strengths than at "zero" ionic strength.

4. Often the solution mixture will contain other solute species that react through proton transfer, complexation, etc., to reduce the actual molarities of the ions forming a precipitate. This must be taken into account in making the prediction, as discussed in more detail in Chapter 17.

COMPLETENESS OF PRECIPITATION: THE COMMON-ION EFFECT

In gravimetric determinations we would like to arrange conditions so that nearly all of the desired ion is precipitated as one of its salts. Given the reversible nature of solubility equilibria, it is, of course, impossible to make *all* of the ion precipitate. For quantitative purposes we typically settle for 99.9% completeness in precipitation. Some salts, such as silver iodide, are so nearly insoluble that, if we were able to mix stoichiometrically equal amounts of the ions, the precipitation would be quantitative. However, many other salts are sufficiently soluble that it is necessary to drive the precipitation reaction to a suitable point through the addition of an excess of the precipitating ion.

For example, consider the case of calcium fluoride CaF_2. The solubility of this salt in pure water is $2.2 \cdot 10^{-4}$ mol/L. This means that precipitation would be only 99% complete if Ca^{2+} and F^- were mixed together in a $1:2$ ratio with initial concentrations (before precipitation starts) of 0.02 and 0.04 molar, respectively.

If we adopt the strong electrolyte model for CaF_2, we may calculate the value for the solubility-product constant from the above-stated solubility in water:

$$[Ca^{2+}] = 2.2 \cdot 10^{-4}, \qquad [F^-] = 2[Ca^{2+}] = 4.4 \cdot 10^{-4}.$$

Therefore,

$$CaF_2(s) \rightleftarrows Ca^{2+} + 2F^-, \quad K_{sp} = [Ca^{2+}][F^-]^2 f_2 f_1^2.$$

If, for simplicity, we ignore activity coefficients in the following discussion, then

$$K_{sp} = (2.2 \cdot 10^{-4})(4.4 \cdot 10^{-4})^2 = 4 \cdot 10^{-11}.$$

It is important to realize that this equilibrium expression will apply to all mixtures of calcium and fluoride ions, provided solid CaF_2 is at equilibrium with the solution.

It is instructive to consider what happens to the solubility of CaF_2 if additional calcium ion, or additional fluoride ion is added to the equilibrium mixture. First, consider the effect of adding fluoride ion. Suppose this is done so that the final equilibrium molarity of fluoride ion is $0.10M$, to pick a nice round figure. According to the K_{sp} expression, the corresponding equilibrium molarity of calcium ion is:

$$[Ca^{2+}] = \frac{K_{sp}}{[F^-]^2} = \frac{4 \cdot 10^{-11}}{0.1^2} = 4 \cdot 10^{-9}.$$

This is a remarkable reduction in solubility: from $2.2 \cdot 10^{-4}$ in pure water to $4 \cdot 10^{-9}$ with $0.1M$ F^- present. This is called the **common-ion effect** because the solution has been made to contain an excess of an ion which is *in common* with an ion in the precipitate.

Now, what if the experiment is repeated but this time with the addition of excess Ca^{2+} as the common ion? Again, suppose the added Ca^{2+} has an equilibrium molarity of $0.10M$. From the K_{sp} expression we may calculate the corresponding molarity of the fluoride ion:

$$[F^-] = \left(\frac{K_{sp}}{[Ca^{2+}]}\right)^{1/2} = \left(\frac{4 \cdot 10^{-11}}{0.1}\right)^{1/2} = 2.0 \cdot 10^{-5}.$$

Since each molecule of CaF_2 yields two fluoride ions, the solubility of CaF_2 is therefore $1.0 \cdot 10^{-5}$ mol/L. Note the big difference: the addition of $0.1M$ F^- lowered the solubility more than 50 000 times, while the addition of an equal molarity of Ca^{2+} lowered the solubility only about 20 times. This is due to the square term for $[F^-]$ in the K_{sp} expression. For symmetrical salts, such as $AgCl$, both common ions have the same effect in reducing the solubility. The behavior of CaF_2 with each of the common ions is summarized in Fig. 6.1.

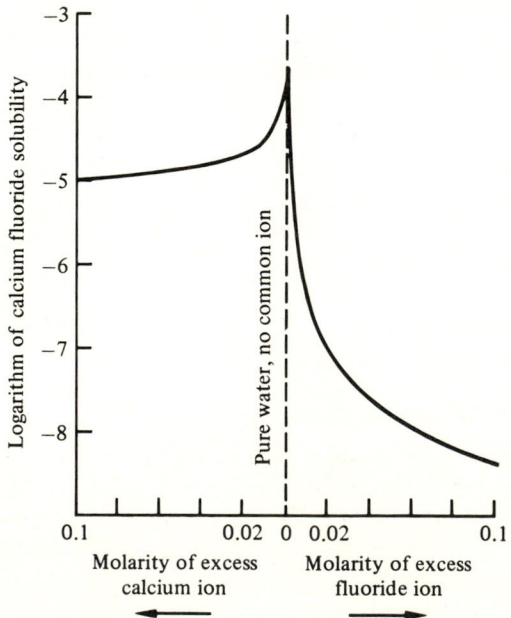

Figure 6.1

Example 1 It is desired to effect a quantitative precipitation of oxalate ion $C_2O_4^{2-}$ from a solution containing $0.02M$ oxalate by formation of silver oxalate. What is the minimum required concentration of excess silver ion?

From Appendix 5 we find the solubility product constant for silver oxalate:

$$Ag_2C_2O_4(s) \rightleftarrows 2Ag^+ + C_2O_4^{2-}, \qquad K_{sp} = 10^{-10.5} = 3 \cdot 10^{-11}.$$

If at least 99.9% of the oxalate is to be precipitated, its final molarity must be no more that $0.02 \cdot 0.001 = 2 \cdot 10^{-5}$. Therefore, the equilibrium molarity of Ag^+ may be found:

$$[Ag^+] = \left(\frac{K_{sp}}{[C_2O_4]}\right)^{1/2} = \left(\frac{3 \cdot 10^{-11}}{2 \cdot 10^{-5}}\right)^{1/2} = 1.2 \cdot 10^{-3}M.$$

Therefore, enough silver nitrate should be added to precipitate the oxalate and to have enough left over to make the silver-ion molarity about $1.2 \cdot 10^{-3}$ or higher.

Example 2 Is it reasonable to try to precipitate bromide ion quantitatively from a solution containing $0.03M$ KBr by adding an excess of lead nitrate?

The solubility product for lead bromide is found in Appendix 5:

$$PbBr_2(s) \rightleftarrows Pb^{2+} + 2Br^-, \qquad K_{sp} = 10^{-4.4} = 4 \cdot 10^{-5}.$$

If the bromide-ion molarity is to be reduced to 0.1% of its initial value, i.e., to $3 \cdot 10^{-5}M$, then the equilibrium lead-ion molarity must be:

$$[Pb^{2+}] = \frac{K_{sp}}{[Br^-]^2} = \frac{4 \cdot 10^{-5}}{(3 \cdot 10^{-5})^2} = 4.4 \cdot 10^4 M.$$

It is obviously impossible to attain such a high equilibrium concentration of lead ion (or of any other ion for that matter). Lead bromide is not a good candidate for a precipitation form.

At the more reasonable lead-ion concentration of $0.1M$, we may calculate the corresponding molarity of bromide ion:

$$[Br^-] = \left(\frac{K_{sp}}{[Pb^{2+}]}\right)^{1/2} = \left(\frac{4 \cdot 10^{-5}}{0.1}\right)^{1/2} = 2 \cdot 10^{-2}M.$$

Thus, approximately one-third of the Br^- in the original $0.03M$ solution would be precipitated under these conditions.

PRECIPITATION OF METAL HYDROXIDES

Many metal ions form slightly soluble hydroxides when the pH of their solutions is raised sufficiently. This may be desired in some cases, when the particular metal is to be separated from the solution either simply to remove it as an interference or to determine it by weighing the dried hydroxide precipitate. In many chemical studies the important goal is to *prevent* the formation of the hydroxide precipitate, so that the metal in question will *stay* in the solution. In either case we may make use of the equilibrium theory, admittedly in a rather oversimplified way, to understand the control of metal-hydroxide precipitation.

Although the subject of solution pH is covered in detail in Chapter 9 in terms of the acid–base equilibrium theory, the reader has already learned the principles essential for this discussion. To review very briefly, we begin with the reversible dissociation of water:

$$2H_2O \rightleftarrows H_3O^+ + OH^-, \qquad K_w = [H_3O^+][OH^-]f_i^2 = 1.0 \cdot 10^{-14} \quad \text{at } 25°C.$$

That is, in *any* aqueous solution the product of the hydrogen-ion and hydroxide-ion activities must be 10^{-14}. In absolutely pure water the molarity of each ion is 10^{-7} and $pH = pOH = 7$. In acidic solutions ($pH < 7$) the molarity of hydroxide ion is controlled (via the water-dissociation equilibrium) by whatever molarity of H^+ (abbreviation for H_3O^+) happens to be present, and $pH < 7$, while $pOH > 7$. In basic solutions, e.g., a solution of sodium hydroxide, it is the molarity of H^+ that is controlled by whatever molarity of OH^- is present, and $pH > 7$, while $pOH < 7$.

The reversible reaction between a metal ion and a hydroxide ion is described in a simple way by the equilibrium constant known as the solubility product K_{sp}. If the activity coefficients are ignored, we get:

$$M(OH)_n(s) \rightleftarrows M^{n+} + nOH^-, \qquad K_{sp} = [M^{n+}][OH^-]^n.$$

Thus, the key idea in causing (or preventing) the formation of the metal-hydroxide precipitate is control of the hydroxide-ion molarity, which in turn is directly related to the pH of the solution. The best way to illustrate these relationships is by some sample calculations. To avoid complication, we will ignore activity coefficients.

Example 1 A solution contains $0.050M$ magnesium chloride. If sodium hydroxide is added gradually to this solution, what hydroxide concentration must be attained before the magnesium hydroxide just begins to precipitate?

From Appendix 5 we find that the solubility product for $Mg(OH)_2$ is $10^{-11.2}$. When the precipitate *begins* to form, the solution still contains Mg^{2+} at a molarity of 0.05. Therefore, the hydroxide molarity at that point must be:

$$[OH^-] = \left(\frac{K_{sp}}{[Mg^{2+}]}\right)^{1/2} = \left(\frac{10^{-11.2}}{0.05}\right)^{1/2} = 1 \cdot 10^{-5}.$$

This corresponds to a $pOH = 5$ or to a $pH = 14 - 5 = 9$.

Example 2 If it is desired to precipitate nearly all the magnesium from the above solution, what should be the final pH of the solution?

First it is necessary to define *nearly all*, and we may arbitrarily choose $5 \cdot 10^{-5}$ as a desirable final molarity of Mg^{2+}. This corresponds to 99.9% precipitation. To reach this low molarity of Mg^{2+}, we must have a hydroxide molarity of:

$$[OH^-] = \left(\frac{10^{-11.2}}{5 \cdot 10^{-5}}\right)^{1/2} = 3.6 \cdot 10^{-4}.$$

This corresponts to pOH = 3.5 or pH = 10.5.

Example 3 What will be the equilibrium molarity of Mg^{2+} in the solution if the pH is raised to 12.5 by the addition of excess sodium hydroxide?

At this pH, we have pOH = 1.5 or $[OH^-] = 0.03M$. The solubility-product expression is arranged to give the magnesium-ion molarity:

$$[Mg^{2+}] = \frac{K_{sp}}{[OH^-]^2} = \frac{10^{-11.2}}{0.03^2} = 7 \cdot 10^{-9},$$

which is a very low molarity indeed.

Example 4 A solution contains $0.10M$ calcium nitrate and $0.02M$ magnesium nitrate. It is desired to remove as much magnesium as possible by precipitating it as the hydroxide, while not causing any precipitation of calcium hydroxide. From Appendix 5 we find that $K_{sp} = 10^{-11.2}$ for $Mg(OH)_2$ and $K_{sp} = 10^{-5.2}$ for $Ca(OH)_2$.

First we find the maximum $[OH^-]$ that will *not* cause precipitation of the calcium; by using $0.1M$ as the calcium ion molarity we get

$$[OH^-] = \left(\frac{10^{-5.2}}{0.1}\right)^{1/2} = 8 \cdot 10^{-3} \quad \text{or} \quad pOH = 2.1.$$

Thus, at pH = $14 - 2.1 = 11.9$, the calcium will not be precipitated but the equilibrium molarity of magnesium will be:

$$[Mg^{2+}] = \frac{10^{-11.2}}{(8 \cdot 10^{-3})^2} = 1.0 \cdot 10^{-7},$$

which corresponds to nearly complete precipitation. We conclude that these two elements may be separated from one another by control of solution pH.

Example 5 In the Volhard titration it is necessary to have a strongly acidic solution to prevent the formation of ferric hydroxide. If the solution is to contain $0.01M$ ferric ion as indicator, what is the highest permissible value for the solution pH?

The solubility product for ferric hydroxide is very small:

$$Fe(OH)_3(s) \rightleftarrows Fe^{3+} + 3OH^-, \quad K_{sp} = 10^{-38.8}.$$

With $0.01M$ Fe^{3+} we would expect precipitation to begin when the $[OH^-]$ becomes

$$[OH^-] = \left(\frac{10^{-38.8}}{0.01}\right)^{1/3} = 5 \cdot 10^{-13} \quad \text{or} \quad pOH = 12.3.$$

This means that the pH should be 1.7 or lower, and it is usually kept at about 0.5 by addition of a strong acid.

6.2   A CLOSER LOOK AT PRECIPITATE PARTICLES

There are many precipitates that are hard to work with. Some, such as aluminum or ferric hydroxides, are gelatinous or "gunky" because they strongly adsorb water. These are difficult to wash and transfer to the filter, which they also clog. Others have a tendency to "creep up" the film of water on the sides of the beaker, again making transfer tedious. Depending on the cleanliness of the beaker, some precipitates stick tightly to the sides, requiring diligent work with the rubber policeman. Colloidal precipitates are common and can run right through most filters, and all precipitates carry a small load of impurities adsorbed on the surface.

Most of these problems can be attributed to the very small size and highly irregular shape of individual particles of a freshly formed precipitate. Through a microscope we can see that the particles can be intricate, with tufts like tassels of corn or, in the case of barium oxalate, large branching aggregates in fibrous bundles or sheaf-like masses of fibrous drusy dendrites (Fig. 6.2). Not only do such particles present an enormous surface area for adsorption of ions, but the structures provide countless protected channels where solution can be trapped and retained in spite of washing

Nucleation

Imagine an *instantaneous and thorough mixing* of two solutions: potassium chloride and silver nitrate. The mixture is unstable because of the strong tendency for the reaction

$$Cl^- + Ag^+ \rightleftarrows AgCl(s).$$

In less than 0.1 sec the precipitation will be nearly complete and the beaker will be filled with a colloidal suspension of silver chloride, giving it the appearance of milk. But what happens in the first millisecond or in the first microsecond of the process? What is the *mechanism* of particle formation? How do particles grow and change shape? These questions have inspired active research for decades. The problems prove to be theoretically complex and experimentally difficult, with each precipitate system having its unique aspects. However, a few simple generalizations can be made.

First, it is inevitable in regular practice that water and reagents both contain traces of insoluble matter (dust, fragments of glass, etc.) in the form of *millions* of submicroscopic particles suspended throughout the solutions used. Each of these provides a site, a focal point, so to speak, for the adsorption of solute species from the solution. To continue our example, when a few ions of silver and chloride have been adsorbed, they encourage the addition of more ions, and a crystal **nucleus** has been born. Because of the role of the initial insoluble particle of dust or other solid particle, this mechanism is called **heterogeneous nucleation** and is probably important in all precipitations.

 At the same time, however, it is kinetically possible for **homogeneous nucleation** to occur. Without the aid of a dust particle, the hydrated ions of

Platinum precipitated as cupric platino-tetrammine hydrochloride. (100×)

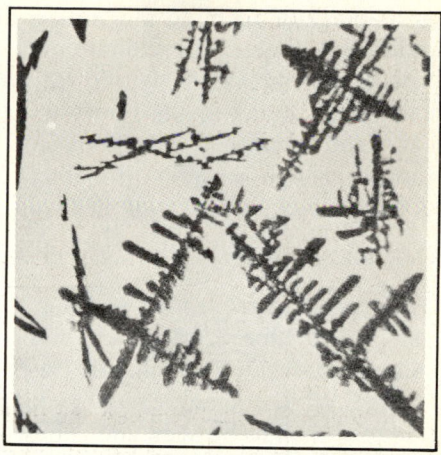

Molybdenum with quinoline and ammonium thiocyanate. (200×)

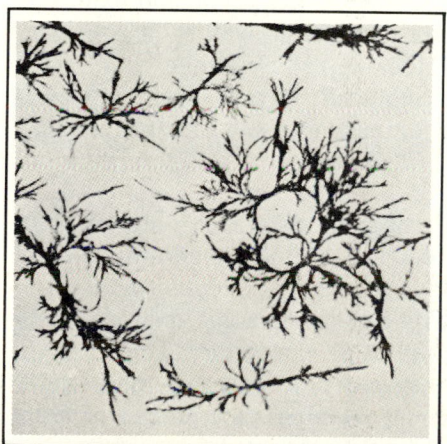

Barium with oxalic acid. (100×)

Gold with potassium mercuric thiocyanate. (100×)

Fig. 6.2 Examples of irregular shapes of precipitate particles formed by rapid mixing of reagents. (Photomicrographs from *Handbook of Chemical Microscopy*, by E. Chamot and C. Mason, John Wiley and Sons, 2nd ed., 1938.)

silver and chloride may collide, forming a silver–chloride bond as the water of the hydration sheath is forced aside. It may be that the ion pair thus formed is short-lived because of a collision with a water molecule that breaks the bond and reestablishes the separate ions. These types of collisions, occuring with blinding speed and frequency throughout the solution, have a statistical probability of forming ion triplets and even higher aggregates. Research has indicated that each precipitate may have a critical size for such an aggregate, perhaps as small as 8–10 ions. Below this size, collisions with water molecules can cause break-up and return to individual ions. But when this size is exceeded through the random collision and bonding of the precipitate ions, further growth is highly probable. It is at this critical size that nucleation has occurred and precipitation has truly begun.

Colloidal Particles

The transition from a crystal nucleus to a precipitate particle requires only a very short time, during which the ions still in solution "pile on" through random collisions in chaotic fashion, usually leading to irregular particle shapes. While the aggregate is young and contains only a few dozen ions, it is still too small to be seen and is best regarded as homogeneously dispersed like other solutes of molecular dimensions. It requires something like a cluster of 1000 atoms for the aggregate to qualify as a **colloidal particle** having an approximate diameter of 10^{-7} cm. Even with a diameter of 10^{-4} cm, containing about a billion atoms, the particle is still classified as colloidal and cannot be seen without a microscope. However, there is a simple way to inspect what appears to be a clear solution to see whether it really contains a colloid in suspension: in a darkened room a beam of light, as from a flashlight, is viewed from the side as it passes through the solution contained in a beaker. If a colloidal suspension is present, the light will be reflected and the beam will "glow".

The physical and chemical properties of colloids are so different from those of bulk matter that colloid science becomes a vitally important research field. The unique behavior of colloids is due to the tremendous surface area they present for a given mass of material. To help visualize this, imagine a cube of gold, with a 4-cm edge and a surface area of $6 \cdot 4^2 = 96$ cm². This could be sliced into eight smaller cubes with 2-cm edges; when separated these would present an exposed surface area of 192 cm². If we repeat the process, making smaller and smaller cubes, the surface area will be doubled for each division into cubes of one-half the previous edge size. To subdivide the original 4-cm cube into particles of a typical colloidal size (10^{-5} cm), there would have to be 18 successive subdivisions and the final exposed surface area would be a surprising $40 \cdot 10^6$ cm². For this large area to be exposed by a single cube, the edge of the cube would have to be about 26 m (85 feet).

The surface of a crystal is far from being chemically inert. The internal ions (or atoms or molecules) are fully bonded by surrounding ions, but those on the surface have one free side for further interaction with whatever comes their way. When the crystal is suspended in the aqueous solution from which it has just been precipitated, the interactions will be rapid and varied. Water molecules collide most frequently and, with some precipitates, particularly hydroxides, water will be strongly adsorbed on the precipitate surface.

As the precipitation is taking place, the ions comprising the precipitate will continue to form new nuclei through homogeneous nucleation, but the probability increases that they will collide with the surfaces of colloidal particles already formed, causing further growth.

This leads to particles that grow large enough to be seen and settle out of the suspension. Of course, this is an important goal as far as gravimetric determination is concerned. Often such growth involves improvement of crystal shape as well as size.

At this stage it is also possible for impurities to be incorporated into the crystal structure through **mixed-crystal formation**: if the solution contains species that normally would not precipitate but do have an ionic charge and/or radius that matches the makeup of the regular crystal lattice, such ions may enter the lattice as substitutes and remain there. An interesting demonstration of this effect is as follows. To a solution of $1.5M$ potassium nitrate containing a little potassium permanganate add a solution of $1.5M$ perchloric acid. The precipitate of potassium perchlorate will not be white (its normal color) but purple because of the substitution of permanganate ion for perchlorate ion in the crystal lattice. This can be readily seen without filtering if the excess permanganate is reduced with a little sodium bisulfite. From an analytical viewpoint, mixed-crystal formation leads to error due to increased mass of the precipitate.

It is especially important to consider the interactions of precipitate particles with whichever ion of the precipitate happens to be present in excess in the solution. Imagine that a solution of $0.01M$ silver nitrate is added slowly, with constant stirring, to a solution of $0.01M$ sodium bromide. The precipitation reaction is

$$Ag^+ + Br^- \rightleftarrows AgBr(s)$$

and a colloidal suspension will be apparent almost immediately. When only a little silver nitrate has been added, the silver bromide colloid is suspended in a solution containing sodium ions, bromide ions, and the nitrate ions left over from the added silver nitrate. The silver ions are virtually completely incorporated into the precipitate, so that their solution concentration is negligible. Collisions of water, sodium ion, and nitrate ion with the surfaces of the suspended AgBr particles are likely to be relatively elastic because both sodium bromide and silver nitrate are quite soluble. However, the collisions of the bromide ions with the particle surface can lead to the formation of stable bonds with the silver ions exposed on the outside of the crystal lattice. This leads to the formation of a **primary adsorption layer** that gives the precipitate particle an *overall negative charge* (Fig. 6.3).

This negative charge, distributed more or less evenly over the particle, has important consequences. First, because all the particles are similarly charged, they repel each other according to the laws of electrostatics. *It is this repulsion that is the chief cause for stability of a colloidal suspension* for,

$$
\begin{array}{c}
\text{Br}^- \qquad \text{Br}^- \\
\text{BrAgBrAgBrAg} \\
\text{Br}^- \ \text{AgBrAgBrAgBrAgBrAgBrAgBr}^- \\
\text{AgBrAgBrAgBrAgBrAgBrAgBr} \\
\text{AgBrAgBrAgBrAgBrAgBrAgBrAgBr}^- \\
\text{Br}^- \ \text{AgBrAgBrAgBrAgBrAgBrAgBrAgBr} \\
\text{BrAgBrAgBrAg} \qquad \text{Br Ag} \\
\text{Br}^- \qquad \text{Br}^- \qquad \text{Br}^-
\end{array}
$$

Crystal lattice of silver bromide with
surface adsorption of bromide ions.

Figure 6.3

without it, the particles would settle to the bottom of the beaker because of gravity. Although the gravitational force is present, the tendency for the particles to move together toward the bottom is offset by the electrostatic repulsion. Of course, if the particles grow to a sufficient size, the force of gravity can take over.

Secondly, the layer of chemically bound ions (bromide ions in the above example) exerts an attractive force on all the oppositely charged ions in the surrounding solution. In the above example it is the sodium ion that is attracted to form a secondary adsorption layer (Fig. 6.4).

The secondary layer is loosely structured and is due only to rather weak electrostatic attractions, not to chemical-bond formation. To some extent it screens the repulsive forces between precipitate particles. The combination of the primary and secondary adsorption layers is called the **electrical double layer**.

Continuing the imaginary addition of $0.01M$ silver nitrate to $0.01M$ sodium bromide, let us suppose that an *excess* of silver is added, so that virtually all the bromide ion is precipitated as small particles of solid silver bromide and the surrounding solution contains silver ion, sodium ion, and nitrate ion. In this situation the ion that can form stable bonds with the surfaces of the particles is *silver ion* Ag^+, and so all the colloidal particles will

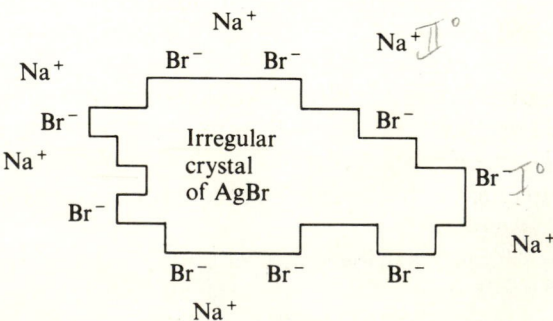

Figure 6.4

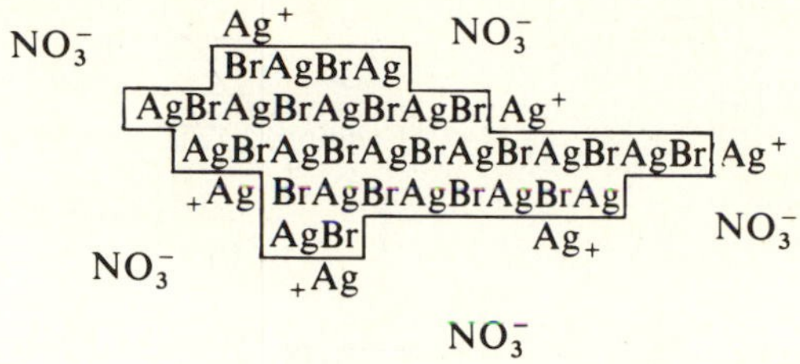

$$NO_3^- \quad \boxed{\begin{array}{l} Ag^+ \\ \hline \text{BrAgBrAg} \\ \hline \end{array}} \quad NO_3^-$$

Figure 6.5

carry a *positively charged* primary adsorption layer, while the loosely held secondary layer will be the negative nitrate ions (Fig. 6.5).

Again the individual particles will repel each other electrostatically because they carry charges of the same sign, and such a positively charged colloid will also remain in fairly stable suspension rather than settling out.

Coagulation of a Colloid

What if the amount of $0.01M$ silver nitrate added is *equal* to the amount of $0.01M$ sodium bromide? Virtually all of the silver and bromide ions will be in the precipitate, and the surrounding solution will contain only sodium and nitrate ions in significant concentration. Since neither of these ions form slightly soluble compounds with either silver or bromide ion, there will be very little adsorption to form a primary layer, and so the individual particles will be essentially neutral, or uncharged. Because the particles may be very small, there will be some tendency for thermal energy to keep the colloid in suspension, but in the absence of the electrostatic repelling effect the particles can bump into each other and coagulate into clumps of higher mass that will settle out.

Actually the formation of an uncharged colloid does not occur precisely when *equal* amounts of silver ion and bromide ion are mixed, because these two ions are not adsorbed on the surface with quite equal tendencies, bromide being slightly favored. Therefore it requires a very small excess of silver ion in solution to make the surfaces electrically neutral, when the colloid is said to be at its **isoelectric point**.

These phenomena can serve as the basis for a fairly accurate titration of silver ion: a sodium bromide solution is added from a buret into a constantly stirred silver-ion solution. Before the endpoint, the colloid is positively charged and stays in suspension, but when the isoelectric point is reached (which is very close to the stoichiometric point), there is a sudden coagulation (also called **flocculation**).

It is not necessary to be at the isoelectric point to achieve the coagulation of a colloid if the solution contains enough strong electrolyte to provide a high concentration of ions for the secondary adsorption layer. For example, a negative colloid of silver bromide is rather stable in the presence of $0.01M$

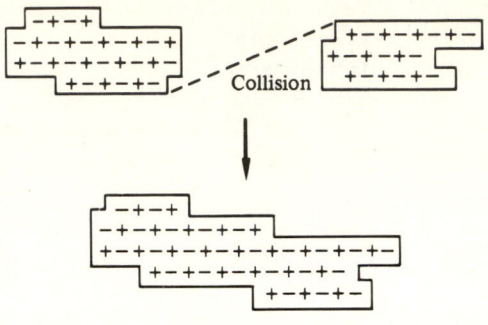

Figure 6.6

sodium bromide but will coagulate if some $1M$ potassium nitrate is added. Apparently the high concentration of potassium ions can *saturate* the secondary layer and effectively shield the repulsive forces between particles. This allows the particles to approach more closely, to collide, and to "cement" together by merging their lattice structures (Fig. 6.6).

The ionic charge of the flocculating ion has a striking effect: a negative colloid will be coagulated by much lower concentrations of calcium ion Ca^{2+} than by singly charged potassium ions, and triply charged aluminum ions are still much more effective in causing coagulation. For positive colloids, similar statements hold for comparisons of, say, nitrate ions (singly charged) and sulfate ions SO_4^{2-}.

Coprecipitation and Aging

In a gravimetric determination it is essential to be sure that a desired constituent is completely precipitated, and this makes it necessary to have a certain excess of precipitating agent in the solution. Thus, a precipitate inevitably becomes contaminated through adsorption of some of these excess ions in the primary layer. When the precipitate is coagulated and filtered, it retains some fraction of both the primary and secondary adsorption layers in spite of the usual washing.

Further, it is not only the ions of the excess precipitating agent that can be incorporated into the primary adsorption layer. In fact, all ions present in the solution will be adsorbed to *some* extent depending on their tendencies to form insoluble compounds with the ions of the precipitate. This is often called the **Paneth–Fajans–Hahn Rule**. For example, we have seen that a silver bromide colloid will preferentially adsorb either silver ion or bromide ion if one of these is present in excess in the solution. If a neutral silver bromide colloid were formed in the presence of, say, sodium acetate and potassium nitrate, the primary layer would be chiefly acetate ion because silver acetate is less soluble that the other three possibilities: silver nitrate, sodium bromide, and potassium bromide. The secondary layer would be a mixture of sodium and potassium ions, and the final precipitate would therefore be contaminated with **coprecipitated** sodium and potassium acetate.

To minimize the coprecipitation effect it is desirable to encourage the precipitate to develop crystals of more regular shapes and larger size, so that the surface area is decreased. This can be accomplished (more effectively with some precipitates than with others) by the **digestion step**. This is simply a

Magnesium with disodium phosphate and ammonium hydroxide. (50×)

Magnesium with disodium phosphate and ammonium hydroxide. (100×)

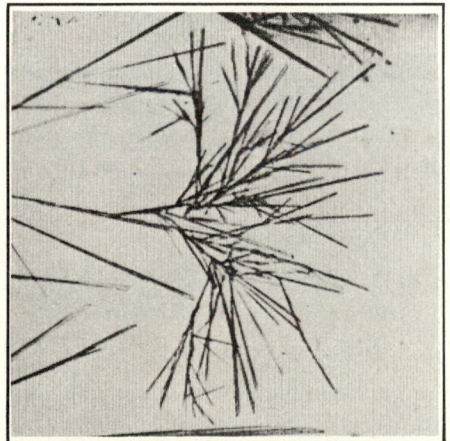

Cesium with hexanitrodiphenylamine. (100 ×)

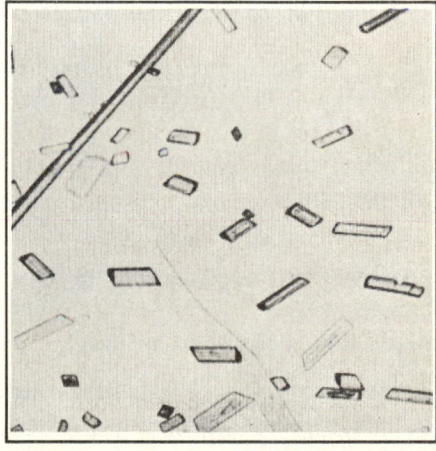

Cesium with hexanitrodiphenylamine. (100 ×)

Fig. 6.7 Effect of aging. Left photomicrographs show freshly formed precipitate, while the right ones show same material after 1 to 2 hours. (From the same source as Fig. 6.2.)

waiting period while the precipitate is allowed to interact with the hot solution with the help of stirring. During this time the smaller particles tend to redissolve and to reprecipitate on the surfaces of larger particles, thus releasing the adsorbed impurities and forming more nearly perfect crystals. The precipitate is said to **age** in the process and the result is a product that is not only less contaminated but easier to wash and filter.

A striking example of the digestion effect is the precipitate of magnesium ammonium phosphate. It precipitates in a granular condition, but soon changes into dendritic forms and large feathery stars and crosses, which in turn develop into plates and tabular forms and finally rectangular ortho-rhombic prisms (Fig. 6.7).

Practical experience in working with precipitation methods has led to a few general rules for obtaining optimal results in the purity and handling properties of precipitates. They are as follows:

1. The precipitant is added slowly, with vigorous stirring, so that the small particles have more time to grow, rather than precipitate "all at once."

2. Both the constituent solution and the precipitant solution are made rela-tively dilute, so that particles are not formed as quickly, thereby again having more chance to grow.

3. The precipitation is carried out in hot solution to increase the solubility of the precipitate temporarily and to speed up the process of crystal growth.

4. After the precipitant has been added, the mixture is digested, which means that the precipitate is left in contact with the hot solution for a time, often overnight, to allow the particles to improve their size and shape.

A superb alternative to these rules is sometimes possible in the form of a technique called **precipitation from homogeneous solution**, or PFHS. This involves chemical generation of the precipitant *in situ* in a homogeneous solution rather than the conventional heterogeneous addition of a solution of the precipitant.

6.3   PRECIPITATION FROM HOMOGENEOUS SOLUTION (PFHS)

Precipitation of Lead Chromate

Before getting into PFHS, let us identify a fundamental problem in carrying out an ordinary (heterogeneous) precipitation by using a specific illustration. Suppose a beaker contains a solution, the lead content of which is to be determined by precipitation and weighing of lead chromate $PbCrO_4$. Potassium chromate solution may be used as the precipitation solution, and when this is poured into the lead-containing solution, there is a contact zone in which nuclei and colloidal particles are quickly formed (Fig. 6.8).

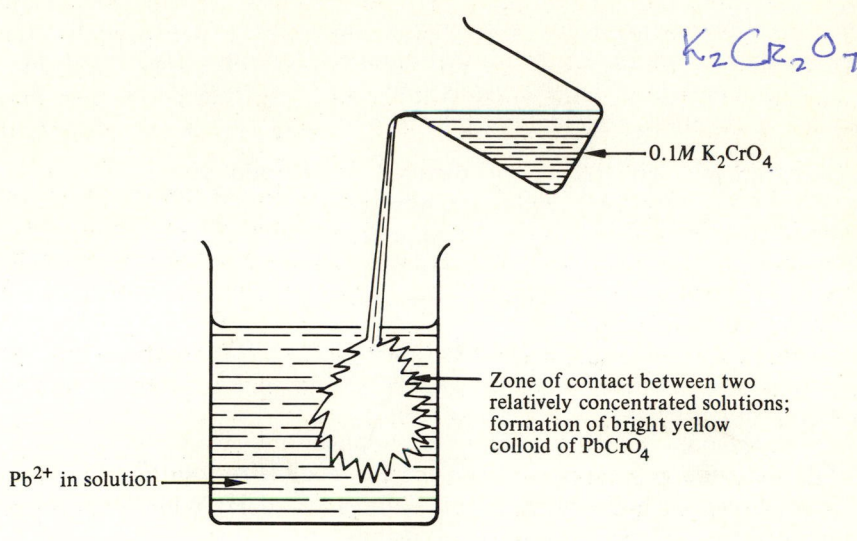

K₂Cr₂O₇

Figure 6.8

Demonstration 1 To 100 mL of distilled water add 10 mL of 0.1M lead nitrate, stir, and then add 10 mL of 0.1M potassium chromate.

Even with slow addition and rapid stirring it is not possible to avoid forming colloidal and contaminated particles. A better precipitate could be obtained if extremely dilute solutions were mixed very slowly, with constant vigorous stirring, but this would require tedious attention and large volumes of solution.

The problem can be solved, with striking benefits, by the clever application of chemical kinetics. First, a solution of chromium nitrate $Cr(NO_3)_3$ is added to the solution containing the lead ions. No reaction occurs because Cr^{3+} ions cannot react with Pb^{2+} ions in any way. Next, some potassium bromate $KBrO_3$ is added and the solution is stirred to make it *homogeneous*, but still no reaction occurs. The ions in the solution are: Cr^{3+}, Pb^{2+}, BrO_3^-, NO_3^-, H^+, and K^+.

When this mixture is heated to near boiling, a slow oxidation–reduction reaction occurs:

$$BrO_3^- + Cr^{3+} \rightarrow Br_2 \text{ (vapor)} + HCrO_4^- \text{ (hydrogen chromate ion)},$$

and after a few minutes the *slow and homogeneous* buildup of the $HCrO_4^-$ ion reaches the point where the solubility of lead chromate is exceeded and the first nuclei of solid lead chromate are formed:

$$HCrO_4^- + Pb^{2+} \rightleftharpoons PbCrO_4(s) + H^+.$$

However, these nuclei are formed under conditions of virtually no excess chromate ion, so that (a) few nuclei are formed and (b) the continuing slow generation of chromate ion encourages the *growth* of the nuclei and not merely the formation of additional nuclei. Consequently, the precipitate eventually consists of large crystals that can be seen without a microscope, instead of countless colloidal coagulated particles.

Demonstration 2 To 100 mL of distilled water add 10 mL of $0.1M$ lead nitrate, 10 mL of $0.1M$ chromium nitrate, and 10 mL of $0.1M$ potassium bromate. Stir, heat to near-boiling, and observe the formation of crystals (deep orange-red) of lead chromate. Compare these crystals with the yellow colloidal precipitate using a microscope at low magnification.

Because of the formation of large and nearly perfect crystals, this technique gives a product that is easily washed and transferred to the filter crucible.

There have been many ingenious applications of chemical kinetics for PFHS, and the reader is referred to the book *Precipitation from Homogeneous Solution*, by L. Gordon, M. Salutsky, and H. Willard (Wiley, 1959). Only two more brief illustrations will follow.

PFHS by Slowly Raising the Solution pH.

Much improved precipitates of hydroxides and basic salts are obtained when the solution is gradually made basic through the slow hydrolysis of urea:

$$NH_2CONH_2 + H_2O \xrightarrow{\text{heat}} CO_2 + 2NH_3.$$

The ammonia formed by this reaction neutralizes whatever acids might be present and, for example, iron hydroxide is precipitated. The precipitate is much more compact and easier to handle than when a solution of ammonia is added directly and heterogeneously. This technique may also be used to liberate precipitating anions, such as chromate and oxalate, from their conjugate acid forms.

PFHS by Slowly Lowering the Solution pH.

The ester 2-hydroxyethyl acetate slowly hydrolyzes at room temperature:

$$HOCH_2CH_2OCOCH_3 + H_2O \rightarrow HOCH_2CH_2OH + CH_3COOH.$$

The acetic acid thus generated will react with whatever bases are present. For example, one may begin with a solution of silver in the form of the silver–ammonia complex. When chloride ion is added, there is no precipitate of silver chloride because of the high stability of the complex. However, if the ester is added and the solution is allowed to stand, the ammonia is gradually protonated, releasing the silver ion and allowing AgCl to form:

$$Ag(NH_3)_2^+ + 2CH_3COOH \overset{\text{slow}}{\rightleftharpoons} Ag^+ + 2NH_4^+ + 2CH_3COO^-.$$

$$+$$

$$Cl^- \rightleftharpoons AgCl(s)$$

This technique results in relatively large cubic crystals of silver chloride.

6.4   CASE STUDY: GRAVIMETRIC STANDARDIZATION OF HYDROCHLORIC ACID

In many research applications it is necessary to have a precisely standardized solution of hydrochloric acid. The molarity of a solution of HCl can be determined by titration of a weighed sample of "tris" or of sodium carbonate, but for highest accuracy it is best to do a precipitation of silver chloride in a gravimetric determination. In the following discussion we assume that a solution of about $0.1M$ is to be prepared from ordinary commercial concentrated acid and that it is desired to know the precise molarity, molality, and percentage by weight of the prepared solution.

The Commercial Stock Solution

Hydrochloric acid is so widely used that several chemical firms offer it. For example in a typical catalog we find this listing:

HYDROCHLORIC ACID, 37%, REAGENT (ACS)

HCl   FW 36.46

Assay—36.5–38.0% HCl
Appearance—free of sediment or suspended matter
Color (APHA) 10 max
Residue after ignition 0.0005%
Bromide 0.005%
Sulfate 0.0001%
Sulfite 0.0001%
Extractable organic substances—pass test
Free chlorine—pass test
Ammonium 0.0003%
Arsenic 0.000001%
Heavy metals (Pb) 0.0001%
Iron 0.00002%

1-pint bottle . . . $3.50
5-pint bottle . . . $8.50

It is evident from the ACS specs that this is a high-quality reagent with very low impurity levels. Also evident is the economic advantage of buying the larger-size bottle.

The molarity of this solution may be calculated from the percent composition (37%) and the density (1.19 g/mL) which is not given in the catalog but can be found in a handbook. Thus, one liter of the solution weighs 1190 g and the amount (moles) of HCl in a liter is:

$$n = 1190 \text{ g(solution)} \cdot \frac{37 \text{ g HCl}}{100 \text{ g(solution)}} \cdot \frac{1}{36.46 \text{ g/mol}} = 12.1.$$

At this concentration a sizable fraction of the dissolved HCl is not dissociated into its ions and exists as the volatile gaseous species. Your eyes, nose, and throat will quickly warn you of this if you open a bottle outside of a fume hood.

Further Purification by Distillation

The already small concentration of impurities can be easily brought to lower levels when desired. With water, hydrogen chloride forms a constant–boiling azeotrope, whose concentration is about $6M$. Therefore it is a simple procedure to dilute the $12M$ stock with an equal volume of distilled water, and then to do a distillation. It is good practice to reject the first portion of the distillate and to collect the large middle portion, rejecting also the residue that remains in the pot. The distillate will have a concentration close to $6M$ and can be stored indefinitely without appreciable change. Suppose that such a purification resulted in a solution that proved to be $6.09M$ according to titration with standardized sodium hydroxide solution.

Preparation of the Approximately 0.1M Working Solution

Suppose that 2 liters are desired. The total required HCl is:

$$2000 \text{ mL} \cdot 0.1 \text{ mmol/mL} = 200 \text{ mmol}.$$

To obtain this quantity from the $6.09M$ distilled stock solution, we must take

$$\frac{200 \text{ mmol}}{6.09 \text{ mmol/mL}} = 32.84 \text{ mL},$$

so we settle for 33 mL as measured in a graduated cylinder and dilute it to 2 liters in a volumetric flask, expecting that the diluted solution will be within 1–2% of the desired value of $0.1M$.

Measurement of the Sample

With a good analytical balance, the imprecision of a single weighing is typically about 0.1 mg. We would like to keep the uncertainty in the hydrochloric-acid concentration at a low level, say, $\pm 0.01\%$, or 1 part in 10,000; and therefore it is necessary for the precipitate of silver chloride to weigh at least 1 gram. The GFW for $AgCl$ is 143.321, so the required amount is

$$n = \frac{1\,\text{g}}{0.14\,\text{g/mmol}} = 7\,\text{mmol}.$$

Therefore the minimum sample size for portions of the $0.1M$ solution is

$$v = \frac{7\,\text{mmol}}{0.1\,\text{mmol/mL}} = 70\,\text{mL},$$

and for convenience we decide to take 100-mL portions using a calibrated pipet that delivers 100.01 mL according to standardization with pure water. For increased confidence in the result we decide to carry three samples through the procedure, but in the following discussion only the data for one sample will be discussed.

By pipetting the 100-mL sample into a previously weighed beaker we find the apparent weight to be 99.836 g. At the same time the temperature of the $0.1M$ solution is found to be 23.0°. The true mass of the sample is calculated by applying the buoyancy factor of 1.00105 (for sample of density 1):

$$\text{True mass} = 99.836 \cdot 1.00105 = 99.941\,\text{g}.$$

Therefore, for the record, the true density of the $0.1M$ solution at 23° is

$$d^{23} = \frac{99.941\,\text{g}}{100.01\,\text{mL}} = 0.9993\,\text{g/mL}.$$

We may assume that for this solution, as for pure water, the temperature coefficient for the density is -0.00026 per degree.

Precipitation of the Silver Chloride

Knowing that the sample contains about 10 mmol of HCl and that a slight excess of silver nitrate must be added, we decide to add 11 mmol of $AgNO_3$. The GFW is 169.87, and so the weight is

$$w = 11\,\text{mmol} \cdot 0.170\,\text{g/mmol} = 1.87\,\text{gram}.$$

Of course, this does not have to be very precise; 1.9 ± 0.1 g will do. The silver nitrate is dissolved in about 50 mL of distilled water and is added to the hot solution of HCl slowly and with constant stirring. The white colloidal precipitate that first forms coagulates nicely after a few minutes, but the solution still contains a good deal of finely divided material. Further, the white solid begins to turn light purple because of photodecomposition:

$$AgCl(s) + \text{Light} \rightarrow Ag(s) + \tfrac{1}{2}Cl_2(aq).$$

For this reason it is best to carry out the precipitation without bright light, especially without sunlight. However, even though the precipitate will eventually be rather dark purple on its surface, the loss in mass (due to the escape of Cl_2) will not cause a serious error.

The precipitate is left in contact with the hot solution in the dark for a few hours to allow time for aging and complete coagulation. At the end of this time the supernatant solution is quite clear; to make sure that we really did add a sufficient amount of silver, we add a drop or two of a silver solution. If any new precipitate should form in this test, we would have to add additional silver nitrate until we were sure a slight excess existed, and the digestion would have to be repeated.

Washing and Filtration

A previously weighed filter crucible is used with a vacuum filtration flask, and the supernatant liquid is decanted through the filter, typically carrying with it a small fraction of the precipitate which is, of course, caught on the filter. About 20 mL of $0.01 M$ nitric acid is added to the precipitate in the beaker, swirled around as a rinse, and then decanted through the filter. Then, with the aid of a wash bottle containing $0.01 M$ HNO_3 and a rubber policeman on a stirring rod, the precipitate is transferred *quantitatively* to the filter. There should be no trace of it left in the beaker. The accumulated precipitate on the filter is washed with two or three small portions of water.

Drying the Precipitate and Crucible

Silver chloride does not melt below 455°, so the temperature of the drying oven is not critical. Typically a setting of about 120° is used. The important thing is that the drying treatment for the crucible *with* the AgCl in it should be the same as when the clean and empty crucible was dried. This is because the empty crucible weight must be subtracted and crucible weight will vary a few tenths of a milligram with different drying and cooling techniques. For most accurate results it is good practice to carry a "dummy", or empty crucible, through the entire procedure, to have an internal check on variations in crucible weight that can be caused by atmospheric humidity, for example. When this is done it may also be better practice *not* to use a desiccator to store the crucibles after removal from the oven, but rather to let them cool in a beaker placed in the vicinity of the analytical balance so that they equilibrate with the surrounding air.

The Final Weighing

Suppose that the final, constant weight of the precipitate (apparent mass) turns out to be 1.4559 grams. The density of silver chloride is 5.56 g/mL, so the buoyancy factor is quite small (1.00007) but it might as well be applied anyway:

$$\text{True mass of AgCl} = 1.4559 \cdot 1.00007 = 1.4560 \text{ g.}$$

Calculation of Concentration

Let us consider four ways of expressing the concentration of the $0.1M$ solution:

Molarity The amount (moles) of HCl in the sample is the same as that of AgCl in the precipitate:

$$n = \frac{1.4560 \text{ g}}{143.321 \text{ g/mol}} = 0.010159 \text{ mol}.$$

This was contained in the pipetted volume of 100.01 mL, so the molarity at the time of pipetting was:

$$M = \frac{0.010159 \text{ mol}}{0.10001 \text{ L}} = 0.10158 \text{ mol/L}.$$

This is the value for a solution temperature of 23°. The molarity at other nearby temperatures depends on thermal expansion, for water expands at 0.00026 per degree. Thus,

$$M \text{ (at } t°) = \frac{0.10158}{1 + 0.00026(t - 23)}.$$

For example, if we should use this solution on a day when the lab temperature was up to 28°, the corrected value for molarity is 0.10145.

Molality This method of expressing concentration has the advantage of not varying with temperature. The true mass of the sample was 99.941 grams. The mass of HCl contained in it was

$$w = 0.010159 \text{ mol} \cdot 36.461 \text{ g/mol} = 0.37041 \text{ g}.$$

Therefore the mass of water in the sample was

$$w_{H_2O} = 99.941 - 0.370 = 99.571 \text{ g}$$

and the molality of HCl in the solution was therefore

$$m_{HCl} = \frac{0.010159 \text{ mol HCl}}{0.099571 \text{ kg } H_2O} = 0.10203 \text{ mol/kg}.$$

Moles/kg of solution These units for concentration have no accepted name but are useful. As with molality, there is no temperature coefficient. The result is:

$$\frac{0.010159 \text{ mol HCl}}{0.099941 \text{ kg of solution}} = 0.10165.$$

Weight percent This is simply the ratio of solute mass to solution mass, multiplied by 100 to convert to percentage. For the solution under discussion,

$$\%HCl = \frac{0.37041 \text{ g HCl}}{99.941 \text{ g of solution}} \cdot 100 = 0.37063\%.$$

Averaging the Triplicate Determinations

Suppose that the three samples carried simultaneously through the procedure gave the following values for molarity:

Sample number	Molarity calculated at 23°
1	0.10158
2	0.10157
3	0.10153
Average:	0.10156

Whereas samples 1 and 2 gave results that agree to within 1 part in 10,000 (which is the precision we can realistically attain in this determination), sample 3 gives a value that is 5 parts in 10,000 lower. Should we reject sample 3 and simply take the average of the other two? Not unless we have some previously noted reason to expect a low result for that sample! For example, if in the digestion step there had been a careless moment when the solution boiled and spattered a little, it would be legitimate to ignore the third result. But if there is no explanation, then the result should be included when it agrees this closely, preferably without after-the-fact rationalization.

The individual deviations of the calculated molarities from their average are $+0.00002$, $+0.00001$, and -0.00003. The average absolute deviation is 0.00002 and so the label on the bottle should be:

$$\text{HYDROCHLORIC ACID}$$
$$M = 0.10156 \pm 0.00002 \ (23°C)$$

Final Comment

In most applications it is not necessary to strive for the high accuracy demonstrated in the above discussion. An uncertainty of 1 part in 1000 is generally regarded as excellent for most purposes. Smaller samples can be taken, for example, to conserve both solution and the expensive reagent, silver nitrate. Also, duplicate determinations often suffice. However, it does not require very much additional effort to aim for high accuracy, buoyancy corrections and all.

The procedure outlined above works equally well for other insoluble silver salts, such as bromide, iodide, and thiocyanate.

6.5   PRECIPITATION TITRATIONS

In this section we will be concerned only with three classical titration methods, all involving the precipitation of silver salts.* The three titrations discussed below are:

1. **The Mohr titration** (1856). Solutions of chloride, bromide, iodide, and thiocyanate may be titrated with a standard solution of silver nitrate, using potassium chromate as the endpoint indicator.

2. **The Fajans titration** (1924). An adsorption indicator such as dichlorofluorescein shows when the silver precipitation is complete.

3. **The Volhard titration** (1878). A solution of silver ion is titrated with a standard solution of potassium thiocyanate, using ferric ion as indicator.

The Mohr Titration

We begin with a neutral or slightly basic solution containing chloride (or bromide, iodide, or thiocyanate) ion. About 1–2 mL of 5% potassium chromate solution is added and the mixture is then titrated with a solution of silver nitrate until the white precipitate of AgCl becomes very slightly tinged with the red-orange color of silver chromate Ag_2CrO_4.

In the first stages of the titration the only reaction is the precipitation of silver chloride:

$$Cl^- + Ag^+ \rightleftarrows AgCl(s), \qquad K_{sp} = 10^{-10}.$$
$$\text{still in} \qquad\qquad\quad \text{white}$$
$$\text{excess}$$

As the equivalence point is approached, the concentration of excess chloride ion decreases and therefore the equilibrium concentration of Ag^+ continuously increases. At the equivalence point exactly the solution will contain equal concentrations (both about $10^{-5}M$) of chloride ion and silver ion. With the recommended amount of potassium chromate present, the next small addition of silver nitrate is sufficient to cause a perceptible precipitation of silver chromate:

$$CrO_4^{2-} + 2Ag^+ \rightleftarrows \quad Ag_2CrO_4(s), \qquad K_{sp} = 10^{-12}.$$
$$\text{yellow} \qquad\qquad \text{deep red–orange}$$

Thus this titration depends upon the fact that silver chromate is more soluble than silver chloride, so that the latter is completely precipitated before the color change occurs.

---

*For a more comprehensive treatment see *Volumetric Analysis*, Vol. II, by I.M. Kolthoff and V. Stenger (Interscience Publishers, 1947).

It has been determined that the excess silver-ion concentration required to form enough silver chromate to be visible corresponds to about 0.004 mmol of Ag^+. That is, there is a slight positive titration error, but a correction for this effect can be applied in the calculations.

Because of the tendency for the chromate ion to take on protons,

$$HCrO_4^- \rightleftarrows H^+ + CrO_4^{2-}, \qquad K_a = 10^{-7},$$

it is necessary to perform the Mohr titration in neutral or slightly basic solutions. The presence of appreciable acid will increase the solubility of silver chromate (see Chapter 17) and cause a delay in the formation of the red precipitate. On the other hand, the solution may not be very basic because there is a tendency for silver ion to precipitate as silver hydroxide. The pH range that proves suitable is from 6.5 to 10.5, and this can be achieved (if the initial sample is acidic instead of neutral) by adding borax or sodium bicarbonate or excess solid calcium carbonate to form buffers (see Chapter 9).

The Mohr titration should be performed at room temperature, because at high temperatures the solubility of silver chromate becomes too high.

The Fajans Titration

An understanding of this titration method depends strongly upon the earlier discussions in this chapter of the primary layers adsorbed on the surfaces of precipitates. When chloride (or bromide, iodide, or thiocyanate) ion is titrated with silver nitrate, there is a change in the primary layer as the equivalence point is passed.

Before equivalence: $\quad Cl^- \quad + \quad AgCl(s) \quad \rightarrow \quad AgCl/Cl^-$
in excess $\qquad$ solid in $\qquad$ a negative primary
$\qquad\qquad$ suspension $\qquad$ adsorption layer

After equivalence: $\quad Ag^+ \quad + \quad AgCl(s) \quad \rightarrow \quad AgCl/Ag^+$
in excess $\qquad\qquad\qquad$ a positive primary
$\qquad\qquad\qquad\qquad$ adsorption layer

If this titration is performed with no indicator present, there is no visible difference in the silver-chloride precipitate before and after the equivalence point. But if the solution contains a small concentration of fluorescein or dichlorofluorescein, a striking color change occurs as the precipitate makes its transition from a negative to a positive form. In neutral or very weakly acidic solution the indicator is present as a yellow anion with a green fluorescence. While the silver precipitate has a negative primary layer, the anion D^- is repelled from the surface of the particles and so keeps its normal color. However, as soon as silver becomes excessive, causing the precipitate surface to become positive, the D^- is strongly adsorbed in the *secondary* adsorption layer, in effect forming a red precipitate of AgD *on the surface*:

Before equivalence: $AgCl/Cl^-$ ... D^-
                                      white   yellow–green

After equivalence: $AgCl/Ag^+D^-$
                                   white   red

A suitable indicator solution is prepared by dissolving 0.1 gram of di-chlorofluorescein in 100 mL of alcohol. Only two or three drops are necessary in the titration.

As with the Mohr titration, the Fajans titration cannot be done in acid solutions. In this case it is important that the indicator be present as an anion, rather than as its uncharged protonated species.

The titration should be carried out in subdued light because the precipitate is photosensitive, especially when the indicator is present.

At or near the equivalence point in silver-halide titrations the precipitate tends to coagulate and to settle out. Of course, this corresponds to the condition of nearly uncharged precipitate particles, when neither chloride nor silver ions are in excess. With the Fajans titration in particular it is better to keep the precipitate in colloidal suspension, so the color change is more obvious. This may be accomplished by the addition of 5 mL of 2% dextrin, which serves as a protective colloid.

The Volhard Titration

In contrast to the Mohr and Fajans titrations, the Volhard titration does not use a standard solution of silver ion as the titrant. Instead, the titration reaction is the precipitation of silver thiocyanate upon adding a standard solution of potassium thiocyanate KSCN as the titrant:

$$Ag^+ + SCN^- \rightleftarrows AgSCN(s), \qquad K_{sp} = 10^{-12}.$$
$$\text{white}$$

As an endpoint indicator, one adds 1–2 mL of a saturated solution (about 40%) of ferric ammonium sulfate. This provides ferric ions that react with the first excess of thiocyanate to form a red–orange complex ion:

$$SCN^- + Fe^{3+} \rightleftarrows FeSCN^{2+}, \qquad K_f = 1.5 \cdot 10^2.$$
$$\text{red–orange}$$

Also in contrast to the Mohr and Fajans titrations, this titration *must* be carried out in rather strongly acidic solution ($[H^+] > 0.3$) to minimize the formation of ferric-hydroxide species, which would diminish the ability of ferric ion to react with the thiocyanate ion.

The color of the complex is a very light shade of tan if the titration is performed well, meaning that a minimum of excess KSCN is added. It requires only about 0.001 mmol of excess thiocyanate to give a perceptible color. There is actually more of a problem of *under*titration because, as the silver thiocyanate precipitate is being formed in the presence of the excess (as yet untitrated) silver ion, there is primary-layer adsorption of some of the silver ion. To ensure that all the silver ion is made available to the thiocyanate, it is necessary to swirl the suspension vigorously as the endpoint is approached and to wait until the tan tint becomes permanent instead of fading slowly.

Although this titration may certainly be used for the determination of the amount of silver in a sample, it is more commonly used for the indirect determination of chloride, bromide, or iodide by applying the method of **back titration**. This is what is usually meant by the **Volhard method**. It works as follows:

To an acidic solution of the unknown halide ion we add an excess of standard silver nitrate solution over that required to precipitate the silver halide. Now the sample has been transformed from a halide unknown to a silver-ion unknown, and the titration of the excess silver is carried out with potassium thiocyanate.

This method causes little problem with bromide and iodide because the solubilities of silver bromide and silver iodide are much lower than that of silver thiocyanate. With chloride, however, there is a tendency for the first excess of thiocyanate to displace chloride ion from the silver-chloride precipitate:

$$SCN^- + AgCl(s) \rightleftarrows Cl^- + AgSCN(s).$$

This means that a larger excess of thiocyanate must be added before the color of the ferric-thiocyanate complex becomes visible. This problem can be handled in two ways: one may digest the silver-chloride precipitate by boiling it for a few minutes and then filter off the coagulated solid before doing the Volhard titration. An alternative is to add 1 mL of nitrobenzene, which is immiscible with water. This coats the silver chloride particles and slows down the rate at which they can react with the thiocyanate.

There are many variations on these precipitation titrations involving silver; they depend upon the concentrations of the unknowns and upon the particular mixtures of halides and various interfering ions present. The purpose of the above discussion is to give the essential equilibrium concepts of the titrations; should the reader have a need for applying these methods it would be well to consult the above-mentioned work by Kolthoff and Stenger.

Example calculation for Volhard method

A soluble sample known to contain bromide (and no interferences) and having a mass of 0.4234 g is dissolved in 50 mL of $0.5M$ nitric acid. A pipet is used to add 25.00 mL of $0.1046M$ silver nitrate, and the resulting precipitate is allowed to age for a few minutes with vigorous swirling. Then 2 mL of saturated ferric ammonium sulfate is added, and the mixture is titrated with $0.0513M$ potassium thiocyanate, with 15.43 mL required to reach the pale tan endpoint.

Thus, there were

$$(0.0513M)(15.43 \text{ mL}) = 0.792 \text{ mmol of KSCN},$$

corresponding to this amount of excess silver ion. Since the total silver added was equal to

$$(25.00 \text{ mL})(0.1046M) = 2.615 \text{ mmol},$$

by subtracting we find that 1.823 mmol of silver (and hence bromide) were precipitated. The atomic weight of bromine is 79.91, and therefore there must have been

$$(1.823 \text{ mmol})(0.07991 \text{ g/mmol}) = 0.1457 \text{ g}$$

of bromide in the sample. The %Br was therefore

$$\frac{100 \cdot 0.1457}{0.4234} = 34.4\%.$$

6.6 DEMONSTRATION: SILVER-BROMIDE COLLOIDAL PRECIPITATION

1. In a 600-mL beaker, add 250 mL of $0.01M$ KBr. Then add about 1 mL of $0.01M$ $AgNO_3$. The precipitate will be nearly invisible, but use a beam of light to show the Tyndall effect.

2. Add more Ag^+ (up to 200 mL) to form a clearly visible quantity of colloidal precipitate. Filter through paper, showing that the particles are small.

3. In a 600-mL beaker, add 250 mL of $0.01M$ $AgNO_3$. Then add 200 mL of $0.01M$ KBr, showing that the precipitate is comparable to that formed in step (2).

4. To each of the solutions (100-mL portions poured into 250-mL beakers) add a little 0.1% dichlorofluorescein. The difference in color shows that there must be something different about the two colloids. Discuss the formation of AgD on the surface due to the presence of the primary layer.

5. Show that dichlorofluorescein will not form a pink precipitate with a solution of $0.001M$ (or so) silver nitrate, proving that the pink color on the colloidal suspension must be related to the precipitate and not to the excess Ag^+ in the solution.

6. Show the changeover of the primary adsorption layer by (a) adding excess $AgNO_3$ to the $AgBr$—Br^- colloid, and (b) adding excess KBr to the $AgBr$—Ag^+ colloid.

7. Coagulate the two colloids by adding $1M$ $MgSO_4$ to each, with stirring. Point out that the great increase in turbidity is due not to more precipitate, but to coagulation.

6.7   DEMONSTRATION: MIXED CRYSTAL FORMATION

1. To 100 mL of $2M$ KNO_3 add 25 mL of concentrated $HClO_4$ to show white precipitate.
2. To 100 mL of $2M$ KNO_3 add a crystal of $KMnO_4$ and stir to dissolve. Then add 25 mL of concentrated $HClO_4$ to form a pink precipitate.
3. Filter a portion of the pink precipitate and wash with distilled water to show that it is the solid that is pink and not the surrounding solution.
4. Add $NaHSO_3$ to the mixture of pink precipitate and excess acidic solution of $KMnO_4$ to decolorize the solution, leaving the pink precipitate. This proves that the purple color of the precipitate is not merely a surface adsorption effect.

PROBLEMS

Gravimetric determinations

1. A sample of barium chloride dihydrate is examined to find its precise composition. A sample weighing 2.7287 g is dried at a temperature high enough to drive off all the water, and the residue weighs 2.3297 g. What is the precise value of x in the formula $BaCl_2 \cdot xH_2O$?
2. Predict the approximate percentage of potassium ion that will precipitate when 50 mL of concentrated perchloric acid is added to 100 mL of $1.0M$ potassium chloride (ignore activity coefficients). The solubility-product constant for $KClO_4$ is about 0.01.
3. The equilibrium molarity of the hydroxide ion can easily be controlled at any desired value by using acid–base buffers. Suppose it is desired to precipitate 99.9% of the ferric ion as $Fe(OH)_3$ from a solution that initially contains $0.02M$ Fe^{3+} and $0.08M$ Mg^{2+}.

    a)   How high must the hydroxide ion molarity be to precipitate the iron (99.9%)?

    b)   At what $[OH^-]$ will the $Mg(OH)_2$ begin to precipitate?

4. A physical chemist, working with zinc-amalgam electrodes, hires you to find the percentage of mercury in one of his preparations. You weigh S grams of the sample, dissolve it in nitric acid, and dilute to 25 mL in a volumetric flask. A pipetted 5-mL portion is treated appropriately in order to precipitate all the mercury as calomel (mercury(I) chloride Hg_2Cl_2, formula weight 472.09), which after drying has a mass of P grams. Given the atomic weight of mercury (200.59), set up the equation you would use for calculating the percentage of mercury in the sample.
5. The percentage of phosphorus in a detergent product may be determined by sample treatment including the precipitation of magnesium ammonium phosphate $MgNH_4PO_4 \cdot 6H_2O$ (MW = 245.41), using an excess of both magnesium and ammonium ions. The separated and washed precipitate is then heated strongly to decompose it to magnesium pyrophosphate $Mg_2P_2O_7$ (MW = 222.57), which is weighed.

a) What would you predict to be the primary adsorption layer on the precipitate and why?

b) If a 5.00-g sample of detergent yielded 0.255 gram of the weighing form, calculate the percentage of phosphorus (AW $= 30.97$) in the sample.

Gravimetric stoichiometry

6. Table 6.1 indicates the precipitation and weighing forms that are useful for determining a number of elements. Use these data and atomic weights (from Appendix 2) to calculate (a) the gravimetric factor for each method and (b) the percentage of the specified element in each sample.

Table 6.1

	Sample mass (g)	Element sought	Precipitate	Weighing form		
				Formula		Mass (g)
a)	3.6191	Ag	AgCl	AgCl	143.32	1.1492
b)	2.1651	Al	Al(OH)$_3 \cdot x$H$_2$O	Al$_2$O$_3$	101.96	0.7791
c)	1.3214	As	As$_2$S$_3$	As$_2$S$_3$	213.97	0.5236
d)	0.4622	Ba	BaSO$_4$	BaSO$_4$	233.40	0.2092
e)	3.8938	Bi	Bi$_2$(CO$_3$)$_3$	Bi$_2$O$_3$	465.96	1.2122
f)	1.7618	Br	AgBr	AgBr	187.78	0.6619
g)	4.3827	Ca	CaC$_2$O$_4 \cdot x$H$_2$O	CaO	56.08	1.3201
h)	2.9220	Cl	AgCl	AgCl	143.32	0.9804
i)	3.5947	Co	nitrosonaphthol salt	Co$_3$O$_4$	240.80	1.1435
j)	6.0317	Cr	BaCrO$_4$	BaCrO$_4$	253.33	1.6517
k)	4.4936	F	PbClF	PbClF	261.64	1.3438
l)	5.2230	Fe	Fe(OH)$_3 \cdot x$H$_2$O	Fe$_2$O$_3$	159.69	1.4946
m)	5.5770	Hg	HgS	HgS	232.65	1.5646
n)	2.7902	I	AgI	AgI	234.77	0.9468
o)	6.7853	K	K$_2$PtCl$_6$	K$_2$PtCl$_6$	415.11	1.7902
p)	0.6561	Mg	MgNH$_4$PO$_4 \cdot 6$H$_2$O	Mg$_2$P$_2$O$_7$	222.57	0.2869
q)	2.0564	Na	NaZn(UO$_2$)$_3$(OAc)$_9$ $\cdot 9$H$_2$O	Same $- 9$H$_2$O	1429.84	0.7483
r)	8.5782	Ni	Ni(C$_4$H$_7$O$_2$N$_2$)$_2$	Ni(C$_4$H$_7$O$_2$N$_2$)$_2$	288.94	2.0949
s)	5.8337	P	MgNH$_4$PO$_4 \cdot 6$H$_2$O	Mg$_2$P$_2$O$_7$	222.57	1.6141
t)	5.2343	Pb	PbSO$_4$	PbSO$_4$	303.25	1.4968
u)	8.1279	S	BaSO$_4$	BaSO$_4$	233.40	2.0212
v)	5.1424	Th	Th(C$_2$O$_4$)$_2$	ThO$_2$	264.04	1.4784
w)	8.2541	U	(NH$_4$)$_2$U$_2$O$_7$	U$_3$O$_8$	842.09	2.0421
x)	3.4078	Zn	ZnS	ZnS	97.43	1.0995

The use of the solubility-product constant*

7. The solubility-product constant for calcium sulfate is $10^{-5.9}$.

a) If equal volumes of 0.10M calcium chloride and 0.20M sodium sulfate are mixed, will a precipitate of calcium sulfate form?

---

*Use Appendix 5 for K_{sp} data.

    b)   Predict the solubility of calcium sulfate in pure water.

    c)   Predict the solubility in $0.50M$ calcium nitrate.

8. If a solution of $0.10M$ silver nitrate is slowly added to a solution containing $0.010M$ potassium iodate and $0.20M$ sodium bromate, what precipitate will form first? Will this precipitate form nearly completely (99%) before the second precipitate starts to form?

9. How much solid sodium oxalate must be added to 100 mL of $0.10M$ zinc nitrate to cause 99.9% precipitation of the zinc as zinc oxalate?

10. Which is more soluble, lead sulfate or lead iodide?

11. What is the highest pH allowed in a solution of $0.05M$ chromic nitrate if no precipitate of chromic hydroxide is desired?

BASIC THERMODYNAMIC RELATIONSHIPS

7

In a way science might be described as paranoid thinking applied to Nature: we are looking for natural conspiracies, for connections among apparently disparate data.

Carl Sagan
The Dragons of Eden

7.1 GIBBS FREE-ENERGY CHANGE ΔG

We have seen in Chapter 4 that the ratio r/Q may be used to predict the spontaneous direction of a chemical reaction in a nonequilibrium mixture. For the special case of $r/Q = 1$, the mixture is already at chemical equilibrium and there will be no net change. For various mixtures the more the r/Q ratio differs from 1.00, that is, the farther away the system is from equilibrium, the greater will be the reaction tendency.

This text cannot include the development of the logical arguments of chemical thermodynamics, but some of the conclusions of that elegant discipline cannot be ignored in our study of chemical equilibria. One of these is that the reaction tendency, or driving force, in a nonequilibrium mixture may be expressed in terms of the energy that, in principle, can be obtained to do useful work. The work might be mechanical (expanding gases pushing pistons in a car engine) or electrical (a digital calculator driven by a chemical battery). It can be shown that this useful energy, called the **Gibbs free energy** and symbolized by ΔG, is directly related to the r/Q ratio as follows:

$$\Delta G = RT \ln \frac{r}{Q}, \tag{7.1}$$

where R is the gas constant (8.314 joules/deg or 1.987 cal/deg) and T is the thermodynamic temperature (298.15 K corresponds to 25°C).

It is important to understand what is meant by ΔG. It is *not* the total Gibbs energy change that would be observed if the nonequilibrium mixture were allowed to react until it had reached equilibrium. Instead we must imagine a large quantity of the nonequilibrium mixture, so large that when one stoichiometric unit of reaction takes place, the overall composition of the mixture is virtually unchanged. To illustrate, consider this hypothetical situation: a 10-billion-liter tank is filled with a solution containing a billion moles of sodium hydroxide and also half a billion moles of hydrochloric acid. Naturally, this would be impossible to achieve in practice even on a one-liter scale because of the very fast reaction:

$$H_3O^+ + OH^- \rightarrow 2H_2O.$$

But the value of hypothetical arguments is that we are allowed to postulate the unlikely and examine the results. Now, with the tank filled and all those moles dispersed uniformly throughout the solution, we will "let" just one mole of H_3O^+ react with just one mole of OH^- while we measure in some way the quantity of the Gibbs free energy involved. The chemical change is represented as follows:

$$H_3O^+(0.05 \text{ molar}) + OH^-(0.1 \text{ molar}) \rightarrow 2H_2O(\text{liquid}).$$

This equation not only defines what we mean by the **stoichiometric unit**, but it also shows the *states* of the reacting species and products. Because of the huge amount of material present, these states are essentially the same after the reaction as they were before it. Because the solution is dilute, we approximate $X_{H_2O} = 1.00$.

The value for the equilibrium quotient Q for this reaction under the particular ionic-strength conditions specified above is (from a reference table)

$$Q = 5.8 \cdot 10^{13} = \frac{X^2_{H_2O}}{[H_3O^+][OH^-]} \qquad \text{(equilibrium values)}.$$

In the hypothetical solution there is a certain value for r:

$$r = \frac{X^2_{H_2O}}{[H_3O^+][OH^-]} = \frac{1^2}{(0.05)(0.1)} = 200 \qquad \text{(nonequilibrium values)}.$$

Therefore, we may calculate the value for the Gibbs free-energy change:

$$\Delta G = RT \ln \frac{r}{Q} = 1.987 \cdot 298.15 \ln \frac{200}{5.8 \cdot 10^{13}}$$

$$= -1.56 \cdot 10^4 \text{ cal/stoich. unit}.$$

This is the quantity we would have observed in the hypothetical measurement; the negative sign means that the Gibbs energy is released rather than absorbed during the reaction. When the Gibbs energy change is negative, the reaction will proceed to the right in the equation used to define the stoichiometric unit. We say the reaction is spontaneous under the existing conditions. A positive value for the Gibbs energy change means that the reaction is nonspontaneous as written and would actually proceed to the left, kinetics permitting.

The Reaction Isotherm

Equation (7.1) may be easily transformed to a more commonly used form by introducing the activity coefficients for the reacting species.

We must distinguish between the equilibrium quotient Q written in terms of concentrations and the equilibrium constant K written in terms of activities. The relationship is simple:

$$K = Q \cdot F,$$

where F is the appropriate cluster of activity coefficients. By introducing the cluster F into Eq. (7.1) we obtain the expression for **the reaction isotherm**

$$\Delta G = RT \ln \frac{r}{Q} = RT \ln \frac{r \cdot F}{Q \cdot F} = RT \ln \frac{r \cdot F}{K},$$

or

$$\boxed{\Delta G = -RT \ln K + RT \ln (rF).} \qquad (7.2)$$

This relationship is one of the more important conclusions of chemical thermodynamics, and we will find it very useful. By inserting the numerical quantities

$$R = 1.987 \text{ cal/deg}, \quad T = 298.15\text{K} \text{ (or 25°C)}, \quad \ln x = 2.3026 \log_{10} x,$$

into Eq. (7.2), we find

$$\Delta G(\text{cal}) = -1364 \log K + 1364 \log(rF).$$

If the value of R is expressed in joules, $R = 8.314$ joules/deg, then

$$\Delta G(\text{joules}) = -5708 \log K + 5708 \log(rF).$$

Standard States and the Standard Gibbs Energy Change $\Delta G°$

From the reaction isotherm we see that for a given reaction there is an infinite number of possible ΔG values corresponding to an infinite number of possible concentration states of the reactants. In order to compare different chemical reactions on a consistent basis, we define the **standard Gibbs energy change** as the ΔG value corresponding to the hypothetical case of all reacting species and products being present in their **Standard states** of unit activity. These states are defined as follows:

A *gas* is in standard state when its partial pressure is exactly 1 atm at 25°C.

A *solid and a liquid* are in standard state when present in pure form, with mole fraction equal to unity.

A *solute* is in standard state when in a 1 molar (or 1 molal) solution under the hypothetical condition that there are no environmental effects causing nonideal behavior. In other words, although the concentration is 1.000 and there may be other solutes present as well, the behavior is the same as would be found in an infinitely dilute solution. This means, of course, that the activity coefficient is also unity.

If a mixture is created with all the species present in their standard states, it follows that the nonequilibrium ratio r would be unity and the activity coefficient cluster would also be unity. The reaction isotherm would take the form

$$\Delta G \equiv \Delta G° = -RT \ln K + RT \ln 1.000,$$

or, since $\ln 1 = 0$,

$$\Delta G° = -RT \ln K = -1364 \log K \text{ cal} = -5708 \log K \text{ joules.} \qquad (7.3)$$

Therefore the reaction isotherm may be written in terms of Gibbs energy changes:

$$\Delta G = \Delta G° + RT \ln(rF).$$

Gibbs Free-Energy Change for an Equilibrium System

When a chemical system is at equilibrium, it must be true that the equilibrium ratio r is exactly equal to the equilibrium quotient Q. It follows from Eq. (7.1) that ΔG must be zero:

$$\Delta G = -RT \ln K + RT \ln(rF) = 0,$$

since $rF = K$ at equilibrium.

This is another way of saying that for a system at equilibrium there is no net driving force to make it move away from equilibrium.

Calculation of K from Gibbs Energies of Formation

Sometimes it will not be possible to find a value in the reference tables for a particular equilibrium constant. One should not overlook the possibility of turning to thermodynamic tables to find values for Gibbs energy changes. We have seen that

$$\log K = \frac{-\Delta G^\circ}{1364} \text{ cal},$$

where ΔG° is the standard Gibbs energy change for the reaction in question. It is also true that this change can be calculated from the standard Gibbs energies of formation for the species involved in the equilibrium. The standard Gibbs energy change for formation is defined in terms of the hypothetical reaction for the formation of a species from its elements:

$$\text{Elements in standard state} \xrightarrow{\Delta G_f^\circ} \text{Species.}$$

One may deduce that the standard Gibbs energy change for any reaction must be the following:

$$\Delta G^\circ \text{(reaction)} = \sum \Delta G_f^\circ \text{(product species)} - \sum \Delta G_f^\circ \text{(reactant species)}.$$

Values for Gibbs energies of formation for a large variety of chemical species are available.*

Example The use of formic acid to remove an excess of bromine from a solution is a reaction of analytical significance:

$$Br_2 + HCOOH = 2Br^- + CO_2 + 2H^+$$

---

*National Bureau of Standards, Technical Note 270, Selected Values of Chemical Thermodynamic Properties, 440 pp., 1968.

From the NBS reference we find:

Species	ΔG_f^0, cal
$Br_2(aq)$	+940
$HCOOH(aq)$	−89,000
$Br^-(aq)$	−24,850
$CO_2(aq)$	−92,260
$H^+(aq)$	0

From these data we may calculate the standard Gibbs energy change for the reaction:

$$\Delta G^\circ = [2(-24,850) - 92,260 + 2(0.0)] - [+940 - 89,000] = -53,900 \text{ cal.}$$

Since this is large and negative we conclude that the bromine–formic acid reaction has a large spontaneous tendency. The equilibrium constant follows:

$$\log K = \frac{-\Delta G^\circ}{1364} = -\frac{-53,900}{1364} = 39.5,$$

whence

$$K = 10^{39.5} = 3 \cdot 10^{39}.$$

Clearly, one would expect this reaction to go to completion.

7.2   EFFECT OF TEMPERATURE ON K VALUES

Tables of numerical values for equilibrium constants typically list only the values for a temperature of 25°C (Appendixes 4 and 5). An example is given in Table 7.1.

Table 7.1

Reaction	K (25°)	pK
$H_2O = H^+ + OH^-$	$1.00 \cdot 10^{-14}$	14.00
$NH_4^+ = H^+ + NH_3$	$5.5 \cdot 10^{-10}$	9.24
$CH_3COOH = H^+ + CH_3COO^-$	$1.75 \cdot 10^{-5}$	4.76

Indeed, it has become customary over the years to choose 25°C (298.15K) as a standard reference temperature for the comparison of thermodynamic data. However, a great deal (perhaps most) of practical laboratory work is carried out at temperatures higher or lower than 25°, and it is important to realize that for all chemical reactions the values of equilibrium constants are dependent on temperature. If one performs calculations dealing with an equilibrium system, it is necessary to account for the temperature effect unless approximate calculations are adequate for the case at hand.

To illustrate the temperature effect, Table 7.2 gives the pK values for the above reactions as determined by direct experiment over the range of 5–45°C.

Table 7.2

t, °C	pK_w	$pK(NH_4^+)$	$pK(CH_3COOH)$
5	14.734	9.903	4.770
15	14.346	9.564	4.758
25	13.997	9.246	4.756
35	13.680	8.947	4.763
45	13.396	8.671	4.777

Plots of the pK-values versus temperature are shown in Fig. 7.1; we note that the variations are great and the curves differ in appearance. The acetic-acid data are relatively constant and go through a minimum at about 25°; the value for K_w (not pK_w) varies about 20-fold over this temperature range.

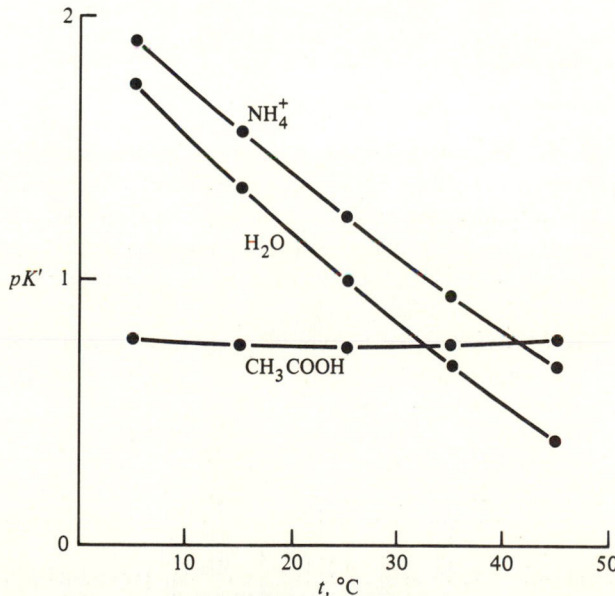

Fig. 7.1 Effect of temperature on pK values. To bring the three plots into the same graph, constants of 13, 8, and 4 have been subtracted from the pK values for water, ammonium ion, and acetic acid, respectively.

The "A, B, C Function"

The variation of pK with temperature might be precisely described by various algebraic functions. However, there are good reasons* for selecting the following three-term equation:

$$pK = \frac{A}{T} + B + C \log T,$$

where A, B, and C are numerical constants that may be determined for each individual case, and T is the thermodynamic temperature.

Once the A, B, and C values are known, it is a simple matter to use the above $pK(T)$ equation to find the appropriate value for pK at any desired temperature, provided it is within the range for which the numerical coefficients have been shown to be valid or at least within a few degrees of that range. For the dissociation of water into its ions, we have

$$A = 6034.623, \qquad B = -65.1356, \qquad C = 23.8.$$

Suppose we desire to know the value of pK_w at the normal temperature of the human body (37°C or 310.15K). Then,

$$pK(37°) = \frac{6034.62}{310.15} - 65.1356 + 23.8 \cdot \log 310.15 = 13.621.$$

The reader should show that these coefficients are consistent with the individual experimental values for pK_w listed in Table 7.2.

Determination of the A, B, and C Values

Once a set of $pK(T)$ values is known from experimental work, it becomes a problem in numerical analysis to find the best set of coefficients for the equation. If a special computer program is available, it is a simple matter to fit the data to the equation by the least-squares method. However, there is an alternative approach that is not only simpler mathematically but also has perceptual advantages.

Consider the following three values of pK_a for benzoic acid in water.

t °C	pK_a	T
15	4.214	288.15
25	4.212	298.15
35	4.221	308.15

---

*See R. W. Ramette, *Journal of Chemical Education*, 54, 280 (1977).

Obviously, this particular equilibrium constant is not very sensitive to temperature changes within the range shown. At some point in the range it goes through a minimum value. Suppose that for a precise research application it is necessary to predict a precise value for this pK_a at an intermediate temperature, such as 28°. We need an equation that realistically describes the $pK(T)$ variation, and so it is important to deduce the values of the coefficients, A, B, and C.

STEP 1   DETERMINATION OF THE VALUE FOR C

By differentiation of the $pK(T)$ equation, we find

$$\frac{dpK}{dT} = -\frac{A}{T^2} + \frac{C}{LT},$$

where $L = \ln 10 = 2.30259$. Let us multiply this derivative by the quantity LT^2, calling the product D.

$$D = LT^2\frac{dpK}{dT} = -LA + CT.$$

The derivative dpK/dT is simply the slope of the plot of pK versus temperature. We may make a good approximation of its value by using finite differences:

$$\frac{dpK}{dT} \simeq \frac{\Delta pK}{\Delta T} = \frac{pK_2 - pK_1}{T_2 - T_1}.$$

Example For the first two values in the benzoic-acid data we have

$$\frac{\Delta pK}{\Delta T} = \frac{4.212 - 4.214}{298.15 - 288.15} = -0.0002.$$

This is the value for the slope at the average temperature of 20°C or 293.15K. The average temperature must be used in calculating the value for D:

$$T_{av} = \frac{T_2 + T_1}{2} \quad \text{and} \quad D = LT_{av}^2\frac{\Delta pK}{\Delta T}.$$

For the first two points, then,

$$D = 2.30259 \cdot 293.15^2(-0.0002) = -40 \quad \text{(at } T = 293.15\text{)},$$

and for the second and third points the reader can show that

$$D = +190 \quad \text{(at } T = 303.15\text{)}.$$

Now, since D is a linear function of T, it follows that a *second* difference calculation yields the value for C:

$$\frac{\Delta D}{\Delta T} = C = \frac{190 - (-40)}{303.15 - 293.15} = 23.0.$$

Thus the value for C has been calculated with ease. If data for additional temperatures are available, the several individual values for C may simply be averaged to find the proper value for the next step. Alternatively, one might choose to use a linear least-squares fit (D-values versus T-values) in which case C would be returned as the slope.

STEP 2 DETERMINATION OF THE VALUES FOR A AND B

Once the value for C is known, we may calculate the values for Σ (sigma) defined as follows by rearrangement of the $pK(T)$ equation:

$$\Sigma = [pK - C \log T] = \frac{A}{T} + B.$$

The values of A and B are readily found by a linear least-squares fit (Σ versus $1/T$) that has A as the slope and B as the intercept. This is the recommended procedure when the data set contains, say, more than four values of pK and T. However, in the rather common situation of only three data points, as in the above benzoic-acid example, it is quite easy to calculate a value for A using each pair of points. Once again, difference calculations are the key (prove this):

$$A = \frac{(\Sigma_2 - \Sigma_1)T_1 T_2}{T_1 - T_2}.$$

Once a value for A has been calculated (preferably an average value of all successive pairs of data points), individual values of B may be found by the simple equation

$$B = \Sigma - \frac{A}{T}.$$

Example Continuing with the benzoic-acid data, we find the data of Table 7.3.

Table 7.3

t	T	Σ	$\Delta\Sigma$	A	B
15	288.15	−52.35723			−62.5771
20	293.15		−0.34277	2944.82	
25	298.15	−52.70000			−62.5771
30	303.15		−0.32053	2944.86	
35	308.15	−53.02053			−62.5771

The average A value is 2944.84, while $B = -62.577$. We may now write the $pK(T)$ expression insofar as it is determined by this particular set of data:

$$pK = \frac{2944.84}{T} - 62.577 + 23.0 \log T.$$

The value of pK for any desired temperature may be calculated by using the appropriate value for T in this equation. Caution must be taken when calculating values that lie outside the temperature range of the data set used to deduce the A, B, and C values.

THE TERM C LOG T MAY BE SUPERFLUOUS

For many chemical reactions the variation of the equilibrium constant with temperature is adequately described by the equation

$$pK = \frac{A}{T} + B.$$

A prominent example is the acid dissociation constant for the ammonium ion, given in Table 7.1. When the data set consists of only three values, as in the above benzoic-acid example, only one value of C may be calculated. However, when more data are available we may examine several values for C to judge whether their average is significantly different from zero. For example, with the pK data on the ammonium ion, the successive calculations for C give values of -3.4, -3.5, and $+9.5$. The average is $+0.87$, but in comparison with the individual values this is not significantly different from zero. Therefore, Σ becomes identical with pK in the calculations of A and B.

7.3   CALCULATING THERMODYNAMIC FUNCTIONS FOR THE CHEMICAL REACTION

Once the relationship between pK and T has been found by the foregoing procedure

$$pK = \frac{A}{T} + B + C \log T,$$

where C may have a zero value in some cases, it is a straightforward operation to find the standard values for the changes in free energy, enthalpy, and entropy by using the following relationships.

Gibbs Free-Energy Change

The relationship between free-energy change and K is

$$\Delta G^\circ = -RT \ln K = -RLT \log K = RLT \cdot pK.$$

It is merely necessary to multiply the p$K(T)$ function by RLT to obtain a function showing the temperature dependence of the free-energy change:

$$\Delta G^\circ = RL(A + BT + TC \log T).$$

Enthalpy Change

From the fundamental thermodynamic relationship

$$\frac{d \ln K}{dT} = \frac{\Delta H^\circ}{RT^2} \quad \text{or} \quad \frac{dpK}{dT} = -\frac{\Delta H^\circ}{RLT^2},$$

we see that it is necessary to take the first derivative of the $pK(T)$ function and to multiply it by $-RLT^2$ to find the function for the temperature dependence of the enthalpy change:

$$\Delta H^\circ = R(LA - CT).$$

Note that $\Delta H^\circ = \text{constant} = RLA$ for the case where $C = 0$.

Entropy Change

The change in entropy is the (negative) derivative of the free-energy change with respect to temperature:

$$\Delta S^\circ = -\frac{d\Delta G^\circ}{dT}.$$

It follows that

$$\Delta S^\circ = -R(LB + C + LC \log T).$$

Note that ΔS° is constant if $C = 0$.

Heat Capacity Change

We define the heat capacity change as the temperature coefficient of the enthalpy change $d\Delta H^\circ/dT$; then we find

$$\Delta C_p^\circ = -RC.$$

PROBLEMS

Using the reaction isotherm

1. Boric acid has only a very slight tendency to donate protons to water:

$$H_3BO_3 + H_2O = H_2BO_3^- + H_3O^+, \quad K = 10^{-9.23}.$$

   a) Use Eq. (7.2) to calculate the Gibbs energy change for the reaction when $[H_3BO_3] = 0.10$, $[H_2BO_3^-] = 0.01$, and $[H_3O^+] = 0.05$. From the sign of ΔG, state whether the above reaction will proceed from left to right or vice versa. Assume the ionic strength to be 0.06.

   b) Suppose the above mixture is perturbed by the addition of some solid sodium hydroxide, which reacts with the H_3O^+ ion and momentarily reduces its molarity to only $1 \cdot 10^{-11}$ mol. Assume that the ionic strength remains at 0.06 and calculate ΔG for the new conditions. What would happen?

2. For each of the reactions given below, calculate the value for K or the value for the standard Gibbs energy change $\Delta G°$ using Eq. (7.3):

a) $AgCl(s) = Ag^+ + Cl^-$, $K = 1.77 \cdot 10^{-10}$, $\Delta G° = ?$ (cal)

b) $H_3O^+ + OH^- = 2H_2O$, $\Delta G° = -1.9 \cdot 10^4$ cal, $K = ?$

c) $Cd^{2+} + SCN^- = CdSCN^+$, $K = 10$, $\Delta G° = ?$ (joules)

d) $I_3^- = I_2(aq) + I^-$, $\Delta G° = +1.6 \cdot 10^4$ joules, $K = ?$

Using standard Gibbs free energies of formation

3. Acetic acid is a very stable compound in a kinetic sense, for its solutions seem unchanged after years. Let us examine its thermodynamic stability. From NBS Technical Note 270 we find the following standard Gibbs energies of formation.

Species	$\Delta G_f°$, cal
$CH_3COOH(aq)$	$-94{,}780$
C(graphite)	0
H_2O(liquid)	$-56{,}687$

Calculate the equilibrium constant for the decomposition reaction

$$CH_3COOH(aq) = 2C(graphite) + 2H_2O(liquid).$$

What do you predict for the equilibrium molarity of acetic acid in a water solution?

4. We welcome the dissolving of air in lakes because oxygen is good for the fish and other living things. Since nitrogen is rather inert, it causes no effects. In short, we do not worry about the reaction

$$\frac{1}{2}N_2(g) + \frac{5}{4}O_2(g) + \frac{1}{2}H_2O(l) = H^+ + NO_3^-.$$

As long as kinetic barriers exist for this reaction, we may continue to ignore it. Is it thermodynamically feasible? Here are some data:

Species	$\Delta G_f°$, cal
N_2(gas)	0
O_2(gas)	0
H_2O(liquid)	$-56{,}687$
H^+	0
NO_3^-	$-26{,}410$

a) Calculate the standard Gibbs energy change for the above reaction. We see that it is positive, and so the reaction does not have a strong forward tendency. However, let us continue.

b) Show that the equilibrium constant for the reaction is about 0.04, which is small but not *very* small. Write the equilibrium constant expression.

c) Taking the partial pressure of nitrogen and oxygen in air as 0.8 and 0.2 atm, respectively, and setting the equilibrium value $[H^+] = [NO_3^-]$, show that there would be an undesirable concentration of nitric acid in our lakes if a catalyst for the reaction were inadvertently released into the environment.

d) Write a short science fiction story based on this idea. Let the hero(ine) be an undergraduate student of analytical chemistry. Send me a copy.

GALVANIC CELLS AND POTENTIOMETRIC DETERMINATIONS

8

One day I didn't eat for a whole day and night, and went into deep speculation. However, nothing came of it.

Confucius

8.1 ELECTRON-TRANSFER REACTIONS

Many of the chemical reactions classified as oxidation–reduction are believed to occur by collision mechanisms that involve one or more transfers of electrons between species. An example is the oxidation of iodide ion by ferric ion. When a solution of ferric perchlorate is mixed with one of potassium iodide, there is a rapid formation of a deep yellow-brown color due to the triiodide ion:

$$2Fe^{3+} + 3I^- \rightarrow 2Fe^{2+} + I_3^-.$$

How do the five ions on the left collide and react to form the three ions on the right? We assume that the mechanism must involve a series of bimolecular collisions like the possible scheme in Fig. 8.1. The two collisions marked "e" involve electron transfer, the ferric ion taking an electron from I^- or I_2^-.

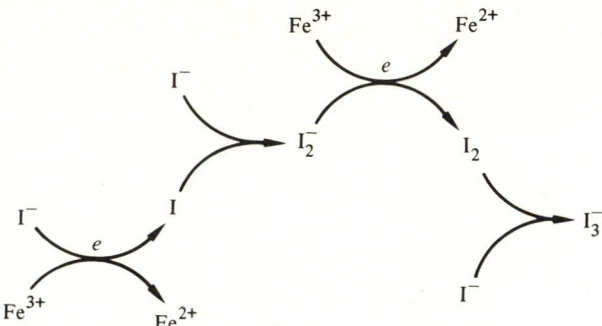

Figure 8.1

We know that electric currents are caused by the flow of electrons in conductors. Therefore it seems reasonable to suggest that reactions with electron-transfer mechanisms might be carried out by keeping the two redox couples in separate beakers, letting the electrons flow from one beaker to the other through a metal wire. Let us examine this idea for the reaction above.

Now, instead of mixing the ferric and iodide ions in the same beaker, let us have separate beakers, each with a platinum wire (chosen because Pt is a good conductor of electrons as well as being chemically inert). We will include a sensitive ammeter in the circuit as shown in Fig. 8.2.

The hypothesis is that the iodide ions will give up electrons to the electrode, according to the half-reaction:

$$3I^- \rightarrow I_3^- + 2e.$$

These electrons will pass through the wire and the ammeter, which detects low rates of electron flow, and will be accepted by the ferric ions at the surface of the other electrode:

$$2e + 2Fe^{3+} \rightarrow 2Fe^{2+}.$$

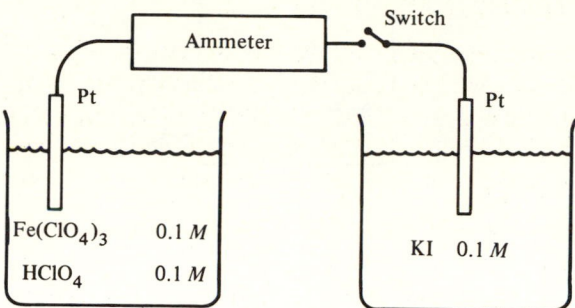

Figure 8.2

However, when the switch is closed the results are disappointing. At best there is a momentary and very tiny pulse of electrical current, after which the zero reading on the ammeter shows that electron flow is not taking place.

The failure can be explained in terms of electrostatic forces: if the two half-reactions *do* take place, there will be an imbalance of electrical charge in the solutions. The iodide half-reaction produces one singly charged ion in place of three singly charged ions; the solution then has two extra K^+ ions. This excess positive charge accumulates around the platinum wire, and electrostatic attraction makes it more difficult for electrons to move away.

At the other electrode the half-reaction produces four positive charges in place of the original six charges (on the two ferric ions). This solution accumulates excess perchlorate ions ClO_4^- which will repel the incoming electrons. Therefore the process comes rapidly to a standstill.

The experiment may be improved by providing a path for the migration of the ions in the solutions. This will let the excess K^+ ions and the excess ClO_4^- ions move toward each other, thereby relieving the buildup of excess charge at each electrode. We will use a special glass vessel that has separate compartments for the ferric and iodide solutions, but also has porous glass connections to a central compartment, into which we may place some inert electrolyte such as sodium perchlorate. This time we will include a voltmeter as well as an ammeter, as shown in Fig. 8.3.

With the switch as shown in Fig. 8.3, the ammeter is out of the circuit and the electrodes are connected only through the very-high-resistance voltmeter. We note with some satisfaction that the voltmeter reads a few tenths of a volt, with the +/− polarity, showing that there definitely is a tendency for current (electron) flow.

When the switch is connected to the ammeter, we find that there is a steady current of a few milliamperes. It is obvious that electron transfer is taking place, especially as we observe buildup of a yellow color at the surface of the platinum wire in the KI solution.

Given sufficient time, the reaction will proceed until equilibrium is reached; the current and the voltage will eventually become zero.

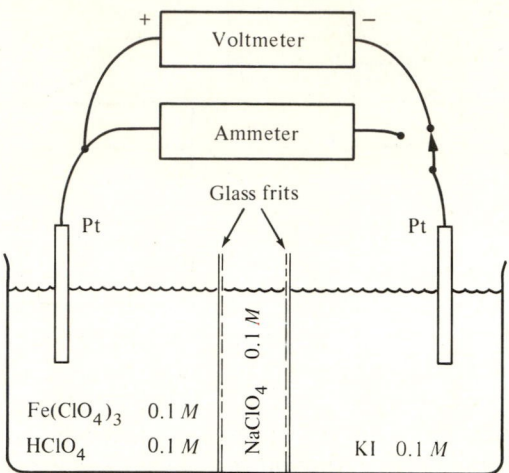

Figure 8.3

Within the three solutions there will be migration of ions as required to maintain electroneutrality. All positive ions (cations) in the system will migrate from right to left in Fig. 8.3, and all negative ions (anions) will migrate from left to right.

8.2   THE CONCEPT OF ELECTRODE POTENTIAL

When the voltmeter was connected to the cell in the example above, it was noted that the measured voltage had a definite polarity, with the potassium iodide side being more negative than the ferric perchlorate side. We could just as correctly say that the potassium iodide side was less positive, since the measurement tells only the *difference* in the two electrode potentials but gives no information about their absolute values. Any attempt to determine the absolute potential of a single electrode electrolyte half-cell runs immediately into the following difficulty. As soon as the voltmeter is connected to the metal electrode and to the solution, the wire connected to the solution automatically becomes a second electrode, so that once again it is only the difference in potentials that is observed.

Yet it is logical that a certain electrostatic potential does exist for any single electrode. The origin of this potential is in the transfer of electrons between the metal electrode and the redox couple in the solution. For example, suppose a platinum wire is immersed in a solution that contains both ferric ions and ferrous ions. The ions are in rapid motion due to thermal diffusion and are constantly colliding with the platinum surface in huge numbers each second.

Whenever a ferric ion collides with the electrode there is some probability that it will remove an electron from the relatively loose **sea of electrons**

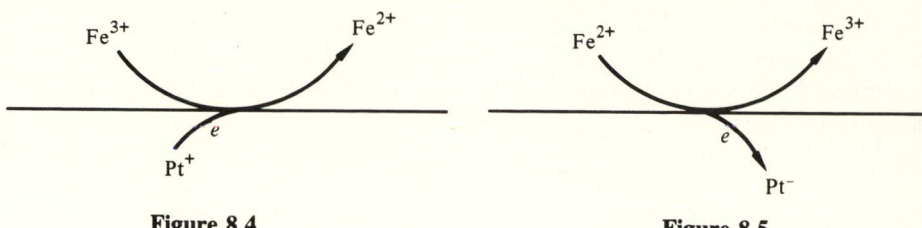

Figure 8.4   **Figure 8.5**

characteristic of a metal-lattice structure. If this happens, the ferric ion is reduced to the ferrous state, and the platinum metal is left with a deficiency of one electron. If this process occurs for a large number of collisions, the metal electrode will accumulate a significant positive charge (Fig. 8.4).

However, the solution also contains a great number of ferrous ions that are constantly colliding with the platinum electrode surface. With each collision there is some probability that the ferrous ion will *release* an electron into the sea of electrons in the metal lattice. To the extent that this process occurs, the platinum electrode will accumulate an excess of negative charge due to the added electrons (Fig. 8.5).

Clearly these two electron transfer processes are in competition, one tending to make the electrode positive, the other tending to make it negative. The net result depends on which process is dominant, and this in turn depends on two factors. First, each chemical species has its own intrinsic tendency either to donate or to accept an electron from the platinum metal; this is a characteristic based on the nature of the electron bonding structure in the species. Second, when the two members of the redox couple are not present in equal concentrations, the one in higher concentration obviously has a kinetic advantage in that it will undergo more collisions per second with the electrode surface than the member with lower concentration.

In summary, a half-cell consists of a metal electrode that is in direct contact with both members of a redox couple. The electrode will acquire a certain electrostatic charge called the **electrode potential**. The sign (+ or −) and the magnitude of this potential is characteristic of the particular redox couple in the half-cell. However, we may not measure the magnitude directly, but only in comparison with the potential for some other half-cell.

8.3   REQUIREMENTS FOR PRACTICAL HALF-CELLS

In using galvanic cells for experiments in equilibrium and analysis our main interest is in systems that quickly establish reproducible potentials at the electrode/electrolyte interface. The following requirements must be met:

1. A metal (wire, foil, or rod) electrode must be immersed in the electrolyte solution. The electrode provides the sea of electrons at the interface and also is a point of contact for use by the experimenter.

2. One (and only one) active redox couple must be present at the interface with all its species present in finite amounts. By "all its species" we mean those that appear in the balanced half-equation for the redox couple. For example, if the permanganate/manganous redox couple is used, the half-equation is

$$MnO_4^- + 8H^+ + 5e \rightleftarrows Mn^{2+} + H_2O.$$

Therefore the solution must contain finite concentrations of permanganate ions, manganous ions, hydrogen ions, and water. We will deal only with aqueous solutions in which the water is present at a mole fraction nearly equal to 1.00.

3. In the case of a redox couple that has a metal as the reduced species, the metal is used for the electrode. Otherwise, as with the permanganate/manganous couple, the electrode is typically platinum metal, chosen because of its ability to withstand direct chemical attack.

4. The redox couple must be able to establish equilibrium fairly quickly with the sea of electrons in the electrode. Otherwise the electrode potential will drift and/or not be reproducible. The term **exchange current** refers to the rate at which the electrons flow back and forth between the electrode and the couple (the net flow being zero, of course). Systems with a high exchange current show steady, reproducible potentials that are in accord with the Nernst equation (see Section 8.4). This is what we mean by "active redox couple" in requirement (2) above.

8.4 THE NERNST EQUATION

The potential of an electrode depends not only on which particular redox couple is present but also on the concentrations of the species involved in the half-reaction. The fundamental equation relating these variables is named in honor of German electrochemist Walter Nernst, who in 1889 derived it from thermodynamic arguments.

The Nernst equation for an electrode is always written by reference to a balanced half-reaction written in the *reduction* direction. Its basic form is:

$$E = E^\circ - \frac{RT}{nF} \ln \frac{\text{Activities of half-reaction products}}{\text{Activities of half-reaction reactants}},$$

where E is the potential difference (volts) across the electrode/electrolyte interface; E° is the **standard reduction potential** and is the value of E under the hypothetical conditions of all species being present in their standard states at activities of unity; R is the gas constant, $8.314 \text{ joule/K} \cdot \text{mol}$; T is the absolute (Kelvin) temperature (e.g., 298.15K at 25°C); n is the number of electrons in the balanced half-reaction; and F is the Faraday constant, 96,487 coulombs/mol. The natural logarithm may be replaced by $\log_{10}$ by introducing the factor 2.3026. Therefore the Nernst equation for a temperature of 25°C is

$$E = E° - \frac{0.05916}{n} \log \frac{\text{Activities of products}}{\text{Activities of reactants}}.$$

For approximate calculations the constant may be rounded to 0.059 or even to 0.06.

A few examples follow that show how the Nernst equation is written for typical half-cells.

a) A platinum electrode is immersed in a solution containing $0.03M$ ferric ion and $0.02M$ ferrous ion, the solution having an ionic strength of 0.2. The half-reaction is

$$Fe^{3+} + e \rightleftarrows Fe^{2+},$$

and the Nernst equation is

$$E = E° - 0.059 \log \frac{[Fe^{2+}]f_2}{[Fe^{3+}]f_3} = E° - 0.059 \log \frac{0.02 \cdot 0.31}{0.03 \cdot 0.07}.$$

b) A copper electrode is immersed in a solution containing $0.005M$ copper nitrate (ionic strength 0.015):

$$Cu^{2+} + 2e \rightleftarrows Cu(s) \qquad \text{(Note that } n = 2\text{).}$$

In this case the reduced species is solid copper metal, and we take its activity to be its mole fraction equal to unity:

$$E = E° - \frac{0.059}{2} \log \frac{X_{Cu}}{[Cu^{2+}]f_2} = E° - \frac{0.059}{2} \log \frac{1}{0.005 \cdot 0.62}.$$

c) A platinum electrode is immersed in a solution of $0.0100M$ hydrochloric acid, and hydrogen gas is slowly bubbled through the solution. The partial pressure of H_2 above the solution happens to be 0.944 atm, and we take this to be its activity:

$$H^+ + e \rightleftarrows \tfrac{1}{2}H_2(g),$$

$$E = E° - 0.059 \log \frac{P^{1/2}}{[H^+]f_1} = E° - 0.059 \log \frac{0.944^{1/2}}{0.0100 \cdot 0.902}.$$

Important point   When the activity of a species is used in the log term of the Nernst equation, it is raised to the power corresponding to its stoichiometric coefficient in the balanced half-reaction. Thus we use the square root of the hydrogen gas pressure.

It is perfectly legitimate to write the half-reaction with doubled coefficients:

$$2H^+ + 2e \rightleftarrows H_2(g).$$

The Nernst equation corresponding to this half-reaction is

$$E = E° - \frac{0.059}{2} \log \frac{P}{[H^+]^2 f_1^2}.$$

This causes no change in the implied value for E because of the properties of logarithms: $\log a^b = b \log a$.

d) A platinum electrode is in a solution containing $0.002M$ potassium permanganate, $0.005M$ manganous nitrate, and $0.01M$ perchloric acid. The ionic strength of this solution is

$$I = 0.002 + 3 \cdot 0.005 + 0.01 = 0.027.$$

The balanced half-reaction is

$$MnO_4^- + 8H^+ + 5e \rightleftarrows Mn^{2+} + 4H_2O(l).$$

The Nernst equation must include *all* species involved in the half-reaction, and so we write

$$E = E^\circ - \frac{0.059}{5} \log \frac{[Mn^{2+}]f_2 \cdot X_{H_2O}^4}{[MnO_4^-]f_1 \cdot [H^+]^8 f_1^8}.$$

The mole fraction of water in dilute aqueous solutions is taken to be unity, and therefore

$$E = E^\circ - \frac{0.059}{5} \log \frac{0.005 \cdot 0.53}{0.002 \cdot 0.86 \cdot 0.01^8 \cdot 0.86^8}.$$

Important point   Note that the potential of this electrode depends very strongly on the pH of the solution, since hydrogen-ion molarity is raised to the eighth power.

A Comment about Activity Coefficients

Given the marked variation of activity coefficients with ionic strength, it is obvious that they must be included in the Nernst equation whenever we want to do realistic calculations involving electrode potentials. Frequently, however, we need only the roughest of estimates and will omit the activity coefficients for simplicity.

A special circumstance appears when experiments are carried out at high and nearly constant ionic strength. Then the Davies equation does not permit reliable calculations of activity coefficients, but at least we may presume them to be relatively constant throughout the experiment. They may be factored out of the log term and be combined with the E° value, giving what is called a **formal potential** symbolized by $E^{\circ\prime}$. The modified Nernst equation may then be written as

$$E = E^{\circ\prime} - \frac{0.059}{n} \log (\text{Concentrations}).$$

8.5   SIMPLIFIED CELL DIAGRAMS

To eliminate the need for artistic talent in describing experimental galvanic cells, it is customary to use a symbolic diagram that not only simplifies the appearance of the cell but, as shown later, helps in making thermodynamic deductions about the cell potential.

Any actual galvanic cell is made up of two half-cells, each with an electrode immersed in a solution. Either half-cell may be designated as the "left" one in the diagram, which is patterned as follows:

Left electrode | left electrolyte ‖ Right electrolyte | right electrode.

A vertical line indicates the surface of the electrode, that is, the interface between the electrode and the solution; it is at this interface that the electrode potential has its origin.

The double vertical line is used to show the zone of contact between the two electrolyte solutions. Although there may be a small electrical potential at this interface, due to differences in diffusion rates of ions, we will usually ignore it.

In writing cell diagrams it is possible to give complete information about the cell composition by showing all substances present. Here is an example:

$$\text{Ag} \left| \begin{array}{l} \text{AgNO}_3 \ (0.010M) \\ \text{KNO}_3 \ \ \ (0.050M) \end{array} \right\| \begin{array}{l} \text{Fe(ClO}_4)_3 \ (0.005M) \\ \text{Fe(ClO}_4)_2 \ (0.003M) \ | \ \text{Pt} \\ \text{HClO}_4 \ \ \ \ \ (0.250M) \end{array}$$

It is always understood that the substances are dissolved in water, unless it is indicated that they are present in solid, liquid, or gaseous form. Thus the diagram above leaves no doubt about the chemical composition of the cell. However, it is frequently convenient to emphasize only those species that are of direct significance in establishing the electrode potential. For the cell above, the potassium ions, nitrate ions, perchlorate ions, and hydrogen ions are not part of the redox couples, and we may write a simpler diagram as follows:

$$\text{Ag} \ | \ \text{Ag}^+ (0.010M) \ \| \ \text{Fe}^{3+} \ (0.005M), \ \text{Fe}^{2+}(0.003M \ | \ \text{Pt}.$$

The cell diagram above is instructive because it illustrates two of the important types of half-cells. At the right we see the use of an inert metal, nearly always platinum (although gold and palladium are sometimes used), placed in contact with a solution that contains two *soluble* members of a redox couple. The purpose of the platinum is to provide the interface for the sea of electrons.

At the left is a silver metal electrode immersed in a solution that contains silver ions. In this half-cell the electrode has two roles: one is to provide the sea of electrons and the other is to serve as the reduced form of the redox couple. The mechanism for the establishment of a potential is different in this case: one may picture the silver ions Ag^+ breaking free from the metal lattice and leaving electrons behind, making the electrode negative. But the competing reaction is the entry of silver ions from the solution *into* the lattice, causing an accumulation of positive charge. Depending on the concentration of silver ions in the solution, the equilibrium at the interface will establish a definite electrode potential. Some other metals that can also serve as electrodes in contact with solutions of their ions are copper, mercury, lead, zinc, and cadmium.

When one member of the redox couple is a gas, that fact is shown in the cell diagram. For example, a half-cell based on the hydrogen ion (hydrogen gas couple) is as follows:

$$\| H^+(0.02M), H_2(gas, 0.9 \text{ atm}) \mid Pt.$$

This diagram represents a solution of an acid, with hydrogen gas being bubbled through it at a pressure of 0.9 atm. When gases are bubbled through solutions in the laboratory, we do not try to control the pressure but simply accept whatever atmospheric conditions exist. As the gas goes through the solution it becomes saturated with water vapor in accordance with the vapor pressure of the solvent. The total pressure, which is the sum of the partial pressure of the gas and the partial pressure of the water vapor, is determined by reading a barometer. Then the actual partial pressure of the gas is calculated

$$P_{gas} = P_{barometric} - P_{H_2O}.$$

The vapor pressure of water P_{H_2O} is found in reference tables for whatever temperature is being used for the cell.

There are two very important things to remember when considering galvanic cells. One is that each half-cell must have a well-defined redox couple present. In looking at a cell diagram it is necessary only to look at all the oxidation states of the elements present: the redox couple can be deduced by finding what element is present in two different oxidation states.

The second point to keep in mind is that not all redox couples actually function in a half-cell. Many couples that look all right on paper have kinetic barriers that prevent them from establishing a predictable potential through interaction with the sea of electrons in the electrode. For example, it would be useless to construct a half-cell based on the nitrogen gas/nitrate ion couple:

$$\| HNO_3 (0.1M), N_2 (gas, 0.9 \text{ atm}) \mid Pt,$$

even though one may write a hypothetical half-reaction:

$$NO_3^- + 6H^+ + 5e = \tfrac{1}{2}N_2 + 3H_2O.$$

This couple does not reach equilibrium within a reasonable time, if ever.

Because of the importance of perchlorate ion in aqueous equilibrium studies, it is worth mentioning that it shows no kinetic tendency to reach equilibrium with other species. The half-cell

$$\| NaClO_4\,(0.2M),\ NaCl\,(0.1M) \,|\, Pt$$

would not display a useful potential, in spite of the fact that there are two soluble members of a redox couple:

$$ClO_4^- + 8H^+ + 8e = Cl^- + 4H_2O.$$

However, the following half-cell may be used reliably:

$$\| NaClO_4\,(0.2M),\ Cl_2\,(gas, 0.9\ atm),\ NaCl\,(0.1M)\,|\,Pt.$$

There are three oxidation states of chlorine present but, as pointed out above, the perchlorate ion can be discounted. In this half-cell the electrode potential is established by the chlorine/chloride couple

$$Cl_2 + 2e = 2Cl^-.$$

8.6   VOLTAGE AND POLARITY FOR A GALVANIC CELL

Every galvanic cell is a combination of two half-cells and, in general, the two electrode potentials will differ, causing the cell to have a finite voltage with a $+/-$ or a $-/+$ polarity in the left/right sense. For example, we may construct a cell comprising a chlorine/chloride ion electrode for one half-cell and a hydrogen/hydrogen ion electrode for the other, using hydrochloric acid as the electrolyte. The cell diagram may be written as follows:

$$Pt\,|\,Cl_2\,(gas,\ 1\ atm),\ HCl\,(0.1M)\,|\,HCl\,(0.1M),\ H_2\,(gas,\ 1\ atm)\,|\,Pt.$$

We find experimentally that the voltage of the cell is 1.49 volt and that the chlorine electrode is the positive pole. These experimental facts are *not* changed by writing the cell diagram oppositely, with the chlorine electrode on the right. The potentials of the electrodes are invariant and the chlorine electrode is always positive with respect to the hydrogen electrode. It is equally correct to state that the hydrogen electrode is negative with respect to the chlorine electrode.

Should we say that the voltage of this cell is $+1.49$ volt or -1.49 volt? The accepted convention is as follows: given the cell diagram, with its left/right orientation of the two half-cells, the cell voltage is defined by the equation

$$E_{cell} = E_{right\ electrode} - E_{left\ electrode}.$$

This is equivalent to saying that the algebraic sign for E_{cell} is the same as the relative sign of the right electrode. Thus for the cell diagrammed above we have

$$E_{cell} = -1.49\ volt,$$

and the right electrode (the hydrogen electrode) is the negative pole.

If the cell diagram above had been written in the opposite direction, with the chlorine electrode on the right, E_{cell} would be $+1.49$ volt. It is very important to adhere to this convention that the sign of the cell voltage is directly related to the cell diagram.

8.7 NUMERICAL SCALE OF ELECTRODE POTENTIALS

It is easy to construct two half-cells experimentally, to connect them to form a whole galvanic cell, and to measure the voltage of that cell. The measurement can give only the *difference* in the electrode potentials and there is no way we can determine the absolute potentials of the individual electrodes. Yet it is very useful to establish a scale of relative electrode potentials so that we may do various numerical calculations using cell data. The logical and experimental sequence for establishing the scale is as follows:

Step 1 We arbitrarily *assign* a value of 0.000 volt to the standard potential $E°$ of the hydrogen ion/hydrogen gas electrode. The standard hydrogen electrode, usually abbreviated as SHE, is hypothetical because it is based on the couple in standard-state conditions:

$[H^+] = 1.000 M$, but ideal behavior with $f_H = 1.000$,

$P_{H_2} = 1.000$ atm, behaving in accordance with the ideal-gas law.

Step 2 The assignment of $E° = 0$ now permits us to *calculate* (not to measure) the electrode potential for a *real* electrode based on the H^+/H_2 couple. For example, suppose we have a platinum electrode in a solution of $0.005 M$ perchloric acid (with pure hydrogen gas bubbling through the solution) and that, after correction for the vapor pressure of water, the partial pressure of hydrogen gas above the solution is 0.962 atm. Then we calculate for this electrode:

$$E = E° - 0.05916 \log \frac{P^{1/2}}{[H^+]f_1}$$

$$= 0 - 0.05916 \log \frac{0.962^{1/2}}{0.005 \cdot 0.927}$$

$$= -0.1376 \text{ volt (at 25°C).}$$

This gives us a numerical value for an actual, working electrode that can be used as one electrode in a galvanic cell.

Step 3 A galvanic cell is constructed by using the real hydrogen electrode in combination with any other electrode. For example, we may use the silver ion/silver couple for the other half-cell:

$$Pt \mid H_2 \, (0.962 \text{ atm}), \, HClO_4 \, (0.005 M) \parallel AgNO_3 \, (0.002 M) \mid Ag \, (s).$$

A precise high-impedance voltmeter is used to find that this cell has a voltage of 0.7757 volt, with the silver electrode being the positive pole. Therefore, by the convention that the sign of E_{cell} is the same as that of the right electrode, we have the relationship

$$E_{cell} = +0.7757 = E_{right} - E_{left} = E_{right} - (-0.1376);$$

where

$$E_{right} = +0.6381 \text{ volts.}$$

This measurement and calculation has established a value for the electrode potential for the particular silver ion/silver electrode.

Step 4　The standard potential for the silver ion/silver electrode may now be calculated on the assumption that the Nernst equation is applicable to the electrode:

$$E_{Ag} = E^{\circ}_{Ag} - 0.05916 \log \frac{1}{[Ag^+]f_1}.$$

Table 8.1　Standard reduction potentials of analytical interest
(acidic aqueous solution, 25°)

Reduction half-reaction	E°_{red}, volts
$Na^+ + e = Na(s)$	-2.71
$Zn^{2+} + 2e = Zn(s)$	-0.76
$Fe^{2+} + 2e = Fe(s)$	-0.44
$Cr^{3+} + e = Cr^{2+}$	-0.41
$2H^+ + 2e = H_2(g)$	0.00^*
$S_4O_6^{2-} + 2e = 2S_2O_3^{2-}$	$+0.08$
$Sn(IV) + 2e = Sn^{2+}$	$+0.15$
$Cu^{2+} + 2e = Cu(s)$	$+0.34$
$I_3^-(aq) + 2e = 3I^-$	$+0.54$
$H_3AsO_4 + 2H^+ + 2e = HAsO_2 + 2H_2O$	$+0.56$
$Fe^{3+} + e = Fe^{2+}$	$+0.77$
$Hg_2^{2+} + 2e = 2Hg(l)$	$+0.79$
$Ag^+ + e = Ag(s)$	$+0.80$
$2Hg^{2+} + 2e = Hg_2^{2+}$	$+0.92$
$NO_3^- + 4H^+ + 3e = NO(g) + 2H_2O$	$+0.96$
$Br_2(aq) + 2e = 2Br^-$	$+1.10$
$IO_3^- + 6H^+ + 5e = \frac{1}{2}I_2(aq) + 3H_2O$	$+1.20$
$O_2(g) + 4H^+ + 4e = 2H_2O$	$+1.23$
$Cr_2O_7^{2-} + 14H^+ + 6e = 2Cr^{3+} + 7H_2O$	$+1.33$
$Cl_2(g) + 2e = 2Cl^-$	$+1.36$
$BrO_3^- + 6H^+ + 6e = Br^- + 3H_2O$	$+1.44$
$MnO_4^- + 8H^+ + 5e = Mn^{2+} + 4H_2O$	$+1.52$
$Ce^{4+} + e = Ce^{3+}$	$+1.70$
$H_2O_2(aq) + 2H^+ + 2e = 2H_2O$	$+1.77$
$F_2(g) + 2e = 2F^-$	$+2.87$

*Reference couple.

Taking the ionic strength of the silver nitrate solution to be 0.002, making $f_1 = 0.952$, we get

$$+0.6381 = E^\circ_{Ag} - 0.05916 \log \frac{1}{0.002 \cdot 0.952};$$

therefore

$$E^\circ_{Ag} = +0.7990 \text{ volt.}$$

By repeating Steps 3 and 4 with other electrode systems, we may build up a set of E° values for redox couples that meet the requirements for applicability of the Nernst equation. Unfortunately most systems of redox couples do not fall in this category because of kinetic barriers in their electron-transfer mechanisms at the electrode surface. Later we will show how such couples may be assigned values of E° by using standard Gibbs energies of formation. Table 8.1 contains a brief list of E° values.

Finally it should be admitted that in practice the cell voltage experiments illustrated above have an inescapable uncertainty due to the **liquid-junction potential** at the boundary between the different electrolytes in the two half-cells. In most cases this potential remains unknown and is perhaps a few millivolts. In Chapter 11 we shall see the application of cells that are especially designed to avoid the error of liquid-junction potentials.

Comments on Appendix 5: Equilibria by Element

For elements that can exist as species with a variety of oxidation states it is useful to group the standard potential data in a single diagram for each element. This provides an easy reference when data are needed for calculations. For example, in Fig. 8.6 we find the diagram for bromine.

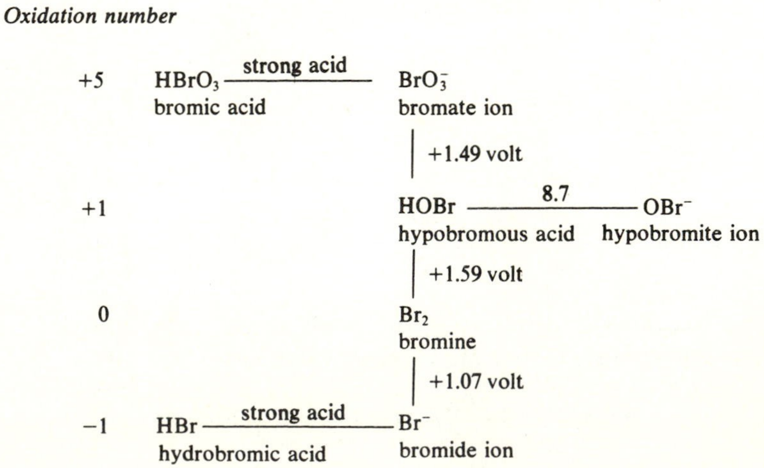

Oxidation number

+5	$HBrO_3$ $\xrightarrow{\text{strong acid}}$ bromic acid	BrO_3^- bromate ion	
		$\Big|$ +1.49 volt	
+1		HOBr $\xrightarrow{\;\;8.7\;\;}$ OBr$^-$ hypobromous acid hypobromite ion	
		$\Big|$ +1.59 volt	
0		Br_2 bromine	
		$\Big|$ +1.07 volt	
-1	HBr $\xrightarrow{\text{strong acid}}$ hydrobromic acid	Br^- bromide ion	

Figure 8.6

The number next to the vertical line connecting two species is the standard reduction potential for that particular redox couple. The balanced half-reaction for the reduction may be determined by the usual rules. Thus,

$$BrO_3^- + 5H^+ + 4e = HOBr + 2H_2O \qquad E_{red}^\circ = +1.49 \text{ volt,}$$

$$HOBr + H^+ + e = \tfrac{1}{2}Br_2 + H_2O \qquad E_{red}^\circ = +1.59 \text{ volt,}$$

$$Br_2 + 2e = 2Br^- \qquad E_{red}^\circ = +1.07 \text{ volt,}$$

Note that the value of E_{red}° pertains to the half-reaction in acidic solution. Later we will consider the way to find the value for E° when the reduction occurs in basic solution.

From the above diagram we find not only the values for E°, but also the acid–base properties of the species involved in the system; the numbers above the horizontal lines connecting the acid–base conjugate pairs are the pK_a values for the acid member of the pair. Thus from the above diagram we see that the dissociation constant for hypobromous acid is $10^{-8.7} = 2 \cdot 10^{-9}$.

8.8   PREDICTION OF CELL VOLTAGE

Given a cell diagram, including actual concentrations for the species in the active redox couples, we may use the Nernst equation to calculate a predicted value for the cell voltage. For example, consider the cell

$$Zn \,|Zn(ClO_4)_2 \,(0.04M) \,\|\, KI_3 \,(0.01M), \; KI \,(0.11M) \,|\, Pt.$$

Assuming the salts to be strong electrolytes, the ionic strength will be the same in both half-cells, 0.12. The steps to find E_{cell} are as follows:

Step 1   For each electrode write the Nernst equation corresponding to the active redox couple. At the left we have:

$$Zn^{2+} + 2e \rightleftarrows Zn, \qquad E_{left} = E_{Zn}^\circ - \frac{0.059}{2} \log \frac{X_{Zn}}{[Zn^{2+}]f_2}.$$

At the right:

$$I_3^- + 2e \rightleftarrows 3I^-, \qquad E_{right} = E_I^\circ - \frac{0.059}{2} \log \frac{[I^-]^3 f_1^3}{[I_3^-]f_1}.$$

Step 2   Look under zinc and under iodine in Appendix 5 (or in other tables) to find the standard potentials for the two couples: $E_{Zn}^\circ = -0.76$ volt and $E_I^\circ = +0.54$ volt.

Step 3   Calculate the potentials of both electrodes using the given molarities and the appropriate values for activity coefficients:

$$E_{left} = -0.76 - \frac{0.059}{2} \log \frac{1}{0.04 \cdot 0.36} = -0.81 \text{ volt,}$$

$$E_{right} = +0.54 - \frac{0.059}{2} \log \frac{0.11^3 \cdot 0.77^3}{0.01 \cdot 0.77} = +0.57 \text{ volt.}$$

Step 4   Use the convention that $E_{cell} = E_{right} - E_{left}$:

$$E_{cell} = +0.57 - (-0.81) = +1.38 \text{ volt.}$$

From this result we conclude that the iodine electrode is the positive pole of this particular cell.

Concentration Cells

When the same active redox couple is present at both electrodes of a galvanic cell, it follows that the same $E°$ value will apply to each side. The cell voltage will then depend only on the difference in concentration (activity) on the two sides. For example, suppose the following cell is constructed:

$$\text{Ag} \mid \text{AgNO}_3 \, (0.01M) \parallel \text{AgNO}_3 \, (0.10M) \mid \text{Ag.}$$

By using the Nernst equation we find for $n = 1$:

$$E_{right} = E_{Ag}° - 0.059 \log \frac{1}{0.10 \cdot 0.78} \qquad \text{when } I = 0.1.$$

$$E_{left} = E_{Ag}° - 0.059 \log \frac{1}{0.01 \cdot 0.90} \qquad \text{when } I = 0.01.$$

In calculating E_{cell}, the $E°$ values cancel:

$$E_{cell} = E_{right} - E_{left} = -0.059 \log \frac{0.01 \cdot 0.90}{0.1 \cdot 0.78} = +0.055 \text{ volt.}$$

If the ionic strength is the same in both half-cells, the activity coefficients will cancel:

$$E_{cell} = -\frac{0.059}{n} \log \frac{\text{Molarity at left}}{\text{Molarity at right}}.$$

Example What must be the molarity of Cu^{2+} in the left compartment of the following cell if the cell potential is found to be -0.033 volt? Assume $I = 0.1$ in both sides and remember that $n = 2$. Then

$$\text{Cu} \mid Cu^{2+} \, (xM) \parallel Cu^{2+} \, (0.01M) \mid \text{Cu,}$$

$$E_{cell} = -0.033 = -\frac{0.059}{2} \log \frac{x}{0.01},$$

whence

$$\log x = \frac{2(-0.033)}{-0.059} + \log(0.01) = -0.88,$$

$$x = 10^{-0.88} = 0.13M.$$

8.9 THE SPONTANEOUS CELL REACTION

If a galvanic cell has anything other than zero as its value for E_{cell}, then a chemical reaction will occur when the electrodes are connected together by a metal wire. The negative pole will have an accumulation of electrons greater than that at the positive pole, and these electrons will move spontaneously through the wire from the negative pole to the positive pole. Upon entering the electrode/electrolyte interface at the positive pole, the electrons will not simply flow into the solution but rather will be taken up by the oxidized species of the active redox couple. Thus a reduction half-reaction will occur spontaneously at the positive pole. At the negative pole, because electrons are being withdrawn from the interface to replace those that moved through the wire, there must be oxidation of the reduced species of the active redox couple.

Example The observed voltage for the following cell is -0.67 volt. What is the spontaneous cell reaction?

$$\text{Pt} \,|\, Cl_2(0.2 \text{ atm}), \text{HCl}\,(0.1M) \,\|\, \text{AgNO}_3\,(0.1M) \,|\, \text{Ag}.$$

From the value of E_{cell} we conclude that the silver electrode is the negative pole. Electrons will move through a wire *from* the silver *to* the platinum, causing Cl_2 to be reduced to Cl^- in the left half-cell. The loss of electrons from the silver interface will cause the oxidation of silver metal to silver ions in the right half-cell. Therefore the spontaneous cell reaction is

$$2\text{Ag(s)} + Cl_2\text{(g)} \rightarrow 2\text{Ag}^+ + 2Cl^-.$$

8.10 CONVENTIONAL CELL REACTION

It is very useful for thermodynamic calculations to define a **conventional cell reaction** that does not depend on the polarity of the cell. This reaction *may* be the same as the spontaneous cell reaction or it may be just the opposite. The rule for definition is simple: Given the cell diagram with its left/right orientation, we define the conventional cell reaction as that which involves oxidation at the left and reduction at the right.

Example Write the conventional cell reaction for the cell in the previous section.

Step 1 Write a balanced *oxidation* half-reaction for the active redox couple at the left electrode:

$$2Cl^- \rightarrow Cl_2\text{(g)} + 2e.$$

Step 2 Write a balanced *reduction* half-reaction for the active redox couple at the right electrode:

$$\text{Ag}^+ + e \rightarrow \text{Ag(s)}.$$

Step 3 Add the two half-reactions to get the whole, conventional reaction, taking care to use proper multiples so that the electrons cancel. The result in the present case is

$$2Cl^- + 2Ag^+ \rightarrow Cl_2(g) + 2Ag(s).$$

We note that for this cell the conventional cell reaction is the reverse of the spontaneous cell reaction. This will always be the case for cells with negative values of E_{cell}. If E_{cell} is positive, the conventional reaction is spontaneous (has a tendency to proceed as written).

Note The reader may be surprised to see a reaction showing the formation of chlorine gas and silver metal from silver and chloride ions, knowing that these ions normally react to form an insoluble precipitate of silver chloride. However, in the galvanic cell these ions are not in contact with each other because they are in separate half-cells. One of the unique characteristics of galvanic cells is that they allow us to study reactions that cannot be carried out by direct mixing of the reactants.

8.11 THE NERNST EQUATION FOR THE WHOLE CELL

In Section 8.4 we showed how the Nernst equation may be applied to a cell when the electrodes are considered separately. This approach is probably easiest for the beginning student, and it is often useful to deal with a single electrode in various situations. However, now that the conventional cell reaction has been defined, it is worthwhile to see how it relates to the Nernst equation for the cell as a whole. To illustrate, let us examine the following cell:

$$Pt\,|\,O_2(gas, 0.9\,atm),\ HCl\,(0.3M)\,\|\,CuCl_2\,(0.1M)\,|\,Cu.$$

Step 1 Identify the active redox couple for each electrode and write the conventional cell reaction:

Oxidation at the left:	$H_2O \rightarrow \frac{1}{2}O_2 + 2H^+ + 2e$
Reduction at the right:	$Cu^{2+} + 2e \rightarrow Cu$

Conventional cell reaction: $Cu^{2+} + H_2O \rightarrow \frac{1}{2}O_2 + Cu + 2H^+.$

Step 2 Look up the standard reduction potentials for the active couples:

$$E^{\circ}_{Cu} = +0.34\,volt;\qquad E^{\circ}_{O} = +1.22\,volt.$$

Step 3 Write the Nernst equation for the cell as follows:

$$E_{cell} = E^{\circ}_{cell} - \frac{0.0592}{n}\log(rF),$$

where E°_{cell} is the hypothetical value of the cell voltage under the condition that all reacting species are present in their standard states at unit activity. The term (rF) has the same meaning as in Chapter 7, that is, it is the nonequilibrium ratio of activities written in the same format as the equilibrium constant expression. For the present case, using the conventional cell reaction and the values as they exist in the cell, we can write

$$rF = \frac{P_{O_2}^{1/2}[H^+]^2 f_1^2}{[Cu^{2+}]f_2}$$

Step 4   Taking $E_{cell}^{\circ} = E_{right}^{\circ} - E_{left}^{\circ}$ we may now insert all the known numerical quantities into the equation ($I = 0.3$ on each side):

$$E_{cell} = +0.34 - (+1.22) - \frac{0.0592}{2} \log \frac{0.9^{1/2} \cdot 0.3^2 \cdot 0.73^2}{0.1 \cdot 0.29} = -0.89 \text{ volt.}$$

The reader should confirm this result by using the Nernst equation to calculate the separate electrode potentials.

To illustrate a common variation on this problem, consider this question: To what value would the Cu^{2+} activity in the right half-cell have to be changed in order to make the cell potential equal to -0.94 volt, assuming that the left half-cell remains unchanged? Again the Nernst equation is written:

$$-0.94 = +0.34 - (+1.22) - \frac{0.0592}{2} \log \frac{0.9^{1/2} \cdot 0.3^2 \cdot 0.73^2}{[Cu^{2+}]f_2}.$$

First find the value of

$$\log (rF) = \frac{2 \cdot (-0.94 - 0.34 + 1.22)}{-0.0592} = 2.027.$$

Then use the 10^x key on a calculator to find

$$(rF) = 10^{2.027} = 106.4.$$

Finally, solve for the copper ion activity:

$$[Cu^{2+}]f_2 = \frac{0.9^{1/2} \cdot 0.3^2 \cdot 0.73^2}{106.4} = 4.3 \cdot 10^{-4}$$

8.12   SECONDARY STANDARD REFERENCE ELECTRODES

Although the hydrogen ion/hydrogen gas electrode serves well as the primary standard half-cell for precise research, it certainly is a bit of trouble to use in practice. The hydrogen gas must be scrupulously purified, barometric pressures must be monitored, and the platinum electrodes themselves must be carefully prepared by depositing an electrolytic surface of "platinum black," a finely divided platinum metal that helps to catalyze the equilibrium at the electrode surface.

Many alternatives have been explored for reference electrodes of known potential that may be used more readily. The requirements for a satisfactory reference include the following:

a) The half-cell should be simple in its chemical composition.

b) When constructed by different workers using slight variations in technique and purity of materials, the half-cell should show a reproducible potential. The construction should not be too fussy.

c) Once prepared, the half-cell should remain stable over a long period of time. Its chemical constituents should not decompose on exposure to light, or warm temperatures, or oxygen (in air).

d) If the half-cell is used in experiments that involve a flow of electrons in the circuit, thus forcing slight changes in composition in the half-cell, its potential should nevertheless remain essentially constant.

We will consider only two examples, both widely used.

Silver Chloride/Silver Reference Electrodes

This electrode consists of a silver wire in contact with a solution that (a) contains a certain substantial concentration of chloride ion and (b) is saturated with the relatively insoluble salt, silver chloride. Typically the silver wire is coated with a thin layer of solid silver chloride, produced by electrolysis when the silver wire serves as the anode in a solution of HCl. This ensures that the solution at the electrode surface is indeed saturated with AgCl.

To write the Nernst equation for this electrode, we first must recognize that the reduction half-reaction is

$$AgCl(s) + e = Ag(s) + Cl^-.$$

This is the *net* half-reaction, even though one might imagine a two-step process:

Step 1　　　　$Ag^+ + e = Ag(s).$

(A small concentration is present in equilibrium.)

Step 2　　　　$AgCl(s) = Ag^+ + Cl^-$

(A little AgCl dissolves to replace the Ag^+.)

It is necessary to consider only the net chemical change when writing the Nernst equation, and therefore

$$E_{AgCl,Ag} = E^\circ_{AgCl,Ag} - 0.0592 \log [Cl^-]f_{Cl}.$$

(The AgCl and Ag are present at unit activity.)

The standard potential for this important electrode has been very accurately determined over a wide range of temperatures by Bates and Bower.* At 25°C, the value of $E^\circ_{AgCl,Ag}$ is +0.2224 volt, with variations of only about ±0.0001 volt in repeated preparations by different workers using standard techniques.

In making this half-cell for use as a secondary reference electrode, one may immerse the silver-chloride-coated silver wire in whatever chloride solution one desires. The equilibrium reference potential $E_{AgCl,Ag}$ (not E°) is seen from the Nernst equation to vary only with the activity of chloride ion in the solution.

Accepted values for E are as follows ($t = 25°C$):

Half-cell solution	$E_{AgCl,Ag}$, volt
0.0100M HCl, KCl, NaCl, etc.	+0.343
Saturated KCl	+0.199

Electrodes Based on Mercury Redox Couples

The applications of mercury-based electrodes is so common and important in analytical chemistry that we should review some fundamental observations. We will restrict the following discussion to the three simplest species:

Hg^{2+}	Hg_2^{2+}	Hg
mercuric ion,	mercurous ion,	mercury metal,
oxidation state +2	oxidation state +1	oxidation state 0

The mercurous ion has a strong tendency to form the insoluble species, mercurous chloride (calomel):

$$Hg_2Cl_2(s) = Hg_2^{2+} + 2Cl^-, \quad K_{sp} = [Hg_2^{2+}][Cl^-]^2 f_2 f_1^2 = 10^{-17.9}$$

THE FORMULA FOR MERCUROUS ION

Among the metals of periodic table group 2b (zinc, cadmium, and mercury), only mercury can be formed in a +1 oxidation state. The dimeric (two-atom) formula for mercurous ion Hg_2^{2+} (instead of the simpler Hg^+) was proved by Linhart† who used galvanic cells of the type

Pt, H_2(gas, 1 atm) | $HClO_4(c_1)$ ‖ $HClO_4(c_1)$, mercurous salt (c_2) | Hg.

Suppose we prepare a mercurous salt e.g., the nitrate or perchlorate, and dissolve it in the acidic solution. The question is: Does it dissociate to form

---

*R. G. Bates, V. Bower, *Journal of Research*, National Bureau of Standards, *53*, 283 (1954).
†G. Linhart, *Journal of the American Chemical Society*, 38, 2356 (1916).

monomeric mercurous ions

$$HgNO_3 \rightarrow Hg^+ + NO_3^-$$

or is the mercurous ion dimeric, so that the reaction would be properly written as

$$Hg_2(NO_3)_2 \rightarrow Hg_2^{2+} + 2NO_3^-?$$

In general, if the formula for the mercurous ion in solution is Hg_n^{n+}, we would expect the electrode potential for the right electrode in the cell above to be based on the half-reaction

$$Hg_n^{n+} + ne = n\,Hg(\text{liquid}).$$

The corresponding Nernst equation for the electrode would be written as

$$E = E^\circ - \frac{0.0592}{n} \log \frac{1}{[Hg_n^{n+}]f_n} = E^\circ - \frac{0.0592}{n} \log \frac{1}{c_2 f_n}.$$

Linhart prepared cells with varying amounts of mercurous salt present and measured the potentials using a constant composition for the left electrode. By considering the *difference* in potentials between two cells, one with $c_2 = 0.001$ and the other with $c_2 = 0.01$, he found that the tenfold concentration change resulted in a potential change of 0.0296 volt. This proved that the value for n is 2, and therefore we use the formula Hg_2^{2+} for the mercurous ion, which consists of two atoms joined by a covalent bond. Apparently there is no appreciable tendency for this ion to dissociate into separate Hg^+ species in solution.

E° FOR THE MERCUROUS ION/MERCURY METAL COUPLE

The standard reduction potential for the mercurous ion/mercury metal couple was determined by El Wakkad and Salem* by using cells of the same type as those used by Linhart. The active redox couples are

Right electrode: $Hg_2^{2+}(c_2) + 2e \rightarrow 2Hg(\text{liquid})$,

Left electrode: $2H^+(c_1) + 2e \rightarrow H_2(\text{gas})$.

We may write the expression for cell potential using the Nernst equation:

$$E = E_{Hg} - E_H = \left\{ E_{Hg}^\circ - 0.0296 \log \frac{1}{[Hg_2^{2+}]f_2} \right\} - \left\{ 0.00 - 0.0296 \log \frac{P}{[H^+]^2 f_1^2} \right\}.$$

In a typical measurement we get $E = 0.7997$ volt when $c_1 = 0.0100$, $c_2 = 1.70 \cdot 10^{-4}$, and $P = 1$. With activity coefficients from the Davies equation ($I = 0.01$ for the left half-cell and $I = 0.0105$ for the right), the reader should show that $E_{Hg}^\circ = +0.796$ volt.

---

*S. El Wakkad, T. Salem, *Journal of Physical Chemistry*, *54*, 1371 (1950).

$E°$ FOR THE MERCURIC/MERCUROUS COUPLE

In the study above the mercury metal could serve both as the reduced species in the redox couple and as the metal electrode material needed for the transport of electrons. When we consider the determination of the standard potential for the mercuric ion/mercurous ion couple, it is necessary to use a cell that employs a platinum electrode. Such a study was reported by Hietanen and Sillen* who used cells of the type

$$\text{Ag} \left| \begin{array}{c} \text{AgClO}_4 \,(0.0100M) \\ \text{HClO}_4 \,(0.0100M) \end{array} \right\| \begin{array}{l} \text{Hg}_2(\text{ClO}_4)_2 \,(c_1) \\ \text{Hg}(\text{ClO}_4)_2 \,(c_2) \\ \text{HClO}_4 \qquad (c_3) \\ \text{NaClO}_4 \quad\;\; (c_4) \end{array} \right| \text{Pt.}$$

In this cell the left electrode serves as a constant reference electrode, having a potential that may be calculated from the known standard reduction potential for the silver ion/silver metal couple. In the right-electrode compartment, we see the two oxidation states of mercury, each present in some definite concentration, and presumably at equilibrium with the inert platinum electrode. The purpose of the perchloric acid is to decrease the tendency for the ions to form complexes with hydroxide ion, and the purpose of the sodium perchlorate is simply to make the right electrode have the same ionic strength (0.020) as the left electrode.

    The active redox couples for this cell are

        Left electrode: $\text{Ag}^+(0.0100M) + e = \text{Ag},$

        Right electrode: $2\text{Hg}^{2+}(c_2) + 2e = \text{Hg}_2^{2+}(c_1).$

The corresponding Nernst equation (at 25°) is

$$E = \left\{ E_{\text{Hg}}^\circ - 0.0296 \log \frac{c_1 f_2}{c_2^2 f_2^2} \right\} - \left\{ E_{\text{Ag}}^\circ - 0.0592 \log \frac{1}{0.01 \cdot f_1} \right\},$$

as the reader should verify. (Note that there is some canceling of f_2 values.)

    In a typical measurement Hietanen and Sillen found $E_{\text{cell}} = +0.1373$ when $c_1 = c_2 = 0.00100$, $c_3 = 0.0100$, and $c_4 = 0.00400$. The reader should show that the calculated value for the standard potential of the mercury(II)/mercury(I) couple is +0.911 if it is assumed that $E_{\text{Ag}^+, \text{Ag}}^\circ = +0.799$.

CALOMEL REFERENCE ELECTRODES

The most widely used reference electrodes are those based on the mercurous chloride/mercury metal couple:

$$\tfrac{1}{2}\text{Hg}_2\text{Cl}_2(\text{s}) + e = \text{Hg}(1) + \text{Cl}^-.$$

The term **calomel** is an old name for mercurous chloride and is still commonly and widely used. A half-cell based on this redox couple has the following

---

*S. Hietanen, L. G. Sillen, *Arkiv för Kemi*, **10**:2, 103 (1956). (In English.)

diagram:

$$\| \, Hg_2Cl_2(\text{solid}), \, KCl(CM) \, | \, Hg(\text{liquid}).$$

Since the mercury metal is present as a pure liquid while the calomel is a pure solid, the only variable in the half-cell system is the concentration of potassium chloride. By far the most common choice is to use a solution that is saturated with potassium chloride. This makes the molarity of chloride ion about $4.17M$ at 25°C and, with such a large concentration, the activity of chloride ion remains virtually constant even when small electric currents are allowed to pass in the cell. Therefore, the *saturated calomel electrode* (SCE) maintains a constant potential.

Some workers prefer to use lower concentrations of potassium chloride in *un*saturated calomel electrodes, with the advantage that such half-cells are less dependent on temperature changes. (In the SCE the activity of chloride ion depends directly on the solubility of KCl, which varies with temperature.) The reduction potentials of the calomel electrodes are summarized as follows:

Calomel electrode	KCl molarity	E (25°C)
Saturated	4.17	+0.244
"Normal"	1.00	+0.280
"Tenth-normal"	0.100	+0.336
Standard		+0.268*

* This is the $E°$ value for chloride ion activity of unity.

Calomel (and also silver chloride) electrodes are referred to as **electrodes of the second kind** because their potentials are controlled by a secondary reaction. The mercury metal in the half-cell is primarily in equilibrium with whatever activity of mercurous ions Hg_2^{2+} exists in the solution. Because of the very low solubility of calomel, this is controlled at low levels by the large excess of chloride ion.

$E°$ FOR THE CALOMEL/MERCURY METAL COUPLE

As noted earlier, a calomel/mercury electrode is based on the half-reaction:

$$\tfrac{1}{2}Hg_2Cl_2(s) + e = Hg(\text{liquid}) + Cl^-.$$

The standard reduction potential for this couple was determined through very accurate measurements by Hills and Ives* who used cells having no liquid junction potential:

$$Pt \, | \, H_2(P \text{ atm}), \, HCl \, (CM), \, Hg_2Cl_2(s) \, | \, Hg.$$

---

*G. Hills, D. Ives, *Journal of the Chemical Society, 318* (1951).

Thus the electrolyte solution contains CM of each hydrogen ion and chloride ion (due to complete dissociation of HCl) and is also saturated with calomel. The concentration of mercurous ion is controlled by the solubility product

$$[Hg_2^{2+}] = K_{sp}/[Cl^-]^2 f_2 f_1^2.$$

The amount of chloride ion that enters the solution through the solubility of calomel is negligible compared to the amount present from the HCl.

The cell potential may be expressed in terms of the Nernst equation as follows:

$$E = E_{right} - E_{left} = \left\{ E_{Hg}^{\circ} - 0.05916 \log \frac{[Cl^-]f_1}{1} \right\} - \left\{ 0.00 - 0.05916 \log \frac{P^{1/2}}{[H^+]f_1} \right\},$$

where both $[H^+]$ and $[Cl^-]$ have the value of c.

In a typical measurement, Hills and Ives found $E = 0.60033$ when $P = 732.5$ mm of Hg (corrected for the vapor pressure of water) and $C = 1.6077 \cdot 10^{-3}$. Taking $f_1 = 0.956$ (Davies equation), the reader should show that the calculated value of E° for the calomel electrode is $+0.2680$ volt.

EXPERIMENTAL DETERMINATION OF E FOR THE
SATURATED CALOMEL ELECTRODE

When a half-cell is prepared with an electrolyte solution of saturated potassium chloride (about $4.17M$), it is not possible to calculate the activity coefficients of the ions because the ionic strength is far too high for the Davies equation to be useful. The potential of the SCE must be directly determined by comparison with an electrode of known potential. Measurements were reported by Hitchcock and Taylor* who used cells of the type

$$Pt \,|\, H_2(1 \text{ atm}), \text{HCl}\,(CM) \,\|\, \text{SCE}.$$

When the concentration of HCl was $0.02000M$, the cell showed a potential (at 25°) of 0.3483. From the Nernst equation we have:

$$E = E_{SCE} - \left\{ 0 - 0.0592 \log \frac{P^{1/2}}{[H^+]f_1} \right\}.$$

The reader should show that the potential for the SCE is $+0.244$ volt, relative to SHE.

8.13   POTENTIOMETRIC DETERMINATION OF AN UNKNOWN CONCENTRATION

From an analytical viewpoint the most important application of galvanic cells is the use of an electrode as a sensitive probe for finding the concentration of a species in solution. The most common example is in the measurement of pH, which is discussed in detail in Chapter 11.

---

*D. Hitchcock, A. Taylor, *Journal of the American Chemical Society*, *59*, 1812 (1937).

The Nernst equation provides the mathematical basis for potentiometric determinations. A typical cell for this application uses a reference electrode together with an electrode that responds to the concentration (activity) of the species to be determined. For example, suppose it is desired to find the concentration of copper ions in a solution. The cell would be as follows:

$$SCE \parallel Cu^{2+}(\text{unknown molarity}) \mid Cu.$$

(Instead of a copper metal electrode, which does not work well because of a layer of copper oxide on its surface, an **ion-selective electrode** made of copper and silver sulfides is used. Its behavior is, however, mathematically similar to that of a properly functioning copper metal electrode.)

For this cell the Nernst equation would have the form:

$$E = k - S \cdot \log \frac{1}{[Cu^{2+}]} = k + S \cdot \log [Cu^{2+}],$$

where k is a collection of constants, including the potential of the reference SCE, the standard potential for the copper electrode, the activity coefficient for the Cu^{2+} ion, and the liquid-junction potential that exists at the interface between the two half-cells.

In a typical application, the cell would first be calibrated by measuring a series of *known* concentrations of Cu^{2+}, using (as closely as possible) the same ionic medium as the one containing the unknown sample. By obtaining a set of E-values for these standards, we are able to determine the values for k and S because they are the intercept and slope, respectively, of a plot of E versus $\log [Cu^{2+}]$.

One hopes that the value for S will be close to 0.0592/2 for this case, but it is not uncommon to find significant deviations from this ideal. As long as the value of S is constant over the range of concentrations studied, the plot will be linear.

Once the values for k and S are known for a particular set of electrodes, the cell potential is determined when the unknown sample is placed in the half-cell. Let this potential be E_x. Then the unknown $[Cu^{2+}]_x$ can be calculated:

$$[Cu^{2+}]_x = 10^{(E_x - k)/S}.$$

The Method of Standard Additions

If one is willing to assume a certain value for S for a given electrode, perhaps because it has been determined by previous calibrations, it is possible to use the unknown sample solution itself for the addition of standards. Suppose the sample consists of V_0 mL of solution which contains the unknown concentration C_x of Cu^{2+}. This is placed in the half-cell and the potential is measured. A series of small volumes of a known standard solution of Cu^{2+} is added, with

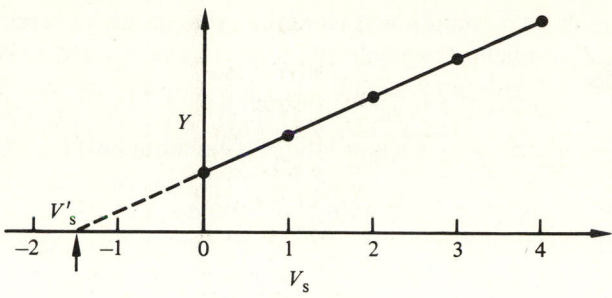

Figure 8.7

the potential measured after each addition. At any point, let the total volume of added standard be V_s mL and let the concentration of the standard solution be C_s. The total concentration of copper ion at any point is simply

$$[Cu^{2+}] = \frac{V_0 C_x + V_s C_s}{V_0 + V_s} = 10^{(E-k)/S} = 10^{E/S} \cdot 10^{-k/S}$$

This may be rearranged to the form

$$Y \equiv (V_0 + V_s) \cdot 10^{E/S} = 10^{k/S}(V_0 C_x + V_s C_s).$$

Using the several values for E that correspond to the added volume V_s, we plot the quantity Y versus V_s. The equation suggests that this plot will be linear since k and S are constant. (See Fig. 8.7.) Let V'_s be the point where the extrapolated line intersects the V_s-axis. At this point $Y = 0$, and therefore $(V_0 C_x + V'_s C_s) = 0$. The unknown concentration is then found:

$$C_x = -\frac{V'_s C_s}{V_0}.$$

8.14 BASIC CELL THERMODYNAMICS

We have seen that the algebraic sign of E_{cell} tells whether the conventional cell reaction has a positive or negative tendency to proceed as written. Such a qualitative conclusion may be useful, but it is even more valuable to make a quantitative statement about the reaction tendency. Consider the cell

$$Cd \,|\, Cd^{2+} (c_1) \,\|\, Fe^{3+} (c_3),\, Fe^{2+} (c_2) \,|Pt.$$

Suppose that the concentrations are such that $E_{cell} = +0.92$ volt. This means that there is a spontaneous tendency for the conventional cell reaction to proceed, i.e., that cadmium metal will be oxidized and ferric ion will be reduced if the cell is allowed to discharge:

$$Cd + 2Fe^{3+} \rightarrow Cd^{2+} + 2Fe^{2+} \qquad \text{(conventional reaction)}.$$

The Gibbs free-energy change accompanying one stoichiometric unit of this reaction (one mole of cadmium metal reduces two moles of ferric ion) is given simply by the relationship (remembering that 1 joule = 1 volt · coulomb)

$$\Delta G = -nFE = -2 \cdot 96487 \cdot (+0.92) = -1.78 \cdot 10^5 \text{ joules.}$$

The minus sign in the expression unites two conventions: one is that a plus-value of E indicates a spontaneous reaction and the other is that a minus-value of ΔG is associated with a spontaneous change.

This calculation holds only for the conditions of c_1, c_2, and c_3 being such that $E_{cell} = +0.92$ volt. It is much more important for us to calculate the *standard* Gibbs energy change for the reaction. To do this we must first find the value of $E_{cell}^\circ = E_{right}^\circ - E_{left}^\circ$. From Appendix 5 we find:

$$Fe^{3+} + e \rightleftarrows Fe^{2+}: \qquad E_{Fe}^\circ = +0.77,$$
$$Cd^{2+} + 2e \rightleftarrows Cd: \qquad E_{Cd}^\circ = -0.40.$$

Therefore, $E_{cell}^\circ = +0.77 - (-0.40) = +1.17$ volt. The standard Gibbs energy change for this reaction is

$$\Delta G^\circ = -nFE^\circ = -2 \cdot 96487 \cdot (+1.17) = -2.6 \cdot 10^5 \text{ joules.} \qquad (8.1)$$

The special significance of ΔG° is that we may now calculate the equilibrium constant for the conventional cell reaction. From Chapter 7 we know that $\Delta G^\circ = -RT \ln (K_{eq})$, and so now we may write

$$nFE^\circ = RT \ln (K_{eq}). \qquad (8.2)$$

When the numerical values for F, R, T, and the conversion factor for changing to base-10 logs are introduced, we find two useful forms of the equation (using $T = 298.15$ K):

$$nE^\circ = 0.059 \log (K_{eq}) \qquad \text{and} \qquad K_{eq} = 10^{nE^\circ/0.059}.$$

Example Calculate the equilibrium constant for the reduction of ferric ion by cadmium metal:

$$K = 10^{2(+1.17)/0.059} = 4.6 \cdot 10^{39} = \frac{[Cd^{2+}][Fe^{2+}]^2 f_2^3}{[Fe^{3+}]^2 f_3^3}.$$

From this large value of K we conclude that the reaction would go virtually to completion in reaching equilibrium.

If an excess of cadmium metal is left in contact with a solution that initially contained $0.01M$ ferric ion, what would be the approximate equilibrium molarity of ferric ion? Assuming that the reaction is nearly complete, we may assume that the final molarity of ferrous ion is 0.01, i.e., virtually all the ferric ion has been reduced, and that the Cd^{2+} molarity is 0.005 (only one Cd^{2+} for each two ferrous ions produced). Ignoring activity coefficients for simplicity, we find, by rearrangement of the equilibrium expression

$$[Fe^{3+}]_{eq}^2 = \frac{0.005 \cdot 0.01^2}{4.6 \cdot 10^{39}},$$

that

$$[Fe^{3+}] = 1.0 \cdot 10^{-23}.$$

Given the number of particles in a mole ($6.02 \cdot 10^{23}$), the reader may well question whether this result has any statistical validity.

Values of $E°$ for Half-Reactions that Cannot Be Used in Cells

In the discussion and examples of how the Nernst equation may be used to predict cell potential, or how to find $E°$ for a couple, it was assumed that the cell would actually function reversibly, i.e., that each electrode would establish equilibrium at the electrode surface. For many redox reactions this is impossible because of kinetic barriers.

An example is the hypothetical reduction of carbon dioxide to the hydrogen oxalate ion:

$$2CO_2(g) + H^+ + 2e \rightarrow HC_2O_4^-.$$

This reaction will not proceed directly because of activation energy barriers in the mechanism. Perhaps a student of organic chemistry can propose a series of reactions for the synthesis of $HC_2O_4^-$ using CO_2 as a starting material, but this would involve several other chemical reagents and certainly less than 100% yield. The fact remains that the half-reaction above will not reach reversible equilibrium at a platinum electrode surface and therefore will not yield a meaningful potential.

Yet in the tables of standard reduction potentials we see many entries for half-reactions that are similarly irreversible. They are derived from Gibbs free-energy relationships. We have seen in Eq. (8.1) that for a reversible couple there is a simple relationship between the potential and the Gibbs energy change, hence,

$$\Delta G° = -nFE° \qquad \text{or} \qquad E° = \frac{-\Delta G°}{nF}.$$

Thus if the value for $\Delta G°$ can be calculated from the Gibbs energies of formation (taking values from published tables such as *NBS Technical Note 270*), then it is easy to calculate a value for $E°$. For reversible couples this calculation has experimental significance because the couples function in actual electrodes. For the irreversible couples the calculation is more a matter of analogy, giving a hypothetical $E°$ value that can be used in theoretical calculations but will not be observed in practice.

For the reaction above we find the following Gibbs energies of formation:

$$CO_2(gas), \quad -92,310 \text{ cal}; \qquad HC_2O_4^-, \quad -167,100 \text{ cal}.$$

The Gibbs energies of formation for H^+ and the electron (which is an abbreviation for the $H_2(gas)$, H^+ couple) are assigned zero values. Therefore, for the CO_2, $HC_2O_4^-$ half-reaction we calculate:

$$
\begin{array}{ccccc}
\Delta G^\circ & = & \Delta G^\circ & - & 2\Delta G^\circ \\
\text{reaction} & & \text{formation} & & \text{formation} \\
& & \text{for } HC_2O_4^- & & \text{for } CO_2
\end{array}
$$

$$= (-167,100) - 2(-92,310) = +17,520 \text{ cal.}$$

It follows that the hypothetical E° value for this couple is

$$E^\circ = -\frac{17,520 \text{ cal}}{2(23,060 \text{ cal/volt})} = -0.38 \text{ volt.}$$

(Note that it is necessary to use the value of the faraday in units of cal/volt when the Gibbs energy change is in calories. If the Gibbs energy were in units of joules, which are volt · coulombs, the faraday would be inserted as 96,487 coulombs.)

Some of the E° values given in Appendix 5 have been determined by galvanic-cell experiments, but many others have been deduced from the Gibbs free-energy data.

8.15 ADDITIVITY RULES FOR COMBINING EQUATIONS AND GIBBS ENERGY CHANGES

It is frequently useful to add together two or more chemical equations to derive a balanced equation for a composite reaction. In all cases the separate equations will have some species in common that cancel out upon making the addition.

The important principle to stress at this point is that *the value for the standard Gibbs energy change for the composite reaction is simply the sum of the Gibbs energy changes for the separate reactions.* For example, consider the following reactions where the C species cancels:

1. $A + B = C$ $\Delta G_1^\circ = 250$ joules
2. $C + D = E$ $\Delta G_2^\circ = -80$ joules
3. $A + B + D = E$ $\Delta G_3^\circ = 170$ joules

Since ΔG° values are additive in this way, it follows that we may also add the individual values for $\log(K)$ and nE° since, by Eqs. (8.1) and (8.2) these quantities are directly proportional to ΔG°. It is important to note that nE° is the additive quantity, not merely E° itself. Thus for the reactions symbolized above we get

$$
\begin{aligned}
\log(K_3) &= \log(K_1) + \log(K_2), \\
n_3 E_3^\circ &= n_1 E_1^\circ + n_2 E_2^\circ, \\
\Delta G_3^\circ &= \Delta G_1^\circ + \Delta G_2^\circ.
\end{aligned}
$$

Many equations, with their corresponding numerical quantities, may be combined in this way. It is important to realize that each equation in the set must have a chemical species in it that will cancel out during the addition. Otherwise the composite equation would be merely the sum of two independent equations and there would be no meaningful conclusion.

Applications of the Additivity Rules

1. ADDING TWO HALF-REACTIONS TO FIND A WHOLE REACTION

Most whole reactions formally can be regarded as the sum of two half-reactions, one being an oxidation and the other a reduction. For example, the reduction of chromic ion by zinc metal is the sum of:

an oxidation:	$Zn \rightarrow Zn^{2+} + 2e$
and a reduction:	$2(Cr^{3+} + e \rightarrow Cr^{2+})$
with the result:	$Zn + 2Cr^{3+} \rightarrow Zn^{2+} + 2Cr^{2+}.$

According to the additivity rules, we may add the $nE°$ values for these half-reactions to find the value of $nE°$ for the whole reaction. It is vital to remember that the algebraic sign of $nE°$ (or $\Delta G°$, or log K) depends on the direction in which the reaction is written. In dealing with half-reactions we will follow these simple rules: *The value for $E°$ will always be that of the standard reduction potential*, e.g., as given in Appendix 5, *but the value for n will carry a "+" sign if the half-reaction is written as a reduction (n electrons are gained) and a "−" sign in the case of oxidation (n electrons are lost).*

Thus, for the above case we have

	$E°_{red}$	n	$nE°$
$Zn \rightarrow Zn^{2+} + 2e$	−0.76	−2	+1.52
$2(Cr^{3+} + e \rightarrow Cr^{2+})$	−0.41	+2	−0.82
$Zn + 2Cr^{3+} \rightarrow Zn^{2+} + 2Cr^{2+}$			+0.70

Once the value of $nE°$ has been found for the whole reaction, it is a simple matter to calculate the value of the equilibrium constant for the reaction:

$$K = 10^{nE°/0.059} = 10^{+0.70/0.059} = 7 \cdot 10^{11}.$$

This fairly large value indicates that the reaction will go nearly to completion.

As another example consider the reduction of ferric ion by iron metal

	$E°$	n	$nE°$
$Fe \rightarrow Fe^{2+} + 2e$	−0.44	−2	+0.88
$2(Fe^{3+} + e \rightarrow Fe^{2+})$	+0.77	+2	+1.54
$Fe + 2Fe^{3+} \rightarrow 3Fe^{2+}$			+2.42

In this reaction the equilibrium constant is

$$K = 10^{2.42/0.059} = 10^{41}.$$

In the foregoing examples we added two half-reactions to find the equilibrium constant for a whole reaction that is an oxidation–reduction type. But addition of half-reactions does not always give a redox reaction. For example, suppose it is desired to calculate the value of the solubility-product constant for silver chloride, given the potential diagram for silver:

$$Ag^{2+} \xrightarrow{+2.0} Ag^+ \xrightarrow{+0.80} Ag(s)$$
$$AgCl(s) \xrightarrow{+0.22}$$

$$AgCl(s) = Ag^+ + Cl^-; \qquad K_{sp} = [Ag^+][Cl^-]f_1^2 = ?$$

The solution to the problem is to add the half reactions and nE° values as follows:

	nE°	
$AgCl(s) + e \rightarrow Ag(s) + Cl^-$	+0.22	(reduction)
$Ag(s) \rightarrow Ag^+ + e$	−0.80	(oxidation)
$AgCl(s) \rightarrow Ag^+ + Cl^-$	−0.58	

Thus a value for nE° is obtained for the reaction, which is not itself a redox reaction even though it can be viewed as a sum of two electron-transfer half-reactions.

The value for the solubility product constant is then calculated:

$$K_{sp} = 10^{(-0.58)/0.059} = 1.5 \cdot 10^{-10}.$$

2. ADDING TWO HALF-REACTIONS TO FIND A THIRD HALF-REACTION

Since bromate ion BrO_3^- is often used in applications that involve its reduction to bromide ion, it would be helpful to have a value for the standard reduction potential for the half-reaction:

$$BrO_3^- + 6H^+ + 6e \rightarrow Br^- + 3H_2O$$
$$E^\circ_{red} = ?$$

From the reduction potential diagram in Appendix 5 we find

$$BrO_3^- \xrightarrow{+1.51} Br_2 \xrightarrow{+1.07} Br^-$$

but no entry for the overall reduction of bromate to bromide. However, the desired half-reaction can be viewed as the sum of the two half-reactions for which E°_{red} values *are* listed. Writing these, with the nE° values, we find:

	E°	n	nE°
$BrO_3^- + 6H^+ + 5e \rightarrow \frac{1}{2}Br_2 + 3H_2O$	+1.51	+5	+7.55
$\frac{1}{2}Br_2 + e \rightarrow Br^-$	+1.07	+1	+1.07
$BrO_3^- + 6H^+ + 6e \rightarrow Br^- + 3H_2O$			+8.62

As shown above, the sum contains six electrons.

Once the value for nE° for the composite half-reaction has been found, it is only necessary to divide by n to find the E°_{red} value:

$$E^\circ_{red} = \frac{(nE^\circ)}{n} = \frac{+8.62}{6} = +1.44 \text{ volt.}$$

Important note   When adding half-reactions in this way, it is vital to take proper account of the number of electrons in each half-reaction. It is *not* valid to add the E° values themselves.

3. ADDING A HALF-REACTION TO A WHOLE REACTION

This application is valuable for finding the effect on a reduction potential when one or both members of the redox couple can be precipitated or complexed. For example, silver(I) is a moderately strong oxidizing agent when it is present as the aqueous ion Ag^+, as judged by the positive value for E°_{red}:

$$Ag^+ + e = Ag(s); \qquad E^\circ_{red} = +0.80 \text{ volt.}$$

Another way of stating this is to say that silver metal is a rather poor reducing agent under conditions in which its oxidation product is Ag^+.

We ask what happens to the redox properties of this couple when it is placed in a solution of potassium iodide. Silver iodide has a very low solubility, as we judge by the small value of its solubility-product constant:

$$AgI(s) = Ag^+ + I^-; \qquad K_{sp} = 10^{-16} \quad \text{or} \quad \log(K_{sp}) = -16.$$

This value for $\log(K_{sp})$ is first converted to the corresponding nE° value:

$$nE^\circ = 0.059 \log(K_{sp}) = 0.059(-16) = -0.94.$$

Now the two reactions may be added, as may the nE° values:

	nE°
$Ag^+ + e \rightarrow Ag(s)$	+0.80
$AgI(s) \rightarrow Ag^+ + I^-$	−0.94
$AgI(s) + e \rightarrow Ag(s) + I^-$	$-0.14 = E^\circ_{red}$ (since $n = 1$).

We conclude that the iodide has greatly lowered the oxidizing capability of silver(I). In fact, silver metal in the presence of iodide ion can serve as an effective reducing agent in some applications.

4. FINDING A VALUE FOR E°_{red} IN BASIC SOLUTION

For half-reactions that include hydrogen ion as reactant or product, the values for E_{red} are greatly affected by the solution pH. For work carried out in strongly basic solution, for example $1M$ NaOH, it is convenient to define standard-state conditions with the activity of OH^- being unity. For example, in acidic solution we usually regard the reduction of hydrogen(I) as follows:

$$H^+ + e = \tfrac{1}{2}H_2; \qquad E^{\circ}(\text{activity of } H^+ = 1) = 0.000.$$

When hydrogen(I) undergoes reduction in basic solution it is more realistic to ignore H^+ as a kinetic entity and to write the half-reaction as follows:

$$H_2O + e = \tfrac{1}{2}H_2 + OH^-; \qquad E^{\circ}(\text{activity of } OH^- = 1) = ?$$

To find the value for the standard potential in basic solution we must add the equation for the dissociation of water into its ions:

$$
\begin{array}{lll}
H^+ + e \rightarrow \tfrac{1}{2}H_2 & & nE^{\circ} = 0.00 \\
\underline{H_2O \rightarrow H^+ + OH^-} & \underline{\log(K_w) = -14,} & \underline{\text{so } nE^{\circ} = -0.83} \\
H_2O + e \rightarrow \tfrac{1}{2}H_2 + OH^- & & E^{\circ}_{basic} = -0.83
\end{array}
$$

In general, if a half-reaction in acid solution involves h hydrogen ions, it will be necessary to add h units of the water dissociation, which means $h \cdot (-0.83)$ for the nE° quantity, to convert to basic conditions. Typically the switch to basic conditions also causes changes in the reacting species (weak acids form their conjugate bases, some metal ions precipitate as hydroxides), and those effects must also be taken into account by adding in the appropriate nE° values as calculated from the corresponding $\log(K)$ values.

Another example: Refer to Appendix 5 to find the equilibrium diagram for bromine, showing its several oxidation states. We find the value for E° in acidic solution for the $BrO_3^-/HOBr$ couple:

$$BrO_3^- + 5H^+ + 4e = HOBr + 2H_2O; \qquad E^{\circ} = +1.49 \text{ volt.}$$

But suppose we would like to have the E° value for this couple under different standard-state conditions, when it is the activity of OH^- that is unity, rather than at unit activity of H^+. This means that we must rebalance the equation, not only showing hydroxide ions instead of hydrogen ions but also accounting for the change in chemical form of hypobromous acid, which would be dissociated to form hypobromite ions OBr^-. The half-reaction in basic solution is then

$$BrO_3^- + 2H_2O + 4e = OBr^- + 4OH^-.$$

To obtain the $E°_{basic}$ value for this half-reaction, given only the diagram data, we may use the additivity rules as follows:

	$E°$	$nE°$	$\log K$
$BrO_3^- + 5H^+ + 4e \rightarrow HOBr + 2H_2O$	$+1.49 \rightarrow$	$+5.96$	
$HOBr \rightarrow H^+ + OBr^-$		$-0.52 \leftarrow$	-8.7
$4H_2O \rightarrow 4H^+ + 4OH^-$		$-3.30 \leftarrow$	-56.0

Sum: The desired reaction $+2.14$

Since the sum of the $nE°$ values, $+2.14$, refers to a half-reaction involving four electrons, the corresponding $E°$ value is $+2.14/4 = +0.53$ volt.

In the above calculations, the pK_a for HOBr $(+8.7)$ was obtained from the diagram, changed to $\log K$, and multiplied by 0.059 to find the $nE°$ value. The well-known value for pK_w $(+14.00)$ was used to obtain the $\log K$ (-56.00) for the fourfold multiple of the water-dissociation reaction needed to make the equation balance properly.

The result $(+0.53$ volt) shows that bromate ion is not so strong an oxidizing agent in basic solution as it is in acidic solution.

8.16 SUMMARY: MAIN CONCEPTS OF GALVANIC CELLS

1. A galvanic cell must have two half-cells, each with an active redox couple and a suitable electrode. To identify the redox couples in a given cell, simply figure out which element is present in two different oxidation states at each electrode.

2. The conventional cell diagram uses a left–right concept. Either half-cell may be designated as "left," but this must be consistent with the conventional cell reaction.

3. The conventional cell reaction involves oxidation at the left electrode and reduction at the right electrode.

4. The Nernst equation for the cell
    a) has $E°_{cell} = E°_{red}(\text{right electrode}) - E°_{red}(\text{left electrode})$;
    b) uses the conventional cell reaction as the basis for the rF expression in the log term;
    c) reflects the right-electrode polarity by the sign of E_{cell}.

5. The spontaneous cell reaction corresponds to oxidation at the negative electrode.

6. There are three important types of calculations based on cells:
    a) From a known chemical composition of both half-cells, so that rF is known, and from $E°$ values found from a table one may calculate the expected value for E_{cell}.

b) Using a measured value for E_{cell}, with known chemical composition of both half-cells, one may calculate the value of E_{cell}°. This is the basis for finding values of equilibrium constants for cell reactions.

c) Using a measured value for E_{cell}, and E° values from a table, one may calculate the value of rF in the log term. If only one constituent of one half-cell is present in unknown concentration, then its concentration (or activity) may be found.

PROBLEMS

Galvanic cells

The problems in this group are actually very similar but are presented in different ways. For each problem you are asked to do several of the following operations:

a) Write a description of the cell.

b) Write the conventional cell diagram.

c) Write the half-reactions (oxidation at the left, reduction at the right) and add them properly to find the conventional cell reaction.

d) Write the Nernst equation for the cell, but without numerical substitution for the E° values or the molarities.

e) Since the cell composition is fully specified, look up E° (or $E_{reference}$) for the half-reactions and use the Nernst equation to find E_{cell}.

f) Since the value for E_{cell} is given and the composition is specified, use the Nernst equation to calculate E_{cell}°, or $E_{half\text{-}cell}^{\circ}$ if a reference electrode is being used.

g) Since E_{cell} is given, use the tables to find E° values and then solve the Nernst equation to find the unknown concentration of the species.

h) Calculate the equilibrium constant for conventional cell reaction using E° value.

i) State which electrode is more positive and whether the spontaneous cell reaction is the same as the conventional reaction or not.

1. A cell is constructed with a cadmium-metal electrode in a solution of $0.010M$ cadmium nitrate and a zinc-metal electrode in a solution of $0.050M$ zinc nitrate. Do: (b), (c), (d), (e), (h), (i).

2. Given the cell diagram: $Pt \,|\, H_2(0.9\ atm),\ HClO_4(0.2M) \,\left\|\begin{array}{l} AgNO_3(0.05M) \\ KNO_3(0.15M) \end{array}\right.\,|\ Ag.$

    Note that the ionic strength is the same on each side. Do: (a), (c), (d), (e), (h), (i).

3. We wish to design a cell that has the following as its conventional reaction:

$$Cl_2(g) + 2Br^- = 2Cl^- + Br_2(aq).$$

    Do: (a), (b), (d).

4. For a certain galvanic cell the Nernst equation may be written as follows:

$$E = E^{\circ} - \frac{0.059}{2} \log \frac{[Fe^{2+}]^2[Pb^{2+}]f_2^3}{[Fe^{3+}]^2 X_{Pb} f_3^2}.$$

    Do: (a), (b), (c).

5. The potential for the following cell is $+1.58$ volt:

$$Pt\,|\,H_2(0.9\,atm),\, \begin{array}{c} HClO_4(xM) \\ NaClO_4(0.02-x) \end{array} \,\|\, HCl(0.02M),\, Cl_2(0.9\,atm)\,|\,Pt.$$

Taking the ionic strength to be 0.02 on each side, do: (a), (c), (d), (g).

6. A cell is prepared with a saturated calomel electrode on the left and, on the right, a platinum wire dipping into a solution containing $0.01M$ I_3^- and $0.1M$ I^-. Do: (b), (c), (d), (e).

7. A cell is prepared with two copper-metal electrodes. One dips into a solution of $0.015M$ Cu^{2+} and the other dips into a solution of unknown Cu^{2+} molarity. The cell potential is 0.0200 volt, with the electrode in the unknown solution being more positive. Ignoring activity coefficients for simplicity, do: (b), (c), (d), (g).

8. Given the cell diagram

$$Hg(l)\,|\,Hg_2^{2+}(0.001M)\,\|\,Hg^{2+}(0.002M),\, Hg_2^{2+}(0.1M)\,|\,Pt,$$

ignore activity coefficients for simplicity and do: (c), (d), (e), (h), (i). (*Hint*: The mercurous ion is present in different concentrations. Keep them separate.)

9. A cell is prepared with the left electrode being a silver wire coated with solid silver bromide and dipping into a solution of $0.040M$ sodium bromide. The right electrode is a silver wire coated with silver chloride and dipping into $0.040M$ sodium chloride. Do: (b), (c), (d), (e), (i).

10. We wish to construct a cell that has the following conventional reaction:

$$Zn(s) + 2VO^{2+}(0.03M) + 4H^+(0.5M) = Zn^{2+}(0.1M) + 2V^{3+}(0.02M) + 2H_2O.$$

Do: (b), (c), (d), (e), (h), (i).

The additivity rules
For each of the problems in this group consult Appendix 5 for standard reduction potentials.

11. Calculate $E°$ for the phosphoric acid/phosphine couple.

12. Calculate the $E°$ value for the H_3AsO_4/As couple in acid solution.

13. Calculate the $E°$ value for the half-reaction: $Fe^{3+} + 3e = Fe(s)$.

14. Calculate the equilibrium constant for the disproportionation of hypobromous acid:

$$HOBr = BrO_3^- + Br_2.$$

(Balance this first.)

15. The solubility product for lead sulfate is $1.6 \cdot 10^{-8}$. Calculate the $E°$ value for the half-reaction:
$$PbSO_4(s) + 2e = Pb(s) + SO_4^{2-}.$$

16. The $E°$ value for the $AgBr(s)/Ag(s)$ couple is often given as $+0.095$ volt. Is this value consistent with the solubility-product constant for silver bromide, $K_{sp} = 5 \cdot 10^{-13}$, and the $E°$ value for the Ag^+/Ag couple?

17. Calculate the standard reduction potential for the silver iodide/silver metal couple, given the solubility product $K_{sp} = 7.9 \cdot 10^{-17}$ for AgI.

18. Calculate the equilibrium constant for the decomposition of hydrogen peroxide:

$$H_2O_2 = \tfrac{1}{2}O_2 + H_2O.$$

19. Calculate the equilibrium constant for the reduction of vanadium(III) by metallic zinc:

$$2V^{3+} + Zn = 2V^{2+} + Zn^{2+}.$$

If an excess of zinc metal is stirred with a solution that initially contains $0.01M$ V^{3+}, what will be the equilibrium molarity of V^{3+}? Ignore activity coefficients.

20. Will cadmium metal be useful as a reagent to reduce uranium(IV) to U(III)?

21. Some tabulated Gibbs free energies of formation (in calories) are as follows:

Pb(s)	Pb²⁺(aq)	I⁻(aq)	PbI₂(s)	PbI₂(aq)
0	−5,830	−12,330	−41,500	−34,300

a) Calculate $E°$ for the Pb^{2+}/Pb couple and compare the result with that in Appendix 5.
b) Calculate the solubility product for lead iodide and compare with the value in Appendix 5.
c) Calculate the $E°$ value for the half-reaction: $PbI_2(s) + 2e = Pb + 2I^-$.
d) Calculate the molarity of PbI_2 molecules in a solution saturated with solid PbI_2. This is the intrinsic solubility, the value for the equilibrium constant for the reaction $PbI_2(s) = PbI_2(aq)$.

22. Some standard Gibbs energies of formation relating to the element gallium are as follows (in calories/mol):

Ga(s)	Ga²⁺	GaF₃(s)	F⁻
0	−21,000	−259,400	−66,640

Also, we find the standard reduction potential:

$$Ga^{3+} \xrightarrow{\;-0.55\;} Ga(s).$$

a) Calculate $E°$ for the Ga^{3+}/Ga^{2+} couple.
b) Calculate the solubility-product constant for gallic fluoride.
c) If metallic gallium were placed in contact with a Ga^{3+} solution, would you expect formation of Ga^{2+} to be extensive?

23. Ethyl alcohol can be oxidized to acetic acid in aqueous solution. Calculate the oxidation potential for this change. Also, show that further oxidation of acetic acid to carbon dioxide is considerably more difficult, in thermodynamic terms. The following Gibbs energies of formation are needed (in calories/mol):

H₂O(liquid)	C₂H₅OH(aq)	CH₃COOH(aq)	CO₂(aq)
−56,687	−43,440	−94,780	−92,260

The use of potentiometric data

24. The following cell is to be used for the potentiometric determination of the concentration of lead ion in the right compartment:

$$SCE \parallel \text{Solution containing } Pb^{2+} \mid Pb(s).$$

a) When the solution is $1.00 \cdot 10^{-3}M$ $Pb(NO_3)_2$ in $7 \cdot 10^{-3}M$ HNO_3, what would you predict for E_{cell}?

b) An unknown solution containing lead ion is placed in the compartment, and the cell potential is found to be -0.523 volt. Assuming that the ionic strength is 0.01, what is the molarity of Pb^{2+}?

c) A saturated solution of lead fluoride PbF_2 in $0.0100M$ sodium fluoride is placed in the cell. The observed potential is -0.486 volt. Calculate the value for the solubility-product constant of lead fluoride.

25. The following cell has a potential of $+0.370$ volt.

$$Hg(\text{liquid}) \mid Hg_2Cl_2(s), KCl(0.10M) \left\| \begin{array}{c} Fe^{3+}(1.2 \cdot 10^{-3}M) \\ Fe^{2+}(\text{unknown}) \\ I = 0.1 \end{array} \right| Pt.$$

Consult Appendix 5 for the potential of the reference electrode. Calculate the molarity of the ferrous ion in the solution.

26. Consider the following cell:

$$Ag(s) \mid AgBr(s), KBr(\text{variable}) \mid SCE.$$

Would it be possible to make a solution of potassium bromide so that the cell would have zero potential?

27. The potential of the following cell is -0.549 volt.

$$\text{Reference electrode} \mid HCl(0.0500M), H_2(0.92 \text{ atm}) \mid Pt.$$

a) What is the reduction potential for the reference electrode?

b) If the solution in the right compartment is replaced with a certain buffer, the cell potential is observed to be -0.834 volt. What is the activity of H^+ in the buffer solution, assuming the hydrogen pressure stays at 0.92 atm?

28. A solution containing an unknown molarity of silver ion is to be examined by the method of standard additions. A 100-mL portion is placed in the cell:

$$\text{Reference electrode} \mid \text{Solution of } Ag^+ \mid Ag$$

and the potential is found to be $+0.325$ volt. Small increments of $1.00M$ silver nitrate are added, with the following results:

mL added:	0.0200	0.0400	0.0600
E observed:	0.347	0.359	0.367

Make a standard additions plot and determine the molarity of Ag^+ in the original solution. Assume that the ionic strength is constant and that the silver electrode responds according to theory, with the log term coefficient of 0.0592.

29. When a solution of potassium iodide is allowed to stand for a few days there is usually some yellow color due to air oxidation:

$$I^- + H^+ + O_2(0.21\ atm) = H_2O + I_3^-.$$

   a) Balance the equation and draw a diagram for a cell having this as its conventional reaction. Use reasonable concentrations of constituents.
   b) Write the Nernst equation for the cell and rearrange it to show how, in principle, the potential of the cell may be used to indicate the molarity of triiodide ion in the solution.

30. When an electrode is used for direct potentiometric measurement of a molarity, we depend on a rearrangement of the Nernst equation as follows: $[x] = 10^{(E-k)/S}$, where k and S are constants for a given cell system. Show, for the case of $S = 0.0592$, what percentage error in the calculated molarity of x is caused by a two-millivolt error in the determination of the cell potential E.

Case study
31. The standard potential for the reduction of aqueous bromine to bromide ion

$$Br_2(aq) + 2e = 2Br^-$$

was determined over 60 years ago by G. N. Lewis and H. Storch (*Journal of the American Chemical Society*, *39*, 2544, 1917). Although this research antedated the Debye–Hückel theory of ionic activity coefficients (1923), the authors successfully treated the nonideal behavior of ions in terms of an apparent incomplete dissociation of HBr, which we now recognize as a strong electrolyte. In the following problems we will review some of the essential ideas of their work.

A. The goal is to find the $E°$ value for the galvanic cell:

$$Pt\,|\,H_2(P\ atm),\ HBr\ (C_1)\,\|\,HBr\ (C_1),\ Br_2\ (C_2)\,|\,Pt.$$

   a) For each half-cell identify the redox couple and, using the appropriate half-reactions, write the conventional reaction for the cell.
   b) If the measured potential for the cell is +1.2 volt, what chemical change would occur if the two platinum electrodes were connected by a copper wire? Explain your answer in terms of electron flow.
   c) Write the Nernst equation for the cell, including activity coefficients.

B. For the preparation of the cell solutions it was necessary to have an accurately standardized solution of hydrobromic acid. A good method is to use a gravimetric determination of bromide via precipitation of silver bromide with a K_{sp} value of $5 \cdot 10^{-13}$.

   a) Without going into detail, list the steps that would be followed in the gravimetric determination, using a solution of silver nitrate as precipitant.
   b) Briefly explain why the initial precipitate is colloidal and why the primary adsorption layer is different at the end of the precipitation.
   c) If an excess of precipitant is added so that the final molarity of Ag^+ is 0.001, what is the equilibrium molarity of unprecipitated bromide ion in the solution? Ignore activity coefficients.

d) If a 10.02-mL portion of the HBr solution yielded 0.2623 gram of dried AgBr (MW = 187.77), what is the precise molarity of the acid? (Find the correct number of significant digits.)

C. It is also necessary to have a standardized solution of aqueous bromine. A stock solution was prepared by dissolving an approximate quantity of purified liquid bromine in redistilled water. A 19.98-mL portion was pipetted directly into a titration flask containing 100 mL of 0.5M potassium iodide, and the bromine was rapidly reduced:

$$Br_2(aq) + 3I^- = I_3^- + 2Br^-.$$

Titration with a standard 0.08223M solution of arsenic(III) required 30.81 mL to reach the endpoint. The titration reaction is

$$I_3^- + As(III) = 3I^- + As(V).$$

a) Describe the color change in the titration, with and without the appropriate indicator present.

b) Calculate the precise molarity of the aqueous bromine solution.

D. When the solution is prepared for the right half-cell it is merely necessary to dilute the chosen volumes of the standardized HBr and Br$_2$ solutions. However, there is a complication that must be considered. In the Nernst equation it is essential to use the actual equilibrium molarities of Br$^-$ and Br$_2$(aq), but these will not be the same as the initial concentrations because of the complexation equilibrium that forms tribromide ion:

$$Br^- + Br_2(aq) = Br_3^-; \qquad K = 16.$$

If a solution is prepared having initial concentrations of 0.0300M HBr and 0.0100M Br$_2$, what will the equilibrium molarities be for the three species?

E. In one experiment Lewis and Storch used initial concentrations of 0.100M HBr and 0.03021M Br$_2$ and calculated the equilibrium molarities to be

$$[Br^-] = 0.0827; \qquad [Br_2] = 0.01291.$$

With the barometric pressure at 755.5 mm of Hg and a temperature of 25° (at which the vapor pressure of water is 23.8 mm of Hg), they found the cell potential to be +1.1639 volt. Using 0.05915 as the constant in the Nernst equation and taking the activity coefficients for singly charged ions to be 0.781, calculate the value for $E°$ for the Br$_2$(aq)/Br$^-$ redox couple.

F. The equilibrium constant for the disproportionation of aqueous bromine has been determined:

$$Br_2(aq) + H_2O = \underset{\substack{\text{hypobromous}\\\text{acid}}}{HOBr} + H^+ + Br^-; \qquad K = 2.5 \cdot 10^{-9}.$$

a) Calculate the value of Q for this reaction for the conditions in the cell of part E.

b) For the H^+ and Br^- molarities present in the cell, determine whether the concentration of hypobromous acid formed would be negligible compared with that of Br_2.

c) Using your $E°$ value from Problem 5 and the above K-value for the disproportionation reaction, calculate the standard reduction potential for the following half-reaction:

$$HOBr + H^+ + 2e = Br^- + H_2O.$$

ACID–BASE EQUILIBRIA AND pH CALCULATIONS

9

Liquor fixus and spiritus acidus nitri are in their nature unlike, foes and adversaries of each other, and when the two are brought together, and the one part has overcome and killed the other, neither a fiery liquor nor a spiritus acidus can be found in their dead bodies, but the same has again been made as both were before and from which they were derived, namely, ordinary saltpeter.

Rudolph Glauber, 17th century

$$KOH + HNO_3 = KNO_3 + H_2O$$

(20th-century translation)

In comparison with the other main categories of solution equilibria (complex formation, solubility, oxidation–reduction), acid–base systems have few complications. Therefore, reliable acid–base calculations may be done with rigorous models and with high precision. Because solution pH is a very important variable in experimental chemistry, biology, and geology, the control of pH by acids and bases is a truly fundamental topic.

9.1 REVIEW OF DEFINITIONS

Water Autoprotolysis

A discussion of acid–base behavior must start with consideration of the solvent itself. In pure water, or in a solution of a solute that has no acid–base properties (e.g., NaCl), the ions of water are produced in small, equal concentrations by the reaction

$$H_2O + H_2O \quad = \quad H_3O^+ \quad + \quad OH^-. \tag{9.1}$$
$$\text{hydronium} \qquad\qquad \text{hydroxide}$$
$$\text{ion} \qquad\qquad\qquad \text{ion}$$

In pure water the equilibrium molarities of these ions are only $1.00 \cdot 10^{-7}$, and at this low ionic strength the activity coefficients are 1.00. Therefore the *equilibrium constant* for the reaction may be written as

$$K_w = [H_3O^+][OH^-]\, f_i^2 = (1 \cdot 10^{-7})^2 1^2 = 1.00 \cdot 10^{-14}. \tag{9.2}$$

Note that the mole fraction of water, X_{H_2O}, is omitted from the denominator by convention, for it remains essentially constant at a value of 1.0.

If the water contains a solute such as sodium chloride, which dissociates into inert ions that change the ionic strength but not the equality of the water-ion concentrations, then reaction (9.1) still describes the acid–base model correctly, but it will proceed to a greater extent. For example, consider a solution of $0.030M$ calcium chloride, a completely dissociated electrolyte:

$$CaCl_2 \overset{100\%}{=} Ca^{2+} + 2Cl^- \quad \text{(inert ions)}.$$

The ionic strength of this solution is 0.090 and therefore the activity coefficient for singly charged ions is 0.79. By applying Eq. (9.2) we find

$$K_w = 1.00 \cdot 10^{-14} = [H_3O^+][OH^-](0.79)^2;$$

and, since the water ions are present in equal amounts, we get

$$[H_3O^+] = [OH^-] = Q_w^{1/2} = \left(\frac{1.00 \cdot 10^{-14}}{0.79^2}\right)^{1/2} = 1.3 \cdot 10^{-7}.$$

The inert solute has, through the effect of ionic strength, caused a 30% increase in the hydronium-ion molarity. It is worth noting, however, that the *activity* of H_3O^+ stays unchanged and equal to $1.00 \cdot 10^{-7}$. Therefore, the equilibrium constant is always $1.00 \cdot 10^{-14}$, but the equilibrium *quotient* in terms of molarities

$$Q_w = [H_3O^+][OH^-],$$

will depend on the ionic strength. In the above example, Q_w is 70% greater than K_w. The relationship between K_w and Q_w will be used frequently:

$$K_w = Q_w f_1^2 \qquad \text{or} \qquad Q_w = K_w / f_1^2.$$

The pH Notation

Any numerical quantity X may of course be converted to its logarithmic value $\log X$. The notation pX simply refers to the *negative* of the logarithm:

$$pX = -\log X.$$

When this notation is applied to the hydronium ion there are two important quantities. One is the negative log of the hydronium-ion *molarity*:

$$pcH = -\log [H_3O^+],$$

where the "c" is a reminder that concentration is used. The other quantity uses the *activity*:

$$paH = -\log ([H_3O^+]f_H),$$

where the "a" for activity is usually omitted and just pH is written.

In the example above of $0.030M$ $CaCl_2$, the pH of the solution is 7.00 and the pcH is 6.90.

The "p" notation is also used for numerical values of equilibrium constants and quotients. Thus

$$pK_w = -\log K_w = -\log(1.00 \cdot 10^{-14}) = -(-14.00) = 14.00$$

and

$$pQ_w = -\log Q_w = 13.80 \qquad \text{(for the above example having } I = 0.090\text{)}.$$

Acids and Acid Dissociation Constants

The equality of the water ion molarities is destroyed if the water contains a solute that transfers protons to the solvent molecules,

$$HA + H_2O = H_3O^+ + A \qquad \text{(ignoring charges on HA and A),} \qquad (9.3)$$

where HA is the formula symbol for any monoprotic acid and A is its conjugate base. Two species, HA and A, that differ only by a proton are called a **conjugate pair**.

The equilibrium constant for reaction (9.3) is called the **acid dissociation constant**. It is written as follows:

$$K_a = \frac{[H_3O^+][A]}{[HA]} \cdot \frac{f_H f_A}{f_{HA}}. \qquad (9.4)$$

The corresponding equilibrium quotient, in terms of molarities, is

$$Q_a = \frac{[H_3O^+][A]}{[HA]}. \tag{9.5}$$

Of course, the ionic charge on the conjugate base species will always be one unit less than the charge on the conjugate acid species, due to the loss of the positive proton. Some examples of different charge types are given in Table 9.1. (See Appendix 4 for tables of pK values.)

Table 9.1

Charge on acid species	Acid formula and name	Conjugate base formula and name	Value of pK_a
$+3$	$Al(H_2O)_6^{3+}$ aluminum ion	$AlOH(H_2O)_5^{2+}$ hydroxyaluminum ion	5.0
$+2$	$^+H_3NCH_2CH_2NH_3^+$	$H_2NCH_2CH_2NH_3^+$ ethylenediammonium ion	7.1
$+1$	NH_4^+ ammonium ion	NH_3 ammonia	9.2
0	$HCOOH$ formic acid	$HCOO^-$ formate ion	3.8
-1	HSO_4^- bisulfate ion	SO_4^{2-} sulfate ion	2.0
-2	HPO_4^{2-} monohydrogen-phosphate ion	PO_4^{3-} phosphate ion	12.3

Bases and Base Dissociation Constants

In contrast to *acids*, which upset the balance of water ions by *donating* protons to the solvent molecules, *bases* are substances that *accept* protons from the solvent molecules.

The terms "donor" and "acceptor" are in a sense misnomers. The acid's donation is a forced one and the base's acceptance is rather active. It is like saying that a man "accepts" a stone which the ground "gives" him as he lifts it. More appropriate terms might be "proton loser" and "proton winner." For the base, "proton snatcher" would be more descriptive and striking.*

The proton-transfer reaction between base and water is as follows:

$$A + H_2O = HA + OH^-, \tag{9.6}$$

*T. Hazlehurst, *Journal of Chemical Education*, 17, 466 (1940).

where changes on HA and A are again omitted and the **base dissociation constant** is the equilibrium constant for this reaction:

$$K_b = \frac{[HA][OH^-]}{[A]} \cdot \frac{f_{HA}\, f_{OH}}{f_A}. \tag{9.7}$$

Note that K_b is *not* the reciprocal of K_a.

In terms of molarities the **base dissociation quotient** is

$$Q_b = \frac{[HA][OH^-]}{[A]}. \tag{9.8}$$

The Conjugate Relationship

A very useful relationship exists between the numerical values of K_a and K_b for a given conjugate pair. It is quickly derived by multiplying Eq. (9.4) by Eq. (9.7):

$$K_a K_b = \frac{[H_3O^+][A]\, f_H f_A}{[HA]\, f_{HA}} \cdot \frac{[HA][OH^-]\, f_{HA}\, f_{OH}}{[A]\, f_A}.$$

We note that all the terms involving HA and A can be cancelled; the result is:

$$K_a K_b = [H_3O^+][OH^-]\, f_H f_{OH} = K_w = 1.00 \cdot 10^{-14} \quad \text{(at 25°C)}. \tag{9.11}$$

In logarithmic form, this becomes

$$pK_a + pK_b = pK_w = 14.00, \tag{9.9}$$

and so it is quite easy to calculate a pK_b value for a base, given the pK_a value for its conjugate acid. For example, at 25°C acetic acid has a pK_a value of 4.756. Therefore, pK_b for the acetate ion (the conjugate base of acetic acid) is

$$pK_b = 14.000 - 4.756 = 9.244.$$

The "strength" of an acid or base refers to the extent to which it undergoes proton transfer with the solvent, by reaction (9.3) or (9.6). A **strong acid** such as HCl is virtually 100% dissociated, while a typical **weak acid** such as acetic acid will be only about 1% dissociated according to reaction (9.3). Of course, there is a correlation between strength and pK value, as shown in Table 9.2, which indicates the continuous spectrum of acid and base strengths, with a few specific examples at intervals. Note that as the strength of an acid (or base) increases, the strength of its conjugate species decreases. In the middle region of the spectrum it is correct to say that both members of the conjugate pair are weakly dissociated in water.

As a general principle, we may say that any acid in the spectrum will react (i.e., transfer its proton) with any base that is above it in the chart. The farther apart the acid and base are in Table 9.2 the greater the extent to which the transfer will occur.

Table 9.2

Acid class	Acid	pK_a	pK_b	Base	Base class
	K^+	20	−6	KOH	
Nonacids		19	−5		Strong bases
		18	−4		
		17	−3		
		16	−2		
		15	−1		
	H_2O	14	0	OH^-	
	HS^-	13	1	S^{2-}	
Feeble acids		12	2		Moderately strong bases
		11	3		
		10	4		
	NH_4^+	9	5	NH_3	
		8	6		
Weak acids	$H_2PO_4^-$	7	7	HPO_4^{2-}	Weak bases
		6	8		
		5	9		
	HOAc	4	10	OAc^-	
		3	11		
Moderately strong acids	HSO_4^-	2	12	SO_4^{2-}	Feeble bases
		1	13		
		0	14		
	H_3O^+	−1	15		
		−2	16	H_2O	
		−3	17		
		−4	18		
Strong acids		−5	19		Nonbases
		−6	20		
	HCl	−7	21	Cl^-	
		−8	22		
	HI	−9	23	I^-	
		−10	24		
		pK_a	pK_b		

On Using Appendix 4 of pK_a Values

In the sections that follow we will illustrate a variety of calculations involving acid–base reactions and their equilibrium states. These require knowledge of (1) whether a given species is an acid, a base, or simply an inert substance, (2) the numerical values for acid and/or base dissociation constants, and (3) the amounts of all species present in the mixture.

Appendix 4 certainly is less than comprehensive, but it does contain information on most of the acid–base species commonly encountered in chemical analysis. For each entry the formulas, names, and pK_a values are given. For example, we find:

HCl   <u>strong</u> Cl^-
hydrochloric acid   chloride ion,

HOCl   <u>7.5</u> OCl^-
hypochlorous acid   hypochlorite ion.

Interpretation   The conjugate base of hydrochloric acid (hydrogen chloride) is the chloride ion. HCl is a strong acid, meaning that in dilute aqueous solutions it is *completely* dissociated into its ions H^+ and Cl^-. This means that in dilute aqueous solutions Cl^- has *no* tendency to accept a proton to form molecular HCl. Thus we often refer to the conjugate bases of strong acids as **inert ions** without appreciable basic properties.

Hypochlorous acid has the hypochlorite ion as its conjugate base. The number above the connecting line is the pK_a value for the acid. In this case the meaning is as follows. Hypochlorous acid dissociates by reaction with water:

$$HOCl + H_2O = H_3O^+ + OCl^-,$$

which we usually abbreviate as

$$HOCl = H^+ + OCl^-.$$

The equilibrium constant for this dissociation reaction is expressed as K_a, which is numerically equal to 10^{-pK_a}:

$$K_a = \frac{[H^+][OCl^-]f_1^2}{[HOCl]} = 10^{-7.5} = 3.2 \cdot 10^{-8}.$$

The fact that K_a is so small, tells us that hypochlorous acid is quite weak, having little tendency to dissociate in water. We also may derive information about the basic properties of the OCl^- ion. From the conjugate relationship we obtain:

$$pK_b = 14.00 - pK_a = 6.5; \qquad K_b = 10^{-6.5} = 3.2 \cdot 10^{-7}.$$

This means that there is only a slight tendency for the conjugate base to react with water by taking a proton:

$$OCl^- + H_2O = HOCl + OH^-; \qquad K_b = \frac{[HOCl][OH^-]}{[OCl^-]} = 3.2 \cdot 10^{-7}.$$

Thus the entry in the table gives full information about the acid–base properties of the conjugate pair.

At this point the reader should refer to Chapter 2 on aqueous solution stoichiometry and review the listing of strong electrolytes. In carrying out equilibrium calculations on acid–base systems it is vital to recognize what species are present in the ionic form. In particular the following should be remembered:

a) Sodium, potassium, and ammonium salts are completely dissociated.

b) Perchlorate and nitrate salts are completely dissociated.

c) The common strong acids are: perchloric $HClO_4$, hydrochloric HCl, nitric HNO_3, hydrobromic HBr hydroiodic HI (also, hydriodic),sulfuric H_2SO_4 (first step).

d) The common strong bases are: sodium hydroxide NaOH, potassium hydroxide KOH.

9.2 SOLUTIONS CONTAINING ONE ACID

Derivation of Equations

A chemist may add an acid to an aqueous solution for the direct purpose of increasing the molarity of hydronium ion, thus decreasing the pH. Or the chemist may already have a solution containing a known amount of an acid and may need to calculate the value of the existing pH. In either case it is necessary to understand the fundamental relationship that will now be derived.

To keep the model simple, consider a solution of C moles per liter of the acid HA, and let its acid dissociation quotient be Q_a.

The proton transfer reaction with water produces equal concentrations of H_3O^+ and the conjugate base A. **The symbols $[H^+]$ and $[A]$ will stand for the concentrations of the hydronium ion H_3O^+ and the conjugate base A, respectively.** Therefore, in this solution we have:

$$[H^+] = [A].$$

It is realistic to ignore the small amount of H_3O^+ contributed by dissociation of water. The concentration of undissociated acid is therefore:

$$[HA] = C - [A] = C - [H^+].$$

By substitution for $[A]$ and $[HA]$ in Eq. (9.5) for Q_a, a quadratic equation is obtained:

$$Q_a = \frac{[H^+][A]}{[HA]} = \frac{[H^+]^2}{C - [H^+]} \quad \text{or} \quad [H^+]^2 + Q_a[H^+] - CQ_a = 0. \quad (9.10)$$

Given a value for C and a value for Q_a, the quadratic formula readily yields a value for the corresponding equilibrium molarity of hydronium ion:

$$[H^+] = \frac{-Q_a + (Q_a^2 + 4CQ_a)^{1/2}}{2}. \quad (9.11)$$

The advantage of Eq. (9.11) over the simplified form, which will be shown below for a weak acid, is that it immediately gives accurate answers for all cases when both C and Q_a are known. However, what if the solute is a strong acid, such as HCl, for which Q_a is very large and not accurately known? Nothing could be simpler because, if dissociation is complete, then

$$[H^+] = [A] = C.$$

Instead of solving the quadratic expression in Eq. (9.10) it is often sufficient to obtain an *approximate* result for $[H^+]$ provided that the dissolved acid is so weak (i.e., Q_a is smaller than about 10^{-4}) that the degree of dissociation is small. In that case,

$$[HA] \approx C$$

and

$$Q_a \approx [H^+]^2/C \quad \text{or} \quad \boxed{[H^+] \approx (CQ_a)^{1/2}.} \tag{9.12}$$

Examples of problems with one acid solute

1. Calculate the pH of a solution that contains $0.010M$ phenol and $0.05M$ sodium chloride, taking 10.00 as the value for pK_a for phenol.

Solution The NaCl is merely an inert electrolyte which establishes the ionic strength 0.05. The value for Q_a can be calculated first:

$$K_a = 1.00 \cdot 10^{-10} = \frac{[H^+][C_6H_5O^-]}{[C_6H_5OH]} \cdot \frac{f_1^2}{f_0}.$$

We assume that the activity coefficient for the neutral molecule, phenol, is 1.00. From the Davies equation we calculate that $f_1 = 0.82$ at ionic strength 0.05, and therefore,

$$Q_a = \frac{K_a}{f_1^2} = \frac{1.00 \cdot 10^{-10}}{0.82^2} = 1.5 \cdot 10^{-10}.$$

Since phenol is very weak, it is appropriate to use the approximate Eq. (9.12)

$$[H^+] \approx (0.010 \cdot 1.5 \cdot 10^{-10})^{1/2} = 1.2 \cdot 10^{-6}.$$

This result immediately justifies the approximation that the degree of dissociation is negligible. The quadratic equation would give the same answer in this case.

The activity of hydronium ion and the pH are now calculated:

$$a_H = [H^+]f_1 = 1.2 \cdot 10^{-6} \cdot 0.82 = 1.0 \cdot 10^{-6}, \qquad pH = -\log(1.0 \cdot 10^{-6}) = 6.00.$$

2. How much nitric acid must be added to 20 mmol of potassium nitrate if, upon dilution to one liter, the resulting solution is desired to have a pcH of 2.37? What will be the value of pH? Nitric acid is a strong acid.

Solution Given pcH = 2.37, $[H^+] = 10^{-2.37} = 0.0043$ mol/L. Therefore, 4.3 mmol of nitric acid must be added. The ionic strength of the resulting solution will be

$$I = 0.020 \text{ (due to } KNO_3) + 0.0043 \text{ (due to } HNO_3) = 0.0243.$$

At this ionic strength, $f_1 = 0.86$. Therefore, $a_H = 0.0043 \cdot 0.86 = 0.0037$ and pH = 2.43.

3. Calculate the pH of a solution of $0.010M$ salicylic acid (HSal), given that $K_a = 1.0 \cdot 10^{-3}$.

Solution The dissociation constant is large enough to require use of the quadratic equation. However, a problem is that the necessary value of Q_a is to be obtained by taking ionic strength into account, and the statement of the problem gives no clue for the value of ionic strength. It is necessary to use successive approximations: first use the approximate Eq. (9.12) with K_a in place of Q_a, to get an approximate value for $[H^+]$:

$$[H^+] \approx (0.01 \cdot 1.0 \cdot 10^{-3})^{1/2} = 0.0032.$$

Because the dissociation of salicylic acid is the only source of ions in this solution (ignoring the infinitesimal dissociation of water), this calculation serves as an approximation for ionic strength. At $I = 0.0032$, $f_1 = 0.94$, and therefore

$$Q_a = K_a/f_1^2 = 0.0010/0.94^2 = 1.13 \cdot 10^{-3}.$$

Now this value can be used in the quadratic expression of Eq. (9.11):

$$[H^+] = \frac{-0.00113 + (0.00113^2 + 4 \cdot 0.01 \cdot 0.00113)^{1/2}}{2} = 0.0028.$$

This corresponds to $pH = 2.55$. If a more accurate answer is desired, the value of $[H^+] = 0.0028$ could be taken as a second approximation for the ionic strength, a refined Q_a value obtained, and the quadratic equation solved once again.

The degree of dissociation for salicylic acid in this solution is seen to be

$$\frac{[Sal^-]}{C} = \frac{[H^+]}{C} = \frac{0.0028}{0.0100} = 0.28 \qquad \text{or} \qquad 28\%.$$

4. What must be the concentration of acetic acid in water if the value of $[H^+]$ is to be about the same as that in a solution of $0.0012M$ hydrochloric acid? The ionic strength is 0.0012, and so $f_1 = 0.96$.

Solution   Hydrochloric acid is completely dissociated, so $[H^+] = 0.0012$. The dissociation constant for acetic acid is

$$K_a = 10^{-4.76} = 1.74 \cdot 10^{-5}$$

and

$$Q_a = K_a/f_1^2 = 1.89 \cdot 10^{-5}$$

So

$$1.89 \cdot 10^{-5} = \frac{[H^+]^2}{C} = \frac{0.0012^2}{C}.$$

Therefore, solving for C,

$$C = \frac{0.0012^2}{1.89 \cdot 10^{-3}} = 0.076 \text{ mol/L}$$

5. What is the Q_a value for an acid if its 1.0 molar solution in water has a pH of 6? Assume $f_1 = 1.00$ at this low ionic strength.

Solution   If $[H^+] = 1.0 \cdot 10^{-6}$ and $[HA] = 1.0$, it follows that

$$Q_a = \frac{(1 \cdot 10^{-6})^2}{1} = 1 \cdot 10^{-12}.$$

6. If a certain acid solution has a pH of 4.0, what is the contribution of $[H^+]$ due to the dissociation of the solvent, water? Assume $f_1 = 1.00$.

Solution   The water dissociation is shown by the hydroxide-ion molarity, since there is no other source of OH^-:

$$[OH^-] \approx \frac{K_w}{[H^+]} = \frac{1 \cdot 10^{-14}}{1 \cdot 10^{-4}} = 1 \cdot 10^{-10}.$$

9.3   SOLUTIONS CONTAINING ONE BASE

Derivations of Equations

There is an attractive symmetry between the Brönsted definitions of acid and base, and the expressions for K_w, K_a, and K_b. The conjugate relationships make the algebraic treatment of solutions of bases exactly analogous to that already discussed for acids.

Consider a solution of C moles per liter of a base A, and let its base dissociation quotient be Q_b.

The proton-transfer reaction with water produces equal concentrations of OH^- and the conjugate acid HA. Therefore,

$$[OH^-] = [HA] \quad \text{(ignoring the contribution from water).}$$

The concentration of conjugate base that remains as the base species is therefore

$$[A] = C - [HA] = C - [OH^-].$$

By substitution for $[HA]$ and $[A]$ in the expression for Q_b in Eq. (9.8), a quadratic equation is obtained:

$$Q_b = \frac{[OH^-][HA]}{[A]} = \frac{[OH^-]^2}{C - [OH^-]} \quad \text{or} \quad [OH^-]^2 + Q_b[OH^-] - CQ_b = 0. \quad (9.13)$$

As with acid solutions, the quadratic formula may be used to solve the equation for solutions of bases when both C and Q_b are known. However, for solutions of strong bases, such as NaOH and KOH, the hydroxide-ion molarity is simply equal to the base concentration:

$$[OH^-] = C.$$

Also in analogy to the acid solutions, an approximate result for weak bases may be obtained if the degree of dissociation of the base is quite small:

$$[\mathrm{OH}^-] \approx (CQ_b)^{1/2}. \qquad (9.14)$$

Examples of problems with one base solute

1. Calculate the pH of $0.005M$ sodium hydroxide.

Solution   NaOH is a strong base, completely dissociated into OH^- and the hydrated sodium ion. Therefore,

$$[\mathrm{OH}^-] = C = 0.005.$$

Since NaOH is a 1:1 electrolyte, the molarity is equal to the ionic strength, and $f_1 = 0.93$ at $I = 0.005$. Therefore,

$$a_{\mathrm{OH}} = 0.005 \cdot 0.93 = 0.0047.$$

Finally, $\mathrm{pOH} = -\log 0.0047 = 2.33$ and $\mathrm{pH} = 14.00 - 2.33 = 11.67$.

2. Calculate the pH of $0.03M$ calcium acetate, given K_a for acetic acid equal to $1.8 \cdot 10^{-5}$.

Solution   First, since $\mathrm{Ca(OAc)_2}$ is a 2:1 electrolyte, $I = 0.090$ and $f_1 = 0.79$. Each calcium ion is accompanied by two acetate ions, and so the nominal concentration of the weak base, acetate ion, is 0.060.
   The value of K_b for acetate ion is obtained by the conjugate relationship:

$$K_b = \frac{K_w}{K_a} = \frac{1 \cdot 10^{-14}}{1.8 \cdot 10^{-5}} = 5.6 \cdot 10^{-10}.$$

The value of Q_b turns out to be the same as that for K_b because of the cancellation of activity coefficients for this charge-type:

$$Q_b = K_b \frac{f_{\text{acetate ion}}}{f_{\mathrm{OH}} f_{\text{acetic acid}}} = K_b \frac{f_+}{f_+ \cdot 1.00} = K_b = 5.6 \cdot 10^{-10}.$$

Using Eq. (9.14) because the base is so weak, we get

$$[\mathrm{OH}^-] = (0.060 \cdot 5.6 \cdot 10^{-10})^{1/2} = 5.8 \cdot 10^{-6}.$$

Finally,

$$a_{\mathrm{H}} = \frac{K_w}{[\mathrm{OH}^-] f_1} = \frac{1.0 \cdot 10^{-14}}{5.8 \cdot 10^{-6} \cdot 0.79} = 2.2 \cdot 10^{-9} \qquad \text{and} \qquad \mathrm{pH} = 8.66.$$

3. Approximately what concentration of ammonia must be dissolved in water to make the $\mathrm{pH} = 11$? The dissociation quotient may be taken as $1.8 \cdot 10^{-5}$.

Solution   If $\mathrm{pH} = 11$, then $\mathrm{pOH} = 3$ and $[\mathrm{OH}^-] \approx 1 \cdot 10^{-3}$. Thus,

$$Q_b = \frac{[\mathrm{OH}^-][\mathrm{NH_4^+}]}{[\mathrm{NH_3}]} = \frac{(1 \cdot 10^{-3})^2}{C} = 1.8 \cdot 10^{-5}.$$

Solving for C,

$$C = \frac{1 \cdot 10^{-6}}{1.8 \cdot 10^{-5}} = 5.6 \cdot 10^{-2} \, \text{mol/L}.$$

4. A $0.10M$ solution of a certain base is found to have a pH of 10.0. What is the approximate value for pQ_b?

Solution If pH $= 10.0$, then pOH $= 4.0$ and $[OH^-] \approx 1.0 \cdot 10^{-4}$. This is small compared to the total concentration of 0.10, and therefore the molarity of undissociated base is about 0.10. Now Q_b is readily calculated:

$$Q_b = \frac{[HA][OH^-]}{[A]} = \frac{[OH^-]^2}{C} = \frac{1 \cdot 10^{-8}}{0.1} = 1.0 \cdot 10^{-7},$$

and

$$pQ_b = 7.0.$$

9.4   REPRESSION OF A WEAK DISSOCIATION

The foregoing discussions have dealt with single-solute solutions: strong acid, weak acid, strong base, weak base. Suppose a solution is prepared by mixing a strong acid with a weak acid. The strong acid will be completely dissociated into H^+ ions and its anions. The extent to which the weak acid will dissociate in such a solution will be much less than when it was in water by itself. This is because the dissociation equilibrium is forced to the left by the additional source (strong acid) of H^+ ions:

$$HA(weak) + H_2O \rightleftarrows H_3O^+ + A.$$

The equilibrium molarity of H^+ will be essentially the same as if the strong acid were present alone. Similar considerations hold for the repression of a weak-base dissociation when a strong base, such as NaOH, is also present.

9.5   SOLUTIONS CONTAINING A CONJUGATE PAIR

Buffer Action and pH Control

The practical range of values for solution pH is from -1 (e.g., $10M$ HCl) to 15 (e.g., $10M$ NaOH). In other words, the hydronium-ion molarity can range from a high of about 10 to a low of about $1 \cdot 10^{-15}$, or sixteen orders of magnitude. Because so many chemical and biological processes are vitally dependent upon solution pH, it is important in laboratory work to be able to adjust the pH of a system to any desired value *and* to ensure that the pH will be maintained close to that value for the duration of the experiment.

    We have seen that solutions of a desired low pH may be easily prepared by dissolving the proper amount of a strong acid. If one wants a solution of pH $= 2$, then $0.01M$ hydrochloric acid will do nicely. But what if an experiment must be carried out under conditions of pH = 6? It is possible, using careful successive dilutions, to prepare a solution of $1 \cdot 10^{-6}M$ HCl, but there would be no point in doing so because it would not be stable. Even very small amounts of other acid or base substances in the system (in fact, even absorption of carbon dioxide from the air) would change this low level of hydronium ion. Yet it is essential to be able to adjust and maintain pH in the middle range of values and this brings us to the subject of **buffer** solutions.

According to the dictionary, a buffer is a device for lessening the shock of concussion. Boxing gloves, auto shock absorbers, and politically neutral countries located between hostile nations all are examples of buffers. When the term is applied to aqueous solutions it usually refers to a solution whose pH is not very sensitive to the addition of small quantities of either strong acid or strong base.

The practical way to achieve this is to make a solution that contains substantial concentrations of an acid *and* of the conjugate base of that acid. An example would be acetic acid $(0.3M)$ and sodium acetate $(0.2M)$, both in the same solution. This solution contains:

CH_3COOH, undissociated molecules $0.3M$,

Na^+, CH_3COO^-, dissociated salt $0.2M$.

Here, the ionic strength due to the salt is 0.2, and Q_a for acetic acid at $I = 0.2$ is $3.1 \cdot 10^{-5}$.

The hydronium-ion concentration is easily calculated by direct substitution of the molarities into the equilibrium quotient expression:

$$Q_a = \frac{[H^+][OAc^-]}{[HOAc]} \qquad so \qquad [H^+] = 3.1 \cdot 10^{-5} \frac{0.3}{0.2} = 4.65 \cdot 10^{-5}.$$

This solution tends to maintain its value of pcH = 4.333 because (1) if a small amount of strong acid is added, the $0.2M$ acetate ion will react with it to form undissociated acetic acid, and (2) if a small amount of strong base is added, the $0.3M$ acetic acid will react with it to form acetate ion. But because *both* acetic acid and acetate ion are already present in substantial amounts, the ratio of their molarities will not change greatly. And the calculation above showed that it is this ratio, multiplied by Q_a, that determines the value of $[H^+]$ and hence of pH.

A comparative calculation is fully convincing: consider two solutions having a pcH of 4.333. One is the buffer solution described above, and the other is a carefully prepared solution of $4.7 \cdot 10^{-5}M$ hydrochloric acid. Let 1.00 mmol of NaOH be added to one liter of each of these solutions. In the case of the dilute hydrochloric acid solution, there is only

$$1000 \text{ mL} \cdot 4.7 \cdot 10^{-5}M = 0.047 \text{ mmol}$$

of HCl present. This will react with the NaOH to form NaCl, of course, but there will be

$$1.00 - 0.047 = 0.953 \text{ mmol}$$

of NaOH left over, and this means that the hydroxide-ion concentration will then be

$$\frac{0.953 \text{ mmol}}{1000 \text{ mL}} = 9.53 \cdot 10^{-4} M.$$

This corresponds to a pcOH of 3.02 and a pcH of about 11. A mere millimole of strong base has made the pH jump from 4.33 to 11, a huge change of about seven units.

Now, in the case of the buffer solution, the 1.00 mmol of NaOH will react with the acetic acid, present in the amount of

$$1000 \text{ mL} \cdot 0.3 M = 300 \text{ mmol}.$$

Because of the reaction

$$HOAc + NaOH \rightarrow NaOAc + H_2O,$$

there will be 299 mmol of HOAc left, and the new amount of NaOAc will be

$$\underset{\text{originally there}}{1000 \text{ mL} \cdot 0.2 M} + \underset{\text{from reaction}}{1.00 \text{ mmol}} = \underset{\text{total}}{201 \text{ mmol}.}$$

Therefore, the new molarities of the conjugate-pair species are

$$[HOAc] = 0.299, \qquad [OAc^-] = 0.201.$$

The new value for pcH is calculated to be

$$[H^+] = 3.1 \cdot 10^{-5} \cdot \frac{0.299}{0.201} = 4.61 \cdot 10^{-5}, \qquad pcH = 4.336.$$

What a contrast—the change in pcH was only 0.003 unit instead of about seven pH units! If the calculation had been done with the addition of 1.00 mmol of strong acid, the buffer ratio would have become 0.301/0.199 with similar favorable results for the buffer solution, but the pcH of the unbuffered HCl solution would have dropped to about 3.0.

The ability of a solution to resist attempts to change its pH is called the **buffer capacity**. The dilute HCl solution has a higher buffer capacity than does pure water, but is very poorly buffered compared to the acetic acid sodium acetate solution. Buffer capacity is subject to precise algebraic treatment using differential calculus, but the simple rule is: *The higher the molarities of the conjugate pair, the better the buffer capacity.* And, for a certain total concentration of HA and A, the buffer capacity will be at a maximum when [HA] = [A].

Effect of Dilution

In addition to showing resistance to chemical threats to its pH, a conjugate-pair buffer is relatively immune to dilution by water. Again taking the two solutions of pcH = 4.333, compare the effects of doubling their volumes by adding pure water. In the case of $4.7 \cdot 10^{-5} M$ HCl, the dilution will change the molarity to $2.35 \cdot 10^{-5}$ and the new pcH will be 4.633. (The change of 0.30 pH unit is simply the logarithm of the dilution factor—a tenfold dilution would change the pcH by 1.0 unit.)

By contrast, the twofold dilution of the acetic acid sodium acetate buffer would result in the new molarities:

$$CH_3COOH: \qquad 0.3/2 = 0.15 M,$$
$$Na^+, CH_3COO^-: \qquad 0.2/2 = 0.10 M.$$

There has been no change in the **buffer ratio**. There is a small effect on the Q_a value because the ionic strength has dropped from 0.2 to 0.1, making $Q_a = 2.9 \cdot 10^{-5}$. The new [H$^+$] and pcH are calculated in the same manner as before:

$$[H^+] = 2.9 \cdot 10^{-5} \frac{0.15}{0.10} = 4.35 \cdot 10^{-5}; \qquad pcH = 4.362.$$

The change is only 0.03 pH unit, compared with 0.30 for the HCl solution. Of course, a price has been paid in that the new buffer capacity is only half as great.

These properties of conjugate-pair buffers are extremely valuable in analysis, in chemical and biological research, in physiology, and in geochemistry. Therefore we will take a closer and harder look at the precise algebraic basis for buffers and will examine the methods for laboratory preparation of buffers with desired characteristics.

Chemical Reactions and the Equilibrium Buffer pH

Buffers based on different conjugate pairs may differ greatly in the pH of their solutions. Some buffers are rather acidic (low pH) such as the bisulfate/sulfate case, while others are decidedly basic (high pH) such as the ammonium ion/ammonia case. To explain this variation we must consider the several chemical reactions that occur very rapidly and simultaneously in any buffer solution.

To keep the discussion general, let us use the symbol HA for the conjugate acid and A for the conjugate base, without bothering to show the ionic charge(s). The acid species is constantly colliding with water molecules and if the orientation of the collision is favorable, there will be proton transfer:

$$\text{Acid dissociation:} \qquad HA + H_2O \rightleftharpoons H_3O^+ + A.$$

To the extent that this reaction occurs, the solution will tend to become acidic (low pH) due to the accumulation of hydronium ions.

However, the conjugate base species is also undergoing continuous collision with water molecules according to the proton-transfer reaction:

$$\text{Base dissociation:} \qquad A + H_2O \rightleftharpoons OH^- + HA.$$

This reaction tends to make the solution basic because the buildup of hydroxide ions will depress the dissociation of water

$$\text{Water dissociation:} \qquad 2\,H_2O \rightleftharpoons H_3O^+ + OH^-,$$

and thereby decrease the molarity of hydronium ions, resulting in a higher pH.

All three of these reactions occur in every buffer. Whether the equilibrium value for the pH is low or high depends on which of the first two reactions is the dominant one. If the acid dissociation is more extensive than the base dissociation, then the solution will become acidic, and vice versa. In the case of equal reaction tendencies, the solution will simply be neutral (pH = 7.00) because equal amounts of H_3O^+ and OH^- will be produced per second and, through combination to form water, they will exist only at the low level of about $10^{-7}M$.

We may characterize the intrinsic tendency for the acid-dissociation reaction by the value for K_a, and that for the base dissociation reaction by the value for K_b. Thus if K_a is greater than K_b and the values for C_a and C_b are equal we should expect the pH to be below 7. If K_a is smaller, the buffer will be basic. These conclusions are supported very convincingly by examining the expression for the acid-dissociation constant. For the acid-dissociation reaction

$$HA + H_2O \rightleftharpoons H_3O^+ + A,$$

we write the K_a expression:

$$K_a = [H^+] f_1 \frac{[A] f_A}{[HA] f_{HA}}.$$

By taking logarithms and rearranging, we obtain the pH in terms of the other quantities:

$$pH = -\log([H^+] f_1) = pK_a + \log \frac{[A]}{[HA]} + \log \frac{f_A}{f_{HA}}. \qquad (9.15)$$

This equation shows clearly the several factors that influence the pH value of a buffer. Of greatest importance is the first term on the right, the pK_a value. Since pK_a values range from about 1.0 to 13 for various acids, we see that the greatest consideration in choosing a conjugate pair to make a buffer is the pK_a value.

If the pK_a might be regarded as the "coarse-tuning" control for solution pH, then the base/acid ratio [A]/[HA] is the "fine-tuning" device. The ratio can be made to vary from 0.01 (e.g., using 0.01M A and 1.0M HA) to 100 (using 1.0M A and 0.01M HA). Thus the logarithm of the ratio can vary from -2 to $+2$ and, unlike the pK_a value, which is constant and single-valued, the ratio may be varied continuously.

The term involving the activity coefficients is significant (i.e., not negligible) but is not used for pH control purposes in the way that pK_a and the ratio are used.

Therefore, considering the first two terms only, we see that the pH of a series of buffers based on a given conjugate pair will be within the following:

$$pH = pK_a \pm 2.$$

When we discuss the preparation of buffers of desired pH, we will first choose a conjugate pair in accordance with its pK value and then will calculate the precise concentration ratio that is needed.

Derivation of Equations for Calculation of Buffer pH

Although Eq. (9.15) is exact and includes all the variables in a buffer solution, it cannot be used directly for precise calculation of pH. Suppose we prepare a buffer solution, measure out C_a moles of the conjugate acid and C_b moles of the conjugate base, mix them, and dilute to one liter. Thus, the *initial* concentrations are:

$$[HA] = C_a; \qquad [A] = C_b.$$

But the acid and base dissociation reactions occur immediately and, within the time it takes to dilute and mix the solution, these concentrations have shifted to their equilibrium values, which will differ from C_a and C_b because of the production of H^+ (or OH^-).

A direct way to state the equilibrium relationship between [HA] and C_a, and between [A] and C_b, is to express the proton balance.

Proton balance In any solution in which proton-transfer reactions have occurred, *the sum of the molarities of species that have gained protons must be exactly equal to the sum of the molarities of the species that have lost protons.*

In pure water, for example, application of the proton balance is simple:

$$[H_3O^+] = [OH^-],$$

where $[H_3O^+]$ is the only species to gain a proton and $[OH^-]$ is the only species to lose a proton.

In a solution of a weak acid, such as HCN, we have

$$[H_3O^+] = [OH^-] + [CN^-],$$

where both species on the right have lost protons.

In a solution of a weak base, such as SO_4^{2-} (e.g., a solution of sodium sulfate), we have

$$[H_3O^+] + [HSO_4^-] = [OH^-],$$

where both species on the left have gained protons.

The application of proton balance to a buffer solution is a little harder to visualize, but is helpful. We begin with C_a molar HA and C_b molar A. Let us suppose that the acid dissociation dominates, so that there is a net production of H_3O^+ and A. The proton-balance condition is:

$$[H_3O^+] = [OH^-] + ([A] - C_b). \tag{9.16}$$

Since there was already C_b molar A present, we use $[A] - C_b$ as a measure of the proton loss. That is, the equilibrium concentration of A is greater than the original value, C_b, and it is this increase that corresponds to proton loss from HA.

Equation (9.16) is easily rearranged to show the equilibrium molarity of conjugate base A:

$$[A] = C_b + [H^+] - [OH^-]. \tag{9.17}$$

We also will need a similar expression for the equilibrium molarity of HA. By noting that the total concentration of the conjugate pair can be expressed in two ways:

$$C = C_a + C_b = [HA] + [A],$$
$$\text{total} \quad \text{original} \quad \text{equilibrium}$$

we see that

$$[HA] = C_a + C_b - [A].$$

Then, by using Eq. (9.17) for [A], we get

$$[HA] = C_a - [H^+] + [OH^-]. \tag{9.18}$$

Finally, Eqs. (9.17) and (9.18) may be substituted into Eq. (9.15) to obtain a useful relationship:

$$pH = pK_a + \log \frac{C_b + [H^+] - [OH^-]}{C_a - [H^+] + [OH^-]} + \log \frac{f_A}{f_{HA}} \tag{9.19}$$

or, in nonlog form

$$[H^+] f_1 = aH = K_a \frac{(C_a - [H^+] + [OH^-]) f_{HA}}{(C_b + [H^+] - [OH^-]) f_A}. \tag{9.20}$$

In Chapter 11 we will refer to this exact relationship because it provides the key to the experimental determination of pK_a values through the potentiometric measurement of the pH of prepared buffer solutions. However, for our present purpose (the calculation of theoretical pH) it will be more practical to modify Eq. (9.20) slightly. By remembering that the relationship between a K value and a Q value is

$$K_a = Q_a \frac{f_1 f_A}{f_{HA}},$$

we note that Eq. (9.20) can be rewritten without activity coefficients in a form sometimes called the **Charlot equation**:

$$[H^+] = Q_a \frac{C_a - [H^+] + [OH^-]}{C_b + [H^+] - [OH^-]} = Q_a \frac{C_a - [H^+] + Q_w/[H^+]}{C_b + [H^+] - Q_w/[H^+]}. \qquad (9.21)$$

This *could* be rearranged to a cubic polynomial and, given values for Q_a, C_a, and C_b, solved for the value of $[H^+]$ by some method of numerical analysis. But for all practical situations such effort would be a waste of time because the equation can be simplified according to the type of buffer:

a) For **midrange buffers** (pH = 4 to 10) it will usually be true that both H and $[OH^-]$ will be negligible compared to C_a and C_b, so that the very simple relationship given below holds quite well:

$$[H^+] \approx Q_a \frac{C_a}{C_b}. \qquad (9.22)$$

b) For **acidic buffers** (pH below 4) the $[H^+]$ is not negligible:

$$[H^+] = Q_a \frac{C_a - [H^+]}{C_b + [H^+]} \quad \text{or} \quad [H^+]^2 + (C_b + Q_a)[H^+] - Q_a C_a = 0. \qquad (9.23)$$

A reliable value for [H] can be obtained simply by using the quadratic formula.

c) For **basic buffers** (pH above 10) the $[OH^-]$ is not negligible:

$$[H^+] = Q_a \frac{C_a + Q_w/[H^+]}{C_b - Q_w/[H^+]} \quad \text{or} \quad C_b[H^+]^2 - (Q_w + Q_a C_a)[H^+] - Q_a Q_w = 0.$$

$$\qquad (9.24)$$

Again, the quadratic formula readily gives a value for $[H^+]$.

The practical rule for doing calculations on buffers is this: first simply use the easiest expression, Eq. (9.22) to find a value for $[H^+]$. If this value is not negligible compared to C_a or C_b, then perform the exact calculation using Eq. (9.23). But if the value of $[OH^-]$, that is Q_w/H, is not negligible compared to C_a or C_b, then use Eq. (9.24) for the accurate calculation of $[H^+]$.

Examples of pH calculation, given buffer-solution composition

The steps in each calculation are as follows:

a) calculate ionic strength,

b) look up or calculate activity coefficient values,

c) look up K_a and convert to Q_a,

d) convert K_w to Q_w,

e) use Eq. (9.22) or, if necessary, Eq. (9.23) or Eq. (9.24) for $[H^+]$,

f) convert $[H^+]$ to activity of hydronium ion aH

g) convert aH to pH.

Example 1 One of the standard reference buffers established by the National Bureau of Standards consists of $0.025M$ sodium bicarbonate and $0.025M$ sodium carbonate. It has been assigned a pH value of 10.012 based on electrochemical cell measurements; we will see if the calculated pH agrees with this.

First, the ionic strength of the buffer solution is 0.100 (0.025 contributed by the 1:1 salt and 0.075 by the 2:1 salt). At this ionic strength, the activity coefficients are $f_1 = 0.78$ and $f_2 = 0.37$, according to the Davies equation.

From a table of pK_a values we find that the bicarbonate ion has $pK_a = 10.33$ or $K_a = 4.68 \cdot 10^{-11}$. The conversion to Q_a gives:

$$Q_a = K_a \frac{f_{HCO_3^-}}{f_H f_{CO_3^{2-}}} = 4.68 \cdot 10^{-11} \frac{0.78}{0.78 \cdot 0.37} = 1.26 \cdot 10^{-10}.$$

The conversion of K_w gives

$$Q_w = \frac{K_w}{f_H f_{OH}} = \frac{1.00 \cdot 10^{-14}}{(0.78)^2} = 1.64 \cdot 10^{-14}.$$

The simple Eq (9.22) then gives

$$[H^+] = Q_a \frac{[HCO_3^-]}{[CO_3^{2-}]} = 1.26 \cdot 10^{-10} \frac{0.025}{0.025} = 1.26 \cdot 10^{-10},$$

and conversion to aH and pH gives

$$aH = [H^+] \cdot f_H = 1.26 \cdot 10^{-10} \cdot 0.78 = 9.83 \cdot 10^{-11},$$

$$pH = -\log(9.83 \cdot 10^{-11}) = 10.01.$$

The agreement with the experimentally determined value of 10.012 is excellent. In this case it was not necessary to use the more exact Eq. (9.24) for basic buffers because the molarity of hydroxide

$$[OH^-] = \frac{Q_w}{[H^+]} = \frac{1.64 \cdot 10^{-14}}{1.26 \cdot 10^{-10}} = 1.3 \cdot 10^{-5}$$

is quite small compared to $C_a = C_b = 0.025$.

Example 2 Another NBS reference buffer, assigned an experimental value of 1.679 at 25°C, is a mixture of $0.0500M$ oxalic acid and $0.0500M$ potassium hydrogen oxalate. The acid-dissociation constant for oxalic acid is 0.0536:

$$H_2C_2O_4 = H^+ + HC_2O_4^-; \qquad K_a = 0.0536.$$

Because oxalic acid is an uncharged molecule, its only contribution to ionic strength is through whatever part of it dissociates to establish the equilibrium. At this point we don't know what this amounts to, and we'll therefore estimate the ionic strength as due merely to the 1:1 salt KHC_2O_4:

$$I \approx 1 \cdot C_b = 0.050.$$

By using activity coefficients at this ionic strength, $f_1 = 0.82$, we find the value of Q_a:

$$Q_a = K_a \frac{f_0}{f_1 f_1} = \frac{0.0536}{0.82^2} = 0.0797.$$

Since this buffer is acidic, there is no point in dealing with K_w and Q_w. Now, the approximate Eq. (9.22) brings us bad news in this case:

$$[H^+] \overset{?}{\approx} Q_a \frac{C_a}{C_b} = 0.0797 \frac{0.050}{0.050} = 0.0797.$$

Obviously this is impossible because, even if *all* the oxalic acid dissociated, the value of $[H^+]$ could not exceed 0.050. Clearly, this is a case for the quadratic Eq. (9.23). Without changing our rough estimate for ionic strength, we find

$$[H^+]^2 + (0.050 + 0.0797)[H^+] - 0.0797 \cdot 0.050 = 0,$$

and, using the quadratic formula,

$$[H^+] = \frac{-0.1297 + (0.1297^2 + 4.0 \cdot 003986)^{1/2}}{2} = 0.0257.$$

This is certainly more reasonable, showing that about half the oxalic acid has dissociated. However, this also shows that the approximate value for the ionic strength was not too good. In addition to the 0.0500 from the potassium-hydrogen oxalate, there is a contribution of 0.0257 due to the dissociation of the acid.

With a better value of ionic strength, $0.050 + 0.026 = 0.076$, a refined conversion of K_a gives

$$Q_a = 0.0536/0.79^2 = 0.0859.$$

The quadratic formula is solved again, using this value for Q_a, with the result

$$[H^+] = 0.0265,$$

which is appreciably different from the previous value of 0.0257.

Finally, $aH = 0.0265 \cdot 0.79 = 0.0209$ and $pH = 1.679$. In view of the uncertainties in the activity coefficients, this precise agreement with the experimental value is remarkable, and shows that calculations of this sort do have validity.

Example 3 Here is a buffer that has a high pH, making it necessary to take account of $[OH^-]$. The solution contains $0.0194M$ sodium phosphate Na_3PO_4 and $0.0056M$ sodium monohydrogen phosphate Na_2HPO_4.

The ionic strength is calculated:

due to Na_3PO_4 (a $3:1$ type): $6 \cdot 0.0194 = 0.116$
due to Na_2HPO_4 (a $2:1$ type): $3 \cdot 0.0056 = 0.017$

Total: 0.133

Activity coefficients at this ionic strength are $f_1 = 0.77$, $f_2 = 0.35$, and $f_3 = 0.10$. Even though there is considerable uncertainty in f_2 and f_3, we will use these values.

The dissociation constant for the HPO_4^{2-} ion is $4.79 \cdot 10^{-13}$, so conversion to Q_a gives

$$Q_a = K_a \frac{f_{HPO_4^{2-}}}{f_H f_{PO_4^{3-}}} = 4.79 \cdot 10^{-13} \cdot \frac{0.35}{0.77 \cdot 0.10} = 2.18 \cdot 10^{-12},$$

and

$$Q_w \doteq 1.00 \cdot 10^{-14}/0.77^2 = 1.69 \cdot 10^{-14}.$$

Substitution of these quantities into Eq. (9.24) yields

$$0.0194\,[H^+]^2 - (1.69 \cdot 10^{-14} + 2.18 \cdot 10^{-12} \cdot 0.0056)[H^+] - 2.18 \cdot 10^{-12} \cdot 1.69 \cdot 10^{-14} = 0,$$

and the quadratic formula gives as a solution

$$[H^+] = 2.32 \cdot 10^{-12}.$$

Therefore,

$$aH = 2.32 \cdot 10^{-12} \cdot 0.77 = 1.79 \cdot 10^{-12} \quad \text{and} \quad pH = 11.75.$$

This isn't bad, considering that the experimental value for this particular buffer is 11.80.

9.6   ACID–BASE MIXTURES THAT REACT EXTENSIVELY

The foregoing discussions dealt with solutions of acids and/or bases that reacted only with the solvent to produce equilibrium concentrations of H^+ or OH^-. There were six fundamental types of such solutions:

1. *Inert salt*, such as NaCl. Neither Na^+ nor Cl^- has acid–base properties. The solution is neutral, pH $= 7$, due only to water dissociation.

2. *Strong acid*, such as HCl. Completely dissociated into H^+ and Cl^-. The solution is acidic, pH below 7, due to this added hydrogen ion.

3. *Strong base*, such as NaOH. Completely dissociated into Na^+ and OH^-. The solution is basic, pH above 7, due to the repression of water dissociation by this added OH^-.

4. *Weak acid*, such as acetic acid, CH_3COOH. Only partly dissociated into H^+ and its conjugate base, acetate ion. The solution is acidic, pH below 7, but not as acidic as a solution of the same concentration of a strong acid.

5. *Weak base*, such as ammonia, NH_3. Partly reacts with solvent to form OH^- and its conjugate acid, ammonium ion. The solution is basic, pH above 7, but not as basic as a solution of the same concentration of a strong base.

6. *Conjugate-pair buffer*, such as a mixture of the weak boric acid H_3BO_3 (pK_a 9.23) and its conjugate-base borate ion $H_2BO_3^-$ (pK_b 4.77), the latter being present in the form of its completely dissociated sodium salt. The pH of a buffer solution is within about two units of the pK_a value for the conjugate-acid species, and therefore may be acidic for some buffers and neutral or basic for others.

   Solutions of these types may be prepared simply by dissolving the desired materials in water. For example, a solution of $0.10M$ ammonium chloride (NH_4^+ is a weak acid, $pK_a = 9.24$) may be made by dissolving 5.35 grams of the solid salt in a liter of water. However, we could prepare an identical solution by mixing 100 mL of $1.0M$ hydrochloric acid with 100 mL of $1.0M$ ammonia, followed by dilution to one liter. This procedure would depend on the reaction:

$$H^+ \quad + NH_3 = NH_4^+.$$
$$\text{from the HCl}$$

We certainly would expect this mixture to have exactly the same properties, including its pH, as the solution prepared by direct dissolving of NH_4Cl. This assumes, of course, that we mixed *exactly* equimolar amounts of HCl and NH_3, which is easier to do on paper than in the laboratory.

   To continue the example, if we deliberately added more ammonia than HCl, the latter would be converted to $NH_4^+Cl^-$ (completely dissociated) and there would be some NH_3 left over. This would result in a buffer solution, with both members of the NH_4^+/NH_3 conjugate pair present. It would be identical in its properties, including pH, to a buffer solution prepared by mixing ammonium chloride and ammonia in the same ratio.

   This example illustrates one reason why it is so important to understand the reactions of acids and bases, and how to calculate the pH of the solutions that result from their mixtures. It is very often necessary to adjust the pH of a solution to some desired value, as a step in an analytical procedure. By adding the right acid (or base) in the right amount, the solution pH can be changed to any desired value.

   There are three main steps to follow in dealing with a given mixture of acids and bases, prepared by adding one solution to another:

1. First examine the compositions of the original solutions before they are mixed. Identify all species present and their molarities. This requires know-

ledge of what electrolytes are completely dissociated and whether there are any weak acids or bases. Look in Appendix 4 for pK_a values for any acids present and deduce pK_b values for any bases present. This information tells what proton-transfer reactions are at least possible with the species in the solutions.

2. Imagine that the solutions are now mixed, so that the proton-transfer reactions may occur. On paper, carry out the following reactions:

a) If one solution contains a strong acid and the other solution contains a strong base, let H^+ react with OH^- to form H_2O, until whichever one was present in lesser amount is used up. There may then be an excess of H^+ or of OH^-, that can react further as follows.

b) If there is H^+ from a strong acid and also a weak base present, let them react to form the conjugate acid of the weak base until whichever one was present in lesser amount is used up. There may then be an excess of H^+ or of weak base.

c) If there is OH^- from a strong base and also a weak acid present, let them react to form water and the conjugate base of the weak acid until whichever one was present in lesser amount is used up. There may then be an excess of OH^- or of the weak acid.

3. Once these reactions, all of which proceed virtually to completion, have been carried out on paper, it will be possible to list the chemical species present in the final mixture. By using Appendix 4 it will be possible to classify each species according to its acid–base properties and pK value. At this point *any mixture that has been made can be classified as one of the six fundamental types given earlier.* Therefore it can be treated by the calculation models that have already been developed. (There is also a seventh solution type, that of the amphiprotic species, which will be discussed later in the chapter.)

The following examples and Table 9.3 show how these ideas can be applied to a variety of cases.

Sample Calculations Dealing with Mixtures

Example 1 (SA + Equal SB)
With this type of problem it is always helpful to work with the number of millimoles of each reacting substance.

Let 400 mL of 0.0300M HCl be added to 600 mL of 0.0200M NaOH. There are 12.0 mmol of strong acid and 12.0 mmol of strong base. They react to form 12.0 mmol of sodium chloride, dissolved in a total volume of 1000 mL, so the concentration of NaCl is 0.012. A solution of such an inert salt is neutral, although it would be virtually impossible to achieve this in the laboratory, because the slightest mismatch in quantities would change the pH.

In this hypothetical case, the ionic strength is 0.012, so $f_1 = 0.89$ and

$$Q_w = 1 \cdot 10^{-14}/0.89^2 = 1.26 \cdot 10^{-14},$$
$$[H^+] = (Q_w)^{1/2} = 1.12 \cdot 10^{-7},$$
$$aH = 1.00 \cdot 10^{-7}, \qquad pH = 7.00.$$

(Of course, all this arithmetic wasn't necessary—the pH is 7.00 in any neutral solution, regardless of ionic strength.)

Example 2 (Excess SA + SB)

If 55 mL of $0.1M$ HCl are mixed with 45 mL of $0.1M$ NaOH, there are 5.5 mmol of acid and 4.5 mmol of base. It is often helpful to set up a before-and-after diagram as follows:

Reaction: $HCl + NaOH \rightarrow NaCl + H_2O$
mmoles at start: 5.5   4.5   0
↓   ↓   ↓
mmoles at end: 1.0   "0"   4.5

The purpose of quotes is to remind the reader that the reaction does not go literally to completion; there is some very low amount of reactant left at equilibrium.

The equilibrium solution contains NaCl at a molarity of $4.5/100 = 0.045$, and also HCl at a molarity of $1.0/100 = 0.010$. Thus, $I = 0.055$, $[H^+] = 0.010$, $f_1 = 0.815$, aH $= 0.00815$, and pH $= 2.09$.

Example 3 (SA + Excess SB)

If 100 mL of $0.05M$ HCl are added to 100 mL of $0.06M$ NaOH, the strong base is in excess.

Reaction: $HCl + NaOH \rightarrow NaCl + H_2O$
mmoles at start: 5.0   6.0   0
↓   ↓   ↓
mmoles at end: "0"   1.0   5.0

At equilibrium, there is NaCl at $0.025M$ and NaOH at $0.0050M$. The ionic strength is 0.030, $f_1 = 0.85$, and so

$$aOH = 0.0050 \cdot 0.85 = 0.00425,$$
$$aH = 1 \cdot 10^{-14}/0.00425 = 2.35 \cdot 10^{-12},$$
$$pH = 11.63.$$

Example 4 (SA + Equal WB)

When 50 mL of $0.08M$ HCl are mixed with 40 mL of $0.1M$ NH₃ the reaction goes essentially to completion, forming 4.0 mmol of ammonium chloride, with neither reactant in excess. The resulting solution is indistinguishable from one formed by dissolving 4.0 mmol of NH_4Cl in 90 mL of water, and is regarded simply as a solution of a weak acid ($pK_a = 9.26$ for ammonium ion). The ionic strength is the same as the molarity, $4.0/90 = 0.0444$, and $f_1 = 0.83$.

For a cationic acid, activity coefficients cancel, so that

$$Q_a = K_a \frac{f_1}{f_1 f_0} = K_a = 5.5 \cdot 10^{-10}.$$

By using the approximate formula for a weak acid we find

$$[H^+] = (0.0444 \cdot 5.5 \cdot 10^{-10})^{1/2} = 4.9 \cdot 10^{-6},$$
$$aH = [H]f_1 = 4.1 \cdot 10^{-6},$$
$$pH = 5.39.$$

Example 5 (Excess SA + WB)

Let $200\,mL$ of $0.1M$ HCl be added to $50\,mL$ of $0.1M$ NH_3. The $5.0\,mmol$ of NH_3 are converted to ammonium chloride, but there are still $15.0\,mmol$ of HCl left over. At equilibrium,

$$C_{HCl} = 15/250 = 0.060,$$
$$C_{NH_4Cl} = 5/250 = 0.020,$$
$$I = 0.080 \quad \text{and} \quad f_1 = 0.79.$$

Therefore, $aH = 0.060 \cdot 0.79 = 0.0474$ and $pH = 1.32$

Example 6 (SA + Excess WB)

In this case, *part* of the weak base is converted to its conjugate acid and the excess is unchanged, and so a conjugate-pair buffer is created. Suppose $100\,mL$ of $0.040M$ HCl is mixed with $100\,mL$ of $0.070M$ NH_3.

$$
\begin{array}{lcccc}
\text{Reaction:} & HCl & + & NH_3 = NH_4Cl & + H_2O \\
\text{mmoles at start:} & 4 & & 7 & 0 \\
& \downarrow & & \downarrow & \downarrow \\
\text{mmoles at end:} & \text{``0''} & & 3 & 4
\end{array}
$$

The equilibrium molarity of NH_3 is $3/200 = 0.0150$, and that of NH_4^+ is $4/200 = 0.0200$, which is also the ionic strength. For this buffer, taking pK_a for NH_4^+ as 9.26, we have

$$[H^+] \approx Q_a \frac{C_a}{C_b} = 5.5 \cdot 10^{-10} \frac{0.0200}{0.0150} = 7.3 \cdot 10^{-10},$$
$$aH = 7.3 \cdot 10^{-10} \cdot 0.87 = 6.4 \cdot 10^{-10},$$
$$pH = 9.20.$$

Example 7 (WA + Equal SB)

If $100\,mL$ of $0.025M$ KOH and $100\,mL$ of $0.025M$ acetic acid (HOAc) are mixed, each supplies $2.5\,mmol$, and therefore $2.5\,mmol$ of potassium acetate are formed in a total volume of $200\,mL$. The solution is no different from one prepared by dissolving $2.5\,mmol$ of solid potassium acetate and diluting to $200\,mL$.

Potassium ion is inert, and acetate ion is a weak base. The pK_a for acetic acid is 4.76, so pK_b for acetate ion is $14.00 - 4.76 = 9.24$. For an anion base, acitivity coefficients cancel, so that

$$Q_b = K_b = 5.8 \cdot 10^{-10}.$$

According to the approximate formula for a weak base solution,

$$[OH^-] = (C_b Q_b)^{1/2} = \left(\frac{2.5}{200} \cdot 5.8 \cdot 10^{-10}\right)^{1/2} = 2.7 \cdot 10^{-6}.$$

Therefore,

$$aOH = 2.7 \cdot 10^{-6} \cdot 0.89 = 2.4 \cdot 10^{-6},$$
$$pOH = 5.62,$$
$$pH = 8.38.$$

Example 8 (WA + Excess SB)
In this case the acetic acid is completely converted to potassium acetate, and there is some KOH left over. The strong base represses the dissociation of the weak base (acetate ion), and so the solution may be treated as follows:

$$[OH^-] = C_{excess\ KOH},$$
$$I = C_{excess\ KOH} + C_{KOAc}.$$

Example 9 (Excess WA + SB)
This illustrates another approach to preparing a buffer solution. Suppose 25 mL of $0.5M$ KOH is mixed with 100 mL of $0.200M$ HOAc and then diluted to one liter.

	Reaction:	HOAc + KOH = KOAc + H_2O		
mmol at start:		20.0	12.5	0
		↓	↓	↓
mmol at end:		7.5	"0"	12.5

The equilibrium molarities are $[HOAc] = 0.0075$ and $[OAc^-] = 0.0125$, and the ionic strength is also 0.0125, due to the KOAc. Using the simple buffer equation, we get

$$[H^+] = \frac{1.75 \cdot 10^{-5}}{0.89^2} \cdot \frac{0.0075}{0.0125} = 1.33 \cdot 10^{-5},$$
$$aH = 1.33 \cdot 10^{-5} \cdot 0.89 = 1.18 \cdot 10^{-5},$$
$$pH = 4.93.$$

Example 10 (WA + Equal WB)
If a weak acid is mixed with an *unequal* amount of weak base, the result is a buffer solution, but of a type not commonly used. Of more interest is the equimolar mixture, because it exists automatically when certain salts are dissolved in water. For example, when ammonium acetate NH_4OAc is dissolved, the solution may be regarded as an equimolar mixture of ammonia and

acetic acid. It is just as realistic to regard it as an equimolar mixture of the weak acid, ammonium ion, and the weak base, acetate ion. From either viewpoint, there is a reversible reaction

$$NH_4^+ + OAc^- = NH_3 + HOAc.$$

The extent of this reaction depends on the values of the equilibrium quotients

$$Q_a = \frac{[H^+][NH_3]}{[NH_4^+]}, \qquad Q_b = \frac{[OH^-][HOAc]}{[OAc^-]}.$$

From the stoichiometry of the reaction it appears that

$$[NH_3] = [HOAc] \qquad \text{and} \qquad [NH_4^+] = [OAc^-].$$

These are not exact equalities because of reactions with water, but it turns out that there is very little error in this approximation. Therefore, when the *ratio* of the equilibrium quotients is written, molarities may be cancelled.

$$\frac{Q_a}{Q_b} = \frac{[H^+][\cancel{NH_3}]}{[\cancel{NH_4^+}][OH^-]} \frac{[\cancel{OAc^-}]}{[\cancel{HOAc}]} = \frac{[H^+]}{[OH^-]} = \frac{[H^+]^2}{Q_w}.$$

Finally, a little rearrangement gives an expression for $[H^+]$ in the solution:

$$[H^+] = (Q_a Q_w / Q_b)^{1/2}.$$

Now it is clear why a solution of ammonium acetate is nearly neutral. Since

$$Q_a \text{ (for ammonium ion)} \approx Q_b \text{ (for acetate ion)} \approx 5.5 \cdot 10^{-10},$$

we have

$$[H^+] = \left(\frac{5.5 \cdot 10^{-10} \cdot 1.00 \cdot 10^{-14}/f_1^2}{5.5 \cdot 10^{-10}}\right)^{1/2} = \frac{1.00 \cdot 10^{-7}}{f_1}$$

or

$$aH = 1.0 \cdot 10^{-7}.$$

But in a solution of ammonium cyanide (for cyanide ion, $Q_b = 1.6 \cdot 10^{-5}$), the weak base is stronger than the weak acid. For a $0.1M$ solution, $I = 0.1$, $f_1 = 0.78$, and the pH turns out to be 9.24. (Check this.)

Table 9.3 summarizes the stoichiometric deductions and the calculation formulas that are appropriate for each of the possible situations of monoprotic systems.

9.7 CASE STUDY: SPECTROPHOTOMETRIC DETERMINATION OF THE DISSOCIATION CONSTANT OF 2,4-DINITROPHENOL*

The classic study by von Halban and Kortum is impressive, even in comparison with modern spectrophotometric studies of equilibria, because of the required extreme purity of reagents used and the extraordinarily high

---

*After H. von Halban, G. Kortum (University of Zurich), *Zeitschrift fur physikalische Chemie*, 170, A, 351 (1934).

Table 9.3 An equilibrium view of mixtures of acids and bases

Type of mixture	Result when the acid is in stoichiometric excess	Result when the acid and base are equimolar	Result when the base is in stoichiometric excess
Strong acid + strong base $HCl + NaOH \rightarrow NaCl + H_2O$	NaCl and HCl (excess) Treat as a solution of strong acid in the presence of inert salt. $[H^+] = C_{SA}$	NaCl A neutral solution of an inert salt. $[H^+] = (Q_w)^{1/2}$	NaCl and NaOH (excess) Treat as a solution of strong base in the presence of inert salt. $[OH^-] = C_{SB}$
Strong acid + weak base $HCl + NH_3 \rightleftharpoons NH_4^+ + Cl^-$	NH$_4$Cl and HCl (excess) The excess HCl represses dissociation of the weak acid NH_4^+. $[H^+] \approx C_{SA}$	NH$_4$Cl A solution of the weak acid, ammonium ion. $[H^+] \approx (C_{WA}Q_a)^{1/2}$	NH$_4$Cl and NH$_3$ (excess) A buffer solution with ammonium ion and free ammonia. $[H^+] = Q_a \dfrac{[NH_4^+]}{[NH_3]}$
Weak acid + strong base $HOAc + NaOH \rightleftharpoons NaOAc + H_2O$	NaOAc and HOAc (excess) A buffer solution with acetic acid and acetate ion. $[H^+] = Q_a \dfrac{[HOAc]}{[OAc^-]}$	NaOAc A solution of the weak base, acetate ion. $[OH^-] \approx (C_{WB}Q_b)^{1/2}$	NaOAc and NaOH (excess) The excess NaOH represses dissociation of the weak base, OAc$^-$. $[OH^-] \approx C_{SB}$
Weak acid + weak base $HOAc + NH_3 \rightleftharpoons NH_4OAc$	NH$_4$OAc and HOAc (excess) A buffer solution with acetic acid and acetate ion. $[H^+] = Q_a \dfrac{[HOAc]}{[OAc^-]}$	NH$_4$OAc A solution of a salt that is amphiprotic. $[H^+] = \left(\dfrac{Q_a Q_w}{Q_b}\right)^{1/2}$	NH$_4$OAc and NH$_3$ (excess) A buffer solution with ammonium ion and free ammonia. $[H^+] = Q_a \dfrac{[NH_4^+]}{[NH_3]}$

accuracy of the photometric measurements. When it is further realized that commercial spectrophotometers were not yet available in 1934, the success of the author's "homemade" instrumentation is remarkable.

The Chemical System

2,4-dinitrophenol is a weak acid ($pK \approx 4$) but is much stronger than the parent compound, phenol ($pK \approx 10$), because of the inductive effect of the two nitro groups.

abbreviated HD   abbreviated D⁻
(colorless)   (yellow)

The above dissociation reaction occurs only partially when dilute solutions of dinitrophenol are prepared in pure water. The object of the spectrophotometric work is to determine the extent of this dissociation and hence to be able to calculate very precise values of the dissociation constant:

$$K_a = \frac{[H^+][D^-]}{[HD]} \cdot \frac{f_1^2}{f_0}.$$

In the simple solution with no other solutes, it may be assumed that $[H^+] = [D^-]$ and that activity coefficients may be accurately calculated by the Debye–Hückel equation or the Davies equation.

Because the D⁻ species is a resonance hybrid with a structure quite different from that of the acid HD, it absorbs ultraviolet and visible radiation differently. Being yellow, the anion absorbs light at the lower wavelengths, and von Halban and Kortum found that when they used the blue 436-nm emission line from a mercury-vapor lamp as the radiation source, the D⁻ species absorbed strongly while the HD species absorbed only negligibly.

Thus given a solution that contains a mixture of HD and D⁻, the molarity of D⁻ can be determined by the spectrophotometric measurement.

Preparation of Materials and Solutions

Samples of supposedly pure dinitrophenol obtained from various sources proved to contain significant levels of impurities. Obvious possibilities are the isomeric 2,3- and 3,4-dinitrophenols, and the 2-, 3-, and 4-nitrophenols. The authors took exacting care in synthesis and purification of a suitable product, using repeated recrystallizations from ethanol and sublimations. Finally a sample was obtained that showed no further changes in optical properties with further purification steps. The authors described "dim yellowish small leaves, melting at 112.6°C, giving a nearly white powder when finely ground."

To minimize the introduction of basic impurities from the glassware used for the solutions, all glass vessels were "steamed for many days" and then allowed to dry in a dust-free area.

Water was purified by repeated distillation until it showed a specific conductivity of less than $8 \cdot 10^{-7}$. This was used to make a stock solution of dinitrophenol, which was weighed on a sensitive balance to the nearest microgram. The solutions used for spectrophotometric measurements were then prepared by dilution.

Spectrophotometric Measurements

The molar absorptivity of the anion D^- was determined by using slightly basic solutions to ensure complete dissociation. For example, a solution was prepared that contained

$$\begin{aligned} \text{dinitrophenol:} \quad & 2.127 \cdot 10^{-4} M, \\ \text{potassium hydroxide:} \quad & 0.0050 M. \end{aligned}$$

Since the hydrogen-ion concentration in this solution is about $2 \cdot 10^{-12}$ and Q_a is about $1 \cdot 10^{-4}$, the ratio of dinitrophenol in the ionized form to that in the dissociated form is

$$\frac{[D^-]}{[HD]} = \frac{Q}{[H^+]} = \frac{1 \cdot 10^{-4}}{2 \cdot 10^{-12}} = 5 \cdot 10^7.$$

In other words, ionization is "complete" and it can be assumed that the measured absorbance is due only to the D^- series. The absorbance, using a 1.000-cm cell, was 0.945, so the molar absorptivity is,

$$\varepsilon_{D^-} = \frac{A}{b \cdot C} = \frac{0.945}{1 \cdot 2.127 \cdot 10^{-4}} = 4443 \text{ L/mol} \cdot \text{cm}.$$

To check on the assumption that the HD series is "nonabsorbing" at the 436-nm wavelength, a more concentrated solution was prepared with

$$\begin{aligned} \text{dinitrophenol:} \quad & 2.376 \cdot 10^{-3} M, \\ \text{hydrochloric acid:} \quad & 1.0 M. \end{aligned}$$

With this high hydrogen-ion concentration, the dissociation of dinitrophenol is nearly completely repressed; but because the anion absorbs light so strongly it is necessary to evaluate its contribution to the measured absorbance of 0.0313. At this high ionic strength it was known from other measurements that Q_a was about $1.3 \cdot 10^{-4}$, and so the concentration of the anion in the solution must be

$$[D^-] = [HD]\frac{Q_a}{[H^+]} = 2.376 \cdot 10^{-3} \frac{1.3 \cdot 10^{-4}}{1} = 3.09 \cdot 10^{-7}.$$

Since the absorbance of 0.0313 was measured with the solution in a long cell of 8.57 cm, the contribution of D^- to the absorbance was

$$A_{\text{due to } D^-} = \varepsilon_{D^-} b[D^-] = 4800 \cdot 8.57 \cdot 3.09 \cdot 10^{-7} = 0.0127.$$

(The value of molar absorptivity is 4800 at ionic strength 1.0, according to separate measurements in basic NaCl solutions.)

Now, this means that the absorbance due to the HD species is

$$A_{\text{due to HD}} = A_{\text{obs}} - A_{\text{due to D}^-} = 0.0313 - 0.0127 = 0.0186.$$

Therefore, the molar absorptivity of HD can be calculated

$$\varepsilon_{\text{HD}} = \frac{A_{\text{due to HD}}}{b\,[\text{HD}]} = \frac{0.0186}{8.57 \cdot 2.376 \cdot 10^{-3}} = 0.91.$$

This is small enough to be neglected in comparison with the high molar absorptivity of the D^- ion, and so it was assumed that the HD present in the equilibrium solutions made no contribution to the absorbance.

The following is an example of a typical equilibrium measurement. A solution of $3.991 \cdot 10^{-4} M$ dinitrophenol in purified water was placed in a calibrated 1.000-cm absorption cell, and the absorbance was determined to be 0.6473, by repeated and immediate comparisons with a basic solution of dinitrophenol having nearly the same absorbance. Therefore, the concentration of anion in this solution is:

$$[\text{D}^-] = \frac{0.6473}{1.000 \cdot 4443} = 1.457 \cdot 10^{-4} M.$$

By difference, then,

$$[\text{HD}] = (3.991 - 1.457) \cdot 10^{-4} = 2.534 \cdot 10^{-4} M.$$

Because of the extreme purity of the water and dinitrophenol and the rigorous care in protecting the solutions from outside contamination, it is an accurate assumption that

$$[\text{H}^+] = [\text{D}^-] = 1.457 \cdot 10^{-4} M.$$

Therefore, the value of the dissociation quotient can be calculated with reliability

$$Q_a = \frac{(1.457 \cdot 10^{-4})^2}{2.534 \cdot 10^{-4}} = 8.38 \cdot 10^{-5}.$$

To calculate the thermodynamic value for the acid-dissociation constant, it is necessary only to estimate the activity coefficient for the singly charged ions. Since the ionic strength is very low, due only to the dissociated dinitrophenol concentration of $1.457 \cdot 10^{-4}$, the Davies equation should be quite accurate.

$$\log f_1 = \frac{-0.51 I^{1/2}}{1 + I^{1/2}} + 0.15 I = -0.00697,$$

whence

$$f_1 = 0.9861.$$

Therefore, combining this result with that for Q_a, we get

$$K_a = Q_a f_1^2 = 8.38 \cdot 10^{-5} \cdot 0.9861^2 = 8.14 \cdot 10^{-5}.$$

Von Halban and Kortum carried out this sort of determination for a large variety of concentrations; the results for the lower values of concentrations (where the activity coefficients should be most reliable) are summarized in Table 9.4.

Table 9.4

Total concentration of dinitrophenol	[D⁻] from absorbance	[HD] by difference	$Q_a \cdot 10^5$	f_1 from Davies eq.	$K_a \cdot 10^5$
$7.018 \cdot 10^{-4}$	$2.049 \cdot 10^{-4}$	$4.969 \cdot 10^{-4}$	8.449	0.9836	8.175
5.505	1.774	3.731	8.435	0.9847	8.179
3.991	1.457	2.534	8.377	0.9861	8.142
2.395	1.056	1.339	8.325	0.9882	8.129
1.5364	0.7883	0.7481	8.310	0.9897	8.140
0.9245	0.5541	0.3704	8.289	0.9914	8.147

The average value for K_a in this set of data is $8.15 \cdot 10^{-5}$, or $pK = 4.089$.

Now, more than 40 years later, there still are no substitutes for quality, attention to detail, and painstaking care.

9.8 POLYPROTIC ACIDS AND BASES

When a molecular structure contains more than one acidic proton that may be transferred to water, the substance is called **diprotic** (two protons), **triprotic** (three protons), etc. The general term for these molecular structures is **polyprotic**. The tendency of the species to transfer a second proton is always weaker than the tendency to transfer the first one, and the third proton is still harder to remove, etc. This may be due largely to the different chemical nature of the molecular function groups containing the protons, but even for acids that have the same functional acidic groups it is more difficult to remove successive protons because the molecule gains one unit of negative charge for each proton lost, so that the successive protons are increasingly attracted by the electrostatic forces. This may be illustrated by considering phosphoric acid—a triprotic acid which has the structure:

$$
\begin{array}{c}
\quad\quad O \\
\quad\quad \| \\
H\!-\!O\!-\!P\!-\!O\!-\!H \\
\quad\quad | \\
\quad\quad O \\
\quad\quad | \\
\quad\quad H
\end{array}
$$

All three protons are acidic, and they are all exactly equivalent in the structure. But the equilibrium constants for the successive losses of these protons show a steady trend:

$$H_3PO_4 + H_2O = H_3O^+ + H_2PO_4^-, \qquad K_1 = 7.6 \cdot 10^{-3},$$

$$H_2PO_4^- + H_2O = H_3O^+ + HPO_4^{2-}, \qquad K_2 = 6.2 \cdot 10^{-8},$$

$$HPO_4^{2-} + H_2O = H_3O^+ + PO_4^{3-}, \qquad K_3 = 4.8 \cdot 10^{-13}.$$

This information is presented in Appendix 4 in the following form:

$$H_3PO_4 \xrightarrow{2.12} H_2PO_4^- \xrightarrow{7.21} HPO_4^{2-} \xrightarrow{12.32} PO_4^{3-}.$$

Other examples of diprotic and triprotic inorganic acids are arsenic acid H_3AsO_4, carbonic acid H_2CO_3, sulfurous acid H_2SO_3, sulfuric acid H_2SO_4, hydrosulfuric acid or hydrogen sulfide H_2S.

Types of Acid–Base Functional Groups

Polyprotic acids are much more common among organic molecular structures, partly because a variety of acidic functional groups are available and partly because the larger molecular structures can accommodate many groups. Some examples of acidic groups and their conjugate base counterparts are as shown in Table 9.5.

The groups in Table 9.5 differ intrinsically from each other in their tendency to donate a proton; a given group will show different tendencies according to the particular structural environment in the molecule. Phenomena such as hydrogen bonding and resonance are very important in influencing acid strength. For example, just to consider the effect of relative location of acidic groups on a molecule, we may examine acids of the type

$$\underset{O}{\overset{HO}{>}}C-(CH_2)_n-C\overset{OH}{\underset{O}{<}}$$

where the value of N may be as low as zero (oxalic acid) or as high as desired.

First, compare oxalic acid with adipic acid ($n = 4$):

$$HOOC\text{---}COOH: \qquad pK_1 = 1.27, \qquad pK_2 = 4.27,$$
$$HOOC\text{---}(CH_2)_4\text{---}COOH: \qquad pK_1 = 4.43, \qquad pK_2 = 5.42.$$

There are interesting questions: (1) Why is pK_1 for oxalic acid so much lower than pK_1 for adipic acid? After all, in each case the proton is being trans-

Table 9.5

Acidic group		Conjugate-base group	
Carboxyl:	$\equiv\!C\!-\!C\!-\!O\!-\!H$ with $\parallel O$ below	Carboxylate:	$\equiv\!C\!-\!C\!-\!O^-$ with $\parallel O$ below
Sulfonic:	$-\!C\!-\!S\!-\!O\!-\!H$ with O above and below	Sulfonate:	$\equiv\!C\!-\!S\!-\!O^-$ with O above and below
Ammonium:	$\equiv\!C\!-\!N\!-\!H^+$ with H above and below	Amine:	$\equiv\!C\!-\!N\!:$ with H above and below
Sulfhydryl:	$-\!S\!-\!H$	Sulfide:	$-\!S^-$
Phenolic:	$Ar\!-\!O\!-\!H$	Phenolate:	$Ar^*\!-\!O^-$
Hydroxyl:	$\equiv\!C\!-\!O\!-\!H$	Alkoxide:	$\equiv\!C\!-\!O^-$
Amidinium:	$\equiv\!C\!-\!C\!-\!NH_3^+$ with $\parallel O$ below	Amide:	$\equiv\!C\!-\!C\!-\!NH_2$ with $\parallel O$ below
Oxime:	$>\!C\!=\!N\!-\!O\!-\!H$	Oximate:	$>\!C\!=\!N\!-\!O^-$

*Ar is an aromatic ring.

ferred from the same acidic group, the carboxyl group. (2) Why is there such a difference in the ΔpK value? For oxalic acid pK_2 is three log units higher, but for adipic acid it is only one log unit higher than pK_1.

The Significance of Charge Dispersal

The key idea in acid strength is really the ability of the conjugate base, through use of its molecular structure, to spread out the negative charge resulting from the proton loss. The more this charge is dispersed, the easier it is for the proton to break away.

Now, with oxalic acid the closely neighboring oxygen atoms of the other carboxyl group are able, due to their high electronegativity, to share some of the negative charge and thereby to delocalize it.

Compare this with the relatively distant location of the other carboxyl group in adipic acid. When one of the protons is transferred, the negative charge remains localized arount the one carboxyl group. The other is simply too far away to share in charge dispersal. Therefore, it is reasonable to expect a large difference in the pK_1 values.

This discussion paves the way for answering the second question as well: In oxalic acid the second proton responds to the negative charge already present because it is so close. In adipic acid the localized charge is so remote that the second dissociation is almost as strong as the first.

An interesting question for the reader   If the value of n in the formula $HOOC—(CH_2)_n—COOH$ were made very large, would pK_2 be equal to pK_1?

To illustrate the importance of hydrogen bonding on acid strength, consider this pair of geometrical isomers:

maleic acid, the cis-isomer with both acid groups on the same side; $pK_1 = 1.92$, $pK_2 = 6.22$.

fumaric acid, the trans-isomer, with acid groups on opposite sides $pK_1 = 3.02$, $pK_2 = 4.39$.

Bear in mind that there is no rotation around the double bond, so that the spatial location of the groups is fixed. The pK values for fumaric acid are in accord with the idea that the two carboxyl groups are relatively isolated from each other and noninteracting. A comparison with succinic acid,

$$pK_1 = 4.21, \quad pK_2 = 5.64$$

shows that ΔpK is about the same. The higher strength (lower pK_1 value) of fumaric acid compared to succinic acid may be explained by the help given to charge dispersal by the pi-cloud of the double bond.

With maleic acid, the lower value of pK_1 is explained by the additional charge dispersal provided by hydrogen bonding in the conjugate base:

The final comparison is between the pK_2 values. Although maleic acid was stronger than fumaric acid in the first dissociation step, it is decidedly weaker in the second. The rationale is simply that additional energy is required to break the hydrogen bond formed in the first step.

Still another important example of the charge-dispersal concept is found in structures that are resonance hybrids. For example, consider the two carboxylic acids:

$$\text{acetic acid, } CH_3COOH: \qquad pK_a = 4.8,$$

benzoic acid, ⬡—COOH: $pK_a = 4.4.$

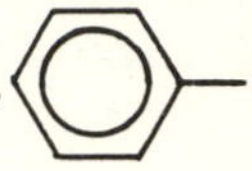

There is not much difference in the pK values, for in each case the negative charge resulting from proton loss remains localized in the same way, on the carboxyl group.

However, there is a striking difference in acidity between methanol and phenol:

$$\text{methanol, } CH_3—OH: \qquad pK_a = 15.5,$$

phenol, ⬡—OH: $pK_a = 10.0.$

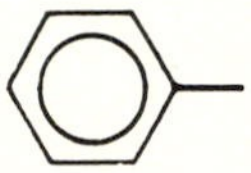

The conjugate base of methanol is the methoxide ion CH_3O^-, and the negative charge is highly localized. But the conjugate base of phenol, the phenolate ion, may be formally written in the variety of canonical forms:

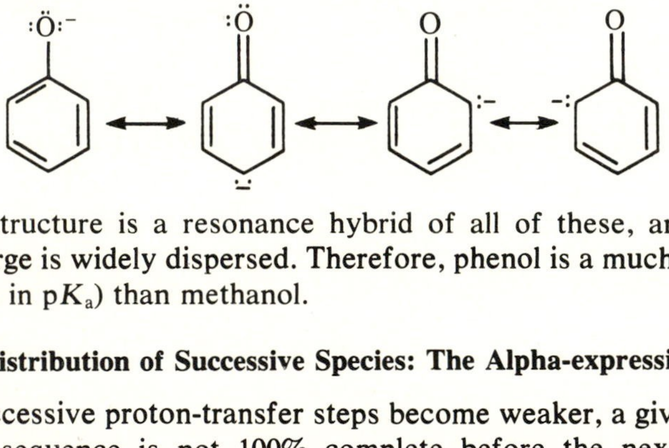

The actual structure is a resonance hybrid of all of these, and clearly the negative charge is widely dispersed. Therefore, phenol is a much stronger acid (5.5 log units in pK_a) than methanol.

Percentage Distribution of Successive Species: The Alpha-expressions

Although successive proton-transfer steps become weaker, a given step in the dissociation sequence is not 100% complete before the next one begins. Particularly with acids such as adipic, where the pK values are not very different, there is an overlapping of steps and all three species H_2A, HA^-, and A^{2-} are simultaneously present to *some* degree. The best algebraic approach to calculating the relative amounts of each species is by using alpha-expressions.

To illustrate the derivation of alpha-expressions we will use a triprotic acid H_3A. The three dissociation steps and their dissociation quotients are as follows (charges on the species are omitted for simplicity):

$$H_3A = H^+ + H_2A, \qquad Q_1 = \frac{[H^+][H_2A]}{[H_3A]}, \tag{9.25}$$

$$\begin{array}{cc} \text{parent} & \text{intermediate} \\ \text{acid} & \text{species} \end{array}$$

$$H_2A = H^+ + HA, \qquad Q_2 = \frac{[H^+][HA]}{[H_2A]}, \qquad (9.26)$$

$$HA = H^+ + A, \qquad Q_3 = \frac{[H^+][A]}{[HA]}, \qquad (9.27)$$

where A is a fully deprotonated species.

Suppose that an aqueous solution contains C moles per liter of the four species of this acid. **The material-balance expression** is:

$$C = [H_3A] + [H_2A] + [HA] + [A]. \qquad (9.28)$$

Now suppose that, by adding various amounts of some strong acid and/or some strong base, the pH of the solution is varied considerably. At higher molarities of H^+, reactions (9.25), (9.26), and (9.27) would be forced to the left, forming protonated species at the expense of the molarities of the deprotonated species. But at low values of $[H^+]$ (e.g., after addition of NaOH in excess) all reactions must shift to the right in order to establish equilibrium. At any given value of $[H^+]$ there will be a certain percentage of the acid in each of its four forms. We may derive expressions to show this as follows.

Each molarity may be written in terms of its conjugate species simply by rearranging the Q values. Thus,

$$[H_2A] = Q_1[H_3A]/[H^+],$$
$$[HA] = Q_2[H_2A]/[H^+] = Q_1Q_2[H_3A]/[H^+]^2,$$
$$[A] = Q_3[HA]/[H^+] = Q_1Q_2Q_3[H_3A]/[H^+]^3.$$

Each of the deprotonated species has now been expressed as a function of $[H_3A]$. These expressions are substituted into Eq. (9.28)

$$C = [H_3A] + \frac{Q_1[H_3A]}{[H^+]} + \frac{Q_1Q_2\,[H_3A]}{[H^+]^2} + \frac{Q_1Q_2Q_3\,[H_3A]}{[H^+]^3}$$

By dividing by C and factoring out $[H_3A]/C$, we find:

$$1 = \frac{[H_3A]}{C}\left(1 + \frac{Q_1}{[H]} + \frac{Q_1Q_2}{[H]^2} + \frac{Q_1Q_2Q_3}{H^3}\right),$$

which may be rearranged to an especially convenient form:

$$\frac{[H_3A]}{C} = \frac{[H^+]^3}{[H^+]^3 + Q_1[H^+]^2 + Q_1Q_2[H^+] + Q_1Q_2Q_3}.$$

Obviously, $[H_3A]/C$ is the fraction (not percentage) of the material present in solution as the parent acid. We call this fraction "alpha 3" and use the symbol α_3. (The subscript indicates the number of acidic protons on the species.)

By a similar series of substitutions that focus on the other species instead of on H_3A, the entire set of alpha-expressions is found:

$$\alpha_3 = \frac{[H_3A]}{C} = \frac{[H^+]^3}{[H^+]^3 + Q_1[H^+]^2 + Q_1Q_2[H^+] + Q_1Q_2Q_3},\tag{9.29}$$

$$\alpha_2 = \frac{[H_2A]}{C} = \frac{Q_1[H^+]^2}{\text{Same denominator}},\tag{9.30}$$

$$\alpha_1 = \frac{[HA]}{C} = \frac{Q_1Q_2[H^+]}{\text{Same denominator}},\tag{9.31}$$

$$\alpha_0 = \frac{[A]}{C} = \frac{Q_1Q_2Q_3}{\text{Same denominator}}.\tag{9.32}$$

Note that the numerator term in each expression is simply one of the successive terms in the denominator, which itself is a regular progression easy to remember. The first term is merely $[H^+]$ raised to a power equal to the number of acidic protons in the parent acid. Each successive term drops one power in $[H^+]$ and includes one more Q value.

For a diprotic-acid system there are three alpha-expressions, one for each species:

$$\alpha_2 = \frac{\text{Fraction present as}}{H_2A \text{ at a given } [H]} = \frac{[H_2A]}{C} = \frac{[H^+]^2}{[H^+]^2 + Q_1[H^+] + Q_1Q_2},$$

$$\alpha_1 = \text{Fraction present as } HA = \frac{[HA]}{C} = \frac{Q_1[H^+]}{\text{Same denominator}}.$$

$$\alpha_0 = \text{Fraction present as } A = \frac{[A]}{C} = \frac{Q_1Q_2}{\text{Same denominator}}.$$

The alpha-expressions for a monoprotic system are quite simple:

$$\alpha_1 = \text{Fraction present as } HA = \frac{[HA]}{C} = \frac{[H^+]}{[H^+] + Q},$$

$$\alpha_0 = \text{Fraction present as } A = \frac{[A]}{C} = \frac{Q}{[H^+] + Q}.$$

The reader should derive the alpha-expressions for the monoprotic system, beginning with the material-balance expression $C = [HA] + [A]$ and introducing the Q expression.

Degree of Dissociation as a Function of pH

Consider a solution of $0.001 M$ acid HA which has a Q_a value of $1 \cdot 10^{-8}$, and let the solution pcH be varied by addition of either hydrochloric acid or sodium hydroxide. Table 9.6 shows the results for the alpha-functions.

The only reason for showing the large number of decimal places is to emphasize that even at the pcH extremes of 0 and 14, the alpha-functions cannot become exactly 1 or 0: there is always a little of the minor form in the equilibrium system. Note, however, that beginning with $pcH = pQ = 8$ (where the dis-

Table 9.6

pcH	[H^+]	α_1	α_0
0	$1 \cdot 10^0$	0.99999999	0.00000001
1	$1 \cdot 10^{-1}$	0.9999999	0.0000001
2	$1 \cdot 10^{-2}$	0.999999	0.000001
3	$1 \cdot 10^{-3}$	0.99999	0.00001
4	$1 \cdot 10^{-4}$	0.9999	0.0001
5	$1 \cdot 10^{-5}$	0.999	0.001
6	$1 \cdot 10^{-6}$	0.99	0.01
7	$1 \cdot 10^{-7}$	0.91	0.09
$pQ_a = 8$	$1 \cdot 10^{-8}$	0.5	0.5
9	$1 \cdot 10^{-9}$	0.09	0.91
10	$1 \cdot 10^{-10}$	0.01	0.99
11	$1 \cdot 10^{-11}$	0.001	0.999
12	$1 \cdot 10^{-12}$	0.0001	0.9999
13	$1 \cdot 10^{-13}$	0.00001	0.99999
14	$1 \cdot 10^{-14}$	0.000001	0.999999

tribution is equal between the two species), each unit change in pcH is accompanied by a tenfold shift in the ratio of one form to the other.

When the pcH is three units lower than pQ, the dominant form HA accounts for 99.9% of the total. When the pcH is three units higher than pQ, the same can be said for species A. Thus variation in pH exerts tremendous control over the chemical state of the acid–base pair in the solution and this effect is exploited to advantage in separations, in analytical procedures, and in the study of equilibrium systems.

Applications
The chief use of an alpha-expression is to calculate the molarity of a species in a complex solution.

Example 1 What is the molarity of sulfide ion in a $0.1M$ solution of H_2S if the pH has been adjusted to 5.0 with a little hydrochloric acid? Assume the ionic strength is low enough to be ignored.

Given K_1 and K_2 for H_2S, $1 \cdot 10^{-7}$ and $1 \cdot 10^{-13}$ respectively, the answer is quickly found.

$$[S^{2-}] = C \cdot \alpha_0 = C \cdot \frac{Q_1 Q_2}{[H^+]^2 + Q_1[H^+] + Q_1 Q_2}$$

$$= 0.1 \cdot \frac{1 \cdot 10^{-7} \cdot 1 \cdot 10^{-13}}{(1 \cdot 10^{-5})^2 + (1 \cdot 10^{-7})(1 \cdot 10^{-5}) + (1 \cdot 10^{-7})(1 \cdot 10^{-13})}$$

$$= 0.1 \frac{1 \cdot 10^{-20}}{1.0 \cdot 10^{-10}} = 1 \cdot 10^{-11}.$$

Example 2 A solution containing $0.001M$ ammonia, among other solutes, has a pcH of 6.7. Find the molarity of ammonium ion, given its pQ_a value of 9.3.

From the alpha-expression, $\dfrac{[NH_4^+]}{C} = \dfrac{[H^+]}{[H^+] + Q_a}$, and so

$$[NH_4^+] = 0.001 \cdot \frac{2 \cdot 10^{-7}}{2 \cdot 10^{-7} + 5 \cdot 10^{-10}} = 9.975 \cdot 10^{-4} \approx 0.001,$$

whereas

$$[NH_3] = 0.001 \cdot \frac{5 \cdot 10^{-10}}{2 \cdot 10^{-7} + 5 \cdot 10^{-10}} = 2.5 \cdot 10^{-6}.$$

This calculation approach coordinates very nicely with practical laboratory work. Usually we know a value for C simply because we have mixed known molar quantities together. Also, measurement of pH is a fairly simple lab procedure. Thus, since the required Q values are available by correction of already tabulated K values, alpha calculations are quite convenient to carry out. They are indispensable in treating the principles of metal–chelate equilibrium systems, as will be seen later. Alpha-expressions are also the key to accurate calculation of theoretical titration curves.

Alpha-expressions may be used to produce a diagram that summarizes the percentages of the various forms of an acid as a function of pH at a constant ionic strength. For example, we consider the variation of pH in a solution of tetraprotic acid EDTA (ethylenediaminetetraacetic acid). At ionic strength equal to 0.1, the successive pQ_a values are 2.0, 2.67, 6.16, and 10.26. By letting these pQ values remain constant, while plugging various values of $[H^+]$ into the five alpha-expressions, we generate Table 9.7 (blanks indicate that α is less than 0.01).

Table 9.7 Alpha values for EDTA

pcH	α_4	α_3	α_2	α_1	α_0
0	0.990				
1	0.907	0.0907			
2	0.452	0.452	0.0966		
3	0.0309	0.309	0.660		
4		0.0044	0.949		
5			0.931	0.0644	
6			0.591	0.409	
7			0.126	0.873	
8			0.0142	0.980	
9				0.947	0.0520
10				0.645	0.355
11				0.154	0.846
12				0.0179	0.982
13					0.998
14					1.00

The results are seen more clearly in graphical form in Fig. 9.1. In such a distribution diagram the relative significance of all species at a given pcH may be seen at a glance.

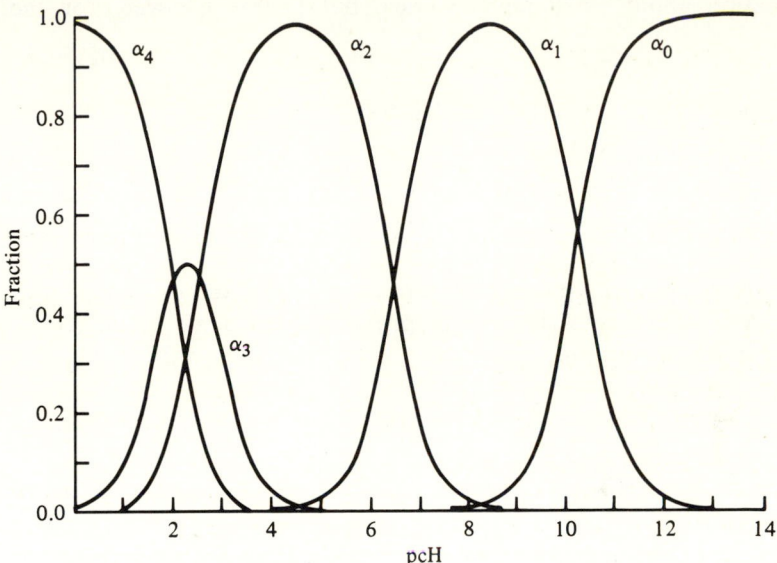

Fig. 9.1 Distribution of EDTA species as a function of pcH.

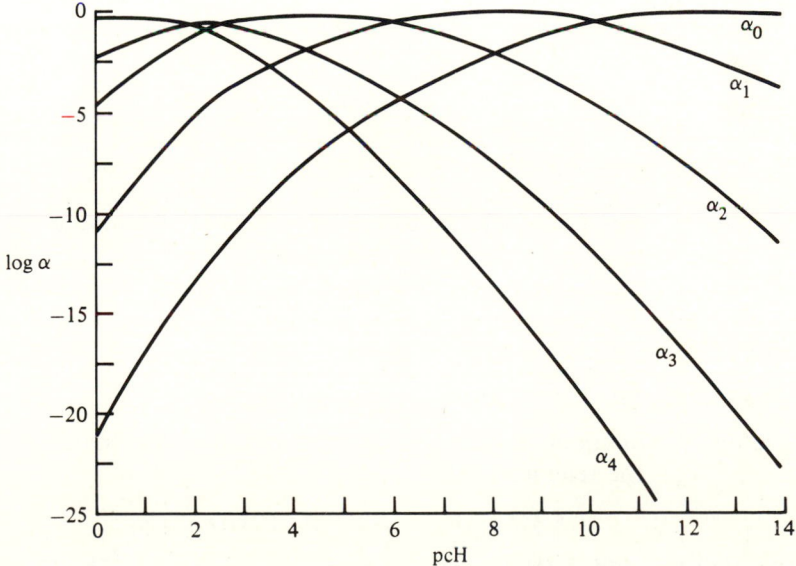

Fig. 9.2 Logarithmic version of Fig. 9.1.

However, with the linear alpha axis it is not possible to read values for very low alphas, and in some applications this is important information. Therefore, we may adopt an alternative mode of plotting, using the logarithm of the alpha values, as shown in Fig. 9.2. This form transmits much more information about small alpha values, but is less accurate for the higher values.

9.9 CALCULATION OF pH IN SOLUTIONS INVOLVING POLYPROTIC SYSTEMS

Nearly all the calculation approaches discussed for monoprotic acids and bases may be applied with only minor changes to polyprotic systems. However there are two important new aspects to consider: (1) the reaction stoichiometry is more complex, with more than one proton loss, or gain, possible; (2) intermediate species in the polyprotic sequence are **amphiprotic**, having both acid *and* base properties.

Stoichiometry of Diprotic Acid with Addition of Sodium Hydroxide

A discussion of diprotic systems will include the essential ideas, and the reader may easily extend these considerations to triprotic and higher systems. First we will consider the addition of certain amounts of sodium hydroxide to a solution of a diprotic acid H_2A.

NONE ADDED

A solution of H_2A in water dissociates primarily according to its first dissociation constant:

$$H_2A \overset{K_1}{=} H^+ + HA^-.$$

It is always true that the second dissociation is weaker than the first:

$$HA^- \overset{K_2}{=} H^+ + A^{2-},$$

Therefore it is usually quite sufficient to calculate the pH of a solution of H_2A using the same model as was derived for the monoprotic acid. Typically, the dissociation of the first step produces enough H^+ to repress the second step nearly completely. Sulfuric acid is exceptional in that K_2 is not so small, being equal to 0.01.

FORMATION OF H_2A, HA^- BUFFERS

When the amount of sodium hydroxide added is less than the amount of H_2A originally present, the reaction

$$H_2A + Na^+OH^- = Na^+HA^- + H_2O$$

proceeds until the added OH^- is almost completely used up. The HA^- thus formed, together with the H_2A that was present in excess, is a buffer mixture. The pH of this solution may be calculated using the monoprotic model:

$$[H^+] = Q_a \frac{[H_2A]}{[HA^-]}.$$

AMPHIPROTIC SUBSTANCE

When the amount of added NaOH is exactly equal to the amount of the H_2A originally present, the reaction goes nearly to completion, so that the solution is identical to the solution that would be prepared by dissolving the compound Na^+HA^- in water. This is different from the monoprotic case, which produces a solution of a weak base A^- when equal amounts of HA and NaOH are mixed. The HA^- species is called an **ampholyte**, or an **amphiprotic** species, meaning that it is both acidic and basic.

For example, the hydrogen oxalate ion $HC_2O_4^-$ may either donate a proton in a collision with a water molecule:

$$HC_2O_4^- + H_2O \rightleftharpoons C_2O_4^{2-} + H_3O^+ \tag{9.33}$$

$$\text{oxalate ion,}$$
$$\text{the conjugate}$$
$$\text{base of } HC_2O_4^-$$

or, with a different collision orientation, it may take a proton away from the water molecule:

$$HC_2O_4^- + H_2O \rightleftharpoons H_2C_2O_4 + OH^-. \tag{9.34}$$

$$\text{oxalic acid,}$$
$$\text{the conjugate}$$
$$\text{acid of } HC_2O_4^-$$

The equilibrium quotient for reaction (9.33) is the acid dissociation quotient for $HC_2O_4^-$ and is identical to what we call Q_2 for oxalic acid:

$$Q_a \text{ for } HC_2O_4^- = \frac{[H^+][C_2O_4^{2-}]}{[HC_2O_4^-]} = Q_2 \text{ for } H_2C_2O_4.$$

For reaction (9.34) the equilibrium quotient is the *base* dissociation quotient for $HC_2O_4^-$;

$$Q_b \text{ for } HC_2O_4^- = \frac{[OH^-][H_2C_2O_4]}{[HC_2O_4^-]}.$$

This is related to Q_1 for oxalic acid by the conjugate relationship:

$$Q_b \text{ for } HC_2O_4^- = \frac{Q_w}{Q_1 \text{ for } H_2C_2O_4}.$$

Noting that reaction (9.33) tends to make the solution acidic, while reaction (9.34) tends to make it basic, we have an apparent ambiguity to resolve. The key is in the *relative values* of Q_a and Q_b. Whichever is greater determines which of the two reactions dominates. For the amphiprotic species $HC_2O_4^-$, the Q_a value is much larger than Q_b, so the solution is acidic. By comparison, a solution of sodium bicarbonate is basic because Q_b for HCO_3^- is larger than its Q_a value.

To calculate the pH of a solution of an amphiprotic substance we must derive an equation that takes account of its molarity C and the fact that the substance has both acidic and basic properties. To keep the derivation general let us ignore ionic charges and consider the species HA with H_2A as its conjugate acid and A as its conjugate base. There are two acid-dissociation quotients to define:

$$H_2A = H^+ + HA, \qquad Q_1 = \frac{[H^+][HA]}{[H_2A]},$$

$$HA \ = H^+ + A, \qquad Q_2 = \frac{[H^+][A]}{[HA]};$$

there is also the omnipresent dissociation of water:

$$H_2O = H^+ + OH^-, \qquad Q_w = [H^+][OH^-].$$

Suppose a solution is prepared by dissolving C moles of HA per liter. Part is converted to H_2A and part is converted to A. Some of the water is converted to H_3O^+ and some is converted to OH^-. Let us use the principle of **proton balance**. When substances are mixed and proton-transfer reactions occur as a result, it must be true that the sum of the molarities of the species that have gained protons will be equal to the sum of the molarities of the species that have lost protons. In the solution of HA, the species that have gained protons are H_3O^+ and H_2A; those that have lost protons are OH^- and A. Therefore, according to proton balance at equilibrium, we have:

$$[H^+] + [H_2A] = [OH^-] + [A].$$

By substituting from the three equilibrium quotient expressions given above, we find:

$$[H^+] + \frac{[H^+][HA]}{Q_1} = \frac{Q_w}{[H^+]} + \frac{Q_2[HA]}{[H^+]}.$$

When we multiply all terms by $[H^+]$ and rearrange to solve for $[H^+]$, the result is an exact expression for the hydrogen-ion molarity in the solution:

$$[H^+] = \sqrt{\frac{Q_w + Q_2[HA]}{1 + [HA]/Q_1}}. \qquad (9.35)$$

Sample calculation The question is: What is the pH of a solution of $0.20M$ potassium hydrogen oxalate? Realizing that potassium salts are fully dissociated, we can write the nominal concentrations of ions as

$$[K^+] = [HC_2O_4^-] = 0.20.$$

Therefore, we take 0.20 to be the ionic strength also. The $HC_2O_4^-$ is amphoteric, we conclude, because a table of equilibrium constants shows for oxalic acid that

$$pK_1 = 1.27, \quad K_1 = 5.37 \cdot 10^{-2},$$
$$pK_2 = 4.27, \quad K_2 = 5.37 \cdot 10^{-5}.$$

At $I = 0.2$, we have $f_1 = 0.75$ and $f_2 = 0.31$, and therefore

$$Q_1 = K_1 \frac{1}{f_1^2} = 9.55 \cdot 10^{-2},$$

$$Q_2 = K_2 \frac{f_1}{f_1 f_2} = 1.73 \cdot 10^{-4},$$

$$Q_w = \frac{1 \cdot 10^{-14}}{f_1^2} = 1.78 \cdot 10^{-14}.$$

Using Eq. (9.35) we get

$$[H^+] = \sqrt{\frac{1.78 \cdot 10^{-14} + 1.73 \cdot 10^{-4} \cdot 0.200}{1 + 0.200/(9.55 \cdot 10^{-2})}}$$

$$= \sqrt{\frac{1.78 \cdot 10^{-14} + 3.46 \cdot 10^{-5}}{1 + 2.09}} = 3.35 \cdot 10^{-3},$$

$$aH = 2.51 \cdot 10^{-3}, \quad pH = 2.60.$$

Note that the value of Q_w in the numerator was totally insignificant. This will be true nearly always, except for cases when Q_2 is very small (for example, less than 10^{-11}) and the concentration is also small.

Quite often the $[HA]/Q_1$ term in the denominator is *much* larger than 1, although this obviously wasn't true in the above example.

If both of these conditions are met, then the equation takes a very simple form:

$$[H^+] = \sqrt{\frac{Q_2[HA]}{[HA]/Q_1}} = \sqrt{Q_1 Q_2} \quad \text{([HA] cancels)} \quad\quad (9.36)$$

or

$$pcH = \frac{pQ_1 + pQ_2}{2}, \quad\quad (9.37)$$

which is a convenient logarithmic form of Eq. (9.36).

The simple form works for all cases when both pQ_1 and pQ_2 are in the range from, say, 4 to 11. For example, the case of $0.1M$ sodium bicarbonate may be considered: carbonic acid has pK_1 and pK_2 values of 6.35 and 10.33, respectively. At $I = 0.1$, we have $f_1 = 0.78$ and $f_2 = 0.37$. Therefore, we may calculate

$$Q_1 = 7.34 \cdot 10^{-7}, \quad Q_2 = 1.26 \cdot 10^{-10}.$$

The exact Eq. (9.35) gives $[H^+] = 9.62 \cdot 10^{-9}$, and so does Eq. (9.36).

With a little perception you may show that Eq. (9.35) is not limited to amphiprotic cases, but embodies all that is needed for neutral solutes ($Q_1 = \infty$ and $Q_2 = 0$), for weak acids ($Q_1 = \infty$, $Q_2 = Q_a$ for the acid), and for weak bases ($Q_1 = Q_w/Q_b$, $Q_2 = 0$).

AMINO ACIDS: IMPORTANT EXAMPLES OF AMPHIPROTIC SUBSTANCES

Appendix 4 includes a number of amino acids important in biological systems. These are substances with molecular structures containing an amine group —NH_2 (whose basic properties are akin to those of ammonia) and one or more acidic groups (such as carboxyl —COOH or sulfhydryl —SH). The simplest amino acid is glycine H_2N—CH_2—COOH. In aqueous solution the amino acids undergo *intra*molecular proton transfer from the acidic group to the amine group; the resulting structure is called a **dipolar ion**, for it exhibits a separation of charge within the molecule even though it is neutral overall. For glycine the dipolar-ion structure is written as follows:

$$^+H_3N—CH_2—COO^-.$$

If a strong acid is added to a solution of an amino acid, the zwitter ion can accept a proton (i.e., it can act as a base):

$$^+H_3N—CH_2—COO^- + H^+ \rightleftarrows {}^+H_3N—CH_2—COOH,$$
glycinium ion

where the glycinium ion is the conjugate acid of glycine. And in basic solution the glycine can give up a proton (i.e., it can act as an acid).

$$^+H_3N—CH_2—COO^- \rightleftarrows H^+ + H_2N—CH_2—COO^-,$$
glycinate ion

where the glycinate ion is the conjugate base of glycine.

It is perfectly reasonable to regard the conjugate acid of glycine as a diprotic acid and to refer to the successive acid dissociations in terms of pK_1 and pK_2. The pK values in Appendix 4 follow this convention. Thus, the dipolar-ion form of glycine (and of other amino acids) is amphoteric. The pH of a solution of glycine, with no added acid or base from another source, is given by eqs. (9.35)–(9.37).

The **isoelectric point** of an amino acid is defined as the pH at which the solution contains equal concentrations of the cation form (conjugate acid) and the anion form (conjugate base). The relationship of this pH to the pK values is easily derived:

$$K_1 = \frac{[H^+]f_1[\text{Dipolar ion}]}{[\text{Cation}]f_1}, \qquad K_2 = \frac{[H^+]f_1[\text{Anion}]f_1}{[\text{Dipolar ion}]},$$

assuming the activity coefficient for the dipolar ion to be unity. By multiplying the K expressions and solving for the activity of hydrogen ion under the

Table 9.8 Equilibrium view of mixtures of a diprotic acid with a strong base: malonic acid + sodium hydroxide.

First step: $H_2Mal + NaOH = NaHMal + H_2O$;   *second step*: $NaHMal + NaOH = Na_2Mal + H_2O$.

Acid dissociation quotients: $Q_{a1} = \dfrac{[H^+][HMal^-]}{[H_2Mal]}$, $Q_{a2} = \dfrac{[H^+]Mal^{2-}]}{[HMal^-]}$.

$n_b/n_a = 0.1$ to 0.9	1.0	1.1 to 1.9	2.0	>2.0
First step incomplete so we have buffer mixture of NaHMal and excess H_2Mal. $[H^+] = Q_{a1} \dfrac{[H_2Mal]}{[HMal^-]}$	A stoichiometric mixture, solution of NaHMal only, an ampholyte. $[H^+] = (Q_{a1}Q_{a2})^{1/2}$	Second step incomplete so we have buffer mixture of Na_2Mal and NaHMal. $[H^+] = Q_{a2}\dfrac{[HMal^-]}{[Mmal^{2-}]}$	A stoichiometric mixture, solution of Mal^{2-} only, a weak base. $[OH^-] = \left(\dfrac{Q_w}{Q_{a2}}[Mal^{2-}]\right)^{1/2}$	With excess (xs) NaOH, the dissociation of Mal^{2-} is repressed. $[OH^-] = C_{xs\ NaOH}$

Table 9.9 Equilibrium view of mixtures of a diprotic base with a strong acid: ethylenediamine + hydrochloric acid

First step: $en + HCl = enH^+Cl^-$;   *second step*: $enH^+Cl^- + HCl = enH_2^{2+} + 2Cl^-$.

Base dissociation quotients: $Q_{b1} = \dfrac{[OH^-][enH^+]}{[en]}$, $Q_{b2} = \dfrac{[OH^-][enH_2^+]}{[enH^+]}$.

$n_a/n_b = 0.1$ to 0.9	1.0	1.1 to 1.9	2.0	>2.0
First step incomplete so we have buffer mixture of enH^+ and excess en. $[H^+] = \dfrac{Q_w}{Q_{b1}}\dfrac{[enH^+]}{[en]}$	A stoichiometric mixture, solution of enHCl only, an ampholyte. $[H^+] = \left(\dfrac{Q_w}{Q_{b1}}\dfrac{Q_w}{Q_{b2}}\right)^{1/2}$	Second step incomplete so we have buffer mixture of enHCl and enH_2Cl_2. $[H^+] = \dfrac{Q_w}{Q_{b2}}\dfrac{[enH_2^{2+}]}{[enH^+]}$	A stoichiometric mixture, solution of enH_2Cl_2 only, a weak acid. $[H^+] = \left(\dfrac{Q_w}{Q_{b2}}[enH_2^{2+}]\right)^{1/2}$	With excess (xs) HCl, the dissociation of enH_2^{2+} is repressed. $[H^+] = C_{xs\ HCl}$

condition that [cation] = [anion], we find:

$$[H^+] f_1 = (K_1 K_2)^{1/2} \qquad \text{or} \qquad pH = \frac{pK_1 + pK_2}{2}.$$

FORMATION OF HA⁻, A²⁻ BUFFERS

When the amount of sodium hydroxide is larger than that required to form the amphiprotic species but less than enough to remove all the second protons from the diprotic acid, the result is a mixture of HA^- and A^{2-}, a conjugate-pair buffer. For the calculation of pH, this type of buffer may be treated just as if it were a monoprotic case. One must be careful to take proper account of the stoichiometry, however. For example, if 50 mmol of H_2A and 65 mmol of NaOH are mixed and diluted to one liter, the buffer mixture contains $0.015 M$ Na_2A and $0.035 M$ NaHA.

FORMATION OF THE WEAK BASE A²⁻

If the amount of added NaOH is exactly twice the amount of H_2A originally present, the result is a solution that could not be distinguished from one prepared by dissolving a sample of Na_2A:

$$H_2A + 2Na^+OH^- \rightarrow Na_2A + 2H_2O.$$
$$\text{both "used up"}$$

Just as the pH of a solution of H_2A may be treated as if it were merely monoprotic, so may a solution of the diprotic base A^{2-} be treated as if it were a monoprotic weak base.

Stoichiometry of Diprotic Base Treated with Various Amounts of Strong Acid

When substance B, which can form BH^+ and BH_2^{2+}, is treated with a strong acid, the sequence of steps is simply the reverse of those just explained for H_2A plus NaOH. We begin with a solution of merely B, which may be treated as if it were a monoprotic base. Then a series of buffer solutions B, BH^+ are formed. When the amounts of strong acid and B are exactly equal, the amphiprotic species BH^+ is formed and the pH of the solution is calculated as described above for the species HA^-. Further addition of strong acid produces a series of BH_2^{2+}, BH^+ buffers, and finally the weak conjugate acid BH_2^{2+} is formed.

The stoichiometric deductions, along with the appropriate calculation formulas, are summarized in Tables 9.8 and 9.9.

9.10   PREPARATION OF BUFFERS WITH SPECIFIC pH AND IONIC STRENGTH

Part 1. Using the Monoprotic Model

It is frequently necessary in research to prepare a buffer solution that is uniquely designed for the experiment of the moment. This requires concern for the chemical and biological properties of the substances used, such as stability, toxicity, side reactions with other species in the mixture, etc. It may be important to specify the ionic strength, and the value of pH is often critical. Designing a buffer solution obviously is more difficult than merely calculating the pH of a given mixture, but the underlying principles and equations are very much the same. This will be shown by a few typical examples.

The steps to be followed are:

a) specify the pH, convert to aH;

b) choose a conjugate pair that is chemically suitable and with pK_a within about one unit of the desired pH;

c) specify the ionic strength and find the activity coefficients;

d) set up simultaneous equations using K_a and ionic strength;

e) calculate the individual values of molarities for the conjugate acid and base species;

f) decide what stock solutions will be mixed or what substances will be weighed out;

g) specify the volume of solution to be prepared and calculate the actual quantities of reagents to be used.

Example 1 Suppose a biochemist wants to study some enzyme behaviors in a variety of media, one of which is a solution with a reliable pH of 5.00 and an ionic strength of 0.200. The conjugate acid should have a pK_a value between 4 and 6, therefore, and inspection of Table 9.10 of acid-dissociation constants reveals a number of candidates.

Table 9.10

Acid	pK_a	Comment
Acetic	4.756	stable, pure, often used
Propionic	4.87	also OK
Butyric	4.817	odor of rotting flesh
Pyridinium ion	5.17	also has a repulsive odor
Anilinium ion	4.596	unstable to air oxidation
Hydroxylammonium ion	5.98	a reducing agent

We will choose propionic acid and use HPr as an abbreviated formula. At the ionic strength of 0.200 the activity coefficient is $f_1 = 0.75$. Therefore the *equilibrium ratio* of the conjugate-pair species can be calculated by rearranging the expression for K_a:

$$K_a = \frac{aH\,[Pr^-]f_1}{[HPr]f_0} \quad \text{and} \quad \frac{[Pr^-]}{[HPr]} = \frac{K_a}{aH} \cdot \frac{1}{f_1} = \frac{1.35 \cdot 10^{-5} \cdot 1}{1.00 \cdot 10^{-5} \cdot 0.75} = 1.80.$$

Of course, there is an infinite set of values of [Pr⁻] and [HPr] that would satisfy the ratio requirement of 1.80. What is needed is another relationship to help pin down the individual values appropriate for this buffer; and the answer is found by considering the ionic-strength requirement:

$$I(\text{total}) = I(\text{due to conj. acid}) + I(\text{due to conj. base}) + I(\text{due to any inert salts}).$$

In the present example, the conjugate acid is a neutral molecular species and so contributes nothing to ionic strength. The conjugate base is a singly charged anion and, if accompanied by a singly charged cation such as K^+, the contribution to the ionic strength is equal to its molarity. If the buffer contains no inert salt, such as sodium chloride, then the expression for ionic strength is

$$C_{KPr} = I = [Pr^-] = 0.200.$$

In this simple case, then

$$[HPr] = \frac{[Pr^-]}{1.80} = \frac{0.200}{1.80} = 0.111.$$

Finally, assuming that one liter of the buffer is to be prepared by weighing out pure propionic acid and potassium propionate, the quantities needed are:

propionic acid:	111 mmol (8.22 grams),
potassium propionate:	200 mmol (22.4 grams).

THREE CHEMICAL APPROACHES TO MAKING A CONJUGATE-PAIR BUFFER

In the example above it was assumed that pure propionic acid and potassium propionate were both available for weighing the desired quantities. This would certainly be a simple and direct approach and can be recommended when possible. However, in some cases it may be that the pure conjugate acid *or* the pure conjugate base may be available, but not both.

Example 1 Suppose that pure propionic acid is available but that the stockroom contains no salt of this acid. An easy solution to the problem is to use the reaction between propionic acid and sodium hydroxide:

$$HPr + NaOH \rightarrow NaPr + H_2O.$$
(in excess)

If 311 mmol of HPr are mixed with 200 mmol of NaOH and diluted to one liter, the result is a buffer mixture with the same properties as specified in the above example (pH = 5.00, $I = 0.200$), because the reaction will produce 200 mmol of NaPr (in solution) and the excess HPr will be 111 mmol.

On the other hand, suppose that some pure potassium propionate is available, but no propionic acid. The buffer may be produced through the reaction between propionate and a strong acid, such as hydrochloric acid:

$$KPr + HCl \rightarrow HPr + KCl.$$
(in excess)

But in contrast to the first two approaches, this method produces some inert salt (KCl) that will have an effect on the ionic strength:

$$I = C_{KPr} + C_{KCl},$$

where both KPr and KCl are 1:1 types and

$$0.200 = [Pr^-] + [HPr]$$

because the reaction produces equimolar HPr and KCl.

From the calculated ratio, $[Pr^-] = 1.80[HPr]$, and so

$$0.200 = 1.80[HPr] + [HPr],$$

or

$$[HPr] = \frac{0.200}{2.80} = 0.0714 \qquad \text{and} \qquad [Pr^-] = 1.80 \cdot 0.0714 = 0.1285.$$

For one liter of the buffer, then, we require $0.0714 + 0.1285 = 0.200$ mol of potassium propionate and 0.0714 mol of hydrochloric acid (e.g., 71.4 mL of $1.00M$ solution).

Example 2 A case of low pH

Using the pure solids (sodium bisulfate and sodium sulfate) prepare five liters of a buffer solution that has a pH of 1.80 and an ionic strength of 0.200.

For HSO_4^-, we have $K_a = 0.0100$, and at $I = 0.20$ the activity coefficients are $f_1 = 0.75$ and $f_2 = 0.31$. Therefore,

$$\frac{[SO_4^{2-}]}{[HSO_4^-]} = \frac{K_a f_1}{aH f_2} = \frac{0.0100 \cdot 0.75}{1.59 \cdot 10^{-2} \cdot 0.31} = 1.52.$$

Also,

$$I = 0.20 = 1 \cdot [HSO_4^-] + 3[SO_4^{2-}].$$

Therefore, by solving these simultaneous equations we find

$$[HSO_4^-] = 0.0360; \qquad [SO_4^{2-}] = 0.0547.$$

These are the equilibrium concentrations, but it is the stoichiometric concentrations that are needed to make the buffer. From Eqs. (9.17) and (9.18) it follows that (ignoring $[OH^-]$):

$$C_a = [HSO_4^-] + [H^+]; \qquad C_b = [SO_4^{2-}] - [H^+].$$

The value of $[H^+]$ is equal to $aH/f_1 = 1.59 \cdot 10^{-2}/0.75 = 2.12 \cdot 10^{-2}$. Therefore,

$$C_a = 0.0572; \qquad C_b = 0.0335.$$

The buffer may be prepared by using 286 mmol of $NaHSO_4$ and 168 mmol of Na_2SO_4, dissolved and diluted to five liters.

Example 3 The same low pH case approached differently

In Example 2 the buffer was prepared using a conjugate-pair system. When the pH is lower than about 2, an alternative is to use a solution of a strong acid. Suppose the desired pH (1.80) and ionic strength (0.20) are still specified. Since $f_1 = 0.75$ at $I = 0.20$, and since $aH = 1.59 \cdot 10^{-2}$, then

$$[H^+] = \frac{aH}{f_1} = \frac{1.59 \cdot 10^{-2}}{0.75} = 2.12 \cdot 10^{-2}.$$

This means that the desired concentration of hydronium ion can be achieved simply by having the solution contain $0.0212M$ strong acid (e.g., perchloric or hydrochloric acid), provided that the ionic strength is made to be 0.20 by the addition of some inert salt, such as sodium perchlorate or sodium chloride.

To prepare five liters of the buffer, one could use

$$5000 \cdot 0.0212 = 106.0 \text{ mmol of } HClO_4$$

and

$$5000 \cdot 0.1788 = 894.0 \text{ mmol of } NaClO_4.$$

Example 4 A case of high pH

A buffer with $pH = 12.00$ and ionic strength 0.150 is needed. At such a high pH it is valid to use a strong base such as potassium hydroxide. To find the necessary molarity of hydroxide ion, the following steps are used:

Since $pH = 12.00$,

$$pOH = 2.00 \qquad \text{and} \qquad aOH = 0.0100;$$

since $I = 0.15$,

$$f_1 = 0.76 \qquad \text{and} \qquad [OH^-] = \frac{0.0100}{0.76} = 0.0132.$$

Therefore, a reasonable procedure is to make a solution that contains

$$\begin{array}{l} 0.0132M \text{ KOH} \\ \underline{0.1368M \text{ KCl}} \quad \text{(inert salt to adjust ionic strength)} \\ 0.1500 = I. \end{array}$$

Example 5 Polyprotic acid plus sodium hydroxide

Given a supply of phosphoric acid H_3PO_4 and sodium hydroxide, we want to prepare one liter of a buffer of pH 6.50 and ionic strength 0.100. The three pK_a values for phosphoric acid are:

$$pK_1 = 2.12, \ pK_2 = 7.21, \ pK_3 = 12.32.$$

Since the specified pH is fairly close to the pK_2 value, we conclude that the buffer must be based on the conjugate pair $H_2PO_4^-$ and HPO_4^{2-}.

At $I = 0.1$, the activity coefficients are $f_1 = 0.78$ and $f_2 = 0.37$. Therefore,

$$K_2 = \frac{aH[HPO_4^{2-}]f_2}{[H_2PO_4^-]f_1}.$$

We may now find the ratio of concentrations using the specified data:

$$\frac{[HPO_4^{2-}]}{[H_2PO_4^-]} = \frac{K_a f_1}{aH f_2} = \frac{10^{-7.21} \cdot 0.78}{10^{-6.50} \cdot 0.37} = 0.410.$$

A second relationship of these concentrations follows from the ionic strength. Since the solution will consist of a mixture of NaH_2PO_4 (a 1:1 salt) and Na_2HPO_4 (a 1:2 salt), we have:

$$I = [H_2PO_4^-] + 3 \cdot [HPO_4^{2-}] = 0.100.$$

Solution of these simultaneous equations gives:

$$[H_2PO_4^-] = 0.0448; \qquad [HPO_4^{2-}] = 0.0184.$$

The sum of these values is 0.0632, and therefore this is the amount in (moles) of phosphoric acid needed. To find the amount of sodium hydroxide needed per liter we realize that the first proton must be removed from all the phosphoric acid, so 0.0632 mol of NaOH is needed for that purpose. Then another 0.0184 mol is needed to convert part of the $H_2PO_4^-$ to HPO_4^{2-}. The total amount of NaOH is therefore 0.0816 mol.

The buffer may be prepared by taking the following quantities and diluting to one liter:

$$\frac{0.0632 \text{ mol of } H_3PO_4}{14.8 \text{ mol/L}} = 0.00427 \text{ L } (4.27 \text{ mL}) \text{ of } 85\% \text{ } H_3PO_4$$

and

$$\frac{0.0816 \text{ mol of NaOH}}{19.4 \text{ mol/L}} = 0.00421 \text{ L } (4.21 \text{ mL}) \text{ of } 50\% \text{ NaOH.}$$

BEING REALISTIC: THE USE OF A pH-METER FOR FINAL ADJUSTMENT

The calculations discussed above are quite valuable for determining the approximate quantities of reagents needed to prepare buffers of desired pH and ionic strength. However, there are usually significant uncertainties in the pK values used and in the values of activity coefficients calculated by the Davies equation. Perhaps even the reagents are not as pure as one would like.

Therefore, once the buffer has been prepared with the calculated quantities, it is wise to measure the pH experimentally with a pH-meter, an instrument that can be found in any good laboratory. Typically the measured value is within 0.1–0.2 pH unit of the expected value and, if greater precision is desired, it is a simple matter to make some judicious additions of small quantities of a strong acid or strong base solution until the reading on the meter is as desired.

9.11   PREPARATION OF BUFFER SOLUTIONS WITH ROUNDED pH VALUES*

Specified amounts (in millimoles) of the reagents listed in Table 9.11 are mixed and diluted to 100 milliliters. (For a more complete listing see the references or the *Handbook of Chemistry and Physics*, Chemical Rubber Publishing Company.)

Table 9.11

pH	Reagents	pH	Reagents
1.00	5.0 KCl, 13.4 HCl	7.50	5.0 KH_2PO_4, 4.11 NaOH
1.50	5.0 KCl, 4.14 HCl	8.00	5.0 KH_2PO_4, 4.67 NaOH
2.00	5.0 KCl, 1.30 HCl	8.00	1.25 borax, 2.05 HCl
2.50	5.0 KHphthalate, 3.88 HCl	8.50	1.25 borax, 1.52 HCl
3.00	5.0 KHphthalate, 2.23 HCl	9.00	1.25 borax, 0.46 HCl
3.50	5.0 KHphthalate, 0.82 HCl	9.50	1.25 borax, 0.88 NaOH
4.00	5.0 KHphthalate, 0.01 HCl	10.00	1.25 borax, 1.83 NaOH
4.50	5.0 KHphthalate, 0.87 NaOH	10.50	1.25 borax, 2.27 NaOH
5.00	5.0 KHphthalate, 2.26 NaOH	11.00	2.5 Na_2HPO_4, 0.41 NaOH
5.50	5.0 KHphthalate, 3.66 NaOH	11.50	2.5 Na_2HPO_4, 1.11 NaOH
6.00	5.0 KH_2PO_4, 0.56 NaOH	12.00	2.5 Na_2HPO_4, 2.69 NaOH
6.50	5.0 KH_2PO_4, 1.39 NaOH	12.00	5.0 KCl, 1.20 NaOH
7.00	5.0 KH_2PO_4, 2.91 NaOH	12.50	5.0 KCl, 4.08 NaOH
		13.00	5.0 KCl, 13.2 NaOH

If a series of buffers is to be prepared, it is convenient to have stock solutions of the reagents as follows:

$0.200M$ KCl (15.11 g/L)
$0.100M$ KHphthalate (20.423 g/L)
$0.100M$ KH_2PO_4 (13.609 g/L)
$0.0250M$ borax, $Na_2B_4O_7 \cdot 10H_2O$ (9.534 g/L)
$0.0500M$ Na_2HPO_4 (7.098 g of anhydrous/L)
$0.200M$ HCl and $0.200M$ NaOH

---

*After reports by R. G. Bates and V. Bower, *Journal of Research, National Bureau of Standards*, *55*, 197 (1955) and *Analytical Chemistry*, *28*, 1322 (1956).

PROBLEMS

Practice in conversions
1. Fill in the following table:

	I	f_1	pQ_w	$[H^+]$	pcH	aH	pH	$[OH^-]$	pOH
a)	0.01			$1 \cdot 10^{-6}$					
b)		0.89			5.3				
c)	0.03						10.52		
d)		0.80				$1.3 \cdot 10^{-7}$			
e)			13.80				4.00		
f)	0.10								2.25
g)								0.0050	2.35
h)					12.70		12.82		

Practice in converting K values to Q values
2. Fill in the Q values: $Q = K/F$

Equilibrium reaction		pK	I	Q	Answers (as pQ)
a)	$H_2O = H^+ + OH^-$	14.00	0.060		13.82
b)	$HCN = H^+ + CN^-$	9.4	0.15		9.16
c)	$NH_4^+ = H^+ + NH_3$	9.24	0.02		9.24
d)	$HSO_4^- = H^+ + SO_4^{2-}$	2.0	0.25		1.47
e)	$Al^{3+} + H_2O = H^+ + AlOH^{2+}$	5.0	0.09		5.42
f)	$Cu^{2+} + H_2O = H^+ + CuOH^+$	8.3	0.002		8.34
g)	$HCOO^- + H_2O = OH^- + HCOOH$	10.25	0.04		10.25
h)	$PO_4^{3-} + H_2O = OH^- + HPO_4^{2-}$	1.7	0.11		2.14
i)	$SO_3^{2-} + H_2O = HSO_3^- + OH^-$	6.8	0.005		6.87

Recognizing and characterizing acids and bases

3. For each of the cases (a) through (t) refer to Appendix 4 on acid-dissociation constants. Characterize the substance given, identify the conjugate acid and conjugate base, and give their pK values. Write the K_a and K_b expressions. State whether a solution of the substance will be acidic or basic.

a) ammonium chloride
b) potassium acetate
c) hydriodic acid
d) sodium cyanide
e) potassium bromide
f) hydrogen fluoride
g) formic acid
h) hydroxylamine
i) potassium hydroxide
j) sodium nitrate

k) ethylammonium perchlorate
l) pyridine
m) potassium benzoate
n) "tris"
o) sodium chloride
p) potassium borate
q) phenol
r) propionic acid
s) potassium chlorite
t) hydrazine

Example Sodium nitrite.

Answer Sodium salts are completely dissociated in water solution, and therefore sodium nitrite will produce Na^+ and NO_2^- ions. The nitrite ion is the conjugate *base* of nitrous acid HNO_2. From Appendix 4 we find that the pK_a for nitrous acid is 3.3. This is the equilibrium constant for the acid-dissociation reaction:

$$HNO_2 + H_2O = H_3O^+ + NO_2^-, \qquad \text{or (abbreviated)} \quad HNO_2 = H^+ + NO_2^-.$$

The expression for K_a is

$$K_a = \frac{[H^+][NO_2^-]f_1^2}{[HNO_2]} = 10^{-3.3} = 5 \cdot 10^{-4}.$$

The expression for the base-dissociation constant for NO_2^- is written for the reaction

$$NO_2^- + H_2O = HNO_2 + OH^-,$$

$$K_b = \frac{[HNO_2][OH^-]f_1}{[NO_2^-]f_1} = \frac{1 \cdot 10^{-14}}{5 \cdot 10^{-4}} = 2 \cdot 10^{-11} \quad \text{(very small)}.$$

Thus since the solution contains sodium ion, which is neither acidic nor basic, and nitrite ion, which is a very weak base, the solution of sodium nitrite will be slightly basic.

Fundamental pH calculations

4. Two of the most commonly used acids are hydrochloric acid (strong) and acetic acid ($pK = 4.756$).

   a) Calculate the pH of the $0.10M$ solution for each acid.

   b) A certain solution of acetic acid has the same pH as that of $5.5 \cdot 10^{-4}M$ hydrochloric acid. What is the acetic-acid molarity? Ignore ionic strength.

   c) Calculate the pH of the $0.050M$ solutions of the sodium salts of the two acids.

   d) What molarity of HCl must be present in a solution containing $0.080M$ acetic acid if only 0.01% of the latter is dissociated? Ignore ionic strength.

5. Solutions of ammonia (a weak base, $pK_b = 4.76$) and of ammonium salts are used often in analysis.

   a) Calculate the pH of the $0.10M$ solutions of ammonia and of ammonium chloride.

   b) What is the pH of a buffer solution containing $0.20M$ NH_3 and $0.12M$ NH_4NO_3?

   c) A solution of $0.050M$ ammonium sulfate has the same ammonium-ion molarity as a solution of $0.10M$ ammonium bromide, but the ionic strengths are different. Calculate the pH of each solution, ignoring the very weak basic properties of the sulfate ion.

6. Sodium hydrogen carbonate $NaHCO_3$ dissociates to give an amphiprotic anion, but sodium carbonate Na_2CO_3 dissociates to give an anion that is basic only.

   a) Calculate the pH of the $0.10M$ solutions of these two salts.

b) What is the pH of a buffer solution containing $0.060M$ sodium bicarbonate and $0.020M$ sodium carbonate?

7. A technician prepares $0.1M$ solutions of potassium dihydrogen phosphate and sodium monohydrogen phosphate, but before the bottles are labeled he forgets which is which. Explain why a strip of litmus paper (which turns red in acidic solution, blue in basic solution) would solve the problem. What would you recommend if the second solution were sodium dihydrogen phosphate?

8. a) By using a pH-meter it is found that a $0.100M$ solution of a certain weak acid has a pH of 4.93. Ignoring activity coefficients, calculate the approximate pK value.

   b) Another solution of this acid is found to have a pH of 4.44. What is its approximate molarity?

9. Although solutions of pure sodium chloride in pure water are neutral, they do not have a pcH of 7.00 because of the ionic-strength effect. Derive an equation that relates the solution pcH to the NaCl molarity.

10. Will the pH of a $1M$ solution of a base be 7 pH units higher than a $1M$ solution of its conjugate acid, regardless of what the pK values are for the conjugate pair?

11. Examine the derivation for the equation relating the $[H^+]$ to the molarity for solutions containing only one weak acid. Suppose a solution is prepared with two (or even three) different weak acids, each having a different pK_a. Discuss.

12. On page 312 there is a list of practical buffers with rounded pH values. For each of the buffers calculate the pH (i.e., predict the pH on the basis of the pK values in Appendix 4) and compare the results with the actual values. (Well, that's too much work. Just do the even integers, pH $= 2.00$, 4.00, etc.)

13. There is a concept called **geologically pure rain**. This refers to rain that fell before the atmosphere became polluted. The dissolved species in such rain would be only nitrogen gas, oxygen gas, and carbon dioxide at a molarity of about $9.2 \cdot 10^{-6}$. Taking the acid-dissociation constant for CO_2 (which is partly hydrated to carbonic acid H_2CO_3) as $5.0 \cdot 10^{-7}$, calculate the pH of geologically pure rain.

14. A technician would like to have a neutral solution of $0.10M$ lithium perchlorate, but the stockroom does not have this salt. He places 100.00 mL (standardized pipet) of $0.2000M$ perchloric acid in a beaker and adds 100.00 mL of a solution of LiOH (a strong base) that is labeled $0.2000 \pm 0.0006M$. What is the approximate range of pH values that may be obtained for the mixture?

15. Given a pH-meter, how could you quickly tell which solution is which in the following pairs?

   a) $0.010M$ nitric acid and $0.20M$ nitrous acid
   b) $0.10M$ sodium acetate and $0.01M$ ammonia
   c) $0.10M$ sodium hydrogen adipate and $0.10M$ potassium monohydrogen citrate
   d) $0.05M$ alanine and $0.15M$ histidine
   e) $0.01M$ sulfuric acid and $0.01M$ perchloric acid

Calculation of equilibrium pH

Table 9.12 contains 130 brief descriptions of mixtures prepared with two solutes. In most cases proton transfer occurs, so that a weaker acid and base are formed. The object is to find the equilibrium pH-value (i.e., after the main reactions have occurred). Answers are given in Table 9.13. In each problem the equilibrium state will be classified as one of the seven fundamental types:

Code number for type	Equilibrium state
1	Inert ions only, a neutral solution
2	A strong acid (or a strong plus a weak acid)
3	A strong base (or a strong plus a weak base)
4	A weak acid, with or without inert ions present
5	A weak base, with or without inert ions present
6	A conjugate-pair buffer, both members present in stable amounts
7	An amphiprotic species, with or without inert ions present

To solve problems like these well, taking account of ionic-strength effects, a sense of organization is needed. Here are some logical steps to consider:

1. Note what solutes are in the initial mixture and state their acid–base characteristics (strong or weak, acid or base, inert or amphoteric). If you do not know these properties, use Appendix 4 to look up pK_a values for the species involved.

2. Let the proton-transfer reactions take place (on paper) until the species that is the limiting reactant is essentially used up.

3. Classify the resulting equilibrium mixture as one of the above seven types, according to what you know about the acid–base characteristics of the products.

4. Write the equilibrium-constant expression (including activity coefficients) that relates the conjugate pair to be considered in the equilibrium mixture.

5. Look up the necessary K value in Appendix 4. (If only rough calculations are needed, go to Step 8 and use K in place of Q.)

6. Calculate the ionic strength of the equilibrium solution, neglecting the minor effects of equilibria.

7. Find the activity-coefficient values needed and convert K to Q.

8. Calculate the hydrogen-ion molarity, using appropriate reasoning.

9. Convert $[H^+]$ to pH.

16. Table 9.12 Problems on pH of acid–base mixtures

Assume that the indicated numbers of millimoles are mixed and diluted to 100 mL. In Problem Aa, for example, 4 mmol of acetic acid plus 5 mmol of NaCl are diluted to 100 mL. For each problem, calculate the pH, including the effects of ionic strength.

	a) NaCl		b) HClO$_4$		c) KOH		d) HCl		e) NaOH	
A. KI	3	0	10	1	6	0.1	0	10	0	1
B. HBr	50	8	5	4	3	2	7	0	12	12
C. CH$_3$COOH	4	5	4	0	7	3	2	5	6	6
D. NH$_3$	9	10	8	3	10	0	3	3	5	7
E. HCOOH	8	15	1	3	5	5	10	0	15	4
F. HNO$_3$	4	20	4	0	4	4	4	4	4	2
G. NaCN	12	6	11	2	4	9	5	5	20	0
H. HF	13	7	8	0	2	6	4	3	7	7
I. NH$_2$OH	8	8	6	6	10	0	10	7	3	9
J. Phenol	7	9	3	3	3	2	4	0	8	8
K. "Tris"	2	11	6	3	5	6	4	4	10	0
L. HNO$_2$	6	12	20	0	8	8	3	2	3	5
M. H$_3$BO$_3$	5	13	4	6	9	4	12	0	11	7
N. Ethanolamine	4	14	10	10	20	0	10	8	5	15
O. H$_2$SO$_4$	10	16	20	0	12	12	10	20	10	20
P. Pyridine	13	17	7	4	2	2	11	11	30	0
Q. H$_3$PO$_4$	10	18	10	30	10	5	30	0	10	30
R. KH$_2$PO$_4$	15	0	9	9	8	8	5	2	5	2
S. K$_2$HPO$_4$	12	0	6	2	6	2	6	6	7	7
T. K$_3$PO$_4$	10	2	5	5	20	0	5	10	4	3
U. H$_2$CO$_3$	1	3	0.01	0	4	4	0.2	0.1	0.1	0.2
V. Ethylenediamine	1	4	2	1	1	1	2	2	3	0
W. Adipic acid	10	5	2	2	3	1	1	0	0.5	0.5
X. Citric acid	10	6	2	0	6	9	3	2	7	17
Y. Alanine	5	0	4	4	4	4	3	2	3	2
Z. Glycine	2	20	7	2	8	3	9	9	10	10

Buffer solutions (consult Appendix 4 for pK values).

17. Solution A is prepared by accurate dilution of a standardized HCl solution and contains nothing but $1.000 \cdot 10^{-4} M$ HCl. Solution B is prepared by mixing 64.58 mmol of benzoic acid with 49.88 mmol of sodium benzoate, dissolving and diluting to one liter.

a) Show that these solutions have the same pH.

b) Calculate the pH of the solution formed by diluting 100 mL of solution A to one liter and do the same for solution B. Why does one dilution change pH more than the other?

Table 9.13 Answers to problems on pH of acid–base mixtures (code for type, ionic strength, pH)
These answers were obtained by computer, with iteration, and therefore may differ somewhat from those obtained with slight approximations and round-offs.

	a) NaCl			b) HClO₄			c) KOH			d) HCl			e) NaOH		
A. KI	1	0.03	7.00	2	0.11	2.111	3	0.061	10.908	2	0.1	1.108	3	0.01	11.955
B. HBr	2	0.58	0.435	2	0.09	1.150	2	0.03	2.071	2	0.07	1.251	1	0.12	7.00
C. CH_3COOH	4	0.051	3.083	4	0.0009	3.082	6	0.03	4.561	2	0.05	1.387	5	0.06	8.676
D. NH_3	5	0.102	11.093	6	0.03	9.532	5	0.0014	11.117	4	0.03	5.452	3	0.07	12.749
E. HCOOH	4	0.155	2.437	2	0.03	1.594	5	0.05	8.14	4	0.0044	2.385	6	0.040	3.242
F. HNO_3	2	0.24	1.53	2	0.04	1.477	1	0.04	7.00	2	0.08	1.197	2	0.04	1.778
G. NaCN	5	0.18	11.112	6	0.11	9.94	3	0.13	12.839	4	0.05	5.35	5	0.2	11.22
H. HF	4	0.0809	2.062	4	0.0074	2.17	3	0.06	12.511	2	0.03	1.594	5	0.07	7.929
I. NH_2OH	5	0.08	9.451	4	0.06	3.703	5	"0"	9.5	6	0.07	5.728	3	0.09	12.85
J. Phenol	4	0.09	5.567	2	0.03	1.594	6	0.02	10.208	4	"0"	5.688	5	0.08	11.333
K. "Tris"	5	0.1102	10.18	6	0.03	8.14	3	0.06	12.687	4	0.04	4.813	5	0.0003	10.534
L. HNO_2	4	0.1267	2.287	4	0.0108	2.012	5	0.08	8.003	2	0.02	1.759	3	0.05	12.215
M. H_3BO_3	4	0.13	5.265	2	0.06	1.313	6	0.04	9.054	4	"0"	5.075	6	0.07	9.376
N. Ethanolamine	5	0.14	11.043	4	0.1	5.357	5	0.003	11.398	6	0.08	8.998	3	0.15	13.056
O. H_2SO_4	2	0.26	1.133	2	0.2	0.827	4	0.217	1.443	2	0.3	0.658	5	0.3	7.294
P. Pyridine	5	0.17	9.142	6	0.04	5.124	3	0.02	12.241	4	0.11	3.177	5	"0"	9.324
Q. H_3PO_4	4	0.211	1.64	2	0.30	0.66	6	0.0582	2.174	4	0.053	1.364	5	0.557	12.203
R. KH_2PO_4	7	0.1506	4.443	4	0.1181	1.664	7	0.2403	9.222	6	0.0549	2.396	6	0.09	6.721
S. K_2HPO_4	7	0.360	9.205	6	0.16	7.146	6	0.235	11.28	7	0.1203	4.481	5	0.385	12.107
T. K_3PO_4	5	0.5763	12.206	7	0.200	9.228	5	1.124	12.484	7	0.150	4.476	3	0.27	12.343
U. H_2CO_3	4	0.030	4.202	4	"0"	5.214	7	0.041	8.205	2	0.001	3.015.	5	0.0026	10.52
V. Ethylene-diamine	5	0.041	10.932	6	0.0101	9.947	3	0.01	11.955	7	0.0209	8.641	5	0.0016	11.182
W. Adipic acid	4	0.0524	2.714	2	0.02	1.759	6	0.0101	4.078	4	0.0006	3.223	7	0.006	4.847
X. Citric acid	4	0.0692	2.131	4	0.0034	2.5	6	0.120	4.498	2	0.02	1.759	6	0.3	5.591
Y. Alanine	7	"0"	6.107	4	0.04	2.024	5	0.04	11.128	6	0.02	2.39	6	0.02	10.062
Z. Glycine	7	0.2	6.108	6	0.02	2.859	6	0.03	9.486	4	0.09	1.851	5	0.1	11.277

c) Calculate the pH of the solution formed when 1.0 mmol of HCl (for example, 0.1 mL of $10M$ stock) is added to 100 mL of solution A, and do the same for solution B. Explain the difference in behavior.

d) Calculate the pH of the solution formed when 1.0 mmol of NaOH is added to 100 mL of solution A, and do the same for solution B. Explain.

18. Using stock solutions of $17.4M$ glacial acetic acid and $2.00M$ sodium hydroxide, how would you prepare one liter of a buffer that has a pH of 6.00 and an ionic strength of 0.0300?

19. A solution formed by mixing 184.7 mmol of propionic acid with 99.96 mmol of sodium propionate and diluting to one liter is found to have a pH of 4.500. Calculate the pK value for propionic acid.

20. What amount of ethylenediamine should be added to 200 mmol of HCl and diluted to one liter to produce a buffer that has a pH of 10.50?

21. A buffer is to be made with the following specifications: $pH = 10.00$, $I = 0.09$, concentration of NaCl 0.04. How would you prepare this, using stock solutions of $1.5M$ NaCl, $1.0M$ phenol, and $0.2M$ NaOH? Assume that one liter is required.

22. A certain buffer contains $0.100M$ HA and $0.070M$ NaA and has a pH of 6.43. What amount of NaOH should be added to two liters to adjust the pH to precisely 6.50? Assume that the ionic strength stays the same.

ACID–BASE TITRATIONS

10

Science is quantitative; alone of the humanities it possesses techniques of measurement from which derive the critical faculties of man—the faculties of discriminations between truth and error. While science is completely impersonal and gives no direct light on what is good or what is bad, it does tell us what is true and what is false in the material world. Is not this gift of science an incalculable one to the humanities?

L. M. Gould, 1945

The principal concepts of titration were discussed in Chapter 2 on aqueous solution stoichiometry, but this classical technique is important in so many varied applications that we should examine both the practical and the theoretical bases in detail. Later chapters will take up the topics of titrations involving metal–ligand reactions and oxidation–reduction reactions, but for now we will be sufficiently busy with a variety of proton-transfer titrations.

The practical goal of an acid–base titration is simple: to determine the precise volume of a solution of a base (or acid) required to react *stoichiometrically* with a sample of an acid (or base). Typically the sample is contained in a flask or beaker so that stirring is easy, and the titrant solution is delivered from a buret in a controlled, systematic, and cautious manner.

Immediately the question arises: How can we recognize when the correct (i.e., stoichiometric) quantity of titrant solution has been added? This correct volume of titrant solution is called the **equivalence point**, while our practical estimation of its value is called the **endpoint**. There are many possibilities for detection of the titration endpoint, and it is a challenge to find a method that brings the endpoint so close to the equivalence point that the error is negligible.

For acid–base titrations, the two most common methods for determining the endpoints of titrations are (1) use of colored indicators and (2) use of a pH meter to monitor the solution pH as a function of the volume of added titrant. Both of these methods can be applied intelligently only with a sound understanding of the theory of the **titration curve**, which is a plot of solution pH versus the volume of added titrant. Much of the following discussion will be devoted to this topic.

It is helpful to list the main types of titration situations:

a) The sample is a solution of a strong acid and the titrant is a strong base.

b) The reverse of (a).

c) A solution of weak acid is titrated with strong base.

d) A solution of weak base is titrated with strong acid.

e) A di- or polyprotic acid is titrated with strong base.

f) A di- or polyprotic base is titrated with strong acid.

First let us consider a direct and simple approach; then it will be useful to derive a fundamental equation that can be applied to all possible situations.

10.1   DIRECT CALCULATION: A STRONG ACID
TITRATED WITH A STRONG BASE

The essential ideas of the titration technique may be nicely illustrated by the reaction between HCl and NaOH. Suppose that a sample of hydrochloric acid is pipetted into a beaker. If the sample volume is V_0 mL and the molarity of

HCl is A_0, the sample contains

$$n_A = V_0 A_0 \text{ mmol of HCl.}$$

Now, if a buret is filled with B_0 molar NaOH solution that is to be added in measured volumes to the acid sample, the titration reaction is simply

$$H^+ + OH^- = H_2O.$$

Let V_B be the volume (mL) of NaOH solution added from the buret, so that at any point we have

$$n_B = V_B B_0 \text{ mmol of NaOH.}$$

The **equivalence point** is defined as the exact amount of NaOH required to react stoichiometrically with the HCl, so that $n_A = n_B$. We have

$$V_B(\text{e.p.}) = \frac{V_0 A_0}{B_0}; \tag{10.1}$$

in the absence of impurities the resulting solution would be merely a neutral solution of sodium chloride in water with pH = 7.00.

But before the equivalence point is reached, n_B is less than n_A and the molarity of *un*titrated H^+ is given by:

$$[H^+] \text{ (before e.p.)} = \frac{V_0 A_0 - V_B B_0}{V_0 + V_B}. \tag{10.2}$$

Once the equivalence point has been passed (so that n_B is greater than n_A), there will be an excess of hydroxide ion in the solution, the molarity of which is given by

$$[OH^-] \text{ (after e.p.)} = \frac{V_B B_0 - V_0 A_0}{V_0 + V_B}, \tag{10.3}$$

and, according to the equilibrium between hydronium and hydroxide ions,

$$[H^+] \text{ (after e.p.)} = \frac{Q_w}{[OH^-]}. \tag{10.4}$$

Let us apply these equations to a typical case. Suppose that 50 mL of $0.12M$ HCl is titrated with $0.15M$ NaOH. From Eq. (10.1) we find that the volume required for stoichiometric reaction is 40.00 mL. At the equivalence point, then, the molarity of NaCl will be

$$\frac{50 \text{ mL} \cdot 0.12M}{90 \text{ mL}} = 0.0667 \text{ mol/L.}$$

At this ionic strength, f_1 is about 0.80 and $Q_w = 1.56 \cdot 10^{-14}$. The molarity of hydronium ion at the equivalence point is, therefore,

$$[H^+] = (1.56 \cdot 10^{-14})^{1/2} = 1.25 \cdot 10^{-7}, \qquad \text{pcH} = 6.90.$$

This shows that the reaction is quantitative, with virtually all the H^+ used up at the equivalence point. Consider the calculated values of $[H^+]$ before and after the equivalence point, as summarized in Table 10.1. (The reader should verify some of these calculations.)

Table 10.1

V_B	$[H^+]$, Eq. (10.2)	$[OH^-]$, Eq. (10.3)	$[H^+]$, Eq. (10.4)	pcH
0	0.12			0.92
10	0.075			1.12
20	0.0429			1.37
30	0.0188			1.73
35	0.00882			2.05
39	0.00169			2.77
39.9	0.000167			3.78
40			$1.3 \cdot 10^{-7}$	6.90
40.1		0.000166	$9.4 \cdot 10^{-11}$	10.03
41		0.00165	$9.5 \cdot 10^{-12}$	11.02
45		0.00789	$2.0 \cdot 10^{-12}$	11.70

Note that even a slight (0.1 mL) excess of NaOH causes the molarity of hydronium ion to drop to a very low level. It is instructive to view the change in $[H^+]$ in two ways, as shown in Figs. 10.1 and 10.2.

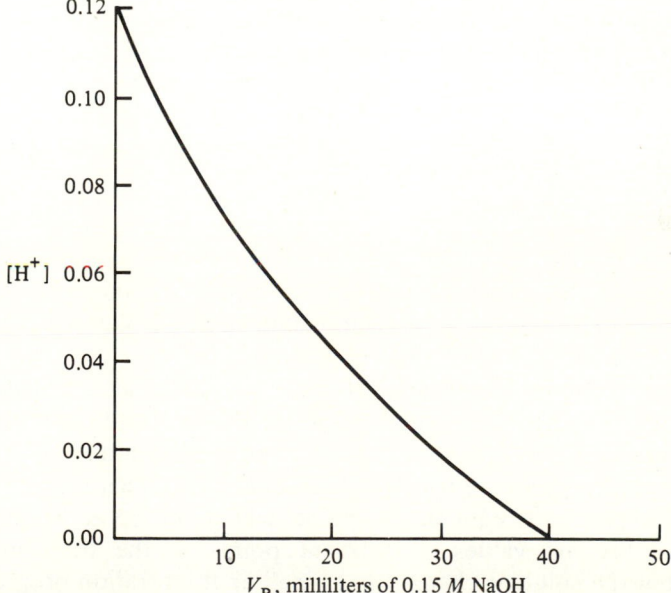

V_B, milliliters of 0.15 M NaOH

Fig. 10.1 If $[H^+]$ is plotted versus V_B, the line seems to intersect the axis at the equivalence point. The curvature is due to the changing volume of the solution. This plot gives no clear information on the way in which $[H^+]$ is changing in the immediate vicinity of the equivalence point.

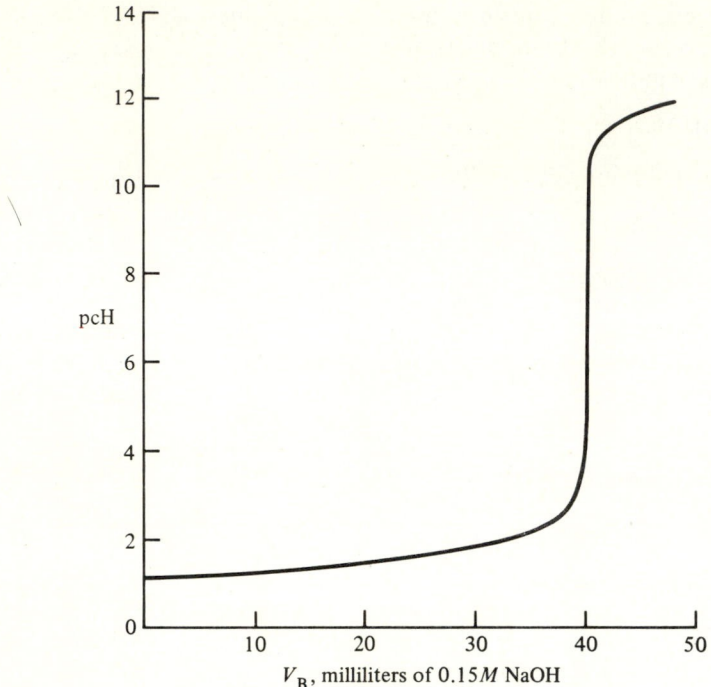

Fig. 10.2 By contrast, a plot of pcH dramatically reveals the huge relative changes in [H⁺] in the vicinity of the equivalence point. Such information is needed for accurate determination of the value of V_B at the stoichiometric point, and so this type of plot is much more valuable than the nonlogarithmic plot.

10.2   DIRECT CALCULATION: A WEAK ACID TITRATED WITH A STRONG BASE

If a monoprotic weak acid is titrated, the stoichiometric relationships in terms of millimoles are the same as those shown above for the strong-acid case. Thus, Eq. (10.1) still holds as the relationship between volumes and concentrations.

However, a different approach is required in the weak-acid case, because the acid is only slightly dissociated. Suppose that a 50-mL portion of $0.12M$ acetic acid (a weak acid with $pK_a = 4.76$) is titrated with a solution of $0.15M$ sodium hydroxide. There are 6 mmol of the acid, and so 6 mmol of NaOH will be required to reach the equivalence point, which will be at 40.00 mL added. To calculate the pH values at different points in the titration we must recognize that the solution changes in "type" as the titration progresses.

1. At the start of the titration, before any NaOH has been added, the solution is merely that of a weak acid, dissociating to give slight and equal concentrations of H⁺ and A⁻.

2. When the amount of added NaOH is less than 40 mL, part of the acid is converted to A⁻ and part remains in the undissociated HA form. This means that a series of conjugate-pair buffers is formed with continuously varying ratios of conjugate acid to conjugate base.

3. At the equivalence point, when exact stoichiometric amounts of the weak acid and the NaOH are mixed, the acid has been stoichiometrically converted to its conjugate base, and so the solution is treated as one containing a weak base only.

4. Once the equivalence point has been passed, the presence of the excess NaOH is sufficient to repress the slight dissociation of the weak conjugate base, and so the solution is essentially a dilute solution of a strong base only.

The calculations for these four "regions" will now be illustrated for the above case. For simplicity, ionic-strength effects will be ignored.

1. Initial pH (see Eq. (9.12)).

For a weak acid solution, $[H^+] = (C_a Q_a)^{1/2}$. With $C_a = 0.12$ and $Q_a = 1.74 \cdot 10^{-5}$,

$$[H^+] = (0.12 \cdot 1.74 \cdot 10^{-5})^{1/2} = 1.44 \cdot 10^{-3};$$
$$pH = 2.84.$$

2. Calculations in the Buffer Region

WHEN 10 mL OF NaOH HAVE BEEN ADDED

The amount of acid originally present is 6.00 mmol, and when $10 \cdot 0.15 = 1.5$ mmol of NaOH have been added, the titration reaction produces 1.5 mmol of Na⁺A⁻, leaving 4.5 mmol of undissociated HA. By using the buffer equation and Eq. (9.22), we find:

$$[H^+] = Q_a \frac{C_a}{C_b} = 1.74 \cdot 10^{-5} \cdot \frac{4.5/60}{1.5/60} = 5.22 \cdot 10^{-5};$$
$$pH = 4.28.$$

Note that the total volume, 60 mL, appears in the numerator and denominator and thus cancels. The millimole ratio may be used instead of the molarity ratio.

WHEN 20 mL OF NaOH HAVE BEEN ADDED

This is just halfway to the equivalence point, and the amounts of A⁻ and HA are equal at 3.00 mmol each. This point is called the **midpoint of the buffer region**, and pH $\approx$ pK = 4.76.

WHEN 30 mL OF NaOH HAVE BEEN ADDED

Now 75% of the weak acid has been converted to its conjugate base. The calculation is:

$$[H^+] = 1.74 \cdot 10^{-5} \cdot \frac{1.5}{4.5} = 5.8 \cdot 10^{-6};$$

$$pH = 5.24.$$

GETTING CLOSE TO THE EQUIVALENCE POINT: WITH 39 mL ADDED

The amount of conjugate base is $39 \cdot 0.15 = 5.85$ mmol, leaving only 0.15 mmol of untitrated weak acid. The buffer equation still holds:

$$[H^+] = 1.74 \cdot 10^{-5} \cdot \frac{0.15}{5.85} = 4.46 \cdot 10^{-7};$$

$$pH = 6.35.$$

This type of calculation is easily repeated for as many points as desired in the buffer region of the titration curve. However, the simple buffer equation becomes less reliable at the two extremes of the region: when only a very slight amount of NaOH has been added and when the point is very close to the equivalence point. The reader should explain why this is true.

3. Calculation of the Equivalence-Point pH

When exactly 40 mL of $0.15M$ NaOH has been added, the weak acid is stoichiometrically converted to its conjugate base. For the case under consideration, the total volume at the equivalence point is 90 mL and the concentration of the weak base is

$$C_b = \frac{6.00 \text{ mmol}}{90 \text{ mL}} = 0.067M.$$

To calculate the pH, we first use the weak-base model to find the $[OH^-]$ (see Eq. (9.14):

$$[OH^-] = (C_b Q_b)^{1/2} = \left(0.067 \cdot \frac{1 \cdot 10^{-14}}{1.74 \cdot 10^{-5}}\right)^{1/2} = 6.2 \cdot 10^{-6};$$

then,

$$pOH = 5.21, \quad \text{and} \quad pH = 14 - 5.21 = 8.79.$$

Important point   The equivalence-point pH in the titration of a weak acid is greater than 7.0. Although we speak of "neutralizing" the weak acid with the strong base, this does not mean that the solution turns out to be neutral in terms of pH. The weaker the acid being titrated, the higher the pH will be above 7.0. Only strong acid–strong base titrations have pH $= 7.0$ at the equivalence point.

4. After the Equivalence Point

The addition of excess NaOH is treated in the same way as when it is added to pure water. The [OH⁻] is taken simply as the molarity of the excess NaOH. For example, when a total of 41 mL have been added, the excess NaOH is

$$C_{sb} = \frac{1.0 \text{ mL} \cdot 0.15M}{91 \text{ mL}} = 0.00165M;$$

$$pOH = 2.78, \quad \text{and} \quad pH = 11.22.$$

When 45 mL have been added, a similar calculation shows that pH = 11.90.

These calculated results are shown as circles in Fig. 10.3. The curve has been drawn to indicate the overall relationship for the titration. Compared with the case of the strong acid–strong base titration, we note that the pH values before the equivalence point are higher but after the equivalence point they are the same.

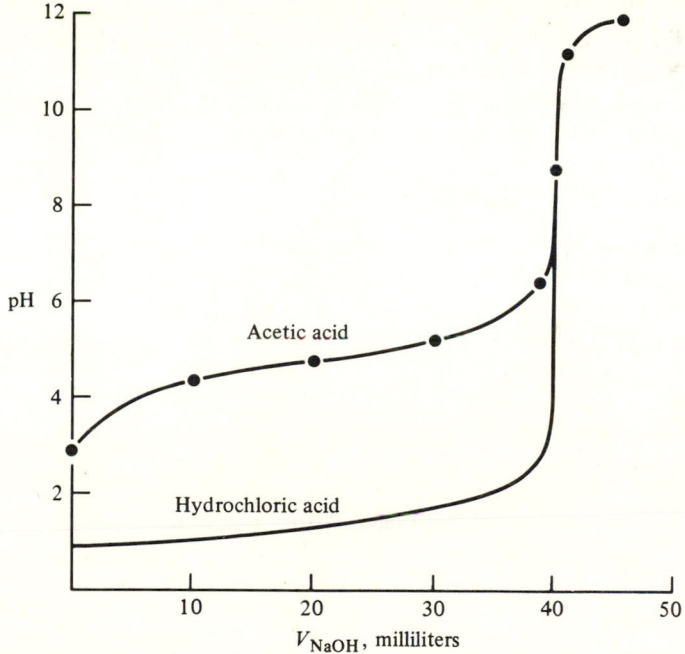

Fig. 10.3 Comparison of titration curves for strong and weak acids. 50 mL of 0.12M acid titrated with 0.15M NaOH.

10.3   EFFECT OF CONCENTRATION ON THE TITRATION CURVES

Strong Acid

When a strong acid is titrated with a strong base it is always true that the equivalence-point pH will be 7.0, since the titration reaction produces an inert salt such as NaCl. However, the initial pH and the values of pH before and after the equivalence point will depend markedly on the concentration of the solutions. For example, when the initial concentration of HCl is 0.1, we know the initial pH = 1. But if the initial HCl concentration is 0.01, the pH = 2 and when the initial concentration is 0.001, the pH = 3, etc. Titration curves calculated for the titrations of 0.1, 0.01, and $0.001M$ HCl with NaOH of equal concentration are shown in Fig. 10.4.

Note particularly that the extent of the change in pH in the vicinity of the equivalence point is markedly less for the dilute solutions. We will see later that this makes the choice of a visual indicator more critical.

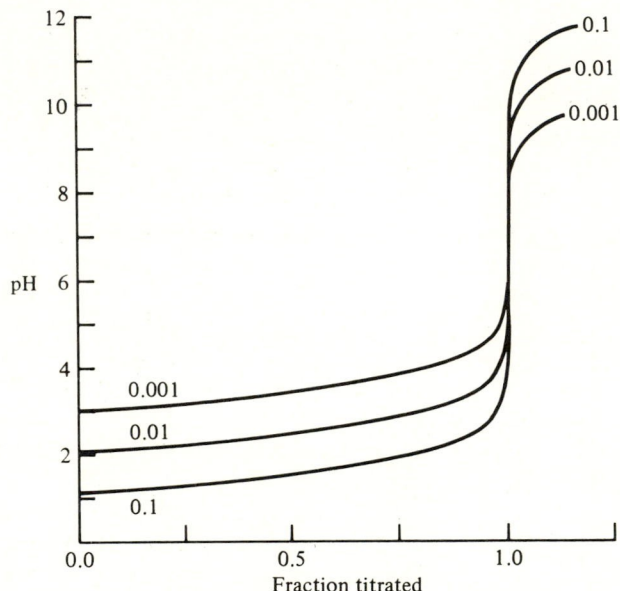

Fig. 10.4 Effect of concentration on strong acid–strong base titration curves. The effect of changing ionic strength is included.

Weak Acid

Because the pH of a conjugate-pair buffer solution is determined chiefly by the acid-to-base ratio, it is not very sensitive to dilution. Therefore, as seen in Fig. 10.5, the titration curve for acetic acid of various concentrations (0.1, 0.01 and $0.001M$) shows much the same form in the buffer regions. Of course, once excess NaOH has been added, the pH follows the same course as in the strong-acid titrations. In the titration of $0.001M$ acetic acid with $0.001M$ NaOH, the equivalence-point pH is only slightly above 7.0.

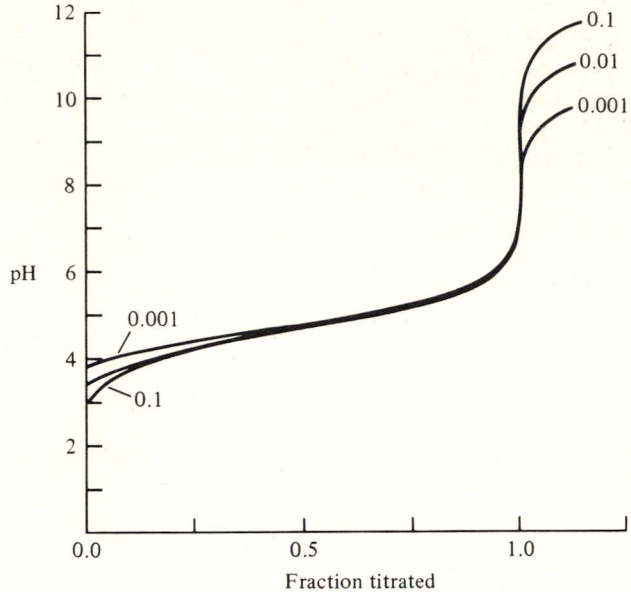

Fig. 10.5 Effect of concentration of weak acid–strong base titration curves. The NaOH titrant solution is of the same molarity as the initial acid molarity.

10.4 HOW TO UNIFY THE CALCULATIONS OF TITRATION CURVES

For any given titration it is possible to calculate the solution pH at any point, given the pertinent volumes and concentrations, by using the methods discussed earlier. However, it is much easier and usually more accurate to use an "inverse" approach proposed by Waser.*

In this approach, we first derive an equation that relates the volume of added titrant to the hydrogen-ion concentration of the solution. Then, by substituting a series of values for $[H^+]$, we calculate the corresponding values of V. It will become clear that this method permits the use of an exact equation (which may involve $[H^+]$ raised to the fourth or fifth power), without the need either for some sophisticated numerical analysis technique or for simplifying approximations.

For illustration, such an equation will be derived for the case of a diprotic acid being titrated with sodium hydroxide. The simplification of the equation to cover the cases of monoprotic acids (strong or weak) will be easy, and the extension of the equation to polyprotic acids will be fairly obvious.

Derivation

To define symbols, let an original volume (V_0 mL) of a solution of a diprotic acid H_2A be titrated with a solution of sodium hydroxide having a molarity B_0.

---

*J. Waser, *Journal of Chemical Education*, **44**, 274 (1967).

Let the original concentration of H_2A be called A_0 and let the two dissociation quotients for the diprotic acid be Q_1 and Q_2.

The chemical species involved are identified by the equations for the titration reactions that take place in two steps:

$$H_2A + Na^+OH^- = Na^+HA^- + H_2O$$

and

$$Na^+HA^- + Na^+OH^- = Na^+A^{2-}Na^+ + H_2O,$$

where, of course, we assume the ions to be completely dissociated in the solution.

As the titration proceeds, the molarities of the various species are changing not only because of the chemical reactions, but also because the solution is increasing in volume. The total volume at any point is $(V_0 + V_b)$, where V_b is the volume of the added titrant. Due to the volume increase, at any point in the titration we have:

$$[H_2A] + [HA^-] + [A^{2-}] = A = \frac{A_0 V_0}{V_0 + V_b}$$

and

$$[Na^+] = B = \frac{B_0 V_b}{V_0 + V_b}.$$

The electroneutrality law must hold at all times, and therefore

$$[Na^+] + [H^+] = [OH^-] + [HA^-] + 2[A^{2-}]. \tag{10.5}$$

To find substitutes for the terms in Eq. (10.5) we remember that

$$[OH^-] = \frac{Q_w}{[H^+]};$$

$$[HA^-] = \alpha_1 A = A \frac{Q_1[H^+]}{[H^+]^2 + Q_1[H^+] + Q_1 Q_2};$$

$$[A^{2-}] = \alpha_0 A = A \frac{Q_1 Q_2}{[H^+]^2 + Q_1[H^+] + Q_1 Q_2}.$$

When all these relationships are substituted into the electroneutrality expression, the result is somewhat intimidating:

$$[H^+] + \frac{B_0 V_b}{V_0 + V_b} = \frac{Q_w}{[H^+]} + \frac{A_0 V_0}{V_0 + V_b} \cdot \frac{Q_1[H^+] + 2Q_1 Q_2}{[H^+]^2 + Q_1[H^+] + Q_1 Q_2}. \tag{10.6}$$

However, Eq. (10.6) does have the desired character that the only variables it contains are $[H^+]$ and V_b. The values of A_0, V_0, B_0, Q_1, and Q_2 are all specified and constant for whatever titration we wish to consider, assuming constant ionic strength.

If a certain numerical value of V_b were inserted into Eq. (10.6) it would be necessary to solve a quartic equation (there would be an $[H^+]^4$ term upon rearrangement) to find the value of $[H^+]$. But if the equation is rearranged, first by multiplying through by $(V_0 + V_b)$, then grouping terms that contain V_b, and finally solving for V_b, the result is:

$$V_b = \frac{V_0\{A_0(\alpha_1 + 2\alpha_0) - [H^+] + Q_w/[H^+]\}}{B_0 + [H^+] - Q_w/[H^+]}. \qquad (10.7)$$

With this form, the entire right side is estimated for a specified value of $[H^+]$, and the result is equal to the volume of added titrant. To use Eq. (10.7) the best approach is to substitute successive values of $[H^+]$ corresponding to pcH values taken in, say, half-pH-unit steps over the desired range. It requires only a simple BASIC computer program to accomplish this. In fact, Eq. (10.7) is easily accommodated in a small, programmable calculator.

Extension of the Equation to Polyprotic Acids

For acids with three or more protons, the same derivation is used, but starting with additional terms in the electroneutrality expression. The form of the final equation is unchanged, but the alpha-term is expanded as follows:

$$V_b = \frac{V_0\{A_0(\alpha_{n-1} + 2\alpha_{n-2} + \cdots + n\alpha_0) - [H^+] + Q_w/[H^+]\}}{B_0 + [H^+] - Q_w/[H^+]}, \qquad (10.8)$$

where n is the number of acidic protons on the species being titrated.

Simplification of the Equation for Monoprotic Acids

In this case only α_0 is needed in the parentheses, so that we have:

$$V_b = \frac{V_0\{A_0 Q/([H^+] + Q) - [H^+] + Q_w/[H^+]\}}{B_0 + [H^+] - Q_w/[H^+]} \qquad (10.9)$$

where Q is the dissociation quotient for the monoprotic acid.
Should the acid be strong with a very large value for Q, then

$$\frac{Q}{[H^+] + Q} \approx 1, \qquad V_b = \frac{V_0(A_0 - [H^+] + Q_w/[H^+])}{B_0 + [H^+] - Q_w/[H^+]}. \qquad (10.10)$$

The calculations then become quite easy. The reader should use Eq. (10.10) to find some of the values tabulated earlier for the titration of HCl with NaOH.
It should be emphasized that Eq. (10.8) and its special forms such as Eqs. (10.7), (10.9), and (10.10) are exact for the entire titration curve, if the ionic-strength effects are neglected. This is an advantage over the approach that must use different equations for each region of the titration curve.

A Similar Equation for the Titration of Base with Acid

When a strong acid, e.g., hydrochloric acid, is used as titrant for a solution of a base, the pH of the solution continuously decreases as the titration progresses. Suppose the base, symbolized by B, is capable of taking on successive protons:

$$B \xrightarrow{\ H^+\ } BH^+ \xrightarrow{\ H^+\ } BH_2^{2+} \xrightarrow{\ H^+\ } etc.$$

The electroneutrality expression for the case of titration with HCl is as follows:

$$[Cl^-] + [OH^-] = [H^+] + [BH^+] + 2[BH_2^{2+}] + 3[BH_3^{3+}], \quad etc.$$

This equation should be compared with Eq. (10.5). Perhaps on some dull, rainy afternoon the reader will take time to prove that the equation for the titration curve is:

$$V_a = \frac{V_0\{B_0(\alpha_1 + 2\alpha_2 + 3\alpha_3 + \cdots) + [H^+] - Q_w/[H^+]\}}{A_0 - [H^+] + Q_w/[H^+]}. \tag{10.11}$$

This equation is the counterpart of Eq. (10.8) and is subject to the same type of simplification for the case of a monoprotic base. Again, the solution of Eq. (10.11) is a mátter best handled by a simple BASIC program or a programmable calculator.

The Problem of Changing Ionic Strength

Because proton transfer always requires a change in ionic charges of the species involved in a titration reaction and because the volume of the solution changes in the process, the ionic strength will vary continuously. If one wishes to calculate an exact titration curve taking these factors into account, a computer program should be used. It is necessary to use repeated iterations (successive approximations) to handle this complex calculation. The curves shown in this chapter were calculated and plotted by computer, and they may be regarded as close representations of the data that would be obtained in the laboratory using a calibrated pH-meter.

However, it should be pointed out that for general discussions of titration curves it is hardly necessary to resort to the exact curves. Plots of titration curves calculated *without* taking account of ionic-strength effects do not differ greatly from the exact versions, as Fig. 10.6 shows.

10.5   EXAMPLES OF ACID–BASE TITRATION CURVES

The curves in Figs. 10.7–10.9 were calculated by a computer program for the following conditions:

1. The initial volume of the solution is 50 mL.
2. The initial molarity of the species being titrated is $0.024/N$, where N is the number of protons that can be given up or accepted.

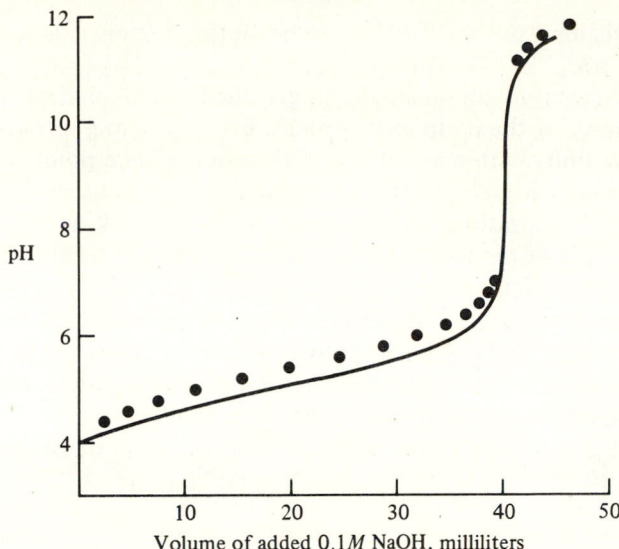

Fig. 10.6 Titration curve: 50 mL of $0.08M$ potassium hydrogen phthalate with $0.1M$ NaOH. *Solid curve*—the exact version with ionic strength taken into account; *dots*—points calculated under the assumption that there is no ionic-strength effect.

3. The molarity of the titrant, which is sodium hydroxide or hydrochloric acid depending on whether an acid or a base is being titrated, is $0.03M$.

4. The temperature is 25°C.

5. The ionic strength is due only to the various species of the substance being titrated and the titrant.

A Group of Curves for Monoprotic Acids

Let us examine four titration curves for acids of differing strength, as shown in Fig. 10.7. The acids are:

$$\begin{array}{rll}
\text{nitric acid, } HONO_2: & pK = -2 & \text{(strong acid),} \\
\text{nitrous acid, } HONO: & pK = 3.3, & \\
\text{hydroxylammonium ion, } HONH_3^+: & pK = 6.0, & \\
\text{ammonium ion, } NH_4^+: & pK = 9.24 &
\end{array}$$

A number of observations should be made about these curves. First, note the differences in the very first portions. As the pK value for the acid becomes larger (weaker acid), the initial pH region becomes much more sensitive to small additions of sodium hydroxide, because the initial dissociation of the weaker acids is so small.

Second, all of the acids display a relatively small slope in the middle portions of the titration curve, which is good reason to refer to these portions as the **buffer regions**. At the middle of the buffer region it is approximately true that $pH = pK_a$.

Third, the stronger the acid, the larger the rate of change of pH in the immediate vicinity of the equivalence point. For the strong nitric acid, the pH changes several units within about 1% of the equivalence point. Later we will see that this makes it easy to choose an acid–base indicator for finding the endpoint of such a titration. By contrast, the very weak acids show such a slight change of pH at the equivalence point that it is not feasible to determine the endpoint with indicators.

Finally, note that in all cases the curves merge to a common set of pH values when excess sodium hydroxide is added. This is understandable in terms of the repression of the dissociation of water, once it is no longer necessary for the added hydroxide to react with the weak acid.

Table 10.2 gives the pH and volume data for some of the points on the curves of Fig. 10.7.

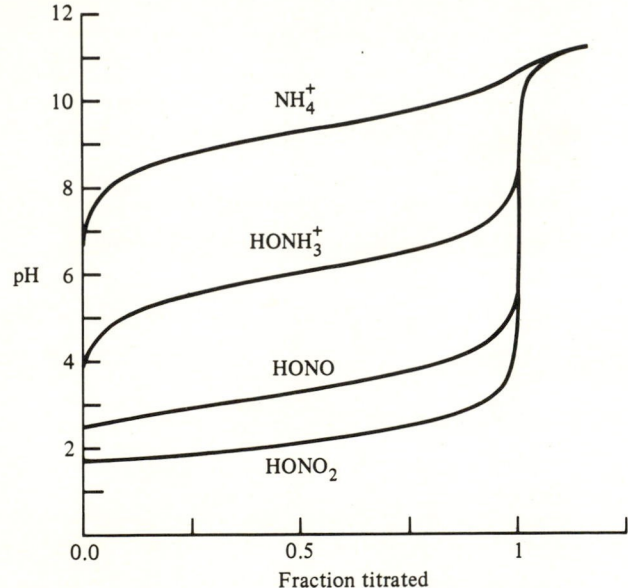

Fig. 10.7 Effect of pK value on weak acid–strong base titration curves.

The values for fraction of acid titrated given in Table 10.2 support the observations made on Fig. 10.7. Clearly, the stronger the acid, the greater the rate of pH change in the vicinity of the endpoint. For nitric acid the pH changes a full four units between 99.9% and 100.1% titrated. Taking the equivalence point as the center of the quickly changing pH range, we note that the pH at the equivalence point becomes higher and higher as the acid becomes weaker. Note also that the curves merge to common values once the equivalence point has been passed.

Table 10.2   Data for Titration of 50 mL of $0.024M$ acids with a solution of $0.030M$ sodium hydroxide. At the equivalence point, 40.00 mL of NaOH would be required. Fraction $= V_b/40.00$.

	Fraction of acid titrated			
pH	$HONO_2$	$HONO$	$HONH_3^+$	NH_4^+
2	0.379			
2.5	0.760	0.004		
3	0.919	0.297		
3.5	0.974	0.616		
4	0.992	0.841	0.004	
4.5	0.997	0.944	0.025	
5	0.999	0.982	0.079	
5.5	1	0.944	0.215	
6	1	0.998	0.467	
6.5	1	0.999	0.737	0.002
7	1	1	0.899	0.005
7.5	1	1	0.966	0.015
8	1	1	0.989	0.047
8.5	1	1	0.997	0.136
9	1.001	1.001	1	0.335
9.5	1.003	1.003	1.002	0.618
10	1.008	1.008	1.008	0.844
10.5	1.027	1.027	1.027	0.968
11	1.088	1.088	1.088	1.068

Titration Curves for Diprotic Acids

When a diprotic acid is titrated with NaOH, there are two equivalence points. The first corresponds to the addition of one millimole of NaOH for each millimole of H_2A originally present in the sample. The second equivalence point corresponds to the addition of two millimoles of NaOH per one millimole of H_2A originally present. Clearly it will take exactly twice the volume of NaOH solution to reach the second equivalence point as it takes to reach the first. The shape of the titration curve will depend markedly on the absolute values of pK_1 and pK_2 and also on their relative values. There are two buffer regions, one for the H_2A/HA buffer and the second for the HA/A buffer. At the midpoints of these buffer regions, the pH will be approximately equal to pK_1 and pK_2, respectively. If pK_1 and pK_2 do not differ by more than a couple of log units, the first equivalence point will not show a "break." This is illustrated by the curves of Fig. 10.8, calculated with Eq (10.7), for the titration of maleic and fumaric acids, which are cis- and trans-isomers. For these curves, the original molarity of the acid was 0.012, and the NaOH molarity was 0.03.

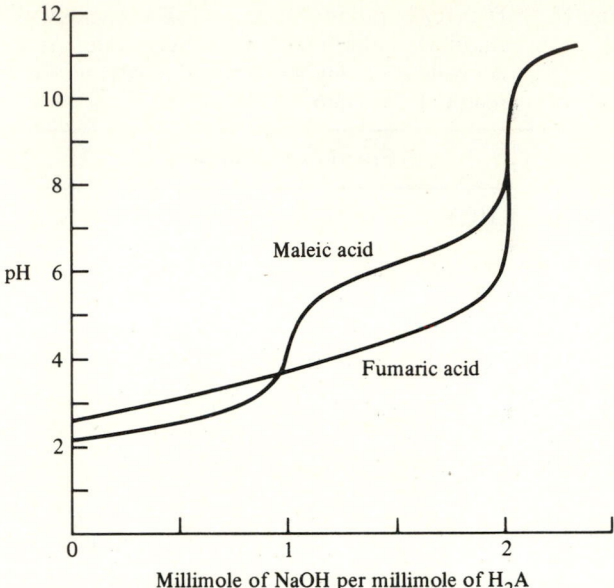

Fig. 10.8 Diprotic acid–strong base titration curves. Maleic acid: $pK_1 = 1.91$; $pK_2 = 6.33$ (first equivalence point easily visible). Fumaric acid: $pK_1 = 3.10$; $pK_2 = 4.60$ (values too close for the first equivalence point to be visible).

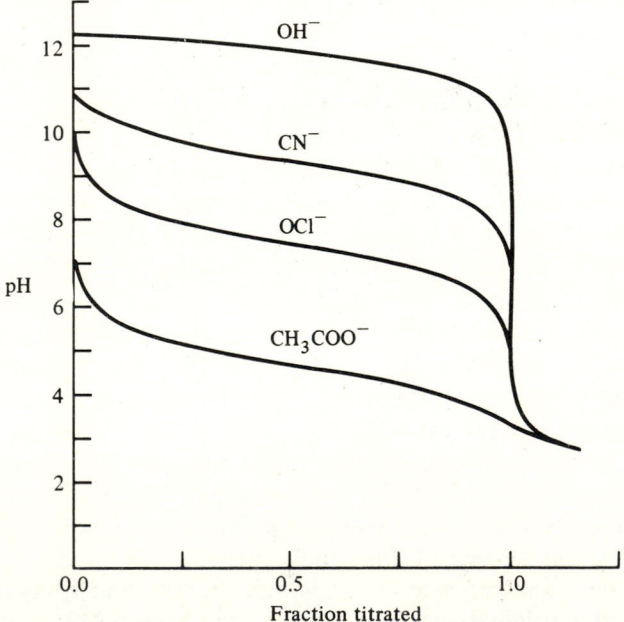

Fig. 10.9 Effect of pK on weak base–strong acid titration curve.

A Group of Curves for Monoprotic Bases

Figure 10.9 shows titration curves for the four anion bases titrated with HCl:

$$\begin{array}{lll}
\text{hydroxide ion, OH}^-: & \text{strong base,} \\
\text{cyanide ion, CN}^-: & pK_b = 4.6, \\
\text{hypochlorite ion, OCl}^-: & pK_b = 6.5, \\
\text{acetate ion, CH}_3\text{COO}^-: & pK_b = 9.24.
\end{array}$$

Note the similarity of the main features of these curves to those of the set for acids being titrated with sodium hydroxide. The midpoints of the buffer regions are approximately where $pH = pK_a$ for the conjugate acid of the base being titrated. The reader should substantiate this statement.

10.6   HOW TO SKETCH A TITRATION CURVE WITH ONLY ONE POINT

Computer-calculated curves are great for accuracy, but it's nice to be able to sketch a "quick and dirty" curve at times. All one needs is the pK value for the acid. Set up the pH and volume (fraction titrated) axes as in Fig. 10.10, with the equivalence point marked with a vertical dashed line, and put an x at the point over the midpoint of the buffer region, where $pH = pK$. (For strong acids, just use $pH = 1$ or 2.)

Knowing that this x marks the middle of the buffer region, sketch a segment with a small (but not flat) slope, indicating that the pH does not change quickly in the buffer region (Fig. 10.11).

Now sketch a short line in the equivalence-point region, recognizing that this is where the pH should change most rapidly. This line should be steep, but not vertical. The endpoint pH is above 7 for weak acids (Fig. 10.12).

The pH will always level off at about 12 once the titration is complete, so sketch a short segment showing this. Also, the pH shows a slight rise at the very start of the titration, but this is less evident as the acid strength increases (Fig. 10.13).

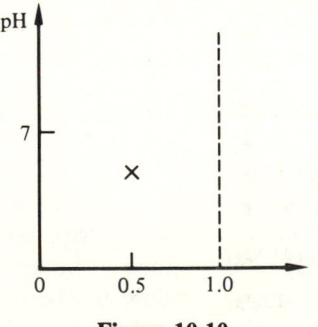

Figure 10.10

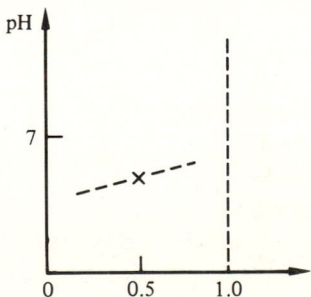

Figure 10.11

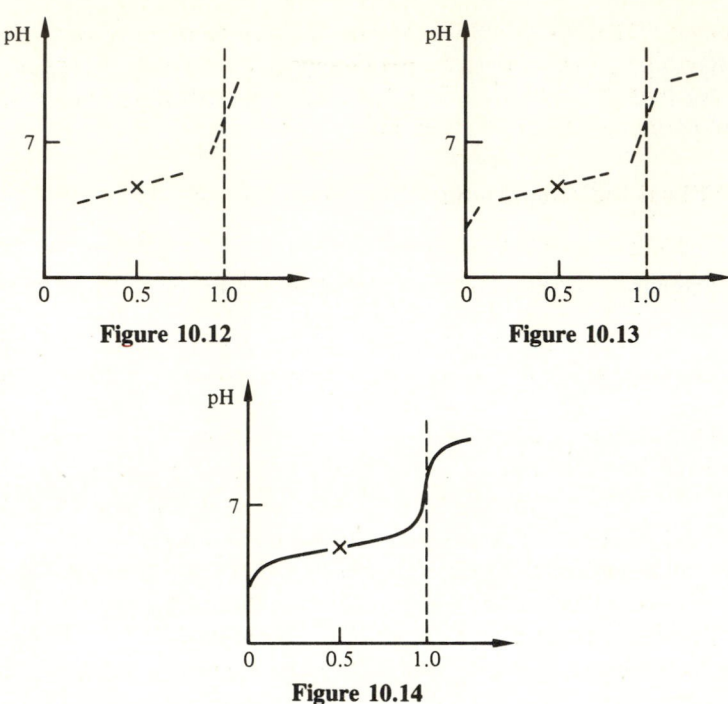

Figure 10.12   Figure 10.13

Figure 10.14

Finally, simply connect these four regions with a neat, smooth curve, and the result is a sketch that is very useful for discussion, for consideration of the choice of indicator, etc. (Fig. 10.14).

10.7 EXPERIMENTAL DETECTION OF THE TITRATION EQUIVALENCE POINT

So far we have dealt only with theoretical calculation of titration curves, using predetermined values for equilibrium constants and concentrations. But the practical importance of titrations is in the determination of an unknown concentration, either of the sample being titrated or of the titrant in the buret. In this case, how is the analyst to know just when the stoichiometric amount of titrant has been added? The two most important procedures are as follows.

1. Use of a pH Meter to Measure Solution pH During the Titration

In this procedure, often called **potentiometric titration**, one uses an electrode made of pH-sensitive glass (see Chapter 11) along with a precise millivolt-meter to obtain a series of values of pH and volume. The experimental titration curve is plotted, and the volume corresponding to the equivalence point is estimated by finding the point on the curve where the pH is rising most sharply. This method is capable of very high precision and can be applied effectively even when the pH change in the vicinity of the equivalence point is not particularly steep. The technique of finding the point of steepest slope will be discussed in Chapter 11.

2. Use of Visual Indicators to Signal the Endpoint

The use of acid–base indicators is convenient, fast, and simple in practice. To the original solution to be titrated, one adds a few drops of a dilute solution of an indicator. If the choice of indicator has been correct and if the titration involves a steep change in pH at the equivalence point, the indicator will undergo a marked change in color in the immediate vicinity of the equivalence point.

To understand the principles of the indicator color change and the limitations of this method of determining the endpoint for titrations, it is important to examine the equilibrium theory and the molecular structure of indicators and to relate this to the theory of the acid–base titration curves. These topics will be discussed in some detail in the next few sections.

ALGEBRAIC BASIS FOR ACID–BASE INDICATORS

No new principles are needed to explain the action of acid–base indicators, which are merely weak acids. It is their strong color that makes them special and useful. Letting HIn stand for the conjugate acid form, we have

$$HIn \rightleftarrows H^+ + In^-; \qquad Q_a = \frac{[H^+][In^-]}{[HIn]} = \frac{[H^+][\text{Blue form}]}{[\text{Yellow form}]}$$

yellow   blue
color   color

Now imagine a solution of this indicator, perhaps $1 \cdot 10^{-5} M$, in which the concentration of hydrogen ion can vary. Assuming that Q_a has a constant value of $1 \cdot 10^{-8}$, for example, we can calculate the equilibrium ratio of blue-to-yellow forms:

$$\frac{[\text{Blue}]}{[\text{Yellow}]} = \frac{Q_a}{[H^+]} = \frac{1 \cdot 10^{-8}}{[H^+]}.$$

What we "see" as the color of the solution will depend on the color ratio and is summarized in Table 10.3

Because the reaction is reversible there will never be a complete conversion to either form, regardless of pH extremes; but, due to our human visual limitations, the color change "seems" complete when one form of the indicator makes up about 90% of the total. The pH range over which we can perceive the changes that occur when the pH is varied is called the **transition range**, or the **color change interval**. An approximate rule is that the transition range is $pH(\text{range}) = pQ_a \pm 1$.

Table 10.3

pcH		[Blue]/[Yellow]	Visual effect
3		0.00001	pure yellow
4		0.0001	pure yellow
5		0.001	pure yellow
6		0.01	nearly pure yellow
7		0.1	greenish yellow
8	Transition range	1	green
9		10	greenish blue
10		100	nearly pure blue
11		1000	pure blue
12		10000	pure blue

The middle of the range is the pH at which both forms are present in equal concentrations, when $pH = pQ$. In the above example this occurs at $pH = 8$, and the green color is not due to the formation of some "green" species, but rather because the blue species absorbs the long wavelengths (red end) while the yellow species absorbs the short wavelengths (blue-violet end), letting the middle wavelengths (green) come through the solution relatively freely.

Table 10.4 of selected acid–base indicators summarizes the characteristics of a few important indicators. The transition ranges are those that have been subjectively determined by placing the indicators in a series of buffers of known pH; they vary from 1.2 to 2.0 pH units.

The semi-chemical names of indicators are merely a convenience. Who wants to say "sodium salt of para-para-anilinophenylazobenzenesulfonic

Table 10.4 Selected acid–base indicators

Indicator name	Acidic charge type	Acidic color	pH range	Basic color
Picric acid	neutral	colorless	0.0–1.3	yellow
o-cresol red	zwitter ion	red	0.2–1.8	yellow
Thymol blue	zwitter ion	red	1.2–2.8	yellow
2,4-dinitrophenol	neutral	colorless	2.4–4.0	yellow
Bromophenol blue	anion	yellow	3.0–4.6	blue–violet
Bromocresol green	anion	yellow	4.0–5.6	blue
Methyl red	neutral	red	4.4–6.2	yellow
Bromocresol purple	anion	yellow	5.2–6.8	purple
Bromothymol blue	anion	yellow	6.2–7.6	blue
o-cresol red	anion	yellow	7.2–8.8	red
Thymol blue	anion	yellow	8.0–9.6	blue
Phenolphthalein	anion	colorless	8.0–10.0	red
Thymolphthalein	anion	colorless	9.4–10.6	blue
Alizarin yellow	anion	yellow	10.0–12.0	lilac
Trinitrobenzoic acid	neutral	colorless	12.0–13.4	orange–red

acid" when a simple "Orange IV" will mean the same thing to the stockroom manager? Some names were adopted before the chemical structures were known; for example, the indicator "malachite green" was given its name because the color resembled that of the copper mineral, malachite.

The entries under *Acidic charge type* show the ionic or molecular form of the conjugate-acid species. This is significant because of the effect of ionic strength on the pQ value (and hence the pH range) of the indicator; but generally the effect is not an obstacle in titration applications. Using the symbol HIn again for the acid form, we can write the following charge types:

$$\text{neutral:} \quad HIn = H^+ + In^-,$$
$$\text{anion:} \quad HIn^- = H^+ + In^{2-},$$
$$\text{cation:} \quad HIn^+ = H^+ + In^0,$$
$$\text{zwitter ion:} \quad {}^+HIn^- = H^+ + In^-.$$

MOLECULAR STRUCTURE OF ACID–BASE INDICATORS

The foregoing algebraic model shows how the color change of an indicator is influenced by solution pH but gives no information on why we might expect such a change at all. We believe that color is due to selective absorption of some wavelength bands, while other bands are relatively free to pass through the solution. For electromagnetic radiation in the visible and ultraviolet regions of the spectrum it is accepted that the mechanism of photon absorption is the excitation of electrons to higher energy levels within the molecular structure of the colored species. Therefore it is reasonable to expect that any major change in the electronic bonding pattern will present a very different set of possible energy transitions; hence photons of different energies (wavelength) will be absorbed.

Three common indicators will be used to illustrate these ideas. The first, called *phenolphthalein* (fee-nol-*thay*-leen), is an example of a one-color indicator and is commonly used for the titration of weak acids with sodium hydroxide because the pH at the equivalence point of such titrations is typically about 9, which is in the middle of the transition range for this indicator. The structures may be drawn as follows.

acid form, colorless

$-2H^+$

base form, deep pink; resonance hybrid

How can the loss of that phenol proton cause such a color change? The basic structure, as shown above, is merely one canonical form of a resonance hybrid. When the proton was in place on the —OH group, conjugated bonding in ring 1 was blocked. However, when the proton is removed, rings 1 and 2 are totally equivalent and there is a major shift of electron density with equal distribution over both rings. This accounts for the drastic change in the selective absorption of visible wavelengths. The electron pattern in ring 3 has relatively little effect, since it is the same for both forms.

The second example is *bromcresol purple*, an indicator that is ideal for titration of strong acids with sodium hydroxide. Although the titration reaction produces a neutral salt (e.g., NaCl), the equivalence-point pH is usually about 6 instead of the theoretical 7 because of contamination with atmospheric carbon dioxide. The CO_2 partially hydrates to carbonic acid, making the solution very slightly acidic.

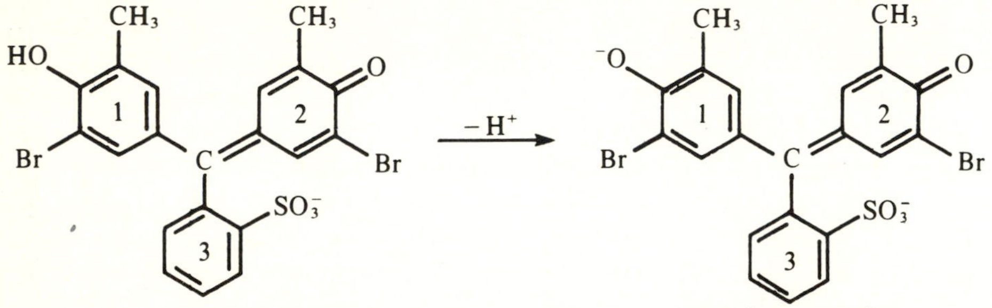

acid form, yellow   base form, purple

The principle of the color change is the same as for phenolphthalein, because the main structural characteristics are identical. However, the presence of the bromine atoms helps to disperse the negative charge of the basic species because of their high electronegativity. This leads to a lower value for pQ (i.e., the yellow form of bromcresol purple is a stronger acid than the colorless form of phenolphthalein), hence to a lower pH transition range (see Table 10.4). The sulfonic acid group also helps to disperse charge.

The final example, *methyl red*, illustrates a different structural type—an **azo dye**. With its color-change interval of 4.4–6.2, this indicator is useful for titrations of weak bases with strong acid.

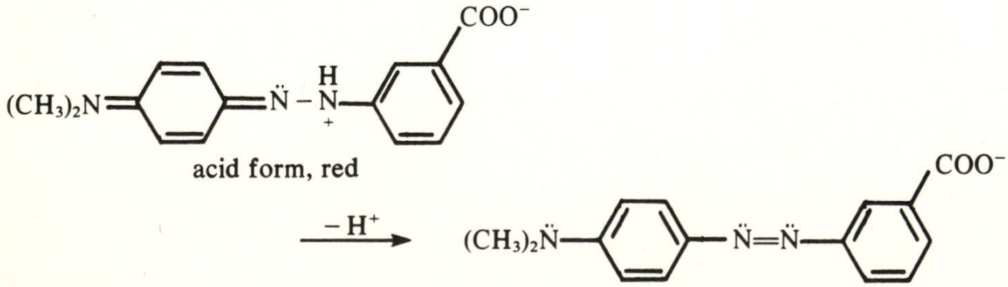

acid form, red

base form, yellow

In this case the strongly colored species is the acidic form. We typically find strong color associated with conjugated double bonds. The loss of the acidic proton causes a shift to the nonconjugated structure, resulting in a change of both color hue and intensity. In still stronger acidic solutions a second proton can bond to the electron pair shown on the other azonitrogen, but this causes only a small shift in electron distribution, and the resulting species is also red.

CHOOSING AN INDICATOR FOR A GIVEN TITRATION APPLICATION

The purpose of an acid–base indicator is to signal, by its change of color, when a certain pH has been reached. Because the color change is spread over a range of about two pH units, there is imprecision in this signal. However, a little thought may convince you that the most reliable signal will be when the pH is close to the middle of the transition range, for it is there that the change of hue is most sensitive to a small change in pH.

With this in mind, the choice of an indicator is simple and direct. First, calculate the value of pH to be expected at the titration equivalence point, and then pick an indicator that has its transition range nearly centered on that pH value.

Example 1 Titration of ammonia with hydrochloric acid

At the equivalence point the titration reaction has produced a solution of ammonium chloride. The ammonium ion is a weak acid with $pK_a = 9.26$. The equivalence-point pH, for a $0.1M$ solution of ammonium ion, is found by the following calculations:

$$[H^+] = (10^{-9.26} \cdot 0.1)^{1/2} = 10^{-5.13} = 7.4 \cdot 10^{-6},$$
$$aH = H \cdot f_1 = 7.4 \cdot 10^{-6} \cdot 0.78 = 5.8 \cdot 10^{-6},$$
$$pH = 5.24.$$

Therefore, either bromcresol green (range 4.0–5.6) or methyl red (range 4.4–6.2) would be a reasonable choice.

Example 2 Titration of a very weak acid with sodium hydroxide

The hydroxylammonium ion NH_3OH^+ has a $pK_a = 6.0$. When a solution of hydroxylammonium chloride is titrated with sodium hydroxide, the solution at the equivalence point contains hydroxylamine (which is a weak base with $pK_b = 14.0 - 6.0 = 8.0$) and the inert salt, sodium chloride. Again, assuming that the molarity is $0.1M$, the equivalence-point pH is calculated as follows:

$$[OH^-] \approx (K_b C_b)^{1/2} = (1 \cdot 10^{-8} \cdot 0.1)^{1/2} = 3 \cdot 10^{-5};$$

therefore,

$$[H^+] \approx \frac{K_w}{[OH^-]} = \frac{1 \cdot 10^{-14}}{3 \cdot 10^{-5}} = 3.3 \cdot 10^{-10},$$

and pH ≈ 9.5. (Activity coefficients have been ignored.)

From Table 10.4 we see that thymol blue will be almost completely converted to its blue form at this pH, while thymophthalein will be barely starting its transition. Phenolphthalein looks like the best bet, although it would be best to titrate until the color change is nearly complete to the red form. Thymolphthalein might work satisfactorily if the titration endpoint is defined by the appearance of the slightest hint of a pale blue color.

10.8 APPLICATIONS OF ACID–BASE TITRATIONS

There are four important applications of acid–base titrations:

1. To standardize (determine the precise molarity of) a solution of a strong base, usually NaOH or KOH. This is accomplished by titration of a known quantity of a reference acid.

2. To standardize a solution of a strong acid, such as H_2SO_4, HCl, $HClO_4$, HNO_3, etc., by titration of a known quantity of a reference base.

3. To use a standardized NaOH solution to titrate an unknown quantity of an acid for the purpose of determining the amount of acid in the sample.

4. To use a standardized strong acid solution to titrate an unknown quantity of a base for the purpose of determining the amount of base in the sample.

Preparation and Standardization of Strong Acid and Strong Base Solutions

Because it is so frequently necessary to have reliable standard solutions of strong acids and bases, the following procedures are of value to all practicing chemists. The object is to produce a solution whose molarity is known with an uncertainty of about 1 ppt.

GENERAL COMMENTS

The principles of acid–base titrations have been discussed earlier. By considering theoretical titration curves we can understand the requirements for accurate titrations, including the choice of the proper indicator for the endpoint. In striving for the highest accuracy in the standardization of acid–base solutions, we must be aware of several sources of error:

a) Volumetric glassware that has not been standardized by weighing the water delivered or that has not been properly cleaned may cause solution volumes to be in error by several parts per thousand.

b) When the standardization is based on a weighed sample of reference material there is usually negligible error in the weighing itself, except for masses lower than about 0.2 gram. However, the reference material should have been dried in an oven to remove surface moisture, and it should be a product of reliable, very high purity (99.9 to 100.1% assay). Considerable care must be taken to transfer the substance quantitatively from vessel to vessel.

c)   When the reference material is first dissolved and diluted to a volume in a volumetric flask and a pipet is then used to transfer a portion to the titration flask, it must be remembered that additional errors are inevitable because of the introduction of the two extra pieces of glassware, even if they have been standardized previously.

d)   A known, highly pure reference standard is called a **primary standard**. A solution that has been standardized by titration of a weighed portion of this pure material is referred to as a **secondary standard**. If a second solution is now standardized by titration of the secondary standard instead of using a procedure based on another primary standard, the second solution cannot be regarded as highly accurate because of cumulative errors.

e)   With skill, the endpoints of the titrations described below can be obtained with a precision of about 0.02 mL in the buret reading. To achieve this it is necessary to approach endpoints with patience and care, adding fractions of drops from the buret tip.

PREPARATION OF A STRONG ACID SOLUTION

The common strong acids are usually readily available in concentrated stock solutions as shown in Chapter 2. Perchloric acid, hydrochloric acid, and sulfuric acid are not affected by air or light and typically may be trusted to be free of impurities. However, concentrated hydrochloric acid will gradually become lower in concentration because of the volatility of gaseous HCl. Concentrated nitric acid, when fresh, is colorless, but on long standing or exposure to light it develops a brown color due to oxides of nitrogen. These may not be harmful once the solution is diluted, but they may be removed by addition of a little sulfamic acid or by boiling the diluted solution. Hydrobromic acid and hydriodic acid are likely to be contaminated with bromine and iodine, respectively, as evidenced by the brown color; the pure acids are colorless.

   To prepare V milliliters of a strong acid with molarity C one simply adds V_s milliliters of the concentrated stock to water and then dilutes to the total volume V. The value of V_s is readily calculated if the concentration C_s of the stock solution is known:

$$V \cdot C = V_s \cdot C_s \qquad \text{and} \qquad V_s = \frac{V \cdot C}{C_s}.$$

Note that dilution does not change the number of millimoles.

Example To prepare two liters of $0.08M$ hydrochloric acid from a stock solution of $12M$, use a graduated cylinder or pipet to take

$$\frac{2000 \cdot (0.08)}{12} = 13.3 \text{ mL}$$

of the stock, add it to a few hundred milliliters of water, dilute to two liters, and mix well.

Caution   When diluting concentrated sulfuric acid (96% by mass), it is important to add the acid slowly, with constant stirring, to a larger volume of water because of the large quantity of heat liberated. It is good practice to follow this rule for the other acids as well.

Generally a dilute (that is, 0.1 to 1.0M) solution of a strong acid will be stable for years if kept in a glass-stoppered bottle. If the solution has been standing for a while, it is often found that drops of water collect on the inside of the bottle above the solution, due to fluctuating room temperature. Therefore, the solution should be remixed each time before using.

STANDARDIZATION OF STRONG ACID BY TITRATION OF SODIUM CARBONATE

A classical approach of long standing is to weigh a sample of pure sodium carbonate Na_2CO_3 and to titrate it with the strong acid to be standardized. The carbonate ion is a diprotic base ($pK_{b1} = 3.67$, $pK_{b2} = 7.6$), and the resulting titration curve is as shown in Fig. 10.15. Because the second step involves the reaction

$$HCO_3^- + H^+ = H_2CO_3 = CO_2(g) + H_2O,$$

the equilibrium constant is not very favorable, and the pH change at the equivalence point is not sharp. However, the recommended practice is to

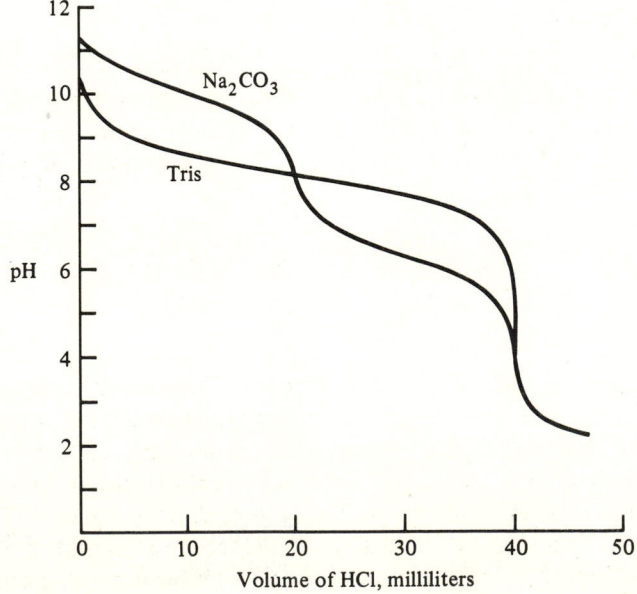

Fig. 10.15 Titration of 50 mL of 0.04M Na_2CO_3, or 0.08M tris, with 0.10M HCl.

continue the titration until the second equivalence point is near and then to boil the solution to remove the accumulated carbon dioxide. Then, when the titration is completed, the pH change is more abrupt. This technique is inconvenient, and so it has become more common to use a different reference base for the standardization.

STANDARDIZATION OF STRONG ACID BY TITRATION OF "TRIS"
(TRADE NAMES: THAM, TRIZMA BASE)

The compound called **tris**, which is an abbreviation for tris-(hydroxymethyl)-aminomethane, serves as a good standard. It has a molecular weight of 121.14, corresponding to the formula

$$(HOCH_2)_3C—\ddot{N}H_2.$$

Tris is thus a monoprotic base ($pK_b = 5.93$) that may be considered a derivative of ammonia. It is nonhydroscopic, does not react with CO_2 from the atmosphere, is stable both as a solid and in aqueous solution, and can be prepared in pure form. Tris reacts quickly and stoichiometrically with hydrogen ion:

$$R—\ddot{N}H_2 + H^+ = RNH_3^+.$$

The titration curve in Fig. 10.15 shows that the equivalence-point pH is 4.8, so that bromcresol green (interval 4.0–5.6) serves well as an indicator.

PROCEDURE USING TRIS

Usually the approximate molarity of the acid solution is known. Let it be called C. In a titration it is desirable to reach the endpoint when about 40 mL have been added from a 50-mL buret. This is large enough to minimize the error in estimating the endpoint reading but is comfortably below the buret capacity, thus making it unlikely that the buret will have to be refilled to complete the titration. Therefore, each titration of tris should require $40 \cdot C$ mmol. The molecular.weight of tris is 121.14, and thus the sample to be titrated should have a mass

$$w = 40 \cdot C \cdot 0.12114 \text{ g}.$$

The required amount of tris (depending on how many replicate titrations will be performed) should be placed in a beaker or weighing bottle and dried in the oven at 100–103°C for one or two hours only. After cooling in a desiccator, the desired samples can be weighed into titration flasks.

Since tris is a weak base, its solutions will absorb carbon dioxide from the atmosphere. Therefore, it is best not to dissolve the sample until it is time to titrate it. Add about 50 mL of water and enough bromcresol green indicator solution to give the solution a distinct blue color. Titrate with the strong acid solution until the color changes from blue to green. Slight overtitration will cause the color to become yellow.

Calculate the concentration of the strong-acid solution, using the average of three (or more) titrations.

Example A solution of $0.1M$ HCl was used to titrate samples of tris by the above procedure, with the following results.

Trial	Mass of tris, g	Endpoint volume, mL
1	0.5221	41.64
2	0.5097	40.58
3	0.5302	42.26

According to the manufacturer's certificate, this particular batch of tris had been assayed at 99.86% purity. (In the absence of such information we are forced to assume a value for purity, for example, 100.0%) To illustrate the calculation for trial 1 we write:

$$M_{HCl} = \frac{n \text{ mmol}}{41.64 \text{ mL}} = \frac{(0.5221 \text{ g})(0.9986)}{(0.12114 \text{ g/mmol})(41.64 \text{ }'} = 0.1034M.$$

The reader should complete the calculations for trials 2 and 3, find the average molarity, and determine the relative average deviation in parts per thousand.

Preparation of a Strong Base Solution

Although potassium hydroxide might be preferred for some special applications, by far the most common choice for a strong base solution is sodium hydroxide. This is readily available in solid form, usually as pellets, but because of exposure to the atmosphere it contains significant amounts of both water and carbon dioxide (which is converted to sodium carbonate). Therefore, it is not recommended that a solution of sodium hydroxide be prepared by weighing and dissolving the solid if the solution is to be used for accurate titrations. Instead, one should use the commercially available 50% (by mass) solution of NaOH, which is about $19.4M$. At this high concentration, sodium carbonate is virtually insoluble and may be filtered out. The clear filtrate is then convenient to use for preparing a dilute solution of desired concentration.

To prepare V milliliters of C-molar NaOH simply use a graduated pipet to measure V_s milliliters of the 50% solution:

$$V_s = \frac{V \cdot C}{19.4}.$$

Example To prepare 500 mL of 0.15M NaOH, given the 50% stock solution that is 19.4M, take

$$\frac{500 \cdot (0.15)}{19.4} = 3.9 \, \text{mL}$$

of the stock, dilute to 500 mL, and mix well. The distilled water used for the dilution may be freed of dissolved carbon dioxide by boiling or by bubbling CO_2-free air through it for an hour.

Since strong base solutions attack glass, they are kept in polyethylene bottles. Glassware (including burets) used for measurement of NaOH solutions should be very thoroughly rinsed after use. Once the solution of NaOH is standardized, it will keep indefinitely in a plastic bottle.

To test for dissolved carbonate, add 1 mL of 1M barium chloride solution to a few milliliters of the NaOH solution. A precipitate of $BaCO_3$ will form. If the barium chloride will not interfere with the intended use of a strong base solution, it may be used to remove carbonate, since the precipitate $BaCO_3$ can be filtered off.

STANDARDIZATION OF NaOH VERSUS
POTASSIUM HYDROGEN SULFOSALICYLATE

Recently, Butler and Bates have proposed* an attractive alternative to the time-honored use of KHP. The double potassium salt of sulfosalicylic acid may be prepared in very pure form. This salt has the formula

$$\text{KHSs} \cdot \text{K}_2\text{Ss}, \qquad \text{MW} = 550.655,$$

where Ss represents the doubly charged anion

The KHSs moiety of this salt contains the singly charged anion in which the carboxyl group carries the proton. The pK_a for this acidic species is 2.85, which makes it a considerably stronger acid than the hydrogen-phthalate anion.

The advantage of using this compound in place of KHP is evident from the titration curve shown in Fig. 10.16. Because of the lower pK value, the change in pH in the immediate vicinity of the equivalence point is considerably sharper. This makes the choice of an indicator less critical and also ensures that the indicator color change is more abrupt.

---

*R. Butler, R. G. Bates, *Analytical Chemistry*, **48**, 1669 (1976).

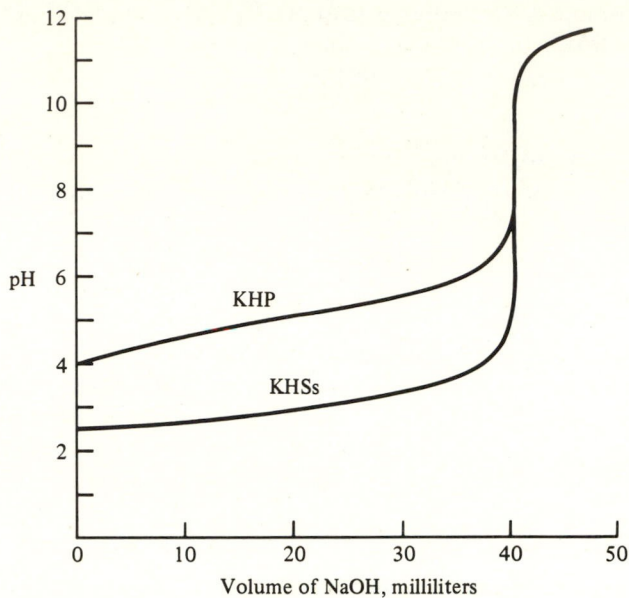

Fig. 10.16 Titration curves for KHP and KHSs with strong base.

STANDARDIZATION OF NaOH VERSUS
POTASSIUM HYDROGEN PHTHALATE (KHP)

The monopotassium salt of phthalic acid (called potassium biphthalate or, more correctly, potassium hydrogen phthalate) is available in very pure form:

$$MW = 204.23, \qquad pK_a = 5.4.$$

When dissolved in water, the salt dissociates completely into its ions, K^+ and HP^-. The acidic hydrogen on the carboxyl group is titrated by sodium hydroxide:

$$HP^- \quad + OH^- = \quad P^{2-} \quad + H_2O.$$
biphthalate ion   phthalate ion

In the titration, a buffer region is observed at about pH = 5.0, and the steep rise in pH at the equivalence point is centered at about pH 8.8. Therefore, either thymol blue (color-change interval 8.0–10.0) or phenolphthalein (8.0–9.6) may be used as a visual indicator. The titration curve in Fig. 10.16 (upper curve) depicts the titration of 50 mL of 0.08M KHP with 0.10M NaOH.

PROCEDURE USING POTASSIUM HYDROGEN PHTHALATE (KHP)

Suppose the approximate concentration of the NaOH solution is C molar. If the titration is designed to require about 40 mL from the buret at the endpoint, then $40 \cdot C$ mmol of KHP are required in the sample. The molecular weight of KHP is 204.23, and thus the sample should weigh

$$w = 40 \cdot C \cdot 0.20423 \text{ g.}$$

The required amount of KHP (depending on the number of replicate titrations to be carried out) should be placed in a beaker or weighing bottle and dried in the oven at 110°C for one or two hours. After cooling in a desiccator the samples may be weighed into a series of titration flasks.

To each sample add about 50 mL of water and two or three drops of phenolphthalein indicator solution (1% in ethanol), and titrate with the NaOH solution until the endpoint is indicated by the permanent appearance of a very pale pink color. On standing, this color will fade because of absorption of carbon dioxide from the air. Slight overtitration will result in an intense red color.

The concentration of the NaOH solution should be calculated from the average of three or more titrations.

Example A solution of $0.05M$ NaOH was used to titrate samples of KHP, with the following results:

Trial	Mass of KHP, g	Endpoint volume, mL
1	0.4097	38.82
2	0.3822	36.19
3	0.4114	39.03

If we assume the KHP to be 100.0% pure, the calculation for trial 1 is as follows:

$$M_{\text{NaOH}} = \frac{n \text{ mmol}}{38.82 \text{ mL}} = \frac{0.4097 \text{ g}}{(0.20423 \text{ g/mmol})(38.82 \text{ mL})} = 0.05168M.$$

The reader should do the calculations for trials 2 and 3, find the average molarity, and determine the precision in parts per thousand.

Comparison of a Strong-Acid Solution with a Strong-Base Solution

If a standardized strong-acid solution is available, it may be used as a secondary standard reference for the standardization of a sodium hydroxide solution. The converse is also true. When highest accuracy is not required, this procedure may be preferred because it is easy to pipet a sample of strong-acid solution instead of weighing out tris or KHP samples.

In performing a strong acid–strong base titration it is always good practice to place the NaOH in the buret, whether it is the standard or the unknown. This is done because the buret helps to protect the basic solution against absorption of carbon dioxide from the air.

The theoretical equivalence-point pH for this titration is 7.00, since the product of the titration reaction is simply an inert salt such as NaCl. However, unless the solution in the titration flask is protected from air contact, for example, by bubbling nitrogen through it during the titration, there will always be some dissolved carbon dioxide, and this will cause the pH to be about 6 at the equivalence point. The best choice for an indicator is thus bromcresol purple, which has a color-change interval of 5.2–6.8.

Procedure   Pipet a portion of the strong acid solution into a titration flask, add enough water to bring the volume to about 50 mL, add enough bromcresol purple indicator solution to give a distinct yellow color, and titrate with the solution of sodium hydroxide. The color will change very abruptly to purple at the endpoint. If particular care is taken to add very small increments of NaOH at the equivalence-point region, the solution will show a gray or nearly colorless hue, since the yellow and purple colors of this indicator are complementary to each other. It is difficult to observe this "ideal" endpoint, and the slightest overtitration will result in a purple color.

Example When 20.00 mL of 0.1035M HCl is titrated with 0.04925M NaOH, how many milliliters should be required? *Answer*: 42.03 mL.

PROBLEMS

Acid–base titrations

1. A sample of pure potassium hydrogen phthalate weighing 0.7663 gram is dissolved in 50 mL of water, phenolphthalein indicator is added, and the solution is titrated with a solution of sodium hydroxide. If 43.28 mL are required to reach the endpoint, what is the precise molarity of the strong base solution? What is the color change at the endpoint?

2. If we desire to standardize a solution of potassium hydroxide that is known to be approximately 0.15M, how much of the double potassium salt of sulfosalicylic acid (MW = 550.66) should be weighed out so that the endpoint will occur at about 45 mL of added base? What indicator should be used and what is the color change at the endpoint?

3. A solution of nitric acid is standardized by titration of 0.2147 gram of tris, 30.44 mL being required. The acid solution is then used to find the concentration of an unknown ammonia solution, 50.00 mL of which requires 42.36 mL of the nitric acid to reach the endpoint. Calculate the molarities of both the nitric acid and the ammonia solutions. What indicator would you use in these titrations and what is the color change at the endpoint?

4. An impure sample of sodium carbonate weighing 0.1322 gram is titrated with 0.1200M perchloric acid, 17.92 mL being required to reach the second endpoint. What is the percentage by weight of Na_2CO_3 in the sample?

5. Faced with the task of having to titrate a large number of samples of strong base solution, ranging in molarity from about 0.2 to 0.3, each sample being precisely 5.00 mL an analyst decides to simplify the calculations. What molarity of HCl

titrant solution should be used so that the buret reading at the endpoint will simply be 100 times the strong-base molarity in the sample?

6. A buffer solution is prepared by mixing 100 mL of $2.5M$ acetic acid with 200 mL of $0.80M$ sodium hydroxide and diluting to one liter. If a 50-mL portion of the buffer is titrated with standardized $0.0935M$ potassium hydroxide, what volume will be required to reach the phenolphthalein endpoint?

7. A sample of a monoprotic carboxylic acid weighing 0.441 gram is titrated with $0.133M$ NaOH, 38.0 mL being required to reach a phenolphthalein endpoint. What is the molecular weight of the acid? Why is it probably correct to use phenol-phthalein?

8. For each of the following titrations consult Appendix 4 for pK values, calculate the approximate equivalence-point pH (assuming $0.1M$ solutions are used), recommend an acid–base indicator, and state its color change.

  Also make a sketch of the theoretical titration curve and comment on the suitability of doing a visual titration (with an indicator) compared with doing a potentiometric titration with the glass electrode:

a) boric acid with potassium hydroxide

b) chlorous acid with sodium hydroxide

c) sodium cyanide with hydrochloric acid (in the hood, of course!)

d) hydroxylammonium chloride with sodium hydroxide

e) potassium chromate with perchloric acid

f) nitrous acid with potassium hydroxide

g) propionic acid with potassium hydroxide

h) ethylamine with nitric acid

i) formic acid with sodium hydroxide

j) phenol with potassium hydroxide

k) ammonium nitrate with sodium hydroxide

l) ethanolamine with hydrobromic acid

m) sodium malonate with perchloric acid

9. For each of the following amino acids consider two titrations, one with a standard solution of sodium hydroxide, and the other with a standard solution of hydrochloric acid. Assume $0.1M$ solutions and sketch the titration curves, using the axis arrangement shown in Fig. 10.17:

a) alanine

b) glutamic acid

c) histidine

d) aspartic acid

e) cysteine

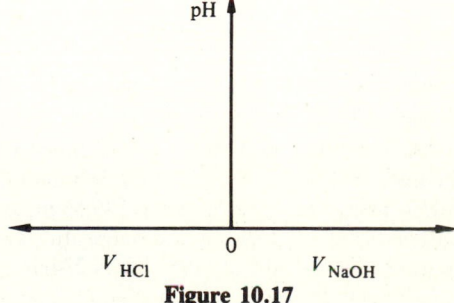

Figure 10.17

10. For each of the following diprotic acids sketch the titration curves and choose an indicator for the endpoint that is best suited for visual titration.

   a) oxalic   b) succinic   c) phthalic   d) sulfuric
   e) carbonic   f) hydrosulfuric   g) sulfurous h) ethylenediammonium ion, enH_2^{2+}

11. The purity of a sample of aspirin may be checked by a simple two-step procedure. First, the sample is boiled with an excess of NaOH, causing hydrolysis.

$$CH_3COO—C_6H_4—COOH + 2OH^- \rightarrow HO—C_6H_4—COO^- + CH_3COO^-.$$
   aspirin   salicylate ion   acetate ion

Then the excess hydroxide is determined by titration with a standard solution of HCl. The quantity of hydroxide that reacts with the aspirin is the difference between the total amount used and the excess.

   A sample of aspirin weighing 1.625 gram was boiled with 50.0 mL of $0.500M$ NaOH for 10 minutes. After cooling, it required 25.24 mL of $0.278M$ HCl to reach the endpoint. Given the molecular weight of aspirin (180.16), what was the purity of the sample?

12. A common technique for determining the percentage of nitrogen in organic material is to use the Kjeldahl method. This begins with a digestion step that converts the nitrogen to ammonium ion.

$$\text{Organic } N \rightarrow NH_4^+.$$

Then an excess of NaOH is added, liberating the ammonia, which is readily distilled out of the mixture and captured in a known volume of standard HCl:

$$NH_4^+ + OH^- (\text{excess}) \xrightarrow[\text{heat}]{} NH_3(g),$$

$$NH_3(g) + H^+Cl^- \rightarrow NH_4^+Cl^- + H^+Cl^- (\text{excess}).$$

The final step is a titration of the remaining HCl to determine the amount of HCl in excess of that required to react with the ammonia.

   In one determination an organic sample weighing 2.00 grams was treated as described above, and the ammonia was captured in 100 mL of $0.0500M$ HCl. Titration required 40.0 mL of $0.0500M$ NaOH.

   a) Calculate the percentage of nitrogen in the sample.
   b) Calculate the (approximate) equivalence-point pH and suggest a suitable acid–base indicator.

13. Lidocaine is a widely used drug for topical anesthesia. According to the U.S. Pharmacopoeia, the chemical assay of a sample of lidocaine is performed by acid–base titration. Since lidocaine is nearly insoluble in water, a 532-mg sample was dissolved in 50 mL of a certain hydrochloric acid solution. The excess HCl was then titrated with $0.1063M$ sodium hydroxide, using bromcresol green as indicator; 18.27 mL were required. When 50.00 mL of the same HCl solution were titrated, using bromcresol purple, 39.55 mL of the NaOH solution were required. Assuming that lidocaine is a monoprotic base, what is the purity of the sample? The molecular weight of lidocaine is 234.34.

POTENTIOMETRIC pH MEASUREMENTS

11

A natural law regulates the advance of science. Where only observation can be made, the growth of knowledge creeps; where laboratory experiments can be carried on, knowledge leaps forward.

Michael Faraday

Perhaps the most common chemical measurement made in all sorts of laboratories—chemical, biological, geological, clinical, waterworks, environmental research, industrial control laboratories, and others—is the determination of the pH of a solution. pH measurements are easy to do, thanks to modern instruments, but to interpret them most usefully we must understand the fundamental limitations of the measurement technique.

11.1 THE HYDROGEN ELECTRODE CELL

The primary standard tool for the determination of pH is the cell

$$\text{Pt} \mid \text{H}_2(g), \text{Solution of unknown pH} \mid \text{Reference electrode,} \qquad (11.1)$$

where the reference electrode is typically the silver/silver chloride half-cell in one of its variations, or the saturated calomel electrode. The hydrogen gas must be highly purified and bubbled through the solution at some known pressure. The platinum electrode must be prepared by electrodeposition of finely divided platinum metal (platinum black) to provide a catalytic surface. The solution of unknown pH must be free of strong oxidizing or reducing agents as well as of certain impurities that poison the platinum surface. Therefore, this electrode is not used for routine pH measurements, particularly under field conditions.

The Nernst equation for cell (11.1) is:

$$E = E_{ref} - 0.0592 \log a_{\text{H}} + E_{\text{j}}, \qquad (11.2)$$

where E_{ref} is the reduction potential for the reference electrode, E_{j} is the (usually) small liquid junction potential between the two half-cell solutions. It is assumed that the pressure of hydrogen is 1 atm and the temperature is 25°. Equation (11.2) is solved for pH:

$$\text{pH} = - \log a_{\text{H}} = \frac{E - E_{ref}}{0.0592} - \frac{E_{\text{j}}}{0.0592}.$$

The experimental measurement gives a value for E. It is assumed that E_{ref} is already known. Because E_{j} is unknown it is necessarily ignored, but with the realization that, for each millivolt in the E_{j} value, the pH calculation is in error by

$$0.001/0.0592 \approx 0.02 \text{ pH unit.}$$

This fundamental limitation in the absolute accuracy of typical pH determinations cannot be overcome by modern instrumentation or by theoretical correction.

11.2 THE GLASS ELECTRODE

It is an interesting discovery that certain compositions of glass, when immersed in water solutions, form a hydrated layer of silicic acid that behaves somewhat like an amphoteric surface, able either to take on or give up small quantities of hydrogen ions to the aqueous solution at the interface. This means that the electrical charge of the thin surface layer will vary according to the activity of hydrogen ion in the solution:

$$^+H_2(glass) \rightarrow H(glass) \rightarrow Glass^-.$$
$$increasing\ pH \rightarrow$$

Even more remarkable is the finding that this surface potential varies nearly in accord with a Nernst-like equation

$$E(\text{of glass surface}) = 0.0592\ pH + k. \tag{11.3}$$

This has led to a great deal of research in attempts to perfect the composition of a glass having the optimum behavior, i.e., as close as possible to Eq. (11.3). A widely used glass for this purpose, manufactured by Corning Glass Works as Corning 015 is 72.2 mol% SiO_2, 6.4% CaO, and 21.4% Na_2O, a composition that also has a relatively low electrical resistance.

Many different manufacturers now market "glass electrodes" fabricated from this and other glass compositions. Figure 11.1 shows the essential physical picture.

When the electrode shown in Fig. 11.1 is dipped into a solution of unknown pH, *both* surfaces of the pH-responsive glass reach equilibrium with their respective solutions and there is no mixing of the solutions (i.e., no diffusion through the glass). It is the *difference* between the two surface potentials that must be sensed by the measuring instrument and interpreted in terms of the pH.

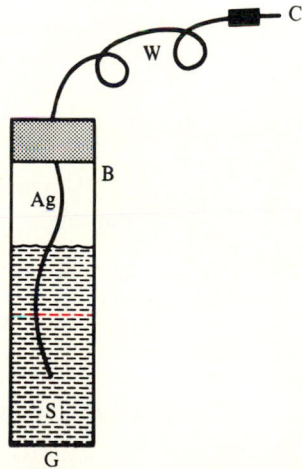

Fig. 11.1 Glass electrode for pH measurements: C—contact for plugging into pH meter; W—shielded cable; B—body of the electrode, completely sealed; Ag—silver wire coated with silver chloride; S—internal reference solution, pH = 7 buffer saturated with silver chloride; G—pH-responsive glass membrane.

11.3   THE REFERENCE ELECTRODE

The cell is completed by using a reference electrode for the other half-cell. A typical example, the saturated KCl version of the calomel half-cell, is constructed as shown in Fig. 11.2.

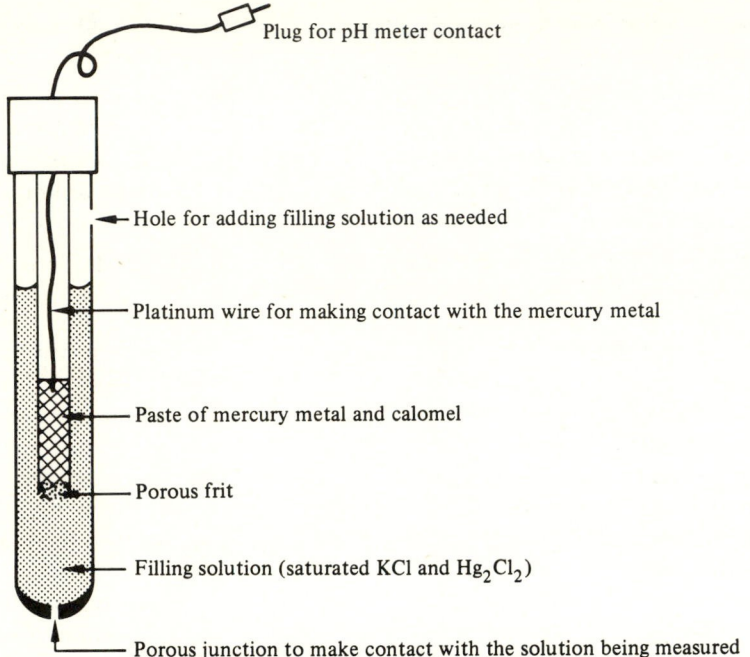

Fig. 11.2 Saturated calomel reference electrode.

11.4   RELATION OF CELL POTENTIAL TO SOLUTION pH

When the two electrodes are dipped into a solution whose pH is to be measured, the following cell has been assembled:

$$\underbrace{Ag \mid AgCl(s), \; Buffer \mid Glass}_{Glass \; electrode} \; \left| \begin{array}{c} Solution \; of \\ unknown \; pH \end{array} \right\| \; \underbrace{KCl(sat), Hg_2Cl_2(s) \mid Hg(liquid)}_{Reference \; electrode}$$

Note that there are five phase boundaries at which electrical potentials are developed; numbering them from left to right, we write the potential for the cell as a whole as:

$$E = E_1 + E_2 + E_3 + E_j + E_5, \tag{11.4}$$

where

E_1 is the potential of the internal silver/silver chloride electrode in the body of the glass electrode assembly. This potential remains constant because the electrode body is sealed, and no mixing occurs with the outer solution.

E_2 is the small potential developed at the inner surface of the pH-responsive glass. It also is a constant because it is always in contact with the fixed inner solution.

E_3 is the potential established at the outer surface of the pH-responsive glass. It varies according to Eq. (11.3) if the glass is in good shape.

E_j is the liquid-junction potential that necessarily is established when the solution of unknown pH makes contact with the inner solution of the reference electrode via the fiber. It is not known with respect either to magnitude or sign and it changes whenever the solution of unknown pH is changed.

E_5 is the potential of the saturated calomel reference electrode. It is constant because the mercury is always in contact only with the same filling solution.

When Eq. (11.3) is substituted for E_3 in Eq. (11.4) and all the constant terms are collected together and called E_k, the result is

$$E_{cell} = E_k + 0.0592 \, pH + E_j. \tag{11.5}$$

Except for the value of the constant term E_k, Eq. (11.5) is the same as Eq. (11.2) as derived for the hydrogen cell. In practice the glass electrode deviates from ideal behavior in its response to changing pH, and the Nernstian slope of 0.0592 must be replaced with a slightly lower value, which we may represent as $\beta \cdot 0.0592$. The correction factor β is known as the **electromotive efficiency** for the particular glass electrode being used. The more practical version of Eq. (11.5) is, then:

$$E_{cell} = E_k + \beta \cdot 0.0592 \, pH + E_j. \tag{11.6}$$

Typical commercial glass electrodes in good condition have values of $\beta = 0.995$ or so in the pH range 1–10. There is usually a falloff in the value of β at higher pH. In the pH range 12–13, the value of β may be as low as 0.90.

Standardization of a Glass Electrode and pH-Meter Assembly

Where the hydrogen cell can be used directly for determination of pH through Eq. (11.3), it is not possible to use a glass electrode in such an "absolute" way because the value for E_k is unknown. First it is necessary to standardize the equipment with reference buffers of known pH and then to measure the pH of the unknown. The usual steps are as follows.

Step 1   Place the glass and reference electrodes in a buffer solution having a pH close to 7 and precisely known. Methods of obtaining such reference buffers are discussed later. Set the temperature control to the proper value. This adjusts the electronic circuit to provide a Nernst constant of whatever value is appropriate for the existing temperature (for example, 0.0592 for 25°). Now adjust the STANDARDIZE control until the reading on the pH-meter agrees precisely with the known pH of the reference buffer. From this point on, the STANDARDIZE control must not be changed.

Step 2   Rinse the electrodes with distilled water and place them into another reference buffer, chosen so that the two reference buffers will bracket the pH-range expected for the unknowns. For example, if the unknowns are expected to be in the pH range 5–6, the second reference buffer might be $0.0500M$ potassium hydrogen phthalate, which has a pH of 4.008 at 25°. Note the reading on the pH-meter once it has stabilized. If it agrees with the known pH of the second reference buffer (rare!), the glass electrode is behaving ideally in that pH range. Suppose it reads 4.19 instead of the expected 4.01. Then the electromotive efficiency for the glass electrode is readily calculated:

$$\beta = \frac{\text{Difference in pH reading}}{\text{Known true pH difference}} = \frac{7.00 - 4.19}{7.00 - 4.01} = 0.940.$$

Step 3   Most modern pH-meters are equipped with a SLOPE adjustment that can be used at this point to compensate for the electromotive efficiency of the electrode. One simply turns the control (typically a screwdriver adjustment) until the pH reading for the second reference buffer agrees with the expected value. This adjustment will have little or no effect on the standardization that was carried out at pH $= 7$, but it is a good idea to check to be sure. The SLOPE control introduces the factor β into the electronic calculations carried out by the circuitry.

Step 4   Once the pH meter assembly has been standardized, it is necessary merely to place the electrodes (after rinsing) into the unknown solution and to read the pH directly from the meter or digital display.

THE PROBLEM OF THE LIQUID-JUNCTION POTENTIAL

When the foregoing procedure is carried out, there are two applications of Eq. (11.6), one for the second standardizing buffer, one for the unknown. Let Eq. (11.6) be written for each step:

$$E_s + E_k + \beta \cdot 0.0592 \, \text{pH}_s + (E_j)_s, \tag{11.7}$$

$$E_x + E_k + \beta \cdot 0.0592 \, \text{pH}_x + (E_j)_x. \tag{11.8}$$

The pH-meter (or the experimenter) has only four pieces of data: E_s, E_x, β, and pH$_s$. These data are used in the following *operational definition* of pH:

$$\boxed{\text{pH}_x = \text{pH}_s + \frac{E_x - E_s}{\beta \cdot 0.0592}.} \tag{11.9}$$

Note that no attempt is made to evaluate the effect of the liquid-junction potential, because it is not possible to do so. The pH-meter simply carries out the arithmetic of Eq. (11.9) via its electronic circuits.

But out of curiosity we can solve Eqs. (11.7) and (11.8) to find an expression for the true value of pH_x. Simply subtract Eq. (11.7) from Eq. (11.8), and solve to find

$$(pH_x)_{true} = pH_s + \frac{E_x - E_s}{\beta \cdot 0.0592} + \frac{\Delta E_j}{\beta \cdot 0.0592}, \tag{11.10}$$

where ΔE_j is the *change* in liquid-junction potential when the unknown solution is substituted in place of the standardizing buffer. The other two terms on the right side are exactly the same as in Eq. (11.9), and so it follows that the operational value of measured pH is in error by about 0.02 pH unit for each millivolt change in the liquid-junction potential. Therefore it is meaningless to carry out pH measurements to the third decimal place, at least so far as absolute meaning is concerned. Under circumstances in which the liquid-junction potential may be presumed constant (e.g., the pH of an essentially constant solution is being watched for small drift), the higher-precision readings are useful and realistic, but only on a *relative* basis. This fundamental limitation in practical pH measurements is not widely appreciated.

11.5   ESTABLISHMENT OF STANDARD REFERENCE BUFFERS

For standardizing the pH assembly in Step 1 of the above procedure it was necessary to have a solution of precisely known pH. Whatever error exists in the standardization will be directly transmitted to the final reading on the unknown. For years there was international confusion on what the term "pH" should really mean and a variety of reference standards were in use, making it difficult to compare results from one laboratory to another. Under the leadership of Roger G. Bates, the U.S. National Bureau of Standards undertook a major research effort to establish internationally uniform procedures and definitions. The above operational definition of pH was one part of this work. A most important aspect was the establishment of reliable buffer mixtures that could be readily used by all laboratories. For a practical reference buffer to be suitable the following four criteria must be met:

1. The solution must be easy to prepare with standard lab techniques.
2. Once prepared, the solution must be chemically stable with an unchanging and highly reproducible value of pH.
3. The chemicals used in preparation must be inexpensive, available in very pure form, and sufficiently soluble.
4. As a group, the set of reference buffers must cover a wide pH range.

Suppose that a certain solution has been prepared that meets the stated criteria and that it is now necessary to establish its precise pH_s value. It will be necessary to go back to the hydrogen electrode, as in cell (11.1), but with an important difference. Recall that the shown cell (11.1) suffers the limitation of having a liquid-junction potential, and for a purpose so important as establishing a reference buffer it is necessary to eliminate this uncertainty. A similar cell, without a liquid-junction potential, can be constructed as follows.

$$Pt \mid H_2(g), \text{ Solution of reference buffer}, NaCl(CM), AgCl(s) \mid Ag.$$

Therefore the two half-reactions may be written as

Oxidation at left: $\frac{1}{2}H_2(g) = H^+(a_H \text{ of some value}) + e,$

Reduction at right: $AgCl(s) + e = Ag(s) + Cl^-(CM).$

The Nernst equation applies as follows:

$$E = E^0_{\text{red, AgCl, Ag}} - 0.0592 \log \frac{[Cl^-]f_{Cl}a_H}{p_{H_2}^{1/2}}.$$

This may be solved for $pH_s = -\log a_H$ (the subscript "s" refers to "standard"):

$$pH_s = \frac{E - E^0}{0.0592} - \log (p_{H_2})^{1/2} + \log C + \log f_{Cl}. \tag{11.10}$$

Everything on the right side is precisely known from experimental measurement except for f_{Cl}:

E is the measured cell voltage, which can be precise to 0.00001 volt.

E^0 has been exceptionally carefully studied for the silver chloride/silver electrode, over a range of temperatures; at 25°, $E^0 = 0.22234$ V.

The pressure of hydrogen is simply the atmospheric pressure (barometer) minus the vapor pressure of water at the temperature used.

The coefficient shown as 0.0592 (for 25°) is a simple function of temperature: $RT \ln 10/F$; the precise value at 25° is 0.059157.

C, the molarity of added sodium chloride, is precisely controlled to whatever value is desired.

The equation can thus be solved for pH once an estimate is made for f_{Cl}. It is conventionally accepted (Bates and Guggenheim) that the activity coefficient for the chloride ion may be calculated by the equation

$$\log f_{Cl} = \frac{-0.51 \, I^{1/2}}{1 + 1.5 \, I^{1/2}},$$

rather than by the Davies equation. The ionic strength of the solution is easily calculated because the precise composition of the buffer solution is known and its ionic-strength contribution can be added to the value of C.

But there's a catch here: First of all, the larger the value of C, the larger the ionic strength and hence the greater the uncertainty in the calculation of f_{Cl}. Second, the sodium chloride has to be present for the cell to function without a liquid-junction potential, but this salt is not a desired part of the reference buffer and, through its contribution to ionic strength, it may exert a small but appreciable effect on the actual pH of the solution. The problem is handled as follows: the cell is set up with varying values of C, say from 0.005 to 0.05 in steps of 0.005, and Eq. (11.10) is solved for each composition. In this way a set of pH' values is determined, one for each value of C, and then a plot of the calculated pH' value is made versus C and extrapolated to $C = 0$ to find pH$_s$. This is the most reliable way to eliminate from the results any effect of the added chloride on the pH of the reference buffer. A complete discussion of the subject is found in the excellent monograph by R. G. Bates, *Determination of pH, Theory and Practice*, John Wiley and Sons, 1973. A summary of the pH values established for the primary and secondary standard buffers, taken from this book, is found in Table 11.1.

11.6   DETERMINATION OF pK_a VALUES FROM BUFFER pH MEASUREMENTS

Because of the importance of acid–base reactions in nearly all areas of chemistry and biochemistry, much attention has been given to the accurate determination of pK_a values. In Chapter 13 we will see how spectrophotometric data may be used for this purpose, and in Chapter 18 examples will be given of indirect determination of K_a by interpreting the effect of pH upon solubility. However, by far the most important procedure for finding pK_a values is that based on use of galvanic cells for measurement of the pH of buffer solutions of known composition. First we will see how this approach is applied to monoprotic acids, and then we will consider the more complicated case of diprotic acids.

Monoprotic Acids and Their Buffers

The object is to determine a value for K_a for the acid HA, which dissociates according to the simple equation (charges omitted, to keep the discussion general):

$$HA + H_2O = H_3O^+ + A.$$

To see what is necessary we write the exact expression for K_a:

$$K_a = [H^+]f_1 \frac{[A]f_A}{[HA]f_{HA}},$$

where the values of the activity coefficients f_A and f_{HA} will depend on the charge type of the particular conjugate pair used.

The experiment to be performed is conceptually simple: One mixes C_a mol of substance HA with C_b mol of substance A, dilutes to one liter, and measures the equilibrium pH. The measurement may be done approximately, using a typical pH-meter with glass and reference electrodes, or it may be

Table 11.1 The pH(s) of NBS primary standards from 0 to 95°C. [m = molality (mol/kg).]*

Temperature °C	KH tartrate (saturated at 25°C)	KH₂ citrate (m = 0.05)	KH phthalate (m = 0.05)	KH₂PO₄ (m = 0.025) Na₂HPO₄ (m = 0.025)	KH₂PO₄ (m = 0.008695), Na₂HPO₄ (m = 0.03043)	Borax (m = 0.01)	NaHCO₃ (m = 0.025), Na₂CO₃ (m = 0.025)	0.05m KH₃(C₂O₄)₂ · 2H₂O	Ca(OH)₂ saturated at 25°C
0	—	3.863	4.003	6.984	7.534	9.464	10.317	1.666	13.423
5	—	3.840	3.999	6.951	7.500	9.395	10.245	1.668	13.207
10	—	3.820	3.998	6.923	7.472	9.332	10.179	1.670	13.003
15	—	3.802	3.999	6.900	7.448	9.276	10.118	1.672	12.810
20	—	3.788	4.002	6.881	7.429	9.225	10.062	1.675	12.627
25	3.557	3.776	4.008	6.865	7.413	9.180	10.012	1.679	12.454
30	3.552	3.766	4.015	6.853	7.400	9.139	9.966	1.683	12.289
35	3.549	3.759	4.024	6.844	7.389	9.102	9.925	1.688	12.133
38	3.548	3.755	4.030	6.840	7.384	9.081	9.903	1.691	12.043
40	3.547	3.753	4.035	6.838	7.380	9.068	9.889	1.694	11.984
45	3.547	3.750	4.047	6.834	7.373	9.038	9.856	1.700	11.841
50	3.549	3.749	4.060	6.833	7.367	9.011	9.828	1.707	11.705
55	3.554	—	4.075	6.834	—	8.985	—	1.715	11.574
60	3.560	—	4.091	6.836	—	8.962	—	1.723	11.449
70	3.580	—	4.126	6.845	—	8.921	—	1.743	—
80	3.609	—	4.164	6.859	—	8.885	—	1.766	—
90	3.650	—	4.205	6.877	—	8.850	—	1.792	—
95	3.674	—	4.227	6.886	—	8.833	—	1.806	—

*Except in precise research studies, there will be negligible error caused by substituting molarities of these values.

done more accurately by using a hydrogen electrode with a silver chloride/silver electrode that does not have a liquid-junction potential. To find the value for K_a from these data we must do the following:

1. From the pH measurement, calculate $([H^+]f_1) = 10^{-pH}$.

2. Calculate the ionic strength of the mixture, taking into account the ionic charge(s) on each member of the conjugate pair. Look up the f_A and f_{HA} values as calculated from the Davies equation.

3. Remember, from the discussion of buffers in Chapter 9, that the equilibrium molarities of HA and A differ somewhat from their initial values of C_a and C_b:

$$[A]_{equil} = C_b + [H^+] - [OH^-],$$
$$[HA]_{equil} = C_a - [H^+] + [OH^-].$$

These equations may be solved by finding the value for $[H^+]$ or $[OH^-]$:

$$[H^+] = \frac{10^{-pH}}{f_1}, \qquad [OH^-] = \frac{K_w}{[H^+]f_1^2} = \frac{1 \cdot 10^{-14}}{10^{-pH}f_1}.$$

If the measured pH is lower than about 6 we may ignore the $[OH^-]$, and if the pH is higher than about 8, we may ignore $[H^+]$. Both $[H^+]$ and $[OH^-]$ will be negligible compared to C_a and C_b if the pH is between 5 and 9. (The reader should show that these assertions are justified by considering numerical quantities.)

Example An organic chemist has prepared a purified sample of a new compound HR, which is a monoprotic acid, and would like to know the pK_a value because of its significance in supporting a proposed molecular structure. The molecular weight has already been determined by an accurate potentiometric titration with standard sodium hydroxide. A sample of precisely 8.00 mmol is weighed out and placed in a 100-mL volumetric flask. From a standard NaOH solution, 5.00 mmol of NaOH are added and the mixture is mixed and diluted to the mark. The pH is determined with a glass electrode that has been standardized using NBS reference buffers, and is found to be 6.52.

Following the steps listed above we find:

1. $[H^+]f_1 = 10^{-6.52} = 3.02 \cdot 10^{-7}$.

2. Since the sodium hydroxide reacted with the acid to form 5.0 mmol of the conjugate base Na^+R^-, the ionic strength of the buffer solution is 0.050. The value of f_1 is therefore 0.821.

3.
$$[R^-]_{equil} = 0.0500 + [H^+] - [OH^-] = 0.0500,$$
$$[HR]_{equil} = 0.0300 - [H^+] + [OH^-] = 0.0300.$$
(both are
negligible)

Then the value for K_a is readily calculated:

$$K_a = 3.02 \cdot 10^{-7} \frac{0.0500 \cdot 0.821}{0.0300} = 4.1 \cdot 10^{-7} \quad \text{or} \quad pK_a = 6.38.$$

Another example, with low pH It is desired to find the pK_a value for the species BH^+, which is the conjugate acid of the very weak base B. A buffer is prepared by taking 12.00 mmol of the base, adding 6.00 mmol of hydrochloric acid from a standard solution, mixing, and diluting to 100 mL in a volumetric flask. Thus, the nominal composition of this buffer is

$$C_a = 0.0600; \quad C_b = 0.0600.$$

The pH is determined to be 2.37. Following the logical steps:

1. $[H^+] = 10^{-2.37} = 4.27 \cdot 10^{-3}$.
2. Since the reaction that formed the buffer is

$$B + H^+Cl^- = BH^+Cl^-,$$

we see that there will be 6.00 mmol of strong 1:1 electrolyte present, and that

$$[H^+] + [BH^+] = 0.0600 = \text{Ionic strength.}$$

Therefore, $f_1 = 0.810$.

3. In this case the pH is low, and $[H^+]$ will not be negligible compared to C_a and C_b. Therefore

$$[B] = C_b + [H^+] = 0.0600 + \frac{4.27 \cdot 10^{-3}}{0.810} = 0.0653,$$

$$[BH^+] = C_a - [H^+] = 0.0600 - \frac{4.27 \cdot 10^{-3}}{0.810} = 0.0547.$$

Finally

$$K_a = [H^+]f_1 \frac{[B]}{[BH^+]f_1} = 4.27 \cdot 10^{-3} \frac{0.0653}{0.0547 \cdot 0.810} = 6.3 \cdot 10^{-3} \quad \text{or} \quad pK_a = 2.20.$$

Diprotic Acids and Their Buffers

As with the discussion of monoprotic acid buffers, we will keep the discussion general by not specifying the ionic charge(s) on the species. Thus we will consider the diprotic system

$$H_2A \xrightarrow{\quad pK_1 \quad} HA \xrightarrow{\quad pK_2 \quad} A.$$

The object of the pH measurements is to find the values for K_1 and for K_2, corresponding to the successive steps of the acid dissociation:

Step 1 $H_2A + H_2O = H_3O^+ + HA,$ $K_1 = [H^+]f_1 \dfrac{[HA]f_{HA}}{[H_2A]f_{H_2A}};$

Step 2 $HA + H_2O = H_3O^+ + A,$ $K_2 = [H^+]f_1 \dfrac{[A]f_A}{[HA]f_{HA}}.$

For some diprotic acids the value of pK_2 is much larger than that of pK_1. For example

$$H_2Gly^+ \xrightarrow{\quad 2.35 \quad} HGly \xrightarrow{\quad 9.78 \quad} Gly^-,$$
$$\text{glycine}$$

$$H_2Ma \xrightarrow{\quad 1.91 \quad} HMa^- \xrightarrow{\quad 6.33 \quad} Ma^{2-}.$$
$$\text{maleic acid}$$

In such cases it is possible to determine the pK values by using the same calculation approach as described for the monoprotic acids. For example, one buffer could be made by mixing some hydrochloric acid with an excess of glycine, resulting in a buffer of H_2Gly^+ and $HGly$. The pH would be so low, about 2–3, that the second step of dissociation to form Gly^- would be completely negligible. A second buffer would be made by mixing some sodium hydroxide with an excess of glycine. In this solution containing $HGly$ and Gly^-, the pH would be high, about 9–10, and there would be negligible formation of H_2Gly^+. Thus both buffers would fit the monoprotic model satisfactorily.

We are more concerned in this section with diprotic acids that have more closely spaced pK_1 and pK_2 values. An example is adipic acid $HOCO(CH_2)_4COOH$:

$$H_2Ad \xrightarrow{\quad 4.42 \quad} HAd^- \xrightarrow{\quad 5.41 \quad} Ad^{2-}$$

In a buffer that contains chiefly H_2Ad and HAd^-, there is significant dissociation to form Ad^{2-}. Similarly, in a buffer of HAd^- and Ad^{2-}, the pH is not high enough to eliminate formation of appreciable H_2Ad. The pK values are too close to allow these buffers to be treated by the simple monoprotic model, and we must derive a more complicated approach that takes account of the overlapping dissociation steps. The details of each buffer will be developed separately, and then the two results will be combined for the final calculation of K_1 and K_2.

THE FIRST BUFFER, H_2A AND HA CHIEFLY, WITH A LITTLE A

Consider a buffer solution that has been prepared by mixing C_a mol of H_2A with C_b mol of HA and diluting to one liter. (This is equivalent to mixing C_b moles of NaOH with $C_a + C_b$ moles of H_2A.) Equilibrium is quickly reached according to the two acid-dissociation steps. As H_2A dissociates in reaching equilibrium there is an increase in the molarity of HA over its initial value of C_b, and there is an increase of the molarity of A over its initial value of zero. Thus both HA and A increase at the expense of H_2A. The proton-balance condition is:

$$[H^+] = [OH^-] + ([HA] - C_b) + 2[A],$$

where the left-hand species is the only one that results from proton gain, while the right-hand species are due to proton loss.

Note that we use only the increase in the [HA] and that it is necessary to use 2[A] because each A species results from the loss of two protons from H_2A. (We cannot say that the A is formed from HA, because HA has actually increased.)

The proton-balance equation is now rearranged:

$$C_b + [H^+] - [OH^-] \equiv L_1 = [HA] + 2[A]. \tag{11.11}$$

The reason for defining the collection of terms on the left as L_1 is simply to make later algebraic manipulations less cumbersome. Note that L_1 will be a known quantity for the buffer, given the value of C_b used to make the buffer, the pH measurement, and the ionic strength of the buffer.

The next step is to introduce the alpha-expressions for HA and A (see Chapter 9):

$$[HA] = (C_a + C_b)\alpha_1 = (C_a + C_b)\frac{Q_1[H^+]}{[H^+]^2 + Q_1[H^+] + Q_1Q_2}.$$

$$[A] = (C_a + C_b)\alpha_0 = (C_a + C_b)\frac{Q_1Q_2}{[H^+]^2 + Q_1[H^+] + Q_1Q_2}.$$

These expressions are substituted into Eq. (11.11):

$$L_1 = (C_a + C_b)\frac{Q_1[H^+] + 2Q_1Q_2}{[H^+]^2 + Q_1[H^+] + Q_1Q_2}.$$

Now, do three careful operations: (a) multiply both sides by $[H^+]^2 + Q_1[H^+] + Q_1Q_2$, (b) collect the terms having $Q_1[H]$ and those having Q_1Q_2, (c) divide all terms by $[H^+]$. With a final rearrangement, the equation becomes the following:

$$\frac{L_1[H^+]}{(C_a + C_b - L_1)} = Q_1 + Q_1Q_2\frac{2(C_a + C_b) - L_1}{(C_a + C_b - L_1)[H^+]}. \tag{11.12}$$

This may look rather complicated to the reader, because it really is. We must be sustained by the hope that we will be able to solve the perplexing difficulty of the overlapping acid dissociations. Having completed the analysis of the first buffer, let us march on to the second buffer.

THE SECOND BUFFER, HA AND A CHIEFLY, WITH A LITTLE H_2A

In this case the buffer solution has been prepared by mixing C_a mol of HA and C_b mol of A and diluting to one liter. (This is equivalent to mixing $C_b + C_a$

moles of NaOH with $2C_a + C_b'$ moles of H_2A.) The shift in concentrations (which is necessary to reach equilibrium) results in an increase in the molarity of A over its initial value of C_b, and an increase in the molarity of H_2A over its initial value of zero.* The proton-balance expression is then

$$[H^+] + [H_2A] = [OH^-] + ([A] - C_b). \tag{11.13}$$

Equation (11.13) includes the molarity of H_2A, and we will want an expression in terms of the chief species, HA and A. To make the transformation we use the material-balance equation:

$$C_a + C_b = [H_2A] + [HA] + [A],$$

whence

$$[H_2A] = C_a + C_b - [HA] - [A].$$

By substituting for $[H_2A]$ in Eq. (11.13) and rearranging, we find

$$C_a + 2C_b + [H^+] - [OH^-] \equiv L_2 = [HA] + 2[A]. \tag{11.14}$$

Again, it is for algebraic convenience that we define the collection of terms on the left as L_2 which, like L_1, can be regarded as an experimental quantity once the pH has been measured.

Since the *form* of Eq. (11.14) is identical to that of Eq. (11.11) the only difference being that we have L_2 instead of L_1, we will omit the details of introducing the alpha-expressions and write directly the final result:

$$\frac{L_2[H^+]}{(C_a + C_b - L_2)} = Q_1 + Q_1 Q_2 \frac{2(C_a + C_b) - L_2}{(C_a + C_b - L_2)[H^+]}. \tag{11.15}$$

Now we have a bewildering equation for *each* of the buffers. However, they are only in terms of the Q_1 and Q_2 values because it was easier to do the derivations without the cluttering effect of activity coefficients. Unless the two buffers happen to have the same ionic strength, the Q values will not be the same in the two solutions. Therefore, we should transform Eqs. (11.12) and (11.15) into equivalent versions with K values instead. This is not difficult to do, because there is a direct relationship between K and Q:

$$K_1 = Q_1 \frac{f_1 f_{HA}}{f_{H_2A}}; \qquad K_2 = Q_2 \frac{f_1 f_A}{f_{HA}}.$$

When these expressions are rearranged, we find

$$Q_1 = K_1 \frac{f_{H_2A}}{f_1 f_{HA}}; \qquad Q_1 Q_2 = K_1 K_2 \frac{f_{H_2A}}{f_1^2 f_A}.$$

---

*The shift in molarities will be as described if the buffer has a pH lower than 7, because HA would have to dissociate to make the solution acidic. However, if pK_2 is higher than 7, the shift would be to form HA and H_2A at the expense of A. Although the proton balance expression would differ, the final conclusion reached will be the same.

Substitution into Eqs. (11.12) and (11.15) gives the following formidable results:

For the first buffer, H_2A and HA,

$$\frac{L_1[H^+]f_1f_{HA}}{(C_a+C_b-L_1)f_{H_2A}} = K_1 + K_1K_2\frac{\{2(C_a+C_b)-L_1\}f_{HA}}{(C_a+C_b-L_1)[H^+]f_1f_A},$$

and for the second buffer, HA and A,

$$\frac{L_2[H^+]f_1f_{HA}}{(C_a+C_b-L_2)f_{H_2A}} = K_1 + K_1K_2\frac{\{2(C_a+C_b)-L_2\}f_{HA}}{(C_a+C_b-L_2)[H^+]f_1f_A}.$$

Simply to make these equations easier to write and to look at, the complicated variables on the left will be called Y and those on the right will be called X. Using subscripts 1 and 2 to distinguish between the first and the second buffers, we then have:

$$Y_1 = K_1 + K_1K_2X_1; \tag{11.16}$$

$$Y_2 = K_1 + K_1K_2X_2. \tag{11.17}$$

At this point it is very important to realize that Y and X are both known from the buffer composition and the pH measurement. This means that Eqs. (11.16) and (11.17) are independent simultaneous equations with only two unknowns, K_1 and K_2. Therefore they may be combined to find the solution for each of the equilibrium constants. The reader should verify the following result

$$K_1 = \frac{Y_2X_1 - Y_1X_2}{X_1 - X_2}; \tag{11.18}$$

$$K_2 = \frac{Y_1 - Y_2}{Y_2X_1 - Y_1X_2}. \tag{11.19}$$

Note that the denominator of Eq. (11.19) is the same as the numerator of Eq. (11.18).

Thus by using the pH data on the two buffers of known composition, the values of K_1 and K_2 may be found, no matter how close they may be. No approximations were made in the foregoing derivations, so that the secondary dissociations are fully accounted for.

EXAMPLE CALCULATION: THE DETERMINATION OF pK_1 AND pK_2

Suppose two buffers are prepared as follows. For the first buffer, 10.00 mmol of a diprotic acid H_2R is mixed with 20.0 mL of 0.100M sodium hydroxide solution and then diluted to 100 mL. The measured pH of the resulting solution is 2.89.

For the second buffer, 10.00 mmol of H_2R is mixed with 15.0 mL of 1.00M sodium hydroxide and then diluted to 100 mL. The measured pH is 4.60.

Buffer 1   We find that

$$C_a = \frac{10.00 - 2.00}{100} = 0.0800\,M, \quad \text{and} \quad C_b = \frac{2.00}{100} = 0.0200\,M.$$

Therefore, the ionic strength is approximately 0.02 (due to the Na^+HR^-) plus $10^{-2.89}$ (due to the H_3O^+) for a total of 0.021. From the Davies equation, $f_1 = 0.87$ and $f_2 = 0.56$.

The next step is to calculate

$$L_1 = 0.020 + \frac{10^{-2.89}}{0.87} = 0.0215.$$

Then

$$Y_1 = \frac{0.0215 \cdot 10^{-2.89} \cdot 0.87}{0.1 - 0.0215} = 3.06 \cdot 10^{-4}$$

and

$$X_1 = \frac{(2 \cdot 0.1 - 0.0215) \cdot 0.87}{(0.1 - 0.0215) \cdot 10^{-2.89} \cdot 0.56} = 2739.$$

Buffer 2   In this case, $C_a = 0.0500$, $C_b = 0.0500$, and $I = C_a + 3C_b = 0.20$. Therefore, $f_1 = 0.75$ and $f_2 = 0.31$. We find

$$L_2 = 0.05 + 2 \cdot 0.05 + \frac{10^{-4.6}}{0.75} = 0.1500.$$

Then

$$Y_2 = \frac{0.150 \cdot 10^{-4.6} \cdot 0.75}{0.1 - 0.150} = -5.644 \cdot 10^{-5}$$

and

$$X_2 = \frac{(2 \cdot 0.1 - 0.15) \cdot 0.75}{(0.1 - 0.150) \cdot 10^{-4.6} \cdot 0.31} = -96316.$$

Now we may calculate K_1 and K_2:

$$K_1 = \frac{-5.644 \cdot 10^{-5} \cdot 2739 - 3.06 \cdot 10^{-4} \cdot (-96316)}{2739 - (-96316)} = 2.96 \cdot 10^{-4} \quad \text{or} \quad pK_1 = 3.53;$$

$$K_2 = \frac{3.06 \cdot 10^{-4} - (-5.644 \cdot 10^{-5})}{-5.644 \cdot 10^{-5} \cdot 2739 - 3.06 \cdot 10^{-4}(-96316)} = 1.24 \cdot 10^{-5} \quad \text{or} \quad pK_2 = 4.91.$$

11.7 POTENTIOMETRIC TITRATIONS OF ACIDS AND BASES

In Chapter 10 we discussed the theoretical calculation of acid–base titration curves and gave several examples. Such curves may be obtained experimentally and rather easily. It is only necessary to use a pH-meter to display the solution pH continuously during the execution of the titration. Typically the titrated sample is placed in a beaker so that the glass and reference electrodes of the pH-meter may be accommodated along with the tip of the buret, and the solution is kept stirred with a magnetic stirrer. The technique of potentiometric titration has several advantages over the use of visual indicators.

1. If the sample is of unknown nature, it is not possible to choose an indicator because the pH of the equivalence point cannot be predicted.
2. The sample may display more than one endpoint (inflection point) that can be observed when the data are plotted (pH versus volume added).
3. When endpoints are not sharp, indicators are generally unsatisfactory because of the too gradual color change. The potentiometric data may be plotted to find the inflection point quite accurately.
4. Because the potentiometric method yields pH data in the buffer regions of the titration curve, it is possible to calculate the pK value(s) of the titrated substance.

Obtaining the Data for the Curve

By referring to several of the theoretical titration curves shown in Chapter 10 we are reminded that the typical shape has a relatively flat buffer region followed by a rise to the inflection point. It can be shown that the position of this inflection point on the volume axis is extremely close to the true equivalence point for the titration reaction (see Fig. 11.3).

Therefore, in the performance of a potentiometric titration it is good practice to take quite a few pH readings in the immediate vicinity of the equivalence point, and this means that rather small volume additions must be made so that this important part of the curve will be well defined by the

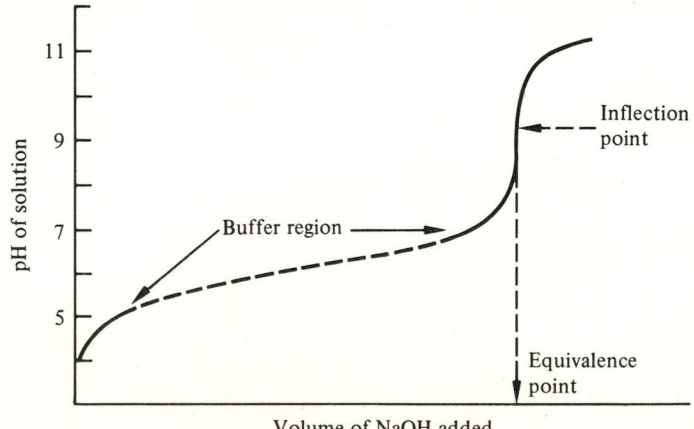

Fig. 11.3 Key points of a titration curve.

experimental data. On the other hand, since the pH does not change quickly throughout the buffer region, it is reasonable to make much larger volume additions during that stage (for example, 2 mL to 5 mL).

If the molarity of the titrated sample is not even approximately known, it is wise to perform an "exploratory" titration to get a rough idea of where the titration endpoint will be. Then a second portion of the sample may be titrated with more care, as follows: First add, say, 2-mL increments of titrant (recording the pH after each addition), until the total added volume is within 1–2 mL of that expected for the endpoint. Let one or two drops of titrant form slowly and drop of their own accord. Then read the meniscus level to the nearest 0.01 mL. From this point, until the titration is finished, the volume additions are made in terms of drops instead of milliliters. The size of slowly forming and self-falling drops is highly reproducible (perhaps to ± 0.001 mL) and so this technique will result in accurate increments. Take a pH reading for each drop, or two drops, or four drops (depending on the sharpness expected for the particular endpoint). After some definite number of drops (say, 20) have fallen, read the meniscus again. The change in meniscus reading, divided by the number of drops, gives the average size per drop and this should be calculated to the nearest 0.0001 mL. This part of the titration is rather slow and painstaking, and will take some time if the analyst has misjudged the position of the titration endpoint and started the dropwise addition too soon.

When the pH readings are clearly higher than what is expected for the endpoint and have started to level off, the titration is finished and the notebook is full of data. It becomes clear at this point why in titration we prefer to use visual indicators that are well suited to that type of endpoint detection. Table 11.2 shows computer-simulated data for the equivalence-point region in a titration of 100 mL of a weak acid solution of unknown molarity. The titrant was $0.1004M$ NaOH. The reason for using the computer is to obtain error-free data for the sake of discussing the principles of the data interpretation. We will also consider the effect of errors in pH values for the real case.

Interpreting the Data To Find the Precise Endpoint

Table 11.2 shows, in the first three columns, the volume and pH data, as they might be obtained by means of a high-precision pH-meter. The reader will recall earlier admonitions about the folly of trying to obtain pH values more accurately than about ±0.02 unit, because of the uncertainty of the liquid-junction potential in the cell. However, in the vicinity of the titration endpoint we may expect that the solution composition is not changing very much, and so the liquid-junction potential is either constant or perhaps undergoing a very slight drift. This means that we may measure *relative* pH values with higher

Table 11.2 Data for the titration of 100.0 mL of xM HA with 0.1004M NaOH, using a pH-meter

Volume NaOH, mL	Drops added	Observed pH	ΔpH, per 2 drops	$\frac{1}{\Delta pH}$
43.00	—	6.523		
			0.046	21.7
	+2	6.569		
			0.051	19.6
	4	6.620		
			0.057	17.5
	6	6.677		
			0.065	15.4
	8	6.742		
			0.077	13.0
	10	6.819		
			0.093	10.75
	12	6.912		
			0.118	8.47
	14	7.030		
			0.161	6.21
	16	7.191		
			0.257	3.89
	18	7.448		
			0.683	1.46
	20	8.131		
			1.597	0.63
	22	9.728		
			0.361	2.77
	24	10.089		
			0.195	5.13
	26	10.284		
			0.134	7.46
	28	10.418		
			0.102	9.80
	30	10.520		
			0.082	12.20
	32	10.602		
			0.069	14.49
	34	10.671		
			0.060	16.67
44.74	36	10.731		

$$\text{Average volume per drop} = \frac{44.74 - 43.00}{36} = 0.0483 \text{ mL/drop.}$$

precision, for example, ±0.001 unit. For the purpose of finding a titration endpoint we are not as concerned about the absolute accuracy of the pH values as we are about their relative precision. This is because we are interested only in finding the position of the inflection point on the curve, and if the entire curve is shifted vertically one way or the other from its true position, this will have no bearing on the position of the inflection point on the volume axis. Therefore, when possible we prefer to use a pH-meter with an expanded scale, so that pH is displayed to three decimal places.

Several graphical techniques have been proposed for finding the position of the inflection point in potentiometric titration data. The most direct method, which is often quite satisfactory, is to make a plot of pH versus volume based only on the data in the vicinity of the endpoint, so that an expanded volume axis may be used. Such a plot is shown for the above data in Fig. 11.4. Note that the use of the dropwise addition technique makes it particularly easy to plot the data because it is not necessary to locate odd volumes on the axis.

Often it is quite sufficient to "eyeball" such a graph to estimate the position of the inflection point. For the above data we would probably be content to choose 21 drops as the endpoint. In terms of milliliters, this would make the endpoint volume $43.00 + 21 \cdot 0.0483 = 44.01$ mL.

A second graphical method, widely suggested in spite of its obvious limitations, is based on the idea that the inflection point corresponds to a

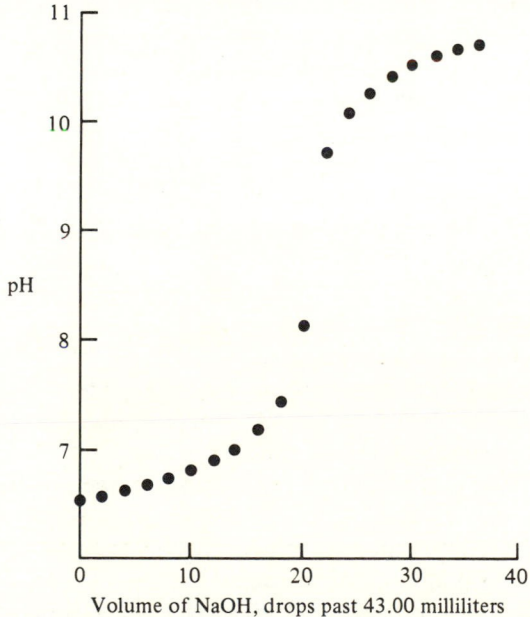

Fig. 11.4 Titration curve: pH versus volume. The volume axis is expanded; it shows the number of drops added after 43.00 mL had been added. A large pH change is observed even with only a two-drop addition. The endpoint is not very clear: it is somewhere between 20 and 22 drops added.

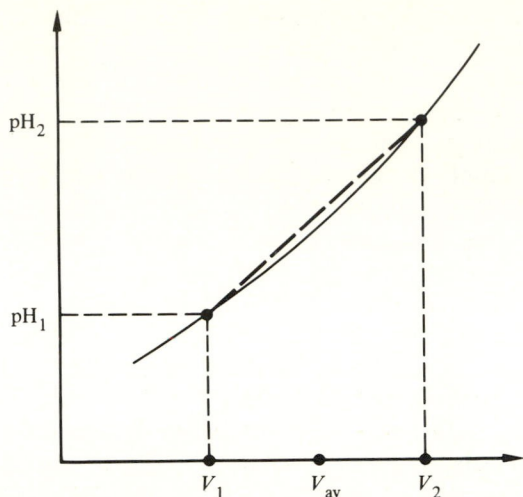

Fig. 11.5 Approximation of the slope of a titration curve.

maximum in the slope of the plot of pH versus volume. Therefore, if we can find the position of maximum slope, we have the endpoint. It is easy to obtain an estimate of the slope of the plot at intervals by using pairs of data points. Figure 11.5 will help to explain the calculation.

In Fig. 11.5, pH_1 and pH_2 represent two successive points on the titration curve, corresponding to volumes V_1 and V_2. The titration curve is shown by the curved line passing through the two points. The points may also be connected by a straight line, as shown by the dashed line in the figure. We may easily calculate the slope of this dashed line:

$$\text{Slope} = \frac{\Delta pH}{\Delta V} = \frac{pH_2 - pH_1}{V_2 - V_1}.$$

Now, this is not the slope of the titration curve at V_1 or at V_2, but there is some place between V_1 and V_2 where the curve does have this slope. (Imagine moving the dashed line parallel to itself until it is just tangent to the titration curve. The point where it touches the curve has that slope.) We make the simple approximation that the slope of the dashed line is equal to the slope of the curve at the point where $V = (V_1 + V_2)/2$, in other words, at the average of the two volumes. Since the points on a titration curve are usually not too far apart, this is a rather good approximation.

The fourth column of Table 11.2 gives the values of the slopes. Because the volume increments were constant, two drops each, it is sufficient to use simply ΔpH as the equivalent of the slope. We note from the table that the slope goes through a sharp maximum at about 20–22 drops, and this is even more obvious from Fig. 11.6. Note that it is necessary to plot the slope versus the *average* of the two volumes used to calculate it. Even with the precise

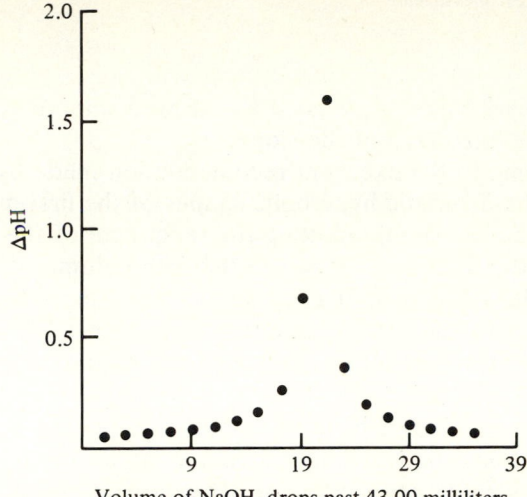

Fig. 11.6 First-derivative plot: ΔpH per two drops versus volume. The volume axis is expanded; it shows the average of the two volumes used to calculate the derivative. The precise position of the maximum is not easy to find: it is somewhere between 19 and 21 drops added.

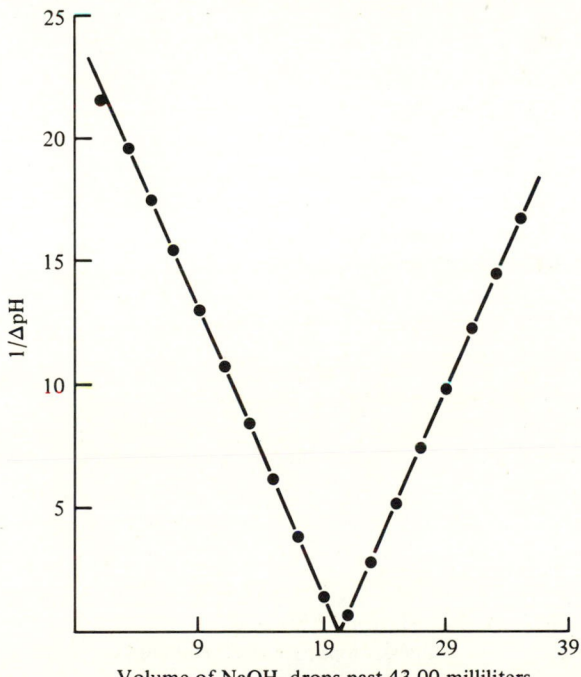

Fig. 11.7 Inverse-derivative plot: $1/\Delta pH$ versus volume. The volume axis is expanded; it shows the average of the two volumes used to calculate the derivative. The precise endpoint is easy to find by drawing straight lines: $V_{ep} = 20.3$ drops past the addition of 43.00 mL.

$$V = 43.00 + 20.3 \cdot 0.0483 = 43.98 \text{ mL.}$$

computer data, we see that it is guesswork to draw a smooth curve through the points to locate the maximum of the slope.

Finally we come to the excellent recomendation made by G. Gran* who showed that the characteristic hyperbolic shapes of the first-derivative curves could be made linear by simply plotting the reciprocal of the derivative. The present data give the quantities shown in the fifth column of Table 11.2, and Fig. 11.7 shows the Gran plot. It is of great importance that this approach allows us to make good use of the points that are somewhat distant from the equivalence point. This was not possible with the nonlinear plots of either of the other graphical methods.

As a final example of data interpretation in finding the endpoint, we consider the potentiometric titration of a very weak acid ($pK_a = 7.5$). Fig. 11.8 shows the conventional titration curve of pH versus V and the Gran plot. The scale for the axes has been kept the same as for the previous figures, so that the less favorable slope at the equivalence point is evident. In this case a visual indicator would undergo a rather gradual color change, and the potentiometric method is much more promising. Because of the relatively small slope, the data were obtained (again, by computer) by using four-drop increments

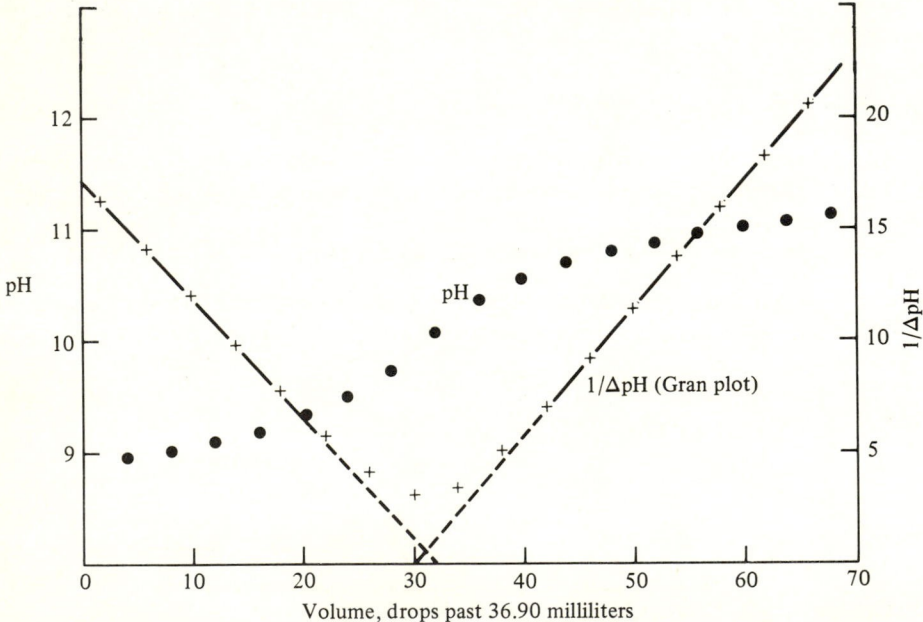

Fig. 11.8 Data for potentiometric titration of a very weak acid ($pK = 7.5$). Endpoint from Gran plot:

$$V = 36.90 + 31 \cdot 0.0483 = 38.40 \text{ mL}.$$

---

*G. Gran, *Acta Chemica Scandinavica*, 4, 559–577 (1950).

instead of two-drop increments. This provides better accuracy for the values of ΔpH.

The value of the Gran plot is evident. By drawing straight lines through the points that are not close to equivalence, we are able to extrapolate and find their intersection, which gives a reliable estimate for the endpoint.

Potentiometric Titration Data in the Real (Noncomputer) World

The foregoing discussion used computer-simulated titration data, so that the ideal nature of the graphical methods would be evident. The real data that we obtain in the laboratory will certainly have more errors, mostly in the pH values. Since we need the *change* in pH for the Gran plot, small errors in pH will lead to relatively large errors in ΔpH. This will cause a considerable scattering of the points on the graph. Nevertheless, the Gran plot may still prove to be useful because it allows averaging when the straight lines are drawn through the scattered points. In general, however, if a typical pH-meter precise to ± 0.01 unit is used for the titration, it is likely that the Gran plot will be somewhat disappointing compared with the examples shown above. It is well to keep in mind the advantage of using volume increments that are large enough to cause significant changes in pH, because then ΔpH values will be more reliable.

Stoichiometric Calculations

Once the endpoint has been determined, the ensuing calculations are similar to those for visual-indicator titrations. The condition for the equivalence point is

$$n(\text{titrant}) = N \cdot n(\text{constituent}).$$

where $N = 1$ for monoprotic titrations, and may be either 1 or 2 depending on which endpoint is used in a diprotic titration. Since

$$n(\text{titrant}) = V(\text{endpoint}) \cdot C(\text{titrant})$$

and

$$n(\text{constituent}) = V_{\text{init}} \cdot C_{\text{init}} \qquad \text{or} \qquad \frac{w(\text{constituent})}{\text{GFW}(\text{constituent})},$$

there are several possible ways to use the titration endpoint data.

Deduction of pK Value(s) from the Titration Curve

Earlier in this chapter it was shown how the determination of the pH of buffers of known composition facilitates the calculation of the pK_a value for a monoprotic acid. Also, for a diprotic acid, the determination of pH values for two buffers, one based on the H_2A/HA conjugate pair and the other based on the HA/A conjugate pair, allows calculation of both pK_1 and pK_2.

Since buffer solutions of the requisite type are automatically created in the course of a titration, we may use the potentiometric data for the buffer region(s) of the titration curve, rather than make up special buffer solutions for the purpose of determining pK value(s). This approach may be used even when the initial molarity of the acid (or base) being titrated is unknown, because the pH data in the equivalence-point region will provide that information, as shown in the previous section.

It is only necessary to do some simple stoichiometric calculations to find the necessary values for C_a and C_b for the buffer corresponding to a chosen point in the titration. Suppose that an initial volume of V_0 milliliters contains C_0-molar monoprotic acid, and that the titration is performed with B_0-molar sodium hydroxide. If a volume of V milliliters of the NaOH has been added to form a buffer, then

$$C_b = \frac{VB_0}{V_0 + V} \qquad \text{(for first buffer)},$$

and

$$C_a = \frac{V_0 C_0 - VB_0}{V_0 + V} \qquad \text{(for first buffer)}.$$

These equations hold also for the first-buffer region in the titration of a diprotic acid. As for the second-buffer region, it is necessary to remember that $(V_0 C_0)$ millimoles of the NaOH were required just to remove the first proton from the molecules of H_2A. Therefore, what we call C_b in the second buffer is due only to the NaOH added past the first equivalence point:

$$C_b = \frac{VB_0 - V_0 C_0}{V_0 + V} \qquad \text{(for second buffer)},$$

and

$$C_a = \frac{V_0 C_0}{V_0 + V} - C_b \qquad \text{(for second buffer)}.$$

Given the values for C_a and C_b, the calculation proceeds as shown earlier.

IDENTIFICATION OF AN UNKNOWN ACID AS MONOPROTIC OR DIPROTIC

Suppose an unknown acid is dissolved and titrated with sodium hydroxide, using the potentiometric technique. Will it always be possible to tell whether the substance is monoprotic or diprotic? At first thought it might seem that a diprotic acid may be characterized by having two inflection points, one for each equivalence point. This is often true, but if the values for pK_1 and pK_2 do not differ by more than about two units, there may be no rising inflection point at the first equivalence point because of the overlapping dissociation. This is well shown in Fig. 10.8, in which the titration curves for maleic and fumaric acids are compared. Maleic acid ($pK_1 = 1.9$, $pK_2 = 6.3$) shows two

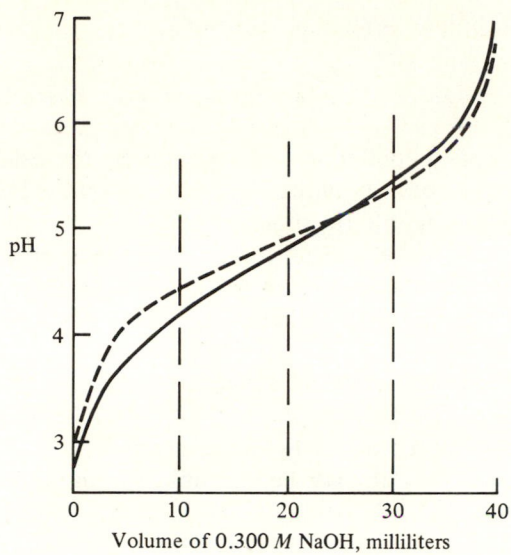

Fig. 11.9 Comparison of titration curves of monoprotic and diprotic acids. *Dashed line—monoprotic acid; solid line—diprotic acid.*

distinct equivalence points, while fumaric, its *trans*isomer ($pK_1 = 3.1$, $pK_2 = 4.6$), shows a good second equivalence point but has no inflection at the first.

Figure 11.9 shows titration curves for a monoprotic acid ($pK = 5$) and a diprotic acid ($pK_1 = 4.3$, $pK_2 = 5.7$). The pK values of diprotic acid average 5.0, the same as the monoprotic acid. To keep the scale the same, the monoprotic acid was present at an initial molarity of 0.12, while the diprotic acid was at 0.06. Therefore, a 100-mL sample of each solution requires 40.00 mL of $0.300M$ NaOH.

The question is: If one obtained either of these curves in the laboratory, by titration of an unknown, would it be possible to state whether the acid is monoprotic or diprotic? Well, the curves are not identical in spite of the fact that they have the same average pK value. The diprotic-acid curve has a steeper slope throughout the buffer region. We may calculate a change in pH by noting the pH values at the points where the titration is 25% and 75% complete (in this case at 10 mL and at 30 mL) since the endpoint is at 40 mL:

$$\Delta pH = pH(75\%) - pH(25\%).$$

It turns out that ΔpH is *always* about 0.95 unit for monoprotic acids, and *usually* is greater than 1.0 for diprotic acids. The proof for monoprotic acids is quite easy: when the acid is 25% titrated, it is 25% in its conjugate base form, while 75% remains as the conjugate-acid form. By the simple buffer equation,

$$[H^+]_{25\%} = Q\frac{C_a}{C_b} = Q\cdot\frac{75}{25} = 3Q.$$

At the 75% titrated point, by similar reasoning,

$$[H^+]_{75\%} = Q \cdot \frac{25}{75} = \frac{Q}{3}.$$

By taking logarithms and combining these equations the reader will have little trouble in proving that for a monoprotic acid $\Delta pH(75\% - 25\%) = 0.95$.

It is theoretically possible for diprotic acids to have such close values of pK_1 and pK_2 that ΔpH will approach the limiting value of 0.95. However, the vast majority of diprotic acids will show a ΔpH greater than 1.0. For example, the diprotic acid of Fig. 11.9, with rather close pK values, shows a ΔpH value of 1.7.

In conclusion, it will usually be possible to classify an unknown acid as either monoprotic or diprotic, but there may be an occasional diprotic acid that will pass as monoprotic as far as the appearance of the titration curve is concerned. In such a case it may be possible to use spectrophotometry to reveal the formation of the species if H_2A, HA^-, and A^{2-} have different absorptivities.

Resolution of Mixtures by Potentiometric Titration

Because the potentiometric titration technique can reveal the entire titration curve, in contrast to an indicator that responds to pH change only in a single region, we often may determine the amounts of acid (or base) constituents in a mixture. Suppose a solution contains two acids, HA and HB, and that a sample is titrated with standard sodium hydroxide. It is the nature of chemical equilibrium that the stronger acid (lower pK value) will react preferentially with the added hydroxide ion. If the acids differ greatly in pK values, the stronger one will react essentially completely before the weaker one reacts appreciably.

However, it is an important property of the titration curve that a rising inflection point will appear precisely at the equivalence point for the stronger acid, while with continued titration a second rising inflection point will appear precisely at the equivalence point for the weaker acid. It does not matter if, in fact, the weaker acid begins to react before the stronger acid has been used up: The position of the inflection point on the volume axis will mark the amount of sodium hydroxide required for the stronger acid alone. It is well to admit at this point that, for the case of a mixture of acids with close pK values, there will be no visible first inflection point. This is similar to the situation for the first equivalence point in the titration of a diprotic acid. However, if the two acids differ by more than about three units in pK, the first endpoint will be visible.

For example, Fig. 11.10 shows the titration curve for a 100-mL sample of a solution containing $0.0200M$ HA ($pK = 3.3$) and $0.0440M$ HB ($pK = 6.0$), the titrant being $1.00M$ NaOH. The stronger acid, HA, is titrated first and we

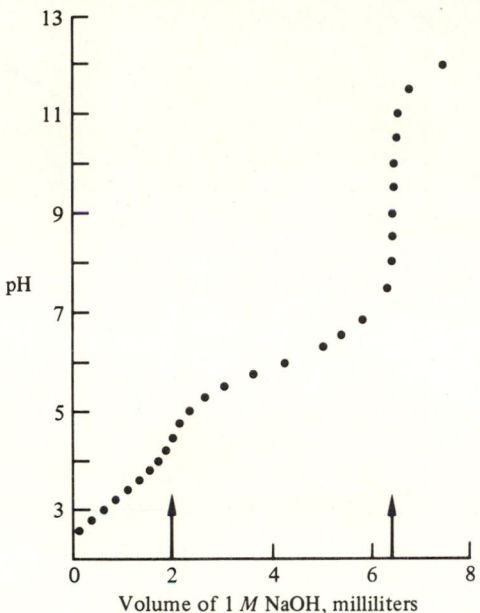

Fig. 11.10 Titration of a mixture of two weak acids. The 100-mL sample contains $0.0200M$ HA ($pK = 3.3$) and $0.0440M$ HB ($pK = 6.0$). Equivalence points are marked by vertical arrows.

note that the first rising inflection point occurs at 2.00 mL, in accord with expectation. Then the buffer region for the HB/B⁻ system is followed by a steep-inflection point that occurs at precisely 6.40 mL. The difference in the two endpoints, 4.40 mL, corresponds precisely to the amount of NaOH needed to titrate the weaker acid.

By noting that the pH is about 3.3 at the middle of the first buffer region, and about 6.0 at the middle of the second buffer region, we can estimate the pK values for the two acids as well.

A curve such as this would not be mistaken for that of a diprotic acid because the endpoints are not evenly spaced, as would be required for the successive steps of a diprotic acid.

Analysis of a Buffer by Potentiometric Titration

Our previous discussions of titration curves have assumed that the acid or base is initially present in pure form. But in a buffer solution there is a mixture of both members of the conjugate pair and, in a sense, we may consider the solution to be already partly titrated. When we begin a titration of a buffer solution we are already more or less near the middle of the fairly flat region of the titration curve. By titrating one portion of the buffer with standard NaOH, we may find the amount of the conjugate acid present. By

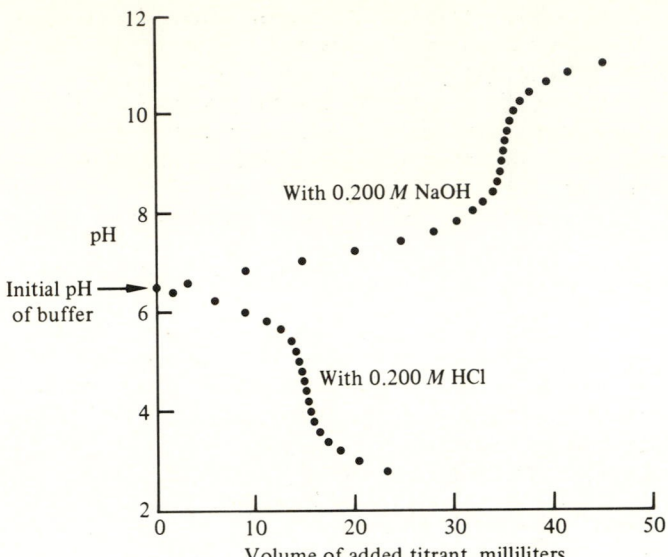

Fig. 11.11 Titration of a buffer.

titrating another portion with standard HCl, we may determine the amount of conjugate base present. This is illustrated by Fig. 11.11, which shows the data for the titration of 100-mL portions of a buffer containing KH_2PO_4 and K_2HPO_4. The NaOH endpoint at 35 mL shows that the buffer contained $0.070M$ conjugate acid, while the HCl endpoint at 15 mL indicates $0.030M$ for the conjugate base. This is a favorable case, because both endpoints are fairly sharp. For buffers with higher pH values, the NaOH endpoint will be more gradual, while the HCl endpoint will be even sharper, and vice versa.

PROBLEMS

1. The following cell may be used for pH determinations:

$$\text{Reference electrode} \mid \text{Solution, } H_2(1 \text{ atm}) \mid \text{Pt.}$$

   a) When a reference buffer known to have a pH of 9.19 is placed in the cell, the observed potential is -0.824 volt. Calculate the potential of the reference electrode.

   b) A solution of unknown pH is substituted for the buffer, and the potential of the cell is -0.786 volt. Calculate the pH of this solution.

2. The precise pH of a proposed buffer is determined by using a cell without a liquid-junction potential:

$$\text{Ag} \mid \text{AgCl(s), NaCl}(0.01M), \text{ Buffer, } H_2(1 \text{ atm}) \mid \text{Pt.}$$

The cell potential is measured to be -0.7255 volt. Taking E^0 for the silver chloride/silver electrode to be 0.2224 volt, the numerical constant in the Nernst equation to be 0.05916, and the ionic strength contribution by the buffer substances to be 0.0100, calculate the pH of the solution.

To find the pH of the buffer without the presence of sodium chloride, three additional measurements were made using $C_{NaCl} = 0.02$, 0.03, and 0.04. The corresponding cell potentials were -0.7084, -0.6987, and -0.6918 volt, respectively. Calculate the pH of each solution, and plot the four pH values versus the molarity of NaCl. Extrapolate to $C_{NaCl} = 0$ to find the pH of the buffer by itself.

3. We want to find the pK_a value for substance HR, using a galvanic cell of the following composition:

$$SCE \parallel HR, \text{ NaR Buffer, } H_2(1 \text{ atm}) \mid Pt.$$

The buffer is prepared by mixing 50 mL of $0.140M$ HR and 20 mL of $0.250M$ NaOH and diluting to 100 mL. The potential of the cell is observed to be -0.706 volt (Pt electrode negative). The reduction potential (E not E^0) for the saturated calomel electrode is $+0.244$ volt.

a) Find the pH of the buffer solution.

b) Calculate the pK value for HR.

4. To establish a reference buffer for use by chemical oceanographers, Bates and co-workers (*Analytical Chemistry 49*, 29 and 867 (1977)) made measurements with cells having no liquid-junction potential:

$$Pt, H_2(1 \text{ atm}) \mid \text{Synthetic seawater with HCl or buffer} \mid AgCl(s), Ag$$

An important assumption is that seawater may be treated as a *constant ionic medium* in which activity coefficients are at least constant, even if not accurately known. This means that the Nernst equation for the cell may be written with a reference potential E^* in place of the usual standard potential E^0. That is, the activity coefficient terms are lumped with E^0 to define the new constant term E^*. In the following calculations, use 0.05916 for the Nernst constant and use molality in place of molarity in all expressions.

a) To determine the precise value for E^* it was necessary to prepare a synthetic-seawater solution containing a known molality of HCl so that both the hydrogen ion and chloride ion molalities would be known for use in the Nernst equation. For this purpose a synthetic-seawater recipe with NaCl, KCl, $CaCl_2$, and $MgCl_2$ was used, containing $0.01000m$ HCl and a total chloride molality of 0.6070. The observed cell potential was 0.3700 V. Calculate the value of E^*.

b) Since natural seawater contains about $0.03m$ sodium sulfate, it would have been desirable to include this salt in the above recipe. The difficulty is that there is a significant tendency for the reaction $H^+ + SO_4^{2-} = HSO_4^-$, to take place; and therefore the precise molality of H^+ would be uncertain. However, once E^* has been determined, as in part (a), it may be used to estimate the value for the acid-dissociation quotient for HSO_4^-. For example, when the cell solution contained synthetic seawater including $0.02924m$ sodium sulfate, $0.01000m$ HCl, and a total Cl^- molality of 0.5691, the observed cell potential was 0.3783 V. Use these data, together with the E^* value, to find the actual H^+

molality in this solution. By difference, calculate the molalities of HSO_4^- and SO_4^{2-} in the equilibrium mixture, and then find the value for Q_a for HSO_4^-.

c) One studied reference buffer was 0.04000m tris/0.04000m trisH$^+$Cl$^-$ in the synthetic seawater. The solution had a total chloride molality of 0.5690, and the cell potential was 0.7383 volt. Calculate the pmH ($-\log$ (molality of H$^+$)) in this buffer, and the pQ_a value for trisH$^+$ in the seawater medium.

5. A 50.00-mL portion of a solution of a fairly strong acid was titrated with 0.1007M NaOH. When the pH reading was about 4, the buret reading was 20.30 mL, and the NaOH solution was added one drop at a time with the results shown in the table. The pH-meter was precise only to ±0.01 unit. The drop size was found to be 0.0441 mL.

Drops past 20.30 mL	Observed pH	Drops past 20.30 mL	Observed pH
1	4.03	6	9.95
2	4.25	7	10.10
3	4.70	8	10.21
4	9.21	9	10.30
5	9.73		

Plot pH versus the number of drops and, on the same graph, plot $1/\Delta$pH, remembering to place the points midway between the drop numbers. Find the titration endpoint and calculate the precise concentration of the acid in the original solution.

6. A 50.00-mL portion of a solution of a very weak acid (pK about 7) was titrated with 0.1000M NaOH. At 23.80 mL the analyst began adding titrant in two-drop increments with the results shown in the table. Find the endpoint and the molarity of the acid solution.

Drops past 23.80 mL	Observed pH	Drops past 23.80 mL	Observed pH
2	8.68	14	10.19
4	8.77	16	10.45
6	8.90	18	10.61
8	9.06	20	10.73
10	9.32	22	10.83
12	9.76	24*	10.91

*The buret reading was 25.00 mL at this point.

7. A mixture of a certain monoprotic acid HA ($pK = 3$) with a certain diprotic acid H_2B ($pK_1 = 4$, $pK_2 = 7$) having a total volume of 100 mL is titrated with 0.100M NaOH. The data obtained with a standardized pH meter are shown in the table:

V NaOH, mL	pH, observed	V NaOH, mL	pH, observed
2.00	2.50	39.50	7.00
10.50	3.00	42.00	7.50
18.40	3.50	43.30	8.00
24.70	4.00	43.80	8.50
29.00	4.50	43.95	9.00
31.00	5.00	44.03	9.50
32.20	5.50	44.17	10.00
33.50	6.00	44.55	10.50
36.00	6.50	45.75	11.00

Find the titration endpoints by plotting pH versus V, and deduce the original molarities of HA and H_2B. *Hint* At the first endpoint, HA has been converted to A^-, and H_2B has been converted to HB^-.

8. You are given 100 mL of water that contains 1.00 gram of a dissolved substance that is the hydrochloride salt of either the amino acid asparagine or of proline. Explain how a potentiometric titration with standard NaOH would enable you to identify the material, in terms of both pK values and molecular weight. Include sketches of the titration curves.

9. An organic chemist has requested your help in assaying a new compound that is a weak base with a molecular weight of 150. By using a pH meter in the titration of 100 mL of water containing 0.300 gram of the compound, with $0.0645M$ HCl as titrant, you find that the pH is 4.30 when the added volume is 11.50 mL. The endpoint, which is not very precise, is judged to occur at 29.7 mL.

   a) Calculate the percentage purity of the sample, assuming that any impurities are inert to reaction with HCl.

   b) Estimate the value of pK_b for the substance that is an uncharged molecular species B forming the conjugate acid BH^+.

10. The titration curve shown in Fig. 11.12 was obtained in the potentiometric titration of 0.473 gram of an amino acid (HCl salt) dissolved in 50 mL of water. The titrant was $0.120M$ NaOH. Interpret the curve to find the molecular weight, pK values, and the identity of the amino acid. *Note* The molecular weight for the amino acid is 36.5 units smaller than that of the titrated hydrochloride salt.

11. A mixture of three monoprotic acids is to be analyzed by titration with a standard solution of sodium hydroxide. In addition to hydrochloric acid, the mixture contains HA, a weak acid with $pK = 4.306$, and HB, which is much weaker.

   It is found that species HB and A^- do not absorb significantly at a wavelength of 473 nm, while both HA and B^- have appreciable molar absorptivities. A photometric titration is carried out by titrating a 100-mL portion of the mixture with $1.00M$ sodium hydroxide. The titration vessel is a beaker with 6.3-cm internal diameter, placed in the light path of the spectrophotometer adapted for that purpose. The results, with absorbance corrected for dilution, are shown in Fig. 11.13.

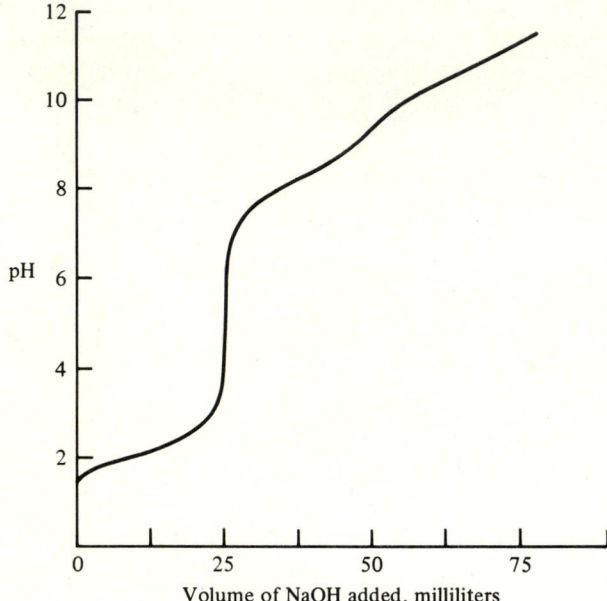

Volume of NaOH added, milliliters

Figure 11.12

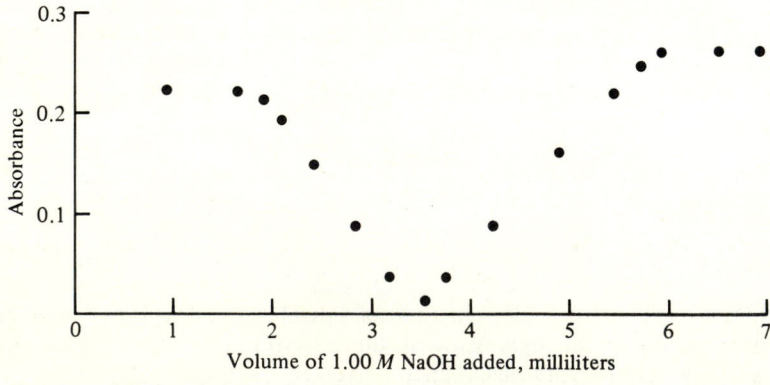

Volume of 1.00 M NaOH added, milliliters

Figure 11.13

a) Extrapolate the *linear* segments of the photometric titration data to find the three titration endpoints and calculate the molarities of the acids in the original sample.

b) Calculate the molar absorptivity of species B⁻ and show the units.

c) It was noted that the solution pH was 8.20 when 5.00 mL of the NaOH had been added. Use the titration data to calculate the pK value for HB, ignoring activity coefficients.

d) This mixture could also be analyzed by potentiometric titration, using a glass electrode. Including a point for the calculated initial pH, make a neat and fairly accurate sketch of the predicted titration curve.

e) To standardize the pH-meter for use in the titration, a buffer was prepared by mixing 25.0 mL of 0.100M perchloric acid with 50.0 mL of 0.0800M NaA, and diluting to 100 mL. Calculate the precise pH, taking $f_1 = 0.79$.

12. A certain diprotic acid has the following properties:

$$H_2A \xrightarrow{pK_1 = 3.30} HA^- \xrightarrow{pK_2 = 5.9} A^{2-}$$

ϵ at 500 nm: 6.5   61.0   23.0

A buffer containing 0.00500M H₂A and 0.0100M NaHA is to be titrated with 0.100M NaOH as a check on its composition, using a 100-mL sample.

a) Calculate the precise pH of the buffer, taking activity coefficients into account and considering the slight dissociation of the conjugate acid.

b) Without using activity coefficients, calculate the approximate pH values to be expected at each endpoint in the titration.

c) Draw the expected potentiometric titration curve, pH versus V_{NaOH} and discuss the feasibility of using acid–base indicators for each endpoint.

d) Noting the molar absorptivities given above, draw the expected photometric titration curve, assuming that the cell path is 1 cm. For this plot use linear line segments; i.e., assume the reactions go virtually to completion one step at a time. Ignore any effect of dilution on the absorbance.

13. Phosphoric acid is triprotic, with pK values of 2.12, 7.21, and 12.32. A low-pH buffer is prepared by dissolving 10 mmol of H₃PO₄ and 15 mmol of KH₂PO₄ and diluting to 100 mL. The entire solution is then titrated with 1.00M NaOH, using a pH meter.

a) Without concern for activity coefficients, calculate the approximate values for pH at the start and at each equivalence point in the titration.

b) Sketch the titration curve, taking care to have the equivalence points correspond to the right points on the volume axis.

c) What characteristics would be desired for an indicator for the second endpoint?

14. A solution of fumaric acid in 0.66M sodium chloride was titrated with 0.0694M NaOH. The initial molarity of the acid was 0.0233M, and the presence of the NaCl

served to keep the ionic strength essentially constant throughout the titration. A pH-meter, standardized with buffers that also contained $0.66M$ NaCl, was used to find the pcH of the solution during the titration. The second endpoint was sharp, at 33.50 mL added. When 4.17 mL of NaOH had been added, the pcH was 2.52, and when 29.22 mL had been added, the pcH was 4.39. Calculate the values of pQ_1 and pQ_2 for fumaric acid in this salt medium, using Eqs. (11.12) and (11.15).

15. For the determination of pK_1 and pK_2 of a certain diprotic acid H_2R^+, two buffers were prepared. For the first, 25.0 mmol of HR were mixed with 15.0 mmol of HCl and diluted to 100 mL; the pH of this solution was 4.44. The second buffer was prepared by mixing 25 mmol of HR with 15.0 mmol of NaOH and diluting to 100 mL; the pH was 5.92. Calculate pK_1 and pK_2.

16. A sample of the hydrochloride salt of an amino acid $H_2Am^+Cl^-$ weighing 114.4 mg was dissolved in 60 mL of water and titrated with $0.0300M$ NaOH, and the titration curve shown in Fig. 11.14 was observed using a glass electrode.

   a)   Calculate the molecular weight of HAm.

   b)   Estimate the pK_a values, ignoring activity coefficients.

   c)   Show that the pH at the start and at each endpoint is consistent with your estimated pK values.

   d)   Comment on the use of a visual acid–base indicator for this titration.

   e)   Sketch the titration curve expected for the titration of a sample of HAm (not H_2AmCl) with hydrochloric acid.

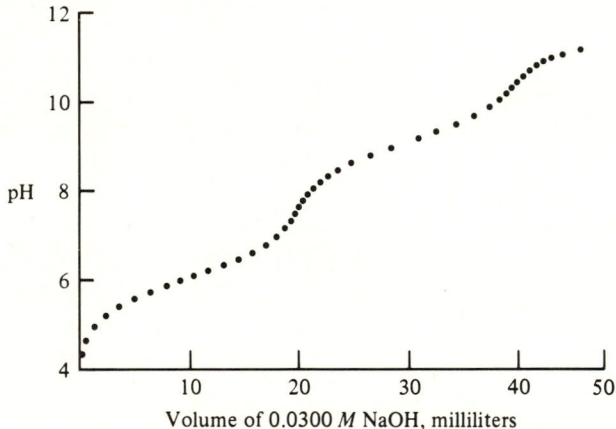

Volume of 0.0300 M NaOH, milliliters

Figure 11.14

REDOX TITRATIONS AND REAGENTS

12

I'd rather balance a redox equation than eat fried chicken.

Frank Quiring

As a category of chemical reactions, those classified as oxidation–reduction, or redox, are exceptionally valuable in analytical chemistry. However, in contrast to the relatively simple category of proton-transfer reactions, they are much less predictable and usually are much more complicated. We will distinguish two types of applications of redox reagents. One is for sample treatment in which mixtures may be heated and then allowed to stand for some time, and where clean stoichiometry is often not a problem because of the deliberate addition of excess reagent. The other type of application is that of titration, and this will be considered first.

12.1   REQUIREMENTS FOR A REDOX TITRATION

From an operational view, redox titrations do not differ from other types, e.g., acid–base titrations. When a titrant is added from a buret to the analyte solution, it is essential that the following criteria be met:

1. The titration reaction must be *fast* for otherwise the analyst would be unable to observe the approach of the endpoint. While proton-transfer reactions are extremely fast, making this criterion no problem for acid–base titrations, the rates of redox reactions vary tremendously. As pointed out in Chapter 4, there is no correlation between thermodynamic tendency and kinetic tendency. Reactions that have very large equilibrium constants may nevertheless be immeasurably slow. Therefore we cannot predict the feasibility of a redox titration, using equilibrium theory, in the way we can for acid–base titrations.

Some redox reactions that are too slow for titration purposes may be greatly speeded up by the addition of a catalyst, thus becoming suitable. Small amounts of substances that can cycle between different oxidation states (silver, iodine, for example) have often proved useful.

2. The titration reaction must proceed with very clean stoichiometry in accord with the balanced equation for the reaction. This criterion is essential. If it is not met, we would not be able to use a precise relationship between the amount of titrant added and the amount of constituent being determined. Again, in acid–base titrations this is never a problem because of the simplicity of the proton–transfer process. But many redox reactions have very complicated mechanistic pathways which, at points, may branch into alternative subpaths. This may involve, for example, the reactivity of an intermediate oxidation state of an element with the formation of products that are not included in the main path. Sometimes such problems may be prevented by the addition of a catalyst that can greatly reduce the average lifetime of the intermediate oxidation state.

Also because of the mechanistic pathway, redox reactions are often highly dependent on solution conditions including pH, temperature, and the presence of species that can form stable complexes with metal ions. In many cases the detailed mechanism of useful redox reactions is not well understood, and often we are not sure of the identity of the intermediate species that are formed as the reaction proceeds. Fundamental studies of reaction mechanism and kinetics are therefore important for the continued development of reliable analytical methods.*

3. Finally, for a reaction to be useful for titration purposes it is necessary to be able to find the endpoint with high precision and in close accord with the theoretical equivalence point for the titration. In some cases a titrant or constituent may be so strongly colored that it can serve as its own indicator. We will also discuss redox indicators, which have a role identical in principle to that of acid–base indicators. For some redox titrations we may make use of a platinum electrode to sense the potential, using the potentiometric titration technique. Other instrumental methods of endpoint detection are often possible as well, including photometric titrations.

12.2 EQUILIBRIUM CONSTANTS FOR REDOX TITRATION REACTIONS

In discussing acid–base titrations and titration curves in Chapter 10 we noted that the most favorable change in pH at the equivalence point was for the strong acid–strong base case in which the equilibrium constant for the reaction was 10^{14}:

$$H^+ + OH^- = H_2O; \qquad K = \frac{1}{K_w} = 1 \cdot 10^{14}.$$

The titration of a weak acid such as acetic acid ($pK \approx 5$) was also quite suitable for endpoint detection with an indicator:

$$HOAc + OH^- = OAc^- + H_2O; \qquad K = \frac{1}{K_b} = 1 \cdot 10^{9}.$$

Many redox reactions have even larger equilibrium constants and, if they meet the criteria listed earlier for titrations, show very sharp changes in potential at the equivalence point. For example, consider the titration of a rather extreme case—the titration of chromous ion with ceric ion:

$$Cr^{2+} + Ce^{4+} = Cr^{3+} + Ce^{3+}.$$

---

*For interesting discussions of a few redox mechanisms of analytical importance the reader should see *Chemical Analysis*, by H. A. Laitinen and W. E. Harris, 2nd ed., McGraw-Hill, New York, 1974.

The equilibrium constant is found by using the standard-reduction potential:

$$Cr^{2+} = Cr^{3+} + e, \qquad E^0_{red} = -0.41, \qquad nE^0 = +0.41 \text{ (for oxidation)}$$
$$Ce^{4+} + e = Ce^{3+}, \qquad E^0_{red} = +1.70, \qquad \underline{nE^0 = +1.70}$$
$$\text{Sum:} \qquad +2.11$$

$$K = 10^{nE^0/0.059} = 10^{2.11/0.059} = 6 \cdot 10^{35} \quad (\textit{very large}).$$

In fact, such equilibrium constants are so large that one should be wary of depending on certain calculations. For example: What is the equilibrium molarity of Cr^{2+} in an equilibrium mixture of $0.01M$ each of the other three ions? The direct approach is as follows:

$$[Cr^{2+}] = \frac{[Ce^{3+}][Cr^{3+}]}{[Ce^{4+}]K} = \frac{0.01 \cdot 0.01}{0.01 \cdot 6 \cdot 10^{35}} = 1.6 \cdot 10^{-38} \text{ mol/L}.$$

Since Avogadro's number is $6 \cdot 10^{23}$, there would be only one Cr^{2+} ion in 100 trillion liters! Since equilibrium depends on statistical behavior, such a conclusion is meaningless.

12.3 BASIC THEORY OF REDOX TITRATIONS

Quite a few oxidation–reduction reactions show very precise stoichiometry and also are sufficiently rapid to be used for titrations.* In some cases the constituent to be determined is oxidized by the titrant, as in the titration of arsenic(III) with a standard solution of potassium permanganate:

$$5HAsO_2 + 2MnO_4^- + 6H^+ + 2H_2O \rightarrow 2Mn^{2+} + 5H_3AsO_4$$

or, in simplified form,

$$5As(III) + 2Mn(VII) \rightarrow 5As(V) + 2Mn(II).$$

When enough permanganate has been added to oxidize all the As(III), the next slight excess gives a pale pink color to the solution, signifying that the endpoint has been reached.

In other cases the titrant is a reducing agent, causing the constituent being determined to undergo a decrease in oxidation state. An example is the titration of chromium(VI) with a standard solution of ferrous ammonium sulfate:

$$HCrO_4^- + 3Fe^{2+} + 7H^+ \rightarrow Cr^{3+} + 3Fe^{3+} + 4H_2O$$

or

$$Cr(VI) + 3Fe(II) \rightarrow Cr(III) + 3Fe(III).$$

---

*An excellent source for specific methods and fundamental critical discussion is *Volumetric Analysis*, Vol. III, by I. M. Kolthoff and R. Belcher, Interscience Publishers, New York, 1957.

The endpoint of this titration can be signaled by a redox indicator, which is a substance that undergoes a striking color change due to change in oxidation state in the immediate vicinity of the equivalence point of the titration.

Although various redox titrations differ in stoichiometry, in practical details, and in methods of endpoint detection, they all share a simple theoretical principle. Before the endpoint is reached, the reduction potential of the redox system is controlled by the redox couple being titrated, and after the endpoint the potential is controlled by the excess titrant and its conjugate species. This is quite analogous to the way in which, for acid–base titrations, the solution pH is controlled before the endpoint by the constituent conjugate pair and after the endpoint by the titrant conjugate pair. Whereas pH is the master variable that is followed for acid–base titrations, it is the reduction potential E that is the master variable for redox titrations.

The Use of Formal Potentials in Titration Theory

Many of the reagents and visual indicators used in redox titrations undergo changes in oxidation state according to the generalized half-reaction:

$$Ox^a + hH^+ + ne = Red^b + wH_2O,$$

where "a" and "b" are the ionic charges on the oxidized and reduced forms of the redox couple. The Nernst equation for such a half-reaction is written as follows:

$$E = E^0 - \frac{0.059}{n} \log \frac{[Red]f_b}{[Ox]f_a[H^+]^h f_1^h}.$$

The activity coefficients for the redox species depend upon the ionic charges, and when the coefficient h is other than zero, the potential will depend on the solution pH. When it is further realized that either Ox or Red, or both species, may be subject to competing equilibria, we see that the calculation of E may be formidable. For example, consider the chromium(VI)/chromium(III) couple:

$$HCrO_4^- + 7H^+ + 3e = Cr^{3+} + 4H_2O.$$

The Nernst equation may be written on the basis of this half-reaction. However, in strongly acid solution, there is some formation of H_2CrO_4, and at pH $= 7$ there is dissociation to CrO_4^{2-}. In addition, Cr^{3+} forms complexes with various anions and other ligands, including OH^-, even in weakly acidic solutions. These reactions would change the $[HCrO_4^-]/[Cr^{3+}]$ ratio and therefore the potential.

In the performance of redox titrations it is usually true that the reaction is carried out in an aqueous medium of high and constant ionic strength, at a relatively fixed pH and with relatively constant concentrations of complexing species. Thus typically we have a constant influence of the many complicating factors.

It is further true that in titration analysis we are chiefly concerned with reliable stoichiometry and less with the exact nature of the reacting species. Therefore it is much to our practical advantage to separate the logarithm term of the Nernst equation into two parts: We retain the part that shows the ratio of the two concentrations, but with the *total* concentration of each oxidation state and without concern for what actual species or mixtures of species may be present. The other part of the log term, which includes all the factors of activity coefficients, pH, complexes, acid dissociations, etc., which are often unknown, *is simply incorporated with the E^0 value.* Thus for the chromium case:

$$E = E^0 - \frac{0.059}{3} \log(\text{of various factors}) - \frac{0.059}{n} \log \frac{[\text{Cr(III)}]}{[\text{Cr(VI)}]},$$

where the first term on the right is a true constant, while the second term is constant only for the particular conditions of the titration.

In this way, the formal potential $E^{0'}$ is defined as:

$$E = E^{0'} - \frac{0.059}{3} \log \frac{[\text{Cr(III)}]}{[\text{Cr(VI)}]}.$$

The formal potential may be regarded as a practical working version of the standard potential for the situation at hand. If we know its value we can use the above equation to predict values of E for specified ratios of the redox couple and vice versa.

Numerical values of formal potentials must be determined experimentally, for the factors influencing them are too complex to allow reliable theoretical prediction.

Example   To determine the formal potential for the chromium couple in the titration medium of $0.5M$ HCl, we might prepare a cell as follows:

$$\text{SCE} \,\|\, 0.01M \text{ Cr(VI)}, 0.005M \text{ Cr(III)}, 0.5M \text{ HCl} \,|\, \text{Pt.}$$

Suppose the observed cell potential is 0.732 V. Calculation then shows the value for the formal potential:

$$E^{0'} \text{ (in } 0.5M \text{ HCl)} = 0.732 + 0.244 + \frac{0.059}{3} \log \frac{0.005}{0.01} = 0.97 \text{ V.}$$

FINAL COMMENT

The use of formal potentials greatly simplifies the theoretical discussion of titration curves. When we use standard potentials instead, we must realize that our predictions will be inaccurate, even if qualitatively useful.

Reduction Potential in the Titration System

The Nernst equation provides the fundamental basis for calculating the values of the reduction potential throughout the course of a redox titration. To illustrate, we will consider the arsenic(III)/manganese(VII) case mentioned above. Before the endpoint is reached the solution contains some As(III) as yet untitrated, and some As(V) formed by the titration reaction. It also contains some Mn(II) formed by the titration reaction, but no appreciable amount of Mn(VII) is present because the reaction goes largely to completion. This is why we may say that the potential is determined by the arsenic couple before the endpoint. In terms of the formal potential for the couple, the Nernst equation may be written:

$$E \text{ (before endpoint)} = E_{As}^{0'} - \frac{0.06}{2} \log \frac{[As(III)]}{[As(V)]}, \tag{12.1}$$

We will let X stand for the fraction of the As(III) that has been oxidized to As(V). For example, when the titration is 50% complete, $X = 0.5$. The reader should prove that before the endpoint the concentration ratio is:

$$\frac{[As(III)]}{[As(V)]} = \frac{1-X}{X}. \tag{12.2}$$

This may then be substituted into the Nernst equation, so that E may be calculated for various values of X.

Once the endpoint has been passed, there is no appreciable amount of As(III) left in the solution, and so the arsenic couple is ineffective in controlling the reduction potential of the system. Instead it is now the manganese couple that is in control because the solution contains the Mn(II) formed by the titration, plus whatever excess Mn(VII) has been added. The appropriate Nernst equation is:

$$E \text{ (after endpoint)} = E_{Mn}^{0'} - \frac{0.06}{5} \log \frac{[Mn(II)]}{[Mn(VII)]}. \tag{12.3}$$

Again letting X stand for the fraction titrated, we may show that after the endpoint

$$\frac{[Mn(II)]}{[Mn(VII)]} = \frac{1}{X-1}. \tag{12.4}$$

Substitution into Eq. (12.3) permits calculation of E values for values of X larger than 1.

To carry out these calculations we will use the standard reduction potentials for the arsenic and manganese couples, as given in Appendix 5. These are +0.56 volt and +1.51 volt, respectively. The results are given in Table 12.1.

Note that the potential before the endpoint does not change rapidly, and is equal to $E_{As}^{0'}$ at the midpoint of the titration ($X = 0.5$). Once a slight excess of permanganate is present, however, the potential jumps markedly to the higher values in the proximity of $E_{Mn}^{0'}$.

Table 12.1 Calculated Values of E_{red} for the Titration of As(III) with Mn(VII)

X	E from Eq. (12.1)	E from Eq. (12.3)
0.01	0.50	
0.1	0.53	
0.3	0.55	
0.5	0.56	
0.7	0.57	
0.9	0.59	
0.95	0.60	
0.99	0.62	
0.999	0.65	
1.001		1.45
1.01		1.47
1.05		1.48
1.1		1.49

Prediction of Potential at the Exact Equivalence Point

We used Eq. (12.1), with substitution from Eq. (12.2), to predict the potential before the equivalence point, while Eqs. (12.3) and (12.4) served for predicting potentials after the equivalence point. Neither would be useful for calculating the equivalence-point potential because As(III) and Mn(VII) are both present at extremely low and unknown molarities at that point. By first finding the equilibrium constant for the reaction, we could go through a calculation to find the equilibrium molarities, which then could be substituted into either Eq. (12.1) or (12.3). However, a more direct approach using the Nernst equation is preferred.

If we assume that the chemical reaction is fast enough to be at equilibrium at any point in the titration, then it is logical to say that both Eq. (12.1) and Eq. (12.3) will always show a correct relationship. In other words, there can be only one equilibrium potential E in the solution (just as there can be only one pH at equilibrium in an acid–base titration system), and each of the redox couples must have a concentration ratio consistent with that potential. Therefore, since E is the same for both couples, we may add Eq. (12.1) to Eq. (12.3). However, for the logarithm terms to be combined, it is necessary that they each have the same multiplier, which will be 0.06 if we first multiply each equation by the number of electrons:

$$2E = 2E^{0\prime}_{As} - 0.06 \log \frac{[\text{As(III)}]}{[\text{As(V)}]},$$

$$5E = 5E^{0\prime}_{Mn} - 0.06 \log \frac{[\text{Mn(II)}]}{[\text{Mn(VII)}]}.$$

Now when the equations are added, we get

$$7E = 2E'_{As} + 5E'_{Mn} - 0.06 \log \frac{[As(III)][Mn(II)]}{[As(V)][Mn(VII)]}. \tag{12.4}$$

At this point we apply stoichiometric reasoning. At the exact equivalence point the molarity of the Mn(II) produced by the reaction will be exactly 2/5 of the molarity of the As(V) produced, given the 2:5 stoichiometry of the balanced equation. Similarly, the concentration of Mn(VII) that remains unreacted at equilibrium will be exactly 2/5 of the molarity of unreacted As(III). When these conclusions are substituted into Eq. (12.4) the concentrations cancel out and the log term becomes zero.

Therefore the potential at the equivalence point is simply

$$E \text{ (endpoint)} = \frac{2E'_{As} + 5E'_{Mn}}{7} = \frac{2 \cdot 0.56 + 5 \cdot 1.51}{7} = +1.24 \text{ V}.$$

This is a weighted average of the two standard-reduction potentials. The endpoint potential will be closer to the $E^{0'}$ of the couple that involves the larger number of electrons in its half-reaction. In general, for a titration of couple (1) that undergoes a change in oxidation state of n_1, with a couple (2) that undergoes a change in oxidation state of n_2 units, the equivalence-point potential will be

$$E \text{ (endpoint)} = \frac{n_1 E'_1 + n_2 E'_2}{n_1 + n_2}. \tag{12.5}$$

This calculation is of theoretical interest but in practice it is highly improbable that a solution can be prepared with this potential. Because of the large equilibrium constant for most titration reactions, the slightest departure from true equivalence will cause the potential to shift significantly, as is suggested by the steep slope of the titration curve shown in Fig. 12.1.

How to Sketch a Redox Titration Curve

In Chapter 10 we saw how to sketch an acid–base titration curve, given only the pK_a for the acid. It is almost as easy to draw an approximate redox titration curve. First set up the axes, E versus percentage titrated, extending the latter to 200%. The E-axis should extend from the E^0 value for the couple being titrated to the E^0 value for the titrant couple. At the 50% point, place a cross (x) at the E^0 value for the titrated couple (Fig. 12.2a). Note the reason for this by examining Eqs. (12.1) and (12.2).

From Eqs. (12.3) and (12.4) we see that $E = E^0$ (titrant) when the added titrant is twice the stoichiometric amount. Place an x at the 200% point at a value of E^0 (titrant) (Fig. 12.2b).

Knowing that the titration curve is relatively flat before and after the endpoint, and steep at the endpoint, sketch the S-shaped curve that connects the two E^0 values. As with the curves for weak acids, there is a slight rise at the start (Fig. 12.2c).

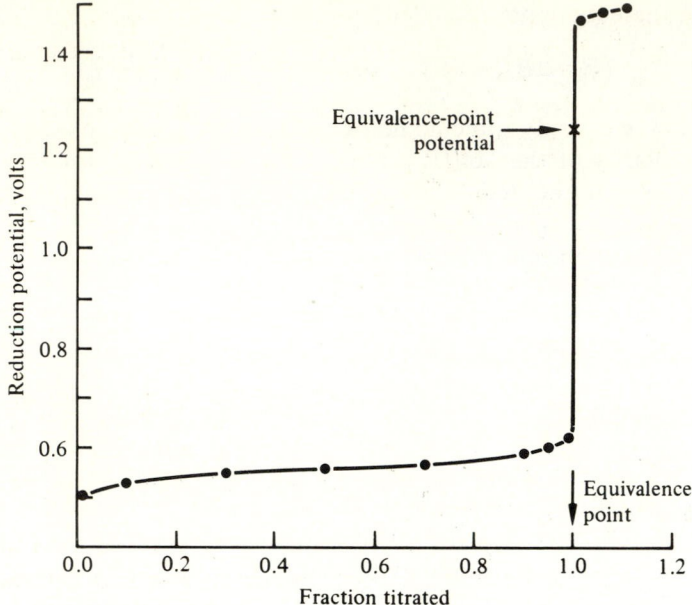

Fig. 12.1 Titration of arsenic(III) with manganese(VII).

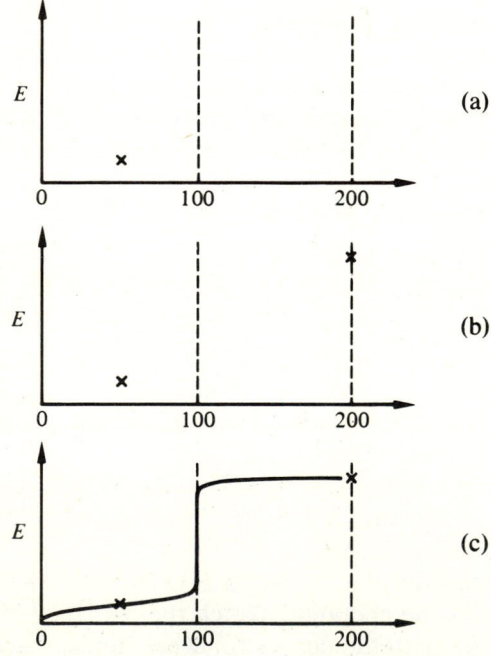

Fig. 12.2 Steps in sketching a redox titration curve.

12.4 SOME IMPORTANT REDOX REAGENTS

This text is not intended to be a reference for the huge variety of chemical reagents and applications related to oxidation–reduction reactions. However, there are certain examples that should be singled out because they are used so widely.*

Potassium Permanganate

In its highest oxidation state (+7) manganese is a powerful oxidizing agent and exists as the permanganate ion MnO_4^-. In acidic solutions this ion is reduced by many different substances to manganous ion:

$$MnO_4^- + 8H^+ + 5e \rightarrow Mn^{2+} + 4H_2O,$$

$$E_{red}^0 = +1.51 \text{ V}.$$

Although solutions of MnO_4^- are capable of oxidizing water:

$$H_2O \rightarrow \tfrac{1}{2}O_2 + 2H^+ + 2e; \qquad E_{ox}^0 = +1.23 \text{ V},$$

the reaction takes place so slowly in neutral solution that the permanganate solution maintains a stable molarity over a period of months provided it is prepared as described below.

The reagent nearly always used for preparation of a solution is potassium permanganate $KMnO_4$; it is available with purities typically higher than 99%. If a solution is needed for immediate and approximate work, it is quite satisfactory to weigh out the desired amount and to dissolve it in distilled water. The solution is intensely purple.

To prepare a solution that will not deteriorate appreciably for several weeks it is necessary to remove the manganese dioxide MnO_2 that is usually present as an impurity in the commercial reagent. Also it is necessary to be concerned about miscellaneous impurities, especially organic matter, that may be present in small amounts in the distilled water.

PREPARATION OF 0.02M $KMnO_4$

Dissolve about 3.2 gram of solid $KMnO_4$ in one liter of distilled water. The solution may be allowed to stand for three days or may be heated to boiling for about one hour. This causes oxidation of various impurities and the formation of a deposit of manganese dioxide. By filtering the solution through a fine sintered-glass crucible, one obtains a stable permanganate solution. For accurate work it is necessary to standardize this solution by titration of a primary standard reducing agent.

---

*For a thorough review, see I. M. Kolthoff and R. Belcher, *Volumetric Analysis*, Vol. III, Interscience Publishers, 1957.

STANDARDIZATION OF 0.02M KMnO$_4$

The best approach is to use the arsenic(III)/manganese(VII) reaction which was discussed earlier in this chapter. As a source of a precise amount of As(III) one weighs a sample of very pure arsenic trioxide As$_2$O$_3$ (MW = 197.84). If the KMnO$_4$ solution is about 0.02M and it is desired to use about 40 mL of it in the standardization titration so that the buret reading will be subject to little error, then 0.8 mmol of KMnO$_4$ will be used. Since the As(III)/Mn(VII) stoichiometry is 5:2, the arsenic sample should contain about 2 mmol of As(III). This corresponds to 1 mmol of As$_2$O$_3$ and therefore the sample should weigh about 0.2 gram and should be placed directly in the titration flask.

The sample of As$_2$O$_3$ must first be dissolved in about 10 mL of 3M sodium hydroxide, which hastens the solubility process through the reaction:

$$As_2O_3 + 2OH^- \rightarrow 2AsO_2^- + H_2O.$$

Then about 15 mL of 6M hydrochloric acid is added, converting the As(III) to the HAsO$_2$ species. The solution is then diluted to about 50 mL with distilled water, a drop of 0.002M potassium iodate is added to serve as a catalyst, and titration with the 0.02M KMnO$_4$ is continued until a very pale, pink color persists, indicating that a slight excess has been added.

Problem A sample of As$_2$O$_3$ weighing 0.2033 gram required 44.62 mL of KMnO$_4$ solution for titration. What is the precise molarity of the permanganate? *Answer* 0.01842M.

APPLICATIONS OF POTASSIUM PERMANGANATE

A common permanganate titration is the determination of iron(II) in acidic solution, using the reaction:

$$5Fe^{2+} + MnO_4^- + 8H^+ \rightarrow 5Fe^{3+} + Mn^{2+} + 4H_2O.$$

The permanganate ion is intensely purple and serves as its own indicator to show that the titration endpoint has been reached. The yellow color due to iron(III) in less acidic solutions can be eliminated by adding a little phosphoric acid, which forms a colorless complex with Fe(III).

Potassium permanganate solutions may also be used to advantage in the titration of oxalic acid (or oxalate salts), hydrogen peroxide, nitrite, etc. (see problems at the end of the chapter).

Examples of titration curves involving permanganate are shown in Figs. 12.3 and 12.4. These curves are computer-calculated and plotted.

Cerium(IV): Ammonium Hexanitratocerate, or Ceric Ammonium Nitrate

Most of the redox titrations that can be performed with potassium permanganate can also be done with a solution of cerium(IV). The reduction

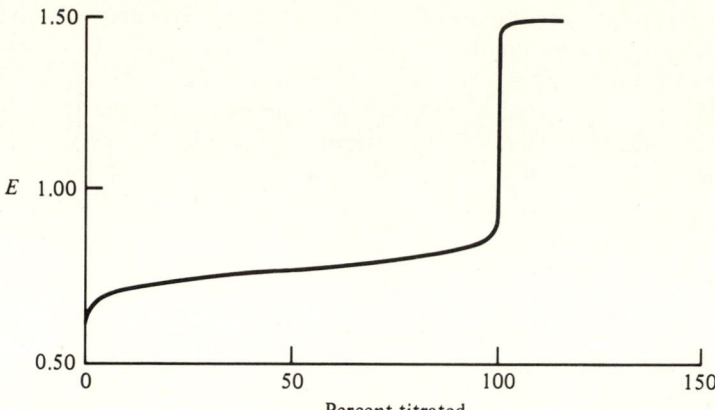

Fig. 12.3 Titration of ferrous ion with permanganate. At the 50% point, the potential is +0.77 volt, E^0 for the ferric/ferrous couple. At the 200% point, which is not shown, the potential would be +1.51 volt, the E^0 value for the MnO_4^-/Mn^{2+} couple. This titration is widely used in chemical analysis.

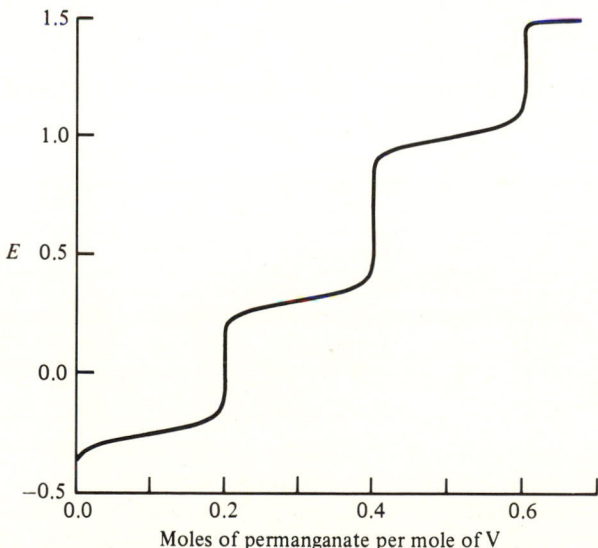

Fig. 12.4 Titration of vanadium (vanadous ion V^{2+}) with permanganate. There are three distinct endpoints. The first corresponds to the formation of vanadic ion V^{3+}. The midpoint of this segment is at $E = -0.26$, which is the E^0 value for the vanadic/vanadous couple. The second segment is for the further oxidation of vanadyl ion VO^{2+}, with $E^0 = +0.36$. The third segment corresponds to the final formation of pervanadyl ion $V(OH)_4^+$, with $E^0 = +1.00$. The endpoints are equally spaced because each step involves one-unit change in the vanadium oxidation state.

potential for the ceric/cerous couple $Ce(IV) + e \rightleftarrows Ce(III)$ depends on the acidic medium used. In $1M$ perchloric acid $E_{red}^0 = +1.7$, making $Ce(IV)$ even stronger than permanganate $(E^0 = 1.51)$. In $1M$ sulfuric acid the value is lower, $E^0 = +1.44$, apparently due to sulfate-ion complexes with the $Ce(IV)$ ion.

An advantage of $Ce(IV)$ over permanganate is the fact that its half-reaction involves only a one-electron change. Because of the five-electron change needed for MnO_4^-, the reactions often involve the formation of intermediate oxidation states of manganese and these sometimes undergo undesired side reactions, changing the apparent stoichiometry.

The most convenient way to prepare a standard solution of $Ce(IV)$ is by using pure ammonium hexanitratocerate $(NH_4)_2Ce(NO_3)_6$ $(MW = 548.23)$. A weighed amount is simply dissolved in perchloric or sulfuric acid and diluted to the desired total volume.

The yellow color of $Ce(IV)$ is not sufficiently intense for the species to serve as its own indicator in titrations, and it is necessary to use a redox indicator such as ferrous phenanthroline.

A good many organic compounds may be determined through their ability to react with $Ce(IV)$. For example, a specimen containing an unknown quantity of glycerol may be treated with $Ce(IV)$ in $4M$ perchloric acid. The reaction that occurs is:

$$\begin{array}{l} CH_2OH \\ | \\ CHOH + 8Ce(IV) + 3H_2O \rightarrow 3HCOOH + 8H^+ + 8Ce(III). \\ | \qquad\qquad\qquad\qquad\qquad\quad \text{formic acid} \\ CH_2OH \end{array}$$

The $Ce(IV)$ is added in precise amount, but in excess of that needed for the reaction. Then, the excess $Ce(IV)$ may be determined by titration with iron(II):

$$Ce(IV) + Fe^{2+} \rightarrow Ce(III) + Fe(III).$$

Potassium Dichromate

This substance, $K_2Cr_2O_7$ $(MW = 294.19)$, is readily available as a very pure product suitable for direct use as a primary standard. One simply weighs the desired quantity and dissolves it in distilled water with dilution to the desired precise volume. The solutions are exceptionally stable and may be kept unchanged for years.

In the water solution there is an equilibrium reaction due to water hydrolysis of the dichromate ion:

$$Cr_2O_7^{2-} + H_2O \rightleftarrows \qquad 2HCrO_4^-.$$
$$\text{hydrogen chromate ion}$$

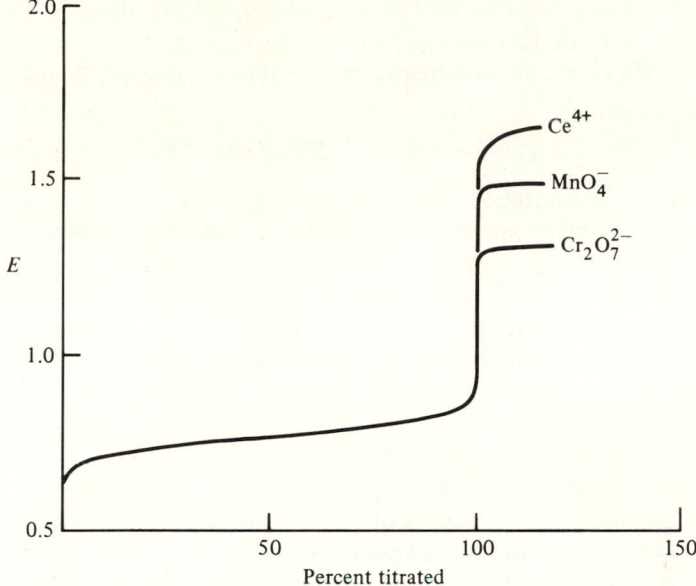

Fig. 12.5 Titration of Fe^{2+} with dichromate, permanganate, and ceric ions.

The equilibrium constant for this reaction is about 0.03, and in the dilute solutions typically used for titration most of the chromium(VI) is present as the monomer $HCrO_4^-$.

An important application for potassium dichromate solutions is in the determination of iron in an ore. This procedure is discussed in detail later.

When Cr(VI) is used as an oxidizing agent in acidic solution the half reaction is:

$$HCrO_4^- + 7H^+ + 3e \rightarrow Cr^{3+} + 4H_2O,$$

and the standard reduction potential is +1.33 volt. However, in the solution used for the iron-ore titration, the potential is lowered to about 1.0 volt, due to the presence of hydrochloric and phosphoric acids.

A comparison of the three oxidizing titrants is shown in Fig. 12.5 for the titration of ferrous iron. In each case the pre-endpoint section is identical, being determined entirely by the ferric/ferrous couple. After the endpoint the relative strengths of the oxidants are evident by the potential at which the curves tend to level off.

Iodine, or Triiodide Ion

Solid elemental iodine I_2 (MW = 253.81) is only slightly soluble in water. However, it is much more soluble in the presence of potassium iodide because of the formation of the triiodide ion.

$$I_2 + I^- = I_3^-; \qquad K = 7.2 \cdot 10^2.$$

Therefore when we refer to an *iodine solution*, we are usually speaking of a solution that contains $K^+I_3^-$ along with excess K^+I^-.

The triiodide ion is not nearly as strong an oxidizing agent as those discussed above:

$$I_3^- + 2e = 3I^-; \qquad E_{red}^0 = +0.54\,\text{V}.$$

Far from being a limitation, this smaller oxidizing power is an advantage because solutions of I_3^- are more selective. In typical concentrations $(0.05M)$ the I_3^- ion appears as a deep red-brown color, but at low concentrations the color is pale yellow, sufficiently intense to permit titration endpoints to be seen without the aid of an indicator. Nevertheless, it is easy and advantageous to use a suspension of starch as an aid in seeing the endpoint, due to the formation of an intense blue complex between starch and triiodide ion.

PREPARATION OF $0.05M$ I_3^- SOLUTION

Weigh accurately about 12.7 grams of pure iodine into a one-liter volumetric flask that contains about 40 grams of potassium iodide dissolved in 500 mL of water. It will take considerable stirring to dissolve the iodine, and this is most conveniently done with the aid of a magnetic stirrer. Then remove the stirring bar, dilute the solution to the mark, and mix thoroughly.

Problem 7.1362 gram of I_2 were dissolved with the aid of KI and diluted to precisely 500 milliliters. What was the molarity of the resulting solution? *Answer* $0.05623M$.

STANDARDIZATION OF AN IODINE SOLUTION

The procedure is similar to that described for the standardization of $KMnO_4$, with arsenic trioxide as a primary standard. The reaction has $1:1$ stoichiometry:

$$As(III) + I_3^- = As(V) + 3I^-.$$

The use of a starch indicator is optional when the iodine solution is not less than about $0.1M$ because the yellow color of excess I_3^- is easily visible if the solution is otherwise colorless. About one drop of $0.1M$ I_3^- added to 100 mL of water will give a distinct color, even though the diluted molarity is only $5 \cdot 10^{-5}$. However, in the usual titration procedure (see below), the molarity of I_3^- at the true equivalence point is much lower than this, about $5 \cdot 10^{-7}M$. This means that the enhanced color due to the formation of the blue starch–triiodide complex offers an advantage in accuracy.

The arsenic(III)/triiodide reaction has been well studied and it is possible to make realistic theoretical calculations. The fully balanced equation, written for acidic conditions, is:

$$HAsO_2 + I_3^- + 2H_2O = H_3AsO_4 + 3I^- + 2H^+.$$

The equilibrium constant for this reaction is small, about $K = 0.2$, and one might think that it therefore would be useless for titration. However, the reaction is chemically reversible and its equilibrium position is pH-dependent, as suggested by the liberation of $2H^+$ ions. If we work at a fairly high pH of about 9, the titration of As(III) by I_3^- becomes quantitative. We will examine this idea in more detail.

When the Nernst equation is written for the arsenic-reduction reaction

$$H_3AsO_4 + 2H^+ + 2e = HAsO_2 + 2H_2O,$$

we have

$$E = E_{red}^0 - \frac{0.059}{2} \log \frac{[HAsO_2]}{[H_3AsO_4][H^+]^2 f_1^2}.$$

Both of the arsenic species are weak acids:

$$H_3AsO_4 \xrightarrow{pK_1 = 2.2} H_2AsO_4^- \xrightarrow{pK_2 = 7.0} HAsO_4^{2-} \xrightarrow{pK_3 = 11.5} AsO_4^{3-};$$

$$HAsO_2 \xrightarrow{pK = 9.2} AsO_2^-.$$

These data indicate that at the usual pH 9 for the titration, As(III) is present as a mixture of $HAsO_2$ and AsO_2^- species, while As(V) is almost entirely present as $HAsO_4^{2-}$. More generally we may use the alpha-expressions to find the molarities of the parent acids in terms of the total concentration and the hydrogen-ion concentration:

$$[HAsO_2] = [As(III)]\frac{[H^+]}{[H^+] + 10^{-9.2}}$$

and

$$[H_3AsO_4] = [As(V)]\frac{[H^+]^3}{[H^+]^3 + 10^{-2.2}[H^+]^2 + 10^{-9.2}[H^+] + 10^{-20.7}}.$$

(We will ignore activity coefficients for simplicity.)

When these expressions are substituted into the Nernst equation, and the logarithm term is split up as discussed earlier to find an expression for the formal potential, we get:

$$E = E^{0'} - \frac{0.059}{2} \log \frac{[As(III)]}{[As(V)]},$$

where the concentrations represent the total amounts of arsenic in each oxidation state, and

$$E^{0'} = E^0 + \log(\text{complex function of } [H^+]).$$

Therefore, the formal potential for the arsenic couple may be calculated as a function of the solution pH. The results are summarized in Fig. 12.6. The E^0 value for the triiodide ion/iodide ion couple is not affected by the solution pH and is shown at its constant value of $+0.54$ volt. From the plot we see that the formal potential for the arsenic couple is equal to the iodine potential at a pH

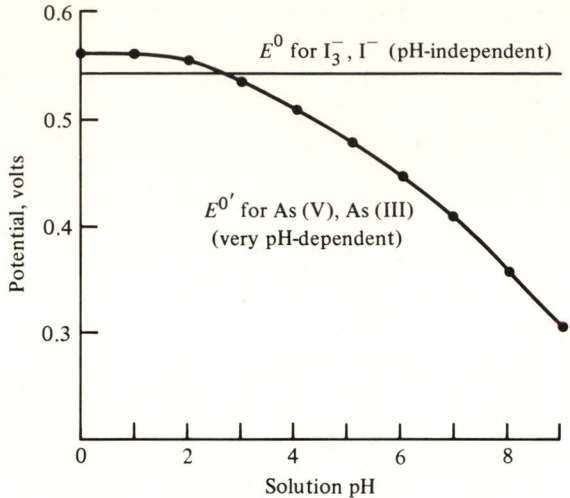

Fig. 12.6 Effect of pH on the As(V)/As(III) formal potential.

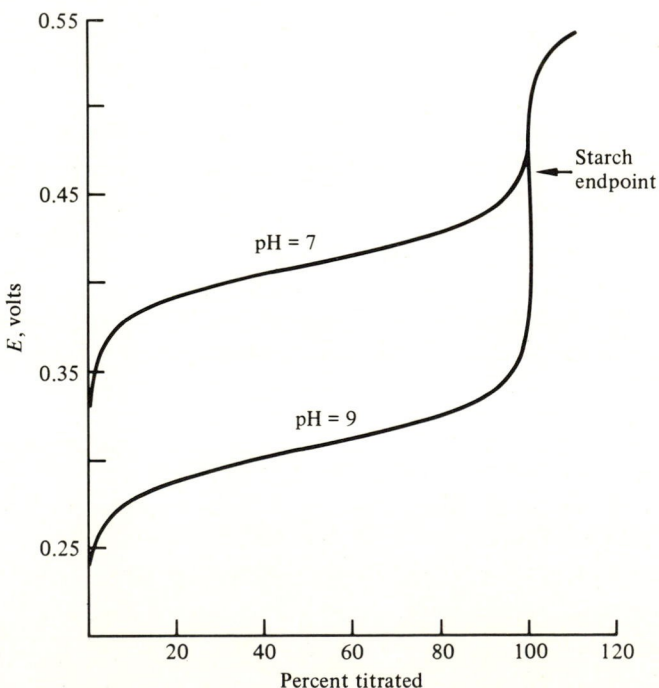

Fig. 12.7 Titration of $0.01M$ As(III) with I_3^- in the presence of $0.1M$ KI.

of about 2.7, and at still lower pH values it is even higher. The decrease in $E^{0'}$ with pH is evident and at pH = 9 the difference in the two potentials is more than sufficient to allow the titration reaction to be quantitative. The routine titration procedure simply uses a gram or two of sodium bicarbonate to adjust the pH to the desired value.

We conclude the discussion of the arsenic–iodine reaction by showing the theoretical titration curves for the titration of $0.01 M$ As(III) with $0.1 M$ I_3^-, using two different pH values. In Fig. 12.7 the lower curve corresponds to the recommended conditions of pH = 9. The slope is very steep at the endpoint, and starch indicator will show a distinct blue color when the solution is only very slightly overtitrated (not enough to cause appreciable error). The less favorable formal potential at pH = 7 causes the slope to be much smaller in the vicinity of the endpoint, and starch would cause the endpoint to appear about 1% early. In fact, at this pH one would be better off not using starch, since then a bit more triiodide would have to be added before the endpoint would be evident (yellow color).

APPLICATIONS OF DIRECT TITRATION WITH IODINE

Because of its relatively low reduction potential, iodine does not compete with permanganate, ceric, and dichromate for titrations of many substances. One particularly useful application is the titration of ascorbic acid (vitamin C, MW = 176.13). The reaction involves 1:1 stoichiometry:

ascorbic acid   dehydroascorbic acid

For example, to check the vitamin C content of a pharmaceutical tablet, it is merely necessary to dissolve the tablet in about 25 mL of distilled water and to titrate immediately with a standard solution of iodine. Usually the tablet will contain considerable insoluble binder material that does not dissolve, but this does not interfere with the visibility of the endpoint.

Problem A tablet containing vitamin C is titrated with a standard solution of $0.0221 M$ iodine. What mass of vitamin C (in milligrams) is contained in the tablet if 25.2 mL of the iodine solution is required to reach the endpoint? *Answer* 98 mg.

Potassium Iodate

A standard solution of potassium iodate KIO_3 (MW = 214.00) may be pre-pared by simply dissolving a weighed sample. Such a solution may be used in place of a standard iodine solution, provided that the solution of the sample being titrated contains some strong acid as well as some potassium iodide. As soon as the KIO_3 solution is added, there is rapid formation of iodine by the redox reaction:

$$IO_3^- + 6H^+ + 5I^- = 3I_2 + 3H_2O.$$

The iodine then can react with, say, ascorbic acid just as if it had been added directly. The iodate–iodide reaction may also be used for convenient prepa-ration of a standard solution of iodine (triiodide ion).

Sodium Thiosulfate

An analytical chemist hardly ever thinks of iodine without also picturing a standard solution of sodium thiosulfate $Na_2S_2O_3 \cdot 5H_2O$ (MW = 248.18). This is because of the highly reliable and useful reaction between thiosulfate ion and triiodide ion:

$$2S_2O_3^{2-} + I_3^- \rightarrow \quad S_4O_6^{2-} \quad + 3I^-$$
$$\text{tetrathionate}$$
$$\text{ion}$$

The structure of the thiosulfate ion is similar to that for sulfate ion:

$$\begin{array}{cc} O \;\; 2- & O \;\; 2- \\ | & | \\ O-S-O & S-S-O \\ | & | \\ O & O \end{array}$$

The oxidation product, tetrathionate ion, is notable for having four sulfur atoms bonded in a chain:

$$\begin{array}{cc} O & O \;\; 2- \\ | & | \\ O-S-S-S-S-O \\ | & | \\ O & O \end{array}$$

Both thiosulfate and tetrathionate ions are colorless. Thus in the titration of a solution containing iodine one sees a gradual disappearance of the yellow color, and the endpoint is taken as the disappearance of the last trace of color. Again, starch indicator may be added to enhance the visibility of the color change, which is then from blue to colorless.

PREPARATION OF 0.1M SODIUM THIOSULFATE

Sodium thiosulfate pentahydrate usually is available as crystalline lumps that are partially opaque due to loss of some of the crystallization water. Therefore it is not accurate to prepare a standard solution by direct weighing of the solid unless one takes the trouble to pick out crystals that are uniformly clear. Simply weigh 25 grams of the solid and dissolve in distilled water to which has been added about 0.2 gram of sodium carbonate. Dilute to one liter and let stand for a day before standardizing. The sodium carbonate serves to raise the solution pH, improving the stability, and to precipitate traces of copper that may be present as an impurity. Copper(II) catalyzes the decomposition of thiosulfate.

STANDARDIZATION OF SODIUM THIOSULFATE SOLUTION

The standardization depends on the titration of a known amount of iodine by using the reaction shown above. If an accurate solution of iodine is already available, it may be used directly. Otherwise it is an easy matter to prepare a known amount of iodine with a weighed sample of either potassium iodate or potassium dichromate, both of which are available in pure form. An accurately weighed sample of one of these salts is added to a strongly acidic solution of potassium iodide, whereupon a stoichiometric amount of iodine is formed by the reaction

$$IO_3^- + 6H^+ + 5I^- \rightarrow 3I_2 + 3H_2O$$

or

$$Cr_2O_7^{2-} + 14H^+ + 6I^- \rightarrow 3I_2 + 7H_2O + 2Cr^{3+}.$$

The "liberated" iodine is then titrated directly with the sodium thiosulfate solution, with or without the starch indicator.

Problem   A sample of pure potassium dichromate (MW = 294.19) weighing 0.1993 gram was added to 50 mL of 0.1M sulfuric acid to which a few grams of potassium iodide had been added. The resulting iodine was titrated with a solution of sodium thiosulfate, 33.65 mL being required to reach the endpoint. What was the molarity of the sodium thiosulfate?

Answer   There was 0.1993/0.29419 = 0.6775 mmol of potassium dichromate. By the reaction with iodide ion, 2.032 mmol of I_2 was produced. This would require 4.065 mmol of thiosulfate (2:1 stoichiometry). Therefore the molarity of the thiosulfate solution must be 4.065/33.65 = 0.1208M.

APPLICATIONS OF THE THIOSULFATE–IODINE TITRATION

The procedure just discussed for the standardization of a thiosulfate solution suggests a wide variety of applications. Because of the relatively low reduction potential of the I_3^-/I^- couple (+0.54 volt), many oxidizing agents have an E_{red}^0 higher than this and are therefore capable of oxidizing iodide ion to iodine:

$$Oxidant + nI^- \rightarrow \frac{n}{2}I_2 + Reductant.$$

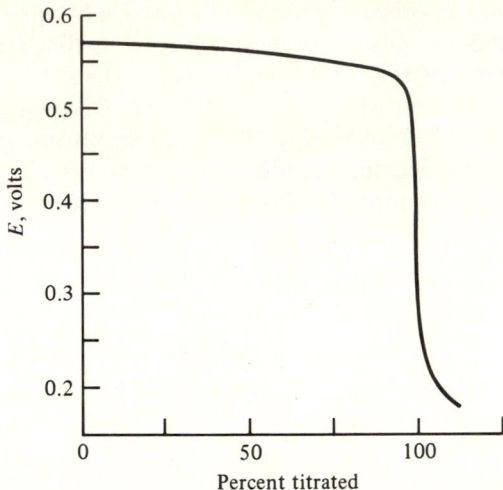

Fig. 12.8 Theoretical curve for the titration of $0.01M$ I_3^- in $0.1M$ KI with sodium thiosulfate solution.

The value of n is simply the change in oxidation state undergone by the oxidant. Thus the oxidant, which probably would not react with thiosulfate with clean stoichiometry, is converted to a precisely equivalent amount of iodine that can be titrated very easily and accurately.

Figure 12.8 shows a titration curve for the triiodide ion/thiosulfate reaction. The calculation in this case required a variation in the form of the Nernst equation before the endpoint, because one member of the redox couple, iodide ion, was essentially constant at $0.1M$. Also, the titrant couple involves a 2:1 ratio, since $2S_2O_3^{2-} \rightarrow S_4O_6^{2-} + 2e$.

Since the E^0 value for the I_3^-/I^- couple is $+0.54$ volt and that for the $S_4O_6^{2-}/S_2O_3^{2-}$ is $+0.08$ volt, the change in potential in the vicinity of the equivalence point is not as great as in many other examples. Nevertheless the slope of the curve at the endpoint is quite steep and allows very precise results to be obtained.

Ferrous Ammonium Sulfate (Mohr's Salt)

Even though it exists as a hexahydrate $Fe(NH_4)_2(SO_4)_2 \cdot 6H_2O$ (MW $= 392.14$), **Mohr's salt**, also referred to as **FAS**, has sufficient purity of composition to be useful as an approximate standard. A solution of ferrous ion, accurate to within a few parts per thousand, may be prepared simply by weighing a sample of the undried salt and dissolving it in $0.1M$ sulfuric acid. The presence of the acid is essential because otherwise there would be rapid air oxidation and formation of ferric hydroxide:

$$Fe^{2+} + O_2(air) \rightarrow Fe(OH)_3.$$

An acidic solution is still subject to air oxidation but the rate of the reaction is much slower. Because of the uncertainty in purity and the gradual change in concentration of Fe^{2+} due to oxidation, solutions of FAS should be standardized before being used for accurate work. Titration with dichromate (or permanganate) is one useful method for accomplishing this.

Oxalic Acid and Sodium Oxalate

These compounds are available in high purity and are especially useful as an alternative to arsenious oxide for the standardization of potassium permanganate solutions. The titration of a weighed sample is carried out in about $0.5M$ sulfuric acid at a temperature of about 70°. The reaction is:

$$5H_2C_2O_4 + 2MnO_4^- + 6H^+ \rightarrow 10CO_2 + 2Mn^{2+} + 8H_2O.$$

The permanganate serves as its own indicator, with about 0.03 mL of $0.02M$ $KMnO_4$ being required in excess to give a perceptible pink color.

12.5   HANDY GUIDE TO STANDARD SOLUTIONS OF REDOX REAGENTS*

Potassium dichromate can serve as the ultimate reference standard for redox, and so can arsenious oxide. However the latter is very expensive and less easy to prepare. The percentages shown in Fig. 12.9 are A.C.S. minimum purities for reagent quality.

12.6   VISUAL INDICATORS FOR REDOX TITRATIONS

Redox indicators are quite analogous to acid–base titration indicators in their operation. The ratio of the two colored forms of an acid–base indicator is controlled by the pH of the solution according to the K_a value for the indicator (Ind);

$$HInd = H^+ + Ind, \qquad K_{HInd} = \frac{[H^+][\text{Basic form}]}{[\text{Acidic form}]}$$

or

$$pH = pK_{HInd} - \log \frac{[\text{Acidic form}]}{[\text{Basic form}]}. \tag{12.6}$$

A redox indicator is a substance, typically an organic dye, that can be oxidized or reduced to a species of different color. We may write a general

---

*For a wealth of practical suggestions and details, consult I. M. Kolthoff and R. Belcher, *Volumetric Analysis*, Vol. III, Interscience Publishers, 1957. These titrations will be illustrated in problems at the end of this chapter.

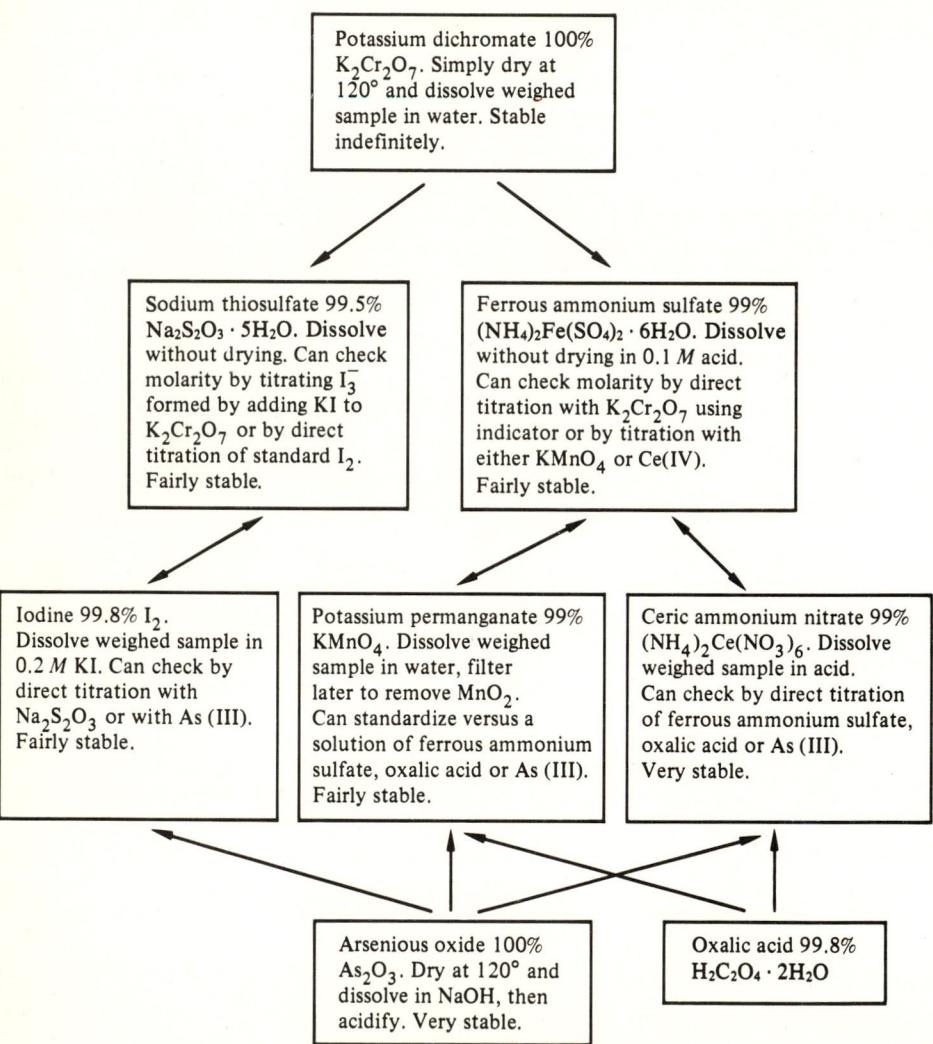

Figure 12.9

half-reaction:

$$\text{Ox} \quad + ne = \quad \text{Red}$$

(e.g., yellow)   (e.g., blue)

and the corresponding Nernst equation

$$E = E^{0'}_{\text{Ind}} - \frac{0.059}{n} \log \frac{[\text{Reduced form}]}{[\text{Oxidized form}]}. \tag{12.7}$$

The color of the acid–base indicator is dictated, according to Eq. (12.6) by the solution pH, which in turn is controlled by the main titration reaction. Similarly, the color of the redox indicator is dictated, according to Eq. (12.7), by the reduction potential of the titration system. In both cases the concentration of indicator is typically too small to have appreciable influence on the pH or potential.

From the examples of redox titration curves we have seen that large changes in potential occur in the vicinity of the endpoints. For a redox indicator to change color as a signal that the endpoint has been reached, it is necessary that the value of $E^{0'}_{\text{Ind}}$ be reasonably close to the value of $E_{\text{eq.p.}}$. This can be shown clearly by using a rearrangement of Eq. (12.7):

$$\frac{[\text{Reduced form}]}{[\text{Oxidized form}]} = 10^{n(E^{0'}_{\text{Ind}} - E)/0.059}. \tag{12.8}$$

Using the typical value of $n = 2$, we find from Eq. (12.8) that the ratio is about 10 when E is 0.03 volt lower than $E^{0'}_{\text{Ind}}$. The ratio is about 1/10 when E is 0.03 volt higher than $E^{0'}_{\text{Ind}}$. The redox indicator is at the center of its color-change interval when the ratio is 1.0, and this will occur when $E = E^{0'}_{\text{Ind}}$.

Therefore, if it is desired that the indicator be in the process of changing color at the same time that the titration is reaching the equivalence point, it is necessary that $E^{0'}_{\text{Ind}}$ be close to $E_{\text{eq.p.}}$. The value for the latter may be predicted by the calculation described earlier, using Eq. (12.5). Since the rate of change of E with volume of titrant is so great (as shown by the steep slope of typical titration curves), we see that it is not necessary to be too fussy about matching $E^{0'}_{\text{Ind}}$ with $E_{\text{eq.p.}}$.

Table 12.2 gives a short list of redox indicators which have proved useful over a range of potentials. Of these, ferroin (the ferrous ion complex with 1,10-phenanthroline) is discussed in Chapter 16 in connection with the spectrophotometric determination of iron, and diphenylamine sulfonate is discussed later in this chapter in the case study of the determination of iron in an ore.

Table 12.2 Redox indicators*

Indicator	Potential[a] at pH = 0, volts	Color	
		Oxidized form	Reduced form
Janus green B	−0.23	—	—
Neutral red	−0.20[b]	red-violet	colorless
Gallocyanine	0.02	violet-blue	colorless
Safranine O	0.24	blue-violet	colorless
Dimethylglyoxime[c]	0.25[d]	colorless	red
5,5′-indigodisulfonic acid, disodium salt	0.29	blue	colorless
Nile blue A	0.41	blue	colorless
Methylene blue	0.53	blue	colorless
Thionin(e)	0.56	violet	colorless
2,6-dichloroindophenol sodium derivative	0.67	red	colorless
Variamine blue B hydrochloride	0.71	blue	colorless
Diphenylamine	0.76	violet	colorless
3,3′-dimethoxybenzidine dihydrochloride	0.76	red	colorless
Barium diphenylaminesulfonate	0.84	violet-red	colorless
o-tolidine	0.87	blue	colorless
2,2-bipyridine[c]	0.97	blue	red
Xylene cyanole FF	1.00	pink	yellow-green
N-phenylanthranilic acid	1.08	violet-red	colorless
1,10-phenanthroline[c]	1.14	pale blue	dark red
5-nitro-1,10-phenanthroline[c]	1.25	pale blue	red
2,2′:6′2″-terpyridine[c]	1.25	pale blue	red

* From the J. T. Baker catalog. [a] Versus standard hydrogen electrode.
[b] At pH = 5. [c] Fe(II) present. [d] At pH = 9.4.

12.7 POTENTIOMETRIC DETECTION OF THE REDOX TITRATION ENDPOINT

There are a number of circumstances that may make it impossible to rely on
visual endpoint indicators for redox titrations. When the solution is highly
colored, for example, the color change of the indicator may be totally
obscured. In some cases it may not be possible to obtain a redox indicator
suitably matched to the equivalence-point potential, so that the endpoint would
be observed either too early or too late. When the E^0 values for constituent
and titrant are not far enough apart, there will be too little change in potential
in the vicinity of the equivalence point to cause the indicator to change
sharply. When we wish to analyze a mixture of constituents, with successive
equivalence points, a single indicator cannot suffice. Finally, when the initial
solution has a high potential, as does a solution of a strong oxidant such as
ceric ion, the added indicator may be irreversibly changed to a useless form
long before the endpoint is approached. Therefore it is fortunate that other
methods of endpoint detection are available. As with any type of titration, one
can examine the possibility of using a spectrophotometer to follow absorbance changes. However, it is the electron-transfer character of many

redox titration reactions that can often be exploited by using a platinum indicator electrode and a reference electrode to follow the potential during the titration. In terms of a conventional cell diagram we have

$$\text{SCE} \,\|\, \text{Solution being titrated} \,|\, \text{Pt.}$$

Before the endpoint the platinum electrode responds to the ratio of the couple being titrated, and after the endpoint it is the titrant couple that establishes the potential. This is the same, in principle, as the theoretical calculation of redox titration curves discussed at the start of this chapter. In many cases the experimental titration curves (observed values of E_{cell} plotted versus volume of added titrant) do indeed resemble the theoretical curves. This requires that the couples reach equilibrium with the platinum surface, which means that the reversible reaction

$$\text{Ox} + ne(\text{Pt surface}) = \text{Red}$$

must be fast. When this condition is not met, the experimental titration curves may be distorted versions of the smooth theoretical shapes, but often they are useful anyway because of the large potential changes at the endpoints.

For the interpretation of the data to find the precise endpoint, one uses the same approach as described in detail in Chapter 11 for potentiometric pH titrations. The difference is that E is used instead of pH.

12.8   CASE STUDY: ANATOMY OF A PROCEDURE— DICHROMATE TITRATION OF IRON IN AN ORE

The classical method of determination of iron in a sample of iron ore involves a good deal of interesting inorganic chemistry and depends for its success on a number of nicely balanced chemical equilibria. In the following discussion we will take one step of the procedure at a time and illustrate how applications of fundamental principles are used to explain the procedure.

Step 1   A sample of the iron ore Fe_2O_3, which is largely the mineral hematite, is weighed and dissolved in concentrated hydrochloric acid with heating.

In this step we face the natural insolubility of ferric oxide, which may be as well regarded as its hydrated form, ferric hydroxide $Fe(OH)_3$. The solubility product is very small, about 10^{-37}. However, the high acidity of the hydrochloric acid provides a strong driving force by reacting with OH^- to form water, and there is also some help in the fact that the ferric ion can form a tetrachloro complex, with $Q_f = 100$. The net solution reaction may thus be considered as the sum of four reversible reactions:

	$\log K$	
$\frac{1}{2}Fe_2O_3(s) + \frac{3}{2}H_2O(\text{liquid}) = Fe(OH)_3(s)$	0	Mole fraction for each substance 1
$Fe(OH)_3 = Fe^{3+} + 3OH^-$	-37	*Very* low solubility
$3(OH^- + H^+ = H_2O)$	$+42$	$(3pK_w)$, major effect
$Fe^{3+} + 4Cl^- = FeCl_4^-$	$+2$	Minor effect
$\frac{1}{2}Fe_2O_3 + 3H^+ + 4Cl^- = FeCL_4^- + \frac{3}{2}H_2O$		Overall, $\log K = +7$ (favorable)

Step 2   Iron(III) in the resulting solution is reduced to Fe(II) species by using a minimal excess of stannous ion. The tin is oxidized to the Sn(IV) state.

In the hydrochloric-acid medium the iron and tin species are present as chloro-complexes, and it is appropriate to use formal reduction potentials to calculate the equilibrium quotient for the reaction. In $1M$ HCl it has been found that $E_{red}^{0'}$ for the Fe(III)/Fe(II) couple is $+0.71$ volt, while it is $+0.14$ volt for the Sn(IV)/Sn(II) couple.

To show that this reduction step is highly favorable, we calculate Q as follows:

		$E^{0'}$	$nE^{0'}$
Reduction half-reaction:	$2(Fe(III) + e = Fe(II))$	$+0.71$	$+1.42$
Oxidation half-reaction:	$Sn(II) = Sn(IV) + 2e$	$+0.14$	-0.28
	$Sn(II) + 2Fe(III) = Sn(IV) + 2Fe(II)$		$+1.14.$

Therefore, the equilibrium quotient has a value of $Q = 10^{(1.14)/0.06} = 10^{19}$, showing that thermodynamically the reduction is highly favorable. It is also kinetically favorable, and one sees the disappearance of the yellow color due to the $FeCl_4^-$ species during the dropwise addition of the Sn(II) reducing solution. One or two drops are added in excess, once the yellow color has vanished, to be quite sure that all the Fe(III) has been reduced.

This means that the solution cannot be titrated immediately with the standard potassium dichromate solution because the slight excess of Sn(II) would be titrated along with the Fe(II), giving high results for the determination. It is necessary to include a step that will selectively get rid of the Sn(II) without changing the Fe(II) concentration. We need a mild oxidizing agent that can oxidize the tin (therefore its $E^{0'}$ value must be higher than $+0.14$ volt), but unable to oxidize the iron (therefore its $E^{0'}$ value must be lower than $+0.71$ volt).

Step 3   An excess of mercuric chloride is added, all at once with rapid mixing. A small amount of white silky precipitate should be formed.

At first glance the use of mercury(II) as the oxidizing agent might seem too drastic, since $E^0 = +0.92$ for the Hg^{2+}/Hg_2^{2+} couple. The oxidizing power of this couple is made still stronger by the tendency of mercurous ion to precipitate in the presence of chloride ion to form calomel (the white silky precipitate). But in the hydrochloric-acid medium the mercuric ion has a strong tendency to form a tetrachloro-complex, which "offsets" the oxidizing effect of the couple. This may be shown by combining the three reactions as follows:

	$E^{0'}$	$nE^{0'}$	$\log K$
$2Hg^{2+} + 2e = Hg_2^{2+}$	$+0.92 \rightarrow$	$+1.84 \rightarrow$	$+31$
$Hg_2^{2+} + 2Cl^- = Hg_2Cl_2(s)$			$+18$ (from K_{sp})
$2(HgCl_4^{2-} = Hg^{2+} + 4Cl^-)$			-30 (from β_4)
$2HgCl_4^{2-} + 2e = Hg_2Cl_2 + 6Cl^-$	$+0.57$	$\leftarrow$	$+19$

The net value for the formal reduction potential, +0.57 volt, meets the criteria mentioned above. In carrying out this step, the directions call for rapid addition of the mercuric chloride solution to prevent the occurrence of a second stage of reduction in which metallic mercury, finely divided, could be produced. If the Sn(II) is quickly "surrounded" by excess Hg(II), it does not have the kinetic opportunity to cause this undesired reaction. If, through mishap, one obtains a grey precipitate instead of the white silky "angel's hair," it is necessary to start over because the mercury metal can react with the dichromate and cause errors in the iron determination.

Step 4   After dilution and the addition of sulfuric and phosphoric acids, a small amount of indicator (diphenylamine sulfonic acid) is added, and the solution is titrated with a standard potassium dichromate solution. The endpoint is easily observed when the indicator changes from colorless to a deep purple color. Because of the background color due to chromium (III), the observed change is from green to purple.

The addition of the phosphoric acid serves two closely related purposes. (1) By forming a stable complex with Fe(III) the phosphate causes a desirable color change, turning the yellow chloro-complex into a colorless phosphate complex. (2) At the same time, because the phosphate complex is more stable, the formal potential for the iron(III)/iron(II) couple is changed considerably, from +0.71 to +0.44 volt. Since the Cr(VI)/Cr(III) couple is not particularly affected by the phosphoric acid, it retains its potential of +1.01 volt. Therefore, the equilibrium quotient for the titration reaction is made much higher:

$$
\begin{array}{lcc}
 & E^{0\prime} & nE^{0\prime} \\
3\{Fe(II) = Fe(III) + e\} & +0.44 & -1.32 \\
HCrO_4^- + 7H^+ + 3e = Cr(III) + 4H_2O & +1.01 & +3.03 \\
 & & \overline{+1.71}
\end{array}
$$

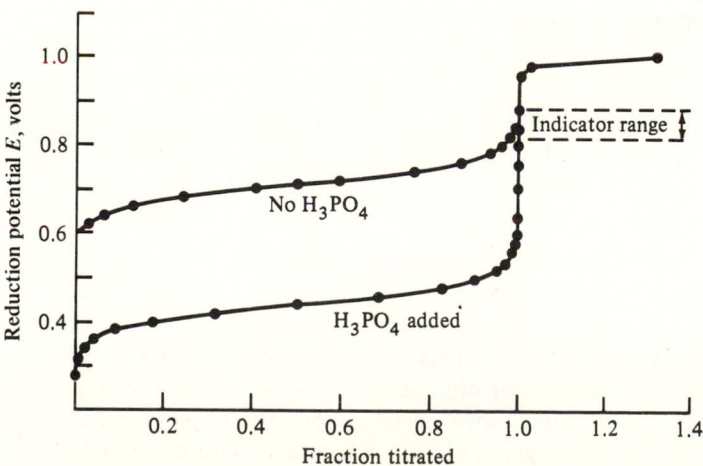

Fig. 12.10 Titration of Fe(II) with dichromate.

Thus the equilibrium quotient for the titration reaction is $Q = 10^{1.71/0.06} = 10^{28.5}$ compared with 10^{15} if the phosphoric acid had not been added.

   This improvement in the equilibrium quotient is very evident in the theoretical titration curves. In Fig. 12.10, the upper curve is the one that would be found in the absence of phosphoric acid, while the lower curve shows the helpful effect of this reagent. In both cases, the value of E at the midpoint of the titration corresponds to the formal potential of the iron couple.

   The formal reduction potential for the indicator, diphenylaminesulfonate, is about +0.85 volt. The change in structure upon oxidation of this indicator is rather drastic and accounts for the striking change in color:

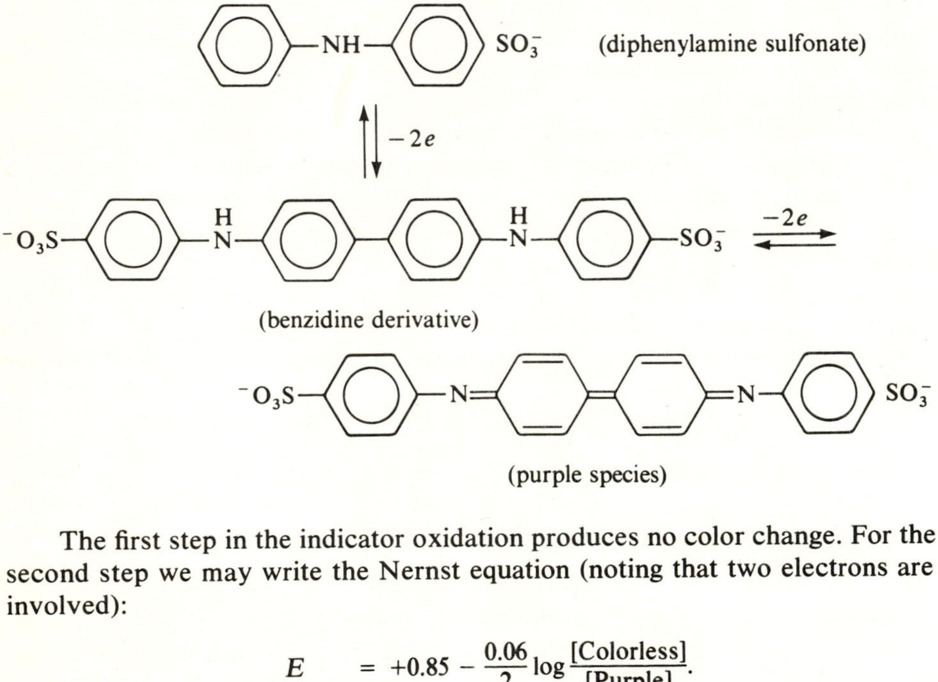

(diphenylamine sulfonate)

(benzidine derivative)

(purple species)

   The first step in the indicator oxidation produces no color change. For the second step we may write the Nernst equation (noting that two electrons are involved):

$$E_{\text{reduction}} = +0.85 - \frac{0.06}{2} \log \frac{[\text{Colorless}]}{[\text{Purple}]}.$$

From this we conclude that the color-change interval will be from +0.82 to +0.88 volt, corresponding to ratios of 10/1 and 1/10, respectively, in the concentrations of the two forms. The reduction potential of the system is established, at any point in the titration, by the iron and chromium couples, and the indicator simply conforms to whatever value of E is established.

   In Fig. 12.10 this color-change interval is shown by horizontal dashed lines, and it is seen even more clearly from the expanded-scale plot in Fig. 12.11, that the indicator gives an inaccurate estimate of the equivalence point when phosphoric acid is not added. But in the presence of phosphoric acid the

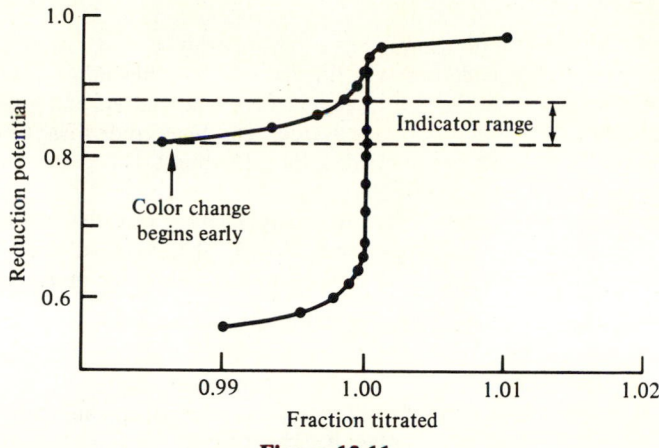

Figure 12.11

potential changes so rapidly in the immediate vicinity of the endpoint that no error can be measured when the indicator completes its color change at the equivalence point.

PROBLEMS

Standard solutions of redox reagents

These problems include realistic statements of conditions for performing the titrations mentioned in the text.

1. A primary standard solution of potassium dichromate is prepared by dissolving 5.9164 g of dried salt in water and diluting to 1.0002 L in a standardized volumetric flask. Calculate the precise molarity.

   What is the approximate size of the sample of ferrous ammonium sulfate that should be used if it is desired to determine its purity by titration with the above dichromate solution?

   In one titration a 1.9423-g sample of ferrous ammonium sulfate hexahydrate was added to 100 mL of $0.5M$ sulfuric acid in a titration flask. About 5 mL of 85% phosphoric acid were added (to decolorize the ferric ion that is formed in the titration), and 6–8 drops of 0.2% sodium diphenylaminesulfonate indicator were added. As the titration proceeded there was a buildup of light green color due to the Cr(III) formed:

$$Cr_2O_7^{2-} + 6Fe^{2+} + 14H^+ \rightarrow 2Cr(III) + 6Fe(III) + 7H_2O.$$

   The color changed sharply to violet at the endpoint, and the buret reading was 40.76 mL. Calculate the purity of the ferrous ammonium sulfate.

   Comment: Once a given batch of FAS has been thus assayed, it may be used as a reliable source of ferrous ions of known purity.

2. A stock solution of sodium thiosulfate is prepared by weighing 12.655 g of the pentahydrate salt of uncertain purity and then dissolving and diluting to 500 mL in a volumetric flask. The molarity of the solution was checked in two ways. First,

25.00 mL of 0.02476M potassium dichromate was added to 25 mL of 0.2M HCl in a titration flask, then 1 g of KI was added and dissolved, and after a few minutes the liberated iodine (triiodide ion) was titrated with the thiosulfate solution. As the endpoint was approached, the color became very pale yellow, and a few milliliters of 0.2% starch indicator solution was added. The blue color disappeared sharply when a total of 36.52 mL had been added from the buret. Calculate the molarity of the thiosulfate (see chemical equations in text).

Second, a sample of pure iodine weighing 0.4108 g was dissolved in 50 mL of water containing 2 g of KI. It required 31.80 mL of the thiosulfate to titrate this sample, again using starch indicator. Calculate the thiosulfate molarity.

3. Two liters of potassium permanganate solution were prepared by dissolving about 6 grams of the reagent grade salt in water. After standing for three days the solution was filtered and ready for standardization. A sample of Mohr's salt weighing 1.6204 grams was transferred to a titration flask and about 50 mL of 0.2M sulfuric acid was added to dissolve it. The titration required 36.86 mL of the permanganate solution, the endpoint being the first permanent tinge of pink. Calculate the precise molarity of the permanganate.

4. A chemist is faced with the task of titrating a large number of samples of arsenic(III), and finds that either permanganate or cerium(IV) can be used as titrant. The price of potassium permanganate is $9 per pound, while ceric ammonium nitrate costs $42 per pound. Taking into account the five-electron change for permanganate compared with the one-electron change for cerium, how many times more expensive is it to do a ceric titration than a permanganate titration?

5. What is the proper size (in grams) for samples of arsenious oxide that are to be used for the standardization of a stock solution of 0.03M potassium permanganate?

6. An inorganic chemist has prepared a solution of chromium(II) by electrolyzing a solution of chromium(III) with a mercury cathode, and desires to find the molarity of Cr(II) in the final solution. Because Cr(II) is quickly air-oxidized, the procedure is as follows: using nitrogen gas from a tank to keep the Cr(II) solution away from air, a pipet is used to take a 25-mL sample. This is delivered directly into an air-free solution of ferric sulfate in 0.1M sulfuric acid, and the immediate reaction is

$$Cr(II) + Fe(III) \rightarrow Cr(III) + Fe(II).$$

Thus for each millimole of Cr(II) originally present, one millimole of the more stable Fe(II) is produced. Titration with 0.01000M potassium permanganate requires 24.48 mL. Calculate the molarity of the Cr(II) in the original sample.

7. A 50.00-mL portion of a solution of ferric ammonium sulfate is added to 50 mL of 0.4M sulfuric acid in a titration flask, and 1 g of KI is added. Titration with 0.0677M sodium thiosulfate requires 47.82 mL. What is the molarity of ferric iron in the original sample?

8. A sample of iron ore weighing 1.548 gram is analyzed by the procedure outlined in the text. It required 29.44 mL of 0.01554M potassium dichromate to reach the endpoint. Calculate the percentage of iron in the ore.

9. It is possible to determine tungsten by direct titration with permanganate. First the tungsten is converted to W^{3+} by reduction with lead amalgam. The W(III) is then oxidized by the permanganate in $2M$ HCl. Suppose a 10-mL portion of $0.0050M$ W^{3+} solution requires 3.0 mL of $0.0100M$ KMnO$_4$. If the oxidation product contains one tungsten atom and two oxygen atoms what is its ionic charge?

10. A research project requires a sample of very pure calcium nitrite Ca(NO$_2$)$_2$, MW = 132.1. The sample on hand is suspected to have undergone some air oxidation to form the nitrate. A portion weighing P grams is dissolved and treated with KI and acetic acid, which causes the reaction

$$4H^+ + 2NO_2^- + 3I^- \rightarrow 2NO + I_3^- + 2H_2O.$$

Titration of the liberated I_3^- requires V milliliters of C-molar sodium thiosulfate. Derive an equation for the purity of the sample (in percent) assuming that no nitrate present would undergo any reaction.

11. A sample of an aqueous solution of glycerol C$_3$H$_8$O$_3$ was treated with 25.0 mL of $0.0200M$ cerium(IV) in a perchloric-acid medium, as described in the text. The leftover Ce(IV) was determined by titration with a standard solution of $0.0100M$ ferrous ammonium sulfate, 16.3 mL being required. What was the mass (in milligrams) of glycerol in the sample?

12. Small amounts of sodium may be determined as follows: The sodium is first separated by precipitation of the triple salt sodium zinc uranyl acetate hexahydrate NaZn(UO$_2$)$_3$(CH$_3$COO)$_9$ · 6H$_2$O (MW = 1538). This precipitate may be weighed or it may be dissolved in acid to liberate the uranyl ion, which can then be reduced to U^{4+} by amalgamated zinc. The U^{4+} may then be titrated with a standard solution of potassium dichromate, the unbalanced equation being

$$U^{4+} + Cr_2O_7^{2-} \rightarrow UO_2^{2+} + Cr^{3+}.$$

If a 4.60-g sample required 12.50 mL of $0.01200M$ dichromate, what is the percentage of sodium in the sample?

13. When sucrose C$_{12}$H$_{22}$O$_{11}$ (MW = 342.3) is treated with an excess of Ce(IV) in $4M$ perchloric-acid solution, it is oxidized to formic acid and carbon dioxide, the partially balanced half-reaction being

$$C_{12}H_{22}O_{11} \rightarrow 11HCOOH + CO_2.$$

The cerium is reduced to the +3-state and the excess cerium(IV) is readily determined by titration with a standard solution of oxalic acid in the balanced reaction

$$2Ce(IV) + H_2C_2O_4 \rightarrow 2Ce(III) + 2CO_2 + 2H^+.$$

A sample containing an unknown amount of sucrose was treated with 50.00 mL of $0.2205M$ Ce(IV), and it required 8.55 mL of $0.1035M$ oxalic acid in the titration, using nitroferroin indicator.
a) Balance the equation for the reaction between sucrose and Ce(IV).
b) Calculate the mass (in milligrams) of sucrose in the sample.
c) Sketch the theoretical titration curve for the titration, assuming that the standard

potentials are as follows:

$$Ce(IV) \xrightarrow{\text{1.7 volt}} Ce(III); \quad CO_2 \xrightarrow{\text{-0.5 volt}} H_2C_2O_4.$$

d) At what potential should the indicator be midway in its color change?

14. The ascorbic-acid content of commercial vitamin C tablets is easily determined by direct titration of the water solution of the tablet with standard iodine (triiodide ion) solution (see the text for the reactions). what should the precise molarity of a triiodide ion solution be if each milliliter is to correspond to 10.0 mg of ascorbic acid?

15. A stock bottle of "30% hydrogen peroxide" is to be checked for its H_2O_2 content. A portion weighing 0.500 gram is transferred to 100 mL of $1M$ sulfuric acid in a titration flask, and it requires 31.45 mL of $0.0514M$ $KMnO_4$ to reach the endpoint. Given the unbalanced titration reaction

$$H_2O_2 + MnO_4^- \rightarrow O_2 + Mn^{2+},$$

calculate the percentage of H_2O_2 in the stock solution.

16. A soluble sample weighing 0.298 g is treated with excess of ammonium oxalate to effect quantitative precipitation of calcium oxalate $CaC_2O_4 \cdot H_2O$. The precipitate is collected on a filter, washed, and then redissolved completely in $2M$ sulfuric acid. After being heated to about $80°$, the solution is titrated with $0.0175M$ potassium permanganate, 26.43 mL being required. Calculate the percentage of calcium in the sample, knowing that $C_2O_4^{2-} \rightarrow 2CO_2 + 2e$.

SPECTROPHOTOMETRIC DETERMINATION OF EQUILIBRIUM CONSTANTS

13

I have yet to see any problem, however complicated, which, when you looked at it in the right way, did not become still more complicated.

Poul Anderson

For decades chemists working toward goals of great variety have shown interest in determining equilibrium constants for reversible reactions of the type

$$D + X \rightleftarrows DX, \tag{13.1}$$

where D and X are solute species capable of independent existence but able to associate to form the complex DX. A few examples will indicate the wide range of chemical changes that correspond to this model.

	D	X	DX
Triiodide ion formation	I^- +	I_2 =	I_3^-
Metal–ligand complexes	Fe^{3+} +	SCN^- =	$FeSCN^{2+}$
Charge-transfer complexes	C_6H_6 +	I_2 =	$C_6H_6 \cdots I_2$
Proton-transfer reactions	H^+ +	CrO_4^{2-} =	$HCrO_4^-$

Another thing these reactions have in common is that X and DX differ in the way they absorb radiation in the UV–visible region. This suggests that absorbance measurements should be helpful in finding the extent to which X has been converted to DX in an equilibrium reaction mixture. A singular advantage of spectrophotometry over other methods of chemical analysis is the ability to "look" at various species in solution through choices of appropriate wavelengths *while not disturbing the system being studied.* In addition, spectrophotometric measurements may be made with excellent precision, over wide ranges of solute concentration, and upon solutions in solvents other than water. It is no wonder that this technique stands supreme in its versatility for research studies of chemical equilibria.

13.1   DEFINING THE SYSTEM

Returning to the reaction shown in Eq. (13.1), we will find it useful to regard substance D as a species that dictates (hence "D") the degree to which the reaction proceeds to the right. In the absence of D, of course, there could be no reaction and any substance X would be unchanged. At high concentrations of D, the reaction would be driven to the right, perhaps approaching 100% conversion of X to DX.

These ideas are described more clearly in connection with the expression for the equilibrium quotient for the reaction:

$$Q = \frac{[DX]}{[D][X]}$$

and therefore

$$\frac{[DX]}{[X]} = Q[D].$$

We see that the ratio of DX to X in the equilibrium mixture depends directly on the equilibrium concentration of D and also on the value of the equilibrium quotient. For strong complexes, having a large value for Q, only a small excess concentration of D is necessary to effect nearly complete conversion of X to DX. For weak complexes, e.g., with $Q = 10$, even with the molarity of D as high as 1.0 the ratio will be only 10, corresponding to about 90% conversion.

Suppose a solution is prepared by mixing C_d mole of substance D and C_x mole of substance X per liter. As the reaction proceeds toward equilibrium there is a decrease in the concentrations of D and X, and an increase in the concentration of DX:

$$
\begin{array}{llccc}
\text{Reaction:} & \text{D} & + & \text{X} & = \text{DX} \\
\text{Initial molarity:} & C_d & & C_x & 0 \\
& \downarrow & & \downarrow & \downarrow \\
\text{Final molarity:} & C_d - [\text{DX}] & & C_x - [\text{DX}] & [\text{DX}].
\end{array}
$$

Clearly, if it were possible by any means to determine the equilibrium concentration of the complex [DX], it would be easy to calculate the equilibrium quotient, given known starting concentrations of D and X:

$$
Q = \frac{[\text{DX}]}{(C_d - [\text{DX}])(C_x - [\text{DX}])}.
$$

For the favorable case where a wavelength can be found at which DX is the only absorbing species, an absorbance measurement may yield the data for calculating [DX], provided that this substance has a known value of molar absorptivity at that wavelength. Alas, such an ideal situation is not common, and we must consider the more general cases.

13.2 HYPOTHETICAL ABSORPTION SPECTRA

Figure 13.1 illustrates the three possibilities for a typical system. At wavelength λ_3 we see the favorable case when only substance DX has a measurable absorptivity. The worst situation corresponds to λ_1 where all three constituents absorb. Generally it is possible to avoid this complication because substance D will not absorb over the entire range. The most typical situation, which will be treated in detail, is that for wavelength λ_2: both X and DX have significant but different molar absorptivities. It does not matter which one has the higher value but, as we shall see, they should differ significantly.

In this figure the solid line shows the spectrum of DX, and the dash–dot line that of X. Suppose the equilibrium mixture has about half-conversion to DX. Then the **apparent** molar absorptivity, defined by the pseudo-Beer's Law,

$$
\epsilon \equiv \frac{A_{\text{mixture}}}{bC_x}, \tag{13.2}
$$

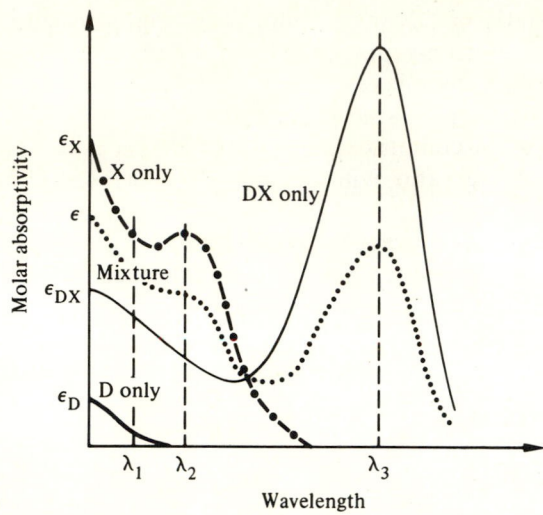

Fig. 13.1 Hypothetical absorption spectra for the system $D + X \rightleftarrows DX$.

would be expected to lie about midway between the two extremes of ϵ_{DX} and ϵ_X, as shown by the dotted line. The greater the conversion of X to DX, the closer the line for the mixture would approach that for DX alone.

13.3   THE BASIC ALGEBRAIC MODEL

Let us assume that the solution under study contains a known *equilibrium* concentration of substance D, such that substance X is only partly converted to DX. The absorbance of the solution, measured at λ_2 using a 1.00-cm cell, would be

$$A = \epsilon_X[X] + \epsilon_{DX}[DX]; \qquad b = 1, \tag{13.3}$$

since absorbances are additive. This is a direct consequence of applying Beer's Law to the two absorbing species in the mixture. However, we also have defined an apparent absorptivity by Eq. (13.2), which is rearranged to give

$$A = \epsilon C_X; \qquad b = 1, \tag{13.4}$$

where C_X is the total concentration of X and DX,

$$C_X = [X] + [DX].$$

By equating the two expressions in Eqs. (13.3) and (13.4) for absorbance, we find

$$A = \epsilon C_X = \epsilon_X[X] + \epsilon_{DX}[DX]. \tag{13.5}$$

By substituting $C_X - [DX]$ in place of $[X]$ and rearranging the result, we find

$$[DX] = C_X \frac{\epsilon - \epsilon_X}{\epsilon_{DX} - \epsilon_X}. \tag{13.6}$$

Similarly, if we substitute into Eq. (13.5) the expression $C_X - [X]$ in place of $[DX]$, the result after rearrangement is:

$$[X] = C_X \frac{\epsilon_{DX} - \epsilon}{\epsilon_{DX} - \epsilon_X}. \tag{13.7}$$

An important result appears when Eq. (13.6) is divided by Eq. (13.7):

$$\frac{[DX]}{[X]} = \frac{\epsilon - \epsilon_X}{\epsilon_{DX} - \epsilon}. \tag{13.8}$$

Equation (13.8) is important because it shows how the *ratio* of molarities may be determined by absorbance measurements alone. Once this ratio is determined, given the above-stated assumption that the equilibrium concentration of substance D is known, the desired equilibrium quotient follows immediately:

$$Q = \frac{1}{[D]} \cdot \frac{[DX]}{[X]} = \frac{1}{[D]} \cdot \frac{\epsilon - \epsilon_X}{\epsilon_{DX} - \epsilon}. \tag{13.9}$$

Example data and calculation

Substance B is a weak base that has an intense blue color. It reacts with hydrogen ion to form a yellow conjugate acid:

$$B + H^+ = HB^+.$$

Three solutions are prepared, each containing the same total concentration of B (equal to $2 \cdot 10^{-5} M$) but with different equilibrium concentrations of H^+. The absorbance of each is measured (1-cm cell) at 640 nm, with the following results:

Solution	Absorbance
1.0M HCl	0.120
0.999M NaCl + 0.001M HCl	0.320
1.0M NaOH	0.660

Each solution has the same ionic strength, so we may presume that activity coefficients are constant. In the first solution, the molarity of hydrogen ion is so large (1.0M) that we will assume complete conversion of the base B to its conjugate acid. Therefore the molar absorptivity of HB^+ may be calculated

$$\epsilon_{HB} = \frac{0.120}{2 \cdot 10^{-5}} = 6000.$$

(Assuming that the conversion is complete, the equilibrium molarity of HB^+ in this solution must be the same as the total concentration of B originally added.)

In the third solution, with sodium hydroxide added, it is safe to assume for such a weak base that it is entirely present in the base form B. The molar absorptivity of B is

$$\epsilon_B = \frac{0.660}{2 \cdot 10^{-5}} = 33,000.$$

Now, in the second solution the observed absorbance is not very close to either of the other two values, showing that there must be a mixture of B and HB^+. The hydrogen ion concentration of 0.001 is large enough to cause partial conversion only. The apparent absorptivity of this mixture is found from Eq. (13.2):

$$\epsilon = \frac{0.320}{2 \cdot 10^{-5}} = 16,000.$$

With these results and Eq. (13.9) we may now calculate the equilibrium constant for the reaction:

$$Q = \frac{1}{0.001} \cdot \frac{16000 - 33000}{6000 - 16000} = 1.7 \cdot 10^3.$$

This example illustrates the fundamental power of the spectrophotometric method: with three simple absorbance measurements one may determine the value of an equilibrium constant. However, to use Eq. (13.9) directly, as illustrated, it is necessary that (a) there be only two absorbing species, X and DX, in the equilibrium, (b) the absorptivities of both X and DX be known, and (c) the equilibrium [D] be known. When these conditions are not met, it is possible to find the value for Q by mathematical techniques discussed in the following sections.

13.4   THE SIGNIFICANCE OF AN ISOSBESTIC POINT

When a substance can exist in two forms of different color it is often true that their absorption spectra cross at one or more wavelengths. This was shown, but not discussed, in the earlier section dealing with hypothetical absorption spectra in which the spectrum of substance X and that of substance DX intersected. The wavelength at which the two species have the same molar absorptivity is called the **isosbestic point** (Greek, *iso-*, equal, and *sbennynai*, to quench). Thus the two species have the same ability to absorb (quench) radiation.

If a series of solutions is prepared in which the substance is variously distributed between its two forms, the total concentration of the substance being constant, then the absorbance–wavelength plots will show this common point, as illustrated in Fig. 13.2 which shows the spectra for the acid–base

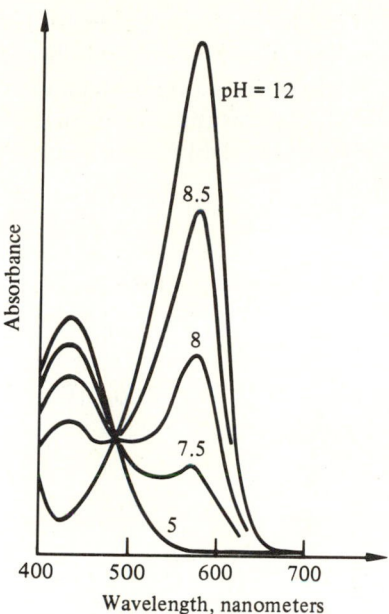

Figure 13.2

indicator m-cresol purple. This substance shows a reversible proton transfer:

$$H^+ + \quad In \quad = \quad HIn.$$
$$\text{purple} \quad \text{yellow}$$

Curve pH = 12 shows the spectrum in $0.01M$ NaOH, where the m-cresol purple was entirely in the purple form. Curve pH = 5 was obtained at the same concentration of indicator, but the protonation was virtually complete and the yellow form dominated. As the indicator was shifted from one form to the other, the other three curves shifted predictably, and all of them went through the common point, the isosbestic point, at 485 nm.

If an isosbestic point is observed for an equilibrium system, it is very strong evidence that there are two, *and only two*, species involved in the system.

If there should happen to be any other significant species, it would be highly improbable that they also would have equal absorptivities at the same point. If an isosbestic point fails to "stay put" when the equilibrium is shifted, it suggests that the system includes more than two components.

13.5   A FEW REALISTIC COMPLICATIONS

1. The Equilibrium Concentration of D Is Not Known

In the example calculation we assumed for convenience that the value of [D] in the equilibrium mixture was known. In practice this is often possible, as when C_d is so large compared to C_x that a negligible amount of D is used up in the formation of the complex. Or sometimes it is possible to make an independent determination of [D] in the mixture, as when D is the hydrogen ion and a pH-meter can be used to find its concentration.

For the case in which a significant portion of D is incorporated in the complex, we must rely on the material-balance expression

$$C_d = [D] + [DX] \tag{13.10}$$

or

$$[D] = C_d - [DX].$$

Knowing C_d, the initial concentration of D, we may calculate the equilibrium concentration by subtracting the concentration of DX. Fortunately, this is not difficult because, as was shown by Eq. (13.6), [DX] can be found from the value of C_x and the absorptivity data:

$$[DX] = C_x \frac{\epsilon - \epsilon_X}{\epsilon_{DX} - \epsilon_X}.$$

2. Direct Determination of ϵ_{DX} May Be Experimentally Impossible

The calculation of the Q value by using Eq. (13.9) requires four experimental quantities: the equilibrium concentration of substance D, the absorptivity of the equilibrium mixture, and the absorptivities for the two species X and DX. In the example calculation it was possible to shift the equilibrium completely to DX (HB^+) by using a high concentration of HCl, so that the value of ϵ_{DX} could be obtained by a single absorbance measurement. Likewise, in the absence of HCl the colored substance was entirely present as X (B), and the absorbance of this solution led to a value for ϵ_X.

There are several problems that might make it impossible to obtain a value for ϵ_{DX} by such a direct measurement: (a) the equilibrium quotient Q may be small so that an unreasonably high concentration of D would be required to drive the reaction quantitatively to the right; (b) even if Q is not very small, perhaps substance D is limited to low concentrations by its own solubility, making it impossible to get enough in solution to drive the reaction to the right; (c) if successive complexes are possible, such as D_2X, D_3X, etc., it may be that whatever concentration of D would be required to convert X completely to DX would also be high enough to cause the formation of one or more of the higher complexes. In each of these cases we find it impossible to obtain a solution containing a known concentration of DX and therefore we would not be able to determine its molar absorptivity by a direct measurement.

Nevertheless it is still possible to use photometric data to determine Q, by making use of rearrangement of Eq. (13.9) in the following form:

$$\epsilon = \epsilon_{DX} - \frac{1}{Q} \frac{\epsilon - \epsilon_X}{[D]}. \tag{13.11}$$

By this rearrangement the unknown quantity ϵ_{DX} is separated from the experimental quantities. The approach is to determine several values of ϵ corresponding to a set of values for [D], and then to plot ϵ versus $(\epsilon - \epsilon_X)/[D]$. Even though ϵ_{DX} is unknown we may at least presume it is a constant. Therefore the plot should be linear with an intercept equal to ϵ_{DX} and a slope equal to $-1/Q$.

Illustrative case study

The acid–base indicator called **methyl red** exists in aqueous solution in three forms (see p. 342):

$$R^- \quad \rightleftarrows \quad HR^{+-} \quad \rightleftarrows \quad H_2R^+$$

singly charged dipolar ion singly charged
yellow anion   red   red cation

At hydrogen ion concentrations lower than about 10^{-7}, methyl red exists almost entirely as the yellow anion. With increasing [H$^+$] there is conversion to the red dipolar ion, but before the transformation is complete there is appreciable formation of the red cation as well. The spectral characteristics of the two red species are somewhat different. It is impossible to prepare a solution that contains only HR^{+-} in known concentration and to determine its molar absorptivity.

To study the equilibrium constant for the first step,

$$H^+ + R^- = HR^{+-}; \qquad Q = \frac{[HR^{+-}]}{[H^+][R^-]},$$

it is therefore necessary to use relatively low acidities and to deal with ϵ_{HR} as an unknown.

Such a study has been reported* and may be summarized as follows: A solution of methyl red with a molarity of $1.00 \cdot 10^{-5}$ was prepared in the presence of $0.0220M$ sodium acetate. This solution has a pH high enough ([H$^+$] low enough) to ensure that all the methyl red is present as the yellow anion. When the absorbance was measured at 420 nm in a 5.00-cm cell it was found to be 1.049. Therefore we may calculate the molar absorptivity of R^-:

$$\epsilon_R = \frac{1.049}{5 \cdot 1.0 \cdot 10^{-5}} = 20,980 \text{ L/mol} \cdot \text{cm},$$

where ϵ_R corresponds to ϵ_X.

Then an ultramicroburet was used to add small volumes of $5.178M$ acetic acid directly to the cell. After mixing, the absorbance was measured for each addition. It is a simple matter (see Chapter 9) to calculate the hydrogen-ion concentrations for each of the series of buffer solutions formed in this series

---

*R. W. Ramette, E. Dratz, P. Kelly, *Journal of Physical Chemistry* **66**, 527 (1962).

of measurements. (The $[H^+]$ corresponds to $[D]$ in this case.) The data are shown in Table 13.1, along with the calculated quantities needed for application of Eq. (13.11).

Table 13.1

$[H^+]$	Corrected absorbance	Apparent absorptivity, ϵ	$\dfrac{\epsilon - \epsilon_x}{[H^+]}$
$1.135 \cdot 10^{-5}$	0.666	13 320	$-6.75 \cdot 10^8$
$1.891 \cdot 10^{-5}$	0.554	11 080	$-5.24 \cdot 10^8$
$3.026 \cdot 10^{-5}$	0.455	9 100	$-3.93 \cdot 10^8$
$3.782 \cdot 10^{-5}$	0.413	8 260	$-3.36 \cdot 10^8$

The absorbances shown have been corrected for the slight dilution that occurred when the acetic acid solution was added to the cell. The values of apparent absorptivity were calculated using Eq. (13.2). The. $[H^+]$ values were calculated using the buffer equation (Chapter 9),

$$[H^+] = 2.33 \cdot 10^{-5} \frac{[\text{Acetic acid}]}{0.022},$$

where $2.33 \cdot 10^{-5}$ is taken as the acid-dissociation quotient for acetic acid at the ionic strength of 0.022 due to the presence of $0.022M$ sodium acetate.

The plot suggested by Eq. (13.11) is shown in Fig. 13.3. By extrapolation

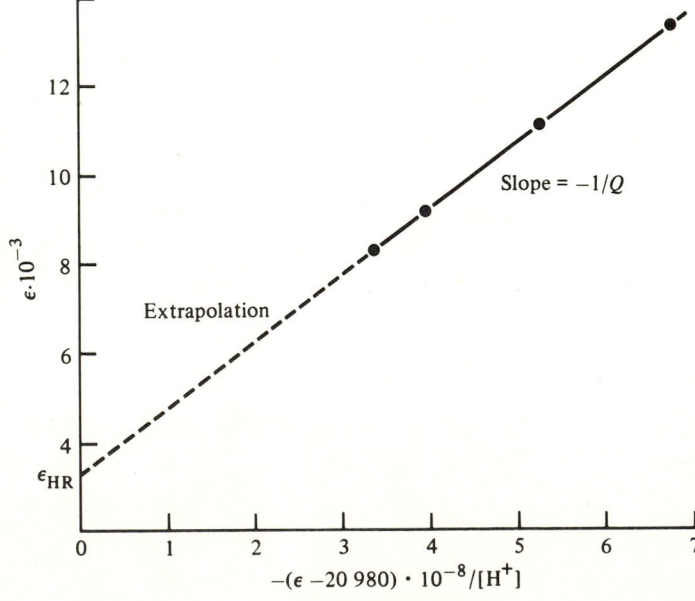

Figure 13.3

of the line connecting the four experimental points, we find the molar absorptivity of HR^{+-} as the intercept of the plot. The value of Q is equal to $-1/\text{Slope}$. A linear least-squares fit of the data, including confidence limits (90%) gives

$$\text{Intercept} = 3235 \pm 65,$$
$$\text{Slope} = -(1.495 \pm 0.013) \cdot 10^{-5},$$

and therefore $Q = (6.69 \pm 0.06) \cdot 10^4$. The precision of these results shows that the inability to measure ϵ_{DX} directly may not be cause for discouragement.

3. Stickier Yet: Neither [D] nor ϵ_{DX} Is Known

As was just discussed, when the absorptivity of DX cannot be directly determined it is helpful to rely on Eq. (13.11), making a plot to find Q from the slope. However, the equilibrium concentrations of D are required in order for the variable to be plotted on the horizontal axis. We recall from Eq. (13.6) and Eq. (13.10) that the value of [D] can be calculated from the optical data, but only if ϵ_{DX} is available.

At this point we are well advised to make use of computer programming. An iterative (successive approximations) approach is easy to program and gets around this whole problem as follows:

a) As a first approximation, just to get started, we simply guess any reasonable value for ϵ_{DX} and use it with Eqs. (13.6) and (13.10) to find tentative values for [D] in each of the solutions.

b) These values of [D] are used with Eq. (13.11), and from the slope and intercept of the plot we find tentative values for Q and ϵ_{DX}.

c) The improved estimate for ϵ_{DX} is then used in part (a) to find better values of [D], which in turn are used in part (b) to find better values of Q and ϵ_{DX}.

After a few iterations around this loop the successive values of Q will become essentially unchanging, showing that further refinement is not necessary. These calculations *may* be done without a computer, but are tedious and time consuming.

4. This Time It Is ϵ_X that Is Unknown

For reasons similar to those discussed in part 2 of Sec. 13.5, it may be impossible to make an independent determination of the molar absorptivity of substance X, although that for DX is available. Again, the problem is solved by rearranging Eq. (13.9), this time to the form

$$\epsilon = \epsilon_X + Q[D](\epsilon_{DX} - \epsilon). \tag{13.12}$$

This suggests a plot of the experimental values of ϵ, obtained for a series of solutions having varying [D], versus the quantity $[D](\epsilon_{DX} - \epsilon)$. The intercept is ϵ_X while the value of the equilibrium quotient is given directly by the slope.

5. Neither ϵ_X nor [D] Are Known

The solution to this difficulty is very similar to that for part 3 of Sec. 13.5. A first approximation is carried out, using C_D in place of [D] in Eq. (13.12). The value of ϵ_X is used in Eq. (13.6) to find [DX], which is used in Eq. (13.10) to find [D]. The process is continued until the results converge to constancy.

A good example in this category is provided by the work of Siddall and Vosburgh, on the "hydrolysis" of ferric ion*:

$$Fe^{3+}(aq) + H_2O = H_3O^+ + FeOH^{2+}.$$

Thus the ferric ion is a weak acid with an acid-dissociation quotient, Q_a:

$$Q_a = \frac{[H^+]\,[FeOH^{2+}]}{[Fe^{3+}]} \approx (4\text{–}5) \cdot 10^{-3}.$$

To make a more accurate determination of this important equilibrium constant, the authors measured the absorbance of a series of solutions containing

Iron(III): $C_X = 1.00 \cdot 10^{-4}$ (ferric perchlorate),
H⁺: $C_D = 0.001$ to 0.015 (perchloric acid).

The ionic strength was maintained equal in all solutions at a value of 0.0166, using sodium perchlorate as required.

By measuring a solution of iron(III) in very strong acidic solution, it was possible to determine that the molar absorptivity of Fe^{3+} is very small, close to $1.0\ L/mol \cdot cm$ at the wavelength used (355 nm). However, it is impossible to make a direct determination of the molar absorptivity of $FeOH^{2+}$ because, when the [H⁺] is made low enough (e.g., $10^{-5}M$) to convert all the Fe^{3+}, there is substantial formation of the higher complexes, $Fe(OH)_2^+$, and even precipitation of $Fe(OH)_3$.

The reaction is cast into the form we have been discussing simply by reversing it:

$$H^+ + FeOH^{2+} = Fe^{3+}.$$
$$\text{D} \qquad \text{X} \qquad \text{DX}$$

When the data are plotted according to Eq. (13.12), the results are as shown in Fig. 13.4. From the intercept of the plot (when the x-axis variable is zero), it is found that the absorptivity of $FeOH^{2+}$ is about 500. The slope yields a value of 227 for Q. Since the acid-dissociation quotient for ferric ion

---

*T. Siddall, W. C. Vosburgh, *Journal of the American Chemical Society*, 73, 4270 (1951).

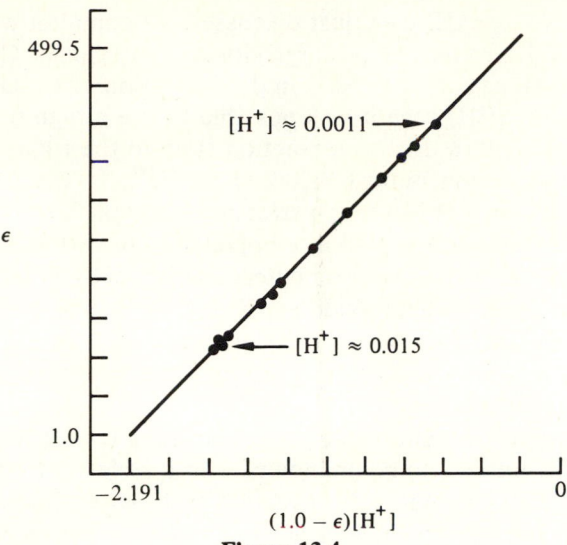

Figure 13.4

is the reciprocal of this value, we have

$$Q_a = \frac{1}{227} = 4.40 \cdot 10^{-3}.$$

To convert to the true equilibrium-constant value, we remember the relationship between K and Q:

$$K = Q\frac{f_1 f_2}{f_3}.$$

At the ionic strength of 0.0166, we find from the Davies equation (see Chapter 4) that $f_1 = 0.880$, $f_2 = 0.599$, and $f_3 = 0.315$. Therefore, $K = 7.36 \cdot 10^{-3}$.

The perceptive reader will realize that in this case Eq. (13.10) cannot be used for the iterative calculation of $[H^+]$. The equilibrium values of $[H^+]$ are actually slightly higher than the initial value C_D because H^+ is released by dissociation of Fe^{3+} as it reaches equilibrium. Therefore,

$$[H^+] = C_D + [FeOH^{2+}].$$

6. The Worst Possible Situation: Neither ϵ_X nor ϵ_{DX} Are Known

At first the reader may feel that this is "too much," that we have already explored all the options for solving difficulties. Please bear with the following. A good chemical example in this difficult category is the hydrolysis of cerium(IV).* Writing the reaction in our now familiar form, we have:

$$H^+ + CeOH^{3+} = Ce^{4+};$$

---

*H. Offner and D. A. Skoog, *Analytical Chemistry*, **38**, 1520 (1966).

this resembles the iron(III) case just discussed. In common with that case, it is impossible to determine the absorptivity of the species $CeOH^{3+}$ because attempts to prepare its solution result in the formation of $Ce(OH)_2^{2+}$, $Ce(OH)_3^+$, etc. Now, in the iron(III) case it was possible to use a high (e.g., $0.5M$) solution of perchloric acid to drive the reaction fully to the right, thus permitting direct observation of the optical behavior of Fe^{3+}. In the cerium(IV) case, however, the Ce^{4+} species is a much stronger acid than Fe^{3+}, and even in $1M$ perchloric acid, cerium(IV) is not completely converted to the $+4$ ion. Therefore, neither ϵ_X nor ϵ_{DX} can be determined directly.

However, at intermediate acidities, in the 0.1 to 1.0M range it appears that $CeOH^{3+}$ and Ce^{4+} are the only cerium species present, and so reliable values of the apparent absorptivity ϵ may be obtained by absorbance measurements at 305 nm.

The best way to extract a value of Q from such meager data is to use Eq. (13.11), which permits ϵ_{DX} to be unknown, *with a series of assumed values for* ϵ_X. Of course, it will be a computer program that makes the desired trials in a systematic way, but the principle is not difficult to understand: If we do find the correct value for ϵ_X by successive trials, we can recognize that fact by observing that the plot is *linear.* If the assumed value of ϵ_X is too low, the plot will be curved in one direction, and if the assumed value is too high, the plot will be curved in the other direction. We need merely to instruct the computer to find the value of ϵ_X that yields the most linear plot, as determined by a linear least-squares fit.

The cerium(IV) data were successfully solved by this approach, and Fig. 13.5 shows three plots corresponding to different assumed vaues of ϵ_X. The

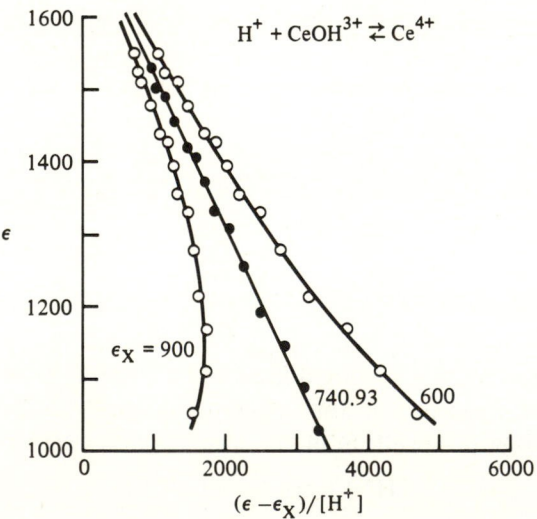

Figure 13.5

computer program found a minimum in the deviations from a straight line when the assumed value was 740.93, which we may certainly round off to 741. For that plot, the intercept was 1739, giving us a value for the absorptivity of Ce^{4+} not otherwise determinable. The equilibrium quotient turned out to be 4.8, making the acid-dissociation quotient (at ionic strength 1.0) equal to 0.21.

In summary, there are many ways to handle seemingly intractable problems if one knows how to apply computer programming.*

A point on which to conclude   In spite of clever and sophisticated data-processing techniques, there is no substitute for accurate raw data. Too often spectrophotometric measurements have been made without regard for instrumental problems such as stray light, without checking Beer's Law, and even without proper definition of the chemical system under study. Temperature and ionic-strength effects are too frequently ignored, and very often the experiments are not repeated sufficiently to establish accuracy and precision.

PROBLEMS

1. A certain very weak base B cannot be quantitatively converted to its conjugate acid BH^+ because of the overlapping formation of the second conjugate acid BH_2^{2+}. Therefore the molar absorptivity of BH^+ is unknown, although that for B has been determined to be $2.6 \cdot 10^4$ L/mol $\cdot$ cm. At acidities lower than about $0.02M$ H^+ there is negligible formation of BH_2^{2+}, and so the following data were obtained in an attempt to determine the equilibrium constant for the reaction

$$B + H^+ = BH^+.$$

The measurements were made at 356 nm by using a total concentration (B plus BH^+) of $2.00 \cdot 10^{-5}M$ and a 2.00-cm cell. Assuming that the given concentrations of hydrogen ion are the equilibrium values, find the value of Q:

$[H^+]$	Absorbance
0.0100	0.473
0.00500	0.598
0.00200	0.765
0.00100	0.876

2. A certain acid–base indicator $HInd \rightleftarrows H^+ + Ind^-$ is added in equal concentrations $(0.0001M)$ to a series of solutions, and the absorbance is measured using a 2-cm cell

---

*For further discussion of the above procedures the reader is referred to an article by the author in the *Journal of Chemical Education*, 44, 647 (1967). A more comprehensive review has been published by W. McBryde, *Talanta*, 21, 979–1004 (1974).

in a spectrophotometer. Data are as follows:

Solution	Absorbance
0.05M HCl	0.096
Buffer, pH = 4.5,	
$I = 0.01$	0.387
0.02M NaOH	0.884

a) Calculate the dissociation constant for the indicator. What is its approximate color-change interval?

b) When 1.00 mL of 0.01M indicator solution is added to 100 mL of a buffer solution of unknown pH, the resulting solution shows an absorbance of 0.450 when measured in a 2-cm cell. What is the pH of the solution? Assumed ionic strength is 0.1.

3. An important example of 1:1 association to form a complex is the reaction of molecular iodine with iodide ion in aqueous solution:

$$I_2(aq) + I^- = I_3^-; \qquad K_f = \frac{[I_3^-]}{[I_2][I^-]} \approx 700.$$

This system has been studied in detail by different techniques, but probably the most precise work was that of Danielle who used ultraviolet spectrophotometry.*

The value of K_f is large enough for the reaction to be forced nearly completely

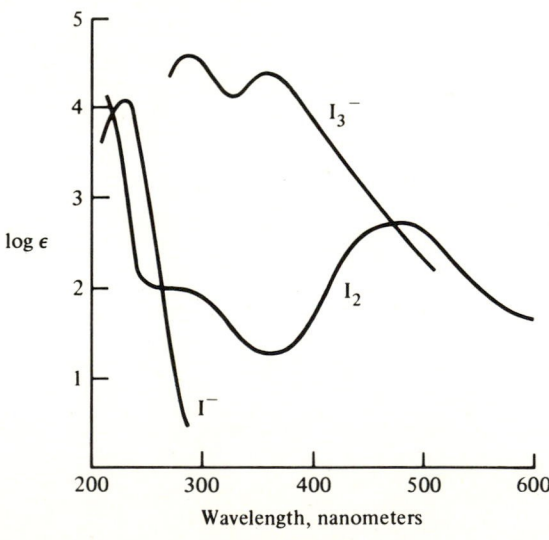

Wavelength, nanometers

Figure 13.6

*G. Danielle, *Gazzetta Chimica Italiana*, **90**, 1068 (1960).

to the right in a $0.2M$ potassium iodide medium. Thus the molar absorptivity of triiodide ion can be directly determined. At the other extreme, it is not difficult to prepare a water solution that contains a known molarity of free iodine, with no iodide ion present, and so it is possible to directly determine the molar absorptivity of molecular iodine.

The experimental absorption spectra reported by Danielle are reproduced in Fig. 13.6. (Note that the ordinate is the *logarithm* of the molar absorptivity, to facilitate showing the wide range of values.)

a) What approximate value would you expect for the absorbance of a solution of $0.001M$ iodine, measured at 490 nm in a 1.0-cm absorption cell?

b) If you were designing a spectrophotometric experiment to determine the value of the formation constant for triiodide, what wavelength would you choose for the precise absorbance measurements? Discuss your reasons.

c) Expecting the value of K_f to be about 700 or so, suppose you prepare a solution that contains (nominally) $1 \cdot 10^{-5}M$ iodine and XM potassium iodide. What should you use for the value of X if you want to force the conversion of iodine to triiodide to the extent of 99%? *Hint*: X is much larger than $1 \cdot 10^{-5}M$.

d) In Danielle's work, at a certain wavelength he found the molar absorptivity of triiodide ion to be 26400, while molecular iodine had a molar absorptivity of only 18 at that wavelength. He prepared a solution containing $3.4 \cdot 10^{-3}M$ potassium iodide and $2.00 \cdot 10^{-5}M$ iodine, and found that the absorbance as measured in a 1.00-cm cell was 0.375. Calculate the value of the formation constant K_f.

METAL–LIGAND COMPLEXES: THEORETICAL EQUILIBRIUM MODELS

14

It seems that most of the basic methods and concepts for treating complex formation equilibria were known already 50 years ago. It is hard to escape the feeling that there was rather a decline than an evolution in this field in the twenties and early thirties. Sometimes I wonder if this was not a side effect of the successes of the Lewis concept of activity, and the Debye–Huckel theory of electrostatic ion interaction. For a while, chemists became used to thinking of most electrolytes as completely dissociated into ions, and to ascribe all deviations from ideal behavior to non-specific electrostatic effects. And there was a period when, in many of the best schools of physical chemistry, the only equilibrium constants that were thought worthwhile determining were those for infinite dilution.

Lars Gunnar Sillen, 1957

14.1 ASSOCIATION OF SOLUTE SPECIES

When the beautiful blue triclinic crystals of copper sulfate pentahydrate are oven-dried, they crumble into a white powder of dehydrated (anhydrous) $CuSO_4$ consisting of partly associated ions Cu^{2+} and SO_4^{2-}. The characteristic color of the hydrated salt, which used to be called **blue vitriol**, is due to the hydrated copper ions that exist in the crystal lattice as $Cu(H_2O)_4^{2+}$, with the four water molecules covalently bound to the central copper ion in a square planar configuration.

When copper sulfate is dissolved in water, the blue solution consists largely of the hydrated ions capable of separate existence *and* of an association species that probably has some covalent character:

$$Cu(H_2O)_4^{2+} + SO_4^{2-}(aq) = CuSO_4(aq).$$

It is not clear whether the sulfate ion replaces one or two water molecules in forming this complex species, which has an absorption spectrum slightly different from that of the tetraaquocopper(II) ion.

A good conceptual definition of a **complex** has been given by Rossotti: a species formed by the association of two or more simpler species each capable of independent existence. In this sense, hydrated ions such as $Cu(H_2O)_4^{2+}$ are properly regarded as complexes (**aquo-complexes** is the usual term) because the simple Cu^{2+} ion is associated with the molecular species H_2O.

There are many varied examples of associations in solution that fit this definition of complex. For example, the familiar formation of triiodide ion

$$I^- + I_2(aq) = I_3^-$$

and the dimerization (through hydrogen bonding) of acetic acid

$$2CH_3COOH = (CH_3COOH)_2,$$

both involve association of species capable of independent existence. However, in this chapter the emphasis will be on associations of metal ions with species that are capable of forming coordinate bonds. Such species are called **ligands** (from the Latin *ligare*, to bind). It is more accurate to say that we will consider **ligand exchange**, whereby the water molecules in the aquo-complex of the metal ion are replaced by other ligands. For example, when a solution of ammonia is added to a solution of copper sulfate, the blue color becomes much more intense because of the stepwise replacement of H_2O by NH_3:

Step 1 $Cu(H_2O)_4^{2+} + NH_3(aq) = CuNH_3(H_2O)_3^{2+} + H_2O.$

Step 2 $CuNH_3(H_2O)_3^{2+} + NH_3(aq) = Cu(NH_3)_2(H_2O)_2^{2+} + H_2O.$

Step 3   Formation of $Cu(NH_3)_3H_2O^{2+}$.

Step 4   Formation of $Cu(NH_3)_4^{2+}$.

Each molecule of ammonia, like each one of water, is capable of forming one coordinate bond with the metal ion. Because it has only one "tooth" (electron pair), it is called a **unidentate ligand** (Latin *dentatus*, toothed).

The **coordination number** N of the central metal ion is the number of coordinate bonds that the metal ion can sustain with the attached ligands (in the above case, $N = 4$). It is often found that the coordination number for a metal ion is equal to twice the ionic charge. Thus continuing the example with NH_3 as the ligand, we find the following complexes:

$$
\begin{array}{lll}
\text{silver, } Ag^+: & Ag(NH_3)_2^+ & N = 2, \\
\text{zinc, } Zn^{2+}: & Zn(NH_3)_4^{2+} & N = 4, \\
\text{chromium, } Cr^{3+}: & Cr(NH_3)_6^{3+} & N = 6.
\end{array}
$$

The *geometry* of the bonds in a metal–ligand complex is closely related to the coordination number:

$$
\begin{array}{lll}
N = 2 & \text{linear structure} & Cu^+, Ag^+, Au^+ \\
N = 4 & \text{square planar} & Ni^{2+}, Cu^{2+}, Pt^{2+}, Pd^{2+} \\
& \text{tetrahedral} & Co^{2+}, Zn^{2+}, Cd^{2+} \\
N = 6 & \text{octahedral} & Co^{3+}, Cr^{3+}, Fe^{3+}
\end{array}
$$

When a ligand can form more than one coordinate bond with the metal ion, it is referred to as **bidentate** (two bonds), **tridentate** (three), **tetradentate** (four), **pentadentate** (five), or **hexadentate** (six bonds). Some examples will make these terms clear.

Example Ammonia is a unidentate ligand with four molecules required to satisfy the copper ion's coordination number $N = 4$, but the same job can be done by only two molecules of ethylenediamine $NH_2CH_2CH_2NH_2$ (commonly abbreviated as "en") which can be visualized as an efficient package of two ammonia molecules in one molecular unit. In the square planar copper(II) complex the structure is

Abbreviated as $Cu(en)_2^{2+}$

In this complex the ethylenediamine behaves as a bidentate ligand that forms two five-membered rings with the central metal ion. Whenever a complex is formed with bidentate (or higher) ligands it is called a **chelate** (Greek *chelos*, crab's claw) because the ligand closes on the metal ion like a pair of pincers.

We can go still further in prepackaging the ammonia molecules. The compound triethylenetetramine $NH_2CH_2CH_2NHCH_2CH_2NHCH_2CH_2NH_2$ (commonly called "trien") contains four nitrogen atoms capable of forming coordinate bonds. The complex formed with copper ion requires only one ligand per copper ion and is a tetradentate chelate. (Sketch the structure, and compare with the above example.)

Triethylenetetramine is not the only way to preconcentrate four molecules of ammonia. Instead of the linear configuration of trien, there is the structure of triaminotriethylamine (abbreviated "tren"):

$$H_2N—CH_2—CH_2—N \Big\langle \begin{matrix} CH_2—CH_2—NH_2 \\ CH_2—CH_2—NH_2 \end{matrix}$$

Like a "tripus" (in analogy to octopus), this ligand can wrap around a metal ion and satisfy four coordination positions, forming three five-membered rings. With zinc ion, which forms tetrahedral complexes, tren is more effective than trien because its geometry is less strained.

Just as the nitrogen atoms in ammonia and these polydentate ligands can form stable coordinate bonds with many metal ions, the oxygen atoms in carboxylic acid groups help ligands to form complexes. A simple example is the moderately strong complexing of lead ion by acetate ion:

$$Pb(aq)^{2+} + CH_3COO^- = PbOCOCH_3^+ \overset{\text{second}}{\underset{\text{step}}{=}} Pb(OCOCH_3)_2.$$

By contrast, lead ion does not form ammonia complexes of appreciable stability. The recognition that some metal ions "prefer" nitrogen bonds while others favor oxygen bonds has led to some ingenious designs of effective ligands. For example, a molecular structure analogous to that of "tren" is found in nitrilotriacetic acid (abbreviated NTA):

$$HOCOCH_2—N \Big\langle \begin{matrix} CH_2COOH \\ CH_2COOH \end{matrix}$$

In basic solution the three protons are dissociated from the acetic acid groups so that a tetradentate ligand is formed, with one nitrogen and three oxygen coordinating sites. The tetrahedral chelate formed with zinc ion has this structure:

Hydrogen atoms omitted for clarity

To avoid cluttering in such structural diagrams it is better not to show the details of the bridging —C—C— groups. Thus a simple curved line can represent this group as follows:

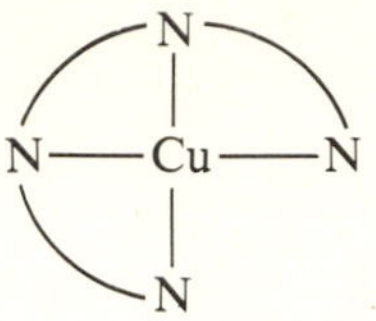

Cu-trien (square planar) 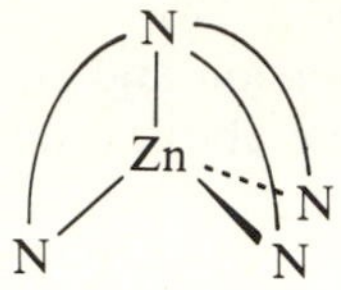   Zn-tren (tetrahedral)

Great advances in metal-ion analysis and in chelate studies quickly followed the introduction of hexadentate ligands, with a mixture of amine and carboxylate groups. By far the most widely used of these is ethylenediamine–tetraacetic acid, (abbreviated EDTA):

With the loss of the carboxyl protons this ligand can form stable 1:1 complexes with metal ions requiring either four or six coordination sites:

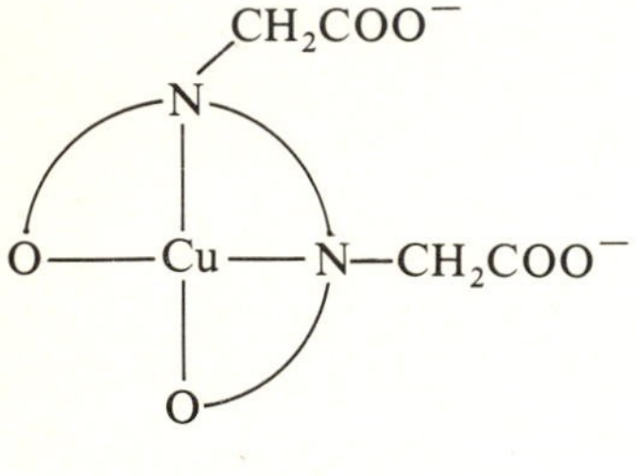

Square planar 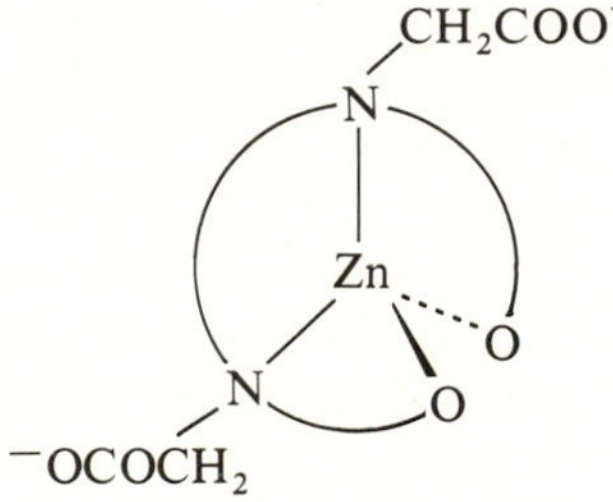   Tetrahedral    Octahedral

14.2   SIMPLIFIED SYMBOLISM FOR COMPLEXES AND CHELATES

Chelate equilibrium systems are so complicated compared with acid–base systems that the algebraic treatment becomes intricate. As an aid to clarity it is useful to adopt the following simplifications:

1. Ionic strength and therefore activity coefficients will be held constant in most cases, so that Q-values in terms of molarities may be used consistently.

2. The ionic charges on metal ions and ligands will be omitted in the equilibrium expressions.

3. The metal ion aquo-complex will be symbolized simply by M, unless the discussion involves more than one metal ion, in which case the regular element symbols will be used.

4. Monodentate and bidentate ligands will be symbolized by L.

5. Polyfunctional chelating ligands will by symbolized by Y.

6. Formulas for complexes and chelates will show the stoichiometry but not the ionic charge. Thus,

$$M \overset{+L}{=} ML \overset{+L}{=} ML_2 \overset{+L}{=} ML_3 \overset{+L}{=} ML_4, \quad \text{etc.};$$

$$M + Y = MY \quad \text{(only one step for polyfunctional ligands)}.$$

7. Chelating ligands that are used as metal-ion indicators by virtue of their intense color will by symbolized by In. The colored metal-indicator chelate is therefore MIn.

8. Protonated forms of ligands will be written as HL, H_2L, etc., or HY, H_2Y, etc., without showing ionic charge.

14.3 SOME COMMENTS ON NOMENCLATURE

The endings *-ous* and *-ic* have been used to distinguish lower and higher oxidation states. However, these suffixes do not identify the oxidation number itself, and we find cuprous ion Cu^+ and ferrous ion Fe^{2+}. Argentic ion is Ag^{2+}, while ceric ion is Ce^{4+}.

    For many of the most commonly used species, chemists tend to stay with the old names (such as ferrous ion) that have lost their ambiguity through repeated and familiar usage. But in general it has proved very useful to adopt a system devised by Stock, whereby the written name of a species carries as much information as the detailed formula.

    In setting forth the ideas of the Stock system, we should consider three types of species: positive ions, negative ions, and neutral compounds. The naming rules differ for each case.

1. Positive ions of the simplest kind, that is, monoatomic species that are deficient in electrons, merely take the name of the element, followed in parentheses by the oxidation number in Roman numerals. Thus we have iron(II) for Fe^{2+}, cerium(III) for Ce^{3+}, etc. When elements are known to exhibit only one oxidation state, the Roman numerals are often omitted. For example, sodium ion is always Na^+ and aluminum ion is always Al^{3+}. The parenthetical indication of oxidation state is especially useful when an element can form different species of the same state.

    Even these simple ions, when in water solutions, are not truly simple but are rather hydrated. A more general view of a simple ion in a water solution is $M(H_2O)_h^{n+}$, where h is the hydration number (the average number of coordinated water molecules per metal ion) and n is the oxidation number of the element M. It is customary to ignore the water of hydration both in names and formulas of dissolved compounds, unless it is of direct concern in the discussion. In the latter case, we refer not to simple ions, but to aquo-ions or aquo-complexes of metal ions. If the hydration number is known, it may be incorporated in the name as follows: The species $Cr(H_2O)_6^{3+}$ would be called the hexaaquochromium(III) ion, an unambiguous name that gives, in order, the number of attached groups (*hexa-*, six), their identity (*aquo-*, water), the identity of the central atom (*chromium*), and its oxidation number (+ 3). Some other examples are diaquosilver(I) for $Ag(H_2O)_2^+$, tetraaquozinc(II) for $Zn(H_2O)_4^{2+}$, and hexaaquocobalt(III) for $Co(H_2O)_6^{3+}$.

In these examples of hydrated ions, the water molecules are coordinated with the metal ion through the previously unshared electron pairs on the oxygen. We say that the water molecules are acting as ligands in the formation of the aquo-complex ion. There are many other chemical species that can act as ligands for metal ions, the only requirement being the possession of an unshared pair of electrons that can form a chemical bond with the metal. A few of the more common examples are listed in Table 14.1.

Table 14.1

Ligand	Ligand stem name used in complex	Example of complex	Name of complex
Ammonia NH_3	ammine-	$Cu(NH_3)_4^{2+}$	tetraamminecopper(II) ion
Hydroxide ion OH^-	hydroxo-	$Fe(OH)_2^+$	dihydroxoiron(III) ion
Chloride ion Cl^-	chloro-	$PbCl^+$	chlorolead(II) ion
Fluoride ion F^-	fluoro-	CaF^+	fluorocalcium ion
Bromide ion Br^-	bromo-	$HgBr^+$	bromomercury(II) ion
Iodide ion I^-	iodo-	BiI_2^+	diiodobismuth(III) ion
Cyanide ion CN^-	cyano-	$CdCN^+$	cyanocadmium ion
Sulfate ion SO_4^{2-}	sulfato-	$CrSO_4^+$	sulfatochromium(III) ion

In most of the samples of Table 14.1, the complex ion in water solution would be hydrated by water molecules in addition to being coordinated by the ligands. If the degree of hydration is known and is of interest in the discussion, it is possible to include it in the name. For example, the sul-fatochromium species at the end of the table actually exists with five water molecules coordinated along with the sulfate ion, and the full name, monosulfatopentaaquochromium(III) ion implies the formula $Cr(H_2O)_5SO_4^+$. In such names, the nonwater ligands are always given first.

2. Neutral compounds containing metal ions can usually be regarded as a combination of positive ion (which may be simple or complex) and negative ion (which also may be simple or complex). Their names are merely the names of the cation and anion, in that order. Therefore, we shall proceed directly to the problem of naming the negative ions.

3. The more common negative ions, such as the halides and the anions of the oxygen acids (like sulfate, nitrate, phosphate, perchlorate, etc.) are named as usual, without recourse to any new rules. We are concerned here with the complex type of anion formed when a metal ion is coordinated with a number of negative ligands sufficient to give the complex an overall negative charge. The rules are different for these cases and may be summarized as follows: (1) the ligands are numbered and identified exactly as in the case of the positive complexes; (2) the central metal element is identified by the (often Latin) stem of the name corresponding to its elemental symbol; (3) all negative complexes end with the suffix -*ate*, followed by the parenthetical Roman numerals for oxidation state. It is easier to see the naming from the examples given in Table 14.2, which include the elements for which Latin stems are used.

Table 14.2

Complex	Element name corresponding to symbol	Name of complex
$CrCl_6^{3-}$	chromium	hexachlorochromate(III) ion
$Fe(CN)_6^{4-}$	ferrum	hexacyanoferrate(II) ion (ferrocyanide)
$Fe(CN)_6^{3-}$	ferrum	hexacyanoferrate(III) ion (ferricyanide)
$CuH_2OCl_3^-$	cuprum	trichloromonoaquocuprate(II) ion
$Ag(IO_3)_2^-$	argentum	diiodatoargentate(I) ion
$Sn(OH)_4^{2-}$	stannum	tetrahydroxostannate(II) ion
$SbOHCl_5^-$	stibium	monohydroxopentachlorostibate(V) ion (but antimonate is also used)
$AuBr_2^-$	aurum	dibromoaurate(I) ion
HgI_4^{2-}	hydrargyrum	titraiodomercurate(II) ion (the form "hydrargyrate" has not caught on)
$PbCl_4^{2-}$	plumbum	tetrachloroplumbate(II) ion

Note that in Table 14.2 the charge on the complex is equal to the oxidation number (given in Roman numerals) minus the number of charges brought in by the coordinating ligands. With a little practice, one can go from name to formula, or vice versa, with no difficulty.

Finally, we give a few examples of names of salts:

K_3CoCl_6—potassium hexachlorocobaltate(III). Note that it is not necessary to say "tripotassium" since the number of potassium ions, each with a $+1$ charge, is determined by the charge on the complex anion.

$Cd(H_2O)_3BrClO_4$—bromotriaquocadmium(II) perchlorate.

$Cr(NH_3)_3Cl_3$—trichlorotriamminechromium(III). This name recognizes that the chlorides are ligands bonded to the chromium. Compare with $Cr(NH_3)_6Cl_3$, with only ammonia ligands, which has the name hexamminechromium(III) chloride.

14.4 EQUILIBRIUM-QUOTIENT EXPRESSIONS

The successive complexing of a metal ion by a ligand is chemically analogous to the successive protonation of a ligand:

	Complex formation	Protonation
Step 1	$M + L = ML,$	$L + H^+ = HL,$
Step 2	$ML + L = ML_2,$	$HL + H^+ = H_2L,$
Step 3	$ML_2 + L = ML_3,$	$H_2L + H^+ = H_3L,$
	etc.	etc.

However it has become customary to treat protonation in terms of *acid-dissociation constants*, and complex formation in terms of *complex-formation constants*. The acid-dissociation constants (or quotients) are numbered with subscripts corresponding to successive losses of protons by whatever species is designated as the "parent acid." For the tetraprotic acid EDTA we have:

Step 1 $\quad H_4Y = H^+ + H_3Y^-,\qquad Q_1 = \dfrac{[H^+][H_3Y]}{[H_4Y]} = 10^{-2.0},$

Step 2 $\quad H_3Y^- = H^+ + H_2Y^{2-},\qquad Q_2 = \dfrac{[H^+][H_2Y]}{[H_3Y]} = 10^{-2.67},$

Step 3 $\quad H_2Y^{2-} = H^+ + HY^{3-},\qquad Q_3 = \dfrac{[H^+][HY]}{[H_2Y]} = 10^{-6.16},$

Step 4 $\quad HY^{3-} = H^+ + Y^{4-},\qquad Q_4 = \dfrac{[H^+][Y]}{[HY]} = 10^{-10.26}.$

For successive complexes, the formation constants (or quotients) are numbered according to the number of ligands that have formed coordinate bonds with the metal ion. For the four-step complexation of copper ion by ammonia we have:

Step 1 $\qquad Cu^{2+} + NH_3 = CuNH_3^{2+},\qquad Q_1 = \dfrac{[ML]}{[M][L]} = 10^{4.0},$

Step 2 $\qquad CuNH_3^{2+} + NH_3 = Cu(NH_3)_2^{2+},\qquad Q_2 = \dfrac{[ML_2]}{[ML][L]} = 10^{3.5},$

Step 3 $\qquad Cu(NH_3)_2^{2+} + NH_3 = Cu(NH_3)_3^{2+},\qquad Q_3 = \dfrac{[ML_3]}{[ML_2][L]} = 10^{2.8},$

Step 4 $\qquad Cu(NH_3)_3^{2+} + NH_3 = Cu(NH_3)_4^{2+},\qquad Q_4 = \dfrac{[ML_4]}{[ML_3][L]} = 10^{1.5}.$

"Overall" Complex-Formation Constants

The relationship between the concentration of the uncomplexed metal ion M and any one of its complexes is easily found by using the product of the

Q-values. These products are called β_i values, with the subscript indicating the number of combined values. To illustrate, we write:

$$\beta_2 = Q_1 Q_2 = \frac{[ML_2]}{[M][L]^2},$$

$$\beta_3 = Q_1 Q_2 Q_3 = \frac{[ML_3]}{[M][L]^3},$$

$$\beta_4 = Q_1 Q_2 Q_3 Q_4 = \frac{[ML_4]}{[M][L]^4}.$$

The beta-values are convenient in dealing with the algebra of complex formation because the products of the Q-values are frequently needed. In some experimental studies it is feasible to determine a beta-value, but not the individual Q-values that it includes.

For the above case of the copper-ammonia complexes, $\beta_2 = 10^{7.5}$, $\beta_3 = 10^{10.3}$, and $\beta_4 = 10^{11.8} = 6 \cdot 10^{11}$. For conformity it is permissible to use the symbol β_1 instead of Q_1.

Formation Constants

In Appendix 5 you will find tabulations of numerical constants for acid–base, oxidation–reduction, and complex-formation equilibria. Arranged in alphabetical order by element, each metal ion is followed by sets of β-values showing complexation tendencies for several selected ligands. For the most part these constants refer to a temperature of 25°C and an ionic strength of about 0.1. The tabulation is not intended to be comprehensive.

14.5 PRECIPITATION: A SPECIAL CASE OF COMPLEX FORMATION

When an uncharged complex is formed by the appropriate combination of metal ion(s) and ligand(s), the tendency to form a stable crystal-lattice structure may overcome the forces of attraction exerted by the water molecules in the solution. Partial (sometimes nearly complete) precipitation is the result. If we assume that an uncharged complex is first formed in the aqueous phase and can exist there up to the concentration called the **intrinsic solubility**, the following equilibrium scheme applies (for the simple case of a 1:1 complex):

$$L + M \rightleftarrows ML(aq) \quad \text{(uncharged species)}$$
$$\updownarrow$$
$$ML(solid)$$

The relationship between the two fundamental equilibrium quotients

$$Q_1 = \frac{[ML(aq)]}{[M][L]} \qquad \text{and} \qquad Q_0 = [ML(aq)]$$

formation quotient   intrinsic solubility is that they both contain the molarity of the aqueous species ML. The **solubility product** for the insoluble substance ML(s) is derived as the combination

$$\frac{Q_0}{Q_1} = Q_{sp} = [M][L];$$

and like Q_0, it holds for a saturated solution *only*, whereas the formation quotient and its reciprocal—the dissociation quotient for ML(aq)

$$Q_d = \frac{[M][L]}{[ML(aq)]}$$

are applicable to unsaturated solutions as well as to saturated solutions.

Since the subjects of precipitation and solubility equilibria are taken up in detail in later chapters, the remainder of this chapter will deal with homogeneous equilibria, i.e., systems not involving precipitates as part of the scheme.

Solubility-Product Constants

Appendix 5 contains, in alphabetical order by element, a table of K_{sp} values for reactions of metal ions with selected ligands. These are mostly the thermodynamic values (zero ionic strength) at 25°C.

14.6   MATERIAL-BALANCE FUNCTION F_0 FOR THE METAL ION

Suppose a solution contains a metal ion M at a concentration of C_M moles per liter. In the absence of a complexing ligand we have

$$C_M = [M].$$

However, if the solution also contains C_L moles per liter of a complexing ligand that can form four successive complexes with the metal ion, then the material-balance expression may be written as

$$C_M = [M] + [ML] + [ML_2] + [ML_3] + [ML_4]. \tag{14.1}$$

Of course, the individual concentrations of these species will depend strongly on the concentration of the added ligand and upon the successive formation constants. A particularly important relationship is the ratio of the total concentration C_M to that of the uncomplexed metal [M].

This can be found by using the beta-expressions to substitute for each of the molarities of the complex species:

$$C_M = [M] + Q_1[M][L] + \beta_2[M][L]^2 + \beta_3[M][L]^3 + \beta_4[M][L]^4. \tag{14.2}$$

When this equation is divided by [M], the result is

$$F_0 = \frac{C_M}{[M]} = 1 + Q_1[L] + \beta_2[L]^2 + \beta_3[L]^3 + \beta_4[L]^4, \tag{14.3}$$

Equation (14.3) defines a **formation function** F_0, which is at the heart of calculations involving successive complexes. F_0 has a value of 1.000 in the absence of the ligand, when $[L] = 0$, and increases steadily in polynomial fashion as the ligand molarity is increased in the system.

In some cases it is possible to determine experimental values for F_0 in a series of solutions of a metal ion, usually in low concentration, containing varying concentrations of ligand. As shown in Chapter 15 (case study of cadmium bromide complexes), a knowledge of such values may make it possible to compute the successive Q-values for the metal–ligand system. Thus, F_0 is a very important function.

Alpha-Functions

Whereas F_0 is used to indicate the ratio of C_M to M, the fraction of the metal that remains *un*complexed is the reciprocal of F_0:

$$\alpha_0 = \frac{[M]}{C_M} = \frac{1}{F_0}. \tag{14.4}$$

By using the Q_1 and beta-expressions it is easy to show that the fractions present as the other individual species are

$$\alpha_1 = \frac{[ML]}{C_M} = Q_1[L]\alpha_0 = \frac{Q_1[L]}{F_0}, \tag{14.5}$$

$$\alpha_2 = \frac{[ML_2]}{C_M} = \beta_2[L]^2\alpha_0 = \frac{\beta_2[L]^2}{F_0}, \tag{14.6}$$

$$\alpha_3 = \frac{[ML_3]}{C_M} = \frac{\beta_3[L]^3}{F_0}, \tag{14.7}$$

$$\alpha_4 = ? \quad \text{(To be completed by the reader: note the pattern.)} \tag{14.8}$$

These expressions will prove useful in plotting diagrams of the relative amounts of species present as a function of the molarity of ligand in the solution. *Note.* Don't confuse these alpha-functions with those derived in Chapter 9 for polyprotic acids, although they certainly are very similar.

Distribution Diagrams

Distribution diagrams are useful for showing the relative concentrations of species involved in successive steps of a complex equilibrium system. The fraction of each species, i.e., its alpha-value, is plotted versus a **master variable** that is a species involved in each step. Such diagrams were introduced in Chapter 9 when we considered the distribution of the several forms of a polyprotic acid as a function of pH. In this case, the hydrogen-ion concentration was the master variable, and it was expressed in terms of pH to make it easier to include a very wide range of values on the plot.

With metal–ligand complex systems we wish to plot the fraction of the metal present as each of the successive complex species, using the molarity of the ligand as the master variable. As with the acid–base case, it will often be best to use a logarithmic scale, plotting fraction versus log[L] to show the distribution clearly over a wide range of ligand concentrations.

Given a set of successive Q-values for a certain metal–ligand system, it is a straightforward procedure (though perhaps a bit tedious) to calculate the alpha-values for a range of ligand molarities. For illustration, consider the complexes of nickel ion with ethylenediamine (en):

$$Ni^{2+} + en = Nien^{2+}, \qquad Q_1 = 4.0 \cdot 10^7;$$
$$Nien^{2+} + en = Ni(en)_2^{2+}, \qquad Q_2 = 3.2 \cdot 10^6;$$
$$Ni(en)_2^{2+} + en = Ni(en)_3^{2+}, \qquad Q_3 = 1.0 \cdot 10^5.$$

There are four species of nickel and we would like to know which one(s) are prevalent at ethylenediamine concentrations of 0.001, 0.1, or whatever. To calculate data for a full distribution diagram, one first has to decide what value will be the lowest ligand concentration represented since, in principle, the log[L] scale goes all the way to $-\infty$, corresponding to zero molarity of ligand. To conserve graph paper we choose as the lowest ligand concentration the value corresponding to a ratio of 100 for the concentration of uncomplexed metal ion to the concentration of the first complex. In the case of Ni–en we simply use the first Q-value to find

$$[en] = \frac{[Nien^{2+}]}{[Ni^{2+}] \cdot 4 \cdot 10^7} = \frac{1}{100} \cdot \frac{1}{4 \cdot 10^7} = 2.5 \cdot 10^{-10}.$$

Therefore, rounding off to the nearest integer log value, we begin the diagram at an equilibrium en concentration of 10^{-10}, or $\log[en] = -10$.

It works out well if the successive values of [L] are in the relative sequence 1, 2, 5, 10, 20, 50, 100, 200, 500, 1000, etc. This will cause the points to be spaced approximately equally along the logarithm axis. In the present case we will use $[en] = 1 \cdot 10^{-10}$, $2 \cdot 10^{-10}$, $5 \cdot 10^{-10}$, $1 \cdot 10^{-9}$, etc.

Now to begin: Take the lowest ligand concentration, and calculate the F_0 function corresponding to it. We find

$$F_0 = 1 + 4 \cdot 10^7 \cdot 1 \cdot 10^{-10} + 4 \cdot 10^7 \cdot 3.2 \cdot 10^6 \cdot (1 \cdot 10^{-10})^2$$
$$+ 4 \cdot 10^7 \cdot 3.2 \cdot 10^6 \cdot 1 \cdot 10^5 \cdot (1 \cdot 10^{-10})^3 = 1.004.$$

Each of the alpha-values can be found by following the simple operations in Eqs. (14.4) to (14.8):

$$\alpha_0 = \frac{1}{F_0} = 0.996 \qquad\qquad = \frac{[M]}{C_M},$$
$$\alpha_1 = \frac{4 \cdot 10^7 \cdot 1 \cdot 10^{-10}}{F_0} = 0.00398 = \frac{[ML]}{C_M},$$

$$\alpha_2 = \frac{4.10^7 \cdot 3.2 \cdot 10^6 (1 \cdot 10^{-10})^2}{F_0} = 1.3 \cdot 10^{-6} \qquad = \frac{[ML_2]}{C_M},$$

$$\alpha_3 = \frac{4.10^7 \cdot 3.2 \cdot 10^6 \cdot 1 \cdot 10^5 (1 \cdot 10^{-10})^3}{F_0} = 1.3 \cdot 10^{-11} = \frac{[ML_3]}{C_M}.$$

There are purposes for which it is desirable to calculate exceptionally small values of alpha, but for the plotting of distribution diagrams there is no point in carrying alpha-values to more than two, or at most three, decimal places. An exception is the decision to plot log(alpha) instead of alpha, in which case a wide range of alpha-values can be represented on the plot.

When the calculation above is repeated for successive values of [en] the series of results includes those in Table 14.3. For simplicity, only the integer values of log[en] are given. Note that the relative concentration of the uncomplexed nickel ion falls steadily and becomes negligible when the free-en molarity is about 10^{-5}. The two intermediate species rise and then fall, taking turns being dominant. The highest complex is not significant until the [en] is about 10^{-6}, and gradually becomes the dominant species.

There is no point in continuing the calculations to ligand concentrations higher than necessary to make the highest complex present to more than about 99%.

The plotted values in Fig. 14.1 show at a glance how the various Ni–en species relate to each other over the wide range of ethylenediamine molarity.

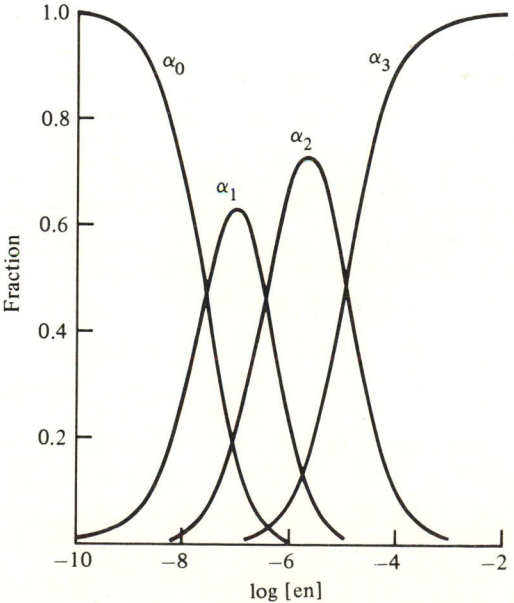

Figure 14.1

Table 14.3   Distribution of nickel–ethylenediamine species at various ligand concentrations

[en]	log[en]	α_0	α_1	α_2	α_3
10^{-10}	-10	0.996	0.004		
10^{-9}	-9	0.961	0.038		
10^{-8}	-8	0.708	0.283	0.009	
10^{-7}	-7	0.159	0.636	0.203	0.002
10^{-6}	-6	0.006	0.220	0.704	0.070
10^{-5}	-5		0.015	0.492	0.492
10^{-4}	-4			0.091	0.909
10^{-3}	-3			0.010	0.990
10^{-2}	-2			0.001	0.999

14.7 SUGGESTED CLASSROOM DEMONSTRATION ON NICKEL–ETHYLENEDIAMINE COMPLEXES

1. Line up four 600-mL beakers, each containing 400 mL of 0.020M NiSO$_4$, which has a light green color due to Ni(H$_2$O)$_6^{2+}$.

2. Leave one beaker as is, and to the others add 2.0, 4.0, and 6.0 mL, respectively, of 4.0M ethylenediamine. Stir. The colors (light blue, dark blue, and purple) will essentially be those of the 1:1, 2:1, and 3:1 complexes because the Q-values are fairly well separated (see Fig. 14.1).

3. Show that cyanide ion forms a more stable complex than does en, by adding solid NaCN to the fourth beaker and stirring its contents. The solution changes to yellow, as the purple octahedral Ni(en)$_3^{2+}$ is converted to the square planar Ni(CN)$_4^{2-}$. *Caution* Be sure no acid is added to this solution because poisonous HCN would be liberated.

4. Show that H$^+$ competes for en by adding some 6M HCl to the third beaker. The color changes back to that of the aquo-complex.

14.8 LIGAND-MATERIAL BALANCE. THE BJERRUM FORMATION FUNCTION FOR AVERAGE LIGAND NUMBER

The material-balance expression for the ligand takes into account the fact that each complex species has a different metal–ligand stoichiometry:

$$C_L = [L] + [ML] + 2[ML_2] + 3[ML_3] + 4[ML_4]. \tag{14.9}$$

Since [L] represents the equilibrium concentration of free ligand (not combined with a metal ion), the total molarity of bound ligand is simple $C_L - [L]$. We define the **average ligand number** $\bar{n}$ by the ratio

$$\frac{\text{Bound ligand}}{\text{Total metal}} = \frac{C_L - [L]}{C_M} = \bar{n}.$$

By introducing Eqs. (14.1) and (14.9) we obtain the result

$$\bar{n} = \frac{[ML] + 2[ML_2] + 3[ML_3] + 4[ML_4]}{[M] + [ML] + [ML_2] + [ML_3] + [ML_4]} = \alpha_1 + 2\alpha_2 + 3\alpha_3 + 4\alpha_4.$$

Introduction of the beta-expressions on p. 451 and rearrangement leads to

$$\bar{n} = (1 - \bar{n})Q_1[L] + (2 - \bar{n})\beta_2[L]^2 + (3 - \bar{n})\beta_3[L]^3 + (4 - \bar{n})\beta_4[L]^4.$$

This equation also proves valuable in experimental studies for the determination of equilibrium constants. This application is discussed in Chapter 15.

14.9 COMPETING EQUILIBRIA

A. The Effect of pH: Protonation of the Ligand

The bonding sites (unshared electron pairs) on ligands that form metal complexes often have affinities for hydrogen ions also. Exceptions are ligands, such as chloride, bromide, iodide, and thiocyanate, which are anions of strong acids. If the pH is high enough, about two units higher than the pQ_a value for the protonated ligand species, this effect will be negligible because too little of the ligand will be converted to HL to make any difference in its concentration. But at lower pH values the ligand will be increasingly forced to form its conjugate acid and therefore will not be as readily available for complexation with the metal ion. The equilibrium scheme is

$$M + L = ML, ML_2, \text{ etc.}$$
$$+$$
$$H^+ = HL, H_2L, \text{ etc.}$$

We want to be able to answer questions such as the following: If a solution containing mercuric ions is made to contain $0.1M$ ammonia, will the mercuric ions be converted to ammine complexes? If the solution has its pH adjusted to about 3, will the mercuric ions still be complexed, or will the protonation of the ammonia force their release?

To treat such a system we must use both the F_0 function for the metal–ligand complexes and the α_0 function for the ligand–proton species. Taking the F_0 part first, we know that

$$C_M = [M] + [ML] + [ML_2] + \cdots,$$

$$C_M - [M] = [M]Q_1[L] + [M]Q_1Q_2[L]^2 + \cdots = \sum [\text{complexes}].$$

Dividing by [M] we find

$$\frac{\sum [\text{complexes}]}{[M]} = Q_1[L] + Q_1Q_2[L]^2 + \cdots = F_0 - 1. \tag{14.10}$$

Thus the function $(F_0 - 1)$ is the ratio of complexed metal ion to uncomplexed metal ion, and hence provides a clear evaluation of the extent of complexing regardless of which individual complex species predominate in a given solution.

When the equilibrium molarity of L is known, Eq. (14.10) may be used immediately. However, if the effect of pH on the molarity of the ligand is significant we must first do a preliminary calculation using the α_0 function. In a solution that contains a mixture of the various protonated species of the ligand, the concentration of the *un*protonated form is given by Eq. (9.32):

$$[L] = \alpha_0 C_L = C_L \frac{Q_{a1} Q_{a2} \cdots Q_{an}}{[H^+]^n + Q_{a1}[H^+]^{n-1} + \cdots Q_{a1} Q_{a2} \cdots Q_{an}}, \qquad (14.11)$$

where C_L is the total ligand molarity in all forms (not counting the one associated with the metal ion) and the Q-values are the successive acid-dissociation quotients for the protonated ligand species. By using Eq. (14.11) with a specified pH, we may calculate the actual concentration of L. This, in turn, is then used in Eq. (14.10) to find the value of $(F_0 - 1)$.

Example calculation Mercury(II) forms successive ammonia complexes, with formation quotients having values of $10^{8.8}$, $10^{8.6}$, 10^{1}, and $10^{0.9}$. Ammonia is a monoprotic ligand, having Q_a (for its conjugate acid, the ammonium ion) equal to $10^{-9.2}$. Suppose a solution contains a small amount of mercury(II) along with $0.10M$ NH_3. At high pH, where there is negligible protonation of the ammonia, $[NH_3] = 0.1$, and according to Eq. (14.10) we get:

$$F_0 - 1 = 10^{8.8} \cdot 0.1 + 10^{17.4} \cdot 0.1^2 + 10^{18.4} \cdot 0.1^3 + 10^{19.3} \cdot 0.1^4$$
$$= 7 \cdot 10^{15}.$$

This huge number shows that under these conditions the mercury is virtually completely complexed. Practically none is left as the simple ion Hg^{2+}.

Will the mercury still be complexed if the pH is adjusted to 3 by addition of acid to react with the ammonia? At this pH the equilibrium concentration of NH_3 is given by Eq. (14.11):

$$[NH_3] = C \frac{Q_a}{[H^+] + Q_a} = 0.1 \frac{10^{-9.2}}{10^{-3} + 10^{-9.2}} = 6.3 \cdot 10^{-8}.$$

Nearly all the ligand has been converted to its conjugate acid. When $6.3 \cdot 10^{-8}$ is substituted into the $(F_0 - 1)$ equation, the result is

$$F_0 - 1 = 1.0 \cdot 10^3.$$

Thus in spite of the strong effect of pH the mercury remains strongly complexed. However, the value of $(F_0 - 1)$ did drop by a factor of more than 10^{12}. The reader should try the calculation at pH $= 0$.

B. Competition between Ligands for the Metal Ion

When a solution contains a metal ion and two ligands, each of which can form complexes, how do we decide which ligand "wins"? Again the key to the answer is found in the $(F_0 - 1)$ values.

Example calculation Copper forms stable complexes both with citrate and with ethylenediamine:

$$Cu^{2+} + Cit^{3-} \overset{10^{14.2}}{=} CuCit^-$$

$$+$$
$$en \overset{10^{10.7}}{=} Cuen^{2+}$$

$$+$$
$$en \overset{10^{9.3}}{=} Cu(en)_2^{2+}.$$

At first glance it appears that the citrate complex is more stable ($Q = 10^{14.2}$), but how important is it that ethylenediamine also forms a second complex? Further, suppose the concentrations of the ligands are unequal, $0.01M$ for en and $0.1M$ for citrate.

By reference to Eq. (14.10) we may write the two $(F_0 - 1)$ expressions:

$$\frac{\sum [en \text{ complexes}]}{[Cu^{2+}]} = (F_{en} - 1) = 10^{10.7} \cdot 0.01 + 10^{20} \cdot 0.01^2 = 1 \cdot 10^{16},$$

$$\frac{[CuCit^-]}{[Cu^{2+}]} = (F_{cit} - 1) = 10^{14.2} \cdot 0.1 = 1.6 \cdot 10^{13}.$$

The results show that the ethylenediamine complexes predominate, since the $(F_0 - 1)$ value is much larger. We may write the ratio of the functions to show the relative amounts of copper in the two complex systems:

$$\frac{\sum [en \text{ complexes}]}{[CuCit^-]} = \frac{F_{en} - 1}{F_{cit} - 1} = \frac{1 \cdot 10^{16}}{1.6 \cdot 10^{13}} = 600.$$

In this solution, the citrate complexing is relatively unimportant compared to that of ethylenediamine. This conclusion could be useful, for example, in predicting the color of the solution for spectrophotometric measurements. In other words, we could conclude that the presence of citrate would not interfere with the photometric determination of copper when ethylenediamine complexes are used as colored species.

C. Conclusion: Combining the pH Effect with Ligand Competition

In the foregoing discussion it was assumed that $[en] = 0.01$ and $[cit^{3-}] = 0.1$. This is equivalent to the assumption that these represent the total concentrations of the ligands and that the solution had a high pH so that protonation was negligible for both ligands. As the pH of the solution is lowered, its effect on the two ligands will not be the same, of course, because the ligands differ in their acid–base properties:

$$H_2en^{2+} \xrightarrow{\ pK_1=7.1\ } Hen^+ \xrightarrow{\ pK_2=9.9\ } en$$

$$H_3Cit \xrightarrow{\ pK_1=3.2\ } H_2Cit^- \xrightarrow{\ pK_2=4.8\ } HCit^{2-} \xrightarrow{\ pK_3=6.4\ } Cit^{3-}.$$

Given a certain pH value, we may first calculate the equilibrium molarities of the ligands, using the total concentrations of 0.01 and 0.1, respectively. The equilibrium molarities can then be used to find the (F_0-1) functions, which serve again as a measure of the complexation tendency.

For example, the use of the α_0 functions at pH $= 6$ gives

$$[en] = 9.3 \cdot 10^{-8}; \qquad [Cit^{3-}] = 0.027.$$

Note that citrate is not as strongly affected by pH because it is less basic than ethylenediamine. When these values are used to find (F_0-1), the values are

$$(F_{en} - 1) = 8.7 \cdot 10^5; \qquad (F_{cit} - 1) = 4.3 \cdot 10^{12}.$$

The tables have been turned by the pH effect. It is now the citrate complex of copper that predominates.

$$\frac{[CuCit^-]}{\Sigma\,[en\ complexes]} = \frac{4.3 \cdot 10^{12}}{8.7 \cdot 10^5} = 5 \cdot 10^6.$$

By now the reader should be thoroughly convinced that an understanding of the pH effect in particular, and competing reactions in general, are important topics in equilibrium and analysis.

PROBLEMS

Metal–ligand complexes (consult Appendix 5 for β values).

1. Name the following ions:

$$FeCl_4^- \qquad Ni(NH_3)_4^{2+} \qquad FeBr^{2+} \qquad Zn(CN)_3^-$$
$$CrOH^{2+} \qquad Cd(SCN)_4^{2-} \qquad CuI_2^- \qquad PbNO_3^+$$

2. Write formulas for the following ions:

dichloroargentate ion   trithiocyanatomercurate(II) ion
monoiodochromium(II) ion   monosulfatopentaaquochromium(III) ion
tetraamminezinc(II) ion   disulfatocuprate(II) ion

3. For each of the metal–ligand combinations in (a)–(f) write the formula, including ionic charge, for the 1:1 complex, and the formula for the highest complex represented by a log β value in Appendix 5. Calculate each of the individual Q values (from the tabulated log β values).

a) aluminum, fluoride   b) cadmium, citrate   c) calcium, acetate
d) copper(II), ethylenediamine   e) silver, ammonia   f) nickel, cyanide

Example Iron(III), oxalate. The $1:1$ complex is $FeC_2O_4^+$ and the highest complex is the $1:3$ species $Fe(C_2O_4)_3^{3-}$. The Q- and beta-values are

$$Q_1 = 10^{7.5}$$

$$\beta_2 = 10^{13.6} \qquad Q_2 = \frac{\beta_2}{Q_1} = 10^{6.1},$$

$$\beta_3 = 10^{18.5} \qquad Q_3 = \frac{\beta_3}{\beta_2} = 10^{4.9}.$$

4. If 1.00 mmol of the indicated ligand is added to a solution that contains 1.00 mmol of *each* of the indicated metals, which metal ion will be preferentially complexed? *Hint*: Compare the log β values.

a) mercury(II), magnesium, ammonia   b) copper(II), lead, citrate
c) calcium, magnesium, EDTA   d) silver, nickel, acetate
e) lead(II), iron(III), oxalate   e) copper(I), silver(I), cyanide

5. Arrange the following metals in decreasing order according to the stability of their EDTA complexes: Ca, Cu, Hg(II), Zn, Cd, Mg, Ag, Fe(III), Fe(II).

6. Assuming that the amount of metal ion is quite small compared to the ligand concentration, for each of the metal ion/ligand systems in (a)–(f) caculate the formation function F_0, the fraction α_0 of the metal ion that is not complexed by the ligand, and the ratio of the concentrations of the two highest complexes.

a) Cd^{2+}, citrate $(0.01M)$ b) Cu^{2+}, ethylenediamine $(1M)$
c) Ag^+, ammonia $(0.005M)$ d) Mg^{2+}, oxalate $(0.2M)$
e) Cd^{2+}, bromide $(0.5M)$ f) Zn^{2+}, chloride $(2M)$

Example Zn^{2+}, pyridine $(0.5M)$. From Appendix 5 we find that zinc forms two pyridine complexes with formation quotients of 10 and 3. Therefore,

$$F_0 = 1 + 10[\text{pyr}] + 10 \cdot 3 \cdot [\text{pyr}]^2 = 1 + 5 + 7.5 = 13.5 = \frac{C_{Zn}}{[Zn^{2+}]}.$$

The fraction of uncomplexed zinc is simply the reciprocal of F_0: $\alpha_0 = 0.074$, or 7.4%. To find the ratio of the two highest complexes we use the highest formation-quotient expression:

$$Q_2 = \frac{[Zn(\text{pyr})_2^{2+}]}{[Zn\text{pyr}^{2+}][\text{pyr}]},$$

whence

$$\frac{[Zn(\text{pyr})_2^{2+}]}{[Zn\text{pyr}^{2+}]} = 3 \cdot 0.5 = 1.5.$$

7. For each of the following metal–ligand systems calculate the value for F_0 and also the percentage of the metal present as each of the various complexes that it forms with the ligand.

a) Cu^{2+}, ammonia (0.04M)   b) Fe^{3+}, fluoride (0.05M)
c) Hg^{2+}, cyanide (0.1M)   d) Zn^{2+}, cyanide (1M)
e) Cd^{2+}, acetate (0.1M)   f) Mg^{2+}, sulfate (0.3M)

8. In each of the following systems a certain metal ion in low concentration is in a solution that contains two ligands, each at a concentration much higher than the metal ion. The concentrations of the ligands are those of the fully deprotonated species (i.e., there is no need to consider a pH effect). Calculate the F_0 and $(F_0 - 1)$ functions and explain whether the metal ion is preferentially complexed by one of the ligands.

a) Ca^{2+}, EDTA (0.01M), citrate (0.1M)
b) Cu^{2+}, oxalate (0.01M), ammonia (1M)
c) Fe^{3+}, citrate (0.01M), thiocyanate (0.1M)
d) Pb^{2+}, acetate (0.1M), citrate (0.1M)
e) Cd^{2+}, tetren (0.01M), EDTA (0.01M)
f) Ag^{+}, ammonia (0.1M), pyridine (1M)

9. In Problem 8 the ligands were assumed to be fully in their deprotonated forms, which is equivalent to saying that the solutions were all at a rather high pH so that the conjugate acids of the ligands were not present in significant amounts. For each of the systems listed in Problem 8, repeat the calculations on the assumption that sufficient strong acid has been added to the solution to make the pH lower. It will be necessary to look up the acid-dissociation constants for the ligand and then to use the alpha-functions to find the free-ligand molarity at the pH specified as follows:

a) $pH = 6$ b) $pH = 6$ c) $pH = 1$
d) $pH = 4$ e) $pH = 8$ f) $pH = 7$

10. A certain metal forms two successive complexes with ligand L, the beta-values being 10^4 and 10^7. This ligand has negligible basic properties. The metal also forms a single complex with ligand Y, the formation quotient being 10^{11}. Ligand Y is the singly charged anion of the acid HY, which has $pK_a = 6.0$. A mixture of $1 \cdot 10^{-5}M$ metal ion, $0.01M$ NaY, and $0.10M$ L is adjusted to various pH values by addition of strong acid. Show that the relative complexing ability of the ligands differs greatly between $pH = 4$ and $pH = 0$.

11. A solution contains $10^{-4}M$ iron(III) and is well buffered at $pH = 4$. How many moles of sodium citrate (per liter) must be added to reduce the $[Fe^{3+}]$ to $10^{-11}M$, assuming no change in pH or volume?

12. A solution contains $10^{-5}M$ $CdCl_2$ and $0.1M$ KI so that the cadmium is largely present as a mixture of iodide complexes. It is desired to add enough sodium cyanide to produce the cadmium/cyanide complexes to the extent that the remaining iodide complexes are only 0.1% of the total. If the pH of the solution is 7.4, will the addition of 5 mmol of NaCN to 100 mL of solution be sufficient to achieve this result?

13. If 0.1 mmol of copper sulfate is added to 100 mL of a buffer containing $0.1 M$ sodium acetate and $0.1 M$ acetic acid, what will the molarity of the uncomplexed Cu^{2+} ion be? If enough potassium iodate is added to make $[IO_3^-] = 0.1$, would you expect to see a precipitate of copper iodate? *Hint:* Show that $[Cu^{2+}] = 1.8 \cdot 10^{-5}$.

14. A solution contains $0.001 M$ silver nitrate and $0.1 M$ ammonia, so that the silver is almost entirely present as the diammine complex. If perchloric acid is added, slowly with stirring, until the pH drops to 7.0, will the silver still be largely complexed with ammonia?

15. Show that in a solution containing $0.10 M$ Zn^{2+} and only $0.001 M$ sodium oxalate, nearly all of the latter is converted to $ZnC_2O_4(aq)$. If enough copper sulfate is added to make the total copper concentration $0.5 M$, will this be sufficient to break up the ZnC_2O_4 complex? Assume that only the first complex CuC_2O_4 is formed due to the large concentration of free metal ion.

16. A solution contains $0.001 M$ each of iron(III) and Zn(II). It is desired to complex the zinc by adding potassium thiocyanate in sufficient amount to make its concentration $1 M$. However, the formation of the red color due to $FeSCN^{2+}$ and $Fe(SCN)_2^+$ is to be avoided. If the solution has a pH of 7, will the presence of $0.10 M$ sodium fluoride be sufficient to convert the iron to the colorless fluoro complexes? What is the equilibrium molarity of Zn^{2+}?

17. If potassium chromate is added to a solution containing $0.01 M$ Pb^{2+} and $0.1 M$ Ag^+, both metals will form insoluble chromates. If it is desired to precipitate the lead, but not the silver, would the presence of $0.1 M$ ammonia be sufficient to lower the $[Ag^+]$ to the point where the K_{sp} for silver chromate would not be exceeded, assuming that the final $[CrO_4^{2-}]$ is $0.01 M$? *Hint:* Show that $[Ag^+]$ becomes 10^{-6}.

METAL–LIGAND COMPLEXES: DETERMINATION OF FORMATION QUOTIENTS

15

The important thing in science is not so much to obtain new facts as to discover new ways of thinking about them.

Sir William Lawrence Bragg

By considering the essential similarity of the chemical model for the titration of a polyprotic base B with a strong acid

$$B \xrightarrow{H^+} BH \xrightarrow{H^+} BH_2 \xrightarrow{H^+}, \quad \text{etc.,}$$

to that of ligand-complexing of a metal ion

$$M \xrightarrow{L} ML \xrightarrow{L} ML_2 \xrightarrow{L}, \quad \text{etc.,}$$

we sense intuitively that the *mathematical* treatment of both systems must be identical. However, the experimental approaches for the study of acid–base equilibria are *all* based on the use of pH as a master variable, prompting the attitude that "if you've seen one, you've seen them all."

With metal–ligand systems, however, we have to consider a large variety of different metal ions and a large variety of master variables (the ligands) as well. Each metal and each ligand show a set of unique chemical properties, and it is up to the ingenuity of the researcher to decide which of these properties may be used to advantage in unraveling the complex system to deduce the equilibrium quotients.

15.1 THE GENERAL EXPERIMENTAL SETUP

If a complex or series of complexes is to be studied in solution, it is obviously necessary to mix the metal ion and the ligand in selected proportions and to make some sort of measurements on the equilibrium solution. Practically all experiments of this sort are the equivalent of carrying out a titration of a known concentration of the metal ion with a solution containing a known concentration of ligand. To generalize, suppose we start with a solution of the metal ion that has a volume of V_0 milliliters and a concentration of M_0 metal ion. To this we will add increments of a ligand solution having a concentration of L_0. Typically L_0 will be much larger than M_0, so that the total volume $V_0 + V$ stays manageable. The cumulative total of the ligand solution added is V milliliters.

At any point in the titration the metal is more or less converted to a mixture of its possible complexes, and the ligand is correspondingly incorporated in the complexes.

Considering the increase in solution volume, the total molarity of the metal is

$$C_M = \frac{M_0 V_0}{V_0 + V} = [M] + [ML] + [ML_2] + \cdots$$

and the total molarity of the ligand is

$$C_L = \frac{C_0 V}{V_0 + V} = [L] + [ML] + 2[ML_2] + 3[ML_3] + \cdots.$$

This exhausts the conclusions we may draw from stoichiometry alone. Now it is time for analytical ingenuity to take over. We have to "get a handle" on what is happening in the solution, either by using some analytical "probe" that can measure the equilibrium concentration of metal ion in its uncomplexed form, and/or by using a probe that can measure the equilibrium molarity of the free ligand, i.e., of species L in its uncomplexed form. Support for this conclusion is found in the two, and only two, expressions we have for material balance in the solution. One is the F_0 function, and the other is the $\bar{n}$ function.

15.2 MAKING USE OF THE F_0 FUNCTION

Since the total metal concentration C_M is known from stoichiometry, if we can find a way to determine the uncomplexed metal-ion concentration [M] in the equilibrium solution we may easily compute the value for F_0:

$$F_0(\text{experimental}) = \frac{C_M(\text{by stoichiometry})}{[M](\text{by determination})}.$$

Suppose this is done for many "points" throughout the titration. Each F_0 value will correspond to a certain value of added ligand concentration C_L.

We also have shown in an earlier section that the functional relationship between F_0, the equilibrium quotients, and the free-ligand molarity is:

$$F_0 = 1 + Q_1[L] + Q_1Q_2[L]^2 + Q_1Q_2Q_3[L]^3 + \cdots.$$

If we wish to use this relationship for deduction of the several Q values, it is necessary to have the values for [L] and not merely for C_L. In some cases it is experimentally feasible to use a ligand concentration that is much larger than the metal-ion concentration (in effect, the solution is greatly "overtitrated"). Then it may be quite reasonable to make the approximation that $[L] = C_L$. Otherwise it is necessary either to have a suitable analytical measurement to find [L] or to use an iterative approach with a computer to find [L] from C_L.

Let us assume that we have indeed obtained the values for [L], one for each experimental value of F_0. The first step is to make a plot of F_0 versus [L], as shown in Fig. 15.1. Typically, this plot will be curved upward as befits a well-behaved polynomial function. If it *should* turn out to be a straight line, this is strong evidence that only the first complex has been formed in the solution. This is equivalent to having zero values for Q_2, Q_3, etc., so that the F_0 equation becomes

$$F_0 = 1 + Q_1[L].$$

The intercept of the plot will be 1.00 in any case. For the straight-line plot the slope would be equal to Q_1, and the study would be finished.

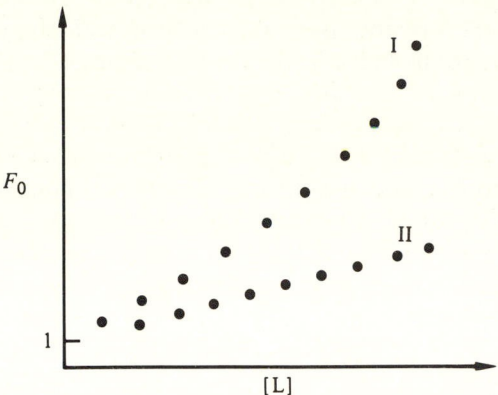

Fig. 15.1 I–typically more than one complex is formed; II–only one complex is formed.

When the plot is curved, the next step is taken. The function F_1 is defined by simple rearrangement of the F_0 function:

$$F_1 = \frac{F_0 - 1}{[L]} = Q_1 + Q_1 Q_2 [L] + Q_1 Q_2 Q_3 [L]^2 + \cdots.$$

This has the effect of decreasing the order of the polynomial by one. A plot of the derived F_1 values versus $[L]$ may be curved or it may be linear. If it *is* a straight line, this is evidence that Q_3, Q_4, etc. are zero (or too small to be determined with the present data), so that the F_1 equation has the form

$$F_1 = Q_1 + Q_1 Q_2 [L].$$

The plot would yield Q_1 as intercept, whether curved *or* straight, and the slope would be $Q_1 Q_2$ for the straight-line case. Even for the curved plot, the limiting slope would be $Q_1 Q_2$.

Once the value of Q_1 has been obtained by extrapolating the F_1 plot to find its intercept at zero ligand concentration, the process continues with the determination of F_2:

$$F_2 = \frac{F_1 - Q_1}{[L]} = Q_1 Q_2 + Q_1 Q_2 Q_3 [L] + Q_1 Q_2 Q_3 Q_4 [L]^2 + \cdots.$$

Again the order of the polynomial has been reduced. A plot of F_2 versus $[L]$ yields β_2 as its intercept and, as the reader doubtlessly has guessed, we define

$$F_3 = \frac{F_2 - \beta_2}{[L]} = \beta_3 + \beta_4 [L].$$

If the system under study involves only four successive complexes, a plot of F_3 versus $[L]$ will give a straight line. The goal of this whole approach is to keep deriving F functions until one of them finally gives a straight-line plot, showing that the polynomial has at last been reduced to a linear equation and that no further analysis is needed.

If the researcher really likes making graphs, then one final test is available: after finding the linear plot, one last F function may be defined:

$$F_4 = \frac{F_3 - \beta_3}{[L]} = \beta_4.$$

If the several F_4 values turn out to be reasonably constant, this is another way of saying that the previous plot was linear. However, this additional work does not contribute much, since a linear F_3 plot is just as informative.

Anyone who understands polynomial least-squares curve-fitting might be a bit outraged at this point. Why not cut out all this graph making and plug the F_0, [L] data into a computer, trying quadratic, cubic, quartic, etc., fits to find which one works best? *Beware of this temptation!* High-order least-squares calculations require intermediate computations using 12 or more digits in the arithmetic, while simple BASIC (for example) uses only seven digits, so that roundoff errors become surprisingly serious. Further, since the experimental data are always complicated by random errors, causing scattering of the points, the parameters (Q values) of the equation are significantly affected when "one grand fit" is attempted. The advantages of the F-function approach are that the errors in individual points become highly visible, that a bit of "sensible subjective extrapolation" is possible, and that each plot can emphasize the points that are most significant for that portion of the "titration."

15.3   CASE STUDY: POTENTIOMETRIC DETERMINATION OF THE FORMATION CONSTANTS FOR THE CADMIUM(II)–BROMIDE SYSTEM*

Recent research by Bond and Hefter illustrates both the feasibility and the limitations of determining the values of a set of successive complex-formation quotients. We will consider the experimental technique, the interpretation of data, and the resulting conclusions about the chemical nature of cadmium(II) in solutions containing bromide ion.

As with many well-designed experiments, the laboratory procedure used by Bond and Hefter is quite simple. An initial solution is prepared to contain $0.975M$ sodium perchlorate and $8.33 \cdot 10^{-3}M$ cadmium perchlorate. The perchlorate ion ClO_4^- is regarded as a noncomplexing ligand, so that the cadmium-ion molarity is taken as precisely $8.33 \cdot 10^{-3}$. The ionic strength of the solution is 1.00. It is important to note that the solution will keep this value of ionic strength throughout the experiment, so that activity coefficients may be regarded as essentially constant.

Two electrodes are placed in the solution: One is a reference electrode of silver/silver chloride connected through a porous junction, and the other is a

---

*After A. M. Bond and G. Hefter (University of Melbourne, Australia), *Journal of Electroanalytical Chemistry*, 68, 203 (1976).

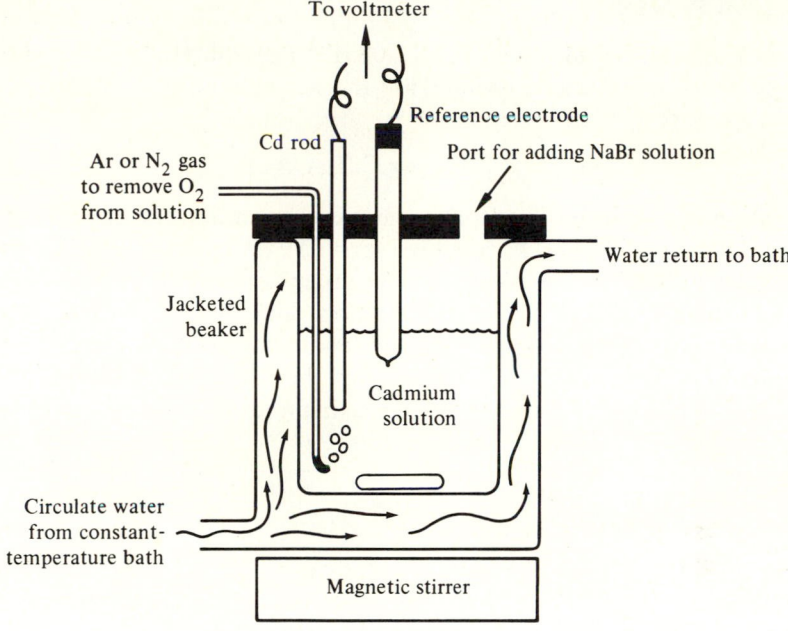

Figure 15.2

rod of pure cadmium metal that has been dipped in mercury to give it a clean amalgamated surface. Provision is made for stirring and the addition of increments of a 1.00M sodium bromide solution, and the experimental setup is as shown in Fig. 15.2.

Analytical Principle

The cadmium metal electrode is a probe that develops an electrical potential depending on the activity of free cadmium ions (the aquo-complex $Cd(H_2O)_4^{2+}$) at its surface. At constant ionic strength the potential measured for the cell can be related to the free molarity of cadmium ions:

$$E(\text{volts}) = a + b \log[Cd^{2+}],$$

where the values of the constants a and b are established for a particular electrode by making several measurements with known concentrations of cadmium ion. The theoretical value for b is 0.0296 at 25°, but this must be checked in practice.

The voltmeter reading therefore gives a continuous indication of the cadmium-ion molarity throughout the course of the experiment. Bond and Hefter reported that these readings were extremely stable and reproducible from day to day.

Equilibrium Principle

With each successive addition of the $1.00M$ NaBr solution, the potential will decrease because the cadmium ions are increasingly converted to a series of bromo-complexes:

$$Cd^{2+} \rightleftarrows CdBr^+ \rightleftarrows CdBr_2 \rightleftarrows etc.??$$

The relationship between the *total* cadmium present and the concentration of the free ion is given by the F_0 function derived earlier in Eq. (14.3). The total, C_M, is known from the original solution composition and whatever dilution has occurred because of the NaBr addition, and [M] is obtained from the potential reading. Thus a series of individual values of F_0 is obtained that corresponds to each equilibrium molarity of the added bromide ion.

These values of F_0 are examined using the F-function approach described earlier.

The Experimental Data

The paper by Bond and Hefter includes a table of a typical set of potentiometric data for the cadmium(II)–bromide system, from which the data in Table 15.1 are taken.

Table 15.1

C_{Cd}(total)	$[Cd^{2+}]_{equil}$	$[Br^-]_{equil}$	F_1
0.00806	0.00387	0.0281	38.5
0.00796	0.00308	0.0393	40.3
0.00757	0.00152	0.0842	47.3
0.00731	0.00103	0.116	52.6
0.00714	0.000810	0.136	57.5
0.00698	0.000665	0.155	61.3
0.00683	0.000557	0.173	65.1
0.00648	0.000368	0.217	76.5
0.00631	0.000300	0.237	84.5
0.00616	0.000257	0.256	89.7
0.00601	0.000219	0.274	96.5
0.00587	0.000187	0.291	104.4
0.00573	0.000161	0.307	112.7
0.00560	0.000143	0.323	118.2
0.00548	0.000128	0.338	123.7
0.00536	0.000116	0.352	128.4
0.00525	0.000102	0.366	137.9
0.00514	0.0000925	0.379	144.0
0.00504	0.0000830	0.392	152.4

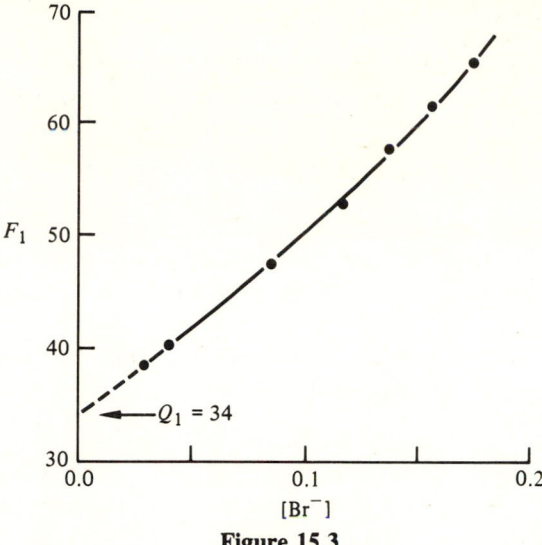

Figure 15.3

Note that when the bromide concentration reaches about 0.4M, less than 2% of the cadmium is present as the aquo-complex. For the plot of F_1 versus [Br⁻] only the first seven points are used (Fig. 15.3). The scatter in the experimental points is not bad. In spite of variability in the way one can choose to extrapolate to the intercept, it seems that the value of $Q_1 = 34$ is reliably determined by this plot.

When the values of $F_2 = (F_1 - 34)/[\text{Br}^-]$ are plotted versus [Br⁻], the effects of experimental errors become more visible, as shown in Fig. 15.4.

In making the extrapolation we ignore the "bad" lower points for a good reason: The accuracy of the F_2 values suffers most when the difference

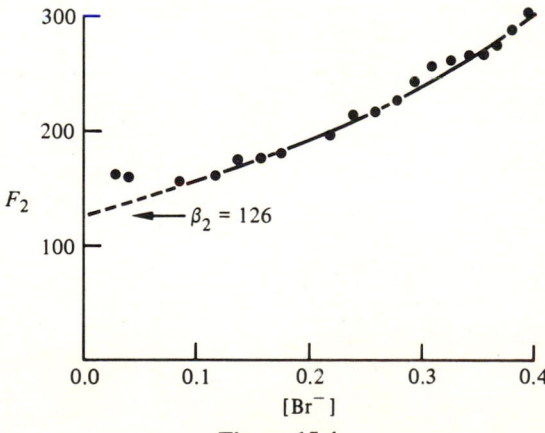

Figure 15.4

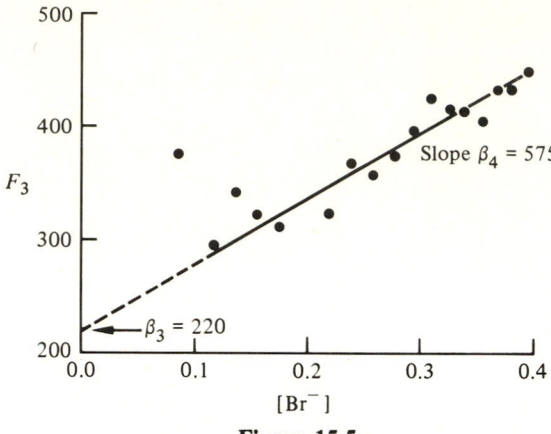

Figure 15.5

$F_1 - 34$ is small, and the two lower F_1 values are only 38.5 and 40.3. Although the intercept is given as 126 in Fig. 15.4 it is obvious that the scatter of the points imposes an imprecision of perhaps 10 or so.

All the data points are used in Fig. 15.4 and, despite the scatter, the curvature seems genuine and not merely an artifact. Therefore it is necessary to go at least one step further, i.e., to make a plot of F_3 versus $[Br^-]$.

Calculation of $F_3 = (F_2 - 126)/[Br^-]$ is equivalent to performing a third differentiation on the original polynomial. The effect of random errors increases with each differentiation step: If the "true" values of F_i are regarded as "signal," while the incoherent errors are "noise," then we may say that the signal-to-noise ratio is deteriorating rapidly. The plot depicted in Fig. 15.5 bears this out.

At this stage the first two data points have "disappeared" from the graph, having meaningless values of 1253 and 885, respectively. Nevertheless, with only a little embarrassment we can accept this plot as linear and draw a straight line by the eyeball method. The results for β_3 (intercept) and β_4 (slope) are 220 and 575, with considerable uncertainties attached.

The quality of this work is typical of what has been achieved to date in the determination of successive formation constants for metal–ligand complexes. It does not compare very well with the high precision that can be obtained in potentiometric studies of acid–base equilibrium systems, where the highly reliable hydrogen electrode can be used. However we may hope and expect that the current technology in designing efficient selective ion electrodes will give chemists the capability of improving the accuracy of low-level metal ion determinations.

The Distribution Diagram: Alpha-Values versus Ligand Concentration

Once the successive equilibrium quotients are determined, they may be used to gain an overall view of the species present in solutions of the metal ion

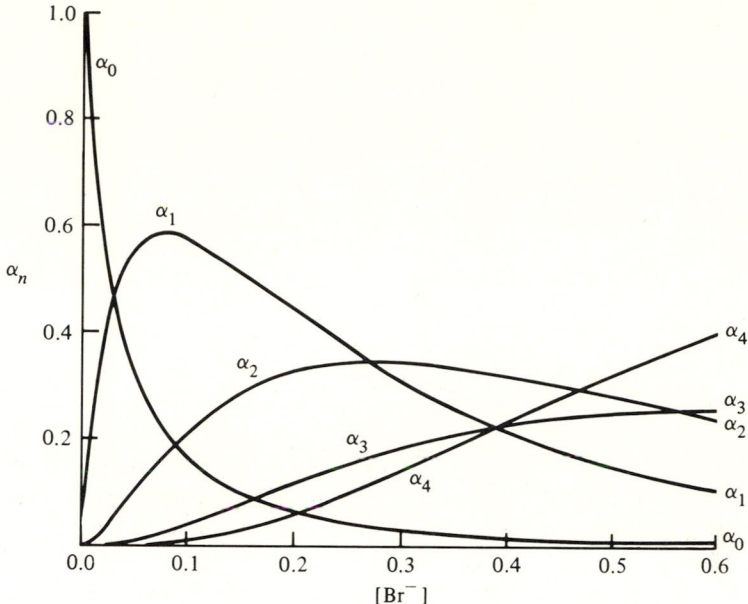

Fig. 15.6 The fraction of cadmium present as $CdBr_n^{2-n}$.

with varying concentrations of the ligand. The way to achieve this is by means of the alpha-expression (14.4) through (14.8). First, values of α_0 are calculated by substituting a series of values for $[Br^-]$ into the expression

$$\alpha_0 = \frac{1}{1 + 34[Br^-] + 126[Br^-]^2 + 220[Br^-]^3 + 575[Br^-]^4}.$$

Then,

$$\alpha_1 = 34[Br^-]\alpha_0, \qquad \alpha_2 = 126[Br^-]^2\alpha_0,$$
$$\alpha_3 = 220[Br^-]^3\alpha_0, \qquad \alpha_4 = 575[Br^-]^4\alpha_0.$$

Note that α_0 is simply multiplied by successive terms of its denominator.

When the values for the ligand concentration cover a wide range of several orders of magnitude, it is best to use a logarithmic scale for the x-axis. In the cadmium(II)–bromide case there is no need for this because complex formation is negligible at very low bromide molarities. Therefore the plot in Fig. 15.6 uses a linear x-axis scale.

Individual Q-Values

The stepwise formation quotients are easily calculated from the experimental beta-values:

$$Q_1 = 34; \qquad\qquad Q_2 = \frac{\beta_2}{Q_1} = \frac{126}{34} = 3.7;$$
$$Q_3 = \frac{\beta_3}{\beta_2} = \frac{220}{126} = 1.8; \qquad Q_4 = \frac{\beta_4}{\beta_3} = \frac{575}{220} = 2.6.$$

These values are relatively small compared, say, to the mercury(II)–bromide complexes, which have Q values of 10^9, $2 \cdot 10^8$, 25, and 20.

In their experimental work, Bond and Hefter did not use bromide-ion concentrations higher than $0.4M$, but the plot has been extended to $0.6M$ on the assumption that, as long as the ionic strength is maintained at 1.0, the values of the equilibrium quotients should hold up.

Note that the molarity of Cd^{2+}, as shown by α_0, drops continuously as the molarity of bromide is increased. However, all the relative concentrations of the intermediate species $CdBr^+$, $CdBr_2$, and $CdBr_3^-$ rise and go through a maximum (not quite reached for the tribromo species at $0.6M$ Br^-), while the highest complex, tetrabromocadmate(II) ion $CdBr_4^{2-}$, eventually becomes the dominant species. But even at the high bromide concentration of 0.6, all four complex species are present in significant amounts. It is this "overlapping" of equilibria that makes this type of system difficult to study with high precision.

It should be possible to make an accurate determination of Q_1 by taking more measurements at $[Br^-]$ in the 0.001 to $0.01M$ range, where there is little contribution of complex species other than $CdBr^+$.

15.4   MAKING USE OF THE BJERRUM FUNCTION $\bar{n}$

This approach was developed and beautifully applied by J. Bjerrum.* It is especially useful in studying systems for which no analytical determination of the free-metal concentration $[M]$ is possible. As with the F-function approach, it *is* necessary to have values for the free ligand concentration $[L]$. This is needed both for the experimental assignment of values to $\bar{n}$:

$$\bar{n} \, (\exp) = \frac{[\text{Bound ligand}]}{C_M} = \frac{C_L - [L]_{det}}{C_M}$$

and for the evaluation of the functional relationship between $\bar{n}$ and the equilibrium quotients (see p. 457):

$$\bar{n} = (1 - \bar{n})Q_1[L] + (2 - \bar{n})Q_1 Q_2 [L]^2 + \cdots.$$

For the calculation of the experimental $\bar{n}$ values it is necessary to obtain accurate values of the difference $C_L - [L]$. This means that, unlike the situation when the F_0 function is to be used, the ligand molarity may not be very large in comparison with that of the metal ion. If it were, then $[L]$ would be practically the same as C_L and the difference calculation would be subject to too much error. In the typical experiment, the titration is carried out in the range of C_L from 0.2 to 5 times C_M so that significant fractions of the ligand become bound to the metal.

---

*J. Bjerrum, *Metal Ammine Formation in Aqueous Solution*. Copenhagen, Haase and Son, 1957.

In certain favorable cases, e.g., when the ligand is fluoride ion, the ligand molarity may be determined by use of an ion-selective electrode. If this is not possible it is necessary to calculate [L] in some way. The classical approach used by Bjerrum, with ammonia as the ligand, is to make use of the acid–base buffer system as follows: The initial metal-ion solution is prepared with a substantial concentration of an ammonium salt, so that the molarity of NH_4^+ is, say, $1.00M$. The titrant solution contains the same concentration of ammonium ion in addition to the L_0-molar ligand (ammonia in this case). Upon addition of increments of the titrant to the metal-ion solution, the ammonium-ion concentration stays constant, and C_L is calculated as shown earlier by simple stoichiometry. Now, some of the ligand reacts with the metal to form a mixture of complexes. *The remaining free ligand concentration is calculated by using a* pH *measurement made in the equilibrium solution.*

Knowing that the solution contains $1.00M$ NH_4^+ regardless of what has happened to the ammonia, the concentration of the remaining free ammonia is obtained from the relationship

$$[NH_3] = Q_a \cdot \frac{[NH_4^+]}{[H^+]},$$

where Q_a is the dissociation quotient for the ammonium ion, determined in separate experiments with solutions of the same general composition, and $[H^+]$ is the hydrogen-ion concentration, which follows from the pH (pcH, really) measurement.

Using this approach we can find fairly accurate values of [L], and we may proceed to a consideration of how to use the $\bar{n}$ function. The technique is quite similar to that already explained for the F functions. The $\bar{n}$ equation is rearranged to give

$$N_1 = \frac{\bar{n}}{(1 - \bar{n})[L]} = Q_1 + Q_1 Q_2 \frac{(2 - \bar{n})[L]}{(1 - \bar{n})} + \cdots.$$

The derived variable N_1 is plotted versus [L] and a value for Q_1 is obtained by extrapolating the plot, which is typically curved, to the intercept. Once Q_1 is thus determined (estimated?), we continue with the definition of N_2:

$$N_2 = \frac{(N_1 - Q_1)(1 - \bar{n})}{Q_1(2 - \bar{n})[L]} = Q_2 + \cdots.$$

Successive plots are made, much in the same manner as discussed for the F functions, until a straight-line function is obtained, signifying that the highest complex present has been identified. These N functions are somewhat more complicated than the F functions, but we must pay the price when it is not analytically possible to include determinations of [M].

Whether the F functions or the N functions are used to study systems of successive complexes, the results for the Q values are typically rather uncertain compared to the precise values which may be obtained for K values for acid–base systems. Studies of complex formations are by necessity often done at high ionic strengths with significantly varying composition, and the analytical procedures for determining [M] and [L] are far less precise than those for measuring pH.

15.5 SPECTROPHOTOMETRIC DETERMINATION
OF THE FORMATION QUOTIENT. THE 1:1 COMPLEX

In Chapter 13 the fundamental equations were derived for the use of spectrophotometry in finding Q values of systems represented by the following model equation:

$$D + X = DX,$$

where D and X are aqueous species, each capable of independent existence, that can associate to form a complex DX. The complex may be weak or fairly stable; i.e., the Q value for the reaction may be small or moderately large.

The important criteria are: (a) there must be a wavelength at which X and DX differ significantly in molar absorptivity; (b) in a series of mixtures of D and X there must be a way to know the equilibrium molarity of D; (c) there must be no other complexes formed.

Using the symbolism of the present chapter, we would recognize that the complex DX would be called ML. Either M *or* L could take the role of D, the other being X. If the criteria listed above are met, we may take advantage of the derived relationship [Eq. (13.9)]:

$$Q = \frac{1}{[D]} \cdot \frac{[DX]}{[X]} = \frac{1}{[D]} \cdot \frac{\epsilon - \epsilon_X}{\epsilon_{DX} - \epsilon}.$$

It was shown that this equation could be used directly if the molar absorptivities of both X and DX were known. This would be possible only if one could prepare a solution containing a *known* concentration of X, so that the measured absorbance would yield ϵ_X by direct application of Beer's Law. It would also be necessary to prepare a solution containing a known concentration of DX, so that the value of ϵ_{DX} would be determined. Then, in a solution that contained a concentration of D such that there was a more or less equal distribution between X and DX, the apparent absorptivity ϵ would be obtained. In conjunction with the known molarity of D in that solution, the three absorptivity values would allow calculation of Q.

In studies of metal–ligand complexes it is usually possible to prepare a solution containing a known concentration of the metal ion by itself, with no ligand present. The absorbance of this solution gives the value of ϵ_M. However, attempts to convert the metal ion "completely" into its complex form ML typically meet with failure for one of the following reasons.

First of all, many metal–ligand complexes are not very stable. That is, they have rather small values of the formation quotient Q. For example, the complexation of copper by chloride has only a weak tendency:

$$Cu^{2+} + Cl^- = CuCl^+, \qquad Q = \frac{[CuCl^+]}{[Cu^{2+}][Cl^-]} \approx 1.3.$$

Suppose that a solution containing $0.001M$ copper is used. Upon addition of a suitable source of chloride (for example, HCl, NaCl, or KCl) there would be some formation of the complex. When the concentration of chloride reached $0.1M$, the ratio of the two species would be only 0.13:

$$\frac{[CuCl^+]}{[Cu^{2+}]} = Q[Cl^-] = 1.3 \cdot 0.1 = 0.13.$$

Even when the chloride concentration is made to be 1-molar, the ratio is only 1.3, meaning that about 60% conversion has been achieved. Going to $10M$ HCl we might predict a ratio of 13. Not only is this still incomplete, but the aqueous medium has changed drastically and has a very high ionic strength. We conclude that it is not feasible to prepare a solution of $CuCl^+$ of known concentration.

15.6   CASE STUDY: THE IRON(III)–THIOCYANATE COMPLEX

The second serious problem is in the successive formation of complexes. For example, suppose we attempt a spectrophotometric determination of the first formation quotient for the iron(III)–thiocyanate system. Considering only the first two possible complexes, we have:

$$\underset{\text{colorless}}{Fe^{3+}} + \underset{\text{colorless}}{SCN^-} = \underset{\text{deep red}}{FeSCN^{2+}}, \qquad Q_1 = \frac{[FeSCN^{2+}]}{[Fe^{3+}][SCN^-]};$$

$$FeSCN^{2+} + SCN^- = \underset{\substack{\text{also} \\ \text{deep red}}}{Fe(SCN)_2^+}, \qquad Q_2 = \frac{[Fe(SCN)_2^+]}{[FeSCN^{2+}][SCN^-]}.$$

At a wavelength of 450 nm the aquo-ferric ion does not absorb appreciably, and so we may set $\epsilon_M = 0$. It is also true that ϵ_L is zero because the thiocyanate ion is colorless. Now, there are two approaches we might use to find the value of ϵ_{ML}, the molar absorptivity of the first complex. By using a high concentration of Fe^{3+} in a solution containing, say, $0.000500M$ SCN^-, we might hope that the reaction for the first step would be driven nearly completely to the right, in which case the concentration of the complex could be taken as 0.000500, and the molar absorptivity would follow from the measured absorbance and Beer's Law.

   The fallacy in this approach is the problem discussed for the copper–chloride system: The value of Q_1 is not particularly large ($Q_1 \approx 125$) although it is much larger than that for $CuCl^+$. With $0.1M$ Fe^{3+}, the ratio of

FeSCN^{2+}/SCN$^-$ is only about 12. It would not be wise to try to use $1M$ Fe^{3+} to achieve a ratio of about 125 (which would be satisfactory) because this would make the ionic strength of the solution quite high, considering that a fairly large concentration of strong acid must also be present to eliminate any significant formation of the hydroxy species FeOH^{2+} that would contribute an unknown amount to the measured absorbance. At high ionic strengths it is often observed that molar absorptivities differ significantly from their values at lower ionic strengths. Therefore, this approach will not do.

The second approach is to use a fairly large concentration of thiocyanate ion in a solution containing, say, 0.000500M iron(III). In this case it would be permissible to use a higher concentration because thiocyanate, being only a singly charged ion, would not boost the ionic strength so high. However, it turns out that step 2 in the complex-formation sequence becomes important well before the conversion of Fe^{3+} to FeSCN^{2+} is complete. This point is worth a fuller discussion and so we will digress to consider a "distribution diagram", and then the problems of the Fe(III)–thiocyanate complex will be taken up again.

Distribution of Successive Complex Species as a Function of Ligand Concentration

In the same way that alpha-functions were used in Chapter 9 to show the dependence of species in polyprotic-acid systems on pH, we may use the alpha-functions for complexes to show the relative amounts of a metal in its successive complex forms. Using the iron(III)–thiocyanate case as an example, we find from the literature that the approximate values for Q_1 and Q_2 are 126 and 20, respectively. When these values are used in the alpha-functions the results are:

$$\alpha_0 = \frac{[\text{Fe}^{3+}]}{C_M} = \frac{1}{1 + 126[\text{SCN}^-] + 2520[\text{SCN}^-]^2};$$

$$\alpha_1 = \frac{[\text{FeSCN}^{2+}]}{C_M} = \frac{126[\text{SCN}^-]}{\text{Same denominator, which is } F_0};$$

$$\alpha_2 = \frac{[\text{Fe(SCN)}_2^+]}{C_M} = \frac{2520[\text{SCN}^-]^2}{\text{Same denominator}}.$$

With the aid of a programmable calculator these equations are readily solved for a series of ligand concentrations. Table 15.2 shows the results for a thiocyanate concentration range from 0.0001 to 0.1 molar.

Inspection of Table 15.2 shows that 10% of the iron is already in the form of the second complex when the thiocyanate concentration is 0.01, at which point only 50% of the iron has been converted to the 1:1 complex. In fact, the *maximum* relative concentration of the 1:1 complex is only about 56% at [SCN$^-$] = 0.02. These results are shown very clearly in Fig. 15.7 as a plot of alpha-values versus the logarithm of the ligand concentration.

Table 15.2

[SCN⁻]	log[SCN⁻]	α_0	α_1	α_2
0.0001	−4.00	0.988	0.012	0.000
0.0002	−3.70	0.975	0.025	0.000
0.0005	−3.30	0.940	0.059	0.001
0.001	−3.00	0.886	0.112	0.002
0.002	−2.70	0.792	0.200	0.008
0.005	−2.30	0.591	0.372	0.037
0.01	−2.00	0.398	0.502	0.100
0.02	−1.70	0.221	0.557	0.223
0.05	−1.30	0.074	0.463	0.463
0.1	−1.00	0.026	0.325	0.649

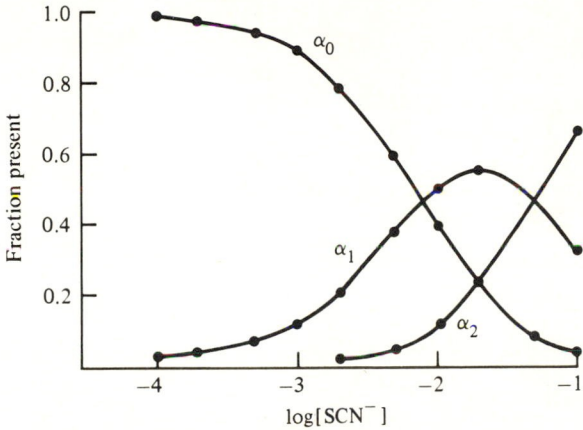

Figure 15.7

Determining Q_1 in Spite of the Problem

From the distribution diagram we note that it is not possible to make a direct determination of the molar absorptivity for the 1:1 complex because we cannot make a solution that contains a precisely known concentration of this species. We further note that it is only when the thiocyanate molarity is kept lower than about 0.001, that is log[SCN⁻] < −3, that the formation of the second complex is truly negligible.

This system was successfully treated by Frank and Oswalt.* These authors used a series of solutions, each containing 0.50M perchloric acid (to

*H. Frank, R. Oswalt, *Journal of American Chemical Society*, **69**, 1321 (1947).

repress formation of $FeOH^{2+}$), $0.000300M$ potassium thiocyanate, and concentrations of ferric perchlorate that ranged from 0.001 to 0.008. The absorbance values were measured at 450 nm, using a cell with the odd path of 1.305 cm. The results they obtained are shown in Table 15.3, where ϵ is the apparent absorptivity (see Chapter 13).

Table 15.3

C_{SCN}	C_{Fe}	A	$\epsilon = \dfrac{A}{1.305 \cdot 0.0003}$
0.0003	0.001	0.1617	413
0.0003	0.002	0.2845	727
0.0003	0.003	0.3900	996
0.0003	0.005	0.5529	1412
0.0003	0.008	0.7163	1830

Knowing that the conversion to $FeSCN^{2+}$ must be quite incomplete even in the solution with the highest iron concentration, we may treat the data according to the graphical method using Eq. 13.11:

$$\epsilon = \epsilon_{ML} - \frac{1}{Q_1} \cdot \frac{(\epsilon - \epsilon_L)}{[Fe^{3+}]}.$$

A plot of ϵ versus $(\epsilon - \epsilon_L)/[Fe^{3+}]$ should be linear and should yield a value of Q_1 as the negative reciprocal of the slope. Since the thiocyanate ion is colorless, $\epsilon_L = 0$ and we merely use $\epsilon/[Fe^{3+}]$ for the variable.

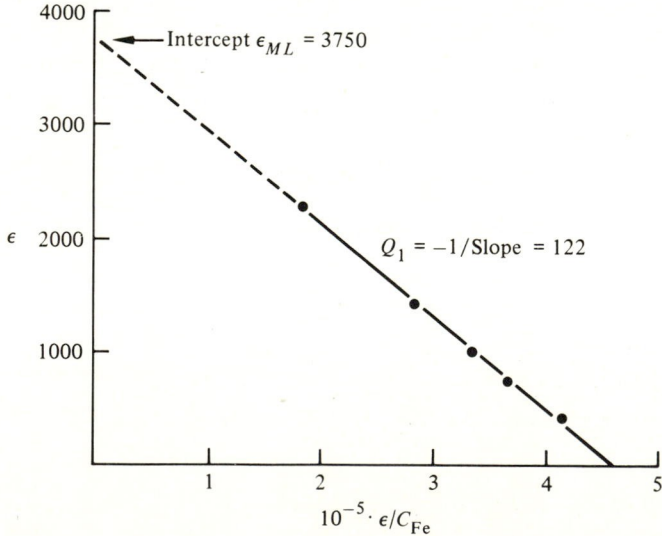

Figure 15.8

An immediate problem arises because Eq. (13.11) requires that the equilibrium values for $[Fe^{3+}]$ be used, and all we have in the data table are the total concentrations of iron, part of which is present as the complex. For a first approximation we will neglect the decrease in $[Fe^{3+}]$ caused by the relatively small concentration of $FeSCN^{2+}$, and make the plot as shown in Fig. 15.8.

Once the value for the molar absorptivity of the complex has been determined it is feasible to make a correction in the $[Fe^{3+}]$ values to account for the decrease due to the complex formation. One uses Eq. (13.6) to find $[FeSCN^{2+}]$ and subtracts this from C_{Fe}. A new plot may then be made to obtain a better estimate of both ϵ_{ML} and Q. The best procedure is to use a simple computer program to carry out the iteration until the results converge to a nearly constant value.

This case study illustrates the typical situation for the study of a $1:1$ complex. Spectrophotometric measurements may also be used in some cases to find values for Q_2, etc., when successive complexes are formed.*

15.7　THE MERCURY ELECTRODE: A PROBE FOR METAL–CHELON EQUILIBRIA

When a metal ion, for example, Cu^{2+}, reacts with a chelon, the reaction typically goes virtually to completion because of the large value of the equilibrium quotient. For this reason it is not feasible to apply the spectrophotometric approach to determining the value for Q; there is not enough reactant at equilibrium to give an appreciable absorbance. An ingenious and fruitful solution to this problem was developed by Schmid and Reilley† who used the potential of a mercury electrode as an indirect sensor for the equilibrium state of the metal–chelon system. Fortunately for this method, the mercuric ion Hg^{2+} typically forms much more stable chelates than do the other metal ions. The success of the method depends on the modest but measurable ability of a metal ion to displace mercuric ion from its chelate:

$$Cu^{2+} + HgY \rightleftarrows CuY + Hg^{2+}.$$

The equilibrium quotient for this displacement reaction is

$$Q = \frac{Q_{CuY}}{Q_{HgY}},$$

where Q_{MY} is the formation quotient for the reaction $M + Y = MY$.

By establishing a solution that contains known molarities of Cu^{2+}, CuY, and HgY, and then using the mercury electrode to sense the (very small)

---

*The interested reader should see L. Newman and D. Hume, *Journal of American Chemical Society*, *79*, 4571 (1957), and W. McBryde, *Talanta*, *21*, 979–1004 (1974).
†R. Schmid, C. Reilley, *Journal of American Chemical Society*, *78*, 5513 (1956).

molarity of Hg^{2+}, we may find the value for Q. Then, if Q_{HgY} is known, we may calculate the value of Q_{CuY}. Therefore, the first matter to be discussed is the use of the mercury electrode to study the mercury–chelon system by itself, for the purpose of finding a value for Q_{HgY}.

Imagine a solution that initially contains mercuric ions at a molarity of 0.00100. Let enough EDTA (or some other chelating agent) be added so that its concentration is $0.00200M$. The mercuric ions become chelated in accordance with the equilibrium model:

$$Hg^{2+} + Y^{4-} \overset{Q_f}{=} HgY^{2-}$$
$$+$$
$$H^+ = HY^{3-} = H_2Y^{2-} = H_3Y^- = H_4Y.$$

In spite of the competition from hydrogen ion, the chelation reaction proceeds very nearly to completion at moderate pH values. Therefore, the equilibrium molarities are:

$$[HgY^{2-}] = 0.00100, \qquad C_Y = 0.00100$$

i.e., virtually all mercury is chelated, and C_Y is the total concentration of the chelon in all of its protonated forms.

The equilibrium molarity of the deprotonated ligand Y^{4-} is highly dependent on the solution pcH, in accordance with the alpha-expression:

$$\frac{[Y^{4-}]}{C_Y} = \frac{Q_1Q_2Q_3Q_4}{[H^+]^4 + Q_1[H^+]^3 + Q_1Q_2[H^+]^2 + Q_1Q_2Q_3[H^+] + Q_1Q_2Q_3Q_4} = \alpha.$$

The equilibrium concentration of mercuric ion must agree with the formation quotient for the mercury–chelon complex:

$$[Hg^{2+}] = \frac{[HgY^{2-}]}{Q_{HgY}[Y^{4-}]} = \frac{[HgY^{2-}]}{Q_{HgY}\alpha C_Y}. \tag{15.1}$$

At higher pH values, the mercuric-ion concentration becomes remarkably small, e.g., 10^{-20} molar.

Now, suppose a mercury metal electrode is placed in contact with this solution. The Hg–Hg_2^{2+}–Hg^{2+} equilibrium system will be quickly established at the electrode surface, but the quantity of mercuric ion that must be reduced to satisfy this equilibrium will be very small indeed since there is such a low molarity of free Hg^{2+}. The chelate HgY^{2-} will dissociate to replace the "lost" mercuric ion, but the molarity of HgY^{2-} will be virtually unchanged by this process.

We may write the Nernst equation in terms of an electrode equilibrium with mercuric ion:

$$E = E^0_{Hg^{2+},Hg} - 0.0296 \log \frac{1}{[Hg^{2+}]f_2}.$$

In the following discussion we will assume that the ionic strength of the solutions used is constant at 0.1. Then the value of f_2 is about 0.37, and since the value for E^0 is +0.854 (relative to SHE), or +0.610 (relative to SCE), the potential for the cell

$$\text{SCE} \,|\, \text{Solution of Hg}^{2+} \text{ at } I = 0.1 \,|\, \text{Hg}$$

is given by:

$$E = 0.597 - 0.0296 \log \frac{1}{[\text{Hg}^{2+}]}. \tag{15.2}$$

This equation is the basic relationship for the applications that follow. As for the solution described above ($0.001\,M$ Hg^{2+} plus $0.002M$ chelon), we note the $[\text{Hg}^{2+}]$ given by Eq. (15.1), and substitute into (15.2):

$$E = 0.597 - 0.0296 \log \alpha - 0.0296 \log \frac{C_Y}{[\text{HgY}^{2-}]} - 0.0296 \log Q_{\text{HgY}}. \tag{15.3}$$

Except at very low pH values, where the hydrogen ions can compete with some success for the chelon, the ratio $C_Y/[\text{HgY}^{2-}]$ is

$$\frac{C_Y}{[\text{HgY}^{2-}]} = \frac{0.001}{0.001} = 1.00,$$

so the term containing it vanishes.

Since Q_{HgY} is a constant, we note that the potential of the cell will vary only with alpha, that is, with pH, since it is only pH that makes alpha vary. Given a numerical value for the formation quotient Q_{HgY} and the successive acid-dissociation quotients for the chelon, it is a simple matter to write a brief BASIC computer program to predict the values of potential over a range of pH values.

When data of this sort are determined experimentally, they may be used to find the value of the formation quotient for the mercury chelonate. Equation (15.3) may be solved for $\log Q_{\text{HgY}}$:

$$\log Q_{\text{HgY}} = \frac{0.597 - E}{0.0296} - \log \alpha, \tag{15.4}$$

assuming that $C_Y = [\text{HgY}^{2-}]$.

Clearly, it is necessary to know the value for alpha before solving this equation. Two possibilities exist: One is to calculate the value for alpha, given the pH of the solution and the values of the acid-dissociation quotients for the chelon. The second approach is much easier when it can be used. One simply adjusts the pH of the solution to values sufficiently high to fully deprotonate the chelon, so that $\alpha = 1.00$.

This method is possible if the last acid-dissociation step for the chelon is not *too* weak, for it is necessary to have the pH about 1.5–2 units higher than the last pQ_a value for the dissociation to be essentially complete.

Determination of Metal–Chelon Formation Quotients

The previous discussion was based on a solution containing a known concentration of mercury chelonate HgY^{2-} and a known excess of chelon C_Y. Now suppose that a known concentration of another metal ion, say Cu^{2+}, is added to the solution. Let this concentration be in excess of that needed to react with the chelon that is not already tied to mercury. For example, the solution might contain the following:

$$[HgY^{2-}] = [CuY^{2-}] = [Cu^{2+}] = 0.00100M.$$

We may write the formation quotient for the copper chelonate:

$$Q_{CuY} = \frac{[CuY^{2-}]}{[Cu^{2+}][Y^{4-}]} = \frac{[CuY^{2-}]}{[Cu^{2+}]C_Y\alpha},$$

and therefore

$$C_Y\alpha = \frac{[CuY^{2-}]}{[Cu^{2+}]Q_{CuY}}. \tag{15.5}$$

When this expression for $C_Y\alpha$ is substituted into Eq. (15.3), the modified Nernst equation becomes:

$$E = 0.597 - 0.0296 \log \frac{[CuY^{2-}]}{[Cu^{2+}][HgY^{2-}]} - 0.0296 \log Q_{HgY} + 0.0296 \log Q_{CuY}. \tag{15.6}$$

Now, if Q_{HgY} has been determined by measurements on the cell without any copper present, using Eq. (15.4), then we may use Eq. (15.6) to find the value for Q_{CuY}, provided that the three molarities are known. They *will* be known within the pH region where the copper and chelon react virtually completely, because they will each be $0.00100M$. Therefore, we have merely to find a pH region in which the values of E are constant.

These ideas will be made clearer by the following calculations, based on measurements for a cell involving the chelon EGTA.

15.8 CASE STUDY: DETERMINATION OF METAL–CHELATE STABILITY CONSTANTS BY USING A MERCURY ELECTRODE*

Holloway and Reilley studied several chelons (aminopolycarboxylate ligands) including EGTA:

$$\begin{array}{l} HOOCCH_2 \\ {>}N-CH_2-CH_2-O-CH_2-CH_2-O-CH_2-CH_2-N{<} \\ HOOCCH_2 \end{array} \begin{array}{l} CH_2COOH \\ \\ CH_2COOH \end{array}$$

(ethyleneglycol bis(β-aminoethylether)-N,N'tetraacetic acid)

The results for this chelon are shown in Fig. 15.9.

---

*After J. H. Holloway, C. N. Reilley, *Analytical Chemistry*, 32, 249 (1960).

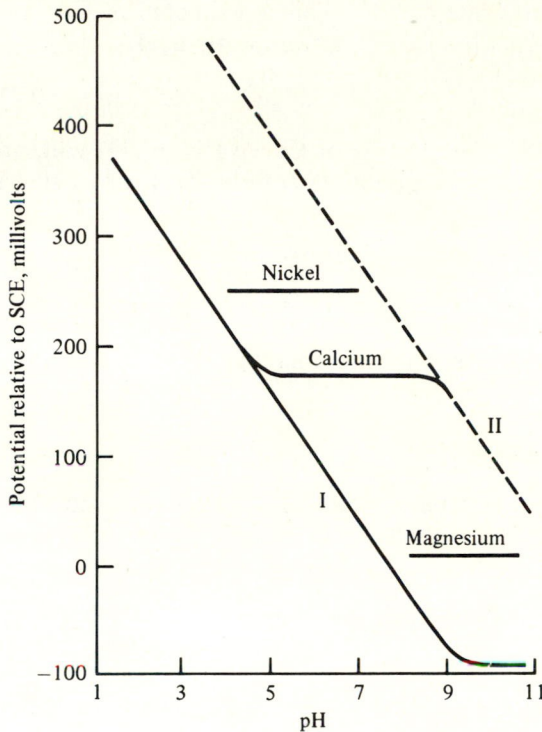

Fig. 15.9 Potential of the Hg electrode relative to SCE as a function of pH.

Curve I

The lowest line in Fig. 15.9 shows the potential of the mercury electrode (relative to that of the saturated calomel electrode as reference) in a solution containing the following:

$$\text{Mercuric perchlorate} \quad 0.0100M,$$
$$\text{EGTA} \qquad\qquad\qquad 0.0200M,$$

where the pH varied from 2 to 11. The Hg–EGTA chelate is so stable that its formation is nearly complete even in the acidic solutions. We may assume that the following equilibrium concentrations exist over the entire pH range:

$$\text{Hg–EGTA chelate} \quad 0.0100M,$$
$$\text{Excess EGTA} \qquad\quad 0.0100M.$$

Equations (15.3) and (15.4) are applicable to this situation, with $C_Y = [\text{HgY}^{2-}] = 0.01$. The decrease in potential as the pH is increased from 2 to about 9 is a consequence of the $\log \alpha$ term in Eq. (15.3). As the pH is increased still further, the excess EGTA is less and less protonated, and the observation of a constant potential at pH 10–11 tells us that alpha must have reached its limiting value of 1.00. This is expected in view of the successive dissociation quotients for EGTA, a tetraprotic acid:

$$pQ_1 = 2.0, \quad pQ_2 = 2.68, \quad pQ_3 = 8.85, \quad pQ_4 = 9.43.$$

The potential at the level portion of Curve I is -0.091 volt, and therefore we may calculate the formation quotient for Hg–EGTA by using Eq. (15.4):

$$\log Q_{\text{HgY}} = \frac{0.597 - (-0.091)}{0.0296} = 23.2,$$

$$Q_{\text{HgY}} = 10^{23.2} = 1.6 \cdot 10^{23}.$$

This large value shows that the mercury–EGTA complex is very stable indeed.

The Metal-Ion Curves

The curves in Fig. 15.9 labeled "Nickel," "Calcium," and "Magnesium" show the electrode potentials obtained in solutions that nominally contained the following (using M^{2+} as a general symbol for the metal ions):

$$\begin{array}{ll}
\text{mercuric perchlorate} & 0.0100M, \\
M^{2+} & 0.0200M, \\
\text{EGTA} & 0.0200M.
\end{array}$$

We have already seen that the mercury–EGTA complex is so stable that it will form quantitatively. Therefore, $[\text{HgY}^{2-}] = 0.0100M$, leaving $0.0100M$ EGTA for reaction with the other metal ion M^{2+}. Since these other metals do not form complexes nearly so stable as that of mercury, a higher pH is required before chelation is quantitative, thus making the following concentrations:

$$[\text{MY}^{2-}] = 0.0100M,$$

$$\text{Excess } [M^{2+}] = 0.0100M,$$

$$[\text{HgY}^{2-}] = 0.0100M.$$

It is useful to view this system as a "buffer" which controls the concentration of the very small equilibrium concentration of free EGTA by way of Eq. (15.5). The free EGTA in turn controls the *very* slight dissociation of the mercury–EGTA complex into mercuric ions and free EGTA, and hence also controls the potential of the electrode, which depends only on the concentration of free Hg^{2+}.

In the portion of the curve that is level, it must be true that the concentrations of MY^{2-} and M^{2+} are equal to 0.0100 because, according to Eq. (15.6), we should be able to observe a constant potential only when the concentration term is constant, and this can be true only in the pH region where the reactions are quantitative.

For the calcium curve, the potential at the level section is $+0.171$ volt. We may now calculate the formation quotient for the Ca–EGTA chelate from Eq. (15.6):

$$0.171 = 0.597 - 0.0296 \log \frac{0.01}{0.01 \cdot 0.01} - 0.0296 \cdot 23.2 + 0.0296 \log Q_{\text{CaY}}.$$

Solution of this relationship gives $Q_{\text{CaY}} = 6 \cdot 10^{10}$, showing that this chelate is fairly stable.

The flat portions of the nickel and magnesium curves are at $+0.249$ and $+0.009$ volt, respectively. Calculate the formation quotients for the two chelates. *Answer* $\log Q = 13.4$ and 5.3.

Curve II

The dashed line in Fig. 15.9 shows an upper limit to the potential that can be observed at each value of pH. It is due to the insolubility of mercuric oxide:

$$HgO(s) + H_2O = Hg^{2+} + 2OH^-,$$
$$Q_{\text{sp}} = [Hg^{2+}][OH^-]^2 = 3 \cdot 10^{-26}. \tag{15.7}$$

During the measurements on curve I, the excess EGTA represses the dissociation of HgY^{2-}, and the equilibrium values of the mercuric-ion activity are unbelievably small. For example, at $pH = 11$, when $E = -0.091$ volt, we find from Eq. (15.2) that the concentration of mercuric ion is only $6 \cdot 10^{-24}$! No HgO can precipitate, according to Eq. (15.7) because at $pH = 11$, $[OH] = 1 \cdot 10^{-3}$, and the solubility product is not exceeded.

However, with calcium (or the other metals) present, the mercuric-ion molarity is much larger because the metal ion is trying to compete with the mercury for the EGTA. In the calcium case, it is at about $pH = 9$ that the HgO begins to precipitate. At all pH values higher than this, the electrode potential is controlled by the mercuric-oxide equilibrium:

$$[Hg^{2+}] = \frac{Q_{\text{sp}}}{[OH^-]^2} = \frac{Q_{\text{sp}}[H^+]^2}{Q_w^2}. \tag{15.8}$$

When Eq. (15.8) is combined with Eq. (15.4), the result is:

$$E = 0.597 - 0.0296 \log \frac{Q_w^2}{Q_{\text{sp}}} - 0.0592 \, \text{pcH}. \tag{15.9}$$

The potential is a linear function of pcH, with a slope of -59.2 mvolts per pcH unit.

Final comment   The use of the mercury electrode for the determination of metal–chelate formation constants has proved exceptionally valuable. These equilibrium constants are very difficult to determine by other techniques. Also, potential–pH diagrams like that shown in Fig. 15.9 are valuable guides for choosing conditions for potentiometric titrations of metal ions with chelons, using the mercury electrode as an indicator electrode.

PROBLEMS

1. In an attempt to confirm the listed value for β_2 for the mercury(II)–ammonia complexes, the following cell was set up and the potential was +0.150 volt. Assuming that the diamminemercury(II) ion is the only significant species, calculate the concentration of Hg^{2+} in the solution and the value for β_2. Note that it is necessary to take account of the decrease in ammonia concentration due to the formation of the complex:

$$SCE \parallel Hg(NO_3)_2(0.00100M), NH_3(0.00500M) \mid Hg.$$

By using the known values for Q_1 and Q_3 (see Appendix 5), show that the assumption that only one complex is present is a good one.

2. If the lead electrodes behaved "properly" (i.e., the Nernst equation was applicable), what would you expect for the potential of the following cell?

$$Pb \mid Pb^{2+}(0.01M) \left\| \begin{array}{c} Pb(NO_3)_2\,(0.01M) \\ Na_3citrate\,(0.1M) \end{array} \right\| Pb$$

3. Suppose the following F_0 values were determined for a metal ion–ligand system at the indicated molarities of uncomplexed ligand. Use the F-function method to find how many complexes exist, and the values of the formation quotients:

F_0:	1.10	1.21	1.58	2.30	4.20	13.52	41.1	141	800	3100
[L]:	0.001	0.002	0.005	0.01	0.02	0.05	0.1	0.2	0.5	1.0

4. The potential of the following cell is −0.017 volt. Assuming that the only complex is the $Ag(en)_2^+$ species, calculate the value of β_2:

$$SCE \parallel AgNO_3(0.0001M), \text{ethylenediamine (en)}(0.1M) \mid Ag.$$

5. *Potentiometric determination of stability constants of metal chelates.*   Figure 15.10 is a potential–pH diagram obtained in the study of a certain chelon, using the mercury electrode versus the saturated calomel electrode. The temperature was 25°C and the ionic strength was held constant with $0.1M$ KNO_3. Cell diagram: $SCE \mid$ Solution $\mid$ Hg(liquid).

Curve I shows the effect of pH on the potential of the electrode in an unbuffered solution containing $0.001M$ mercuric nitrate and $0.002M$ chelon. Some of the data for Curve I are given in Table 15.3.

Table 15.3

pH	E
2.0	0.440
2.65	0.400
4.0	0.277
6.0	0.085
8.0	−0.088
9.0	−0.156
10.0	−0.193
11.0	−0.205
12.0	−0.205

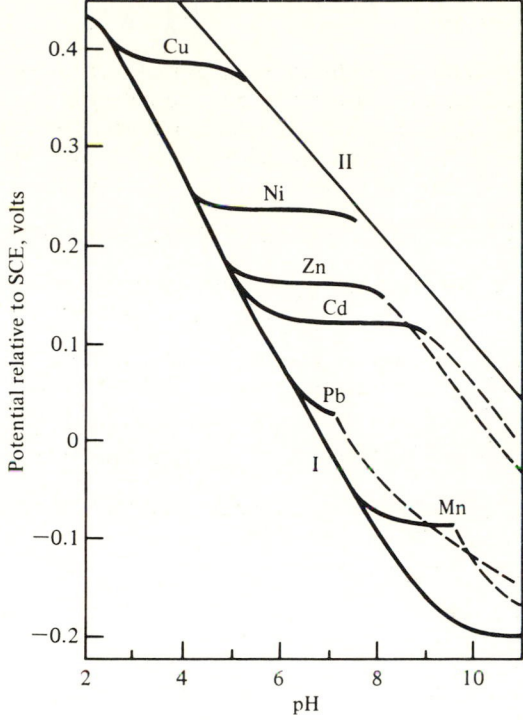

Figure 15.10

Curve II shows the maximum potentials observable with the mercury electrode as a result of the formation of mercuric oxide.

The horizontal curves were determined by varying the pH of solutions that contained $0.001M$ mercuric nitrate, $0.002M$ chelon, and $0.002M$ metal nitrate. The dotted portions show the effect of precipitation of metal hydroxide.

a) Calculate the stability constant for the mercuric chelate. Assume that the standard reduction potential for the mercuric ion/mercury electrode is $+0.612$ volt relative to SCE and that the mercuric chelate is rather stable. *Hint.* Note that the potential levels off at high pH values. What does this mean about the value of α at such pH?

b) Calculate α at each pH and plot log α versus pH. For each pH calculate the molarity of the fully deprotonated ligand.

c) Calculate the logarithms of the stability constants of the copper, nickel, and cadmium chelates, given that the pH-independent portions of the appropriate curves are at potentials of 0.380, 0.230, and 0.125 volt, respectively.

d) Estimate the stability constant of the lead chelate.

METAL–LIGAND COMPLEXES: ANALYTICAL APPLICATIONS

16

It is a capital mistake to theorize before one has data. Insensibly one begins to twist facts to suit theories instead of theories to suit facts.

Sherlock Holmes

16.1   MASKING OF METAL IONS

Consider the reaction of a metal ion M with some analytical reagent R:

$$M + R \rightarrow \text{Products.}$$

In some situations this may be a highly desired reaction but it may be important to prevent its occurrence at other times. For example, in a mixture of two metals M and N, it may be useful to allow only N to react with R. We speak of **masking** metal M against reaction with R, as if to suggest that we use some chemical trick to put M in disguise so that R won't recognize it. Some techniques for achieving this are as follows.

Change in Oxidation State of M

Different oxidation states of metals usually show very different chemical properties and often colors. For example, chromium(III) readily forms an insoluble hydroxide $Cr(OH)_3$, as iron(III) does with $Fe(OH)_3$. If a solution contains a mixture of these metals both hydroxides will form upon raising the pH. If it is desired to prevent the interference of chromic hydroxide while allowing precipitation of ferric hydroxide, the basic mixture may simply be boiled with some hydrogen peroxide present:

$$Cr(OH)_3(s) + H_2O_2 \rightarrow CrO_4^{2-},$$
$$Fe(OH)_3(s) + H_2O_2 \rightarrow \text{No reaction.}$$

By changing the oxidation state of the chromium to $+6$ we can dissolve the chromic hydroxide. The chromium stays in solution while the iron may be collected in the precipitate.

    Both silver(I) and mercury(I) form chloride precipitates, but the mercury may be masked by oxidation to the mercury(II) state, since mercuric chloride is fairly soluble. Silver ion stays in the $+1$ state and precipitates with chloride.

Selective Ligand Complexing of M

This is the most important technique for masking. By using an excess of ligand L, which forms a stable complex with M, we establish a competition between R and L. If the M–L complex is sufficiently stable, the equilibrium concentration of M (uncomplexed) is reduced to very low levels and the reaction with R becomes unfavorable.

    For example, the precipitation of silver as silver chloride has a very favorable equilibrium constant. However, addition of excess ammonia prevents the precipitation from occurring because the silver–ammonia complex is even more stable than silver chloride (at least, when about $3M$ NH_3 is present):

$$Ag^+ + Cl^- = AgCl(s)$$
$$+$$
$$2NH_3 = Ag(NH_3)_2^+$$

Another example. When potassium iodide is added to a solution containing copper(II), a redox reaction occurs:

$$2Cu^{2+} + 5I^- = 2CuI(s) + I_3^-.$$

Cuprous iodide precipitates and the solution becomes colored by the triiodide ion. This may be used to advantage in the determination of copper: the triiodide ion is simply titrated with a standard solution of sodium thiosulfate. However, suppose it was desired to prevent the reaction, perhaps because the solution contained some *other* oxidizing agent to be determined and the copper reaction would thus be an interference. One approach is to add some EDTA to the solution. This chelon forms such a stable complex with copper:

$$Cu^{2+} + Y^{4-} = CuY^{2-}, \qquad K = 10^{18},$$

that reaction with iodide ion is eliminated. The other oxidizing agent(s) in solution, if not affected by EDTA, are free to react, produce triiodide ion, etc.

Selective Precipitation of M

In some cases the metal M may be eliminated as a possible reactant by simple precipitation as an insoluble salt. If the solubility product is quite small, it may not even be necessary to filter the precipitate. This is not much different in principle from the formation of the stable soluble complex.

In a mixture of magnesium, calcium, and barium ions, all three will react with added EDTA in a titration. The barium may be masked by precipitation as the sulfate, and/or the magnesium may be masked at high pH through precipitation as the hydroxide.*

16.2   THE EFFECT OF METAL–LIGAND COMPLEXING ON STANDARD REDUCTION POTENTIALS

The oxidation–reduction tendencies shown by an element in its various oxidation states are summarized in the reduction-potential diagram. Thus, for iron we have:

$$Fe^{3+} \xrightarrow{\ +0.77\ } Fe^{2+} \xrightarrow{\ -0.44\ } Fe.$$

We conclude that ferric ion has a moderately strong tendency to undergo reduction to ferrous ion since E^0 is positive, while it would take a rather strong reducing agent to convert ferrous ion to iron metal. These E^0 values and the conclusions just drawn refer only to acidic solutions containing the aquo-ions, with no ligands that can form complexes.

The reduction potential for iron(III)–iron(II) may be changed greatly when one or the other of the ions forms a more stable ligand complex than the

---

*A thorough discussion may be found in the book by D. D. Perrin, *Masking and Demasking of Chemical Reactions*, Interscience, 1970.

other. If the formation quotients for the complexes are known, it is a simple matter to calculate the effect. For example, both oxidation states of iron can form complexes with EDTA, although the iron(III) complex is quite a bit more stable, as shown by the following quotients:

$$Fe^{3+} + \quad Y^{4-} \quad = FeY^{-}, \quad \log Q = +25.1,$$
$$\text{(EDTA anion)}$$
$$Fe^{2+} + \quad Y^{4-} \quad = FeY^{2-}, \quad \log Q = +14.4.$$

We may add these reactions to the Fe^{3+}, Fe^{2+} half-reaction, and simultaneously add the nE^0 values:

$$\begin{array}{lll}
 & & \underline{\log Q \quad nE^0 = 0.059 \log Q} \\
 & \cancel{Fe^{3+}} + e = \cancel{Fe^{2+}} & +0.77 \\
 & FeY^{-} = \cancel{Fe^{3+}} + \cancel{Y^{4-}} & -25.1 \rightarrow -1.48 \\
 & \cancel{Fe^{2+}} + \cancel{Y^{4-}} = FeY^{2-} & +14.4 \rightarrow +0.85 \\
\text{Sum:} & FeY^{-} + e = FeY^{2-} & +0.14 = E^0 \text{ since } n = 1.
\end{array}$$

Similarly the new E^0 value for Fe(II), Fe(0) may be calculated:

$$\begin{array}{lll}
 & \cancel{Fe^{2+}} + 2e = Fe & -0.88 \\
 & FeY^{2-} = \cancel{Fe^{2+}} + Y^{4-} & -14.4 \rightarrow -0.85 \\
\text{Sum:} & FeY^{2-} + 2e = Fe + Y^{4-} & -1.73 \quad n = 2, \text{ so } E^0 = -0.87.
\end{array}$$

The reduction-potential diagram written in terms of the EDTA complexes of iron therefore becomes:

$$FeY^{-} \xrightarrow{\ +0.14\ } FeY^{2-} \xrightarrow{\ -0.87\ } Fe.$$

Conclusions are that it is even harder to reduce iron(II) to iron metal in the presence of EDTA and that iron(III) has much less tendency to be reduced to iron(II). Both oxidation states have been stabilized by the complexation, the III-state much more than the II-state.

Clearly these ideas suggest that the whole gamut of redox reactions we take for granted in ordinary aqueous solutions is subject to important variations when complexing agents are present. For example, we know that in noncomplexing acidic solutions, potassium iodide will reduce ferric ion, producing triiodide ion. (E^0_{red} for the I^-, I_3^- couple is $+0.54$, compared to E^0_{red} for the iron couple of $+0.77$). However, in the presence of EDTA the iodine couple is unaffected, while the iron couple drops to $E^0 = +0.14$. Iron(III) is *not* reduced by iodide when it is present as an EDTA complex. Of course this is another example of masking.

The change and control of reduction potentials of metal ions through ligand complexing is an important tool in the hands of the informed chemist for it can make possible, or impossible, many valuable reactions.

16.3 SPECTROPHOTOMETRIC DETERMINATION OF METALS, USING COMPLEXES

In many instances the formation of a metal–ligand complex is accompanied by a striking development of color. One example discussed previously is the effect of adding thiocyanate ion in excess to a solution containing a small concentration of iron. The Fe–SCN complexes are deep red and may be used as the basis of a fairly sensitive method for determining small quantities of iron. Similarly, the pale blue cupric ion Cu^{2+} is readily converted to the deep blue complex $Cu(NH_3)_4^{2+}$ by the addition of an excess of ammonia.

Let us consider some criteria for the ideal application of metal–ligand complex formation for spectrophotometric determination of the metal:

1. The metal–ligand complex should be rather stable, so that only a small excess concentration of ligand is necessary to effect complete formation.

2. The complex should have a high molar absorptivity at some wavelength in the ultraviolet–visible range, and other solution species should not absorb.

3. The ligand should react only with the metal to be determined and not with whatever other metals might be present in the same solution. As an alternative it should be possible to prevent the other metals from reacting with the ligand by masking.

4. The system should obey Beer's Law, so that a straight line is obtained for the standard curve.

5. Once formed, the complex (and its absorbance) should be stable with time.

An enormous amount of work has been devoted to the search for excellent spectrophotometric procedures and for color-forming ligands which are highly selective, even specific for individual metals. The majority of the best ligands form stable chelates with the metal ion. The development of such ligands is one of the outstanding chapters in fundamental analytical research in combination with imaginative organic syntheses. It has proved possible in many cases to "tailor" the molecular structure of an organic species so that it becomes an effective ligand for a certain metal ion. An interesting account along this line is the 1966 Detroit Anachem Award Address by Prof. Harvey Diehl* entitled "Development of Metallochromic Indicators."

An important example: determination of iron with 1,10-phenanthroline

One of the better examples of a spectrophotometric procedure based on metal–ligand complexing is the method for determining small concentrations of iron using the chelating reagent 1,10-phenanthroline:

---

*Published in *Analytical Chemistry*, *39*, 30A (1967).

All of the atoms in the phenanthroline molecule are in the same plane, and the two nitrogen atoms, each with a pair of unshared electrons, are in a favorable orientation to form coordinate covalent bonds with the ferrous ion, making a stable five-member ring. The drawing shows only one step in the metal–ligand complexing; the iron is capable of bonding with three phenanthroline molecules, thus showing a coordination number of 6. Abbreviating the reagent as Ph for simplicity, we have for the reversible reaction:

$$\text{Fe}^{2+} + 3\text{Ph} = \text{Fe(Ph)}_3^{2+}, \qquad K = \frac{[\text{Fe(Ph)}_3^{2+}]}{[\text{Fe}^{2+}][\text{Ph}]^3}.$$

The value of this equilibrium constant has been determined as $2 \cdot 10^{21}$, showing that the complex is very stable indeed. It forms quantitatively when an excess of reagent is added to ferrous solutions in the pH range from 2 to 9. The color of the complex is orange-red, and it has a molar absorptivity of $11\,000$ L/cm $\cdot$ mol at the absorption maximum of 508 nm. The sensitivity is judged by the molar absorptivity; if the solution contains only $10^{-5}M$ iron, the measured absorbance will be about

$$A = \epsilon bC = 11\,000 \cdot 1 \cdot 10^{-5} = 0.11$$

when a 1-cm cell is used. With a 10-cm cell it would be reasonable to determine concentrations lower than $10^{-6}M$. The color of the solution after formation of the complex is perfectly stable and highly reproducible. When a good spectrophotometer is used, Beer's Law is very closely followed. Thus, this method easily meets many of the criteria listed earlier for an ideal method.

The difficulty with the phenanthroline method is that various other metal ions *also* react with the reagent and therefore interfere with the iron determination. Nickel, cobalt, and copper give colored complexes, while zinc forms a colorless complex. The latter does not interfere with the color measurement, of course, but it does consume reagent and may prevent the full development of the iron color. Silver and bismuth form precipitates, while cadmium and mercury form slightly soluble complexes which, like zinc, may require the addition of more reagent.

Although the method may be applied successfully in the pH range from 2 to 9, there is a definite effect of pH because the phenanthroline behaves as a monoprotic base. In terms of the dissociation of the conjugate acid HPh$^+$, we have

$$\text{HPh}^+ = \text{H}^+ + \text{Ph}, \qquad K_a = 1.1 \cdot 10^{-5}.$$

At pH values greater than about six, the reagent is present chiefly in its base form Ph. At pH values lower than about four, it exists almost entirely as the conjugate acid HPh^+, and we should consider the fact that the ferrous ion must compete with the hydrogen ion. The net reaction at low pH is as follows:

$$Fe^{2+} + 3HPh^+ = Fe(Ph)_3^{2+} + 3H^+.$$

In strongly acidic solution the proton wins in the competition and the complex cannot be formed.

The procedure for carrying out this determination is relatively simple, as shown in the following description taken from Sandell.*

> Place the sample solution, containing 0.01 to 0.2 mg of iron, in a 25-mL volumetric flask. Add 1 mL of 10% hydroxylamine (to reduce any iron(III) to ferrous ion). Add enough sodium acetate to make the pH 3 to 6. Add 2 mL of 1,10-phenanthroline (0.25% solution in water), mix, dilute to the mark. Wait about 10 minutes for color development, and measure the absorbance in a 1-cm cell at 508 nm. The standards should be treated in the same manner.

The iron(II)-phenanthroline system was thoroughly studied by Lee, Kolthoff, and Leussing.†

16.4 METAL–LIGAND TITRATION CURVES

In analogy to acid–base titration curves, which are plots of pH versus the quantity of base (or acid) added, we may profitably consider the variation of the metal ion concentration in the course of a complex-formation titration. To construct a theoretical titration curve it is necessary to have a set of corresponding values of $pM = -\log M$ and of the total ligand added C_L. To simplify the problem for the sake of illustration let us assume that:

- The initial concentration of metal ion is $0.01M$, and the volume does not change during the titration.
- The ionic strength is constant, so that Q-values may be used as constants.
- The solution pH is high enough for the titrating ligand to be fully deprotonated, so that the pH effect may be ignored, with $\alpha_L = 1$.
- There is no buffer-ligand competition for the metal ion.

Under these admittedly idealized conditions the titration reaction is simply

$$M + nL \rightleftarrows ML_n.$$

---

*Sandell, *Colorimetric Metal Analysis.* 3rd ed., Interscience, 1959, p. 541.
†T. Lee, I. M. Kolthoff, D. Leussing, *Journal of the American Chemical Society*, 70, 2348, 3596 (1948).

To show the importance of chelons in titrations, compared to ligands that form 2:1 or 4:1 complexes with the metal ion, let us consider three titration curves:

1. With ligand L, the metal ion forms four successive complexes, with Q-values equal to 10^5, 10^4, 10^3, and 10^2. The overall value is $\beta_4 = 10^{14}$. This is not very different from the case of titrating a solution of copper(II) with ammonia.

2. With ligand A, the metal ion forms two successive complexes (the ligand is bidentate), with Q-values of 10^9 and 10^5. Again the overall value is $\beta_2 = 10^{14}$.

3. With ligand Y, which is a polydentate chelon, only a 1:1 complex is formed, and it has a formation quotient equal to 10^{14}.

The calculation procedure is as follows:

a) Assume some arbitrary value for the equilibrium molarity of the free ligand and use it in the expression for α_0 (Eq. (14.4)) to calculate the corresponding molarity of the free metal ion M. From Eq. (14.3) we get:

$$[M] = \frac{0.01}{1 + Q_1[L] + \beta_2[L]^2 + \beta_3[L]^3 + \beta_4[L]^4} = \frac{C_M}{F_0},$$

using, of course, only the beta-values appropriate for the specific titration.

b) Use the calculated value of [M] and the assumed value of [L] in the successive beta-expressions to calculate the corresponding equilibrium molarities of each complex.

c) Use the material-balance expression (14.9) for the ligand to calculate the total concentration C_L of the ligand added.

d) Plot pM versus C_L.

e) Repeat the above steps as many times as necessary to obtain enough points to define the titration curve over its entire range, up to a few percent in excess of added ligand.

The results of such calculations for the three titrations listed above are shown in Fig. 16.1.

Even though the overall complex-formation constant is the same in each case, 10^{14}, it is obvious from these curves that the chelon not only complexes the metal ion much more tightly (bringing pM to higher values), but also yields a titration curve with a *much* sharper change in pM at the equivalence point.

Since metallochromic indicators require a change of about 2 pM units to complete their color change, we see that chelometric titrations have a marked advantage over titrations with ligands that are not polydentate. Whereas the titration curve for the monodentate ligand is hopelessly "drawn out" with no

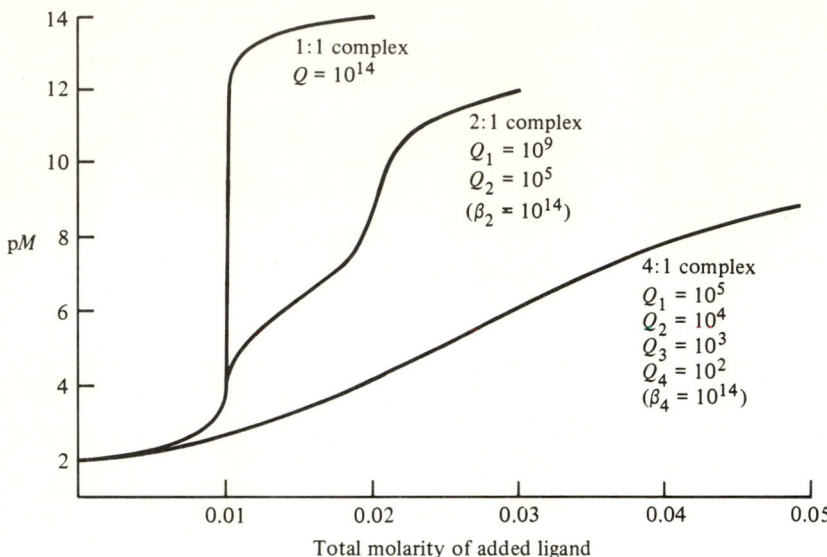

Figure 16.1

sharp change in pM at any point, the curve for the tetradentate ligand shows a change of about eight units in pM in a narrow region very close to the equivalence point.

16.5   CHELOMETRIC TITRATIONS OF METAL IONS

Following the recommendation of Reilley* we will use the term **chelon** for the polydentate ligands that form very stable, soluble, usually $1:1$ chelates and are used as titrants for metal ions according to the reaction:

$$\underset{}{M} + \underset{\text{chelon}}{Y} = \underset{\text{chelate}}{MY} \ .$$

This class of titrations is called **chelometric** because it is the amount of added chelon which is measured in the determinative step. The introduction of polyaminocarboxylate ligands such as EDTA, first by G. Schwarzenbach (Zurich, 1945), and also by A. Ringbom (Finland) and C. N. Reilley (University of North Carolina) among others, has revolutionized the field of metal-ion determinations because it became possible to use rapid and accurate titration techniques in place of slower gravimetric procedures. The new ligands also proved exceptionally valuable for masking and for spectrophotometric methods. In the following sections we will consider the fundamental basis for chelometric titrations.

---

*C. N. Reilley, *Journal of Chemical Education*, **36**, 555 (1959).

The Typical Experimental System

Although each titration method has its unique problems and characteristics, we may identify a few points that most cases have in common:

● The sample consists of a solution of a metal ion at a molarity of about $0.01M$, and this is to be titrated with a standard solution of a chelon having a molarity of about $0.1M$.

● Because the effect of pH is so important in metal–ligand systems, the metal-ion solution is made to contain a buffer that will control the solution pH to a desired fairly narrow range throughout the titration.

● The conjugate base which is part of the necessary buffer system may also interact with the metal ion, forming one or more complexes: ML, ML_2, etc. This competition for the metal ion may be called the F_0 **effect.**

● The equilibrium scheme may be represented by the following diagram:

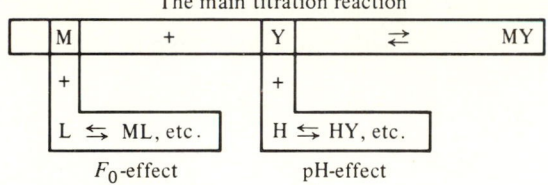

● The equilibrium quotient for the main titration reaction is the formation quotient for the metal–chelon chelate:

$$Q_{MY} = \frac{[MY]}{[M][Y]},$$

where [M] refers to the molarity of the *free* metal ion, i.e., the aquo-ion not complexed in any way by the buffer ligand L. Similarly, [Y] refers to the molarity of the fully deprotonated chelon, not to its proton complexes HY, H_2Y, etc. For the typical case the numerical value of Q_{MY} is rather large: from 10^{10} to 10^{20}.

● To the extent that there is an F_0 effect, the metal ion is hindered in its reaction with the chelon. The shift caused by the buffer ligand L forming a series of ML complexes holds back the main titration reaction to some extent.

● Unless the pH is rather high, there is also a pH effect that tends to interfere with the main titration reaction by forming a series of HY species not able to chelate the metal ion.

Obviously this situation is much more complicated than what we considered for acid–base titrations. Not only is it necessary to consider a number of simultaneous equilibria, but each metal–chelon-buffer system has its own set of variables. This makes it more difficult to generalize, but a fairly typical model system will be demonstrated to show the types of considerations that form the basis for a titration method.

The Model Titration System

The sample to be titrated contains the metal at a molarity of 0.01. The pH is maintained by a buffer that includes ligand L capable of forming four metal–ligand complexes:

$$M + L = ML, \qquad Q_1 = \frac{[ML]}{[M][L]},$$

$$\vdots \qquad\qquad \vdots$$

$$ML_3 + L = ML_4, \qquad Q_4 = \frac{[ML_4]}{[ML_3][L]}.$$

We will consider two sets of values for $Q_1, \ldots, Q_4$. One set will have all values equal to zero (no complexing at all), and the other will have values of 10^3, 10^2, 10, and 1.

For the following discussion we will take the value of the formation quotient for MY to be 10^{14}.

The chelon is tetraprotic, with four successive acid dissociation steps:

$$H_4Y = H^+ + H_3Y, \qquad Q_{A1} = \frac{[H^+][H_3Y]}{[H_4Y]},$$

$$\vdots \qquad \vdots \qquad\qquad \vdots$$

$$HY = H^+ + Y, \qquad Q_{A4} = \frac{[H^+][Y]}{[HY]}.$$

For the present model, let the successive values be 10^{-2}, 10^{-3}, 10^{-6}, and 10^{-10}.

At any point in the titration there is a certain molarity of metal that has *not* been chelated to form MY. We designate this by the symbol C_M.

$$C_M = [M] + [ML] + [ML_2] + \cdots = [M] \cdot F_0.$$

Also, at any point in the titration there is a certain molarity of the chelon that is *not* tied up in the MY chelate, and this is symbolized by C_Y:

$$C_Y = [Y] + [HY] + [H_2Y] + \cdots = [Y]/\alpha_0. \tag{16.1}$$

Systematic Calculation of the Theoretical Titration Curve

By analogy with the titration curves for acid–base reactions, where pH was plotted versus fraction titrated, we define a metal–chelon titration curve as a plot of pM versus the fraction titrated. By *fraction titrated* we mean the ratio

$$\text{Fraction titrated} = \frac{\text{Total molarity of Y, all species}}{\text{Total molarity of M, all species}}.$$

The procedure is as follows, and it is assumed that one has numerical values for the four formation quotients of the M–L reactions and for the four acid dissociation quotients of H_4Y, as well as the formation quotient of the metal chelate Q_{MY}. We assume that $0.01M$ metal ion is used and that volume is constant throughout the titration. Further, we assume that the solution pH is maintained at a constant value by the buffer and that the buffer-ligand concentration is at a constant value of 0.10 throughout the titration.

THE INITIAL VALUE FOR pM

Before any chelon is added to the metal solution there is a mixture of the free metal ion and the buffer-ligand complexes of the metal. The total molarity is 0.01, and from the F_0 relationship we find:

$$[M]_{init} = \frac{0.01}{F_0} = \frac{0.01}{1 + Q_1[L] + Q_1Q_2[L]^2 + Q_1Q_2Q_3[L]^3 + Q_1Q_2Q_3Q_4[L]^4}.$$

(16.2)

Let us consider two cases. First, suppose the buffer ligand has little or no complexing ability toward the particular metal, in which case [M] is simply 0.01 and $pM = 2.00$. This is equivalent to using zero values for the Q's in Eq. (16.2)

It will become clear that this is generally the most desirable situation, and if the necessary pH control can be accomplished with a noncomplexing buffer, one should do so.

For the second case, suppose the Q values are, successively, 10^3, 10^2, 10, and 1. Taking 0.1 as the molarity of the ligand L, we find from Eq. (16.2) that the free-metal-ion molarity has been lowered a great deal:

$$[M] = \frac{0.01}{1 + 10^2 + 10^3 + 10^3 + 10^2} = 4.5 \cdot 10^{-6}, \quad \text{or} \quad pM = 5.3.$$

In general:

$$pM = 2 + \log F_0.$$

(16.3)

THE VALUE OF pM WHEN A 100% EXCESS OF CHELON HAS BEEN ADDED

Of course, in performing a titration one would not end up by adding twice as much titrant as required! However, the "200% titrated" point provides a convenient reference point for sketching a titration curve. By this time, we not only have converted the metal quantitatively to its chelate, so that $[MY] = 0.01$ (ignoring dilution), but the excess chelon is also present at this molarity so that $C_Y = 0.01$.

To find the value for [M] and hence pM at the 200% point we must work with the formation quotient:

$$Q_{MY} = \frac{[MY]}{[M][Y]} \quad \text{or} \quad [M] = \frac{[MY]}{Q_{MY}[Y]}.$$

By incorporating Eq. (16.1) we get

$$[M] = \frac{1}{Q_{MY}\alpha_0} \cdot \frac{[\cancel{MY}]}{\cancel{C_Y}} = \frac{1}{Q_{MY}\alpha_0}$$

or, in logarithmic form,

$$pM_{200\%} = \log Q_{MY} + \log \alpha_0. \tag{16.4}$$

For the present discussion of the model we have assumed that log $Q_{MY} = 14$ and that the successive pQ_a values for the chelon are 2, 3, 6, and 10. Thus, the relationship between α_0 and $[H^+]$ is:

$$\alpha_0 = \frac{10^{-21}}{[H^+]^4 + 10^{-2}[H^+]^3 + 10^{-5}[H^+]^2 + 10^{-11}[H^+] + 10^{-21}}. \tag{16.5}$$

When it is possible to carry out the titration at high pH (for example, 12 or 13), we can see from Eq. (16.5) that α_0 is close to 1.0, that is, the chelon is completely deprotonated. This means that the upper limit to the value for $pM(200\%)$ is simply log Q_{MY}, or 14 in the present case.

However, in a more typical case the titration is performed at a controlled pH value that is lower than the last pQ_a value for the chelon, so that α_0 is much smaller than 1.0. For example, at pH $= 7.00$ we have:

$$\alpha_0 = \frac{10^{-21}}{10^{-28} + 10^{-23} + 10^{-19} + 10^{-18} + 10^{-21}} = 9.1 \cdot 10^{-4}, \quad \text{or} \quad \log \alpha_0 = -3.04,$$

and therefore, $pM = 14 - 3.0 = 11.0$, three log units smaller than the result at high pH.

THE VALUE OF pM AT THE EXACT EQUIVALENCE POINT

By introducing the expressions for F_0 and α_0 into the expression for the chelate formation quotient, we have:

$$Q_{MY} = \frac{[MY]}{[M][Y]} = \frac{[MY]F_0}{C_M C_Y \alpha_0} \quad \text{or} \quad C_M C_Y = \frac{[MY]F_0}{Q_{MY}\alpha_0}$$

Now, when the stoichiometric amount of chelon has been added, there will be close to 0.01M MY, assuming that the titration reaction proceeds nearly to completion, and it is also true that $C_M = C_Y$. Therefore,

$$C_M = \sqrt{\frac{0.01\,F_0}{Q_{MY}\alpha_0}}.$$

But it is also true that $C_M = [M]F_0$, and therefore

$$[M]_{e.p.} = \sqrt{\frac{0.01}{F_0 Q_{MY}\alpha_0}}$$

or, in logarithmic form,

$$pM_{e.p.} = \frac{\log Q_{MY} + \log \alpha_0 - \log 0.01 + \log F_0}{2}. \tag{16.6}$$

By comparing Eqs. (16.3) and (16.4), the reader can see that the value of pM at the equivalence point is exactly midway between the values at the beginning of the titration and the 200% point. Thus, for the model under discussion,

$$pM_{e.p.} = (2 + 14)/2 = 8$$

for the case of a noncomplexing buffer at high pH. For a noncomplexing buffer at $pH = 7$ we have

$$pM_{e.p.} = (2 + 11)/2 = 6.5,$$

and for the complexing buffer at high pH, we get

$$pM = (5.3 + 14)/2 = 9.7.$$

The Shape of the Titration Curve

Given only the three points calculated in the foregoing sections, one might be misled into thinking that a plot of pM versus fraction titrated is merely linear, as shown by the dashed line in Fig. 16.2. However, calculation of pM values

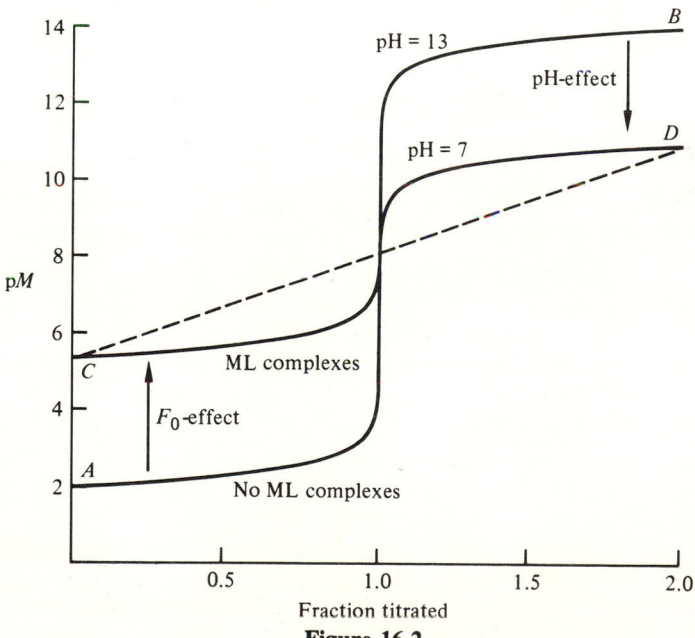

Figure 16.2

over the full range of the titration shows that the titration curve is S-shaped with much the same appearance as a pH-fraction plot for the titration of a strong acid with a strong base. This is not surprising, because both systems are based on models that are chemically and algebraically similar:

$$\text{Strong acid, strong base:} \quad H^+ + OH^- = H_2O, \quad Q = 10^{14},$$

$$\text{Metal ion, chelon:} \quad M + Y = MY, \quad Q = 10^{14}.$$

Figure 16.2 shows the theoretical behavior of the system under four different conditions:

1. The ideal, with high pH and no ML complexing, giving a very large "break" in pM at the endpoint. For more stable chelates the upper section of the curve would be still higher (remember, $pM = \log Q_{MY}$ at the 200% point). Follow curve AB.

2. The case of a complexing buffer, but at high pH. Curve CB.

3. The case of a noncomplexing buffer, but at lower pH. Curve AD.

4. The typical situation, with both the F_0 and pH effects operating. In this case the rate of change of pM at the endpoint may be much smaller than the ideal. Curve CD.

The two halves of the titration curves may be considered separately, since the pH effect is unimportant before the endpoint and the metal–ligand complexing is unimportant after the endpoint. In establishing a procedure for a chelometric titration one should consider each effect and minimize the deterioration of the titration curve as much as possible by choosing weak or noncomplexing buffers, and by operating at as high a pH as may be permitted.

The latter point raises another problem: given a solution of a metal ion, one cannot simply raise the pH indiscriminately by adding, say, sodium hydroxide. This is because many metal ions form insoluble hydroxide precipitates which cannot be titrated very well. And now a dilemma appears: we may prevent the formation of metal hydroxide precipitate by using a *complexing buffer* that masks the metal-ion against precipitation by formation of ML, ML_2, etc. This would enable us to do the titration at a higher pH, but there is a price to pay in the F_0-effect. What we gain in the upper section of the titration curve, we may lose, at least in part, in the lower section of the titration curve. Most practical procedures involve making a compromise between the two effects.

Algorithm for Multipoint Calculation of the Titration Curve

Although the three key points (initial, equivalence, and 200%) are sufficient to enable a useful sketch of the titration curve, it is perhaps worth noting that a very simple computer program may be used for generating pM-fraction values for the entire range of the titration. The stepwise procedure is as follows:

1. From the calculations of the initial and 200% points, decide what range of pM values should be represented in the titration curve. For each of the desired pM values in this range do the following:

2. Convert pM to [M].

3. Using the previously calculated constant value for F_0, find $C_M = [M]F_0$.

4. Find the difference $[MY] = 0.01 - C_M$.

5. Use the value of Q_{MY} to find $[Y] = \dfrac{[MY]}{[M] \cdot Q_{MY}}$.

6. Use the previously calculated value of α_0 (for the particular pH used) to calculate the value of $C_Y = [Y]/\alpha_0$.

7. Find the fraction titrated: $\text{Fraction} = ([MY] + C_Y)/0.01$.

16.6   THE MOST WIDELY USED CHELON: EDTA

The outstanding success of *ethylenediaminetetraacetic* acid, EDTA, as a titrant for metal ions is based on the following properties:

1. The deprotonated form of EDTA contains six sites with pairs of unshared electrons, capable of forming coordinate bonds with a metal ion. Two of these sites are nitrogen atoms, while four are oxygen atoms, making the chelon quite general in its complexing ability. The structure of the anion form is as follows, with each site marked by an asterisk (*):

2. In the reaction with metal ions, EDTA is not only able to satisfy all the coordination needs of the metal, but it can do so by the formation of a set of five-member rings of the types

The EDTA anion can wrap around the metal ion, forming up to five such heterocycles, each having relatively little steric strain.

3. The formation quotients for the metal–EDTA chelates are impressively high, as shown in the following table:

Metal	$\log Q_{MY}$
Magnesium, Mg^{2+}	8.7
Calcium, Ca^{2+}	10.7
Strontium, Sr^{2+}	8.7
Barium, Ba^{2+}	7.9
Iron(II), Fe^{2+}	14.4
Iron(III), Fe^{3+}	25.1
Aluminum, Al^{3+}	16.1
Cobalt, Co^{2+}	16.3
Nickel, Ni^{2+}	18.6
Copper, Cu^{2+}	18.8
Zinc, Zn^{2+}	16.5
Cadmium, Cd^{2+}	16.5
Mercury, Hg^{2+}	22.1
Lead, Pb^{2+}	17.9

4. The acid-dissociation quotients for the protonated form H_4Y have values which permit most of the EDTA to be dissociated at $pH \approx 11$, thus making it possible to avoid a serious pH-effect without going to very high pH values. The successive pQ_a values are as follows (for $25°$ and $I = 0.1$):

$$H_4Y = H^+ + H_3Y^-, \qquad pQ_1 = 2.0,$$
$$H_3Y^- = H^+ + H_2Y^{2-}, \qquad pQ_2 = 2.67,$$
$$H_2Y^{2-} = H^+ + HY^{3-}, \qquad pQ_3 = 6.16,$$
$$HY^{3-} = H^+ + Y^{4-}, \qquad pQ_4 = 10.26.$$

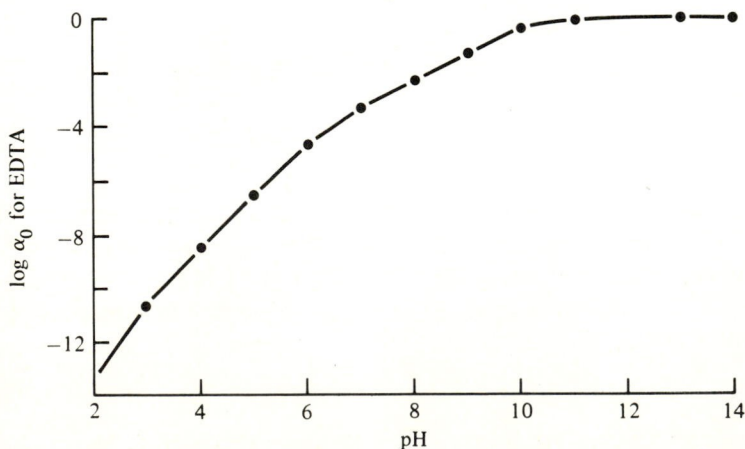

Fig. 16.3 $\alpha_0 = [Y^{4-}]/C_Y$ as a function of pH.

The alpha values for EDTA were shown in Chapter 9. Because of its importance in the present context, a plot of log α_0 is repeated here in Fig. 16.3.

5. Standard solutions of EDTA are easily and accurately prepared by weighing and dissolving a sample of the disodium salt $Na_2H_2Y \cdot 2H_2O$ (MW = 372.24) which is readily available at low cost in highly purified quality. The solutions are stable indefinitely and are nonabsorbing throughout the ultraviolet and visible portions of the spectrum.

6. Since the introduction of EDTA as a titrant some 30 years ago, many researchers throughout the world have developed practical applications for its use in titrations, masking, precipitation from homogeneous solution, and other analytical procedures. These efforts have included the synthesis and examination of quite a few other chelons of varying structure and properties but, except for some special situations, EDTA remains the most preferred reagent for chelometric titrations.* Table 16.1 lists a few of the chelons that have proved useful and Table 16.2 gives the formation constants for a number of common metal ions (see pp. 508, 509).

16.7 VISUAL DETECTION OF THE ENDPOINT: METALLOCHROMIC INDICATORS

Whereas acid–base indicators are strongly colored organic species which change color upon reaction with hydrogen ions, **metallochromic indicators** do the same sort of thing by reaction with metal ions. A metallochromic indicator will usually react with hydrogen ions *also*, so that the equilibrium scheme is similar to that used above for the metal–chelon titration:

$$MInd = M + Ind$$

metal-indicator
chelate, e.g., red   +

$$H^+ = HInd, H_2Ind, \text{etc.}$$

mixture of free indicator
species, e.g., blue at the
pH used for titration

At the beginning of the titration there is a relatively high concentration of the metal ion, and so the indicator is nearly completely present as the red chelate. As the titration proceeds, the molarity of M is steadily decreased by reaction with the chelon titrant, and so the MInd species increasingly dissociates to maintain the above equilibrium system.

---

*The reader is referred to *Handbook of Analytical Chemistry*, ed. L. Meites, McGraw-Hill, New York, 1963, for useful summaries of applications and conditions for titrations.

Table 16.1 Chelons used frequently in metal titrations

Abbreviation	Chemical name
trien	triethylenetetramine $NH_2CH_2CH_2NHCH_2CH_2NHCH_2CH_2NH_2 = Y$ pK values for H_4Y: 3.32, 6.67, 9.20, 9.92
tetren	tetraethylenepentamine $NH_2CH_2CH_2NHCH_2CH_2NHCH_2CH_2NHCH_2CH_2NH_2 = Y$ pK values for H_5Y: 2.6, 4.1, 8.2, 9.2, 10.0
NTA	nitrilotriacetic acid $HOOCCH_2{-}N{<}{\overset{CH_2COOH}{\underset{CH_2COOH}{}}} = H_3Y$ pK values for H_3Y: 1.9, 2.49, 9.73
EDTA	ethylenediaminetetraacetic acid ${\overset{HOOCCH_2}{\underset{HOOCCH_2}{}}}{>}N{-}CH_2CH_2{-}N{<}{\overset{CH_2COOH}{\underset{CH_2COOH}{}}} = H_4Y$ pK values for H_4Y: 2.0, 2.67, 6.16, 10.26
HEDTA	N-hydroxyethylenediaminetriacetic acid ${\overset{HOOCCH_2}{\underset{HOCH_2CH_2}{}}}{>}N{-}CH_2CH_2{-}N{<}{\overset{CH_2COOH}{\underset{CH_2COOH}{}}} = H_3Y$ pK values for H_3Y: 2.64, 5.33, 9.73

Table 16.2 Approximate log values of formation quotients $Q_{MY} = \dfrac{[MY]}{[M][Y]}$ for selected metal chelonates at 25° and $I = 0.1$*

Chelon	Mg^{2+}	Ca^{2+}	Sr^{2+}	Ba^{2+}	Mn^{2+}	Fe^{2+}	Fe^{3+}
trien	negl	negl	negl	negl	4.9	7.8	—
tetren	negl	negl	negl	negl	7.0	—	—
NTA	5.4	6.4	5.0	4.8	7.4	8.8	15.8
EDTA	8.7	10.7	8.7	7.9	13.8	14.4	25.1
HEDTA	7.0	8.0	6.8	6.2	10.7	11.6	—
EEDTA	8.3	10.0	8.6	8.2	13.2	—	—
EGTA	5.4	10.9	8.5	8.4	12.3	—	—
DTPA	9.0	10.7	9.7	8.6	15.5	16.7	27.5
CyDTA	10.3	12.3	10.0	8.0	16.8	—	—

*C. N. Reilley et al., *Journal of Chemical Education*, **36**, 555 (1959).

Table 16.1 (continued)

Abbreviation	Chemical name
EEDTA	ethyletherdiaminetetraacetic acid

$$\text{HOOCCH}_2\diagdown\diagup\text{CH}_2\text{COOH}$$
$$\text{N---CH}_2\text{CH}_2\text{---O---CH}_2\text{CH}_2\text{---N}= \text{H}_4\text{Y}$$
$$\text{HOOCCH}_2\diagup\diagdown\text{CH}_2\text{COOH}$$

pK values for H$_4$Y: 1.90, 2.67, 8.82, 9.49

EGTA — ethyleneglycol-bis-(β-aminoethylether)-N,N'-tetraacetic acid

$$\text{HOOCCH}_2\diagdown\diagup\text{CH}_2\text{COOH}$$
$$\text{NCH}_2\text{CH}_2\text{OCH}_2\text{CH}_2\text{OCH}_2\text{CH}_2\text{N}= \text{H}_4\text{Y}$$
$$\text{HOOCCH}_2\diagup\diagdown\text{CH}_2\text{COOH}$$

pK values for H$_4$Y: 2.0, 2.68, 8.85, 9.43

DTPA — diethylenetriaminepentaacetic acid

$$\text{HOOCCH}_2\diagdown\diagup\text{CH}_2\text{COOH}$$
$$\text{N---CH}_2\text{CH}_2\text{--- N---CH}_2\text{CH}_2\text{---N}= \text{H}_5\text{Y}$$
$$\text{HOOCCH}_2\diagup|\diagdown\text{CH}_2\text{COOH}$$
$$\text{CH}_2\text{COOH}$$

pK values for H$_5$Y: 2.08, 2.41, 4.27, 8.60, 10.55

CyDTA — cyclohexanediaminetetraacetic acid

$$\text{HOOCCH}_2\diagdown\diagup\text{CH}_2\text{COOH}$$
$$\text{N}\text{N}$$
$$\text{HOOCCH}_2\diagup\diagdown\text{CH---CH}\diagup\diagdown\text{CH}_2\text{COOH}= \text{H}_4\text{Y}$$
$$\text{CH}_2\text{CH}_2$$
$$\text{CH}_2\text{---CH}_2$$

pK values for H$_4$Y: 2.40, 3.52, 6.12, 11.70

Table 16.2 (continued)

Co^{2+}	Ni^{2+}	Cu^{2+}	Zn^{2+}	Cd^{2+}	Hg^{2+}	Al^{3+}	Pb^{2+}
11.0	14.0	20.1	11.9	10.8	25.0	negl	10.4
15.1	17.8	22.9	15.4	14.0	27.7	negl	10–11
10.4	11.5	12.6	10.5	9.8	—	—	11.1
16.3	18.6	18.8	16.5	16.5	22.1	16.1	17.9
14.4	17.0	17.4	14.5	13.0	20.1	—	15.5
14.7	14.7	17.8	15.3	16.3	23.1	—	14.4
12.3	13.6	17.8	13.0	16.7	23.8	—	14.6
19.0	20.2	21.0	18.8	19.0	27.0	—	18.6
18.9	19.4	21.3	18.6	19.2	24.4	17.6	19.7

However, if the proper indicator has been chosen, it is only in the immediate vicinity of the equivalence point that the values of pM change quickly through the range which permits complete dissociation of the MInd species, producing the free Ind form, which reacts with H^+ under the pH condition imposed by the buffer to form a certain mixture of the protonated species.

Clearly this is a rather complex business and, unlike the situation with acid–base or redox indicators, we do not have neat tables and simple rules for choosing an indicator for a given application. One relies chiefly on procedures that have been investigated experimentally and have been shown to give good results.

Research is needed to establish the fundamental basis for individual metal-indicator systems. It is necessary to have values for the formation quotients (M + Ind = MInd) and for the acid dissociation quotients for the indicator before a theoretical choice of conditions can be made.

However, as with acid–base titrations, the indicator changes color over a range of about $2\,pM$ units. For best results this range should be centered on the value of pM at the theoretical equivalence point in the titration.

The pM Value at the Center of the Indicator Color-Change Interval

It was shown in Eq. (16.6) that the value for pM at the exact equivalence point in a metal–chelon titration can be calculated if the several equilibrium quotients are known. We now distinguish between this point, which is the ideal place to stop the titration, and the *endpoint*, which is our experimental attempt to find the equivalence point. The difference between the endpoint and the equivalence point is, of course, the **titration error**.

The titration error depends on many things including reagent purity and the skill of the analyst. But at this stage we will consider only the theoretical difference caused by the fact that a metallochromic indicator typically undergoes its color change at pM values that do not coincide exactly with the pM value at the equivalence point.

The starting point is simply the expression for the formation quotient for the metal-indicator complex:

$$M + Ind = MInd, \qquad Q_{MInd} = \frac{[MInd]}{[M][Ind]}, \tag{16.7}$$

where Ind is the fully deprotonated species of the indicator. Let us assume that this indicator is diprotic, so there are two acid-dissociation quotients to consider:

$$H_2Ind = H^+ + HInd^-, \qquad Q_1 = \frac{[H^+][HInd]}{[H_2Ind]},$$

$$HInd = H^+ + Ind, \qquad Q_2 = \frac{[H^+][Ind]}{[HInd]}.$$

Therefore the expression for the fraction of the free (not complexed with M) indicator present as In is given by the alpha-function:

$$\frac{[\text{Ind}]}{C_{\text{Ind}}} = \frac{Q_1 Q_2}{[\text{H}^+]^2 + Q_1[\text{H}^+] + Q_1 Q_2} = \alpha_{\text{Ind}}.$$

By substituting for [Ind] in Eq. (16.7) and solving for [M] we find:

$$[\text{M}] = \frac{[\text{MInd}]}{C_{\text{Ind}}} \frac{1}{Q_{\text{MInd}}\alpha_{\text{Ind}}}. \tag{16.8}$$

The color change corresponds to the transition from the species, MIn, to the mixture of species, C_{Ind}, which is in equilibrium at the pH value controlled by the buffer. At the *middle* of the color change, which we may take to be the visual endpoint in the titration, the conversion is 50% complete, so that [MIn] = C_{In}. Therefore these quantities may be cancelled in Eq. (16.8), and the result may be expressed in logarithmic form as follows:

$$p M(\text{endpoint}) = \log Q_{\text{MInd}} + \log \alpha_{\text{Ind}}. \tag{16.9}$$

Just as the value for $p M$(equivalence point) depends on the solution pH in accord with the properties of the chelon, the value for $p M$(endpoint) depends on the acid–base properties of the indicator.

A Generally Useful Metallochromic Indicator: Methylthymol Blue

An ingenious approach to the design of an indicator was to "insert" a common acid–base indicator (thymol blue) right into the center of the EDTA molecule. The resulting structure combines the strong colors of thymol blue with the strong chelating ability of EDTA. This indicator forms intense blue chelates with many metal ions and can be used both in acidic and in alkaline media; it has been given the trivial name **methylthymol blue**.

Thymol blue

The molecule contains six acidic protons, four from the acetic-acid groups and two phenolic hydrogens. For purposes of describing the effect of pH on the color of the free (no metal present) indicator, it is sufficient to regard methylthymol blue as triprotic:

$$H_3Ind \xrightarrow{pK \approx 7.2} H_2Ind \xrightarrow{pK \approx 11.2} HInd \xrightarrow{pK \approx 13.4} Ind.$$

| yellow | light blue | gray | dark blue |

Since the metal chelates MIn are intense blue, it appears that such chelations affect the light-absorbing properties in a way quite similar to the loss of all the protons in a very basic solution. The use of this indicator in EDTA titrations is practical either at low pH, where the color changes from the intense blue of the MIn species to the yellow color of the free indicator, or in the 11 to 12 pH range, where the color changes from intense blue to gray or very light blue. It cannot be used in a *very* basic solution because there would be no observable color variation when the intense blue of MIn changed to the dark blue of In.

By using nitric acid in the 1 to 3 pH range, methylthymol blue can be applied to metals that form exceptionally stable EDTA complexes: bismuth, scandium, zirconium, and thorium.

When hexamethylenetramine buffer is used in the 5 to 6 pH range, the following metals give good blue-to-yellow endpoints: cadmium, cobalt, iron(II), mercury, manganese, lead, the rare-earth metals, tin(II), and zinc.

At pH ≈ 11 and with ammonia buffers, good endpoints with the blue-to-gray color change are observed with barium, calcium, copper, magnesium, and strontium.

Other Indicators

Myriad substances have been investigated as metallochromic indicators, and many of these have excellent properties for specific applications. A comprehensive and valuable source of information is the book *Indicators*, edited by E. Bishop (Pergamon Press, 1972). This work contains information on the history and theory of titration indicators for acid–base, redox, chelometric, and precipitation systems. For metallochromic indicators it contains extensive summaries of structures, properties and applications.

For an informative review dealing both with metallochromic indicators and other aspects of chelometric titrations, the reader should consult the article by R. Pribil, *Recent Developments in Complexometric Titrimetry*, published in the Chemical Rubber Co. Critical Reviews in Analytical Chemistry, *3*, 113 (1973).

A good source of procedures for chelometric titrations is the *Handbook of Analytical Chemistry*, ed. L. Meites, McGraw-Hill Book Co., New York, 1963.

16.8 PUTTING IT ALL TOGETHER: pM–pH DIAGRAMS

We must consider the effects of pH on the chelon, on the indicator, and on the buffer ligand, in addition to the formation quotients for the metal–chelon and metal–indicator complexes, the buffer–ligand complexes with the metal, and the use of masking agents to reduce effects of interference by other metals. It becomes obvious that typical chelometric titrations are quite complicated, perhaps involving as many as 20 simultaneous equilibria. The best way to achieve an overview of a given metal–chelon indicator system is through construction of a plot of pM versus pH.

To illustrate this we will continue with the model metal–chelon system already discussed, that is:

$$M + Y = MY, \qquad Q_{MY} = 10^{14},$$

$$H_4Y \xrightarrow{\ pQ=2\ } H_3Y \xrightarrow{\ pQ=3\ } H_2Y \xrightarrow{\ pQ=6\ } HY \xrightarrow{\ pQ=10\ } Y.$$

In addition we now define a model metal–indicator system as follows:

$$M + Ind = \underset{\text{red}}{MInd}, \qquad Q_{MInd} = 10^{10},$$

$$\underset{\substack{\text{red-}\\\text{orange}}}{H_2Ind} \xrightarrow{\ pQ=7\ } \underset{\text{red}}{HInd} \xrightarrow{\ pQ=11\ } \underset{\text{blue}}{Ind.}$$

The big question is this: At what pH should the titration be carried out so that (a) the visual endpoint will coincide with the theoretical equivalence point, and (b) the color change at the endpoint will be obvious?

By using Eq. (16.4) to find the values for pM at the 200% point, Eq. (16.6) for the values at the equivalence point, and Eq. (16.9) for the endpoint based on the above model indicator, we may create Table 16.3.

Table 16.3

Solution pH	Values for pM at the			
	start	equiv. point	endpoint	200% point
2	2			
3	2	2.84		3.68
4	2	3.98		5.95
5	2	4.98	2.00	7.95
6	2	5.85	3.96	9.70
7	2	6.48	5.70	10.96
8	2	7.00	6.96	11.99
9	2	7.48	7.99	12.96
10	2	7.85	8.96	13.70
11	2	7.98	9.70	13.96
12	2	8.00	9.96	14.00

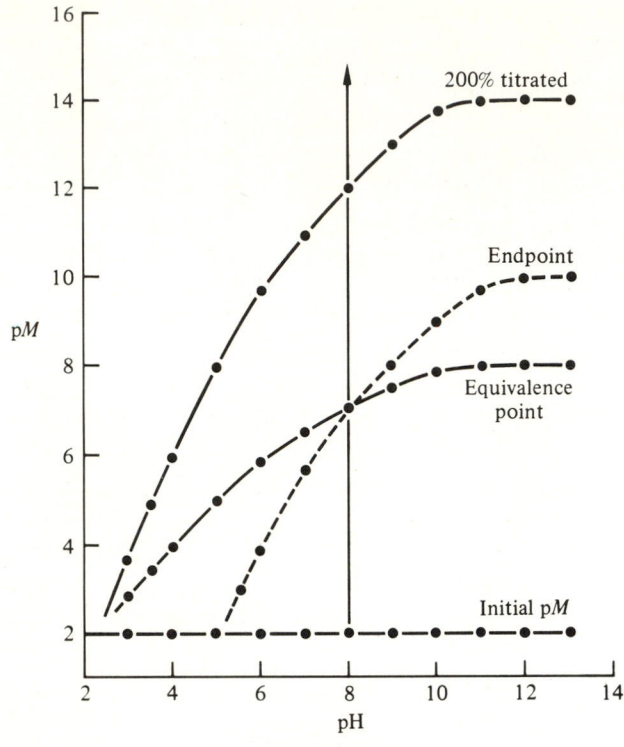

Figure 16.4

Unlike well-designed plots, tables of numbers are always hard to look at and interpret. Figure 16.4 shows the values of pM versus pH, with one line plotted for each of the four situations as labeled. There are several comments that should be made as we contemplate this plot.

• First, it has been assumed that the metal ion, present at an initial molarity of 0.01 ($pM = 2$), is not complexed by whatever buffers are used to vary the pH over the range shown. We are ignoring the F_0-effect. Also, by showing the initial pM constant at 2.0 all the way to high pH values we are implying that this particular metal does not form a hydroxide precipitate.

• Note that the difference between pM(initial) and pM(200%) increases smoothly as the pH increases until it finally levels off at high pH. This is an overview of what we have called the pH-effect and is related to the increasing dissociation of the chelon. The 200% line levels off when α_Y approaches 1.00.

• The pM values at the equivalence point are, as pointed out earlier, exactly midway between pM(initial) and pM(200%).

• The dashed line for the pM values at the endpoint corresponds to the condition of having the indicator 50% in the MInd form and 50% in the free form. Because the alpha-function for the indicator is different from that of the chelon (different pQ_a values) the line has a different shape. It levels off at high pH for the same reason that the value for α_{Ind} approaches 1.00. Just as the 200% line levels off at pM values equal to $\log Q_{MY}$, the indicator-endpoint line levels off at pM values equal to $\log Q_{MInd}$, as required by Eq. (16.9).

• As the titration proceeds, the value of pM steadily increases from its initial value of 2.00 to its ultimate value of $pM(200\%)$. The vertical arrow symbolizes this direction and change.

• If the titration is carried out at pH = 8, where the endpoint line crosses the equivalence-point line, we have the highly desirable situation of negligible titration error. At pH values lower than 8, the endpoint (color change) would occur before the equivalence point is reached, while at pH values higher than 8 the endpoint would be increasingly late. Thus, by inspecting a pM–pH diagram we may make intelligent decisions about what titration conditions to use.

• *However*! In looking back at the model indicator system we note a problem. By considering the pQ_a values for the indicator (7 and 11) we conclude that if the titration is carried out at pH = 8, the *free* indicator will be chiefly in its HIn form. And this is *red*, as is the metal–indicator complex. By operating at the theoretically desirable pH = 8, we would have a color change that cannot be noticed by the human eye.

• The solution to the problem requires a compromise. Since pQ_2 for the indicator is 11, we must operate at a pH of 12 to 13 to make sure that the free indicator is in the In form, which is blue. Then there will be an acceptable color change from MIn(red) to In(blue) at the endpoint. However, inspection of the diagram shows that at pH = 12 the endpoint occurs at $pM = 10$, whereas the equivalence point is at $pM = 8$. The endpoint will be a little late, and a slight positive titration error will occur. However, considering the fast rate at which pM is changing in this titration (in the vicinity of the equivalence point), as shown in Fig. 16.1, we conclude that the titration error is probably quite negligible.

Construction of the pM–pH diagram required complete knowledge of the equilibrium characteristics of the titration system. There is need for research on actual systems involving various metal ions, various chelons, and various indicators. At present the fundamental basis for the theory of chelometric titrations is rather clear* but reliable studies of formation quotients and pQ_a values are needed. There are opportunities here for undergraduate research.

One important system that has been well studied is the titration of calcium ion, or mixtures of calcium and magnesium ions with EDTA. These titrations are important in the analysis of limestone, and especially for

---

*See C. N. Reilley and R. Schmid, *Principles of End Point Detection in Chelometric Titrations using Metallochromic Indicators*, Analytical Chemistry, *31*, 887 (1959).

measurements of the **hardness of water**, a procedure done regularly in nearly all communities and industrial plants. In the following case study we will examine the hardness determination in more detail.

16.9   CASE STUDY: DETERMINATION OF CALCIUM AND MAGNESIUM IN WATER

Water obtained from wells, rivers, and lakes contains calcium ions and magnesium ions, among others, because of the reaction:

$$MCO_3(s) + CO_2 + H_2O = M^{2+} + 2HCO_3^-, \tag{16.10}$$

where M is either calcium or magnesium present in the earth as limestone or dolomite. The carbon dioxide is available from the atmosphere and dissolves slightly in rain water. The presence of Ca^{2+} and Mg^{2+} in municipal water supplies causes both annoyance and expense to consumers for two reasons. First, these ions interfere with the cleansing action of soaps by forming an insoluble curd. This is the source of the "bathtub ring" and the dull sticky deposits in hair when soap is used instead of a shampoo. Soaps contain long chains of $—CH_2—$ groups terminating in a carboxyl group $—COO^-$, and the precipitation reaction is :

$$2CH_3(CH_2)_n COO^- + M^{2+} = M(CH_3(CH_2)_n COO)_2(s). \tag{16.11}$$

Before one can obtain a lather (suds) with soap, the calcium and magnesium ions must be removed. This is the basis for commercial water softeners which exchange sodium ions for the calcium and magnesium before the water is to be used. Otherwise, the soap itself must act as the softening agent first before it can start performing its desired cleaning action.

For many years the determination of the calcium and magnesium content of water was carried out by a "soapimetric" titration, in which a standard solution of a pure soap was used to titrate a sample of the water. As long as there was M^{2+} present, the added titrant would simply precipitate. The endpoint was taken as the point in the titration when vigorous shaking caused the formation of a lather on the surface, showing that the soap was then present in excess over the amount needed for reaction (16.11). This method was not highly accurate, and did not distinguish between calcium and magnesium.

The second problem is that hard water (that is, water containing appreciable concentrations of calcium and magnesium) leaves deposits of $CaCO_3$ and $MgCO_3$ when boiled. This is due to the reversal of the reaction shown in Eq. (16.10), because boiling drives off the CO_2 as a gas and the HCO_3^- ions are reconverted to CO_3^{2-} ions that can form the precipitates. Such deposits are seen in teakettles, water heaters, and boilers. Due to their insulating effect, they decrease the efficiency of heat transfer, and in more serious cases may block the flow of water through pipes. In many industrial processes it is essential to remove the calcium and magnesium before using the water in the boilers. Therefore it is an important analytical problem to determine the original hardness of the water supplies, and also the hardness remaining after water treatment. The discovery that calcium and magnesium may be quickly and accurately determined by a simple titration with EDTA was welcomed with great enthusiasm by water chemists, and probably no other analytical procedure has been so widely adopted in such a short time.*

For some purposes it may be adequate to determine merely the sum of calcium plus magnesium, since they both are the cause of what we call hardness. However, it is still more useful and informative if the individual concentrations of these metals can be determined. Let us first consider the determination of the sum in a sample of water containing appreciable amounts of both calcium and magnesium.

The formation quotients for the two M–EDTA chelates are moderately large:

$$Ca^{2+} + Y^{4-} = CaY^{2-}, \qquad \log Q_{CaY} = 10.7,$$
$$Mg^{2+} + Y^{4-} = MgY^{2-}, \qquad \log Q_{MgY} = 8.7.$$

The larger value for calcium shows that it will be titrated first when EDTA is added. The metallochromic indicator should form a colored chelate with magnesium ions, so that it may change color when enough EDTA has been added to react with both metals. The indicator originally proposed, and still widely used due to inertia on the part of analysts, was **eriochrome black T**, a dihydroxyazo structure:

Eriochrome black T

H_2E^-

---

*The discussions that follow are based on later improvements in the method, taken from the following papers:

F. Lindstrom, H. Diehl, *Indicator for the Titration of Calcium Plus Magnesium with EDTA*, Analytical Chemistry, *32*, 1123 (1960).

G. Hildebrand, C. N. Reilley, *New Indicator for Compleximetric Titration of Calcium in Presence of Magnesium*, Analytical Chemistry, *29*, 258 (1957).

Both papers are recommended for study because they treat not only the analytical problem, but also the equilibrium theory underlying the methods, with determinations of the pertinent equilibrium quotients by methods of spectrophotometry.

This substance behaves both as an acid–base indicator

$$H_2E^- \xrightarrow{pQ=6.3} HE^{2-} \xrightarrow{pQ=11.5} E^{3-},$$

red   blue   orange

where the color changes accompany the successive loss of the hydroxyl protons, and as a metallochromic indicator for magnesium

$$Mg^{2+} + E^{3-} = MgE^-.$$

orange   red

Evidently the magnesium ion forms coordinate bonds with the two hydroxyl oxygens and also with the azo group, forming two five-member rings.

In the EDTA titration of a mixture of calcium and magnesium the solution is buffered at $pH = 10$, which is not high enough to cause precipitation of $Mg(OH)_2$. A small amount of indicator is added, and the red MgE^- complex is formed. As the EDTA is added, the sequence of reactions is as follows:

$$Ca^{2+} + EDTA = CaY^{2-}$$

$$Mg^{2+}(free) + EDTA = MgY^{2-},$$

$$MgE^- + EDTA = MgY^{2-} + HE^{2-}.$$

red   blue

Thus, the visual endpoint does not occur until all the calcium and magnesium is titrated.

An inherent weakness of eriochrome black T is its lack of stability in the indicator stock solution. The decomposition products spoil the quality of the endpoint. Lindstrom and Diehl assumed that this instability was due to the simultaneous presence in the molecule of an oxidizing group $-NO_2$ and reducing groups $-OH$. Since the key part of the indicator structure seemed to be the dihydroxyazo linkage

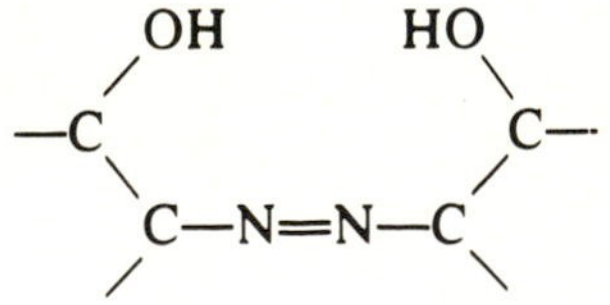

for the coordination of the magnesium ion, they searched through a large number of structures and settled on a dye with the following structure:

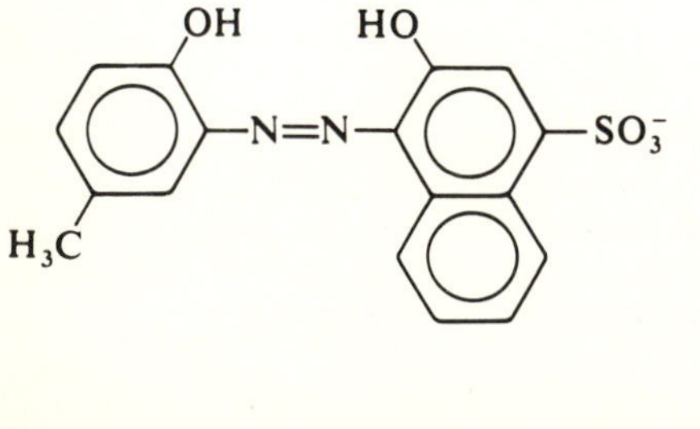

Calmagite:
1-(1-hydroxy-4-methyl-2-phenylazo)-
2-naphthol-4-sulfonate

This substance, which they christened **calmagite**, turned out to be highly stable in solution, with no decomposition or impairment of indicator function over a year in storage. In addition, calmagite gives a somewhat sharper endpoint than does eriochrome black T, and it is now recommended that this indicator be substituted in applications that call for the latter.

The acid dissociation quotients for calmagite were determined by spectrophotometric measurements. The indicator is a weaker acid than eriochrome black T, as might be predicted, since there is no longer a nitro group to help with the dispersal of the anion charge:

$$H_2C^- \xrightarrow{pQ=8.14} HC^{2-} \xrightarrow{pQ=12.35} C^{3-}.$$

$$\begin{array}{ccc} \text{bright} & \text{clear} & \text{reddish-} \\ \text{red} & \text{blue} & \text{orange} \end{array}$$

As with eriochrome black T, the magnesium chelate is red, and the visual endpoint in the EDTA titration (at pH = 10) is a rapid change from red to clear blue.

Taking the formation quotient for MgC^- to be $1 \cdot 10^8$, we get the pM–pH diagram for the titration as depicted in Fig. 16.5. We see a rather good correspondence between the equivalence point and the endpoint at pH = 10. This pH is achieved by adding a few milliliters of a stock buffer solution prepared by mixing 570 mL of concentrated ammonia and 70 g of solid ammonium chloride, and diluting to one liter. The calmagite indicator solution is prepared by dissolving about 50 mg of calmagite in 100 mL of water. When the water sample may contain iron and manganese, it is necessary to add masking agents to prevent these and other metals from "blocking" the indicator by forming such stable chelates that the indicator cannot be released

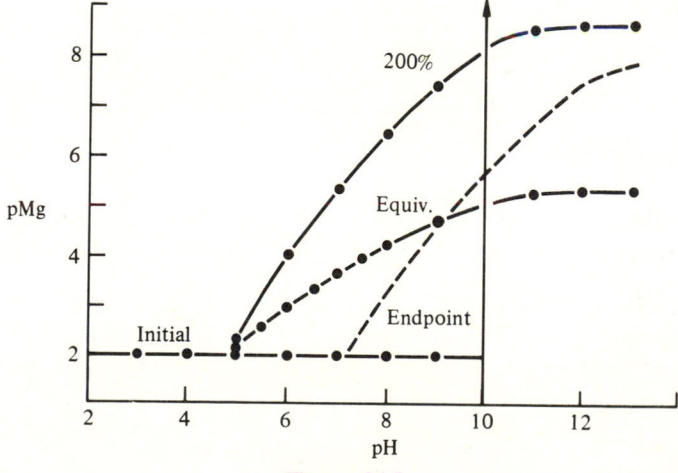

Figure 16.5

at the endpoint. Recently it was shown by Fritz and King* that the interfering metals may be removed by first passing the water sample through a specially prepared ion exchange column, without losing any calcium or magnesium.

Determination of Calcium Only

As discussed above, a water sample containing both calcium and magnesium may be titrated with EDTA, using calmagite indicator, to find the sum of the concentrations of these metals. If either of the metals can be determined by a separate method, the other may be calculated by a difference. A number of possibilities exist for the separation of calcium or magnesium, leaving only one metal to be determined, but the most common practice is simply to adjust the solution pH to a high value (above 12), so that the magnesium hydroxide $(K_{sp} = 2 \cdot 10^{-11})$ is quantitatively precipitated, while the calcium stays in the solution.

Hildebrand and Reilley reported that a dye structure similar to that of calmagite, but with weaker acidic properties, would be desirable as an indicator for calcium at this high pH. They studied the following substance, which they named calcon:

Calcon: H_2Cn^-

The first acid dissociation is stronger than that for calmagite, but the second is weaker:

$$H_2Cn^- \xrightarrow{\text{p}Q=7.4} HCn^{2-} \xrightarrow{\text{p}Q=13.5} Cn^{3-}.$$
$$\text{pink} \qquad\qquad \text{blue} \qquad\qquad \text{pink}$$

The formation quotient for the calcium chelate of calcon is not particularly large:

$$Ca^{2+} + Cn^{3-} = CaCn^-, \qquad Q_{CaCn} = 4 \cdot 10^5.$$
$$\text{red}$$

However, if the titration with EDTA is taken to the pure blue color of HCn^{2-}, which is the dominant form of the indicator at the recommended pH of 12.3, the results for calcium are accurate with the typical uncertainty of about 0.5%.

---

*J. Fritz, J. King, *Analytical Chemistry*, **48**, 570 (1976).

16.10   POTENTIOMETRIC TITRATIONS OF METAL IONS WITH CHELONS

Although the use of visual metallochromic indicators in chelometric titrations can be very convenient and rapid, as would be important in, say, repetitive determinations of water hardness, there are obviously many complicated equilibrium relationships which must be solved before the observed endpoint agrees with the true equivalence point. These difficulties may often be avoided completely by using a mercury electrode as the indicator electrode in a potentiometric titration. The cell may be regarded as follows:

$$\text{SCE} \,|\, \text{Solution being titrated} \,|\, \text{Hg}.$$

To illustrate the important relationships, we will suppose that the initial solution is 100 mL of $0.01 M$ Ca^{2+} (which is to be determined), that it contains $1 \cdot 10^{-4} M$ Hg–EDTA, and that the pH is about 9.4, making $\alpha_0 = 0.1$ for EDTA. This mixture is to be titrated with $0.1 M$ EDTA, so that 10.00 mL will be required to reach the equivalence point.

In the first phase of the titration, before the equivalence point is reached, the addition of EDTA causes the reaction to occur. If V milliliters of $0.1 M$ EDTA ($0.1 V$ millimole) is added to 100 mL of $0.01 M$ Ca^{2+} (1 mmole), we have:

	Ca^{2+}	+	EDTA	$\leftrightarrows$	Ca–EDTA
Initial	1 mmol		$0.1 V$ mmol		0
After reaction	$1 - 0.1 V$		"0"		$0.1 V$ mmol

Of course, the equilibrium molarity of free EDTA is not actually zero because of the reversibility of the reaction. Whatever small amount exists at equilibrium will determine what fraction of the Hg–EDTA is dissociated into Hg^{2+} ions. Hence, the potential of the mercury electrode is directly dependent upon the extent to which the reaction has occurred. The mathematical principle for this situation has been developed in Chapter 15, where Eq. (15.6) is appropriate. Using the data for the above titration (taking $Q_{HgY} = 10^{22.1}$, $Q_{CaY} = 10^{10.7}$) we have:

$$E = 0.597 - 0.0296 \log \frac{[CaY]}{[Ca^{2+}]} - 0.0296 \log \frac{1}{[HgY]} - 0.0296 \log \frac{10^{22.1}}{10^{10.7}}$$

$$= 0.141 - 0.0296 \frac{0.1 V}{1 - 0.1 V}$$

Problem Calculate E for $V = 0.1, 0.5, 1, 2, 4, 6, 8, 9, 9.5$, and 9.9 mL.

Once the equivalence point has been passed, the calcium is quantitatively complexed, and any extra EDTA simply exists in the free state. The appropriate equation for the cell under these conditions is Eq. (15.3):

$$E = 0.597 - 0.0296 \log \alpha_0 - 0.0296 \log \frac{C_Y}{[HgY]} - 0.0296 \log Q_{HgY}.$$

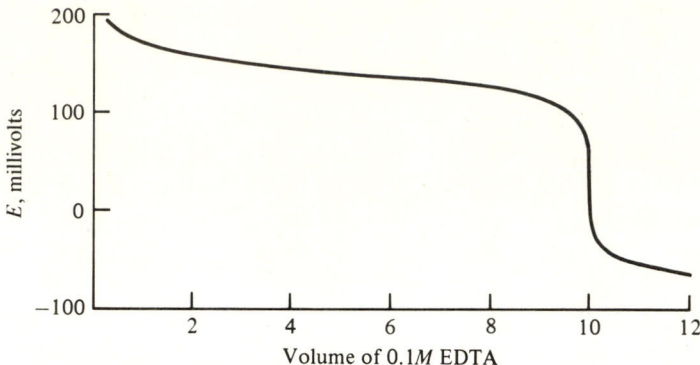

Fig. 16.6 Predicted titration curve for the potentiometric titration of 100 mL of $0.1M$ Ca^{2+} with EDTA at pH = 9.4.

Remembering that $[HgY] = 1 \cdot 10^{-4}$ (ignoring the small dilution effect), we have:

$$E = -0.146 - 0.0296 \log \frac{0.1(V - 10.00)}{110 + V}.$$

Problem Calculate E for $V = 10.1, 10.5, 11, 12$ mL.

When the values for E are plotted versus V, the graph reveals a fairly sharp change in potential in the immediate vicinity of the equivalence point, as shown in Fig. 16.6. The precise endpoint may be found by the techniques explained in Chapter 11. Potentiometric titrations may be slow, but they are more accurate than indicator methods and can also be useful in resolving mixtures of metal ions. A suitable mercury indicator electrode can be a small mercury pool in a J-tube, or a mercury-plated platinum wire.

PROBLEMS

Photometric titrations with EDTA

1. The following spectra in Fig. 16.7 refer to the perchlorates of lead (Pb^{2+}), bismuth (BiO^+), and their chelates with the strongly complexing ligand EDTA.* Neither perchlorate ion nor free EDTA show significant absorption.

    In the following problems assume that the titrations are carried out in a beaker mounted in the light path of a spectrophotometer. The path length is 5 cm and the initial solution volume is 100 mL. Ignore dilution effects.

   a) A solution containing $2 \cdot 10^{-5}M$ lead perchlorate is titrated with $0.001M$ EDTA. Predict the titration curves that would be obtained at the following wavelengths: 220, 240, and 260 nm. Which wavelength is best?

   b) Do the same for the titration of bismuth, using wavelengths of 230 and 265 nm.

---

*R. Wilhite, A. Underwood, *Analytical Chemistry*, **27**, 1334 (1955).

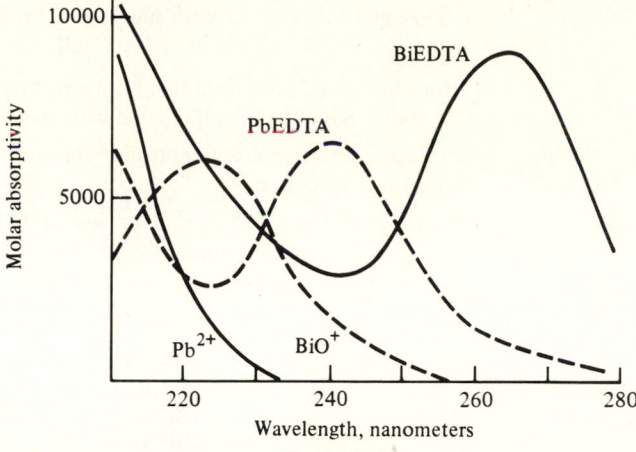

Figure 16.7

c) Suppose it is desired to carry out titrations with more-concentrated solutions, say $0.002M$ metal ions and $0.1M$ EDTA. If it is not practical to use absorbance readings higher than about 0.8, what wavelength(s) would you recommend?

d) Predict the titration curve (both endpoints) for the titration of a mixture containing $3 \cdot 10^{-5}M$ lead and $1 \cdot 10^{-5}M$ bismuth with $0.001M$ EDTA. Consider wavelengths of 220, 230, 240, 250, and 260 nm and choose the most favorable. The bismuth will be titrated before the lead as it forms a more stable chelate.

2. The spectra in Fig. 16.8 show the enhancement of the blue color which occurs when

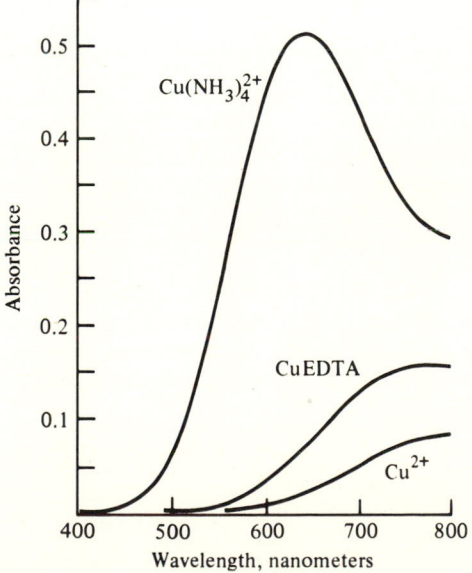

Figure 16.8

copper(II) is chelated with EDTA and complexed with ammonia. Each curve was obtained with a copper concentration of $7.9 \cdot 10^{-3} M$ in a 1-cm cell.

a) Predict the photometric titration curve that would be obtained if $0.002M$ copper is titrated with $0.2M$ EDTA solution in a 5-cm cell using a wavelength of 770 nm.

b) Why would it *not* be advantageous to use a concentrated solution of ammonia as titrant for lower concentrations of copper?

c) What is the molar absorptivity of the $Cu(NH_3)_4^{2+}$ species at 635 nm? What concentration of copper would yield an absorbance of 0.02 at this wavelength if it were present as Cu^{2+}, and if it were present as the tetrammine complex?

3. Figure 16.9 shows the absorption spectra of copper and thorium ions and their complexes with EDTA.* These ultraviolet data were obtained by using 1-cm quartz cells and the following solutions: Curve A: $5 \cdot 10^{-4} M$ CuEDTA; curve B: $0.01M$ EDTA; curve C: $0.01M$ Cu(NO_3)_2$; curve D: $0.01M$ Th(NO_3)_4$.

   A solution of thorium–EDTA showed virtually zero absorbance in this wavelength range. The various protonated forms of EDTA do not absorb significantly.

   a) By reading the graph, estimate the molar absorptivities of the four species: Cu^{2+}, EDTA, CuEDTA, and ThEDTA, at a wavelength of 300 nm.

   b) The authors prepared a standard solution of copper by dissolving 0.7836 gram of pure electrolytic copper metal in about 6 mL of concentrated nitric acid, followed by dilution to precisely one liter. Calculate the copper molarity in the resulting solution.

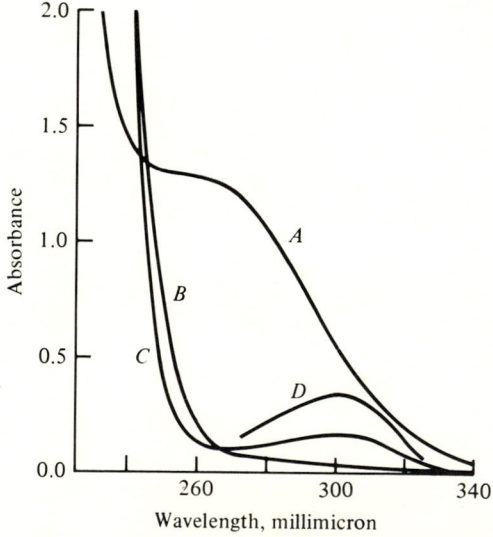

Fig. 16.9 Ultraviolet absorption spectra.

---

*H. Malmstadt, E. Gohrbandt, *Analytical Chemistry*, **26**, 442 (1954).

c) They prepared a solution of thorium by dissolving thorium nitrate in dilute nitric acid. This solution was standardized by precipitation of the oxalate, followed by ignition to form the oxide:

$$Th(IV) \text{ (aqueous)} \rightarrow Th(C_2O_4)_2 \xrightarrow{\Delta} ThO_2$$
$$\text{precipitate} \quad \text{weighing form}$$

If a 50.02-mL portion of the solution yielded 0.1277 gram of ThO_2, what is the molarity of the solution? The molecular weight of ThO_2 is 264.04.

d) To test the photometric-titration technique for thorium, a 5-mL sample of the thorium solution and a 1-mL sample of the copper solution were diluted to 100 mL, and then titrated with 0.00962M EDTA at pH $= 3$ and 300 nm using a cell path of 5 cm. Thorium forms the more stable complex and therefore reacts before the copper. Make an accurate prediction of the photometric titration curve.

4. A solution contains about 0.01M lead and about 0.001M nickel. We desire to determine the precise concentration of each metal by using chelometric titration with standard HEDTA as titrant. This chelon is like EDTA except that one of the carboxyl groups is replaced by a hydroxymethyl group, making it a triprotic acid:

$$H_3Y \xrightarrow{pQ_1 = 2.6} H_2Y^- \xrightarrow{pQ_2 = 5.3} HY^{2-} \xrightarrow{pQ_3 = 9.7} Y^{3-}.$$

Both metals form stable HEDTA chelates:

$$NiY, \qquad \log Q_{Ni} = 17,$$
$$PbY, \qquad \log Q_{Pb} = 15.5.$$

In the first approach to this analysis the technique of photometric titration was used. The pH of a 10.00-mL sample was adjusted to 4.0 and the spectrophotometer wavelength was set to 1000 nm, where only the nickel–HEDTA chelate absorbs. Titration was carried out in a 1-cm absorption cell, using 0.02000M HEDTA in a microburet. The following data were obtained (corrected for dilution):

Volume, mL	Absorbance	Volume, mL	Absorbance
0	0.000	0.4	0.240
0.1	0.062	0.5	0.261
0.2	0.124	0.6	0.268
0.3	0.187	0.7	0.268

The wavelength was then changed to 240 nm, where the Pb chelate absorbs much more strongly than the Ni chelate. The titration was continued on the same sample, but the titrant was changed to a 0.2000M solution. The data were as follows.

Volume, mL	Absorbance	Volume, mL	Absorbance
0	0.054	0.5	0.500
0.1	0.142	0.6	0.535
0.2	0.235	0.7	0.540
0.3	0.324	0.8	0.544
0.4	0.411	0.9	0.544

a) Make very accurate plots of these data and determine the amounts (millimoles) of Ni and Pb in the sample as accurately as you can.

It was decided to attempt a potentiometric titration of the lead–nickel solution, using a lead metal electrode that would (it was hoped) respond to the free-lead-ion concentration throughout the titration with HEDTA. First the nickel would be removed by precipitation with dimethylglyoxime, and then the solution containing the lead only would be placed in the cell:

$$SCE \| Solution \,|\, Pb(s).$$

b) Show how the potential of this cell may be used to calculate the value of pPb at any point in the titration. $E_{SCE} = 0.244$, $E^0_{Pb} = -0.13$.

c) Given the initial concentration of lead, $0.01M$, and the formation quotient for the HEDTA chelate, make a neat sketch of the titration curve: pPb versus volume of HEDTA, that would be expected for the titration of 100 mL of the solution with $0.100M$ titrant. For this sketch assume that there is no buffer complexing of the lead and no protonation of the chelon.

d) Suppose the potentiometric titration is carried out at pH $= 5$ in a buffer containing $0.10M$ acetate ion. There will be formation of lead–acetate complexes:

$$PbOAc^+, \qquad Q_1 = 10^{2.7},$$
$$Pb(OAc)_2, \qquad Q_2 = 10^{1.5}.$$

In addition there will be competition by hydrogen ion for the Y^{3-} ligand (see the pK values given earlier). Calculate F_0 for the lead–acetate system and α_0 for the chelon, and explain how the titration curve will differ from that in part (c). Sketch the modified curve on the same plot as used for part (c).

5. In the chelometric titration of 100 mL of a copper solution, 2.00 mL of $0.050M$ chelon were required to reach the equivalence point. The solution contained a buffer that maintained the pH at 5.0, making the value of $\alpha_{chelon} = 10^{-5}$. The buffer-conjugate base formed copper complexes, making $F_0 = 10^2$. The formation quotient for the copper chelate is 10^{19}. Make a neat and reasonably accurate sketch of the titration curve, pCu versus volume of added chelon, using calculated pCu values for the initial point, the equivalence point, and the 200% titrated point.

SOLUBILITY: THEORETICAL EQUILIBRIUM MODELS

17

Nature is the real antagonist—the friendly enemy who never cheats—always plays fair—but never fails to take advantage of the tiniest oversight or omission.

Arthur C. Clarke
in *The Fountains of Paradise*

17.1 SOLUBILITY: CONCEPTS

Chapter 6 dealt with the formation of precipitates and, for applications such as gravimetric determination, the goal was to drive the precipitation reaction to completion to decrease the loss by solubility as much as possible. In this chapter we will emphasize the reverse process, studying those chemical reactions that cause an "insoluble" substance to enter the aqueous phase and become part of the homogeneous solution. By applying chemical principles already studied in earlier chapters we will discover ways to *encourage* the dissolving of a substance as may be desired for an analytical procedure. We will also explore, in Chapter 18, the use of solubility measurements as a tool, one might say a probe for the indirect study of a number of complexation and acid–base reactions which are, in themselves, not related to solubility.

First consider a simple definition of solubility: the **solubility** S of a substance is the molarity of that substance (counting all its solution species) in a solution that is at chemical equilibrium with an excess of the undissolved substance. This implies that there must also be a uniform temperature throughout the system, because the numerical value of S is typically very dependent upon the temperature.

Substances vary widely in solubility. Consider a series of silver compounds in terms of the amounts (moles) that will dissolve when the solid substance is placed in contact with one liter of water and allowed to reach equilibrium, i.e., allowed to disssolve until the solution is saturated:

Compound	Moles per liter of water
silver perchlorate $AgClO_4$	27.0
silver chlorate $AgClO_3$	1.0
silver chlorite $AgClO_2$	0.03
silver chloride $AgCl$	0.000013

By contrast, potassium perchlorate has a solubility of only about 0.1 mol/L, while potassium chloride is quite soluble, about 4.2 mol/L.

In the early days of the development of analytical chemistry as a science it was particularly important for chemists to have a feeling for the relative solubilities of inorganic salts. It was convenient to group compounds into classes as follows:

soluble:	more than 50 g/L
moderately soluble:	10–50 g/L
slightly soluble:	1–10 g/L
moderately insoluble:	0.01–1 g/L
insoluble:	less than 0.01 g/L.

As discussed in the present and following chapters, solubility equilibria are often complicated. However, it is reasonable to offer a few useful generalizations about the solubility behavior of metal–anion compounds in water:

- Acetates, nitrates, chlorates and perchlorates of the metals are soluble.
- Chlorides are generally soluble, except for AgCl, Hg_2Cl_2, BiOCl, and SbOCl. Lead chloride $PbCl_2$ is slightly soluble.
- Iodides are generally soluble, except for AgI, PbI_2, Hg_2I_2, CuI, BiOI, and SbOI. BiI_3 and SnI_2 are slightly soluble.
- Sulfides are generally insoluble, except that sodium, potassium, and ammonium sulfides are soluble, while barium, strontium, and calcium sulfides are slightly soluble.
- Sulfates are generally soluble, except for $PbSO_4$, Hg_2SO_4, $BaSO_4$, and $SrSO_4$, which are insoluble. Ag_2SO_4, $HgSO_4$, $CaSO_4$ are slightly soluble.
- Carbonates are generally insoluble, except for sodium, potassium, and ammonium carbonates. The insoluble carbonates dissolve in strong acid.
- Oxalates are generally insoluble, except for sodium, potassium, and ammonium.
- Phosphates are generally insoluble, except for sodium, potassium, and ammonium.
- As might be inferred from the above, most sodium, potassium, and ammonium salts are soluble.

Theories of chemical bonding have had only modest success in trying to explain such variations in solubility. Predictions are typically based on correlation and extrapolation of already discovered facts, rather than on first principles. This might be called rationalizing rather than theorizing. However, we may identify a few important factors that pertain to the solubility of a substance.

1. The molecular or ionic forces that hold a substance together must be overcome if the substance is to be dispersed as an aqueous solute. Such forces are quite strong in the crystal lattice of sodium chloride, with positive Na^+ ions and negative Cl^- ions exerting electrostatic attraction. The forces are only moderate in the crystalline structure of solid iodine, where there are no ions but rather I_2 molecules in the lattice. In gaseous helium, interatomic forces are virtually nonexistent. Yet, we find that sodium chloride is much more soluble than iodine, which is much more soluble than helium. Obviously, other factors must also be important.

2. When a substance attempts to enter an aqueous phase, it can do so only by penetrating the hydrogen-bonded structure of water. The hydrogen bonds must be broken before the solute species, ion or molecule, can fit into the liquid structure. In common with the disruption of the crystal structure of the substance, this step requires energy. The lattice attractions and the hydrogen–bonding in water both must be overcome.

3. Substances that can overcome the above two energy barriers by strong chemical interactions with water molecules or with other solutes present in the solution are soluble. Some examples are as follows:

- Why is ammonia NH_3 so much more soluble than its analog phosphine PH_3? Both molecules have an unshared pair of electrons, but ammonia has a very strongly electronegative element nitrogen and is capable of forming hydrogen bonds with water molecules, while phosphine is not. In terms of hydrogen bonding the NH_3 molecules are similar to water molecules.

- Sulfuric acid readily dissociates into ions H_3O^+ and HSO_4^- by direct reaction with water. The ions attract the dipolar water molecules. Sulfuric acid is miscible with water in all proportions. By contrast, sulfurous acid H_2SO_3 is only weakly dissociated ($pK = 1.8$) and is less soluble, being less capable of breaking the hydrogen-bonded water structure.

- Solid iodine has a low solubility, about 0.0013 molar. The molecules of I_2 are nonpolar and do not interact strongly with water molecules. However, if the solution also contains potassium iodide, the solubility of iodine markedly increases because of the formation of triiodide ion I_3^- that interacts with polar water molecules by virtue of its electrostatic charge.

- Calcium carbonate $CaCO_3$ has a strong lattice energy and is insoluble in water. If the solution contains acid, however, the carbonate ion is protonated to form HCO_3^-, then H_2CO_3 and carbon dioxide. Therefore, calcium carbonate is readily soluble in acidic solutions and effervesces, rapidly giving off bubbles of CO_2.

In general we observe that polar solutes tend to dissolve in polar solvents, while nonpolar solutes dissolve in nonpolar solvents. For a long time chemists have summarized this by the maxim: *like dissolves like.*

17.2 CHEMICAL MODELS FOR SOLUBILITY; ALGEBRAIC ACCOMPANIMENTS

It is usually possible to find the solubility S (in moles per liter) of a substance in an aqueous solution. Some type of analysis of a portion of the saturated solution is required to find the total concentration of the substance in all of its possible forms in the solution. For the analysis as such it may not be necessary to have any idea of what those solution species acually are. For example, the solubility of a compound may be determined if a sample of the compound is prepared and labeled with a radioactive isotope. By measuring the radioactivity of a sample of the saturated solution one can tell immediately how much of the compound dissolved, without any need to know what chemical changes it underwent in the solution process. Other analytical procedures, such as atomic absorption spectrophotometry, are likewise nonspecific as far as varieties of solution species are concerned.

If we are to understand a given solubility-equilibrium system, a greater challenge must also be met, and that is the identification of the actual solution species formed by the dissolved substance, and the determination of the values for the equilibrium quotients that describe the interrelationships mathematically. We will see that these two aspects, identification and Q

values, are often very closely intertwined. It is here that research chemists have often fallen far short of the challenge. The deficiency is due partly to human failings of judgment and partly to the inherent intractability of many solubility systems. Researchers have often simply assumed the presence or absence of certain solution species and have then interpreted analytical data with poor accuracy. Too often there has been neglect of activity coefficients, no temperature control, use of reagents that are impure or have undesired effects, and inappropriate techniques of data analysis. One is reminded of the method described by Plato in *Phaedo*:

> This was the method I adopted: I first assumed some principle, which I judged to be the strongest, and then I affirmed as true whatever seemed to agree with this, whether relating to the cause or to anything else; and that which disagreed I regarded as untrue.

Therefore the starting point for building a chemical model (i.e., a scheme of interrelated chemical reactions) for a solubility-equilibrium system is best regarded as a *hypothetical* proposal for the identities of which solution species are present in significant quantities. With such a hypothetical list of species we may write a material-balance expression:

$$S = \Sigma C_i.$$

This merely acknowledges that we may express the total amount of dissolved substance as the summation of the molarities C_i of all the various species formed in the saturated solution.

The second step in making the chemical model is to write equations for the reversible chemical reactions that show the transformations of these species into one another. For each reaction, an expression for the equilibrium quotient may also be written, even if its numerical value may be unknown. The chemical model may be completed by adding judgments about which species are probably dominant under this or that set of solution conditions, such as pH, presence of a ligand, etc.

The algebraic counterpart of the chemical model is derived as a combination of the material-balance expression with the several expressions for equilibrium quotients. The power of the algebraic model is twofold: it enables us to make predictions about the solubility behavior, if we use known or assumed values for the equilibrium quotients. Also, given experimental data for the solubility under controlled solution conditions, the algebraic model may permit us to *test the adequacy of the chemical model* by seeing whether the experimental behavior is really in accord with the algebraic model. If it is in accord, then by methods of numerical analysis we may be able to deduce the values for the equilibrium quotients which are part of the model.

In the following sections we will derive models for a few relatively simple systems of importance to analytical chemistry, and show how a knowledge of the Q values can be valuable in making predictions and thereby controlling solution conditions to our advantage.

In Chapter 18 we will examine the use of the same models for finding the Q values of various systems, using experimental data for the solubility under varying conditions.

17.3   DERIVATION OF ALGEBRAIC MODELS FOR SELECTED SYSTEMS

I. Systems with No Side Effects

Without assuming anything about the actual step-by-step mechanism of the solubility process we may propose the following model for the equilibrium system of a $1:1$ compound in contact with its saturated solution:

$$ML(s) = ML(aq) = M^{z+} + L^{z-}.$$

The intrinsic solubility is simply the value of the equilibrium quotient for the first stage of this scheme, the equilibrium between the solid substance and its molecular counterpart in the saturated solution:

$$Q_0 = \frac{[ML(aq)]}{X_{ML(s)}};$$

if the solid is pure, then $X_{ML} = 1.00$ and

$$Q_0 = [ML(aq)].$$

When the molecular species dissociates into ions, we may consider either a dissociation quotient

$$Q_D = \frac{[M^{z+}][L^{z-}]}{[ML(aq)]} = \frac{[M^{z+}][L^{z-}]}{Q_0}$$

or a formation quotient of the reverse reaction

$$Q_F = \frac{1}{Q_D} = \frac{[ML(aq)]}{[M^{z+}][L^{z-}]} = \frac{Q_0}{[M^{z+}][L^{z-}]}.$$

In those cases where the substance can undergo stepwise dissociation (or formation) we must use an equilibrium quotient for each step, numbering the steps with subscripts.

The solubility product is defined as the equilibrium expression that relates the concentrations of the final, dissociated ions to the solid substance. It is an overall equilibrium quotient that reveals nothing about the concentrations of the intermediate species. For the above simple case, if $ML(s)$ is pure, we have:

$$Q_{sp} = \frac{[M^{z+}][L^{z-}]}{X_{ML(s)}} = [M^{z+}][L^{z-}]. \tag{17.1}$$

Inspection of the above equations will show that the solubility product is directly related to the more fundamental quotients:

$$Q_{sp} = Q_0 Q_D = Q_0/Q_F.$$

When a solute is a weak electrolyte, it is often possible to determine the numerical values for Q_0 and Q_F, in which case Q_{sp} becomes redundant. But for strong electrolytes, that are so nearly completely dissociated that there is no measurable evidence for the existence of a molecular species, we must rely on Q_{sp} expressions and values to make calculations. In other words, it is possible for both Q_0 and Q_F to be too small to be determined, and yet their *ratio* Q_{sp} may be finite and measureable. For example, the solubility of potassium perchlorate is about $0.1M$ at $25°$. There is no evidence for the finite existence of $KClO_4$ molecules (better called *ion pairs*), implying that both Q_0 and Q_F are equal to "zero". Nevertheless,

$$Q_{sp} = [K^+][ClO_4^-] = 0.1 \cdot 0.1 = 0.01.$$

Since the strong-electrolyte case is merely a special example of the more general weak-electrolyte system, we consider the model for the latter. There are three situations to examine:

A) SATURATED SOLUTION OF ML WITH NO OTHER SOURCES OF EITHER ION

The material-balance expression may be written (omitting ionic charges for simplicity) in either of the two ways:

$$S = [ML] + [M] \tag{17.2}$$
$$S = [ML] + [L]. \tag{17.3}$$

Recognizing that $[ML] = Q_0$ and that $Q_{sp} = [M]^2 = [L]^2$, we find

$$S = Q_0 + (Q_{sp})^{1/2}. \tag{17.4}$$

B) SATURATED SOLUTION OF ML IN THE PRESENCE OF ADDITIONAL LIGAND L

When the solution contains some other solute that can contribute L ions to the system, the dissociation of ML will be partly repressed and the concentrations of M and L will not be equal. It would not be correct to use Eq. (17.3) for the material balance because part of [L] would be due to the other source, not to solubility of ML. However, Eq. (17.2) would still be valid because there is no other source of M. Since [M] and [L] are unequal in this case, Eq. (17.4) is not valid for it was derived on the assumption of equal molarities. The solubility-product expression (17.1) itself is still valid, and therefore we may write

$$[M] = \frac{Q_{sp}}{[L]}. \tag{17.5}$$

It is very important to realize that [L] in this expression refers to the equilibrium molarity of L and that this is the total concentration contributed both by the dissociation of ML and by the second source of L. By substituting Eq. (17.5) into Eq. (17.2) of the material balance we get as a result

$$S = [ML] + \frac{Q_{sp}}{[L]} = Q_0 + \frac{Q_{sp}}{[L]}. \tag{17.6}$$

Thus we see that the solubility of the substance is inversely proportional to the equilibrium concentration of the ligand. A prediction of the value for S requires three quantities: Q_0 (which is often assumed to be zero), Q_{sp}, and [L].

Now, how is one supposed to know the value for [L] when L comes from two sources? Even if the extra L from the added source is known, that contributed by the solubility of ML is not obviously apparent. It's time for a supporting derivation that can make Eq. (17.6) more useful.

With L coming into the solution from two sources, we may write

$$[L] = C + [M],$$

where C is the concentration of L from the added source and [M] is a direct measure of how much ML has dissociated and therefore contributed to [L]. But, from Eq. (17.2) we note that $[M] = S - Q_0$, and so

$$[L] = C + S - Q_0.$$

This relationship is substituted into Eq. (17.6) to give

$$S = Q_0 + \frac{Q_{sp}}{C + S - Q_0}. \qquad (17.7)$$

Equation (17.7) is a useful form because it contains only constants (Q values) and variables (S and C) which can be determined experimentally (S can be measured and C is known just by stoichiometry considerations of what was put into the solution).

To use Eq. (17.7) for predictions of solubility, given previously determined values for Q_0 and Q_{sp}, we first recognize that it can be rearranged to the standard quadratic form:

$$S^2 + (C - 2Q_0)S - (Q_0C + Q_{sp} - Q_0^2) = 0. \qquad (17.8)$$

This may be solved by using the quadratic formula, once values for Q_0, Q_{sp}, and C are supplied. Of course, in the case of a strong electrolyte, Q_0 is assigned a value of zero and the equation becomes:

$$S^2 + CS - Q_{sp} = 0. \qquad (17.9)$$

This illustrates an important idea. It is easy to go from Eq. (17.8) to Eq. (17.9) just by letting $Q_0 = 0$. The reverse process would be impossible because, without going through the whole derivation, one could not imagine just how Q_0 would fit in. The important idea is this: It is often better to derive relationships for the more general case and to simplify them as desired for the other situations.

Example calculation using Eq. (17.8)

Silver acetate AgOAc is a weak electrolyte of low solubility. The value for Q_0 has been found to be 0.0100 and the value for Q_F is 3.00, both values referring to ionic strength of 0.50. Therefore, $Q_{sp} = 0.01/3 = 3.33 \cdot 10^{-3}$. To predict the solubility of silver acetate in a solution of $0.500M$ sodium acetate we find the solution to

$$S^2 + (0.500 - 2 \cdot 0.0100)S - (0.0100 \cdot 0.500 + 3.33 \cdot 10^{-3} - 1.0 \cdot 10^{-4}) = 0.$$

The answer is $S = 0.0166M$.

C) SATURATED SOLUTION OF ML IN THE PRESENCE OF ADDITIONAL METAL ION

When it is the M ion that is present from an additional source, the same logic applies. Simply substitute [M] for [L], and vice versa, in the foregoing discussion. Whether extra M, or extra L is added, the solubility of ML is decreased by the common-ion effect. However, it can go no lower than the value of Q_0.

II. Systems Involving Common-Ligand Complexing

In the previous discussion of the simplest system it was assumed that the only effect of the ligand L was to repress the dissociation of ML. This means that the solubility of ML decreases continually as the equilibrium molarity of the ligand increases. Equation (17.6) is in accord with this view. But it is found by experimental measurements on a variety of chemical systems that the solubility may decrease at first, while the ligand concentration is low, but then will pass through a minimum and actually increase as the ligand concentration is made larger. An excellent example of this effect is seen in the report by Jonte and Martin* who used radioactive silver chloride in a series of solutions of sodium chloride; the work covered a wide range of chloride ion concentrations, as shown in Fig. 17.1.

In this figure the molarity of chloride is increasing from right to left, reaching a highest value of about $0.1M$ (pCl $= 1.00$). At the right we see the common-ion effect in action, with the solubility decreasing markedly as the addition of chloride represses the dissociation of AgCl. To account for the minimum in the solubility trend we must postulate a chemical reaction that creates a species of silver which is more soluble than molecular AgCl. The first guess is that a complex ion, dichloroargentate $AgCl_2^-$, is formed. Being an ionic species, it would tend to be stabilized by hydration forces from the polar water molecules. However, this would hold true for other possible complexes, such as $AgCl_3^{2-}$, $AgCl_4^{3-}$, etc. How do we go about proposing a chemical model?

A forceful approach is to postulate the existence of the whole series of silver-chloro complexes. Perhaps some of them will be present in negligible amount, but isn't that just a quantitative conclusion? If we write a material-balance expression for silver, assuming that only mononuclear species are formed (i.e., no species like $Ag_2Cl_4^{2-}$, etc., with more than one silver atom in

---

*J. Jonte, D. Martin, *Journal of the American Chemical Society*, 74, 2052 (1952).

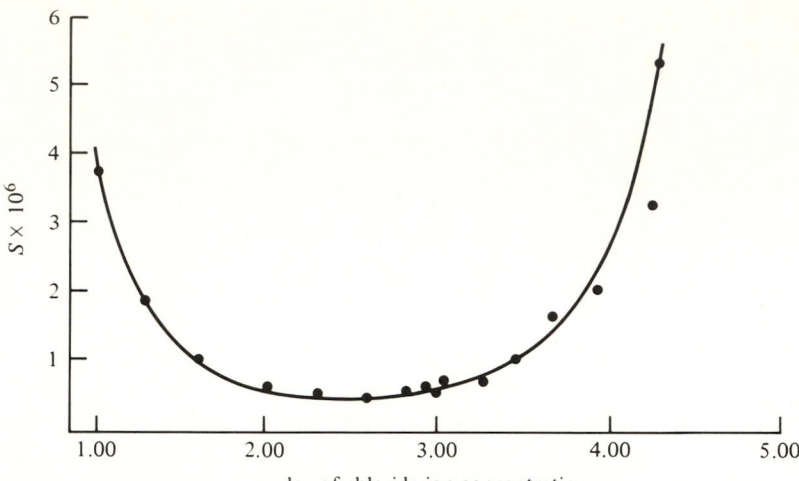

Fig. 17.1 Solubility of AgCl in NaCl solutions.

the structure are present), we come up with the following:

$$S = [Ag^+] + [AgCl(aq)] + [AgCl_2^-] + [AgCl_3^{2-}] + [AgCl_4^{3-}] + \cdots.$$

To make the derivation more manageable we shall leave out species with more than four chloride ions. If later conclusions show that such higher complexes may be present, it will be easy to extend the derived equation.

Now, in accord with the principles of Chapter 14, we may write a series of equilibrium quotients for the formation of the complexes:

$$Q_1 = \frac{[AgCl]}{[Ag^+][Cl^-]}, \qquad Q_2 = \frac{[AgCl_2^-]}{[AgCl][Cl^-]},$$

$$Q_3 = \frac{[AgCl_3^{2-}]}{[AgCl_2^-][Cl^-]}, \qquad Q_4 = \frac{[AgCl_4^{3-}]}{[AgCl_3^{2-}][Cl^-]}.$$

In addition, since we are dealing with saturated solutions of silver chloride,

$$Q_0 = [AgCl] \text{ and } Q_{sp} = \frac{Q_0}{Q_1} = [Ag^+][Cl^-]. \tag{17.10}$$

Note that each complex species may be represented as a function of the next-lowest complex and [Cl]. For example, from the above expression for Q_4, we get

$$[AgCl_4^{3-}] = Q_4[AgCl_3^{2-}][Cl^-].$$

By successive substitutions of this sort into the material-balance expression we find the following:

$$S = [Ag^+]\{1 + Q_1[Cl^-] + Q_1Q_2[Cl^-]^2 + Q_1Q_2Q_3[Cl^-]^3 + Q_1Q_2Q_3Q_4[Cl^-]^4\}. \tag{17.11}$$

We note that the F_0-function has appeared. One more algebraic substitution must be made. From Eq. (17.10), $[Ag^+] = Q_0/Q_1[Cl^-]$, and when this is substituted into Eq. (17.11), the final algebraic model is complete:

$$S = \frac{Q_0}{Q_1[Cl^-]} + Q_0 + Q_0 Q_2[Cl^-] + Q_0 Q_2 Q_3[Cl^-]^2 + Q_0 Q_2 Q_3 Q_4[Cl^-]^3.$$

$$(17.12)$$

If one prefers to use Q_{sp}, it is simple to substitute $Q_{sp}Q_1$ for Q_0:

$$S = \frac{Q_{sp}}{[Cl^-]} + Q_{sp}Q_1 + Q_{sp}Q_1 Q_2[Cl^-] + Q_{sp}Q_1 Q_2 Q_3[Cl^-]^2 + \cdots. \quad (17.13)$$

If the proposed chemical model has merit then this equation should be able to account quantitatively for the results shown in Fig. 17.1. We can see from the algebraic form that Eq. (17.12) has promise, for it includes a term inversely proportional to $[Cl^-]$, as required for the decrease in solubility at low $[Cl^-]$, and it also contains terms with $[Cl^-]$ raised to positive exponents, as required to make the value of S increase at higher chloride molarities.

The trouble is that we need numerical values to plug into Eq. (17.12), values for Q_0, Q_1, Q_2, etc. For the present let us take a recess from silver chloride with the satisfaction that we at least have built a model that is subject to testing. Later, in Chapter 18, we shall return to the actual data and see if it is possible to sharpen our view of the chemical species present and to determine numerical values for the equilibrium quotients.

As a last comment in this section, equations similar in form to (17.12) and (17.13) are applicable to any chemical system involving a 1:1 solute that can take on additional ligands to form higher complexes. Because there are many such systems, the equations are useful. Of course, there are also systems that involve solutes other than the 1:1 type, such as PbI_2, $Ag_2C_2O_4$, etc. The equations will be a bit more complex, but the principles of the derivation are the same.

III. Systems Involving "Foreign" Ligand Complexing

Now we give attention to the equilibrium of a strong electrolyte with the slightly soluble salt having the general formula MA_n. In the case of complete dissociation, the saturated solution involves the equilibrium

$$MA_n(s) = M + nA \quad \text{(charges omitted)}.$$

Since we postulate negligible intrinsic solubility, the only equilibrium expression is that of the solubility product:

$$Q_{sp} = [M][A]^n. \quad (17.14)$$

If there were no other complications, the material-balance expression for the saturated solution of MA_n would be:

$$S = [M] = \frac{[A]}{n}. \quad (17.15)$$

But we wish to consider the effects caused by a ligand L which is different from A and hence is referred to as *foreign* rather than *common*, assuming that M and L can form a series of complexes like those considered in the previous section. By analogy with Eq. (17.11), we may write the material-balance expression in terms of the concentrations of species containing the metal:

$$S = [M]\{1 + Q_1[L] + Q_1 Q_2 [L]^2 + \cdots\}. \tag{17.16}$$

At this point the derivation takes a slightly different tack, because in place of [M] we must substitute from the solubility product for MA_n rather than for ML. From Eq. (17.14),

$$[M] = \frac{Q_{sp}}{[A]^n}.$$

It is also true, from Eq. (17.15), that $[A] = nS$. When these relationships are substituted into Eq. (17.16), the result is:

$$\frac{n^n S^{n+1}}{Q_{sp}} = F_0 = 1 + Q_1[L] + Q_1 Q_2 [L]^2 + \cdots. \tag{17.17}$$

Alternatively, we may write an expression that allows us to calculate S, given the necessary Q values and ligand concentration:

$$S = \left(\frac{F_0 Q_{sp}}{n^n}\right)^{1/(1+n)}. \tag{17.18}$$

For a salt of the $1:1$ type (that is, $n = 1$), Eq. (17.18) takes the form

$$S = (F_0 Q_{sp})^{1/2}. \tag{17.19}$$

Let us apply this equation to an actual case—the effect of ammonia on the solubility of the slightly soluble salt silver iodate $AgIO_3$. First, a few facts from the literature:

- The solubility product K_{sp} for silver iodate has been determined and redetermined for the past 75 years by workers using different techniques. We may take $K_{sp} = 3.09 \cdot 10^{-8}$ as the most probable value.

- Work by Renier and Martin* on the solubility of silver iodate in the presence of added iodate showed that the intrinsic solubility of silver iodate is quite small, about $1 \cdot 10^{-7}$. We will therefore neglect this factor, and consider silver iodate to be a strong electrolyte.

- By using galvanic-cell measurements with a silver electrode it was found by Nasanen† that silver ion forms only two complexes with ammonia, $AgNH_3^+$ and $Ag(NH_3)_2^+$. We adopt his equilibrium constant values, $K_1 = Q_1 =$

---

*J. Renier, D. Martin, *Journal of the American Chemical Society*, 78, 1833 (1956).
†R. Nasanen, *Acta Chemica Scandinavica*, 1, 763 (1947).

$2.04 \cdot 10^3$ and $K_2 = Q_2 = 7.76 \cdot 10^3$. *Interesting note*: It is unusual that the second K value is larger than the first.

- We will use the above information with (17.19) to calculate the solubility of silver iodate for various molarities of the free (uncomplexed) ligand NH_3. A little problem arises: although Q_1 and Q_2 are not dependent upon ionic strength because the activity coefficients cancel in their expressions, the solubility product for silver iodate does include activity coefficients:

$$Q_{sp} = [Ag^+][IO_3^-] = \frac{K_{sp}}{f_1^2}.$$

Therefore, to use Eq. (17.19) for accurate calculations it is necessary to plug in not only the equilibrium $[NH_3]$ but the activity coefficients as well, and this means an assumption about ionic strength.

Rather than picking arbitrary values of $[NH_3]$ and ionic strength just to show that numbers can come out of Eq. (17.19), we will examine the experimental data of Derr, Stockdale and Vosburgh.* They placed pure silver iodate crystals in contact with several solutions containing ammonia and, in some cases, ammonium nitrate to repress the reaction of NH_3 with H_2O. Samples of the saturated solutions were analyzed for the iodate-ion content by treatment with acid and potassium iodide, so that the liberated triiodide ion could be titrated with a standard solution of sodium thiosulfate. The iodate molarity was then simply taken to be the molar solubility of silver iodate.

Example calculation

Excess solid silver iodate was added to a solution containing $0.01277M$ NH_3 and $0.0200M$ NH_4NO_3. After several hours of stirring the solution was filtered to remove the excess $AgIO_3$. A 100-mL portion of the saturated solution was treated with enough nitric acid to protonate all the ammonia, and some potassium iodide was added to convert IO_3^- to I_3^-:

$$IO_3^- + 6H^+ + 8I^- = 3I_3^- + 3H_2O.$$

Titration with $0.0952M$ sodium thiosulfate required 25.22 mL to reach the endpoint. Since it requires two moles of thiosulfate per mole of triiodide, and three moles of triiodide are formed from one mole of iodate, we have:

$$S = [IO_3^-] = \frac{25.22 \cdot 0.0952}{6 \cdot 100} = 0.00400 \text{ mol/L}.$$

The initial solution contained $0.01277M$ NH_3, but part of this had to be used to create the complexes, thereby resulting in the higher solubility. If

---

*P. Derr, R. Stockdale, W. Vosburgh, *Journal of the American Chemical Society*, 63, 2670 (1941).

nearly all the silver in solution is present as the diamine complex $Ag(NH_3)_2^+$, the complexed ammonia amounts to $2 \cdot S = 0.00800$, leaving the equilibrium concentration of free ammonia as

$$[NH_3] = 0.01277 - 0.0080 = 0.00477.$$

Is it reasonable to make this assumption? Knowing that the second formation quotient is $7.76 \cdot 10^3$, we may calculate the ratio of the concentrations of the two complexes at this particular concentration of NH_3 in the solution:

$$\frac{[Ag(NH_3)_2^+]}{[AgNH_3^+]} = Q_2[NH_3] = 7.76 \cdot 10^3 \cdot 0.00477 = 37.$$

Thus, the second complex is expected to be the dominant species in the solution.

Now, what would Eq. (17.19) give as a "prediction" for this solubility value? For the ionic strength of the solution we take the sum of the solubility and the ammonium nitrate concentration:

$$I = C_{Ag} + C_{NH_4^+} = S + 0.020 = 0.0240.$$

With the activity coefficient for a singly charged ion taken as 0.86 at this ionic strength, Eq. (17.19) becomes:

$$S = \left\{ \left(1 + 2.04 \cdot 10^3 \cdot 0.00477 + 2.04 \cdot 10^3 \cdot 7.76 \cdot 10^3 \cdot 0.00477^2 \right) \frac{3.09 \cdot 10^{-8}}{0.86^2} \right\}^{1/2}$$
$$= 0.00394.$$

The agreement with the experimental value 0.00400 is certainly quite satisfactory. We conclude that Eq. (17.19) has merit, *and therefore the proposed chemical model is reasonable.* In other words, an experiment of this sort strengthens our confidence about the nature of the ionic species in solution. But when the process is repeated with data obtained at higher concentrations of ammonia, the agreement is not quite as good, as shown in Fig. 17.2.

The deviations *may* be due to lower reliability of the activity coefficients as the ionic strength becomes higher, or perhaps the values obtained by Nasanen for K_1 and K_2 are not quite correct, or perhaps Derr et al. incurred sampling or analysis errors, or perhaps there are factors at work that we are ignoring here without realizing it, such as the formation of ion pairs between $Ag(NH_3)_2^+$ and IO_3^-, or maybe $Ag(NH_3)_3^+$ is being formed. So don't be too trusting of either experimental data or theoretical calculations because *nature always sides with the hidden flaw.*

Analytical aside   The strong complexing of silver ion by ammonia can be used to advantage in analytical procedures. When the ammonia concentration is fairly high, for example $1M$, salts such as silver chloride, silver iodate, etc., that normally precipitate, will stay in solution. We may say that the silver is masked against precipitation of these salts. However, a salt such as silver

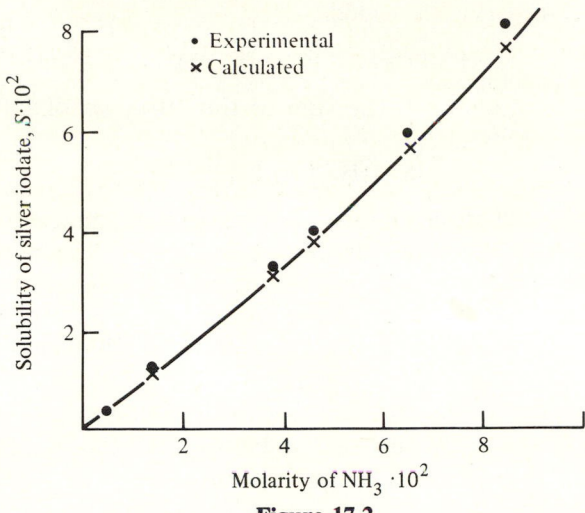

Figure 17.2

iodide is so very insoluble, with $K_{sp} \approx 10^{-16}$, that it will form in spite of the presence of NH_3.

Usually one obtains colloidal or at least very finely divided precipitates of silver chloride or silver iodate. With the aid of ammonia it is possible to greatly improve the crystalline nature of these compounds through precipitation from homogeneous solution: the silver is first complexed with an excess of ammonia, and when the anion (Cl^- or IO_3^-, etc.) is added no precipitate can form. The next step is to add an excess of an ester 2-hydroxyethylacetate. This substance slowly hydrolyzes at room temperature in the aqueous solution to produce acetic acid:

$$HOC_2H_4O\underset{\overset{\|}{O}}{C}CH_3 + H_2O = CH_3COOH + HOC_2H_4OH.$$

The homogeneous production of acetic acid gradually converts the ammonia to ammonium ion, and eventually enough silver ion is freed from the complex to cause nucleation of a precipitate with the chloride (or iodate ion). The precipitate particles grow to large crystals.

IV. Systems Depending upon Solution pH

This is a very important category because so many solutes have either acidic or basic properties. We will consider two main types of solutes: molecular (covalent) species and strong electrolyte salts with basic anions.

A) MOLECULAR SOLUTES

Many organic compounds fit into this class. For example, benzoic acid C_6H_5COOH is a weak acid ($pK = 4.2$) only slightly soluble in water ($S =$

$0.028M$ at $25°$). Considering the model

$$HA(s) = HA(aq) = H^+ + A^-,$$

we realize that this solubility is the sum of the intrinsic solubility [HA(aq)] and the ionic concentration:

$$S = [HA] + [A^-]. \tag{17.20}$$

Theoretically, we would expect the addition of a small amount of strong acid to reduce the solubility, because the dissociation step would be repressed. In fact, the solubility of benzoic acid in the presence of $0.006M$ HCl is 0.0270, and this may be taken as the value of the intrinsic solubility Q_0. The addition of still more HCl produces little effect because the dissociation is already repressed nearly completely.

The solubility of benzoic acid can be greatly increased by the addition of a strong base to the solution for, by combining with the H^+ ions, the base makes it necessary for more acid to dissociate to restore the equilibrium.

By introducing the dissociation quotient for benzoic acid

$$Q_a = \frac{[H^+][A^-]}{[HA]},$$

we find by substitution into Eq. (17.20) that the solubility can be expressed as a function of H:

$$S = Q_0 + Q_0 Q_a/[H^+]. \tag{17.21}$$

Given an equilibrium value for $[H^+]$, we may readily calculate the solubility of benzoic acid in a solution.

Now, consider a molecular solute with basic properties only. For example, 4-bromaniline $BrC_6H_4NH_2$ is listed in the handbook as insoluble in water. That may mean that its solubility is too low to determine or that the person who compiled the table simply was not aware of the experimental value. Anyway, this substance has $pK_b = 10.1$, showing that it is a very weak base. Its water solution would contain the molecular species B up to a maximum concentration equal to the intrinsic solubility Q_0. There would also be some amount of the conjugate acid BH^+, either due to reaction with water or with any acidic substances originally present in the solution. Thus,

$$S = [B] + [BH^+] = Q_0 + \frac{Q_0[H^+]}{Q_a}, \tag{17.22}$$

where Q_a is the *acid* dissociation for the conjugate acid species BH^+. In this case we see that the solubility should increase with increasing $[H^+]$ in contrast to the situation discussed for benzoic acid.

B) AMPHIPROTIC SOLUTES

There are many substances that contain an acidic functional group *and* a basic functional group. All amino acids fall into this category. The chemical model for the solubility of such a substance is as follows:

$$HR(s)$$
$$\updownarrow$$
$$H_2R^+ \rightleftarrows HR(aq) \rightleftarrows R^-.$$

The material-balance expression for the solubility must include all three solution species:

$$S = [R^-] + [HR] + [H_2R^+]. \tag{17.23}$$

We introduce the two acid-dissociation quotients for the protonated species:

$$Q_1 = \frac{[H^+][HR]}{[H_2R^+]} \quad \text{and} \quad Q_2 = \frac{[H^+][R^-]}{[HR]}. \tag{17.24}$$

Also, we have the intrinsic solubility

$$Q_0 = [HR]. \tag{17.25}$$

Combination of Eqs. (17.23), (17.24), and (17.25) gives an expression that shows the hydrogen-ion dependence of the ampholyte solubility:

$$S = \frac{Q_0 Q_2}{[H^+]} + Q_0 + \frac{Q_0[H^+]}{Q_1}. \tag{17.26}$$

The reader probably has already realized that this equation embodies both equations (17.21) and (17.22). In other words, solutes that are only acidic or only basic, are merely special cases of amphiprotic solutes. That is, Q_1 is large for acidic solutes and Q_2 is small for basic solutes, causing one or the other terms of Eq. (17.26) to vanish.

With one term of Eq. (17.26) being inversely dependent upon $[H^+]$ and another being directly dependent upon $[H^+]$, there must be *some* value of $[H^+]$ where the entire function goes through a minimum. Students of calculus should differentiate Eq. (17.26), set the first derivative equal to zero, and find the condition for this minimum in solubility. The answer is:

$$[H^+] = (Q_1 Q_2)^{1/2}, \qquad S_{\min} = Q_0 \left\{ 1 + 2\left(\frac{Q_2}{Q_1}\right)^{1/2} \right\}.$$

This point will occur when the concentration of H_2R^+ is equal to that of R^-. The pH at which this condition is met is often referred to as the **isoelectric point** of the system.

Compare Eq. (17.26) with Eq. (17.12) derived for common-ligand complexing. If the first three terms of Eq. (17.12) are considered, we see that the two systems have the same algebraic properties. The apparent difference is simply that for the metal–ligand system we customarily use formation quotients, while for acid–base systems we use dissociation quotients. Otherwise the equations are quite analogous.

C) STRONG ELECTROLYTE SALTS WITH BASIC ANIONS

There are numerous examples in this category because only the anions of strong acids (perchlorate, chloride, bromide, nitrate, thiocyanate, iodide) have no basic character. Consider the insoluble compound calcium carbonate $CaCO_3$. We may write the equilibrium between this substance and its ions

$$CaCO_3(s) = Ca^{2+} + CO_3^{2-}$$

as well as the solubility product

$$Q_{sp} = [Ca^{2+}][CO_3^{2-}] \approx 5 \cdot 10^{-9}.$$

Because the carbonate ion can readily take on protons to form bicarbonate and then carbonic acid, which in turn can dissociate into water and CO_2, calcium carbonate is quite soluble in the presence of excess strong acid:

$$CaCO_3(s) = Ca^{2+} + CO_3^{2-}$$
$$+$$
$$H^+ = HCO_3^-$$
$$+$$
$$H^+ = H_2CO_3$$
$$\updownarrow$$
$$CO_2(g) + H_2O.$$

Other basic anions that form a number of insoluble salts with metals include iodate, sulfate, chromate, and anions of organic acids (all monoprotic), oxalate, sulfide, sulfite (diprotic), phosphate, arsenate (triprotic). These ions carry charges from -1 to -3, and the insoluble salts involve metals with charges from $+1$ to $+3$. Therefore we might as well proceed to a somewhat general consideration based on the flexible formula $M_m A_a$ for the insoluble substance. On the assumption that this substance is a strong electrolyte, it dissolves in accordance with the equilibrium:

$$M_m A_a(s) = mM + aA, \qquad Q_{sp} = [M]^m[A]^a. \tag{17.27}$$

The molar solubility of $M_m A_a$ may be expressed in terms of the metal as follows:

$$S = [M]/m. \tag{17.28}$$

For saturated solutions in the pH range where protonation of A is *insignificant*, we may also write:

$$S = [A]/a \qquad \text{(at high pH)}.$$

However, the solubility will be increased to the extent that the basic anion can take on protons:

$$S = \frac{1}{a}\{[A] + [HA] + [H_2A] + \cdots\} \quad \text{(at any pH)}. \tag{17.29}$$

In continuing this derivation let us suppose that the anion is diprotic, so that there are only two acid dissociation quotients to consider (if the acid were monoprotic, Q_1 would be very large, i.e., infinity):

$$Q_1 = \frac{[H^+][HA]}{[H_2A]} \quad \text{and} \quad Q_2 = \frac{[H^+][A]}{[HA]}.$$

By substituting these expressions into Eq. (17.29), so that S is expressed in terms of [A] only, we obtain:

$$S = \frac{1}{a}[A]\left\{1 + \frac{[H^+]}{Q_2} + \frac{[H^+]^2}{Q_1 Q_2}\right\}. \tag{17.30}$$

The parenthetical term is simply a version of the F_0 function, when dissociation quotients are used instead of formation quotients.

The next step is to use the solubility-product expression, Eq. (17.27), to introduce the relationship between [A] and [M]:

$$[A] = \left(\frac{Q_{sp}}{[M]^m}\right)^{1/a} = \frac{Q_{sp}^{1/a}}{[M]^{m/a}},$$

and, in view of the solubility expression (17.28), we get

$$[A] = \frac{Q_{sp}^{1/a}}{m^{m/a} \cdot S^{m/a}}.$$

Finally, when this function for [A] is substituted into Eq. (17.30), the result after a little rearranging gives the solubility S in terms of the equilibrium quotients and the hydrogen ion concentration:

$$S = \left\{\frac{Q_{sp}^{1/a}}{a \cdot m^{m/a}}\left(1 + \frac{[H^+]}{Q_2} + \frac{[H^+]^2}{Q_1 Q_2}\right)\right\}^{a/(a+m)}. \tag{17.31}$$

Admittedly, this equation is a bit confusing to look at, but by deriving the relationship for this rather general case we avoid doing separate derivations for each individual case. It is easy to assign various values for a and m to find the equation for simpler cases. For example, with 1:1 salts with monoprotic anions, such as lead sulfate, barium chromate, silver iodate, $a = m = 1$ and $Q_1 = \infty$. The result is:

$$S = \left\{Q_{sp}\left(1 + \frac{[H^+]}{Q_a}\right)\right\}^{1/2}. \tag{17.32}$$

The reader should find the corresponding equations for 2:1 salts (such as silver oxalate), 1:2 salts (such as copper iodate), and other types as desired.

Example calculation
The usefulness of Eq. (17.32) and the like can be illustrated by comparing the pH effect on two salts, barium sulfate and barium chromate, which have quite similar values for their solubility products. We will examine the effect of

adding $0.1M$ strong acid to the solid salts. At ionic strength equal to 0.1 the pertinent equilibrium quotients are:

$BaSO_4$:	$Q_{sp} = 7.3 \cdot 10^{-10}$,	HSO_4^-:	$Q_a = 2.7 \cdot 10^{-2}$,
$BaCrO_4$:	$Q_{sp} = 2.5 \cdot 10^{-9}$,	$HCrO_4^-$:	$Q_a = 8.1 \cdot 10^{-7}$.

Using Eq. (17.32) we may predict the solubility of each salt in the presence of $0.1M$ hydrogen ion:

$$S_{BaSO_4} = \left\{ 7.3 \cdot 10^{-10} \left(1 + \frac{0.1}{0.027} \right) \right\}^{1/2} = 0.00006,$$

$$S_{BaCrO_4} = \left\{ 2.5 \cdot 10^{-9} \left(1 + \frac{0.1}{8.1 \cdot 10^{-7}} \right) \right\}^{1/2} = 0.018.$$

The solubility of the chromate is 300 times higher, in spite of the closeness of the Q_{sp} values, because the chromate ion has a much stronger tendency to take on protons. This result implies that we may successfully precipitate barium sulfate in the presence of chromate ion, simply by using an acidic solution. It also shows that, given the need to clean up some filter crucibles containing such precipitates, it would be foolish to try to dissolve the barium sulfate by pouring acid through, although this would work fairly well for the barium chromate.

Comment on the Derivations

For the sake of giving an overview of the problem, it was efficient to use general models, as in the derivation of Eqs. (17.17) and (17.31). However, one should not memorize such equations.

When you have need for such a function it is far better to examine the specific case you wish to deal with and to derive the appropriate equation on the spot. To this end, learn the concepts of the derivations: begin with a material-balance expression and make substitutions according to the pertinent equilibrium quotients. Keep in mind what you are trying to accomplish, and your background and logic should work for you.

PROBLEMS

Strong electrolyte model with negligible intrinsic solubility

1. For the generalized model of a slightly soluble strong electrolyte

$$A_m B_n(s) = mA + nB \quad \text{(charges omitted for simplicity)}$$

let S be the amount (moles) of the compound that dissolves per liter of water.

a) Assuming no side effects, such as protonation or complex formation, prove the following relationships:

$$Q_{sp} = m^m n^n S^{m+n}, \qquad S = \left(\frac{Q_{sp}}{m^m n^n} \right)^{1/(m+n)}.$$

b) Rederive these expressions for the specific cases of

$AgIO_3$	(a 1:1 electrolyte)
Ag_2SO_4	(a 2:1 electrolyte)
CaF_2	(a 2:1 electrolyte)
$BaSO_4$	(a 2:2 electrolyte)
LaF_3	(a 3:1 electrolyte)

2. The solubility of barium sulfate in $0.030M$ potassium nitrate is $1.82 \cdot 10^{-5}M$. Find Q_{sp} and calculate the solubility of $BaSO_4$ in $0.010M$ potassium sulfate. *Hint:* Consider the common-ion effect.

3. The solubility of lanthanum iodate is $1.7 \, g/100 \, mL$ according to a handbook. In the older literature, the solubility product K_{sp} (in terms of activities) for this salt is given as $pK_{sp} = 11.21$. Are these data mutually consistent?

4. The solubility of potassium perchlorate in water is approximately 0.1 molar. What would you predict for its solubility in $2M$ perchloric acid?

5. Silver bromate has a solubility product of $pK_{sp} = 4.265$. Calculate its solubility in pure water, using successive approximations for the ionic strength. Calculate its solubility in $0.050M$ sodium bromate.

6. The solubility of copper iodate in pure water is 0.00324 molar. Calculate its solubility in $0.090M$ potassium iodate and in $0.090M$ copper sulfate. Which of the two common ions is more effective in repressing the solubility? Why is this so?

Model with negligible intrinsic solubility and significant formation of complexes (*consult Appendix 5 for Q_f values of complexes*)

7. Predict the solubility of mercuric oxide in a solution of $0.01M$ ethylenediamine with $pH = 13$. Assume that $Hgen_2$ is the only significant solution species of mercury.

8. Show that the precipitation of cupric hydroxide $Cu(OH)_2$ would be "complete" at $pH = 10$. Would the presence of $0.10M$ sodium citrate prevent this precipitation at the same pH?

9. Calculate the solubility of cadmium hydroxide in a solution with $pH = 12$ and ammonia concentration of $2M$.

10. A solution is prepared by mixing 10 mmol of silver nitrate, 500 mmol of pyridine, and 10 mmol of sodium iodate, followed by dilution to one liter. Will there be a precipitate?

11. Derive an expression for the solubility of silver thiocyanate in solutions of potassium thiocyanate of varying concentration. Make a graph, plotting solubility versus the equilibrium concentration of SCN^- up to $1.0M$.

12. The molar solubility of a certain strong electrolyte M^+A^- was determined in a series of solutions of constant ionic strength of 0.1, but with varying concentrations of ligand L. With the aid of a computer it was found that the following relationship held over a wide range of ligand concentration:

$$S^2 = 2 \cdot 10^{-8} + 3 \cdot 10^{-6}[L] + 6 \cdot 10^{-5}[L]^2.$$

a) Propose an equilibrium scheme (model) consistent with these findings and deduce the values of the pertinent equilibrium quotients.

b) Predict the solubility of MA in a solution of $0.10M$ NaA in water.

SOLUBILITY: DETERMINATION OF EQUILIBRIUM QUOTIENTS

18

To be successful in unlocking the doors concealing nature's secrets, a person must have ingenuity. If he does not have the key for the lock, he must not hesitate to try to pick it, to climb in a window, or even to kick in a panel. If he succeeds, it is more by ingenuity and determination than by "scientific method".

Joel H. Hildebrand

The previous chapter showed how derived equations could be used to predict solubilities, given facts about solution conditions (e.g., ligand molarities and pH) and the necessary values for equilibrium quotients. We will now explore the possibilities of using the same equations for deducing unknown Q values and for identifying the present solution species, given experimental values for S under controlled conditions. The flowchart shows the essential steps, including feedback loops to correct for poor design, poor analyses, or poor theory. Take a little time to mull this over.

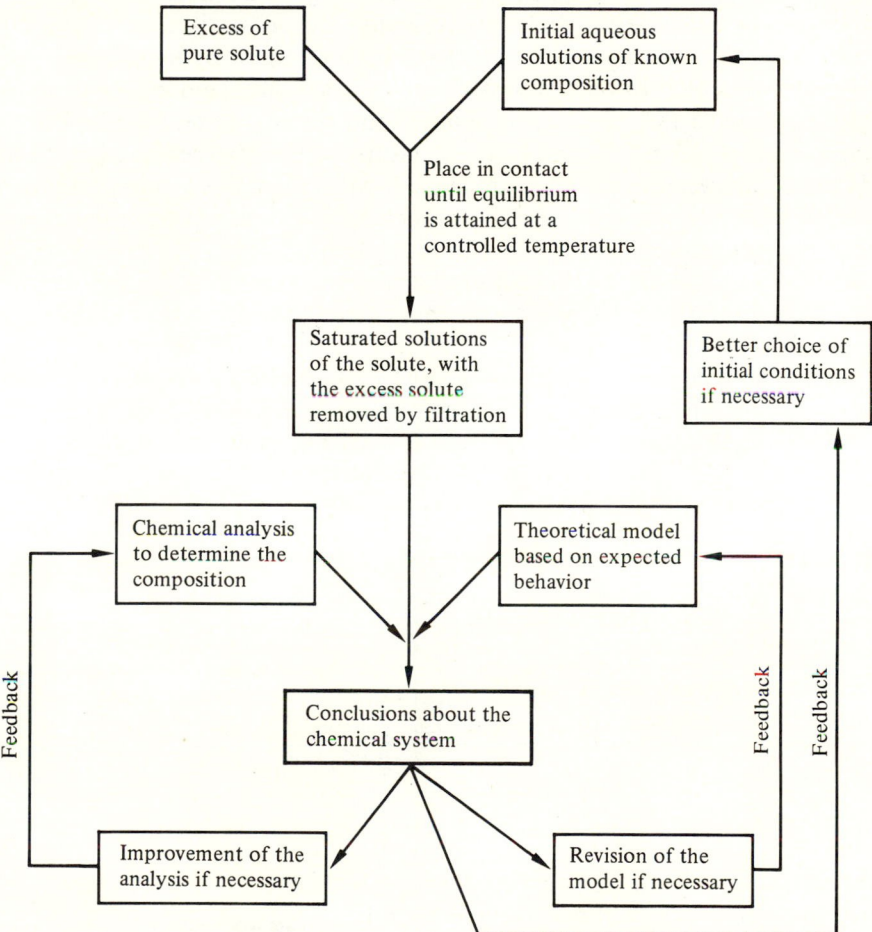

18.1 LABORATORY PREPARATION OF SATURATED SOLUTIONS

To achieve solubility equilibrium it is not sufficient merely to place the solute in contact with the aqueous phase. Substances vary considerably in the *rate* at which they dissolve, and to speed up the process we usually rely on two experimental factors: (1) a large excess of the solid solute is used, so that there will be a greater exposure of surface area to the solution. This also means that saturation will be attained much faster with finely powdered solute than with larger crystals; (2) the solution phase and the solute phase are kept in motion with respect to each other so that the solution in contact with the solid does not become *locally saturated* and thereby slow the process.

There are several inventive techniques for mixing solute and solvent, but we need to consider only two important methods, **batch** and **column**. In the batch method the solute and solvent are sealed up in bottles, immersed in a constant-temperature bath, and mechanically stirred until equilibrium has been reached. The stirring may be by shaking, rotation, or by using a magnetic stirring bar. Samples of the saturated solution are obtained by filtering the mixture to remove the excess solid phase, and this may create difficulties if the equilibration has been done at temperatures rather far removed from the room temperature, because the equilibrium may shift if the temperature is allowed to vary during the sampling process. However, with a little experimental ingenuity we can get around this problem.

The packed-column method offers some advantages. The solute is finely ground and evenly packed in a glass tube surrounded by a water jacket, as shown in Fig. 18.1.

The volume of solution in contact with the solute is relatively small, because most of the column volume is taken up by the solid. This means that the surface-area effect is enormous, and it is often found that equilibrium is

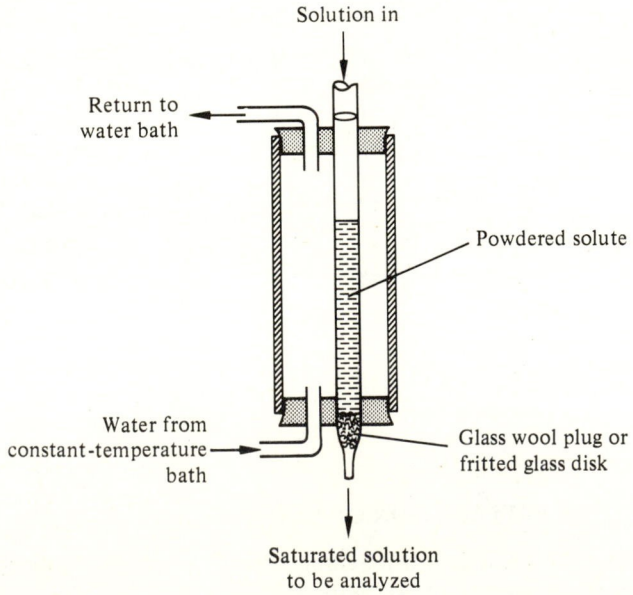

Figure 18.1

reached as quickly as the solution can be forced through the column. Another advantage is that it is a simple matter to change the composition of the input solution without disturbing the column. The first few milliliters are rejected while the new solution flushes out the previous one. Further, the column serves as its own filter, so that the output solution is ready for analysis, and there also is no problem with variation in temperature during the sampling process. As the saturated solution drips from the tip of the column, it may be caught in a preweighed container, and the sample size determined by weight. If small samples can be used for the chemical analysis, this is fast and easy. However, if the analytical procedure requires a large sample (say, 50 mL), then the column method requires patience. The column technique also makes it easy to study the effect of temperature on solubility, for the temperature of the column is changed as quickly as one can adjust the temperature of the water bath used for circulation.

18.2   METHODS OF CHEMICAL ANALYSIS FOR SOLUBILITY DETERMINATIONS

The two essential requirements for the chemical analysis are (1) accuracy and (2) that the total amount of dissolved solute be determined. If the solute dissolves to form a variety of chemical species, care must be taken that the analytical procedure is not "blind" to any of them. For example, a direct spectrophotometric examination may be unsuitable if the solute has more than one colored form in the saturated solution. However, if the sample, once removed from the excess solid, can be chemically treated to convert all the solute to a single colored species, then spectrophotometry may be an excellent choice.

In general there is no restriction as to the type of analytical procedure, provided the above criteria are met. Titrations, gravimetric determination, polarography, potentiometry, spectrophotometry, atomic absorption, gamma counting, and other methods have all been successfully applied. The chemist who understands the advantages and limitations of a wide range of analytical techniques is in the best position for this problem, as for many other research studies.

18.3   GALVANIC-CELL MEASUREMENTS FOR DETERMINING SOLUBILITY PRODUCTS

With the recent development of electrodes which are highly selective for individual metal or ligand species, potentiometric measurements became valuable for obtaining data of interest in solubility studies. Suppose it is desired to determine the solubility product for a strong electrolyte having the simple formula MA:

$$MA(s) = M + A, \qquad Q_{sp} = [M][A].$$

If one can saturate a solution of, say, $0.0100M$ NaA with solid MA and determine the equilibrium concentration of M by means of an electrode, then

$$Q_{sp} = [M] \{0.0100 + [M]\},$$

since it may be assumed that the dissociation of MA contributes just as much A to the solution as it does M. The same system could be studied by saturating a $0.01M$ solution of $MClO_4$ with solid MA, provided there was a means to determine the equilibrium molarity of A, in which case

$$Q_{sp} = \{0.0100 + [A]\}[A].$$

A classical example from the old literature, introduced here because it also illustrates the use of metal-amalgam electrodes, is the study of lead sulfate by Cowperthwaite and LaMer.* They prepared galvanic cells of the following type:

$$Zn(Hg) \,|\, ZnSO_4(C \text{ molar}), \, PbSO_4(s) \,|\, Pb(Hg).$$

The electrodes were solutions of the metals in liquid mercury (such mixtures are called amalgams), consisting of 6 grams of metal to 94 grams of mercury in each case. The reader may verify that these compositions correspond to mole fractions of 0.164 and 0.0582 for zinc and lead, respectively. When the concentration of zinc sulfate is very low, we may assume it is completely dissociated into Zn^{2+} and SO_4^{2-}. The zinc ions establish the left-electrode potential in conjunction with the zinc amalgam. The sulfate ions serve to control the dissociation of lead sulfate through the solubility product, while the equilibrium concentration of lead ions establishes the right-electrode potential.

With the usual convention of assigning oxidation at the left and reduction at the right, the cell reaction is:

$$Zn(Hg) + Pb^{2+}(\text{unknown concentration}) = Zn^{2+}(C\,M) + Pb(Hg),$$

and we may write the Nernst equation by using tabulated values for the standard potentials for the zinc and lead couples:

$$E = +0.76 - 0.13 - \frac{0.0592}{2} \log \frac{[Zn^{2+}]X_{Pb}}{[Pb^{2+}]X_{Zn}}. \tag{18.1}$$

The activity coefficients cancel since both ions are doubly charged. The authors found, for example, that $E = 0.611$ volt when $C = 0.000500$. Solving Eq. (18.1) by using these values and the mole fractions of the metals, we find

$$[Pb^{2+}] = 4.06 \cdot 10^{-5}.$$

---

*I. A. Cowperthwaite, V. K. LaMer, *Journal of the American Chemical Society, 53,* 4333 (1931).

It follows that the total sulfate-ion molarity is $[SO_4^{2-}] = 0.000500 + 0.0000406$. Therefore the solubility product is readily calculated:

$$Q_{sp} = [Pb^{2+}][SO_4^{2-}] = 4.06 \cdot 10^{-5} \cdot 5.41 \cdot 10^{-4} = 2.20 \cdot 10^{-8}.$$

This value may be converted to K_{sp} by noting that the ionic strength of the solution is $4[SO_4^{2-}]$, since the salt types are $2:2$. From the Davies equation we get the activity coefficient $f_2 = 0.81$, and therefore $K_{sp} = 1.46 \cdot 10^{-8}$.

18.4   STUDIES USING COMMON-LIGAND COMPLEXING

Example: Solubility of silver chloride in sodium chloride solutions

Now we return to the question of the solubility of silver chloride in solutions of sodium chloride, introduced earlier in Chapter 17. We will attempt to use Eq. (17.13) in assessing the experimental data by Jonte and Martin, to see whether it is reasonable to account for the minimum in the solubility and, in fact, the precise shape of the solubility curve shown in Fig. 17.1. The data from the original article are reproduced in Table 18.1. Because the several solutions varied in ionic strength, this parameter is included.

Table 18.1   Solubility of radioactive silver chloride in solutions of sodium chloride at 25°C

$[Cl^-]$	S_{AgCl}	I	$[Cl^-]$	S_{AgCl}	I
$5.38 \cdot 10^{-5}$	$5.37 \cdot 10^{-6}$	0.0424	$1.27 \cdot 10^{-3}$	$6.03 \cdot 10^{-7}$	0.0161
$5.92 \cdot 10^{-5}$	$3.31 \cdot 10^{-6}$	$6.4 \cdot 10^{-5}$	$1.59 \cdot 10^{-3}$	$5.62 \cdot 10^{-7}$	0.0123
$1.12 \cdot 10^{-4}$	$2.04 \cdot 10^{-6}$	$1.2 \cdot 10^{-4}$	$2.75 \cdot 10^{-3}$	$4.90 \cdot 10^{-7}$	0.0027
$2.08 \cdot 10^{-4}$	$1.66 \cdot 10^{-6}$	0.0279	$5.50 \cdot 10^{-3}$	$5.75 \cdot 10^{-7}$	0.0055
$3.44 \cdot 10^{-4}$	$1.02 \cdot 10^{-6}$	0.0190	$1.10 \cdot 10^{-2}$	$6.61 \cdot 10^{-7}$	0.0110
$5.51 \cdot 10^{-4}$	$6.92 \cdot 10^{-7}$	$5.8 \cdot 10^{-4}$	$2.75 \cdot 10^{-2}$	$1.10 \cdot 10^{-6}$	0.0276
$9.66 \cdot 10^{-4}$	$6.92 \cdot 10^{-7}$	0.0185	$5.50 \cdot 10^{-2}$	$1.95 \cdot 10^{-6}$	0.0552
$1.10 \cdot 10^{-3}$	$5.25 \cdot 10^{-7}$	0.0011	$1.10 \cdot 10^{-1}$	$3.80 \cdot 10^{-6}$	0.110

For guidance in the interpretation of these data we turn to Eq. (17.13) repeated here:

$$S = \frac{Q_{sp}}{[Cl^-]} + Q_{sp}Q_1 + Q_{sp}Q_1Q_2[Cl^-] + \cdots.$$

Although this equation is quite correct, the effects of activity coefficients are incorporated in the Q-terms. Because the silver-chloride data were obtained with solutions of varying ionic strength, we may not regard the Q values as constants. Equation (17.13) can be converted to a form containing K values by using the relationships:

$$Q_{sp} = \frac{K_{sp}}{f_1^2}, \qquad Q_1 = K_1 f_1^2,$$

$$Q_2 = K_2 \quad \text{(the activity coefficients cancel),} \quad \text{etc.}$$

Considering only the terms shown above, the result is, after rearrangement:

$$S[Cl^-]f_1^2 = K_{sp} + K_{sp}K_1[Cl^-]f_1^2 + K_{sp}K_1K_2[Cl^-]^2 f_1^2. \tag{18.2}$$

This form is useful because it suggests that a plot of the variable $S[Cl^-]f_1^2$ versus the variable $[Cl^-]f_1^2$ should be linear, with an intercept equal to K_{sp} and a slope equal to the product $K_{sp}K_1$, *but only if the third term is negligible.*

The data given in Table 18.1 were used to calculate values for these variables. Since the third term in Eq. (18.2) is directly related to the formation of the complex ion $AgCl_2^-$, only the data at low chloride concentrations were used so that this effect would be small. The plot is shown in Fig. 18.2; we see that the data are so scattered that any curvature would have to be imagined. We assume that the relationship is linear, and when the method of least squares is used the results are as follows:

$$\text{Intercept:} \qquad (1.79 \pm 0.27) \cdot 10^{-10} = K_{sp},$$
$$\text{Slope:} \qquad (3.9 \pm 0.5) \cdot 10^{-7} \quad = K_{sp}K_1 = Q_0.$$

Had the plot been obviously curved, with the points showing far less scatter, we still could have extrapolated to find the value for the intercept K_{sp}.

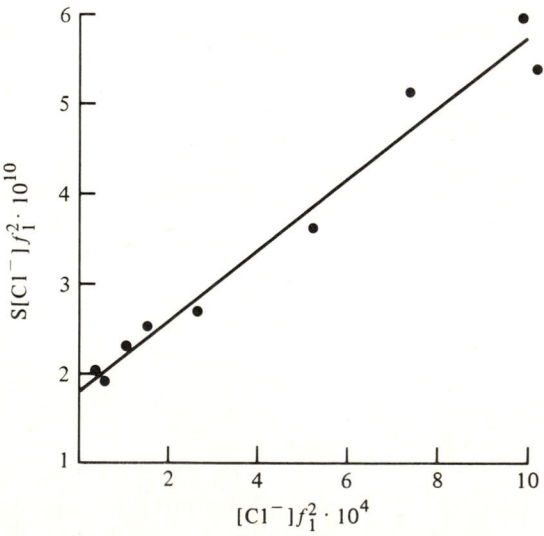

Fig. 18.2 Plot of silver-chloride solubility data.

The next step is to use the determined value for K_{sp} in the following rearranged form of Eq. (18.2). By subtracting K_{sp} from each side and dividing by $K_{sp}[Cl^-]f_1^2$, we find:

$$\frac{S[Cl^-]f_1^2 - K_{sp}}{K_{sp}[Cl^-]f_1^2} = K_1 + K_1 K_2 [Cl^-]. \tag{18.3}$$

This should look familiar, for it is the F_1-function discussed in Chapter 15. All of the quantities at the left of Eq. (18.3) are available from the data at hand, and so the next move is to plot F_1 versus $[Cl^-]$ in accord with the suggestion from Eq. (18.3) that the intercept will be K_1. If the plot is linear, then the slope may be taken as $K_1 K_2$. If it is *not* linear, then we may proceed to a plot of F_2, and then F_3, if necessary. The point is that the solubility data lead us quite naturally into the F-function approach to deducing complex formation quotients. The plot of F_1 is shown in Fig. 18.3.

The results are reasonably linear, and so we apply the least-squares method to find the values for K_1 and $K_1 K_2$:

$$\text{Intercept:} \quad (1.7 \pm 0.5) \cdot 10^3 \quad = K_1,$$

$$\text{Slope:} \quad (1.77 \pm 0.09) \cdot 10^5 = K_1 K_2,$$

$$\text{Slope/intercept:} \quad (1.0 \pm 0.3) \cdot 10^2 \quad = K_2.$$

The linear plot does not mean that silver cannot form complexes with more than two chloride ions, just that under the conditions of these experiments

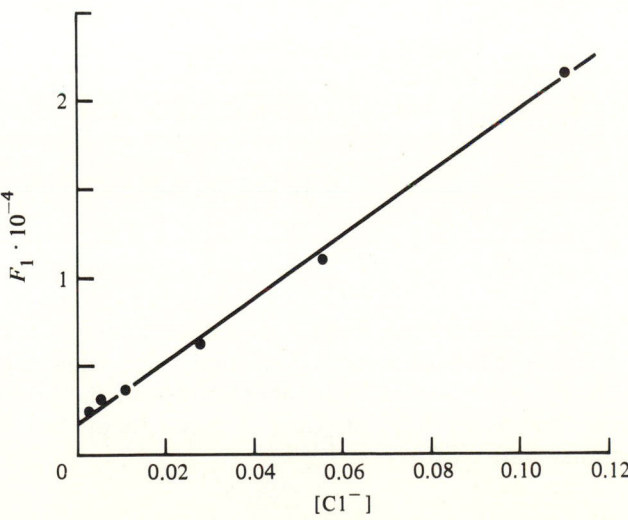

Figure 18.3

such effects are not subject to measurement. There is a hint in the above figure that the plot is slightly curving upward, in compliance with the idea that another term is needed in Eq. (18.3). It would be necessary to carry out measurements at higher chloride molarities to search for the existence of higher complexes.

Analytical comment   This research has shown that silver chloride has a finite intrinsic solubility, about $4 \cdot 10^{-7}$ molar, and that there is a moderate tendency to form at least a dichloro complex that increases the solubility of AgCl in the presence of excess chloride ion. Therefore, for the most precise results in the gravimetric determination of silver it is advisable to keep the excess-chloride concentration low. The solubility minimum is at about $0.003M$ Cl^-.

Other research has shown that a similar complexing effect occurs when silver chloride dissolves in the presence of excess silver ion. Apparently there are complexes with formulas Ag_2Cl^+, Ag_3Cl^{2+}, etc.

Another example: Solubility of copper(I) bromide in sodium bromide solutions

Peters and Caldwell* used a series of ingenious techniques of electroanalytical chemistry both to prepare solid copper(I) bromide and to measure its solubility in a series of sodium bromide solutions. The solutions were prepared to have a constant ionic strength of 2.00, and we may presume that activity coefficients remain constant, so that Q values may be treated as constants.

The object was to identify the species formed in the solutions and to determine values for equilibrium quotients if possible. The overall chemical model for the system is as follows:

$$
\begin{array}{c}
\text{in contact with} \\
\text{NaBr solutions}
\end{array}
$$

$$
\text{CuBr(s)} \quad \rightleftharpoons \quad \text{CuBr(aq)} \rightleftharpoons \text{Cu}^+ + \text{Br}^-
$$

$$
+
$$

$$
\text{Br}^- \text{ (from 0.05 to } 1.5M\text{)}
$$

$$
\updownarrow
$$

$$
\text{CuBr}_2^-, \text{CuBr}_3^{2-}, \text{etc.}
$$

Thus, the chemical model is quite similar to that for silver chloride. The main difference in the experimental approach is that only relatively high ligand concentrations were used, so that the concentrations of the higher complexes are much greater than [CuBr(aq)] or [Cu^+]. The results of the analyses of the saturated solutions were as follows:

---

*D. G. Peters, R. L. Caldwell, *Inorganic Chemistry*, **6**, 1478 (1967).

[Br⁻]	S_{CuBr}, molar	[Br⁻]	S_{CuBr}, molar
0.051	0.00035	0.595	0.0151
0.102	0.00079	0.798	0.0282
0.206	0.00223	1.01	0.0456
0.495	0.0104	1.52	0.118

Note that the solubility continuously increases as the bromide ion molarity increases. This is evidence that the dissociation step of the model is not important at these bromide concentrations.

The algebraic model is once again found in Eq. (17.13), written here with the dissociation step omitted:

$$S_{CuBr} = Q_{sp}Q_1 + Q_{sp}Q_1Q_2[Br^-] + Q_{sp}Q_1Q_2Q_3[Br^-]^2 + \cdots. \qquad (18.4)$$

To see whether the intrinsic solubility (which is $Q_0 = Q_{sp}Q_1$) can be determined from the above data, we plot S versus $[Br^-]$, for Eq. (18.4) suggests that Q_0 is the intercept of such a plot (Fig. 18.4).

Clearly it is impossible to distinguish Q_0 from zero on this plot. In fact, as the reader may care to prove, when the lowest three or four points are plotted on an expanded scale, it is still impossible to estimate a value of Q_0. We conclude that the intrinsic solubility is negligible compared to the total solubilities, and so Eq. (18.4) may be further simplified by dropping the $Q_{sp}Q_1$ term.

The next step in handling the data is to try to determine the coefficient of the linear term, that is $Q_{sp}Q_1Q_2$. By dividing the modified Eq. (18.4) by $[Br^-]$, we find,

$$\frac{S}{[Br^-]} = Q_{sp}Q_1Q_2 + Q_{sp}Q_1Q_2Q_3[Br^-] + Q_{sp}Q_1Q_2Q_3Q_4[Br^-]^2 + \cdots. \qquad (18.5)$$

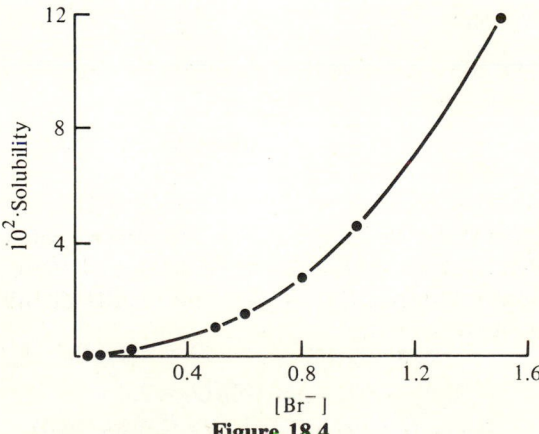

Figure 18.4

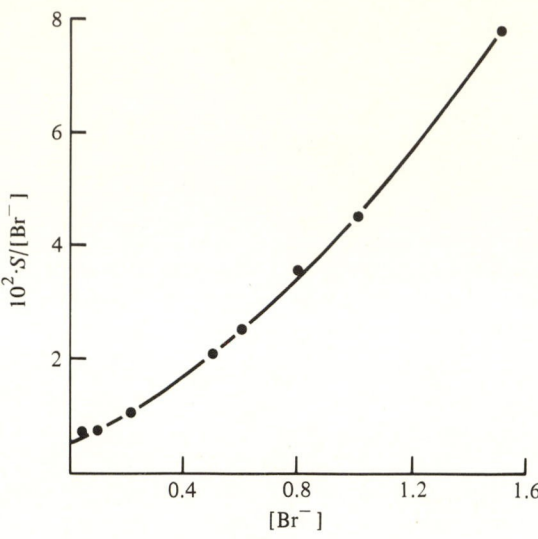

Figure 18.5

This is similar to the F-function approach. Indeed, the concepts are identical: we work with a polynomial function, trying to determine one coefficient at a time. The plot of the calculated $S/[Br^-]$ values versus $[Br^-]$ is shown in Fig. 18.5.

The fact that this plot is definitely curved is positive evidence that Eq. (18.5) is at least a quadratic function. The term involving Q_4 clearly needs to be included in the model. By extrapolating the above curve to the intercept, we estimate the value of $Q_{sp}Q_1Q_2$ to be 0.0055. The accuracy of this estimate is not high, admittedly, but it is the best we can do. For the next step we convert Eq. (18.5) by subtracting the constant term from each side and dividing by $Q_{sp}Q_1Q_2[Br^-]$:

$$\frac{\{S/[Br^-]\} - Q_{sp}Q_1Q_2}{Q_{sp}Q_1Q_2[Br^-]} = Q_3 + Q_3Q_4[Br^-] + Q_3Q_4[Br^-]^2 + \cdots = F_3.$$

It turns out that we have achieved an equation in the F-function sequence, namely the expression for F_3. By using the value of 0.0055 for $Q_{sp}Q_1Q_2$ as determined from the previous plot, a value of F_3 can be readily calculated for each point in the data set. A plot of the F_3 values versus $[Br^-]$ is depicted in Fig. 18.6.

In spite of the scatter, which has to be regarded as normal in this type of experimental study, the plot is satisfactorily linear, and both the intercept and the slope may be estimated by the method of least squares (with 90% confidence):

Intercept:	$Q_3 = 4.2 \pm 0.3$
Slope:	$Q_3Q_4 = 2.9 \pm 0.5$
Slope/intercept:	$Q_4 = 0.7 \pm 0.1$

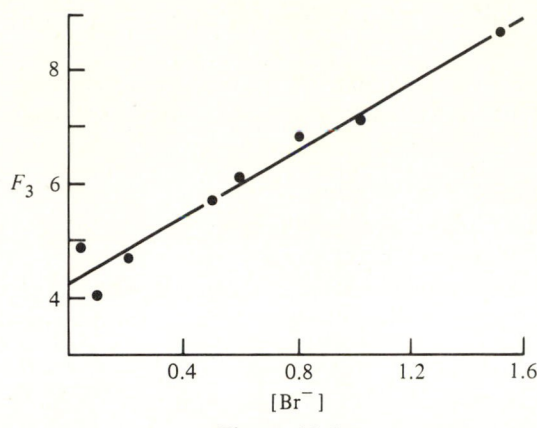

Figure 18.6

Because the F_3 plot was linear, there is no point in continuing the F-function approach. We conclude that the copper(I) bromide system is adequately described by a model consisting of only three complexes: $CuBr_2^-$, $CuBr_3^{2-}$, and $CuBr_4^{3-}$.

Using a very different method of data processing, Peters and Caldwell came to a different conclusion, that $CuBr_4^{3-}$ was absent. This sort of thing is not uncommon in research literature: given the same data set, different workers draw different conclusions.

Suggestion   Use the values estimated above for the equilibrium quotients, and with Eq. (18.4), less the $Q_{sp}Q_1$ term, see how well the original data are reproduced by calculation. This is one way of gaining confidence in the plausibility of one's conclusions.

18.5 STUDIES USING FOREIGN-LIGAND COMPLEXING

A slightly soluble salt will show increased solubility in the presence of a ligand that can form a complex, or several complexes, with the metal ion. The models for this situation were developed earlier, and the situation is described algebraically by Eqs. (17.17) and (17.18). The latter equation is the useful form when one wishes to predict a solubility, given the various equilibrium quotients and the concentration of the complexing ligand. However, we now wish to take a closer look at Eq. (17.17) to see how it can be put to work in determining the formation quotients for the metal–ligand complexes.

First let's identify the logical steps in designing an experiment for this purpose:

a)   We want to study the metal–ligand complexing of a certain metal M with a certain ligand L.

b)   We next try to find an insoluble salt of that metal whose ligand is different from L. Suppose that this salt is MA_n, that it is a strong electrolyte, and that the value for the solubility product Q_{sp} is already known from other experiments. Anion A should not undergo any reactions in the solutions to be used. It will only relate to the molarity of M ions through the value of Q_{sp}.

c)   We prepare a series of solutions containing known and varying concentrations of ligand L. For these solutions to be of constant ionic strength, we use, for example, sodium perchlorate to make the ionic strength adjustment as necessary. The chosen molarities of ligand L should cause the formation of significant amounts of whatever complexes ML_i we hope to study.

d)   To each of the solutions we add a substantial excess of solid MA_n and stir until the solutions are saturated. Alternatively we may use the packed-column technique, with solid MA_n in the column and the solutions of L forced through to obtain samples of the saturated solution.

e)   Samples of the saturated solution are analyzed to determine the total concentration of M, i.e., the solubility. Sometimes this is conveniently done by analyzing M in all its forms, and in other cases it may be better to determine the concentration of the anion A. In that case, $S = [A]/n$.

f)   Given the solubility data and the value of Q_{sp} for MA_n, we use Eq. (17.17) to calculate the F_0 value for each solution. Then the F-function approach is used to unravel the system, one formation quotient at a time.

To illustrate all this let's consider a hypothetical system: it is desired to identify the complex species formed when copper(II) ion reacts with a certain ligand:

$$Cu^{2+} + L = CuL, \qquad Q_1 = \frac{[CuL]}{[Cu^{2+}][L]},$$

$$CuL + L = CuL_2, \qquad Q_2 = \frac{[CuL_2]}{[CuL][L]},$$

etc., as necessary.

Copper iodate $Cu(IO_3)_2$ may serve as the insoluble salt. It has a solubility product value of $1.04 \cdot 10^{-6}$ in $1.00M$ sodium perchlorate, and we will plan to work with solutions of constant ionic strength $I = 1$. Therefore, from Eq. (17.17) we have:

$$F_0 = \frac{4S^3}{1.04 \cdot 10^{-6}} = 1 + Q_1[L] + Q_1Q_2[L]^2 + Q_1Q_2Q_3[L]^3 + \cdots, \qquad (18.6)$$

where S is the molar solubility of copper iodate in the presence of the ligand with molarity [L].

A glass column is packed with solid copper iodate, and solutions of L are passed through at 25°C. Analysis of the resulting saturated solutions yields the following data:

[L]	S, molar	[L]	S, molar
0*	0.00639	0.300	0.0107
0.100	0.0078	0.400	0.0120
0.200	0.0093	0.500	0.0133

*This is where Q_{sp} comes from.

The five values of F_0 may be calculated by using Eq. (18.6). Then, values of F_1 follow immediately in the usual way from $F_1 = (F_0 - 1)/[L]$:

[L]	F_0	F_1	[L]	F_0	F_1
0.1	1.825	8.25	0.4	6.646	14.12
0.2	3.094	10.47	0.5	9.049	16.10
0.3	4.712	12.37			

Since F_1 shows increasing values as the ligand molarity increases, there must be at least one complex beyond CuL. A plot of F_1 versus [L] is shown in Fig. 18.7.

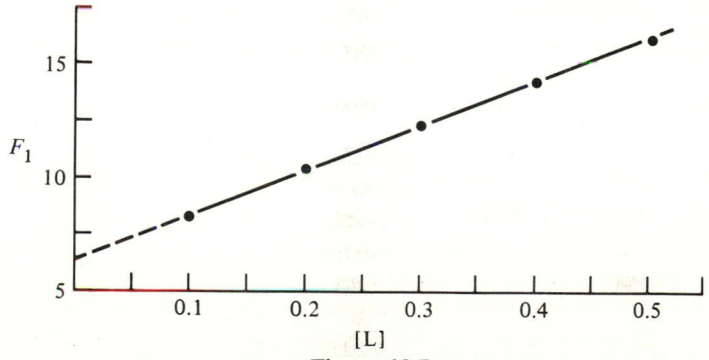

Figure 18.7

The linearity of the plot shows that only two complexes are involved in the equilibrium scheme. The intercept gives the value of $Q_1 = 6.5$, while the slope gives the value of $Q_1 Q_2 = 19.4$. The slope/intercept ratio then is $Q_2 = 3.0$.

18.6 STUDIES OF ACID–BASE EQUILIBRIA THROUGH SOLUBILITY

A. The Use of Insoluble Strong Electrolytes in the Study of Selected Anions

The properties of a basic anion, and hence those of its conjugate acid, may be investigated by finding the effect of varying pH on the solubility of a salt containing that anion. The chemical and algebraic models for these systems were derived earlier, with Eq. (17.31) giving the general relationship. The version of this equation suitable for use with 1:1 salts was Eq. (17.32), repeated here with both sides squared:

$$S^2 = Q_{sp} + \frac{Q_{sp}}{Q_a}[H^+] \quad \text{(for a monoprotic anion).} \tag{18.7}$$

If it is possible to determine a series of values for S at controlled values of $[H^+]$, Eq. (18.7) suggests that a plot of S^2 versus $[H^+]$ should be linear, with an intercept equal to Q_{sp} and a slope equal to Q_{sp}/Q_a. Of course, this assumes that all the solutions have the same ionic strength, so that the Q values are presumed constant. The value for the acid-dissociation quotient is, then, the ratio of intercept to slope.

This approach was used to study the lead-sulfate system in solutions of unit ionic strength containing lithium perchlorate and perchloric acid.* Solid lead sulfate was equilibrated with the solutions by the batch technique, and the samples of saturated solution were analyzed for the dissolved-lead content by a spectrophotometric procedure: after addition of concentrated hydrochloric acid (that converts the dissolved lead to a reproducible mixture of chloroplumbate complexes) the ultraviolet absorption was measured. The results for solubility were as follows:

C_{LiClO_4}	$C_{HClO_4} = [H^+]$	S, molar
0.950	0.050	$0.92 \cdot 10^{-3}$
0.700	0.300	$1.41 \cdot 10^{-3}$
0.400	0.600	$1.84 \cdot 10^{-3}$
0	1.000	$2.28 \cdot 10^{-3}$

By using a plot according to Eq. (18.7), the authors found $Q_{sp} = 6.3 \cdot 10^{-7}$ and Q_2 for sulfuric acid equal to 0.138. The reader should check these results.

B. Investigation of Insoluble Molecular Substances

Many of the reagents and indicators used in analytical chemistry are organic substances with acid–base properties. The neutral (uncharged) species of these reagents are often poorly soluble in aqueous solutions. By determining the effect of pH on the solubility, it may be possible to find the values for the proton-transfer equilibrium quotients of the system. These values, in turn, can be very helpful to the analytical chemist who wants to understand the fundamental behavior of the reagent and hence be able to control it in developing analytical procedures.

---

*R. W. Ramette, R. F. Stewart, *Journal of Physical Chemistry*, *65*, 243 (1961).

The key models were discussed earlier under **amphiprotic solutes**; this automatically covers substances that are only acids, or only bases. Equation (17.26) shows the dependence of solubility upon $[H^+]$, and we will now consider the application of this relationship to an analytical reagent used for the spectrophotometric determination of silver.

In this case study we refer to the work by Sandell and Neumeyer on the ionization constants of p-diethylaminobenzylidenerhodanine.* Called **rhoda-nine** for simplicity, this reagent is amphiprotic and is consistent with the model used earlier:

$$HR(s)$$

$$\updownarrow$$

$$H_2R^+ \rightleftarrows HR(aq) \rightleftarrows R^-,$$

where HR is the symbol for the molecular species

$$
\begin{array}{cc}
HN\!-\!C\!=\!O & \\
| \quad\; | & \\
S\!=\!C \quad C\!=\!CH\!-\!\!\!\bigcirc\!\!\!-\!N(C_2H_5)_2. \\
\backslash / & \\
S &
\end{array}
$$

Its reaction with silver ion (to form a colored complex) depends on displacement of the amine proton by silver ion. This is encouraged at high pH, but at low pH values it is harder for the silver ion to compete for the nitrogen site, and in addition the substance forms the cation H_2R^+ through protonation of the other nitrogen atom. Thus, a knowledge of the acid dissociation quotients is important:

$$Q_1 = \frac{[H^+][HR]}{[H_2R^+]}, \qquad Q_2 = \frac{[H^+][R^-]}{[HR]}.$$

The solubility determinations were made at 20°C, using buffers that contained 20% alcohol to increase the very small solubility of rhodanine. In all cases the ionic strength was 0.050. An excess of solid rhodanine was added to portions of the buffers, which were vigorously shaken for two hours after an initial period of standing for 18 hours. The total rhodanine concentration of filtered portions of the saturated solutions was determined spectrophotometrically: by raising the pH through addition of sodium hydroxide to destroy the buffer, all the dissolved species were converted to the yellow anion form R^-. The absorption maximum is at 470 nm, and Beer's Law is followed.

---

*E. B. Sandell, J. J. Neumeyer, *Journal of the American Chemical Society*, **73**, 654 (1951).

The following data were obtained by Neumeyer:*

$[H^+]$ for buffer	$10^6 \cdot S$	$[H^+]$ for buffer	$10^6 \cdot S$
$1.28 \cdot 10^{-8}$	13.7	$2.22 \cdot 10^{-3}$	2.00
$2.32 \cdot 10^{-8}$	7.59	$3.85 \cdot 10^{-3}$	3.60
$4.03 \cdot 10^{-8}$	4.57	$6.11 \cdot 10^{-3}$	5.40
$5.57 \cdot 10^{-8}$	3.25	$7.51 \cdot 10^{-3}$	6.65
$11.9 \ \cdot 10^{-8}$	1.64	$13.7 \ \cdot 10^{-3}$	11.9
		$17.6 \ \cdot 10^{-3}$	15.3

Note that one group of buffers is at relatively high pH, so that there will be appreciable dissociation of HR to R^-. The other group is in the low-pH region, so that the solutions will contain substantial amounts of the cation H_2R^+. Note also that the solubility goes through a minimum, as would be predicted for an amphiprotic substance.

We will attempt to determine three equilibrium quotients Q_0, Q_1 and Q_2 from these data, using Eq. (17.26) as a guide. The first step is to rewrite that equation, having multiplied all terms by $[H^+]$:

$$S[H^+] = Q_0Q_2 + Q_0[H^+] + \frac{Q_0}{Q_1}[H^+]^2. \tag{18.8}$$

We recognize a familiar picture: a plot of $S[H^+]$ will yield Q_0Q_2 as its intercept. Such a plot is shown in Fig. 18.8, where only the data for the high-pH buffers were used.

The scattering of the points is not as bad as it seems at first glance, because the scale of the vertical axis is expanded. The fact that the plot is linear shows that the $[H^+]^2$ term of Eq. (18.8) is unimportant at these pH

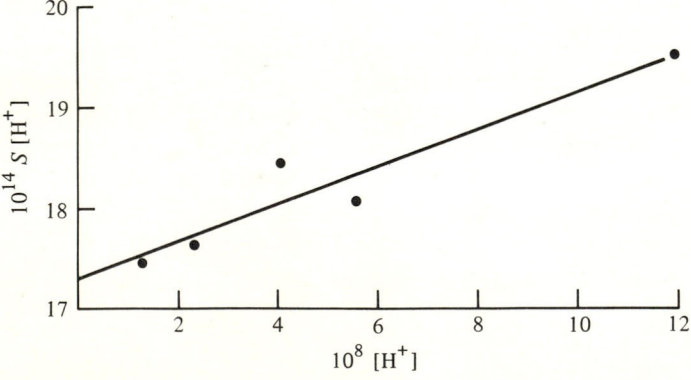

Figure 18.8

*J. J. Neumeyer, MS Thesis, University of Minnesota, 1950.

values. Therefore, we may use a linear least-squares fit to find the following with 90% confidence:

$$\begin{aligned}
\text{Intercept:} &\qquad Q_0 Q_2 = (1.73 \pm 0.05) \cdot 10^{-13}, \\
\text{Slope:} &\qquad Q_0 = (1.9 \pm 0.8) \cdot 10^{-7}, \\
\text{Intercept/slope:} &\qquad Q_2 = (9 \pm 4) \cdot 10^{-7}.
\end{aligned}$$

Once the value for $Q_0 Q_2$ is known, Eq. (18.8) is converted to a linear form:

$$\frac{S[H^+] - Q_0 Q_2}{[H^+]} = Q_0 + \frac{Q_0}{Q_1}[H^+].$$

A plot of the variable on the left versus $[H^+]$ is shown in Fig. 18.9 for the low-pH buffers.

The excellent linearity of this plot implies that a reliable value can be found for the slope Q_0/Q_1. However, the intercept is so close to zero that its value is uncertain. A linear least-squares fit gives the following values:

$$\begin{aligned}
\text{Slope:} &\qquad Q_0/Q_1 = (8.58 \pm 0.12) \cdot 10^{-4}, \\
\text{Intercept:} &\qquad Q_0 = (1.9 \pm 1.2) \cdot 10^{-7}, \\
\text{Intercept/slope:} &\qquad Q_1 = (2.2 \pm 1.3) \cdot 10^{-4}.
\end{aligned}$$

In both plots it is the intrinsic solubility Q_0 that is the factor of low accuracy. Although the values of $Q_0 Q_2$ and Q_0/Q_1 were both fairly reliable, it is not possible to find the individual values of Q_1 and Q_2 without knowing the value of Q_0.

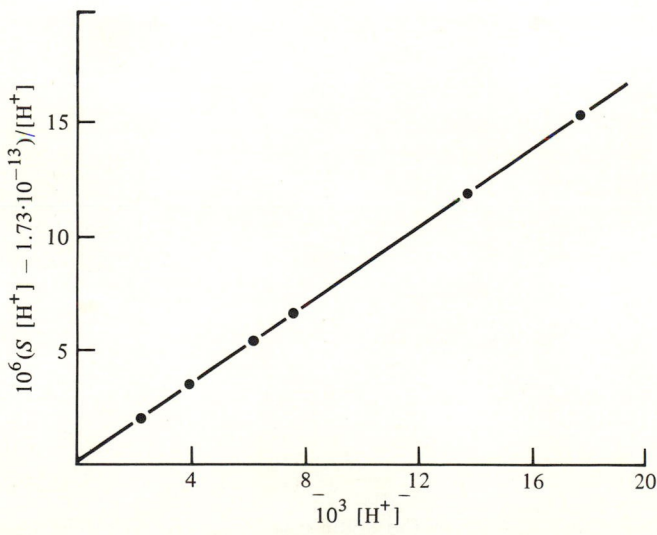

Figure 18.9

This unfortunate situation led Sandell and Neumeyer to attempt a more or less direct determination of Q_0 by measuring the solubility of rhodanine in a buffer whose pH corresponded to the isoelectric point of rhodanine, when the contribution of R^- and H_2R^+ to the solubility was minimal. The hydrogen-ion concentration required for the isoelectric point is, as stated on p. 543,

$$[H^+] = (Q_1Q_2)^{1/2}.$$

The necessary value of Q_1Q_2 is found with fair confidence from the previous values of Q_0Q_2 and Q_0Q_1:

$$Q_1Q_2 = \frac{Q_0Q_2}{Q_0/Q_1} = \frac{1.73 \cdot 10^{-13}}{8.58 \cdot 10^{-4}} = 2.02 \cdot 10^{-10}.$$

Therefore a buffer with $[H^+] = (2.02 \cdot 10^{-10})^{1/2} = 1.42 \cdot 10^{-5}$ was prepared; it contained 20% alcohol and was saturated with rhodanine. A 100-mL portion of the filtered solution was extracted with 10 mL of chloroform, which resulted in the transfer of all the rhodanine into the smaller volume, thus improving the accuracy with which it could be determined spectrophotometrically. The absorption maximum for rhodanine in chloroform is at 415 nm, and a standard curve was established for this determination.

The use of this technique resulted in $Q_0 = (2.1 \pm 0.2) \cdot 10^{-7}$, which is much more reliable than the values determined by the plots. In conclusion, the values for the acid-dissociation quotients are:

$$Q_1 = \frac{2.1 \cdot 10^{-7}}{8.58 \cdot 10^{-4}} = 2.4 \cdot 10^{-4}, \qquad pQ_1 = 3.62,$$

$$Q_2 = \frac{1.73 \cdot 10^{-13}}{2.1 \cdot 10^{-7}} = 8.2 \cdot 10^{-7}, \qquad pQ_2 = 6.09.$$

PROBLEMS

Effect of pH on solubility

1. Equation (17.32) was derived for a 1:1 salt, such as lead sulfate. Derive a similar equation relating solubility to $[H^+]$ for a 2:1 salt, such as copper iodate, using the following model:

$$Cu(IO_3)_2 \cdot H_2O(s) \rightleftarrows Cu^{2+} + 2IO_3^-$$

$$+$$

$$H^+$$

$$\updownarrow$$

$$HIO_3.$$

Interpret the following data on the solubility of copper iodate in constant-ionic-strength solutions of lithium perchlorate and perchloric acid:

C_{LiClO_4}	C_{HClO_4}	$S_{Cu(IO_3)_2}$
0.990	0.010	$5.80 \cdot 10^{-3}$
0.700	0.300	$7.99 \cdot 10^{-3}$
0.300	0.700	$1.06 \cdot 10^{-2}$
0	1.000	$1.23 \cdot 10^{-2}$

The researcher who did this work found $Q_{sp} = 7.6 \cdot 10^{-7}$ and Q_a for iodic acid HIO_3 equal to 0.470. Do you agree?

2. A certain crystalline molecular substance X believed to have basic properties is equilibrated with three acidic solutions by prolonged shaking in sealed bottles. Portions of the saturated solutions are obtained by filtration, and these are analyzed by means of a spectrophotometer. The solubilities thus obtained are as follows:

[H⁺] at equilibrium in the acid solution	Observed molar solubility
0.00100	0.00300
0.00500	0.00900
0.01000	0.01650

a) Derive an expression relating the observed solubility to the intrinsic solubility and the value of Q_b for the original substance using the simple model:

$$X(s) = X(aq), \qquad X(aq) + H^+ = HX^+.$$

b) Interpret the data, either analytically or graphically, to find the value of the intrinsic solubility and the value of the basic dissociation quotient Q_b. Assume the ionic strength constant in all solutions.

c) What experiment might you perform to determine whether substance X is amphiprotic, and if so, how can you find its Q_a value?

3. The solubility of methyl red (a very slightly soluble molecular ampholyte) was determined as a function of [H⁺] in a series of buffers with $\mu = 0.0200$. The buffer compositions and solubility results were as follows:

C_{KCl}	C_{HCl}	C_{NaOAc}	C_{HOAc}	$10^5 \cdot S$, molar
0.019	0.001			0.774
0.016	0.004			1.209
0.010	0.010			2.108
0.007	0.013			2.645
0.004	0.016			3.08
0	0.020			3.68
		0.020	0.00430	2.39
		0.020	0.00472	2.21
		0.020	0.00572	1.92
		0.020	0.00858	1.49
		0.020	0.01715	1.08

Interpret these data to find the intrinsic solubility and the acid-dissociation constants:

$$H_2R^+ = HR = R^-,$$
$$\quad K_1 \quad K_2$$

4. A certain monoprotic acid HA having a molecular weight of 275.6 is only very slightly soluble in water. A sample of the acid is prepared with the radioisotope C-14 incorporated, and portions of the crystalline material are equilibrated with a series of acetic acid sodium acetate buffers of accurately known pH. To determine the concentration of the dissolved acid, the radioactivity caused by C-14 was measured by a scintillation counter. In this technique one simply takes 1.00 mL of the aqueous solution and mixes it with a small volume of a special "cocktail" that generates visible photons when a beta-particle due to the decomposition of C-14 is captured. The photons activate a photomultiplier tube, and the event is recorded as a "count" on the digital readout of the instrument. The number of counts per minute is directly proportional to the molarity of C-14 in the sample.

   The scintillation counter was calibrated by preparing a solution of 9.32 mg (weighed on a microbalance) of the acid in 100 mL of a dilute NaOH solution. When 1.00 mL of this standard solution was placed in the instrument, a count rate of 5 500 counts/min was observed.

   The results for the saturated solutions in the buffers were as follows:

Buffer pH	Counts/min
4.500	402.6
4.726	533.0
4.881	669.9
5.000	815.0

   a) Each buffer had an ionic strength of 0.100. Given that pK for acetic acid is 4.756, explain how you would prepare 100 mL of the pH 4.500 buffer by using stock solutions of 0.774M acetic acid and 0.428M sodium hydroxide; assume $f_1 = 0.79$.

   b) Assuming the following chemical model: $HA(s) = HA(aq) = H^+ + A^-$, derive an equation that relates solubility S to the hydrogen-ion molarity. Explain how the equation may be used to determine the pertinent equilibrium quotients, given a set of S and pH data.

   c) Use the above table of data to deduce the equilibrium quotients (at $I = 0.1$) for the weak acid. Include a neat and precise plot of the data, properly labeled.

Potentiometric titration

5. A chloride-containing sample weighing 0.2507 gram was dissolved in 50 mL of water and was titrated with 0.09860M silver nitrate, using a silver indicator electrode and a saturated calomel reference electrode with a potassium nitrate bridge. The cell voltage was measured as a function of the volume of silver nitrate added, with the following results (the silver electrode was the positive pole):

AgNO₃ added			AgNO₃ added		
drops	mL	E	drops	mL	E
	1.00	0.0677	+1		0.2100
	5.00	0.0735	+1		0.2140
	10.00	0.0816	+1		0.2188
	20.00	0.1011	+1		0.2245
	31.00	0.1409	+1		0.2325
	32.00	0.1485	+1		0.2434
	33.00	0.1591	+1		0.2615
	34.00	0.1760	+1		0.2984
	34.10	0.1786	+1		0.3296
+3		0.1825	+1		0.3455
+3		0.1866	+1		0.3555
+3		0.1920	+2		0.3680
+3		0.1980	+2	35.41	0.3761
+3	34.75	0.2065		36.00	0.4037

a) Interpret these data to find the endpoint of the titration, using a) a large-scale plot of E versus volume, b) a plot of the first derivative versus volume, c) a plot of the reciprocal of the first derivative versus volume. Compare the three methods and finally calculate the percentage of chloride in the sample.

b) Before the endpoint, the cell potential is fixed by the chloride-ion concentration, that controls [Ag$^+$] through the solubility product. After the endpoint, the silver electrode can respond directly to the concentration of excess Ag$^+$. By comparing two potentials, one before and one after the endpoint, and using calculated Cl$^-$ and Ag$^+$ molarities, calculate the solubility product for AgCl. Note that E^0 is not needed.

Weak-electrolyte model with appreciable intrinsic solubility

6. Consider the following results for the solubility of silver chloride in solutions containing sodium chloride and sodium perchlorate with constant ionic strength 0.01:

Molarity of NaCl:	$1 \cdot 10^{-4}$	$2 \cdot 10^{-4}$	$5 \cdot 10^{-4}$
Solubility of AgCl:	$2.4 \cdot 10^{-6}$	$1.4 \cdot 10^{-6}$	$8.0 \cdot 10^{-7}$

Show that the strong electrolyte model is unsatisfactory on the grounds that Q_{sp} would not be constant. Derive the relationship between solubility and [Cl$^-$] according to the model AgCl(s) = AgCl(aq) = Ag$^+$ + Cl$^-$. Show that the relationship predicts a linear plot of S versus $1/[Cl^-]$, and use such a plot to determine the values of Q_0, Q_d, and Q_{sp} for silver chloride.

7. The solubility-product constant for mercuric chloride HgCl₂ is about $4 \cdot 10^{-14}$, and yet the molar solubility of this compound is 0.25. How can this be? Calculate the value for Q_d, i.e., the Q value for the reaction HgCl₂(aq) = Hg^{2+} + 2Cl$^-$. If the formation constant for the reaction Hg^{2+} + Cl$^-$ = HgCl$^+$ is $2 \cdot 10^5$, then what must be the formation constant for the second step HgCl$^+$ + Cl$^-$ = HgCl₂(aq).

DISTRIBUTION EQUILIBRIA AND SOLVENT EXTRACTION

19

There are good scientists who love theory but, in the laboratory, display what an experimentalist might call an inverse Midas Touch.

Eva Menger

Many covalent substances have appreciable vapor pressures even at room temperature. For example, when a stoppered bottle is half-filled with pure water at 25°, the air space at equilibrium contains about $8 \cdot 10^{17}$ molecules of H_2O per milliliter. This corresponds to a gas-phase molarity of $1.3 \cdot 10^{-3}$ and to a partial pressure of 25.7 mm of Hg, or $3.4 \cdot 10^{-2}$ atm. Because of the strength of hydrogen bonding, any attempt to make the water-vapor pressure greater than this (at 25°) results in the formation of liquid water. Thus, the distribution of water molecules between the liquid and vapor phases is perfectly reversible, and we may use a chemical equation and an equilibrium constant to represent the system:

$$H_2O(\text{liquid}) \rightleftarrows H_2O(\text{gas}); \qquad K = \frac{P_{\text{vapor}}}{X_{\text{liquid}}} = \frac{3.4 \cdot 10^{-2}}{1} \quad (\text{at } 25°). \qquad (19.1)$$

The equilibrium partial pressure of water vapor is constant for a given temperature. In the air space above the liquid there is also oxygen gas at a partial pressure of 0.21 atm. Molecules of oxygen also move back and forth between the two phases, and again an equation and equilibrium constant may be used:

$$O_2(\text{gas}) \rightleftarrows O_2(\text{aq}); \qquad K_H = \frac{[O_2(\text{aq})]}{P_{O_2(\text{gas})}}. \qquad (19.2)$$

The value for K_H is $1.4 \cdot 10^{-3}$. According to the equilibrium-constant expression, the molarity of dissolved O_2 is directly proportional to the partial pressure of O_2 in the gas phase:

$$[O_2]_{\text{aq}} = K_H \cdot P_{O_2(\text{gas})}$$

Thus we may predict the concentration of oxygen in water, which is in equilibrium with pure oxygen gas at a pressure of 2.0 atm:

$$[O_2] = 1.4 \cdot 10^{-3} \cdot 2 = 2.8 \cdot 10^{-3} \text{ mol/L}.$$

This simple relationship is known as **Henry's Law**: the solubility of a gas is proportional to the pressure of the gas in contact with the solution. It is not exact, especially for high pressures, but is quite useful for many purposes The **Henry's-Law constant** K_H is simply the numerical value as defined in Eq. (19.2) above.

   Suppose we want to estimate the concentration of dissolved oxygen in water, which is in contact (and at equilibrium) with ordinary air at a total pressure of 1.00 atm. Since according to Eq. (19.1) the water vapor is present at $P_{H_2O} = 0.034$ atm, the partial pressure for the other gaseous species is $1.000 - 0.034 = 0.966$ atm. The percentage of oxygen in dry air is 20.95%, and so the partial oxygen pressure is

$$P_{O_2} = 0.2095 \cdot 0.966 = 0.202 \text{ atm.}$$

Then from Eq. (19.2), we have:

$$[O_2] = 1.4 \cdot 10^{-3} \cdot 0.202 = 2.9 \cdot 10^{-4} \, mol/L.$$

Similar considerations hold for the other air gases: nitrogen, argon, carbon dioxide, etc.

Now let us introduce a second solvent, carbon tetrachloride CCl_4, that does not mix with water. We will see that such **immiscible solvents** are of great utility in separations. Suppose that at 25°C a stoppered bottle contains about 100 mL each of water and carbon tetrachloride, with air space above. The organic solvent, with a density of 1.59 gram/mL, forms the bottom layer (Fig. 19.1 below).

Let the bottle be shaken vigorously for awhile, and suppose that the gaseous pressure above the solvents is 1.000 atm. Several equilibria will be established:

1. A little carbon tetrachloride will dissolve in the water:

$$CCl_4(liquid) = CCl_4(aq); \qquad K_0 = 5.0 \cdot 10^{-3} M.$$

2. A little water will dissolve in the carbon tetrachloride:

$$H_2O(liquid) = H_2O(CCl_4); \qquad K_0 = 4.1 \cdot 10^{-3} M.$$

3. The water will exert its vapor pressure:

$$H_2O(liquid) = H_2O(gas); \qquad K_{eq} = P_{H_2O} = 3.4 \cdot 10^{-2} \, atm.$$

4. The carbon tetrachloride has a higher vapor pressure than water:

$$CCl_4(liquid) = CCl_4(gas); \qquad K_{eq} = P_{CCl_4} = 0.153 \, atm.$$

5. The oxygen in the air will partially dissolve in the water:

$$O_2(gas) = O_2(aq); \qquad K_H = \frac{[O_2]}{P_{O_2}} = 1.4 \cdot 10^{-3}.$$

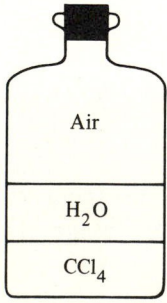

Figure 19.1

When the vapor pressures of the two solvents are subtracted from 1.000 atm, we find 0.813 as the partial pressure for the gases of air. Since air is 20.95% oxygen, we find $P_{O_2} = 0.170$ atm. Therefore,

$$[O_2(aq)] = 1.4 \cdot 10^{-3} \cdot 0.170 = 2.3 \cdot 10^{-4} M.$$

6. The oxygen in the air will also partially dissolve in the carbon tetrachloride:

$$O_2(gas) = O_2(CCl_4); \qquad K_H = \frac{[O_2]}{P_{O_2}} = 1.24 \cdot 10^{-2}.$$

Therefore, the molarity of O_2 dissolved in the CCl_4 is:

$$[O_2(CCl_4)] = 1.24 \cdot 10^{-2} \cdot 0.170 = 2.1 \cdot 10^{-3} M.$$

Note that this is about nine times the concentration in the water phase.

Of course, we may do for the other gases of air what was done for oxygen, using their particular Henry's-Law constants. However, to avoid the extra clutter let us simply summarize what has been calculated above, using a pictorial scheme of the three-phase system:

Air phase:	CCl_4(gas), 0.153 atm	H_2O(gas), 0.034 atm	O_2(gas), 0.17 atm
	⇅	⇅	⇅
Water phase:	CCl_4(aq), $5 \cdot 10^{-3} M$	H_2O(liquid), $X = 1.0$	O_2(aq), $2.3 \cdot 10^{-4} M$
	⇅	⇅	⇅
Organic phase:	CCl_4(liquid), $X = 1.0$	$H_2O(CCl_4)$, $4 \cdot 10^{-3} M$	$O_2(CCl_4)$, $2.1 \cdot 10^{-3} M$

19.1   THE DISTRIBUTION LAW

The above discussion gives a rather complete view of a typical two-solvent system contaminated only by the species of air. In analytical applications of such systems we usually ignore the presence of air and the vapor pressures of the solvents. Further, we typically assume that the mutual solubility of the two solvents is negligible. Instead of considering the distribution of oxygen from air we are interested in the behavior of other substances that are present as solutes in one or both solvents. For example, we may study the way in which iodine I_2 distributes itself between carbon tetrachloride and water. A portion of each solvent is placed in a bottle and varying amounts of iodine are added. After sufficient shaking, the layers separate and may be analyzed for iodine content by using titration with a standard sodium thiosulfate solution. This system was studied by Pearce and Eversole* with the following results (at 25°):

---

*J. N. Pearce, W. G. Eversole, *Journal of Physical Chemistry*, 28, 245 (1924).

Molarity of I_2 in water layer	Molarity of I_2 in CCl$_4$ layer	Equilibrium ratio: organic/water
$2.94 \cdot 10^{-4}$	$2.43 \cdot 10^{-2}$	82.6
$4.03 \cdot 10^{-4}$	$3.34 \cdot 10^{-2}$	82.8
$5.66 \cdot 10^{-4}$	$4.68 \cdot 10^{-2}$	82.7
$9.61 \cdot 10^{-4}$	$7.90 \cdot 10^{-2}$	82.2
		Average: 82.6

Since the distribution of I_2 molecules across the boundary between the two solvents is reversible, we may write an equation and an equilibrium expression:

$$I_2(aq) = I_2(org); \qquad K_D = \frac{[I_2]_{org}}{[I_2]_{aq}}.$$

This is an example of the **distribution law**; it states that a single species will distribute itself between two immiscible solvents in a constant ratio, even though the total amount of the species may change. The equilibrium constant K_D is called the **distribution constant**; it has a particular value for each solute present in the two-phase system. From the above data it is evident that $K_D = 82.6$ for the distribution of iodine between CCl$_4$ and water. For other solutes K_D has quite different values. From the earlier discussion we see that K_D for oxygen is 9.

Relationship between K_D and the Ratio of Intrinsic Solubilities

The solubility of iodine in carbon tetrachloride may be found through titration of a portion of a CCl$_4$ solution that has been shaken with an excess of solid I_2 until equilibrium is reached. The solubility has been reported as 0.115 mol/L:

$$I_2(s) = I_2(CCl_4); \qquad K_0 = 0.115 = \frac{[I_2]_{org}}{1}. \tag{19.3}$$

In water the solubility is much lower, about $1.32 \cdot 10^{-3}$ mol/L:

$$I_2(s) = I_2(aq); \qquad K_0 = 1.32 \cdot 10^{-3} = \frac{[I_2]_{aq}}{1}. \tag{19.4}$$

Imagine a three-phase system of water, carbon tetrachloride, and solid iodine. The distribution law and the two intrinsic solubilities should be simultaneously satisfied. In fact, we may derive the distribution law by dividing Eq. (19.3) by Eq. (19.4):

$$\frac{(K_0)_{org}}{(K_0)_{aq}} = \frac{[I_2]_{org}}{[I_2]_{aq}} = \frac{0.115}{1.32 \cdot 10^{-3}} = 87.1.$$

The fact that this ratio is about 5% higher than that found by the direct distribution study (87.1 compared to 82.6), may be partly due to lack of accuracy in the original data and partly due to our neglect of molecular activity coefficients. In the saturated solutions there perhaps is some deviation from ideal behavior. Therefore, while we may use the ratio of solubilities for estimating the value of K_D, it is always better to rely on direct measurement of the distribution in the two-solvent system.

19.2 THE DISTRIBUTION RATIO D

The distribution law and K_D refer only to the distribution of a single species between the two solvents. However in many situations the species takes part in other reversible reactions in one or both phases, so that at equilibrium it exists as a mixture of chemical forms. It is very useful to define the **distribution ratio** in terms of the total molarity of the substance in each phase:

$$D = \frac{\sum (\text{molarities in the organic phase})}{\sum (\text{molarities in the aqueous phase})} = \frac{C_o}{C_w}. \qquad (19.5)$$

We will use C_o and C_w (often omitting the "w") for total molarities.

Actually it is only D, and not K_D, that can usually be determined experimentally by analysis of the two-phase solvent system. When we have reason to believe that only one molecular species is present in each phase, as in the above example of iodine, we may assume that the value of D is equal to K_D.

If a chemical reaction can occur in the aqueous phase, causing the molecular species to be converted into an ionic species, the distribution will shift in favor of the aqueous phase. This is because ions do not distribute appreciably into the nonpolar organic phase unless they can do so as an associated ion pair which, like a molecule, is uncharged as a whole. A single type of ion cannot cross the phase boundary because this would lead to a separation of charge in the system and electrostatic forces would prevent this from occurring to a measurable extent. Therefore we will consider only uncharged species.

Demonstration

Shake about a gram of solid iodine with 100 mL of distilled water for a few minutes to prepare a dilute solution. It will be pale brown, since the iodine exists in water as the hydrate I_2OH_2. Filter the solution to remove excess solid.

Place the solution in a bottle, add an equal volume of carbon tetra-chloride, and note that the organic layer shows some purple color due to the partial distribution of iodine. The color is purple because in the nonpolar solvent the iodine is present as the unhydrated species I_2. That purple is the color of the free I_2 species is shown by the color of iodine vapor (simply let some solid I_2 stand in a one-liter stoppered bottle). If the CCl_4, H_2O, I_2 system is allowed to stand in the bottle, without shaking, the iodine will eventually diffuse across the boundary until equilibrium is reached with the concentration ratio of 82.6. Such unassisted diffusion would be quite slow, however, and we may greatly speed up the process by vigorous shaking (do this) which breaks up the phases into small droplets. This makes an enormous increase in the area of surface contact between the two solvents and, although the diffusion properties of I_2 are unchanged, the equilibrium is attained much more quickly. At this point the organic layer is purple and the aqueous layer is nearly colorless, because it now contains only about 1/83 of the original amount of iodine.

Now add successive 1-mL portions of $0.1M$ KI solution to the mixture, shaking and allowing the phases to separate every time. With each addition there will be an increase in the brown color of the aqueous phase. Add about 50 mL of $2M$ KI, shake, and the purple color of the organic phase will nearly disappear.

Algebraic Basis for the Demonstration

In this demonstration we are seeing the variation in the distribution ratio D caused by the formation of the triiodide ion in the water phase:

$$\text{Water layer:} \qquad I_2(aq) + I^- \rightleftarrows I_3^-$$
$$\updownarrow$$
$$\text{Organic layer:} \qquad I_2(org)$$

Thus, there are two simultaneous equilibria to consider:

$$I_2(aq) = I_2(org); \qquad K_D = \frac{[I_2]_o}{[I_2]_w} = 82.6, \tag{19.6}$$

$$I_2(aq) + I^- = I_3^-; \qquad K_f = \frac{[I_3^-]}{[I_2]_w[I^-]} = 720. \tag{19.7}$$

It is not difficult to see in which way these equilibria affect the value of the distribution ratio. We write:

$$D = \frac{C_o}{C_w} = \frac{[I_2]_o}{[I_2]_w + [I_3^-]}.$$

By substituting the expression for $[I_3^-]$ obtained from Eq. (19.7), we get:

$$D = \frac{[I_2]_o}{[I_2]_w + K_f[I_2]_w[I^-]}$$

Then, by introducing Eq. (19.6) for the ratio of I_2 concentrations, we have:

$$D = \frac{K_D}{1 + K_f[I^-]} = \frac{82.6}{1 + 720[I^-]}. \tag{19.8}$$

Therefore, if the aqueous phase contains a free-iodide-ion molarity of $1M$, the distribution ratio will be $83/721 = 0.12$, so that most of the iodine will be present (as I_3^-) in the aqueous layer.

19.3   USING DISTRIBUTION DATA TO DETERMINE EQUILIBRIUM CONSTANT VALUES

Once Eq. (19.8) was derived for the iodine triiodide-ion system, it was a simple matter to plug in the known values for K_D, K_f, and for $[I^-]$ to predict the value of the distribution ratio D. But where did those equilibrium constant values come from? We have already seen how the value for K_D was found to be 82.6 by measuring the distribution of iodine in solutions containing no iodide ion. The value for K_f has been determined most accurately by using ultraviolet spectrophotometry, through the methods discussed in Chapter 13. However, it would also be possible to use distribution data and a rearrangement of Eq. (19.8) to find the value of K_f. By taking the reciprocal of Eq. (19.8) we find:

$$\frac{1}{D} = \frac{1}{K_D} + \frac{K_f}{K_D}[I^-]. \tag{19.9}$$

To make use of this relationship we would set up a few bottles containing water, carbon tetrachloride, and a small concentration of iodine. By adding varying amounts of potassium iodide to the water layers, shaking to establish the equilibrium, and then using standard sodium thiosulfate to find the total concentration of iodine in each layer, we would obtain a set of D values, each corresponding to a certain molarity of iodide ion in the water layer.

Then, according to Eq. (19.9), a plot of the $1/D$ values versus $[I^-]$ values should be linear, with an intercept of $1/K_D$ and a slope of K_f/K_D. The desired value of K_f would simply be the slope/intercept ratio.

In fact, fairly good values of K_D and K_f for the iodine system have been determined in this way as well as by interpretations of solubility. The point of this discussion is not merely to learn more about this well-studied system; it is to illustrate the general idea that accurate distribution data may be useful in revealing the nature of interactions between species in one or both solvents. Further examples will follow.

19.4   SOLVENT EXTRACTION AS A TECHNIQUE FOR CHEMICAL SEPARATION

The most typical analytical application of immiscible solvent extraction is the quantitative removal of a constituent from an aqueous phase. Usually it is desired that only one constituent be extracted, thus effecting a separation from impurities and/or from other constituents of the sample. In the optimal situation it is possible to shake a certain volume of the aqueous phase with a significantly smaller volume of the organic phase and yet to extract the constituent to the extent of 99.9% or better. This ideal is rarely achieved, however.

The key to understanding the efficiency of an extraction lies in the distribution ratio D. By controlling the nature of the aqueous phase (pH, addition of ligands, etc.) we may often influence the value of D over a wide range. Of course, if we are interested in efficient extraction, we want the value of D to be as high as possible. This ratio has already been defined by Eq. (19.5):

$$D = \frac{C_o}{C_w},$$

where C_o and C_w are the total concentrations of the constituent, including all its various chemical forms, in each phase. Since we want to use different volumes of the phases at times, we may rewrite the expression:

$$D = \frac{n_o/V_o}{n_w/V_w}, \tag{19.10}$$

where n stands for the number of millimoles of the constituent. It is the ratio of millimoles that gives a simple direct measure of extraction efficiency, and therefore we solve Eq. (19.10) for this ratio:

$$\frac{n_o}{n_w} = D\,\frac{V_o}{V_w}. \tag{19.11}$$

An even more useful function is the **fraction extracted** which is (in terms of millimoles):

$$\text{Fraction extracted} = \frac{n_o}{n_{tot}} = \frac{n_o}{n_o + n_w} = \frac{1}{1 + n_w/n_o}.$$

By inserting the n_w/n_o ratio obtained from Eq. (19.11), we find

$$\text{Fraction extracted} = \frac{n_o}{n_{tot}} = \frac{1}{1 + V_w/DV_o}.$$

Since the fraction left in the aqueous phase must be equal to $(1 - \text{Fraction extracted})$, it follows that:

$$\text{Fraction left} = \frac{V_w/DV_o}{1 + V_w/DV_o} = \frac{1}{1 + DV_o/V_w}. \tag{19.12}$$

It is evident that we have two ways to control the efficiency of an extraction. One is to use chemical means to influence the value of D and the other is to choose particular values for the two volumes V_o and V_w. The larger the value of D, the more efficient the extraction; the larger the ratio of V_o to V_w, the more efficient the extraction.

Example 1 An aqueous solution contains $1 \cdot 10^{-4}$ iodine along with enough potassium iodide to make the value of $D = 5$. If 100 mL of this solution are shaken with 50 mL of CCl_4, what fraction of iodine will remain in the aqueous phase?

$$\text{Fraction left} = \frac{1}{1 + 5 \cdot 50/100} = 0.29 = 29\%.$$

Example 2 If equal volumes of the two phases are used for an extraction, what must be the value for D if 95% of the constituent is extracted?

$$\text{Fraction left} = 0.05 = \frac{1}{1 + D}; \qquad D = 19.$$

Example 3 Under conditions such that $D = 20$, what volume of organic phase should be used to extract 100 mL of aqueous phase if extraction is to be 99% complete?

$$\text{Fraction left} = 0.01 = \frac{1}{1 + 20 \cdot V_o/100}; \qquad V_o = 495 \approx 500 \text{ mL}.$$

In some cases such a calculation may show that an unreasonably large volume of organic phase is required. This leads to the necessity of multiple extractions.

Multiple (Successive) Extractions

Suppose an aqueous phase is extracted with V_o milliliters of an organic solvent, and that 80% of a certain constituent is extracted, leaving 20% in the aqueous phase. If the resulting aqueous phase is then shaken with a fresh portion of solvent (again, V_o milliliters), this second extraction will remove 80% of the 20% that was left, or 16% of the original amount of the constituent. The combined organic-solvent extracts will contain 96% of the constituent and only 4% will now remain in the aqueous phase. Obviously, if this process of successive extractions is continued, we will eventually reduce the remaining constituent to a negligible level in the aqueous phase. However, if N such extractions are performed, the total volume of organic solvent will be $N \cdot V_o$ and may be rather large.

The foregoing idea can be stated in terms of Eq. (19.12). The fraction of the constituent left after N successive extractions is simply:

$$\text{Fraction left} = \frac{n_w}{n_{tot}} = \left(\frac{1}{1 + DV_o/V_w}\right)^N, \tag{19.13}$$

where V_w is the unchanging volume of the aqueous phase and V_o is the volume of the organic solvent used in each of the successive extractions. From this equation we may learn that it is more efficient to use successive extractions when the total volume of organic solvent is specified.

Example The distribution ratio between benzene and water in substance A is 4. A 50-mL portion of a water solution of A is to be extracted with a total volume of 200 mL of benzene. Compare the extraction efficiency using (a) one 200-mL extraction, (b) two 100-mL extractions, (c) four 50-mL extractions. Using Eq. (19.13) for each case, we find:

a) $\dfrac{n_w}{n_{tot}} = \dfrac{1}{1 + 4 \cdot 200/50} = 0.059$ (about 6% left behind);

b) $\dfrac{n_w}{n_{tot}} = \left(\dfrac{1}{1 + 4 \cdot 100/50}\right)^2 = 0.012$ (only about 1% left);

c) $\dfrac{n_w}{n_{tot}} = \left(\dfrac{1}{1 + 4 \cdot 50/50}\right)^4 = 0.0016$ (extraction is nearly complete).

19.5 SOME IMMISCIBLE SOLVENTS AND THEIR PROPERTIES

For an organic solvent to be immiscible (i.e., only slightly soluble) in water it is necessary that it be relatively nonpolar. Otherwise it would be able to disrupt the hydrogen-bonded structure of water and dissolve too extensively. Table 19.1 lists a few commonly used solvents immiscible with water.

Table 19.1 Some Solvents Used For Immiscible-Solvent Extraction

Name	Formula	Density, g/mL	Boiling point, °C
Heptane	$CH_3(CH_2)_5CH_3$	0.684 (20°)	98.5
Benzene	C_6H_6	0.879 (20°)	80.1
Carbon tetrachloride	CCl_4	1.595 (20°)	76.5
Chloroform	$CHCl_3$	1.483 (20°)	61.3
Ethyl ether	$C_2H_5OC_2H_5$	0.719 (15°)	34.5
2,4-pentanedione (acetylacetone, ACAC)	$CH_3COCH_2COCH_3$	0.975 (20°)	140.5
4-methyl-2-pentanone (methyl isobutyl ketone, MIBK)	$(CH_3)_2CHCH_2COCH_3$	0.801 (20°)	115.8

In the above list only two solvents, CCl_4 and $CHCl_3$, form the bottom layer in a two-phase system with water. Ethyl ether boils at a rather low temperature and it is especially important to release the pressure that builds up when ether is shaken with water in a stoppered vessel. When working with any organic solvent it should be assumed that there is a toxic hazard from the vapors or from skin contact. Most solvents also are flammable, and care should be taken to keep all flames out of the area. Portions of solvents should be measured in a fume hood for safety. Waste solvents should not be poured down a drain, but should be placed in an approved recovery vessel.

Chemical manufacturers offer high-purity grades of solvents for analytical purposes. It is often possible to purify a questionable material by distillation, preferably with an efficient column.

19.6 EXTRACTION OF MOLECULAR ACIDS

Organic acids are extracted by organic solvents not very efficiently because acidic functional groups such as carboxyl (—COOH) and phenol (—OH) form hydrogen bonds with water. Values for the distribution constant are in the 0.5 to 5 range. The distribution ratio depends strongly on the pH of the aqueous phase. When the pH is low, forcing the acid into its molecular form, the distribution ratio is most favorable. At high pH the acid is almost completely dissociated into its anion form, which is not extracted at all. This provides the chemist with a useful means of removing an acid from an organic phase that is immiscible with water. When shaken with a basic solution (e.g., sodium bicarbonate), the weak acids in the organic phase are transferred quantitatively to the water layer, while nonacidic substances tend to stay in the organic phase.

Studies of the distribution ratio for acids have revealed an interesting phenomenon: as the concentration of the acid in the system is made larger, an increasing percentage of it is found in the organic phase. In other words, the value of D grows with concentration. This is illustrated by the behavior of acetic acid in a water–ether system, as shown in Table 19.2.*

Table 19.2 Distribution ratio of acetic acid in water–ethyl ether

Equilibrium molarity of acetic acid in the		
water layer	ether layer	$D = \dfrac{C_o}{C_w}$
0.3240	0.1604	0.495
0.6461	0.3406	0.527
1.255	0.7413	0.591

If it were merely simple acetic acid molecules CH_3COOH that existed in each phase, we would expect D to be constant and equal to the distribution constant K_D. As a general principle we may state the following: *when the variation of conditions in an extraction system causes an increase in the tendency of a substance to enter one of the phases, it means that the substance*

*Data taken from A. Seidell, *Solubilities of Organic Compounds*, 3rd ed., Vol. II, D. Van Nostrand, New York, 1941, p. 110. Corrected for slight dissociation in the water phase.

is being changed into other species in that phase. We saw this type of effect earlier, in the discussion of the effect of potassium iodide in the water phase on the distribution of iodine between carbon tetrachloride and the aqueous phase. With increasing KI, the iodine shifted into the aqueous phase, and this was because it was being changed into triiodide ion in that phase. In the present case, it is only the concentration of acetic acid that is being varied. Variation in concentration did not affect the distribution ratio of iodine, but with acetic acid it appears that a new species must be formed in the ether layer.

When this effect was first observed early in the century there were no adequate explanations for it. The principles of molecular structure and bonding were just taking shape with the leadership of G. N. Lewis (at Berkeley) and others. Now any chemistry freshman can understand this phenomenon which is based on hydrogen bonding. In the water phase the acetic-acid molecules are fairly strongly attracted to water molecules through hydrogen bonding (hydrogen bonds shown by dotted lines):

$$
\begin{array}{c}
\quad\quad\quad\quad O \\
\quad\quad\quad / \quad \backslash \\
\quad\quad H \quad\quad H \\
\quad\quad : \\
\quad\quad\; O \\
\quad\quad /\!\!/ \\
CH_3-C \\
\quad\quad \backslash \\
\quad\quad O-H\cdots OH_2 \\
\quad\quad : \\
\quad\quad H \\
\quad\quad | \\
\quad\quad O-H
\end{array}
$$

It is precisely these attractive forces that prevent acids from being efficiently extracted by organic solvents. Since the water solvent contains huge numbers of water molecules for each acetic-acid molecule, the latter exist almost entirely in the above form: single molecules surrounded by water molecules.

Now we consider those acetic acid molecules that do cross the phase boundary and enter the ether (or other organic-solvent layer). There is far less tendency for hydrogen bonding with the solvent. Some solvents, such as CCl_4, are not capable of forming hydrogen bonds. Ethyl ether has an oxygen atom that can bond to the proton of the acetic acid, but it does not have acidic hydrogens that can bond to the oxygens of the acetic acid. Therefore, in the ether layer the acetic acid is not so tightly bound to the solvent and is *free to hydrogen bond to itself*:

$$
\begin{array}{c}
\quad\quad O\cdots H-O \\
\quad\; /\!\!/ \quad\quad\quad \backslash \\
CH_3-C \quad\quad\quad\quad C-CH_3 \\
\quad\; \backslash \quad\quad\quad /\!\!/ \\
\quad\quad O-H\cdots O
\end{array}
$$

This interaction is favored by the lower dielectric constants of organic solvents. Whereas water has a dielectric constant of 78, that for ether is only 4.3, and the values for benzene and carbon tetrachloride are about 2.2.

These *double molecules* of acetic acid are called **dimers** (Greek, two parts), and the reaction between the **monomer** species is called **dimerization**:

$$2CH_3COOH = (CH_3COOH)_2.$$

We refer to the equilibrium constant as a **dimerization constant**. There is no standard symbol for the dimerization constant but we may use K_2, the subscript reminding us that two molecules associate to form the dimer:

$$K_2 = \frac{[\text{Dimer}]}{[\text{Monomer}]^2}. \tag{19.14}$$

Now we may give a qualitative explanation to the increase in the distribution ratio when the concentration of acetic acid in the system is increased. Since the formation of the dimer is proportional to the *square* of the monomer concentration, it follows that the percentage of conversion to dimer will increase with concentration. The explanation (hypothesis) would be more convincing if it were put to a quantitative test. We will therefore derive an equation allowing us to predict the way in which the distribution ratio should change if the dimerization model is correct.

The equilibrium scheme to consider is as follows:

Ether phase: $HA \rightleftharpoons (HA)_2$

$$\updownarrow$$

Water phase: HA

The distribution ratio may be written in terms of the species shown:

$$D = \frac{C_o}{C_w} = \frac{[HA]_o + 2[(HA)_2]_o}{[HA]_w}. \tag{19.15}$$

The factor of 2 is used because dimer molecules contain two monomer molecules. The analytical determination of C_o (e.g., by titration with NaOH) can tell only how many moles of monomer are present in the mixture of species. For example, if 1.00 mmol of NaOH were used to reach the endpoint, there would be no way of knowing whether this was due to 1.00 mmol of HA, 0.50 mmol of $(HA)_2$, or some mixture between these two extremes.

From Eq. (19.14) we may express the dimer concentration in terms of K_2 and the monomer concentration:

$$[(HA)_2] = K_2[HA]^2.$$

When this is substituted into Eq. (19.15) we find:

$$D = \frac{[HA]_o(1 + 2K_2[HA]_o)}{[HA]_w}. \tag{19.16}$$

But, by remembering that the distribution constant is the ratio of the molarities of a single species in the two phase mixture, we have:

$$K_D = \frac{[HA]_o}{[HA]_w} \quad \text{and} \quad [HA]_o = K_D[HA]_w.$$

Therefore we make two final substitutions into Eq. (19.16):

$$D = K_D + 2K_2K_D^2[HA]_w. \tag{19.17}$$

The reason for developing the relationships into the form of Eq. (19.17) is that it provides a means for a quantitative test of the hydrogen-bonding model. Equation (19.17) contains only two variables, D and $[HA]_w$, both known from the experiments (see Table 19.2). If the model is correct, we would expect a plot of D versus $[HA]_w$ to be linear, with an intercept equal to K_D and a slope equal to the product $2K_2K_D^2$. If this is found to be true, then we may calculate the dimerization constant simply by the following:

$$K_2 = \frac{\text{Slope}}{2 \cdot \text{Intercept}^2}.$$

Such a plot is shown in Fig. 19.2, where the three data points of Table 19.2 are used. There is no doubt about its linearity, and we find:

Intercept: $\quad K_D = 0.46,$

Slope: $\quad 2K_2K_D^2 = 0.103,$

$$K_2 = \frac{0.103}{2 \cdot 0.46^2} = 0.24.$$

Thus, not only is the hydrogen-bonding hypothesis supported, but as a bonus we have determined the values for the two equilibrium constants.

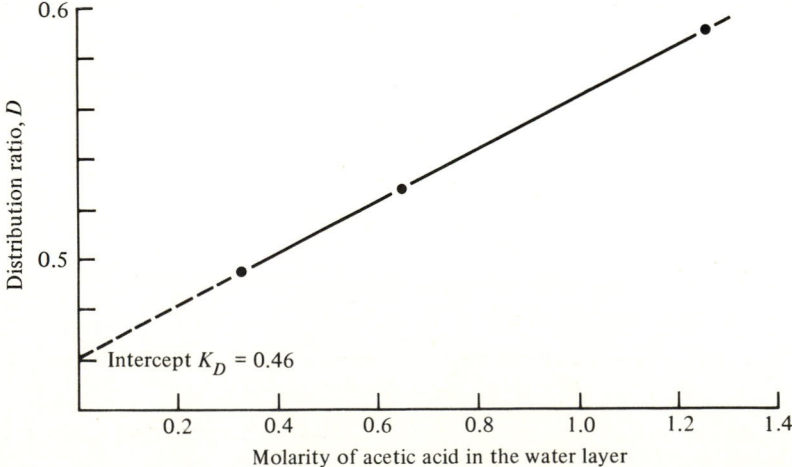

Fig. 19.2 Graphical test of the dimerization hypothesis for acetic acid in ether.

PROBLEMS

1. A certain analytical reagent HB is amphiprotic and is only slightly soluble in water. However it is soluble in chloroform and proves useful as an extractant for various large anions that may be extracted as ion pairs:

$$\text{Water layer} \quad B^- \overset{H^+}{\rightleftarrows} HB \overset{H^+}{\rightleftarrows} H_2B^+ \overset{A^-}{\rightleftarrows} H_2B^+A^-$$

$$\text{CHCl}_3 \text{ layer} \qquad\qquad\qquad HB \qquad\qquad H_2B^+A^-$$

Thus, although the stock solution of HB is in chloroform, when it is used as an extractant we desire that it be transferred to the aqueous phase as its protonated form that can react with the anion and then be reextracted back into the chloroform. The absorbance of the resulting chloroform solution will be proportional to the concentration of the anion, since H_2B^+ is highly colored.

Therefore, to gain a fundamental understanding of the system it was desired to determine the distribution constant for HB and also the acid-dissociation constant for H_2B^+.

To determine K_a for H_2B^+, it was decided to make use of the pH effect on the solubility of HB. Four buffers were prepared, each with $0.100M$ trisH$^+$Cl$^-$ ($pK_a = 8.069$) and with varying concentrations of tris as the conjugate base. An excess of solid HB was added to each solution and the mixture was stirred for two days at a constant temperature of 25°. After removing the excess solid by filtration, the saturated solutions were analyzed for the total dissolved reagent, while $[B^-]$ was considered negligible at the pH values used:

$$S = [HB] + [H_2B^+].$$

The determination of S was quite simple, because it was known that HB and H_2B^+ have an isosbestic point at 558 nm. Thus, the absorbance at this wavelength is directly proportional to the total concentration of the reagent. First, a check on Beer's Law applicability was made by preparing solutions of known concentration and using a 2-cm cell. The results were as follows:

C_{tot}	Absorbance A
0	0.000
$5.00 \cdot 10^{-5}$	0.210
$1.00 \cdot 10^{-4}$	0.440
$1.50 \cdot 10^{-4}$	0.645

a) Plot A versus C and draw what you consider the best straight line. From the slope of the plot calculate the average value of ϵ.

When the saturated solutions in the buffers were examined, again using the 2-cm cell, the spectrophotometric data were as follows:

[H$^+$] in buffer	Absorbance A
$2.00 \cdot 10^{-8}$	0.407
$4.00 \cdot 10^{-8}$	0.543
$6.00 \cdot 10^{-8}$	0.705
$8.00 \cdot 10^{-8}$	0.841

b) What concentration of tris must be present, along with the $0.100M$ trisH$^+$, to make a buffer with $[H^+] = 6.00 \cdot 10^{-8}$?

c) Derive the equation for this system that relates S to $[H^+]$.

d) Make the appropriate plot whose intercept and slope allow the determination of the intrinsic solubility of HB and the acid-dissociation quotient for H_2B^+. Find the values of these equilibrium constants.

e) In a separate experiment it was found that the solubility of HB in pure chloroform is $0.1125M$. What is your estimate for the distribution constant $K_D = [HB]_o/[HB]_w$?

f) Under conditions where the distribution ratio D is 2.5, that is,

$$D = \sum C_o \Big/ \sum C_w = 2.5,$$

suppose that 10 mL of chloroform containing 1.0 mmol of HB is shaken with 50 mL of the aqueous phase. How many millimoles of HB will remain in the chloroform?

2. The following data and questions are based on a computer simulation of a prospective laboratory experiment, the purpose of which is to determine the successive formation quotients for the complexes of silver with 1,10-phenanthroline. The plan is to determine a series of values for the formation function F_0, corresponding to known values for the free-ligand molarity. Then, by plotting F_1, etc., versus [L] it should be possible to deduce the Q values.

Suppose that we start with a solution of $1.00 \cdot 10^{-3}M$ silver nitrate and add small amounts of phenanthroline. By using a silver-wire electrode and a reference electrode we may readily follow the decrease in silver-ion molarity as the complexes are increasingly formed. Therefore it is easy to determine values of $F_0 = C_{Ag}/[Ag^+]$. However, for the F_1-plot method to be successful it is necessary to have reliable values for the *equilibrium molarity* of phenanthroline [L] and not merely for the total molarity of added ligand. We need some sort of probe for L, and the values for [L] must be very low if Q_1 is to be determined.

Since the distribution of L between chloroform and water greatly favors the organic phase, it was decided to use a two-phase system in the research. The equilibrium scheme would be as follows:

Aqueous phase	$HL^+ \rightleftarrows H^+ + \ L + Ag^+ \rightleftarrows AgL^+$, etc.
	$\updownarrow$
Chloroform phase	L

Now, the only solute in the chloroform layer is phenanthroline, present at a much larger concentration than in the aqueous layer because of the large value of the distribution constant K_D. It should be possible to determine the value of $[L]_o$ by ultraviolet spectrophotometry and then to calculate $[L]_{aq}$ by using the known distribution constant.

To minimize the competition from H^+ for the ligand, no acid was added to the aqueous phase, which was essentially neutral.

Experimental procedure

The entire chemical system will be held in a tall glass stoppered silica-absorption cell with a 1-cm path. All work will be done at 25°C.

Initially the cell contains 5 mL of chloroform and 5 mL of $1.00 \cdot 10^{-3}M$ silver-nitrate solution. The phenanthroline is added in very small volumes of a concentrated chloroform solution. After each addition the cell is stoppered, shaken until distribution equilibrium is reached, and then the phases are allowed to separate, perhaps with the help of a centrifuge.

Then the cell is placed in a spectrophotometer with the wavelength set at a chosen value, and the absorbance is measured for the chloroform phase only.

Also, a clean silver wire and a reference electrode are inserted into the aqueous phase. The reference electrode is simply another silver wire immersed in $1.00 \cdot 10^{-3}M$ silver nitrate. Thus, the potential of this cell is related to the equilibrium molarity of free-silver ion.

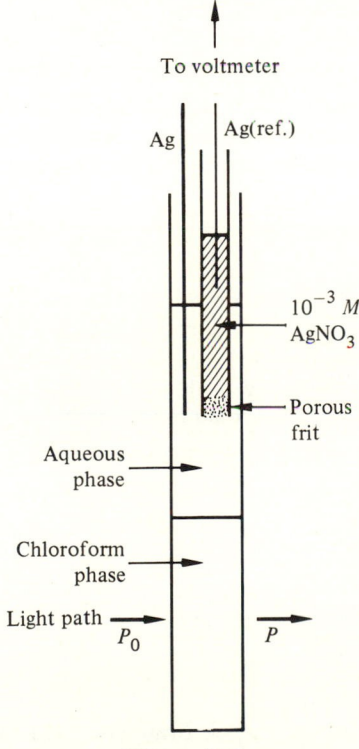

Figure 19.3

Phenanthroline added, mmol	Absorbance of chloroform phase	Cell potential, mV
0.0200	0.160	90.8
0.0400	0.391	120.1
0.0600	0.595	137.9
0.0800	0.844	151.1

Before proceeding with the interpretation of these data it is necessary to consider a few important matters, as pointed out in the following questions.

a)   To interpret the absorbance data for the two-phase system, it was necessary to establish the spectral properties of L in chloroform. Four solutions were prepared by dissolving pure phenanthroline in chloroform, and the absorbances were measured using pure chloroform as a reference solution, with the following results:

Molarity of L	Absorbance ($b = 1$ cm)
0.00400	0.245
0.00800	0.462
0.01200	0.710
0.01600	0.940

b)   Assuming that Beer's Law holds for this case, calculate the four values for molar absorptivity, their average, and the average deviation. From your values does it appear that a plot of A versus C would be linear or is a curvature apparent?

c)   A reliable value for K_D, the distribution constant for phenanthroline between chloroform and water, is required so that the aqueous molarity of L in the two-phase system may be calculated. The experiment was as follows: 50 mL of 0.0600M L in chloroform was shaken with 100 mL of water. The small concentration of L in the aqueous phase was determined by taking advantage of the strong absorption of phenanthroline in the ultraviolet. The absorbance of the aqueous phase was found to be 0.867 when a 1-cm cell was used at a wavelength where the molar absorptivity had previously been determined as 15100. Find the value of the distribution constant K_D.

d)   At last the data for the two-phase system may be interpreted.

Use the silver electrode data to calculate the equilibrium molarities of free-silver ion in each of the four solutions. Knowing that $C_{Ag} = 1.00 \cdot 10^{-3}$, find the four values for F_0.

Use the absorbance data to find the equilibrium molarities of L in the chloroform phase. For this calculation assume the value for ϵ_L to be 59.2, no matter what you found in part (a).

Calculate the equilibrium molarities of L in the aqueous phases using $K_D = 1043$, no matter what you found in part (b).

Make a precise plot of F_1 versus [L]. Draw your conclusions about the formation quotients for the silver–phenanthroline complexes. Comment on whether it is necessary or not to make an F_2 plot, but do not actually make it.

LABORATORY PROGRAM IN CHEMICAL EQUILIBRIUM AND ANALYSIS

The following laboratory program is designed to accompany the theoretical part of this textbook. There are several goals. One is to teach selected techniques for making accurate laboratory measurements. This includes the handling of materials, preparation and standardization of solutions, precise use of volumetric equipment and the analytical balance, measurements in spectrophotometry and potentiometry, etc. Another goal is to demonstrate a good variety of chemical reactions, mostly inorganic, which involve proton transfer, complex formation, oxidation–reduction, and precipitation. Finally, although some of the experiments are chiefly illustrations of chemical analysis for the simple purpose of finding the composition of unknown samples, a number of others are designed to illustrate research in the fundamental study of solution equilibria. The broad purpose is to reveal the close interplay between the theory of chemical equilibrium and the practice of chemical analysis. The laboratory experiments are listed on p. 619.

Note: Each discussion of a laboratory experiment is accompanied by a report form suitable for recording the students' results. This form is intended to summarize, not replace, the regular laboratory notebook record.

General Principles

Preceding the laboratory experiments there is a section dealing with general principles of quantitative laboratory work, including specific recommendations for the use of analytical equipment to a maximum advantage. This section should be read through before beginning the lab program, and will be useful as a reference during the experimental work. The topics discussed in this general section are:

1. SAFETY IN THE LABORATORY

The rules for safe laboratory work do not differ greatly from one type of chemistry to another. We may list a few of the most important rules as follows:

- Always wear shoes (not sandals) in any laboratory. The floors normally have sharp bits of glass and sometimes toxic substances which have spilled. Also, should you spill a chemical solution on your feet, it will not run between your toes if you are wearing shoes.
- Always wear eye protectors, goggles or safety glasses with side shields.
- When using pipets, always use a rubber bulb rather than your mouth.
- When carrying materials, be sure to avoid overloading.
- If sulfuric acid/potassium dichromate cleaning solution is used, take exceptional care because this is a dangerous and corrosive mixture.
- No open fire is allowed when flammable solvents are being used.
- If toxic or irritating vapors are to evolve, use the hood.
- Learn the location and use of the fire extinguisher, safety shower, and eye-wash fountain.
- Report any accidents, especially those involving cuts or burns, immediately to the instructor.
- Always use tongs to remove hot objects from the oven.

2. THE NATURE OF PRECISE WORK

First of all, it is not desirable to strive for maximum accuracy in every step of an analytical procedure. One should not use a precise pipet, for example, when a graduated cylinder or even an unmeasured visual estimate of solution volume is adequate. Nor should one take the time to clean beakers and titration flasks with the same meticulous care that must be given to volumetric flasks, pipets, and burets. Whereas an analytical sample must usually be weighed to the nearest 0.0001 gram, solid reagents added in the procedure often have wide tolerance. The key to making correct judgements is this: understand the experiment before performing it. Know the chemical changes involved, the purpose of each step, and the role of each reagent; also know which steps require special care and which do not. In this way you will not only save time but you will know what is going on. You will be a thinking person and not merely a direction-following machine. As a bonus, you will probably obtain much better results.

3. ANALYTICAL BALANCE

The balance is perhaps the most important instrument in the laboratory. It has been called a *sanctum sanctorum* and must be treated with great care and respect. Because there are many different models of balances, this text will not attempt to explain the details of operation. Generally speaking, the goal is to obtain the weight (mass) of an object with an uncertainty of only ± 0.0001 gram, and this requires attention to a few important rules:

- Operate the balance controls slowly and smoothly. Be gentle.
- Never put the balance in the full-release position unless the weights are within 0.1 gram of the correct value, so that the optical scale can display the final digits. Similarly, never remove an object from the balance pan unless the balance is in the arrested position.
- Always check the zero setting with the pan empty.
- Objects to be weighed should be at the same temperature as the balance, for otherwise air convection currents may cause errors in the readings.
- Handle objects with tongs, or with a strip of paper, unless your fingers are very dry.
- Objects should be fairly well centered on the balance pan.
- Try very hard to avoid spilling any chemical in or around the balance. If a spill occurs, clean it up immediately and very thoroughly.
- Volatile materials must be weighed in covered (stoppered) vessels.
- Chemicals must never be placed directly on the balance pan. Use a beaker, a watch glass, or a square of glossy paper.
- The sliding doors of the balance case must be closed for the final reading.
- When finished with a weighing, leave the balance closed, arrested, and dial all weights to the zero positions.

4. BURETS

Conventional burets are straight glass tubes with uniformly spaced calibration marks and stopcocks to control delivery (Fig. L.1).

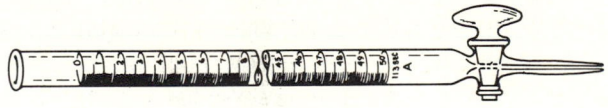

Figure L.1

They are available in sizes ranging from 5 to 1000 mL capacity, with the 50-mL size being most widely used for analytical work. A buret offers the best way to add any desired volume of a solution, as needed in titrations. A 50-mL buret of class A quality has a manufacturing tolerance of ± 0.05 mL, that is, one part per thousand of total capacity. Less expensive burets have a tolerance of ± 0.10 mL. In either case, since the level of liquid in a buret can be estimated to about ± 0.01 mL, it is necessary to standardize a buret if it is to be used to maximum accuracy.

Use of Burets

The first step in preparing a buret for use is to prepare the stopcock. If it is glass, it should be disassembled and wiped clean of old grease. The socket should also be cleaned, and any lumps of grease in the channels should be removed with a bent wire (paper clip). The clean plug should be very lightly covered with petroleum jelly or an approved stopcock grease, care being taken to keep the grease away from the holes. When reinserted into the socket, the plug should be gently rotated to see that the film of grease spreads evenly without plugging the holes.

If the buret has a Teflon stopcock, no lubrication is necessary, but the clamping nut should be adjusted so that the plug turns smoothly but firmly in the socket. If it is too loose, the stopcock will leak. If too tight, it will not be possible to control the flow of liquid easily and accurately.

In using volumetric glassware it is essential that the surface in contact with the solution be exceptionally clean. Otherwise the drainage of the solution will be variable and inaccurate. There is often an invisible film of matter deposited from the atmosphere, and while this has no harmful chemical effect on the solution, it will prevent the solution from wetting the glass uniformly and therefore the buret will not drain properly. The safest way to clean the buret is to use a hot detergent solution that can be drawn into the inverted buret (so that the stopcock area is not affected) and allowed to stand for about 10 minutes. After thorough rinsing with tap water and finally with a small amount of distilled water, the buret is filled with distilled water and placed in its stand. Note, after 1–2 minutes, whether the stopcock is leaking, and make adjustments as needed. Then let the distilled water drain out into the beaker, noting whether any droplets of water bead up on the inside. If drainage is uniform, the film of water left on the surface will be invisible, and the buret is ready for use.

When the buret is to be used for dispensing a chemical solution, it is necessary first to use a little of that solution to rinse the buret. This will remove the small amount of water remaining from the cleaning. Simply pour in about 5–10 mL of the solution, tip the buret so that the solution rinses all surfaces, and then let the solution drain through the stopcock into a waste beaker. Then close the stopcock and fill the buret well above the top mark. At this point the delivery tip, below the stopcock, will probably have an air bubble in it. This is removed by opening the stopcock fully, so that the pressure of the solution sweeps the air out.

The level of the solution is then adjusted to the zero mark or to any position just below the zero mark, and this initial reading is recorded. For accuracy it is useful to have a meniscus reader, which is simply a small card with black and white field. Such a device can be easily made with a pencil and a scrap of paper. When held behind the meniscus with the black edge just below the bottom of the meniscus, the latter is revealed very clearly against the white field.

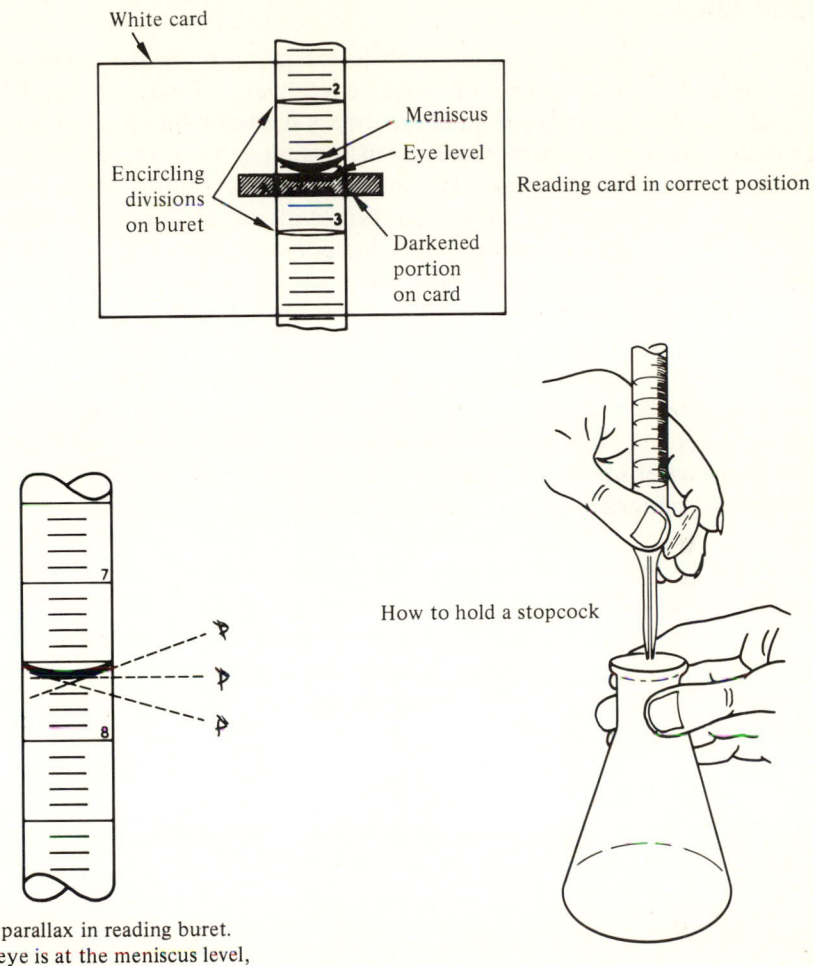

White card

Meniscus

Eye level

Encircling
divisions
on buret

Reading card in correct position

Darkened
portion
on card

How to hold a stopcock

Effect of parallax in reading buret.
Unless the eye is at the meniscus level,
the reading will be in error.

Figure L.2

To deliver solution from the buret, the best technique is to use the left hand for the stopcock manipulation (Fig. L.2). The receiving flask is swirled with the right hand if a titration is being performed. When the delivery is completed, the solution level is read to ±0.01 mL.

When titrations have been completed, the buret should be drained of any remaining solution and then rinsed thoroughly with distilled water before returning it to storage.

Some chemists prefer to store burets filled with distilled water to minimize contact with the atmosphere. With care, it should not be necessary to repeat the cleaning process each time a buret is used.

Standardization

There are two arguments in favor of determining the actual performance of a buret or any other piece of volumetric ware. First, the manufacturing tolerances are always larger than the errors made by an experienced user. We do a disservice to a precise experiment by simply accepting the nominal buret readings, which may be in error by several hundredths of a milliliter. The second argument is that standardization will reveal the reproducibility with which you can use the buret. This includes your skill in setting the initial solution position, in reading the final position, and it also includes the variability in the amount of solution remaining in the undrained surface film due to different rates of delivery.

Since buret tubes are likely to be slightly nonuniform in their internal diameter, the errors in volume delivered, compared to the scale readings, usually fluctuate from place to place on the scale. Therefore it is desirable to check the buret at 10-mL intervals, perhaps even at 5-mL intervals, to obtain the overall pattern. The procedure follows:

a)   Fill the buret with distilled water that has stood long enough in a beaker so that it is at room temperature. The actual temperature must be measured with a thermometer immersed in the water. Adjust the level of the zero mark.

b)   Weigh (to the nearest milligram) a flask fitted with a stopper. It is essential that this flask be dry on the outside and that there be no water film in contact with the stopper.

c)   Touch off the tip of the buret on a beaker and then deliver 10 mL of water into the weighed flask, being sure to touch off the tip on the inner wall below the stopper area. Stopper the flask without delay and record the precise reading from the buret scale.

d)   Reweigh the stoppered flask. The increase in weight is related to the actual volume of water delivered, which may be calculated as discussed in Section 3.7 on the limitations of mass and volume measurements.

e)   Repeat the process for deliveries of 20, 30, 40, and 50 mL, each time refilling the buret to the zero mark.

f)   To check reproducibility, the entire series of measurements should be repeated.

g)   When the calculations of delivered volumes have been completed, summarize the results on a graph as shown in Fig. L.3. It gives the volume, in hundredths of a milliliter, that should be added to the observed buret reading to find the actual volume delivered.

For example, this buret actually delivers $30.00 - 0.06 = 29.94$ mL when the scale reading is 30.00. Interpolation is used for readings between those corresponding to the standardization measurements.

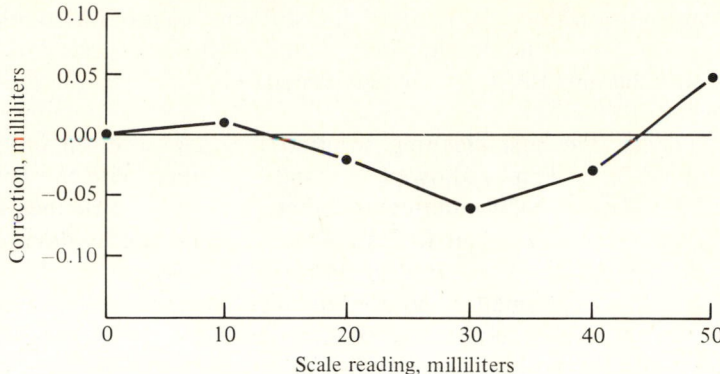

Fig. L.3 Standardization curve for a typical buret.

5. CLEANING GLASSWARE

Analytical experiments may be spoiled because of chemical residues in glassware from previous work. Most of the time, thorough rinsing with tap water, followed by a brief rinse with distilled water, is sufficient to remove these residues. However, there may be some solid clinging rather tightly to the surface, particularly if the glassware has contained a precipitate, and it is necessary to use a test-tube brush with detergent solution to loosen the residue and remove it. In this sense, cleaning of laboratory glassware is not much different from good technique in washing frying pans.

You should not wipe the inside of analytical glassware with a towel because this leaves some lint that may interfere later. It is best to let the beakers and flasks drain and dry in the air. In contrast to common practice in organic chemistry (where water is considered as dirt), it is not advisable to rinse analytical glassware with acetone to speed up drying. The acetone may contain substances that will leave a residue, and generally the residual water will have no harmful effect in an analysis.

More rigorous cleaning procedures are usually needed for volumetric glassware (pipets, volumetric flasks, burets, but not graduated cylinders) designed for measurement of very precise volumes. This is due to an invisible surface film that accumulates upon exposure to the atmosphere and prevents the glass from being wet uniformly by water solutions. In turn, this causes erratic drainage and errors in the measured volumes.

A time-honored cleaning solution is made from concentrated sulfuric acid and potassium dichromate. This is a powerful oxidizing mixture that will clean nearly any glass surface, but it has drawbacks. First of all, it is not recommended for general use in student laboratories because it is so corrosive to clothing and skin. Secondly, it leaves its own residue of chromium(VI) which requires extensive rinsing. Finally, if accidently mixed with water, it can cause boiling and spattering.

Another solution that can be used for stubborn cases is alcoholic potassium hydroxide. This must be kept in a plastic bottle. It is also dangerous to the skin and is flammable. It is not recommended for general use in student laboratories.

For students, the best cleaning solution is a 2% solution of detergent heated to about 50°C. This is allowed to stand in contact with the surfaces to be cleaned for about 10–20 minutes and then is saved for further use. One simply fills the buret or volumetric flask. Pipets can be filled and capped with a rubber policeman. The glassware is then rinsed very thoroughly with tap water and finally with a small amount of distilled water from the wash bottle. We should discourage excessive use of distilled water for rinsing. The only purpose of the distilled water rinse is to remove the tap water, and not much is needed. It is expensive to make distilled water, and it should not be wasted.

Glassware of all sorts should be cleaned before returning it to the locker or stockroom. Otherwise chemical solutions will evaporate and leave deposits that may be difficult to remove.

6. CYLINDER, GRADUATED

Graduated cylinders are available in capacities ranging from 10 to 4000 mL and can be used with an accuracy of about ± 1% of total volume. As a rule, they are not actually used with such care because usually the procedure in an analysis is not so critical. When the directions in a procedure call for *about V milliliters of solution*, it is good practice to use a graduated cylinder and not to attempt to set the meniscus precisely to the specified volume.

Cylinders need not be cleaned with the rigor needed for more precise volumetric ware. Simple scrubbing with detergent and a test-tube brush is excellent, and mere rinsing is usually quite adequate. When a cylinder is to be used for successive additions of different reagents to a mixture, it is usually (not always) unnecessary to rinse it out between additions. This judgment requires chemical knowledge of what is going on.

7. DESICCATOR

The purpose of a desiccator is to provide a relatively dry atmosphere for storage of small vessels and chemicals. For example, a crucible containing a precipitate that has been dried in the oven is typically placed in a desiccator to cool before weighing. It is doubtful that the moist air admitted when the desiccator is opened will actually be dried by the drying agent in the bottom in the short time usually allowed for cooling (30 minutes). Nevertheless, at least a safe storage is assured (Fig. L.4).

Traditional glass desiccators with ground-glass lids and porcelain plates are widely used. They may also be obtained with a stopcock that allows the air to be pumped out so that the material is stored in vacuum. For most work

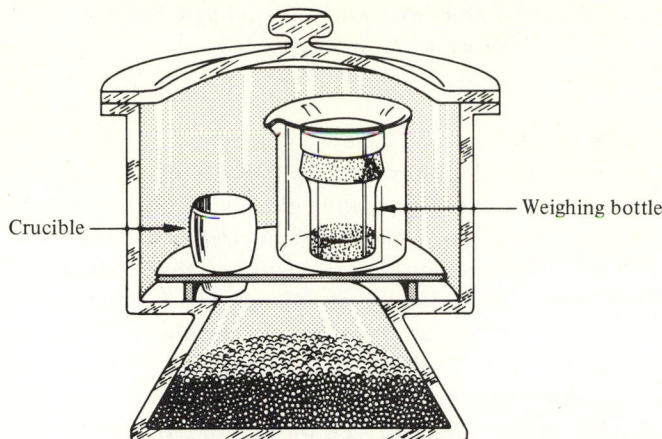

Fig. L.4 Desiccator with crucible and weighing bottle.

these fairly expensive desiccators can be replaced by glass jars with screw tops or metal cans with plastic snap covers.

8. DISTILLED AND DEIONIZED WATER

In quantitative chemistry it is essential to prepare solutions using water that has been purified of the minerals (sodium, calcium, magnesium, sulfate, chloride, carbonate, etc.) found in tap water. Most laboratories have central water-purification systems based either on distillation or deionization. The purified water is stored in large tanks and delivered through tin-lined, glass, or plastic plumbing. The deionizing process is akin to water softening, except that it removes both cations and anions. For some applications the contact with the ion-exchange resin introduces significant amounts of organic substances that may cause errors in certain procedures. Distilled water is excellent for nearly all applications, but it may contain impurities due to volatile substances originally present. Satisfactorily purified water should not give a white precipitate when treated with $0.1M$ silver nitrate. It should have a pH of about 5.7 due to the carbon dioxide that is inevitably present because of exposure to air.

For water used in preparation of sodium-hydroxide solutions it is desirable to remove carbon dioxide either by boiling or bubbling CO_2-free air through it for an hour. The CO_2 can be removed by passing it through an absorption tube of Ascarite (sodium hydroxide on asbestos).

Particularly in research studies it is advisable to further purify the water by distillation, on a small scale, using an all-glass laboratory still.

It is good practice to use distilled water conservatively. General rinsing of glassware should be done first with tap water, and the distilled water rinse should be made with small portions from a wash bottle. Its only purpose is to remove the tap water, and not much is needed for this. It is costly and wasteful to use large amounts of pure water for rinsing.

9. DRYING OVEN

An oven that operates continuously at 105–115°C is essential in accurate analytical work for it is used to remove excess water from materials that are to be weighed on the analytical balance. This undesired water may be present in various forms. Merely because of exposure to the atmosphere, objects and materials acquire a small surface layer of adsorbed moisture, the amount of which is related to the actual surface area exposed. Thus, a fritted glass crucible will adsorb more moisture than a beaker of the same size, and a finely divided powder will adsorb more than the same weight of large crystals. This surface moisture is readily driven off at 110° (Fig. L.5).

Often it is necessary to dry a precipitate that has been formed from an aqueous solution and then rinsed with distilled water. In this case the contaminating water is substantial in amount, and a longer (say, 2–3 hour) drying period is needed.

Many inorganic salts incorporate water into their crystal structures by chemical bonding. For example, copper sulfate forms a fairly stable penta-hydrate $CuSO_4 \cdot 5H_2O$. In many cases this water of crystallization may be removed by heating at 110°, but for some hydrates a much higher temperature is needed.

When placing objects in the oven, such as crucibles, it is best to put them in a protective beaker with a cover (watchglass) to prevent particles from the oven from falling into the precipitate. Label the beaker so that it won't be mistaken for another. When several students open the oven doors the oven temperature will drop to below the set value. A longer drying time is then needed. If space permits, the objects can be left in the oven overnight to dry thoroughly.

When removing objects from the oven, use either beaker tongs or crucible tongs to avoid being burned. It is wise to hold a heavily folded towel in one hand, to lend support at the bottom when tongs are used to remove a beaker. Crucibles and weighing bottles should be placed immediately in the desiccator and allowed to cool for about 30 minutes before weighing.

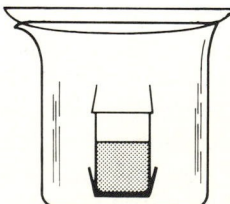

Fig. L.5 Arrangement for drying of samples.

10. FILTER CRUCIBLES

These crucibles are used in a vacuum filtration setup to collect precipitates that are to be dried and weighed. A common type is made of glass with a fritted (porous, sintered) glass disk sealed into the base (Fig. L.6).

In use the crucible is supported in a funnel with a rubber adaptor placed in a side-arm filter flask (Fig. L.7).

After the precipitate has been transferred from its beaker to the filter crucible, the latter is rinsed with distilled water, and then the crucible is placed in a covered beaker for drying in the oven.

The first step in using a filter crucible is to clean it thoroughly, making sure that any precipitate residue from a previous use is dissolved. When the residue is known, use an appropriate chemical treatment (e.g., ammonia for dissolving silver chloride, $6M$ HCl for dissolving lead chromate, etc.). If the past history of the crucible is unknown, then a general cleaning process is advised as follows: the crucible is placed in a beaker and 10 mL of warm $6M$ nitric acid is allowed to trickle through without aid from vacuum. This is followed by a water rinse, and then 10 mL of $6M$ ammonia is trickled through. After another water rinse, use $6M$ HCl. Then place the crucible in

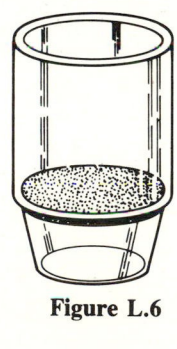

Figure L.6

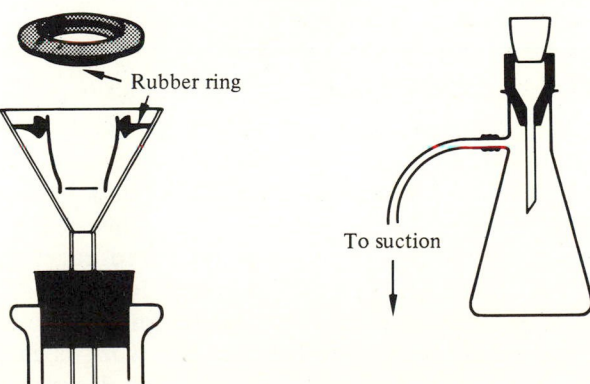

Rubber ring

To suction

Fig. L.7 Adapters for filtering crucibles.

the filter flask setup and, with the aid of vacuum, draw a few hundred milliliters of hot distilled water through the filter. At this point the crucible should at least appear to be clean, and it will probably be quite satisfactory.

When removing a precipitate from a glass (or porcelain) filter crucible, you may use a rubber policeman on a stirring rod to help dislodge the solid. But never use the unprotected stirring rod or a metal spatula to scrape out the precipitate. This will damage the fragile glass frit and spoil the quality of the crucible.

11. FILTER PAPER

Filter paper is available in a wide range of sizes (the diameter of the circular piece) and porosities. It also comes in forms that leave negligible weight in the residue after burning. To separate most precipitates or extraneous matter from a solution, the medium-porosity paper works well. Precipitates consisting of very fine particles require fine-porosity paper, which in turn requires considerable patience because of its slow flow rate.

To use a filter paper, fold it in half and then again in half to form a cone that fits into a funnel. Tear off a small corner (see Fig. L.8). To make it fit snugly in the funnel use a little water from a wash bottle to moisten the paper and press it gently into place. If done correctly, there will be no air leakage between the paper and the funnel.

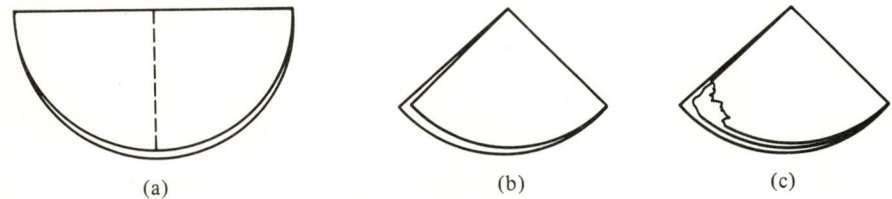

(a)   (b)   (c)

Fig. L.8 Method of folding filter paper.

12. NOTEBOOK

The purpose of the laboratory notebook is to show what was done and what happened. A good notebook may be read by a person with no previous knowledge of the experiment, and the record should enable that person to repeat all of the work in the same way. The notebook record should be complete, it should be clear, and it should be neat. This is no easy task.

The book itself should be bound, with numbered pages, none of which should be torn out. It is well to leave the first few pages for a table of contents, so that specific work may be easily found at a later date. All information should be recorded in ink while the work is in progress. One useful arrangement is to use the right-hand page for descriptive information and data, and to use the left-hand page for later calculations and final results.

Each day's work should begin with the date, followed by a statement about what is about to be done and the purpose of the work. If the procedure to be used is exactly as already written in a reference book, it is not necessary to copy it all down, but the reference should be clearly shown. If there are deviations in the procedure, they should be noted clearly (e.g., "the precipitate did not dissolve when the specified 10 mL of HCl were added, so I added 5 mL more and it did dissolve").

Never use scraps of paper with the good intentions of recording the data later in the notebook. All recordings should go directly and immediately into the book. When you know that you are going to weigh three samples, set up the data as follows:

	1	2	3
Weight of beaker + sample			
Weight of empty beaker			
Weight of sample			

Similarly, when you are doing a series of titrations, set up the data as follows:

	1	2	3
Buret reading at endpoint			
Buret reading at start			
Volume required			

When something goes wrong, do not scribble to obliterate the useless data. Instead, use a single diagonal line to cross it out and add a note of explanation. Be sure that all quantities are identified and that they carry units.

Try to keep notebook pages from becoming cluttered, and certainly keep them from appearing disorganized. If you will keep in mind the goal that the notebook should be forever useful as an unambiguous record of what you did, you will be less likely to write illegibly or vaguely.

Each experiment should be followed by conclusions, and perhaps recommendations about what could be done the next time to make the experiment more reliable. In an instructional laboratory you should tell the instructor what you wish you had been told before starting the experiment.

13. PIPETS

There are two main types of pipets: volumetric (also called transfer) and measuring (also called graduated, or Mohr type).

Volumetric Pipets

These are available in sizes ranging from 0.5 to 200 mL and look like the one shown in Fig. L.9. The circular etched mark shows where the solution meniscus should be set so that the pipet will deliver its nominal volume when used properly.

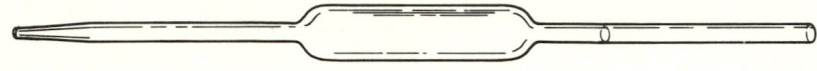

Figure L.9

Since pipets are designed to deliver a certain volume, not to contain that volume, they must be very clean on the inner sufrace, so that drainage will be uniform. Hot detergent solution may be used to fill the pipet, which may then be capped at the tip with a rubber policeman and allowed to stand for 10 minutes. Thorough rinsing with tap water, followed by distilled water rinses from a wash bottle, should prepare the inner surface adequately. If the pipet is used to deliver distilled water, it should be impossible to see the thin film of water remaining on the inner surface. If droplets bead up, the cleaning must be repeated or a more rigorous cleaning solution may be used (see CLEANING GLASSWARE). After cleaning, a pipet will contain some residual water, and this must be removed before using the pipet to deliver an accurate volume of a solution, for otherwise the solution drawn into the pipet will be somewhat diluted. One method of removing the water is to connect the top of the pipet by means of a rubber tube to a house vacuum line or to an aspirator. In a few minutes the water will evaporate and the pipet will be clean and dry. Alternatively, a small portion of the solution to be measured is drawn into the pipet, rolled around to take up the residual water, and then discarded. At this point the pipet is merely wet with the same solution that is to be pipetted, and so no error results.

To obtain the most accurate delivery from a pipet the following steps should be followed:

1. Use a rubber bulb to fill the pipet so that the meniscus is well above the calibration mark.

2. With a deft movement, switch from rubber bulb to your forefinger to keep the level from dropping below the mark (Fig. L.10). If your finger is too dry it will be difficult to maintain a satisfactory seal, but if it is too moist then you will have trouble with precise control of the flow rate. For a well-conditioned forefinger, try rubbing it in your palm. This really works.

3. Still keeping the meniscus a little above the mark, use a tissue or towel to wipe off (one smooth stroke) any droplets of solution that cling to the outside of the tip. This will eliminate the possibility that an extra drop will accidently be included in the delivery.

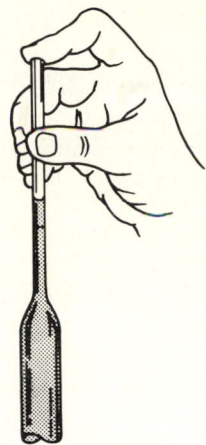

Fig. L.10 Method of holding a pipet.

4. When the pipet tip is in contact with a beaker or with the neck of the vessel from which the sample was taken, let the meniscus approach the calibration mark very slowly. When it is just tangent to the mark, increase finger pressure to stop the flow. Touch off the pipet on the glass surface to remove any partial drop from the outside.

5. Smoothly, to avoid accidental loss from the tip, move the pipet to the receiving vessel. Remove the forefinger completely, letting the solution flow at its natural rate, with the pipet held vertically. During this delivery the tip should be in contact with the inner surface of the vessel for two reasons: one is to prevent any loss by spattering and the other is to maintain a solution "bridge" between the pipet tip and the glass vessel.

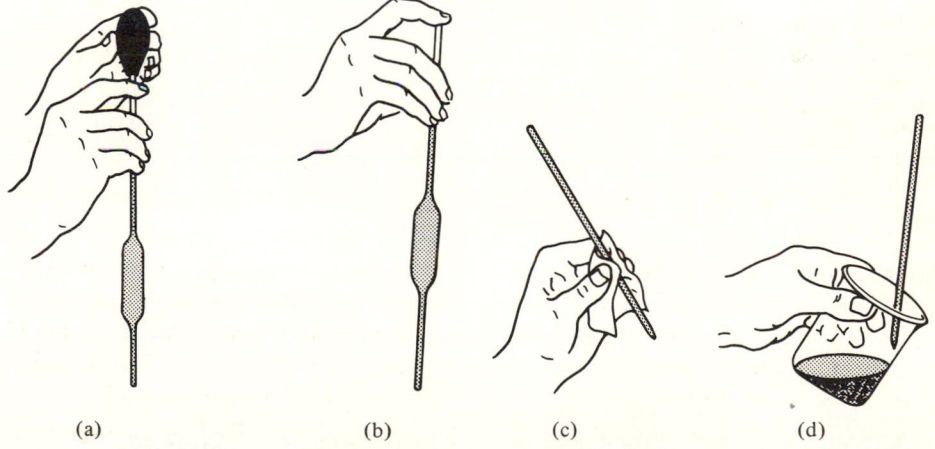

(a)　　　　(b)　　　　(c)　　　　(d)

Fig. L.11 Technique for use of volumetric pipet: (a) draw liquid past graduation mark; (b) use forefinger to maintain liquid level at the graduation mark; (c) tilt pipet slightly and wipe away any drop on outside surface; (d) allow pipet to drain freely.

6. The solution will drain out until about 1 cm or so remains in the tip. Do not blow this part out: it is meant to remain there and the pipet has been calibrated on the assumption that this rule is followed. Wait about 10 seconds to allow the drainage film to catch up with the delivery, and then touch off the tip and remove the pipet.

These steps are shown in Fig. L.11.

Measuring Pipets

Measuring pipets are rather like burets without stopcocks. They are available in sizes ranging from 0.1 to 50 mL, with the two most useful sizes being 1 mL (graduated in 0.01-mL intervals) and 10 mL (graduated in 0.1-mL intervals). The rules for using measuring pipets are identical to those given above for volumetric pipets, except that instead of allowing free flow until drainage stops one must use the forefinger to stop the flow at the desired mark on the graduated scale (Fig. L.12).

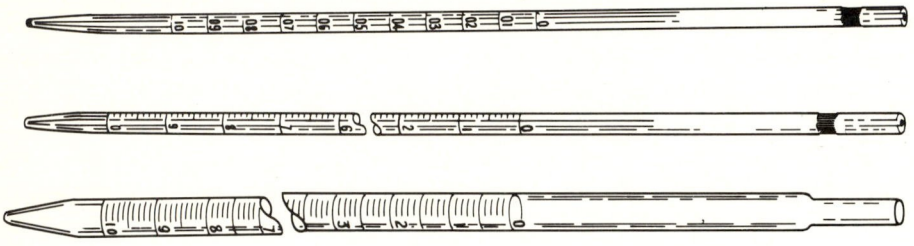

Figure L.12

The 1-mL measuring pipet is valuable for delivering small volumes with an accuracy to about ±0.002 mL if the pipet has been standardized (see below). Since the manufacturing tolerances are greater than those for volumetric pipets (being ±0.02 mL for the 1-mL size and ±0.06 mL for the 10-mL size), standardization is important if highest accuracy is sought. However, for many applications one uses measuring pipets for approximate volumes in much the same way that a graduated cylinder is used.

Standardization of Pipets

After a little practice you will be able to use a volumetric pipet with a reproducibility (e.g., average deviation) of only a few thousandths of a milliliter. The basic procedure for standardization is to find the weight of water delivered from the pipet, using the following steps:

1. Be sure the pipet is clean, as shown by its uniform drainage.

2. Have a beaker of distilled water at room temperature, with a thermometer inserted. Take this beaker to the analytical balance, along with the pipet and other items.

3. Weigh a flask fitted with a stopper to the nearest milligram (there is no point in worrying about the 0.1 mg decimal place when calibrating glassware). This flask must be perfectly dry on the outside, and there must be no water between the stopper and the flask.

4. Use the pipet (according to the rules given earlier) to deliver distilled water into the flask, being sure that the tip is in contact with the flask at a point below that where the stopper will be. Immediately stopper the flask when delivery is complete, to avoid loss by evaporation.

5. Reweigh the flask. The increase is the apparent weight of the water delivered by the pipet. If there is still room in the flask, there is no need to empty it before doing another trial. Simply use the new weight as the starting point for the next trial. When it is necessary to empty the flask, be sure to dry the outside and the stopper area before resuming the standardization.

6. This procedure goes rather quickly, and so you should do 6–8 trials. By noting the successive weights of water you will be able to judge your precision, for each 0.01 gram is equivalent to 0.01 mL. It is likely that you will note considerable improvement in precision after the first 2–3 trials, and your final results may disregard these if they differ appreciably.

7. Use the average of the weights delivered for the calculation of the pipet volume, as discussed in Section 3.7. Once a pipet has been standardized, it should be identified in some way so that it won't be confused with others.

14. TITRATIONS

Using Visual Indicators

In a typical titration one places a sample in an erlenmeyer flask (preferably the wide-mouth type). The solution to be used as titrant is placed in a buret that has been cleaned as described earlier (see BURET). The goal of the titration is to add enough, and only just enough titrant from the buret so that the sample has reacted completely. With some titrations it is possible to determine this endpoint with an uncertainty of only ± 0.01 mL, and so considerable skill is required in the technique of delivery. For the most accurate work the buret must be standardized.

The following steps apply quite generally to visual titrations:

1. Fill the buret with the titrant solution, taking care that the tip is free of air bubbles and that the stopcock does not leak. Adjust the meniscus to the zero mark or to a known position close to zero. Record the reading to the nearest 0.01 mL.

2. Prepare the sample in the erlenmeyer flask as specified in the directions for the titration. When adding the indicator, it is best to use some personal

judgment. The directions may specify "2 or 3 drops of indicator," but this presumes that the indicator solution at hand is of the same concentration as that used by the person who wrote the directions. You should add the indicator solution drop by drop until the solution color is distinct, but do not add more than necessary. Of course, this is not possible with indicators that begin in a colorless form.

3. The most common error in titrations is that caused by accidental over-titration. The analyst does not realize that the endpoint is near, and suddenly it is too late because too much has been added and the determination is spoiled. There is an easy way to avoid this problem: before starting the titration use a 6-inch medicine dropper to remove a little of the sample solution. Lay the dropper sideways on a beaker, with the tip end in the pouring spout of the beaker, so that the solution cannot drain up into the rubber bulb (Fig. L.13). This small part of the sample offers great security against accidental overtitration.

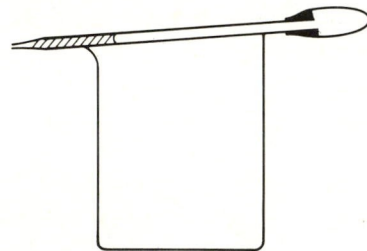

Fig. L.13 This dropper may be called a **titration thief**.

4. Titrate the remaining sample by letting the titrant solution flow from the buret, while simultaneously swirling the flask to mix the solutions. Learn the approved technique for holding and manipulating the buret stopcock with the left hand while swirling with the right. It will take a little practice to swirl rather than simply shake the contents of the flask.

As the endpoint is approached, you will see that the titrant causes a change in the indicator color at the point where the solutions first meet, but that this color disappears upon swirling. When the endpoint is very near, add the titrant drop by drop, swirling after each drop, until the color changes permanently. This is a false endpoint, of course, because of the small amount of solution in the medicine dropper. Because of this "security deposit" it has been possible to reach the false endpoint quickly because there was no threat of overtitrating.

5. Return the solution from the medicine dropper to the flask, being sure to flush out the dropper a couple of times to rinse out the solution (or rinse it out with a wash bottle). You are now ready to approach the true endpoint.

6. Use the wash bottle to rinse down the inside surface of the flask, in case some of the sample spattered. Such rinsing may be done whenever it is desired, for the additional water will not affect the titration.

7. Approach the endpoint with care, first using dropwise additions and judging from the swirls of color when you are getting very close. Then try for highest accuracy by "splitting drops." That is, control the stopcock so that the amount of titrant coming out is not enough to form a full drop. Use a wash bottle to rinse this fraction of a drop into the sample solution. It is possible to deliver as little as 0.005 mL in this way.

8. When you judge from the indicator color that the endpoint has been reached, record the buret reading. The titration volume is, of course, the difference between this and the initial buret reading. With some titrations you will not be sure whether the color change has fully occurred or not. This is typical of titrations that do not have a "sharp endpoint," i.e., a color change that occurs fully with only a very small addition of excess titrant. There are two techniques for dealing with this problem. One is to have, in another flask, a solution for comparison. It should have the same amount of indicator and the volume should be about the same as in the titration flask. This solution might well be the first one from a group of samples, with excess titrant deliberately added to make sure the color change is complete. The second technique is to read the buret accurately when you think the endpoint may have been reached and then to add a drop (or fraction of a drop) to see if there is a definite further change in color. If the color does not change, use the previous reading as the endpoint. If it does change, repeat the process until you are satisfied.

You should realize that some titration endpoints will "fade" after they are reached, due to slow decomposition or reaction with atmospheric O_2 or CO_2.

Using Potentiometric Measurements

A potentiometric titration is somewhat similar to the technique discussed for potentiometric determination, except that now the half-cell containing the unknown becomes a titration vessel. A typical experimental arrangement is depicted in Fig. L.14.

The reference electrode is usually a saturated calomel of the type found with pH meters. The indicator electrode depends upon the type of titration being performed:

a)   for acid–base titrations a glass electrode is used for pH response,

b)   for redox titrations a platinum wire or foil is best,

c)   for titrations involving silver ion a silver wire is used,

d)   for titrations involving mercury a small mercury pool is used,

e)   various ion selective electrodes may be used in many applications.

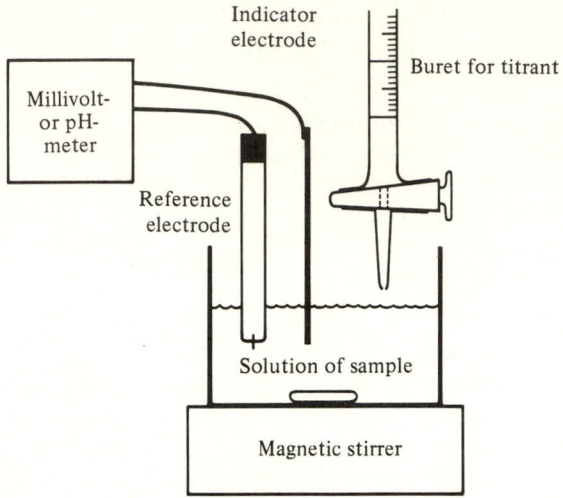

Indicator electrode

Buret for titrant

Millivolt- or pH- meter

Reference electrode

Solution of sample

Magnetic stirrer

Figure L.14

The experimental technique is quite simple: one makes successive additions of titrant and records the value of the cell potential for each addition. The endpoint is found by a plot of potential versus volume of titrant added.

Using Weight Burets

The reader probably has had considerable experience with volumetric titrations by using a buret—a glass tube with regular graduations. At the endpoint one reads the volume (in units of milliliters) and multiplies by the molarity of the titrant solutions (in units of millimoles of solute per milliliter of solution) to find the amount added (in millimoles).

An attractive alternative is to use a plastic "wash bottle" instead of the conventional buret for the delivery of titrant solution. The solution is prepared on a mass basis (in units of millimoles of solute per gram of *solution*), and the amount added in a titration is simply the product of the mass of solution and its concentration. There is no widely accepted term for concentration in units of millimoles per gram, but we might promote **molamity** for this purpose in analogy to the existing terms molarity and molality.

A strong case can be made for performing titrations with weight burets.* One eliminates volumetric glassware entirely, with its attendant problems of cleaning, drainage, standardization, meniscus reading, and high expense. Gravimetric measurements are usually more accurate, and when solutions are prepared on a mass basis there is no concern about change of concentration with temperature.

---

*See *Gravimetric Titrimetry*: *A Neglected Technique*, E. A. Butler, E. H. Swift, *Journal of Chemical Education*, 49, 425 (1972).

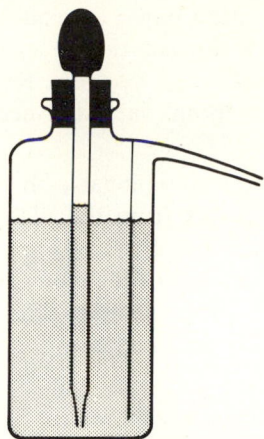

Figure L.15

In performing a volumetric titration one typically adds titrant at a fast rate from the buret until the endpoint is near and then changes to the addition of single drops, and even split drops, to try to hit the endpoint precisely. The same basic approach holds for gravimetric titrations. A convenient procedure is to use a plastic wash bottle with a side spout (e.g., Nalgene Cat. No. 2402-0125) fitted with a medicine dropper, as shown in Fig. L.15.

The bulk of the titrant solution is added to the beaker (or flask) containing the sample by squeezing the bottle to squirt the solution through the side spout. When the endpoint is near, the medicine dropper is used because it allows easy control of drops and split drops. The bottle is weighed at the start of the titration, and again after the endpoint has been reached, to determine the precise mass of the added titrant solution. If the molamity is chosen so that at least 20 grams are used, then a top-loader balance with a precision of ± 0.01 gram is sufficiently accurate.

To realize the best accuracy in a gravimetric titration it is important to observe some precautions as follows:

● When preparing the titrant solution, first weigh a clean dry bottle on a top-loader balance to the nearest 0.01 g. Introduce without loss the titrant substance that has been weighed on an analytical balance to the nearest 0.1 mg. Add water, again using the top-loader balance, until the total solution mass is at about the desired value; avoid wetting the neck of the bottle. Swirl to dissolve the titrant, stopper the bottle, and obtain the final mass of the solution. Calcuate the molamity (moles of solute per kilogram of solution).

● When filling the plastic wash bottle, be sure to rinse it with the solution first. Avoid leaving any solution between the rubber stopper (holding the medicine dropper) and the neck of the bottle.

- In the first phase of the titration (rapid approach to the endpoint) be sure to withhold about 1 mL of the sample solution in a "titration thief," as explained under TITRATIONS, USING VISUAL INDICATORS. Return this "security deposit" to the solution when the endpoint is quite near, and then proceed very cautiously.

- For the final approach to the endpoint use the medicine dropper from the weight buret. When removing this from the plastic bottle, be careful not to inadvertantly squeeze the bottle and thereby expel some solution from the side spout. If the medicine dropper is squeezed gently, it is possible to expel a "split drop" whose volume is even smaller than 0.01 mL (0.01 gram in this context). Such a split drop may be touched off on the stirring rod being used to stir the titrated solution.

- In placing the weight buret on the lab bench, be careful not to let it get wet on the outside, as this will cause problems in determining the correct mass of the solution used.

When you think about it, it seems strange that gravimetric titrations are not widely used. Doubtless this is because the whole art of titration was developed before we had either plastic bottles or top-loader balances. Those of us who look toward the future should encourage the use of this neat technique.

15. TRANSFER, QUANTITATIVE

In analytical procedures it is frequently necessary to transfer, without appreciable loss, a solid or solution from one vessel to another. The term **quantitative transfer** generally means that care is taken to keep any losses negligible.

Transfer of a Solution

Suppose a solution in one beaker is to be transferred to another beaker. If it is simply poured in the usual way, it is likely that a drop or two will run down the outside of the beaker when the pouring is completed. To prevent this, it is wise to use a glass stirring rod held as shown in Fig. L.16.

The stirring rod prevents the loss of a drop, provided the rod touches the beaker spout two or three times as the beaker is returned to its upright position. Of course, the rod must be rinsed with water from a wash bottle. The original beaker still contains some of the solution, since drainage is never complete, and the next step is to rinse down the sides of the beaker with water from a wash bottle. The pouring is repeated, again with the aid of the stirring rod. After about three such rinses, the transfer may be considered to be completed.

If the solution is to be transferred to a volumetric flask, the basic technique is the same as just described, except that a funnel must be used.

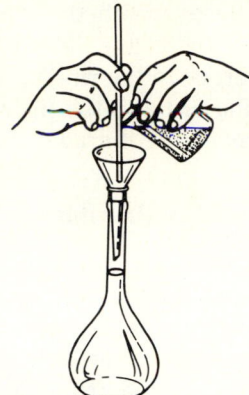

Fig. L.16 Transfer of solution to volumetric flask.

Once the beaker has been well rinsed, a wash bottle is used to rinse the inside of the funnel. Then, as the funnel is withdrawn from the flask, its outside stem surface is also rinsed, to be sure that none of the transferred material is clinging to it.

Transfer of a Solid to a Volumetric Flask

If the solid is a dry crystalline sample in a beaker and if it is to be transferred to a volumetric flask, it is necessary to use a funnel. It is not wise to pour the dry solid into the funnel because crystals can bounce and be lost. Add a little water to the solid from a wash bottle, and then use the same technique as described above for a solution (Fig. L.17). It is not necessary to dissolve the solid completely before the transfer is carried out: one holds the beaker in the pouring position, with the stirring rod in place, and uses the wash bottle to flush the undissolved solid into the funnel. This process is followed by several

Figure L.17

small rinses and pourings, always with the stirring rod, to be sure that the transfer is quantitative.

As with solution transfer, once the rinsing of the beaker is complete, it is important to rinse the stirring rod and the inside of the funnel, and then the outside of the funnel stem as it is withdrawn from the flask.

Transfer of a Precipitate to a Filter Crucible

A precipitate formed in a beaker will not be as easy to transfer as a soluble solid sample. Quite often the precipitate will tend to stick to the beaker surface and must be loosened up with a rubber policeman on a glass rod. Except for this, the transfer is much the same as described above. First it is well to decant the supernatant liquid from the beaker, leaving most of the precipitate behind. This is done with the help of a stirring rod fitted with a rubber policeman. Then, with the beaker fully tipped into the pouring position, a wash bottle is used to flush most of the precipitate into the filter crucible. By touching off the beaker spout with the rubber policeman, loss of any precipitate is avoided. Then a small amount of water is added to the beaker and the rubber policeman is used to thoroughly squeegee the precipitate loose. This is followed by pouring into the crucible, and the process is continued until transfer is complete. Inevitably there will be a very small loss due to precipitate that sticks to the rubber policeman, and this cannot be totally avoided.

As a general comment: if you keep in mind the purpose of a quantitative transfer and are alert to the ways in which losses can occur, you are likely to do it correctly.

16. VOLUMETRIC FLASKS

When it is desired to make a solution of a precisely known volume, a volumetric flask is used. These flasks are available in sizes ranging from 1 to 5000 mL and are designed to contain, not deliver, the nominal volume. They have a circular etching around the neck to indicate the level to which they should be filled (Fig. L.18). It is only the neck of the flask that must be especially cleaned to remove the atmospheric film deposit. Usually this can be done by scrubbing with detergent and a test-tube brush.

Proper technique in using a volumetric flask is as follows: after the sample to be dissolved or diluted has been transferred to the flask, enough distilled water is added to nearly fill the lower part. This is swirled to dissolve the sample and to make the resulting solution homogeneous. Then more water is added until the level is about 1 cm below the mark. At this point the flask is stoppered and inverted, with swirling, about six times. After letting it stand upright for one or two minutes the stopper is removed, the drops that cling to

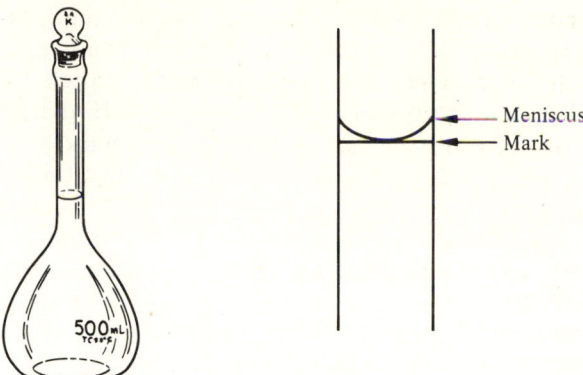

Figure L.18

it are touched off, and the final addition of water is made until the meniscus is tangent to the calibration mark. A medicine dropper is useful at this step. Finally, the stoppered flask is inverted and swirled about 12 more times to guarantee that the solution is homogeneous. When dilute solutions are prepared, negligible error results if the six intermediate inversions are omitted.

Alkaline solutions should not be left in volumetric flasks, because they etch the glass. In general it is good practice to transfer solutions to bottles once they have been made. This releases the volumetric flask for other uses.

For volumetric flasks of sizes greater than 25 mL, the manufacturing tolerances are better than one part per thousand for class A ware, but are somewhat poorer for other grades. If it is desired to standardize a given flask, there are two possible methods. The most accurate method is to fill the flask precisely with water of known temperature and to weigh it. Flasks smaller than 100 mL may be weighed to the nearest milligram on the analytical balance, but it is necessary to use a top-loading balance for 100-mL and larger flasks. A balance accurate to ±0.01 gram is required. By subtracting the weight of the dry flask, one obtains the apparent weight of water contained by the flask. The calculations to find the actual volume are described in Section 3.7.

The second method requires the use of a standardized pipet. Suppose a 250-mL flask is to be standardized: a 50-mL volumetric pipet is used to introduce five measures of distilled water. At this point the water level may, luckily, have its meniscus precisely tangent to the flask calibration mark, in which case the volume of the flask is exactly five times the standardized volume of the pipet.

If the level is not at the mark, which is generally true, one may use transparent tape to mark the meniscus position, thus ignoring the manufacturer's calibration mark and providing a new mark (the top edge of the tape) that yields a flask bearing an integral relationship to the particular pipet used.

If it is desired to use the manufacturer's mark, and the level is below that mark, one uses a 1-mL graduated pipet divided in 0.01-mL units to make the final precise adjustment. The volume of the flask is then the sum of the volumes from the volumetric and measuring pipets. If the level (from the volumetric pipet additions) happens to be above the mark, some water must be withdrawn. Again use the 1-mL measuring pipet, but this time start with the pipet already containing water adjusted to one of its lower graduation marks. Then the pipet is used to withdraw a little more than necessary, followed by return of just enough to fill the flask to the mark. The measuring pipet now contains more than it did at the start, and the extra amount is to be subtracted from the total volume added from the volumetric pipet.

17. WEIGHING

Because of differences in the models and operations of analytical balances used in various laboratories, no attempt at detailed operation will be presented here. The instructor will provide a demonstration and will explain specific problems and local rules. However, the reader should review the earlier section (ANALYTICAL BALANCE) for important general comments.

We are concerned in this section with the handling of materials in the weighing operations. There are several categories to consider:

Direct Weighing

This refers to the case of an object that may be placed directly on the balance pan, when no protection of the pan is required. A piece of glassware, a sample of metal such as a coin, or other objects that cannot possibly damage the pan by contact can be weighed directly. One simply sets the empty balance to zero, places the object as closely as possible to the center of the pan, and dials the appropriate weights. Warnings are often issued against handling such objects with bare fingers. However, if the fingers are dry (wiped on a towel) it is unlikely that the resulting fingerprints will have any noticeable effect on the weight. The reader may test this claim by weighing a 100-mL beaker by using tongs only, then by using fingertips, and finally by unconcerned grasping.

Weighing by Addition

In this technique a beaker (or other vessel) is first accurately weighed. Then the desired quantity of sample is added to the vessel and the increase in weight is the weight of the sample. But how is the *desired quantity* to be added? The safest way is to have the approximate quantity in a small beaker, having already used a top-loading or triple-beam balance. Then this sample can be tapped into the accurately weighed beaker without concern for making a quantitative transfer.

Alternatively, a larger quantity of sample may be ready in a weighing bottle or small beaker. Suppose it is desired to add 0.4 ± 0.05 gram to the accurately weighed beaker without removing the latter from the balance pan.

To illustrate, suppose the empty beaker weighs 34.5832 grams. You want it to weigh 0.4 g more, or 34.98xx grams after the sample is added, where the values of the digits xx are unimportant. To accomplish this, the balance is placed on partial release and the weights are dialed to 34.9 grams. The optical scale will, of course, be off scale. Now, during judicious tapping of small amounts of the sample from the supply beaker into the weighed beaker watch the scale closely, for it will begin to register as soon as the weight reaches 34.9. At this point it is within 0.1 gram of the desired final weight, and the balance may be placed on full release. Next, very careful and small additions (while watching the optical scale) will allow you to tell when the weight has reached 34.95 or so, which will be close enough. The balance doors are closed and the final weight reading is recorded. This procedure requires reaching into the balance case, holding the supply beaker over the weighed beaker, and well controlled tapping without banging the beakers together. It is somewhat tricky, but is a good technique to learn. It has the great advantage of seeing at all times just how much sample has been transferred. Of course, instead of doing the tapping it is reasonable to use a spatula for introducing successive amounts of the solid. In this case it is preferable to use a rounded scoop, for crystalline materials slide off the smooth surfaces of flat spatulas too easily.

Weighing by Difference

This technique is especially useful when a series of samples of similar size are to be weighed because it requires only $(N + 1)$ weighings to obtain N samples, compared to $2N$ weighings for the addition method. It also eliminates the requirement that the receiving vessels be dry; for example, samples may be weighed by difference into a group of titration flasks that have just been cleaned. Further, the technique of weighing by difference is best when the sample should be protected from undue exposure to the atmosphere, e.g., when the material is hygroscopic.

In weighing by difference one uses a capped weighing bottle to contain enough material for all the samples that are to be weighed (Fig. L.19). This

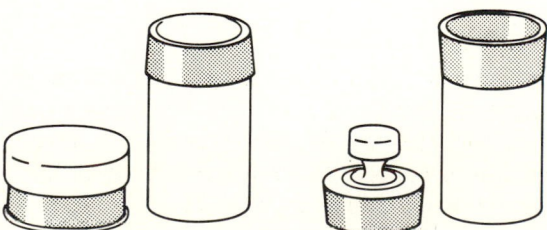

Fig. L.19 Typical weighing bottles.

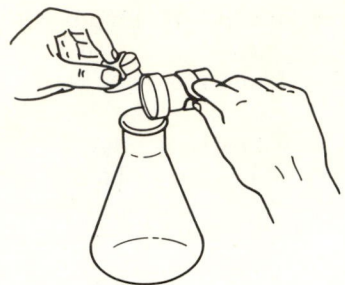

Fig. L.20 Method of transferring sample.

material may be added to the weighing bottle with the aid of a top-loading or triple-beam balance. Then the weighing bottle and contents are weighed accurately on the analytical balance. Next a sample is tapped into the first receiving vessel (Fig. L.20). It is necessary to guess the right amount at this stage, and the best practice is to deliberately add too little at first. The capped weighing bottle is reweighed to see how much actually was transferred. If it is within the desired range for the sample size, then the weight is recorded and the process is repeated for the next sample. If it is too little, the weight is ignored and a little more is tapped into the receiving vessel. Once a satisfactory sample size has been achieved, it becomes easier to judge the quantities to tap out for the successive samples. For each sample the final weight of the weighing bottle becomes the initial weight for the next sample. This is why only $(N + 1)$ weighings are required.

Weighing Liquid Samples

Portions of liquids may be weighed either by addition or difference, the only new factor being the necessity of using a stoppered vessel to minimize errors due to evaporation. In the difference method it is not possible to pour from the supply vessel because of the inevitable wetting of the spout. Perhaps the best (only?) suitable vessel for weighing liquids by difference is a plastic wash bottle that draws the liquid back when delivery is completed.

Laboratory Experiments

1. Analytical weighing and the determination of percentage of water in a hydrated salt
2. Potassium permanganate as a titrant and a stoichiometry study of iron (III)–hydroxylamine reaction
3. Potassium iodate as a titrant and assay for vitamin C in tablets
4. Determination of potassium iodide in commercial table salt
5. Spectrophotometric determination of iron using the bipyridyl complex
6. Photometric titration of copper with EDTA
7. Gravimetric determination of lead content by precipitation of lead chromate
8. Silver-electrode study of equilibria
9. A sequence of acid–base titrations
10. Characterization of an unknown acid by potentiometric pH titration
11. Precise determination of Q_1 for an amino acid
12. Potassium dichromate as a titrant and determination of iron in an ore
13. High-precision assay of Mohr's salt by gravimetric titration
14. Spectrophotometric determination of an acid-dissociation constant
15. Photometric study of the $1:1$ iron(III)–thiocyanate complex
16. The mercury electrode: chelon equilibria
17. The solubility product for copper iodate
18. Distribution of bromine between water and carbon tetrachloride and determination of the formation constant for tribromide ion
19. Titration of calcium with EDTA, and the $1:1$ calcium–acetate complex

LABORATORY EXPERIMENT 1

ANALYTICAL WEIGHING AND THE DETERMINATION OF PERCENTAGE OF WATER IN A HYDRATED SALT

Analytical Weighing

Determination of the mass of an object or analytical sample is the most fundamental laboratory operation in chemistry. Thanks to improvements in the design of analytical balances, weighing is both easy and highly precise. It takes less than a minute to find the mass of an object with an accuracy of about 0.0002 gram. However, it is possible to use a balance incorrectly or carelessly (see ANALYTICAL BALANCE) or to handle materials wrongly. The purpose of this experiment is to teach good balance technique as a basis for many other experiments.

1. The laboratory instructor will explain the details of balance operation and provide an object of known mass. Determine the mass of this object with at least three repeated weighings and compare your results with the known value. If there is a discrepancy of more than about 0.0003 gram, consult the instructor.

2. Using tongs, place a beaker (100–250 mL) on the balance pan and find its precise mass. Remove the beaker with your hand, having dried your fingers on a clean towel, and then determine its mass again. Comment in your notebook on the effect of fingerprints.

3. Place about 1–2 grams of dry sodium chloride in a clean dry weighing bottle or beaker. Take this, a plastic wash bottle with only 30–50 mL of water in it, and a clean dry beaker to the balance. In this part of the experiment you will find the weight of a salt sample and also that of a water sample by two methods of weighing. Carry out the following steps:

    a)   Making sure that the washbottle is dry on the outside, determine its mass. In this and all other weighings, try for the highest precision in weighing.

    b)   Determine the mass of the dry beaker.

    c)   Determine the mass of the weighing bottle and salt.

    d)   Handle the weighing bottle with a strip of paper or tongs and transfer most of the salt to the beaker, being very careful not to lose any. There may still be a little left in the weighing bottle, however. Redetermine the mass of the weighing bottle.

    e)   Redetermine the mass of the beaker, now containing the salt sample.

    f)   Leaving the beaker on the balance, use the wash bottle to add about 5–10 mL of water, trying not to lose any. Immediately reweigh the beaker, now containing salt and the water sample.

g) Reweigh the wash bottle.

The above operations were merely an efficient way to illustrate several ideas. The mass of the salt sample transferred to the beaker may be calculated in two ways: the method of **weighing by addition** involves steps (b) and (e), whereby the weight of the salt sample is taken as the *increase* in mass. The method of **weighing by difference** involves steps (c) and (d), with the mass of the sample being the *decrease* in mass. Use your data to find the two values for the sample mass and report the difference (in tenths of a milligram).

The salt sample is nonvolatile, and it is expected that the two values for its mass will agree rather well. With the water sample we have to consider the evaporation that takes place in the beaker open to the air. By using the data from steps (e) and (f), find the mass of water that apparently was delivered from the wash bottle. By using steps (a) and (g), find the actual mass of the water delivered on the assumption that the closed wash bottle prevents any loss by evaporation. From the difference in the values, calculate the volume of water (in milliliters) that evaporated during the time it took to make the weighing in step (f).

4. How reproducible is a "drop" from the end of a medicine dropper? Obtain a small bottle fitted with a medicine dropper and rubber bulb. Half fill it with distilled water, make sure there is no water on the outside or around the area of the cap, and weigh the bottle accurately. Use the dropper to deliver one drop into a waste beaker, holding the dropper vertically and squeezing the bulb carefully until the drop just falls of its own weight. Then place the dropper back in the bottle and reweigh to determine the precise mass of the drop. Repeat the operation ten times. Calculate the average drop mass, the average deviation, and the range. Then deliver ten drops in rather rapid succession and reweigh the bottle. Compare the average drop mass for the two cases.

5. Check the internal consistency of the optical scale on the balance and the 0.1-gram weight as follows. With the pan empty adjust the scale precisely to zero. Add an object that weighs close to 0.1 gram and read its mass according to the optical scale. This object may be a piece of wire provided by the instructor, or simply some torn bits of paper that by trial and error weigh about 0.1 gram. Then dial the 0.1-gram weight, causing the optical scale to move close to zero. The change in the optical-scale reading should be precisely 0.1000 gram. If it deviates by more than 0.0002 gram consult the instructor.

Determination of Water in a Hydrated Salt

This experiment is a simple example of a gravimetric determination based on loss in mass due to volatilization. Nevertheless, for accurate results it is necessary to take care.

1. Clean two weighing bottles, put them in a labeled beaker covered with a watch glass, and place them in an oven set for about 120°C. After about 2 hours remove them, using tongs, and place them in a desiccator to cool. Find their weights to the nearest 0.1 mg and then repeat the process cutting the heating time to about 30 minutes. Reweigh the bottles and, if their weights differ by more than 0.3 mg from the previous values, again repeat the heating. The object is to bring the bottles to constant weight, showing that all moisture has been removed from their surfaces.

2. Place samples of a hydrated salt (barium chloride dihydrate, copper sulfate pentahydrate, magnesium sulfate heptahydrate, or any other suggested by the instructor) in the weighing bottles. The sample size is not very critical, but should be between 1 and 2 grams. Reweigh the bottles to obtain the precise sample weights. Put the bottles in the oven, again in the covered beaker, for about 2 hours. Let them cool in the desiccator before weighing and again go through the cycles of heating, cooling, and weighing until they have come to constant weight. Record your observations about the change in physical appearance.

3. From the loss in weight for each sample, calculate the percentage of water in the original sample. If you are dealing with a known material, use the table of atomic weights to calculate the theoretical percentage of water in whatever nominal hydrate is used. Find the deviation of the experimental value from the theoretical and express it in parts per thousand.

LABORATORY REPORT 1

ANALYTICAL WEIGHING AND DETERMINATION OF THE PERCENTAGE OF WATER IN A HYDRATED SALT

NAME _____   LAB SECTION _____

DATE _____

Weighing by addition and by difference:

    Weight of NaCl by addition $(e - b)$ _____

                by difference $(c - d)$ _____ $\Delta =$ _____

    Weight of water by addition $(f - e)$ _____

                by difference $(a - g)$ _____ $\Delta =$ _____

Weighing of ten individual drops of water:

    Average mass: _____ Range: _____

    Relative Average Deviation: _____%

Weighing of ten drops of water as a group:

    Average mass: _____

Description of hydrated salt

	Sample 1	Sample 2
Mass of weighing bottle + hydrate		
Mass of empty weighing bottle		
Mass of hydrated salt sample		
Mass of bottle + salt after drying		
Mass of empty weighing bottle		
Mass of dehydrated salt		
Change in mass due to drying		
Volatile material in sample, %		
Theoretical value of water, %		

LABORATORY EXPERIMENT 2

POTASSIUM PERMANGANATE AS A TITRANT AND A STOICHIOMETRY STUDY OF THE IRON(III)–HYDROXYLAMINE REACTION

PURPOSE

1. To standardize a solution of potassium permanganate by using pure arsenious oxide as a primary standard.

2. To determine the stoichiometry of the redox reaction between iron(III) and hydroxylamine.

MATERIALS NEEDED

10- and 25-mL pipets, 50-mL buret, 100- and 250-mL volumetric flasks

Solid potassium permanganate and arsenic trioxide

Solutions supplied by instructor: iron(III) and NH_2OH (see below), $0.002M$ potassium iodate, $6M$ HCl, $1M$ NaOH, $6M$ H_3PO_4.

Potassium Permanganate as a Titrant

PROCEDURE AND COMMENTS

1. Prepare a solution of potassium permanganate $KMnO_4$ as follows: weigh (accurately) a sample of solid $KMnO_4$ in the range of 0.7 to 0.9 gram. Transfer it quantitatively to a 250-mL volumetric flask, add distilled water to the mark, and mix until all the salt has dissolved. This will require several minutes of continuous mixing. Calculate the nominal molarity of the solution, assuming the salt to be 100% pure. The actual purity will be calculated later.

Alternatively, the instructor may provide a sample of more concentrated $KMnO_4$ solution to be treated as an unknown. Transfer this sample quantitatively to a 250-mL volumetric flask, dilute to the mark, and mix thoroughly. In either case, the molarity of the final solution should be approximately 0.02.

2. Prepare a solution of arsenic(III) as follows: weigh (accurately) a sample of pure arsenious oxide (arsenic trioxide As_2O_3) in the range of 0.35 to 0.45 gram. Transfer the sample quantitatively to a 100-mL volumetric flask with the aid of small portions of water. Add about 5 mL of $1M$ sodium hydroxide and swirl the mixture until the solid is completely dissolved. Add 1 drop of phenolphthalein indicator and then add, dropwise, $6M$ hydrochloric acid until the red color just disappears. Dilute to the mark with water and mix thoroughly. Calculate the precise molarity of this solution in terms of As(III), remembering that there are two As atoms per molecule of the original oxide. The molarity should be approximately 0.04.

3. Since the solid potassium permanganate is not of high purity, it is neces-

sary to carry out a titration to standardize (determine the accurate molarity) its solution. Pipet 25-mL portions of the arsenic(III) solution into each of two titration flasks. Add about 10 mL of $6M$ hydrochloric acid and 1 drop of $0.002M$ potassium iodate which serves as a catalyst.

Titrate this solution with the potassium permanganate solution. The endpoint is the first permanent appearance of pale pink color, indicating that a very slight excess of $KMnO_4$ is present. The titration reaction is as follows:

$$HAsO_2 + MnO_4^- \rightarrow H_3AsO_4 + Mn^{2+} \quad (unbalanced).$$

Balance the equation. Use the average of the two titration endpoint volumes to calculate the precise molarity of the potassium permanganate solution (four significant digits).

By comparing this molarity with the nominal value, calculate the percentage purity of the solid $KMnO_4$.

The Iron(III)–Hydroxylamine Reaction

In an acidic solution there is an oxidation–reduction reaction between iron (III) and the hydroxylammonium ion. The iron is reduced to the ferrous state, while the hydroxylamine is oxidized:

$$a\,Fe^{3+} + b\,NH_3OH^+ \rightarrow a\,Fe^{2+} + ?,$$

where a and b are the stoichiometric coefficients to be determined. It will then be possible to deduce the oxidation state of the nitrogen in the unknown reaction product. In hydroxylamine the nitrogen is in the -1 oxidation state, and so there are several species with higher oxidation states: N_2 (oxidation state 0), N_2O (+1), NO (+2), HNO_2 (+3), NO_2 (+4), and NO_3^- (+5). The purpose of this part of the experiment is to find which of these is the most probable according to the observed stoichiometry.

The reaction is carried out by heating a precise amount of hydroxylamine with a substantial excess of iron(III). The hydroxylamine is assumed to be completely used up, and the amount of reacted iron is determined by titration with the standard potassium-permanganate solution.

PROCEDURE

1. The instructor will provide a stock solution of hydroxylamine prepared as follows: 16.41 grams of pure hydroxylamine sulfate $(NH_3OH)_2SO_4$ (MW = 164.14) is dissolved in water and diluted to precisely 2 liters. Alternatively, 13.90 grams of pure hydroxylamine hydrochloride NH_3OHCl (MW = 69.49) may be used. Prove that in either case the molarity of the hydroxylammonium ion NH_3OH^+ is 0.100.

2. The instructor will provide a stock solution of ferric ammonium sulfate prepared as follows: 480 grams of $FeNH_4(SO_4)_2 \cdot 12H_2O$ (MW = 482.19) is dissolved in about 2 liters of water to which 250 mL of concentrated sulfuric

acid has been added. The mixture is then diluted to a final volume of 4 liters. Calculate the final molarity of Fe(III) and that of H_2SO_4.

3. To each of the two titration flasks pipet 10 mL of the hydroxylamine solution. Use a graduated cylinder to add 25 mL of the iron(III) solution. Heat the solutions and boil gently for about 2 minutes to hasten the iron(III)–hydroxylamine reaction. Then add about 50 mL of water and 5 mL of 6M phosphoric acid. The latter serves to decolorize the excess iron (III), so that the titration endpoint will be easier to see.

Titrate the solutions with the standard potassium permanganate solution, again taking the endpoint to be the first permanent appearance of a very pale pink color that indicates a slight excess of $KMnO_4$. The titration reaction is:

$$Fe^{2+} + MnO_4^- \rightarrow Fe^{3+} + Mn^{2+} \quad \text{(unbalanced)}.$$

Balance the equation. Use the average of the two endpoint volumes to calculate the amount (millimoles) of iron(II) formed by the reaction. Of course, this is the amount of iron(III) that reacted with the known amount (1.00 mmol) of hydroxylamine.

Assign values to the stoichiometric coefficients a and b, and deduce the oxidation state of the nitrogen in the reaction product. Propose the identity of the nitrogen species.

LABORATORY REPORT 2

IRON(III)–HYDROXYLAMINE REACTION STOICHIOMETRY

NAME _____   LAB SECTION _____

DATE _____

Preparation of $KMnO_4$ solution

    Mass data: beaker + $KMnO_4$ _____

               empty beaker   _____

               $KMnO_4$ _____

    Calculated molarity of $KMnO_4$ solution (assuming 100% purity):

Preparation of arsenic(III) solution

    Mass data: beaker + As_2O_3 _____

               empty beaker   _____

               As_2O_3 _____

    Calculated molarity of As(III) in the final solution:

Standardization of $KMnO_4$ solution   Trial 1 Trial 2

        Molarity of arsenic(III) reference solution:

        Titration data: buret reading at endpoint   _____ _____

                        buret reading at start   _____ _____

                        volume of $KMnO_4$ used   _____ _____

        Experimental value for $KMnO_4$ molarity   _____

        Apparent purity of the solid $KMnO_4$(%)   _____

Stoichiometry data

      Titrations of iron(II):   buret reading at endpoint   _____ _____

                        buret reading at start   _____ _____

                        volume of $KMnO_4$ used   _____ _____

  Precise amount of hydroxylamine used   _____ _____

  Precise amount (average) of iron(III) reacted   _____

  Precise value of experimental stoichiometric ratio,   _____
  Fe(III)/NH_2OH

  Conclusions (show reasoning on back of this sheet)

LABORATORY EXPERIMENT 3

POTASSIUM IODATE AS A TITRANT AND ASSAY FOR VITAMIN C IN TABLETS

PURPOSE

1. To compare the actual content of vitamin C (ascorbic acid) in commercial tablets with the content declared on the label by the manufacturer.

2. To illustrate the analytical use of the iodate–iodide reaction.

MATERIALS NEEDED

10-mL pipet, 250-mL volumetric flask, 50-mL buret

Solid potassium iodate, solid potassium iodide

Solutions prepared by instructor: $1M$ HCl, ascorbic acid (about 10 mg/mL)

Vitamin C tablets

Potassium Iodate as a Titrant

PROCEDURE AND COMMENTS

1. Prepare a standard solution of potassium iodate KIO_3 (MW = 214.00) as follows: accurately weigh a sample of pure dry KIO_3 in the 0.5–0.6 gram range. Transfer quantitatively to a 250-mL volumetric flask, dilute to the mark with water, and mix thoroughly. Calculate the molarity to four significant digits (it should be about 0.01).

2. To gain experience in titration before attempting the tablet assays, determine the precise concentration of a solution of pure ascorbic acid provided by the instructor. Pipet 10 mL of this solution into a titration flask, add about 50 mL of water, about 5 mL of $1M$ HCl, and about 1 g KI. Titrate immediately with the standard potassium iodate solution, taking the endpoint to be the first permanent appearance of a very pale yellow color, indicating the presence of a slight excess of triiodide ion.

The reactions that take place during the titration are as follows. First, the iodate and iodide ions react in the acidic solution to form triiodide ion:

$$IO_3^- + 8I^- + 6H^+ = 3I_3^- + 3H_2O.$$

As long as the solution contains ascorbic acid, there is a rapid reaction with triiodide ion, during which dehydroascorbic acid and iodide ion are formed:

$$C_6H_8O_6 + I_3^- \rightarrow C_6H_6O_6 + 3I^- + 2H^+.$$

Thus, each millimole of potassium iodate added corresponds to 3 mmol of ascorbic acid in the sample. If a few milliliters of 0.5% potato-starch indicator solution is added to the solution before titrating, the endpoint is more easily seen because of the intense blue color formed by reaction of I_3^- with starch:

$$I_3^- + \text{Starch} = \text{Blue complex}.$$

However, the yellow color of I_3^- is sufficient for an accurate detection of the endpoint.

Repeat this titration until consistent (within 0.1 mL) values are obtained and calculate the concentration of the ascorbic acid solution in units of milligrams per milliliter.

Tablet-Assay Procedure

3. Three individual tablets are to be analyzed. To minimize the error due to air oxidation of the ascorbic acid it is recommended that each tablet be carried through the procedure before the next tablet is dissolved. Weigh the tablet accurately and place it in a titration flask. Add about 50 mL of water, about 5 mL of 1M HCl, and about 1 gram of KI. As the tablet disintegrates, vitamin C will dissolve and the binder material will remain as a finely divided solid. To hasten the process use a stirring rod to break up the tablet. As soon as the disintegration is complete titrate with the standard potassium-iodate solution using the same technique as in step 2.

The binder material of some tablets includes starch, and this may cause the endpoint color to be blue instead of yellow.

For each tablet, calculate the mass of ascorbic acid (MW = 176.13). Also calculate the average deviation for the set of three tablets. The instructor may choose to collect a larger set of results from the class to carry out a statistical analysis of the data.

LABORATORY REPORT 3

ASSAY FOR VITAMIN C IN TABLETS

NAME _____ LAB SECTION _____

DATE _____

Sample identification

Brand: _____

Lot number: _____

Declared vitamin C content (mg): _____

Date purchased: _____

Potassium-iodate-titrant solution

Mass of KIO_3, mg: _____ Total volume, mL: _____

Calculated molarity: _____

Check titration using standard ascorbic-acid solution

Volume of KIO_3 solution required: (1) _____mL (2) _____mL

Average: _____mL

Calculated content of standard ascorbic acid: _____mg/mL

Data and results for tablet assay:

Tablet number	Tablet mass, mg	Volume of KIO_3 used, mL	Vitamin C content, mg	% vitamin C in tablet
1				
2				
3				

Averages

Tablet mass: _____

Vitamin C content, mg: _____

% vitamin C in tablet: _____

Average deviation: _____

LABORATORY EXPERIMENT 4

DETERMINATION OF POTASSIUM IODIDE IN COMMERCIAL TABLE SALT

Table salt is available from most manufacturers in two forms: with or without potassium iodide. This additive is present at a level of about 0.01% in so-called **iodized salt**, which is recommended for persons whose diet would otherwise be too low in iodine, a trace element important in preventing goiter. In this experiment a titration procedure is used to determine the actual KI content of an iodized-salt sample. For regular, noniodized, salt the titration procedure is not useful because the KI level is much lower. Therefore a more sensitive procedure based on the high absorptivity of triiodide ion is used. This involves the technique of standard additions.

MATERIALS NEEDED

> Sodium thiosulfate standard solution, about 0.004M
>
> Commercial iodized salt
>
> Bromine water, saturated
>
> Formic acid and sodium formate solution, 0.5M each
>
> Potassium iodide
>
> Starch indicator
>
> Commercial noniodized salt
>
> Potassium iodate solution, 6.45 mg/L
>
> Potassium iodide solution, 0.5M, freshly prepared
>
> Spectrophotometer

CHEMICAL BASIS FOR THE DETERMINATIONS

First a water solution of the table-salt sample is treated with an excess of aqueous bromine. This results in oxidation of the iodide to iodate:

$$I^- + 3Br_2(aq) + 3H_2O = IO_3^- + 6Br^- + 6H^+.$$

Although the equilibrium constant for this reaction is not large, at the existing low molarities of Br^- and H^+ the reaction goes virtually to completion.

Next, the excess bromine is removed by adding formic acid:

$$HCOOH + Br_2 = CO_2 + 2H^+ + 2Br^-.$$

Fortunately this step does not have an appreciable effect on the iodate ion. At this point the analytical problem has been changed from a determination of iodide ion to a determination of iodate ion in small amount.

By adding an excess of potassium iodide to the solution, the iodate is rapidly converted to triiodide ion by the reaction:

$$IO_3^- + 8I^- + 6H^+ = 3I_3^- + 3H_2O.$$

When a large (say, 20 g) sample of iodized salt is used, there is enough triiodide ion to permit titration by a standard solution of sodium thiosulfate as follows:

$$I_3^- + 2S_2O_3^{2-} = 3I^- + S_4O_6^{2-}.$$

The endpoint is detected by the disappearance of triiodide, with the color enhanced through the presence of starch indicator:

$$\underset{\text{pale yellow}}{I_3^-} + \text{Starch} = \text{Blue complex}.$$

However, for noniodized salt the amount of triiodide formed is too small for accurate titration, and it is better to rely on spectrophotometric measurements. At a wavelength of 350 nm the molar absorptivity of I_3^- is high, about 26000.

The level of KI in the original sample may be found through the method of standard additions by using known amounts of potassium iodate.

PROCEDURE: TITRATION METHOD FOR IODIZED SALT

1. Prepare or obtain from the instructor a standard solution of sodium thiosulfate with a molarity of about $0.004M$. Because the determination of KI in the salt sample need not be highly precise, it is quite satisfactory to weigh a sample of $Na_2S_2O_3 \cdot 5H_2O$ (MW = 248.18), to dissolve it, and dilute the solution to a known volume. Of course, it is also not difficult to standardize this solution by titration of a weighed sample of potassium iodate (see PREPARATION AND STANDARDIZATION OF SODIUM THIOSULFATE in Chapter 12, pp. 410–412).

2. Weigh a sample of iodized salt, about 20 grams, into a titration flask and add about 75 mL of water. Swirl until dissolved. There will be some insoluble material, probably sodium aluminum silicate, which will not interfere with the determination.

3. Add about 1 mL of saturated bromine water and let stand, after swirling, for about 20 minutes. The solution should remain yellow due to excess bromine.

4. Add about 1 mL of formic acid, sodium formate solution, swirl to mix, and let stand for about 1 minute. The yellow color of bromine should disappear.

5. Add about 1 gram of potassium iodide, and swirl to dissolve. The color of I_3^- should be evident.

6. Titrate with $0.004M$ sodium thiosulfate until the color has faded to pale yellow. Then add a few milliliters of starch-indicator solution and continue the titration to the disappearance of the blue color.

7. Calculate the percentage of KI in the sample.

PROCEDURE: SPECTROPHOTOMETRIC METHOD FOR NONIODIZED SALT

1. Weigh (accurately) about 5 g of noniodized salt. Dissolve in about 15 mL of water and filter through medium-porosity paper to remove the insoluble residue of sodium aluminum silicate. The filtration should be done into a 100-mL volumetric flask, and the filter paper should be rinsed with a little water to ensure complete transfer of the soluble part of the sample to the flask.

2. Add about 1 mL of saturated bromine water, swirl to mix, and let stand for about 20 minutes.

3. Add about 1 mL of the formic acid, sodium formate solution, swirl to mix. The yellow color of the bromine should disappear. Dilute to the mark and mix well. At this point the solution contains $1.29 \, \mu g$ of KIO_3 for each $1.00 \, \mu g$ of KI originally present.

4. Set up a row of five small clean and dry beakers and make the additions shown in the following table:

	Beaker no.				
Procedure	1	2	3	4	5
Pipet 10 mL of the salt solution	10	10	10	10	10
Pipet water, mL	8	3	2	1	0
Pipet KIO_3 solution, 6.45 microgram/mL	0	0	1	2	3
Pipet 0.5M KI	0	5	5	5	5
Swirl to mix well					

Note that each beaker ends up with the same total volume of 18 mL. Beaker 1 serves as the blank. Beaker 2 has the sample only. The others have successive standard additions of KIO_3 corresponding to 5, 10, and 15 μg of KI.

Measure the absorbance of each solution at 350 nm, using distilled water in the reference cell and using the same cell (cuvet) for each solution. Subtract the absorbance of the blank from each of the other absorbances.

Plot the corrected absorbance values versus the number of micrograms of added KI. Extrapolate the plot to obtain the intercept on the KI axis (where absorbance is zero). This indicates the mass of KI present in the sample. Remember that each beaker contains only 1/10 of the weighed total sample.

Calculate the percentage of KI in the salt.

LABORATORY REPORT 4

DETERMINATION OF POTASSIUM IODIDE IN TABLE SALT

NAME _____   LAB SECTION _____

DATE _____

Standardization of sodium thiosulfate:

Endpoint volumes: Trial 1 _____ Trial 2 _____

Average _____

Molarity of sodium thiosulfate _____

Analysis of table salt   Iodized   Noniodized

Brand and lot number: _____   _____

Titration method

Mass of table-salt sample: _____

Endpoint volume: _____

Calculated mass of KI in the sample: _____

%KI in the table salt: _____

Spectrophotometric method of standard additions

Mass of table salt used for the 100-mL solution: _____

Beaker No.	Added KI, μg	Observed absorbance	Corrected absorbance
1	0		0
2	0		
3	5		
4	10		
5	15		

(Attach a plot of corrected absorbance versus amount of added KI, in μg)

Intercept of plot, μg of KI in beaker #2: _____

%KI in the table salt: _____

LABORATORY EXPERIMENT 5

SPECTROPHOTOMETRIC DETERMINATION OF IRON USING THE BIPYRIDYL COMPLEX

PURPOSE

1. To find the concentration of iron in a very dilute solution, using the color of the ferrous-dipyridyl complex
2. To illustrate the technique of preparing a standard curve for spectrophotometric determination.

MATERIALS NEEDED

Pure iron wire

$1M$ HCl, 10% hydroxylamine hydrochloride, $2M$ sodium acetate

0.1% bipyridyl (or 1,10-phenanthroline)

Six 100-mL volumetric flasks per student

GENERAL COMMENTS

The principles of spectrophotometric determination by using a standard curve are discussed in Chapter 5. The chemistry of this particular application is discussed in Chapter 16, in terms of 1,10-phenanthroline. This reagent and bipyridyl are very similar in structure and properties and may be used similarly, provided the proper wavelength for measurement is selected.

The sample issued to each student as an unknown will be a water solution of an iron salt. After dilution to a volume of 250 mL the concentration will be anywhere from 1 to 500 parts per million (that is, 1 to 500 μg of iron per 1 mL). Since the standard curve will be prepared only to cover the range of 0 to 5 ppm, some trials must be made before it will be clear what volume of the sample should be taken for the final determination.

The iron in the sample may be in the ferrous, the ferric, or a mixture of these states. Therefore the procedure calls for the addition of hydroxylamine to reduce any ferric iron to ferrous:

$$NH_3OH^+ + 2Fe^{3+} = \tfrac{1}{2}N_2O + 2Fe^{2+} + 3H^+ + \tfrac{1}{2}H_2O.$$

The ferrous ion then reacts with the dipyridyl (abbreviated as L) to form the complex:

$$Fe^{2+} + 3L = FeL_3^{2+}.$$
$$\text{red}$$

This reaction is carried out at a pH of about 5–6 to minimize competition with the hydrogen ion for the ligand. Hence, sodium acetate is added to raise the pH.

PROCEDURE

1. Obtain the sample of unknown solution and dilute it to 250 mL in a volumetric flask.

2. Prepare a standard solution of iron by dissolving pure iron wire in hydrochloric acid. Weigh accurately about 0.05 gram of the wire. Place it in a 1-L volumetric flask and add about 5 mL of $6M$ hydrochloric acid. Allow it to stand until the wire has completely dissolved. *Do not stopper the flask during this step for hydrogen gas is released!* From the precise weight of the wire, calculate the concentration of iron, in parts per million, after diluting to the mark with distilled water. It will be approximately 50 μg/mL.

3. The solutions for the standard curve are prepared as follows: First set up a row of six 100-mL volumetric flasks. Using your 10-mL graduated pipet add 0, 2, 4, 6, 8, and 10 mL of the standard iron solutions to the flasks. The first flask, with no added iron, will serve as the blank. To each flask add 1 mL of $1M$ HCl, 5 mL of 10% hydroxylamine hydrochloride solution, 2 mL of $2M$ sodium acetate solution, and 10 mL of 0.1% bipyridyl solution. Dilute each to the mark and mix well. Wait about 10–20 minutes for the color of the complex to develop fully. Calculate the iron content of each solution, expressed in micrograms.

4. In any spectrophotometric determination the choice of wavelength is of special concern. Generally it is best to make measurements at a wavelength corresponding to a maximum on the absorption spectrum (absorptivity versus wavelength) to achieve maximum sensitivity. The instructor may post an absorption spectrum for the ferrous-dipyridyl complex for you to examine, or you may be asked to determine your own spectrum using either a provided solution or one of the standard solutions you have prepared. In any case, choose a wavelength and use the spectrophotometer to determine the % transmittance or the absorbance for each standard.

5. To prepare the solution of the sample it is necessary to make a trial or two. Noting that the standards cover the range from zero to about 500 μg of iron (present in the 100-mL flasks), it may be necessary for you to use as little as 1 mL or as much as 50 mL of unknown to be within the range. Prepare two different solutions, using the same procedure as specified above for the standards. Measure the $\%T$ or the absorbance of each. If one or both fall within the range of the standard curve, the determination is finished. Otherwise, by noting the absorbance values actually obtained you will be able to choose a different volume of unknown for a final trial. For best accuracy in a spectrophotometric determination the absorbance should be approximately 0.3–0.5.

CALCULATION OF RESULTS

1. Prepare a precise plot of the standard curve, absorbance versus the weight (micrograms) of iron per 100-mL flask. Use graph paper with 10 squares per inch. For the absorbance axis let 1 inch represent 0.1 absorbance unit, and for the weight axis let 1 inch represent 50 μg. Include the point obtained for the blank. Draw the best straight line through all the points, neglecting any point that may deviate seriously from the linear relationship.

2. From the line read the intercept of the plot and calculate the slope of the line. Write the equation for the line:

$$A = A_0 + s \cdot W,$$

where A is absorbance, A_0 is the intercept, s is the slope, and W is the mass of iron in micrograms.

If a computer or programmable calculator is available, use it to find the linear least-squares fit for the data and compare with your visual estimate.

3. Using one or more of the absorbance values determined for the unknown find the weight of iron in the portion taken for the determination. Use the above equation for this purpose. Then calculate the total weight of iron in the original sample.

$$ppm = \frac{mg}{L} \times 10^6$$

LABORATORY REPORT 5

SPECTROPHOTOMETRIC DETERMINATION OF IRON USING THE DIPYRIDYL COMPLEX

NAME _____   LAB SECTION _____

DATE _____

Weight of iron wire used for standard-stock solution: _____

Concentration of this solution after dilution to 250 mL: _____ μg/mL

Data on standard curve (wavelength used: _____ nm)

	Flask 1	Flask 2	Flask 3	Flask 4	Flask 5
Fe stock, mL Fe taken, μg Transmittance, % Absorbance					

(Attach the plot of A versus weight of Fe.)

Intercept of plot: _____ Slope of plot: _____

Least-squares equation for line: _____

Data on the unknown

	Trial 1	Trial 2	Trial 3
Unknown stock, mL Transmittance, % Absorbance Calculated weight of Fe in 100-mL solution, μg			

Weight of iron in the total original sample: _____ μg

LABORATORY EXPERIMENT 6

PHOTOMETRIC TITRATION OF COPPER WITH EDTA

Given an unknown quantity of copper(II) in aqueous solution, it is feasible to determine it by titration with a standard solution of the chelon EDTA, which is ethylenediaminetetraacetic acid, usually prepared as a solution of its disodium salt (MW = 372.24):

$$Na_2[(HOCOCH_2)_2NCH_2CH_2N(CH_2COO)_2] \cdot 2H_2O \qquad \text{or} \qquad Na_2H_2Y \cdot 2H_2O.$$

The technique will be to use a Bausch & Lomb Spectronic 20 spectrophotometer to follow the changes in absorbance as portions of EDTA are added to the sample solution. A 5-mL portion of the sample solution is pipetted into an absorption cell, and the EDTA is added with a precision microburet. (Alternatively, a dispensing pipet may be used.)

At the beginning of the titration the copper is present chiefly as an acetate complex because an acetate buffer is used to control the solution pH at about 5. At a wavelength of 600 nm the initial absorbance will be rather low (see Fig. L.21).

As EDTA is added to the solution, the acetate ligand is displaced from the copper ion because the EDTA complex is much more stable. Thus, the titration reaction is:

$$Cu(OAc)_2 + H_2Y^{2-} = CuY^{2-} + 2HOAc.$$
$$\text{blue-green} \qquad\qquad \text{deep blue}$$

The absorbance will steadily increase because the copper chelate has a higher molar absorptivity.

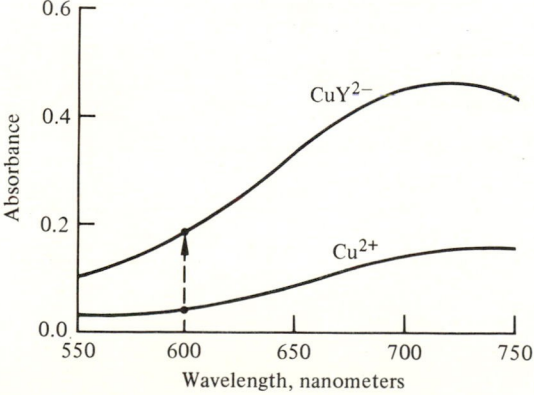

Fig. L.21 Spectra of $7.88 \cdot 10^{-3} M$ Cu^{2+} and $7.88 \cdot 10^{-3} M$ Cu^{2+} EDTA complex CuY^{2-} in pH = 5 acetic acid–acetate buffer. Cell thickness 1 cm.

When enough EDTA has been added to chelate all the copper in the sample, the absorbance will remain constant except for the effect of dilution upon addition of excess (and colorless) EDTA solution.

To interpret the data it is necessary to make a plot of absorbance values versus the volume of added EDTA. Each absorbance value must be corrected for the effect of dilution as follows:

$$A_{cor} = A_{obs} \frac{5 + V}{5},$$

where V is the volume of added EDTA, and 5 is the original volume of the sample.

To understand the nature of this plot we must consider the absorption spectra of Cu^{2+} and Cu–EDTA, as shown in Fig. L.21. Though we would prefer to measure absorbances at the wavelength where Cu–EDTA has maximum absorbance, the B & L Model 20 does not operate above 600 nm unless a special photocell is used. If such a cell is available, the measurements may be made at 720 nm, with the realization that the sample may have to be diluted to keep the final absorbance values in the desirable range, i.e., below $A = 0.5$.

Suppose we choose to work at 600 nm, either because we have only the standard photocell or to enable the titration of somewhat more concentrated solutions. The initial absorbance will correspond to the dot shown on the spectrum for Cu^{2+} and then it will rise to a value corresponding to the dot on the spectrum for CuY^{2-}. The absorbance will then remain constant, no matter how much excess EDTA is added.

Now suppose the unknown contained $0.0220 M$ copper after dilution to 100 mL in the volumetric flask. It will require 1.10 mL of $0.1000\ M$ EDTA to reach the titration endpoint because in the 5.00-mL sample there would be 0.110 mmol of copper. Suppose the EDTA is added in 0.1-mL portions with the microburet, for a total of 2.0 mL. The predicted titration curve would be as shown in Fig. L.22.

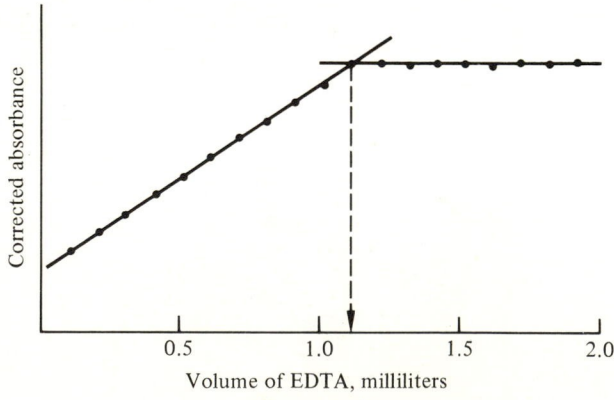

Corrected absorbance

Volume of EDTA, milliliters

Figure L.22

By drawing straight lines through the linear portions of the plot, we find the titration endpoint as the volume corresponding to their intersection. For titrations involving reactions with smaller equilibrium constants there may be curvature in the vicinity of the endpoint, and points deviating from the linear portions are ignored in drawing the lines.

PROCEDURE

1. Two days before performing this experiment give your lab instructor a clean (not necessarily dry) 100-mL volumetric flask. The unknown sample will be placed in it and returned to you. Dilute to the mark with the buffer (see below) and mix very thoroughly.

2. Prepare a standard EDTA solution as follows: Weigh accurately a sample of disodium EDTA dihydrate (MW = 372.24) of about 3.7 grams. Transfer quantitatively to a 100-mL volumetric flask and dilute to the mark with distilled water. This salt is somewhat slow to dissolve, so it is well to do this a day or so before the titration. Be sure the solution is thoroughly mixed before using it.

3. The buffer is prepared as follows: Dissolve 7 grams of sodium acetate in 100 mL of water and add 2 mL of glacial acetic acid. Stir to mix and use this to dilute the unknown sample to the mark in the volumetric flask.

4. Pipet precisely 5 mL of the diluted sample into a clean spectrophotometer cell and measure its absorbance at 600 nm. Your instructor will demonstrate the use of the microburet (Fig. L.23). Carry out the titration, making 0.1-mL

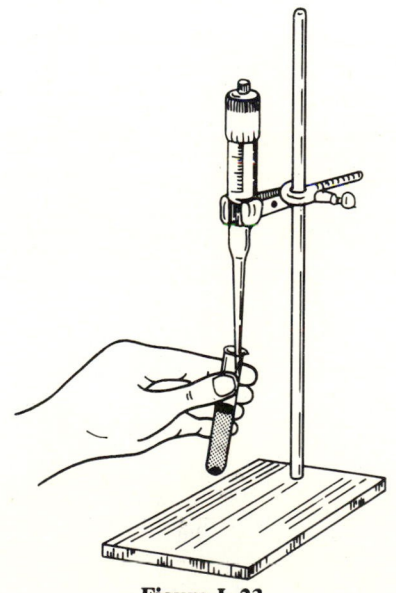

Figure L.23

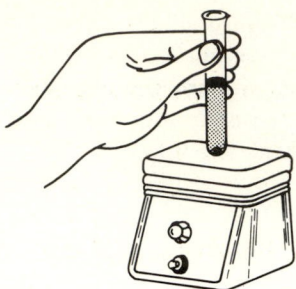

Figure L.24

additions of the EDTA. Mixing is accomplished by having a very small magnetic stirring bar in the cell (Fig. L.24). With each addition the tip of the microburet should be touched to the solution. With each absorbance measurement, you should verify the zero-absorbance setting by using a distilled-water cell. Record the volume added and the measured absorbance for each point.

Interpret the data as shown above, making a precise and careful plot. Report the total quantity (millimoles) of copper in the unknown. Also turn in your titration plot.

LABORATORY REPORT 6

PHOTOMETRIC TITRATION OF COPPER WITH EDTA

NAME _____   LAB SECTION _____

DATE _____

(Attach a graph showing the absorbance, volume data corrected for dilution)

Mass of EDTA used: _____ Molarity of EDTA solution: _____

Volume of EDTA solution required to reach endpoint (from graph): _____

Amount of copper (millimoles) in the total unknown sample:_____

LABORATORY EXPERIMENT 7

GRAVIMETRIC DETERMINATION OF LEAD CONTENT BY PRECIPITATION OF LEAD CHROMATE

PURPOSE

1. To use the standard techniques of gravimetric determination to find the precise amount of lead in a sample.

2. To illustrate the method of precipitation from homogeneous solution in comparison with direct precipitation.

MATERIALS NEEDED

Filter crucibles. Microscope (one for the class), preferably stereo.
$0.1M$ solutions of potassium dichromate and chromic nitrate
$0.2M$ potassium bromate
$1M$ sodium acetate

GENERAL COMMENTS

Refer to Chapter 6 for a discussion of the chemical reactions and general principles of this determination. The procedure as outlined below suggests that the direct precipitation and the PFHS be carried out at the same time, so that comparisons may be made easily. Both approaches are capable of high accuracy in the determination of lead content, and care should be taken in quantitative transfer, drying, and weighing.

The sample for analysis may be provided in various ways. If a solid material such as a lead salt is given, the instructor will provide suggestions on how to dissolve it. An unknown aqueous-solution sample may be diluted to a known volume in a standardized volumetric flask, and then portions may be taken for individual samples using a standardized pipet. To avoid the uncertainties connected with the volumetric glassware, the instructor may provide solutions of lead nitrate in plastic wash bottles. Students may take samples from these using the technique of weighing by difference, thus keeping the experiment entirely gravimetric. The number of samples to be run will be specified by the instructor.

FILTER CRUCIBLES

Prepare in advance and dry to constant weight at 110°.

DIRECT PRECIPITATION BY HETEROGENEOUS MIXING

Place the dissolved sample, which should contain 200–300 mg of Pb, in a 250-mL beaker. Add enough water to make the total volume about 50 mL. Heat the solution to near boiling (avoid actual boiling because of the possibility of loss by spattering) and, while stirring with a glass rod, add slowly about 10 mL of $0.1M$ potassium dichromate solution. Cover with a watch glass, leaving the stirring rod in the beaker, and heat with care so that the mixture boils very gently. Stir occasionally during this process of digestion and aging, and note that the yellow colloidal material gradually coagulates and settles out. When necessary, add a little water from a wash bottle to keep the solution volume from dropping below about 50 mL. When the precipitate has settled out, the supernatant solution should be clear (this does not mean colorless, but free of cloudiness). Its yellow-orange color should indicate that an excess of potassium dichromate has been added. If the solution is color- less, more precipitant is needed, perhaps because the amount of lead in the sample is greater than expected.

Let the mixture cool for about 30 minutes, and then filter through a previously dried and weighed filter crucible, taking care to make the transfer quantitative with the aid of a rubber policeman on a stirring rod. Wash the precipitate on the filter with two or three small portions of distilled water, and place the crucible(s) in a beaker for oven drying at about 110°C. Dry until constant weight is obtained.

PRECIPITATION FROM HOMOGENEOUS SOLUTION

Place the dissolved sample, which should contain 200–300 mg of Pb, in a 250-mL beaker. Add enough water to make the total volume about 30 mL. Add 20 mL of $0.1M$ chromic nitrate and 20 mL of $0.2M$ potassium bromate. Heat rather quickly to near boiling, and then reduce the heat so that very gentle boiling is sustained. During this stage make observations of the occurring changes. The gentle boiling should be continued, with occasional small additions of water from a wash bottle, until the supernatant solution is clear and orange, showing that an excess of chromate has been generated by the redox reaction.

This reaction generates hydrogen ions as well as chromate, and therefore the pH of the solution decreases considerably. Because of the formation of $HCrO_4^-$, the solubility of lead chromate is greater in the acid solution, and significant loss will occur unless the pH is raised. To accomplish this, add 10 mL of $1M$ sodium acetate, which forms an acetic acid–acetate ion buffer of $pH \approx 5$. At this point an additional amount of lead chromate may come out of solution, as indicated by cloudiness. Continue the gentle boiling until the solution is once again clear.

Let the mixture cool for about 30 minutes and then filter through a previously dried and weighed filter crucible, taking care to make a quan- titative transfer with the help of a rubber policeman. Record your obser- vations about the characteristics of this precipitate and the ease of transfer- ring it to the filter, in comparison with the precipitate from heterogeneous solution. Wash the precipitate on the filter with two or three small portions of water and place the crucible(s) in a beaker for oven drying at 110°C. Dry until constant weight is obtained.

CALCULATIONS

1. Balance the equations for the chemical reactions involved in the above procedures, and then answer the following questions:

   a) In the heterogeneous method, how many milligrams of Pb could the sample contain before there is no longer a 10% excess of precipitant (chromate ion)?

   b) In the homogeneous method, what is the percentage excess of bromate ion over that actually required by stoichiometry for the oxidation of all the Cr(III)?

   c) What amount, in millimoles, of H^+ is generated by the redox reaction?

   d) What percentage of the acetate ion from the sodium acetate is used up in reacting with the generated H^+ ion?

2. For each of the samples used, calculate the mass of Pb. Also calculate the percentage of Pb in the sample or the molarity of the sample solution, whichever is appropriate. Calculate the average value and relative average deviation in parts per thousand.

CLEANING THE CRUCIBLES

Never scrape a precipitate from a crucible, except with a rubber policeman, because the porous frit is fragile. First simply tap the inverted crucibles to shake out most of the precipitates. The instructor may request that these be saved in recovery bottles provided for that purpose. A portion of each precipitate may be examined with a low-power microscope, so that the vast differences in crystal forms will be easily visible.

After tapping, flush the crucibles under a stream of tap water. This will still leave some precipitate in the frit. Set the crucible in the vacuum filter assembly and, without turning on the vacuum, add a few milliliters of $6M$ hydrochloric acid. Let it seep through and use a rubber policeman to rub the acid on the insides of the crucible. The lead chromate will dissolve easily because of two reactions:

$$CrO_4^{2-} + H^+ = HCrO_4^-,$$
$$Pb^{2+} + 4Cl^- = PbCl_4^{2-}.$$

Draw the HCl through the filter, repeat with another small portion, and then draw distilled water (wash bottle) through until the frit of the crucible is white. Ideally, if the crucible is now dried again to constant weight, it would have the same empty weight as before, and the student may wish to check this.

LABORATORY REPORT 7

GRAVIMETRIC DETERMINATION OF LEAD BY PRECIPITATION AS LEAD CHROMATE

NAME _____   LAB SECTION _____

DATE _____

Description of sample and its treatment (Tell how it was dissolved or diluted, how the individual samples were measured, and give the standardized volumes of any pipets and volumetric flasks used)

Experimental data

	Sample 1	Sample 2	Sample 3	Sample 4
Sample size, grams of mL				
Method (hetero-geneous or homogeneous)				
Mass of crucible plus precipitate				
Mass of empty crucible				
Mass of $PbCrO_4$				

Calculated results (as specified by the instructor)

LABORATORY EXPERIMENT 8

SILVER-ELECTRODE STUDY OF EQUILIBRIA

In this experiment we will examine the interaction of silver ion with a variety of ligands by measuring the potential of a silver-metal electrode. The three equilibria are as follows:

1. Solubility product for silver chloride:

$$AgCl(s) = Ag^+ + Cl^-, \qquad Q_{sp} = [Ag^+][Cl^-].$$

2. Formation quotient for a silver-ammonia complex:

$$Ag^+ + p\,NH_3 = Ag(NH_3)_p^+, \qquad Q_f = \frac{[Ag(NH_3)_p^+]}{[Ag^+][NH_3]^p},$$

where p is unknown.

3. Solubility product for silver iodide:

$$AgI(s) = Ag^+ + I^-, \qquad Q_{sp} = [Ag^+][I^-].$$

In each part of the experiment the molarity of the ligand (Cl^-, NH_3, I^-) will be present in known excess over the quantity of silver present, and it will be assumed that the reactions *go to completion* to form $AgCl$, $Ag(NH_3)_p^+$, and AgI, respectively. Thus, the only unknown quantity in the above equilibrium expressions is the $[Ag^+]$, and this will be determined in each case by the electrode potential.

The main points of the procedure are as follows: A solution of $1.00 \cdot 10^{-4} M$ silver nitrate will be used to standardize the electrode, in much the same way as a reference buffer is used to standardize a pH-meter. Upon addition of a known and excess quantity of ammonium chloride, precipitation of $AgCl$ will occur and the electrode potential will register the drop in silver-ion concentration in the solution. Then, three additions of a solution of ammonia will be made, and the silver chloride will dissolve in favor of the more stable ammonia complex. Finally, upon addition of potassium iodide the silver–ammonia complex will be converted to a precipitate of silver iodide.

PRINCIPLES OF THE POTENTIAL MEASUREMENT

The reactions described above will be carried out in a beaker that serves as a compartment of a galvanic cell. A silver wire will serve as the indicator electrode for this half-cell, and the other half-cell will consist of a reference electrode (see later for details). The cell diagram may be written as follows:

| Reference electrode | Reaction vessel, with silver ion at various concentrations as the reactions are carried out | Ag |

When this cell is attached to a millivoltmeter, the readings are in accord with the Nernst equation for this cell:

$$E = k - 59.2 \log \frac{1}{[Ag^+]}$$

where

$$k = E^0_{Ag} - E_{ref} + E_{jnc} + 59.2 \log f_1.$$

The standard reduction potential for the silver electrode is $+799 \, mV$, while the liquid-junction potential is unknown but of the order of a few millivolts. Under the conditions described below (ionic strength $I = 0.010$), the activity coefficient for silver ion is $f_1 = 0.90$. The reduction potential for the reference electrode is constant, but naturally depends on the particular electrode used for this purpose. In any case, we may assume that the composite of these values, symbolized by k, is essentially constant throughout the experiment.

At the beginning of the experiment, the cell compartment contains only a solution of $1.00 \cdot 10^{-4}M$ $AgNO_3$ and $0.010M$ NH_4NO_3, the latter mainly to adjust the ionic strength. Let the observed potential of the cell be E_1. Since the molarity of silver ion is precisely known for this solution, it is easy to calculate the value of k:

$$k = E_1 + 59.2 \log \frac{1}{1 \cdot 10^{-4}} = E_1 + 237.$$

Once the value of k has been determined, the Nernst equation will allow calculation of the $[Ag^+]$ in each of the solutions to be examined.* By rearrangement we find:

$$[Ag^+] = 10^{(E-k)/59.2}.$$

PREPARATION OF THE REFERENCE ELECTRODE

Any available reference electrode may be used for this experiment, but it is vital that the solution under study not become contaminated with, for example, chloride ion, which would cause precipitation of silver ion. If a commercial SCE is used, it must be isolated from the silver solution by a potassium nitrate salt bridge (agar gel in U-tube). For instructional purposes it is preferable that the students make a simple reference electrode, an easy and "El Cheapo" version being as follows:

Insert a fine cotton thread into a 6-inch glass tube, wet a rubber septum (Aldrich Chemical Company, 7 mm size, Catalog No. Z100) with a drop of $0.1M$ copper sulfate, and affix it over the end of the tube so that the thread can serve as a slow wick. Fill the tube with $0.1M$ copper sulfate, insert a

---

*For a temperature t other than 25°, instead of 59.2 for the constant use the quantity $59.2(273 + t)/298$.

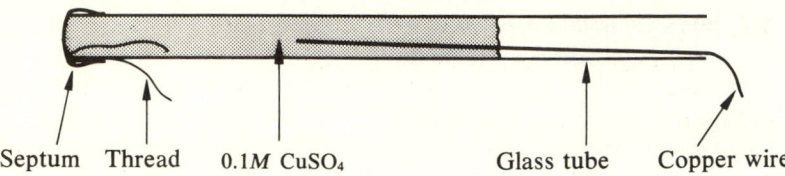

Septum   Thread   0.1M CuSO$_4$ Glass tube   Copper wire

Figure L.25

6-inch length of ordinary copper wire and, *voila*, a reference electrode with a potential about 42 mV higher than that of the SCE (Fig. L.25).

It won't work to use string instead of fine thread, for the leakage rate is then too large. Rinse the outside of the reference electrode thoroughly before inserting it into the silver solution.

PREPARATION OF THE SILVER-INDICATOR ELECTRODE

The silver wire used for the electrode must be first cleaned to remove any tarnish, or else it will not respond in accordance with the Nernst equation. A very simple cleaning method is to immerse the wire in near-boiling sodium bicarbonate solution and to touch it with a piece of aluminum metal. This creates a short-circuited galvanic cell, and the aluminum ($E^0 = -2.3$ V) readily reduces any silver tarnish. This method works well on the family sterling also, but it removes the tarnish from the grooves in the pattern and therefore may not appeal to some. The silver wires may be cleaned by immersion (brief) in 6M nitric acid, but this dissolves some of the wire.

THE SETUP OF THE EXPERIMENT

Mount the two electrodes in a rubber stopper and clamp, so that they may be properly immersed in the solution to be studied. A 250-mL beaker serves as the container, and it is very convenient to have a magnetic stirrer so that

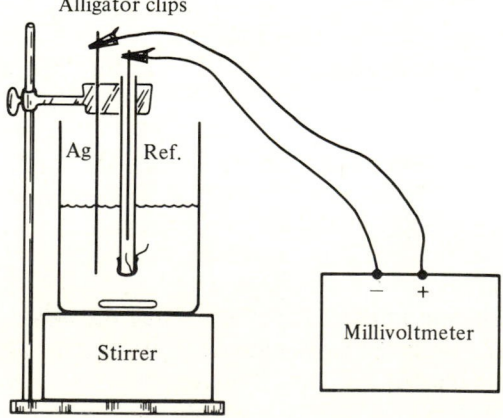

Figure L.26

mixing will be rapid and thorough when the various solutions are added to the cell.

The millivoltmeter may be any high-impedance voltmeter precise to 1 mV, and a regular pH-meter operated in "millivolt mode" is a good choice. At the beginning of the experiment, it is the silver electrode that will have the positive polarity (Fig. L.26).

EXPERIMENTAL PROCEDURE

Initial conditions   Into a 100-mL volumetric flask pipet 10 mL of the stock solution containing $0.001000M$ silver nitrate and $0.1000M$ ammonium nitrate. Dilute to the mark and mix thoroughly. If there is any cloudiness at this stage, your flask is not clean and you must start all over.

With the cell set up, pour *all* of this solution into the silver-electrode compartment. Volumetric flasks will not deliver their nominal volume because of incomplete drainage, but this error will not be serious and you may assume that 100 mL was delivered. Determine the temperature of the solution and use the corresponding value of the *Nernst constant* in your calculations.

Measure the cell voltage with the millivoltmeter, reading carefully to the nearest millivolt. Check this reading with the instructor before proceeding.

Effect of addition of chloride ion   With the magnetic stirrer going, pipet 10 mL of $0.0100M$ ammonium chloride into the cell. Note the formation of a precipitate and the immediate drop in cell voltage. Wait a few minutes for the precipitation to reach equilibrium and for the freshly formed colloidal precipitate to "age" a little. Then read the cell potential to the nearest millivolt. What would you predict for the primary adsorption layer on the surface of the silver chloride particles?

Effect of addition of ammonia   Pipet 10 mL of the stock solution containing ammonia (the instructor will give its molarity, which should be between 0.5 and 0.8) and $0.0100M$ ammonium nitrate. Note that the silver chloride redissolves and that the potential drops still further, and read it to the nearest millivolt. The silver–ammonia complex is more stable than silver chloride:

$$AgCl(s) + p\,NH_3 = Cl^- + Ag(NH_3)_p^+,$$

where p is the coordination number for the complex, to be calculated from your data.

Make two more 10-mL additions of the NH_3 solution and record the potentials.

Effect of addition of iodide ion   Pipet 10 mL of $0.0100M$ potassium-iodide solution into the cell. Note the formation of a precipitate of silver iodide and describe its color compared to that of silver chloride. What is the primary adsorbed layer in this case? Wait a few minutes for the precipitate to age before measuring the potential.

CALCULATIONS (*Note*: Throughout the experiment the ionic strength has been 0.010)

Initial conditions   Calculate k as discussed earlier in this writeup. Taking $+799\,\text{m}V$ for the standard reduction potential of the silver ion–silver metal couple, calculate the potential of your reference electrode, assuming that E_{jnc} is negligible.

Solubility product for silver chloride   When the ammonium chloride was added, the chloride ion was in considerable excess over the silver ions present, and one may assume that precipitation was virtually complete. Using the *excess*-chloride molarity and the silver-ion molarity calculated from the observed potential, calculate the solubility product:

$$K_{\text{sp}} = [\text{Ag}^+][\text{Cl}^-] \cdot f_1^2$$

where f_1 should correspond to the ionic strength of the solution. Compare your value of K_{sp} with the literature value(s) found in your text or other sources. *Note*: Finely divided or colloidal precipitates often show a higher solubility than "fully aged" precipitates. Do your results support this?

Formula and formation constant of the silver–ammonia complex   The equilibrium constant for the reaction

$$\text{Ag}^+ + p\,\text{NH}_3 = \text{Ag}(\text{NH}_3)_p^+$$

is written as

$$K_{\text{f}} = \frac{[\text{Ag}(\text{NH}_3)_p^+]}{[\text{Ag}^+][\text{NH}_3]^p}$$

where the activity coefficients cancel. It is assumed that the activity coefficient of the uncharged ammonia molecule is unity.

   The molarity of the complex can be calculated simply by taking account of the dilution of the cell solution, assuming that virtually all the silver is converted to the complex ion:

$$[\text{Ag}(\text{NH}_3)_p^+] = 0.0001000 \cdot \frac{100}{100 + V},$$

where V is the cumulative volume of the reagents added to the initial 100 mL. The molarity of silver ion is readily calculated as outlined earlier. The equilibrium molarity of ammonia may be calculated simply by considering dilution of the concentrated stock solution:

$$[\text{NH}_3] = M_{\text{stock}} \cdot \frac{V_{\text{NH}_3}}{100 + V},$$

assuming that the small amount used to form the silver complex is negligible. (Justify this assumption by a later calculation, once you have determined the value of p.)

However, the formation constant cannot be calculated directly because p is as yet unknown. First it is necessary to determine p by making a plot. The expression for K_f is rearranged and put into logarithmic form:

$$Y \equiv \log \frac{[Ag(NH_3)_p^+]}{[Ag^+]} = \log K_f + p \log [NH_3].$$

This suggests that a plot of Y versus $\log [NH_3]$ should be linear, with a slope of p and an intercept of $\log K_f$. Make such a plot with your three data points and determine the value of p from the slope. The formula for the complex is thus deduced, by rounding the experimental value of p to the nearest integer.

Although it is possible in principle to find the value of K_f from the intercept of the plot just described, the extrapolation is rather long and the accuracy of K_f thus obtained would be quite sensitive to errors in the data points. It is better to assume that the value of p is the integer obtained by rounding and then to calculate individual K_f values for the three data points by using the last equation.

Average the three K_f values and find the relative average deviation, which is a measure of the precision of your work. Compare your result with that found in the text or other sources.

Solubility product for silver iodide   This calculation is identical in principle to the approach used for silver chloride. Calculate the solubility product and compare with the literature value. Is there evidence for higher solubility of a fresh precipitate?

LABORATORY REPORT 8

SILVER-ELECTRODE STUDY OF EQUILIBRIA

NAME _____   LAB SECTION _____

DATE _____

Condition	Total volume	Ligand molarity*	E_{cell}, mV	$[Ag^+]$	Y	log $[NH_3]$
Start	100					
Cl⁻ added	110					
NH₃ added	120					
NH₃ added	130					
NH₃ added	140					
I⁻ added	150					

Calculated value of k: _____ mV Calculated E_{ref} _____ mV

Temperature of solution: _____ °C   Nernst constant: _____

Molarity of NH₃ stock: _____

Calculated constants

K_{sp} for AgCl: _____

p for $Ag(NH_3)_p^+$: _____

K_f for $Ag(NH_3)_p^+$: _____

K_{sp} for AgI: _____

---

*The concentration of *excess* chloride, or ammonia, or iodide after reaction with silver, taking dilution into account.

LABORATORY EXPERIMENT 9

A SEQUENCE OF ACID–BASE TITRATIONS

PURPOSE

To give experience in three types of acid–base titrations: weak acid with NaOH, strong acid with NaOH, and weak base with strong acid. In terms of laboratory technique the goal is to strive for good accuracy in spite of a steady cumulation of experimental uncertainties.

Primary standard potassium hydrogen phthalate will be used to standardize a solution of sodium hydroxide, which then will be used to standardize a solution of hydrochloric acid. The sodium hydroxide solution may be called a secondary standard and the hydrochloric acid solution is then a tertiary standard. The hydrochloric acid solution will finally be used to determine the molarity of a solution of ammonia.

Before doing this experiment it is important to have studied Chapter 10 on acid–base titrations, and to have read the section entitled PREPARATION AND STANDARDIZATION OF STRONG ACID AND STRONG BASE SOLUTIONS (p. 344). In the general section of the laboratory material the important parts are: Buret, Cleaning Glassware, Distilled Water, Pipets. Titrations Using Indicators, Volumetric Flasks, and Weighing.

MATERIALS NEEDED

The instructor will provide stock solutions of approximate molarities as follows: $1M$ NaOH, $0.8M$ HCl, and $0.1M$ NH_3. The student should use standardized glassware: 25-mL pipet, 250-mL volumetric flask, 50-mL buret. The following indicators are required: bromcresol green, phenolphthalein, bromcresol purple. The potassium hydrogen phthalate should be primary standard grade dried at 110°.

PROCEDURE

In the following directions we have deliberately abstained from giving complete detail, because the student should make use of the above-mentioned background reading.

1. Prepare 250 mL of $0.1M$ NaOH by *accurate* dilution of the $1M$ stock solution. Use a clean dry pipet to avoid any chance of contamination of the stock solution. Transfer this solution to a plastic bottle that has been rinsed with a small portion of this solution.

2. Weigh, by difference from a weighing bottle, two samples of potassium hydrogen phthalate into titration flasks. These samples should be sufficient to require 30–40 mL of the $0.1M$ NaOH for titration. Add about 25 mL of water to each and titrate with the NaOH, using an appropriate indicator. From the titration results calculate the precise molarity of the $0.1M$ NaOH and, using standardized glassware volumes, calculate the precise molarity of the $1M$ stock solution.

3. Prepare 250 mL of $0.08M$ HCl by accurate dilution of the $0.8M$ stock solution. Again use a clean dry pipet and the same volumetric flask as was used for the NaOH dilution.

4. Pipet 25-mL portions of the $0.08M$ HCl into titration flasks. Add an appropriate indicator and titrate with the $0.1M$ NaOH. Use the average endpoint volume to find the precise molarities of the $0.08M$ and the $0.8M$ HCl solutions.

5. Replace the NaOH solution in the buret with the $0.08M$ HCl. Use a 10-mL pipet to place a portion of the $0.1M$ NH$_3$ into a titration flask which already contains about 25 mL of water. Cautiously sniff the flask. The fact that you can smell ammonia shows that there is a steady loss by volatilization. Add an appropriate indicator, and titrate this sample with the HCl. The only purpose of this titration is to give a fairly close idea of where the endpoint of the more accurate titration will be.

6. Knowing that the ammonia-solution sample will be 25 mL, use the buret to add enough HCl to a titration flask so that it will be nearly enough to react with the ammonia. Then pipet 25 mL of the ammonia solution into the flask. In this way the loss of NH$_3$ by volatilization is minimized, since most of it is immediately protonated. Then finish the titration, after adding indicator, with HCl from the buret. Repeat the determination. Calculate, using the average endpoint volume, the precise molarity of the NH$_3$ solution.

7. Report the precise molarities of the $0.1M$ NaOH, $1M$ NaOH, $0.08M$ HCl, $0.8M$ HCl, and $0.1M$ NH$_3$. Use standardized glassware volumes in all calculations. Assuming that the uncertainties of the pipet and volumetric flask are 0.01 and 0.1 mL, respectively, and that the titration endpoints have an uncertainty of 0.04 mL, estimate the confidence limits to be placed on each of the five molarities.

Note to instructor: An alternative sequence of titrations is as follows: Supply stock solutions of $0.8M$ HCl, $1M$ NaOH, and $0.1M$ acetic acid. First prepare $0.08M$ HCl by accurate dilution and standardize this with respect to the primary standard tris(hydroxymethyl)aminomethane. Then prepare $0.1M$ NaOH by accurate dilution and standardize it by titrating the $0.08M$ HCl. Finally, use the $0.1M$ NaOH to titrate the $0.1M$ acetic acid.

LABORATORY REPORT 9

A SEQUENCE OF ACID–BASE TITRATIONS

NAME _____   LAB SECTION _____

DATE _____

Volume of standardized 25-mL pipet: _____

Volume of standardized 250-mL flask: _____

Titration of KHP

	Mass of KHP	Corrected volume of NaOH
Trial 1		
Trial 2		

Titration of HCl

    Corrected volume of NaOH (1) _____ (2) _____   Average: _____

Titration of NH_3

    Corrected volume of HCl (1) _____ (2) _____   Average: _____

Summary of results (the estimates of uncertainty should be based on the assumptions stated in the lab directions):

Solution	Precise molarity	Uncertainty
0.1M NaOH		
1M NaOH		
0.08M HCl		
0.8M HCl		
0.1M NH_3		

LABORATORY EXPERIMENT 10

CHARACTERIZATION OF AN UNKNOWN ACID BY
POTENTIOMETRIC pH TITRATION

PURPOSE

1. To illustrate the standardization and application of the glass electrode for pH determination.
2. To illustrate the technique of potentiometric titration.
3. To apply principles of stoichiometry and acid–base equilibria.
4. To determine the molecular weight and the pK value(s) of an acid.

GENERAL COMMENTS

Although it is very convenient to use a colored acid–base indicator for the detection of a titration endpoint, this technique can be used only when the pK value(s) of the titrated substance are known. Otherwise it is not possible to make a confident choice of indicator, since the equivalence-point pH is a function of the acid–base properties of the material. All an indicator can do is to change color in a certain pH range; it can give no information about the variation of pH during the earlier part of the titration. Yet this is the part of the titration that is most informative about the pK value(s) of the substance.

By contrast, when a pH-meter with a standardized glass-indicator electrode is used to follow the solution pH as a function of the volume of titrant added, we may obtain complete information in a reasonably short time. The theoretical calculation of titration curves is discussed in Chapter 10, and the interpretation of experimental titration curves is covered in Chapter 11.

In this experiment we shall use a solid sample of an acid or a solution containing a known mass of an acid. The acid may be monoprotic or diprotic and, in the latter case, the pK values may be close to each other or quite different.

By using a pH-meter to titrate a known mass of the acid with a standard sodium hydroxide solution, it will be possible to find both the molecular weight (by noting the precise endpoint on the curve) and the pK values (by using pH values taken from the buffer regions for calculations).

PREPARATION OF THE ACID SOLUTION

If the sample of unknown acid is a solid, weigh accurately about 0.5 gram and transfer it to a 250-mL volumetric flask. Dissolve and dilute to the mark, and mix well. If the sample is provided as a solution, use an appropriate pipet to measure a portion which contains approximately 0.5 gram and dilute to 250 mL in a volumetric flask.

STANDARDIZATION OF THE pH-METER

The instructor will provide operating guidelines for the particular pH-meter to be used. The instrument will be standardized by using reference buffer solutions of pH = 7 and pH = 4.

EQUIPMENT SETUP FOR POTENTIOMETRIC TITRATION (see Fig. L.27)

Notes: (a) The electrodes will be supported by a clamp, and care should be taken to keep their tips away from the stirring bar. (b) Until it is desired to add and count drops, the buret tip may touch the solution to ensure complete delivery without rinsing.

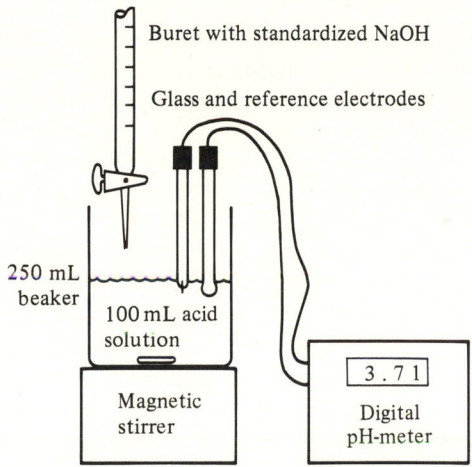

Figure L.27

PERFORMANCE OF THE TITRATION (a standard solution of $0.1M$ NaOH is required)

For this experiment it is important to have a well-defined titration curve with several points in the buffer region(s) and with points closely spaced in the endpoint region(s). Since the acid is treated as an unknown, it is not possible to predict the endpoint. Therefore it is wise to do the titration twice: the first time can be rather casual, with the goal of merely finding out the general shape of the curve and the approximate locations of the endpoint. Then the titration is repeated with a fresh portion of acid solution, using larger (say, 1 mL) increments of NaOH through the buffer region and smaller (say, 0.1 mL) increments to define the endpoint.

    Into a 250-mL beaker pipet 100 mL of the diluted acid solution and assemble the equipment as shown in the diagram. Read the initial pH and prepare to plot the observed pH values obtained during the titration. The data will also be recorded in a table in your notebook, but the simultaneous plot will be very helpful in describing the curve. For the first titration add the NaOH in 1–2 mL increments until the pH-meter shows, by temporary jumps to high pH, that the endpoint is close. Then add smaller increments to find the endpoint without concern for high accuracy. From the plot of the results note whether the acid seems to be monoprotic or diprotic with two distinct endpoint regions. A diprotic acid with pK values which are within one or two log units will not be distinguishable from a monoprotic acid at this point.

Rinse the beaker and electrodes with distilled water and check the pH-meter standardization with the reference buffers. Again rinse the electrodes and repeat the titration with a second 100-mL portion of the acid solution. In this titration take accurate readings of both pH and volume, being sure to obtain several points in each buffer region. When approaching an endpoint region with a rapidly changing pH, use the drop-counting technique: take an accurate buret reading and raise the buret tip above the solution. Then make additions of one or two drops at a time, recording the number of drops added with the corresponding pH readings. After the endpoint has been passed, make another accurate buret reading. The difference in the two buret readings divided by the total number of drops gives the precise average volume per drop. Later, when making an expanded plot of the endpoint region, it will be convenient to plot pH versus the number of drops. With this technique it is possible to define the endpoint region with great precision.

When the titration has been completed, rinse the electrodes with water and return them to their storage vessel containing a pH = 7 buffer. Rinse the buret very thoroughly to remove traces of NaOH which otherwise may attack the glass.

PLOTTING THE TITRATION CURVE

Two plots are desired. One is a plot of the entire titration curve, and the other is an expanded plot of the endpoint region. If the acid shows two endpoints, each should be shown on an expanded plot.

Use the expanded plot(s) to deduce the precise endpoint(s) and calculate the molecular weight of the acid from stoichiometric considerations. A problem arises in the case of a diprotic acid which does not show an inflection point at the first equivalence point. To decide whether a titration curve is for this case or for a monoprotic acid, a simple test may be used: for monoprotic acid the change of pH between 25% and 75% titrated will be close to 0.95 units. For a diprotic acid the change of pH over this range will be close to $pK_2 - pK_1$, and this difference will nearly always be greater than 1.0.

DEDUCTION OF pK VALUE(S)

If the acid is diprotic, it will be necessary to solve simultaneous equations as discussed in Chapter 11. The fundamental relationship for the titration curve is as follows:

$$\frac{L \cdot [H^+]f_1 \cdot f_{HA}}{(A - L)f_{H_2A}} = K_1 + K_1K_2 \frac{(2A - L)f_{HA}}{(A - L)[H^+]f_1 \cdot f_A},$$

where $A = C_a + C_b$. The equation may be symbolized as follows (see Eq. (11.16)):

$$Y = K_1 + K_1K_2 \cdot X.$$

Calculate values for Y and for X, using one point from each buffer region, and solve the equations to find K_1 and K_2. The instructor may provide a computer program as a check on your calculations.

If the acid is monoprotic, the value for K_2 is zero, and the above equation is used with the omission of the term containing K_2.

THEORETICAL TITRATION CURVE FOR THE ACID

Once the pK values have been determined, the instructor may provide a computer program to calculate the theoretical curve for your titration. Data from this program should be plotted on the same graph paper as the experimental curve, for comparison.

LABORATORY REPORT 10

**CHARACTERIZATION OF AN UNKNOWN ACID BY
POTENTIOMETRIC pH TITRATION**

NAME _____   LAB SECTION _____

DATE _____

Instructor's code number for sample: _____

Mass of sample used to make 250 mL of solution: _____ gram

Molarity of sodium hydroxide solution: _____ M

Volume of NaOH for the first endpoint: _____ mL;

for the second endpoint: _____ mL.

The acid is   (   ) monoprotic

(   ) diprotic

Molecular weight: _____

Molarity of the acid in the initial solution: _____ M

Calculation of pK values

Parameter	First buffer region	Second buffer region
pH		
V_{NaOH}		
A		
$[\text{Na}^+]$		
$[\text{H}^+]$		
L		
$A - L$		
$2A - L$		
I		
Y		
X		
K values	$K_1 =$	$K_2 =$
pK values		

LABORATORY EXPERIMENT 11

PRECISE DETERMINATION OF Q_1 FOR AN AMINO ACID

Glycine is the simplest amino acid and its formula is H_2NCH_2COOH. In aqueous solution, glycine exists chiefly as a zwitter ion in equilibrium with its protonated cation and the deprotonated anion:

$$^+H_3NCH_2COOH = H^+ + {}^+H_3NCH_2COO^-, \qquad Q_1 = \frac{[H^+][HGly]}{[H_2Gly^+]},$$
$$ H_2Gly^+ HGly$$

$$^+H_3NCH_2COO^- = H^+ + H_2NCH_2COO^-, \qquad Q_2 = \frac{[H^+][Gly^-]}{[HGly]}.$$
$$ HGly Gly^-$$

The object of this experiment is to determine the value for Q_1 under conditions of constant ionic strength I. To accomplish this, a glass pH-responsive electrode will be calibrated in solutions of known concentration of hydrochloric acid and will then be used to determine the $[H^+]$ in buffer solutions containing known amounts of glycine and hydrochloric acid. All solutions will have the same ionic strength, so that activity coefficients may be presumed constant.

THEORETICAL BASIS FOR THE EXPERIMENT

The galvanic cell comprises a reference electrode (silver wire coated lightly with silver chloride) and a glass pH-responsive electrode immersed in solutions containing constant concentrations of chloride ion and stable concentrations of hydrogen ion:

$$\text{Glass electrode} \,|\, \text{Cell solution} \,|\, \text{Ag(s), AgCl(s).}$$

The potential of the glass electrode varies only with $[H^+]$, and the potential of the reference electrode is constant because it depends only on the $[Cl^-]$, assuming activity coefficients are fixed by the constant ionic strength. Therefore the Nernst equation for the cell can be written in a simple form

$$E = E^* - k \cdot \log [H^+], \qquad (L11.1)$$

where E^* is the formal potential for the cell including the standard potentials for the two electrodes, the activity coefficients for the H^+ and Cl^- ions, and the constant concentration of chloride ion. The coefficient k has the theoretical value of $59.17\,mV$ for a temperature of $25°C$. However, glass electrodes typically show a different response (see the discussion of electromotive efficiency in Chapter 11), and it is necessary to determine the value of k for any individual electrode. This can be done by measurements on solutions having known values of $[H^+]$.

Once the characteristic values for E^* and k have been determined for a particular electrode pair, the cell may be used to find $[H^+]$ for other solutions.

Equation (L11.1) is easily rearranged to

$$[H^+] = 10^{(E^*-E)/k},$$ (L11.2)

where E is the potential measured for the solution in the cell.

PREPARATION OF SOLUTIONS FOR THE EXPERIMENT

Two solutions of hydrochloric acid are needed for the determination of E^* and k, and three glycine buffer solutions will be used for the determination of Q_1. Each buffer solution will have a different ratio of $[H_2Gly^+]/[HGly]$.
   Three stock solutions are needed with the following compositions:

- A standardized solution of hydrochloric acid, approximately $0.1M$.

- A standard solution of sodium chloride, prepared by weighing an accurate mass of the pure salt and then dissolving and diluting it to a known volume. The concentration of this solution may be from $0.2M$ to $2M$, depending on the ionic strength chosen for the experiment.

- A standard solution of glycine, prepared by weighing an accurate mass of the recrystallized material and then dissolving and diluting it to a known volume. The concentration should be approximately $0.3M$.

   The following table lists the amounts of HCl, NaCl, and glycine in terms of millimoles to be taken from the stock solutions by using burets and/or graduated pipets as accurately as possible. These amounts are added to a clean 100-mL volumetric flask, distilled water is added to the mark, and the solution is thoroughly mixed before being transferred to the galvanic-cell vessel. The cell vessel may be a beaker, if the measurements are to be done at room temperature, or a large test tube, if the solutions will be placed in a controlled-temperature water bath. The amount of NaCl in each case is chosen to adjust the ionic strength I to the desired value: $n_{NaCl} = 100I - n_{HCl}$. The lowest value for I is 0.02.

Solution	$[H^+]$	$\dfrac{C_{H_2Gly}}{C_{HGly}}$	Amounts to be used, mmol		
			n_{HCl}	n_{NaCl}	n_{Gly}
A	$1.00 \cdot 10^{-2}$		1.000	$100I - 1$	
B	$1.00 \cdot 10^{-3}$		0.100	$100I - 0.1$	
C	?	$2:1$	2.000	$100I - 2$	3.000
D	?	$1:1$	2.000	$100I - 2$	4.000
E	?	$1:2$	2.000	$100I - 2$	6.000

PREPARATION OF THE REFERENCE ELECTRODE

A silver wire 12–15 cm long is first cleaned by *brief* immersion in $6M$ nitric acid and then rinsed with water. It is then made the anode in a simple electrolysis cell where a copper wire is used as cathode: a 250-mL beaker is filled with $1M$ HCl and a power supply is used to pass a current of 50–100 mA through the circuit for 30–60 seconds. The silver wire will turn dark as silver chloride forms on its surface. Only the lower half need be coated, and then the wire is placed in a beaker of water until it is to be used in the cell. Do not let it dry out.

PREPARATION OF THE GLASS ELECTRODE

The response of the glass electrode will depend upon its age and surface condition. Often it is possible to improve a sluggish glass electrode by letting it stand in $0.1M$ HCl for a few minutes. It is possible to perform this experiment using a combination glass/reference-electrode assembly rather than the silver-wire electrode described above. However, there will be an uncertainty due to the liquid-junction potential between the cell solution and the filling solution in the reference electrode. The use of the silver/silver-chloride wire eliminates this problem.

DETERMINATION OF E^* AND k

Connect the two electrodes to a pH-meter in millivolt mode, preferably one capable of 0.1 mV precision. Immerse the *rinsed* and *tissue-patted-dry* electrodes into solution A. Wait for the potential reading to stabilize and record its value, which we will call E_A. Then rinse and pat-dry the electrodes and immerse them into solution B, recording the potential E_B. Repeat these measurements two more times and find the average of the results.

To find the value for the coefficient k of the cell response to changing $[H^+]$, simply use Eq. (L11.1) to set up two simultaneous equations:

$$E_A = E^* - k \cdot \log 0.01 \ = E^* + 2k,$$
$$E_B = E^* - k \cdot \log 0.001 = E^* + 3k.$$

We see that $k = E_B - E_A$ if the hydrochloric-acid concentrations are prepared as specified earlier. It then follows that $E^* = E_A - 2k$ or $E^* = E_B - 3k$. This completes the standardization of the galvanic cell.

MEASUREMENT OF $[H^+]$ IN THE GLYCINE BUFFER SOLUTIONS

Immerse the electrode pair successively in each of the three glycine solutions, being careful to rinse and pat-dry before changing solutions. Repeat the potential measurements two more times and find the average potentials E_C, E_D, and E_E. For each solution, calculate the hydrogen-ion concentration using Eq. (L11.2).

Store the electrode pair as specified by the lab instructor. Remember that the glass electrode is both expensive and fragile.

CALCULATION OF Q_1

In calculating the values for the dissociation quotient it is necessary to take into account the incomplete reaction between H^+ and HGly. For example, in solution C the stoichiometric (or nominal) concentrations of H_2Gly^+ and HGly are 0.0200 and 0.0100 mol/L, respectively. However the actual equilibrium concentrations are given by the relationships (see the Charlot equation, Chapter 9):

$$[H_2Gly^+] = 0.0200 - [H^+],$$
$$[HGly] = 0.0100 + [H^+],$$

and these corrected values must be used to calculate Q_1.

Report the average of the Q_1 values obtained for the three buffer solutions.

VARIATIONS ON THE EXPERIMENT

Other amino acids (alanine, for example) may be used. Different values of ionic strength may be used by different students, so that the assembled class results will show any trend in the effect of ionic strength. A plot of pQ_1 versus I may be extrapolated to find pK_1 at $I = 0$. If the experiment is carried out over a range of temperatures, it may be possible to infer a value for ΔH^* (see Chapter 7).

LABORATORY REPORT 11

PRECISE DETERMINATION OF Q_1 FOR AN AMINO ACID

NAME _____   LAB SECTION _____

DATE _____

Name and formula of amino acid: _____

Actual concentration of stock solutions

    Hydrochloric acid: _____ Sodium chloride: _____ Amino acid: _____

Volumes of stock solutions used to prepare the five solutions:

Solution	HCl, mL	NaCl, mL	Amino acid, mL
A			
B			
C			
D			
E			

Ionic strength of the solutions: _____

Temperature of solutions at time of measurement of potential: _____°C

Observed potentials (average values), mV

    A: _____ B: _____ C: _____ D: _____ E:_____

Calculated results:

    Value of E^*, mV: _____ Value of k: _____

    [H^+] values

        A: _____ B: _____ C: _____ D: _____ E: _____

    Q_1 values

        C: _____ D: _____ E: _____ Average:_____

LABORATORY EXPERIMENT 12

POTASSIUM DICHROMATE AS A TITRANT AND DETERMINATION OF IRON IN AN ORE

There are several good procedures for the determination of the percentage of iron in samples of iron ore. The present experiment deals with the analysis of hematite, which is largely ferric oxide Fe_2O_3, by using potassium dichromate solution to titrate the iron after it is reduced to the ferrous state. This method is widely used because the titrant can be prepared as a primary standard and because the titration endpoint is particularly sharp and accurate.

The theoretical chemical basis for this experiment is presented in detail as a Case Study in Section 12.8, and the student should read it carefully before performing the experiment. The purpose of the experiment is to illustrate the importance of sample treatment, using several inorganic reactions that are important in their own right, and also to make an accurate determination of the percentage of iron, expressed in terms of Fe_2O_3, in a typical ore sample.

MATERIALS NEEDED

Solid potassium dichromate, primary standard quality

Concentrated hydrochloric acid

$0.25M$ stannous chloride in $1M$ HCl

$0.2M$ mercuric chloride

$3M$ sulfuric acid

Concentrated phosphoric acid

Indicator: sodium diphenylamine sulfonate

EXPERIMENTAL PROCEDURE

1. Prepare a standard potassium dichromate solution. Place 2.5 g of $K_2Cr_2O_7$ in a weighing bottle and dry for 2 hours at 110°. After cooling in a desiccator, weigh the sample by difference into a 500-mL volumetric flask. Dilute to the mark and mix thoroughly. Calculate the molarity of the solution to the nearest part per thousand.

2. Weigh the iron-ore samples. Place about 3 grams of sample in a weighing bottle and dry for 2 hours at 110°. After cooling in a desiccator, weigh by difference three samples of about 0.8 gram each into 500-mL erlenmeyer flasks.

3. Dissolve the samples. To each sample add about 10 mL of concentrated hydrochloric acid. These may be allowed to stand until the next lab period, if they are covered with watch glasses and placed in the hood. Otherwise, use the hot plates in the hood to heat the solutions to hasten solution of the ore. Do not allow the solutions to evaporate to dryness, and add more HCl as necessary. When solution is complete, there may be a small amount of white or gray residue. If the residue is dark add a few drops of the stannous chloride solution and continue heating.

From this point on the samples should be treated one at a time.

4. Reduce the iron with stannous chloride. To the hot hydrochloric acid solution of the ore, which will be yellow due to the ferric chloro-complex, add stannous chloride dropwise until the yellow color has just disappeared. In effect, this is a crude titration. Then add *one* drop of stannous chloride in excess. The solution may have a very faint greenish color.

5. Remove the excess stannous ion with mercuric chloride. Cool the solution to room temperature by swirling the flask under the tap. Add, all at once with immediate swirling, 10 mL of mercuric chloride solution. A small amount of mercurous chloride should appear as a white silky precipitate. If no precipitate forms, this indicates that there is no excess of stannous ion, and the sample must be rejected. If the precipitate is gray, it indicates that metallic mercury, finely divided, has formed due to the presence of too great an excess of stannous ion. Again it is necessary to reject the sample because the mercury will consume some of the titrant.

6. Titrate the sample with dichromate solution as follows: dilute the solution to a volume of about 250 mL, add 10 mL of $3M$ sulfuric acid, 5 mL of concentrated phosphoric acid, and 8–10 drops of indicator. Titrate slowly with the potassium dichromate solution until the green color (due to the chromium(III) formed in the titration reaction) changes to a gray-green. Then titrate with split drops until the indicator changes to purple. The color change is very sharp and will occur with only 0.01 mL excess of the dichromate.

Treat the other two samples in the same way.

CALCULATIONS

Calculate the weight of Fe in each sample, remembering that the stoichiometry is:

$$6Fe^{2+} + Cr_2O_7^{2-} = \text{Products.}$$

Report the percentage of iron, expressed in terms of Fe_2O_3, in each sample. Give the average and the relative average deviation (in parts per thousand).

LABORATORY REPORT 12

DETERMINATION OF IRON IN AN ORE USING DICHROMATE TITRATION

NAME _____   LAB SECTION _____

DATE _____

Weight of $K_2Cr_2O_7$ used to prepare standard solution: _____ gram

Molarity of the solution: _____

	Sample 1	Sample 2	Sample 3
Sample weight Volume of titrant Weight of Fe in the sample Fe_2O_3 in sample, %			

Average % Fe_2O_3: _____

Average deviation (ppt): _____

LABORATORY EXPERIMENT 13

HIGH-PRECISION ASSAY OF MOHR'S SALT BY GRAVIMETRIC TITRATION

PURPOSE

Ferrous ammonium sulfate 6-hydrate (Mohr's salt) can serve well as a reference-standard reducing agent (see Chapter 12) provided the material to be used has a precisely known purity. The commercial reagent cannot be classified as a true primary standard because it cannot be dried to a definite composition. Therefore it must be used *as is* with two or three parts per thousand uncertainty about its water content. But if a given batch is kept in a tightly stoppered bottle, it is fair to assume that it will retain its precise composition. Therefore, by making a very accurate assay it is possible to certify the composition so that the salt can be used as a practical standard in chemical analysis.

GENERAL COMMENTS

In routine chemical analysis we commonly accept uncertainties of 2–5 ppt in the results. This is because of errors in sampling, volumetric glassware calibrations, titration endpoint errors, reagent impurities. less-than-perfect stoichiometry, etc. It usually requires rather special effort to bring the uncertainty down to the nominal value of a mere 1 ppt. In this experiment you will attempt to break the 1 ppt barrier and to determine the purity of a sample of Mohr's salt to within a few parts in 10000.

The key chemical assumptions in this assay are (a) that the potassium dichromate used as titrant is 100.00% pure or is at least of precisely known purity and (b) that the stoichiometry of the iron(II)–dichromate reaction is exactly 6:1. The reactions and the analytical basis for this titration are thoroughly discussed in Chapter 12 in connection with the determination of iron in an ore.

To avoid the myriad errors associated with the use of volumetric glassware, the titration will be done entirely on a mass basis (see TITRATIONS USING WEIGHT BURETS for a discussion of the technique and the important precautions).

MATERIALS NEEDED

Certified potassium dichromate, dried at 120°. For example, the J. T. Baker Chemical Co. offers a primary standard product with a purity between 99.95 and 100.05%, with the actual-lot analysis reported on the label.

Ferrous ammonium sulfate solution, 0.1M. Dissolve 4 grams of Mohr's salt in 100 mL of 1M sulfuric acid.

Reagent-quality ferrous ammonium sulfate 6-hydrate. Alternatively, any other ferrous salt may be used.

Barium (or sodium)-diphenylaminesulfonate indicator solution, prepared by dissolving 0.1 gram of the salt in 50 mL of water. Warm to hasten solution.

Sulfuric acid ($1M$) and phosphoric acid (85%)

Plastic wash bottle to serve as weight buret (e.g., Nalgene Catalog No. 2402–0125) fitted with a medicine dropper in a rubber stopper.

Magnetic stirrer (convenient, but optional, as the solution may simply be stirred with a glass rod)

PREPARATION OF THE POTASSIUM DICHROMATE TITRANT SOLUTION

Unless the instructor provides this solution for class use, proceed as follows: Use an analytical balance to weigh 0.7 gram of pure dry $K_2Cr_2O_7$ to the nearest 0.1 mg, thus keeping the uncertainty at about one part in 10000. Transfer this without the slightest loss to a 250-mL bottle or erlenmeyer flask. This bottle must have been previously thoroughly cleaned, dried, and weighed on a top-loader balance to the nearest 0.01 gram. Add about 100 mL of distilled water and swirl gently until the salt has dissolved. Then add another 100 mL of water (try to avoid wetting the neck, and dry with a tissue if necessary). Stopper the bottle, reweigh to the nearest 0.01 gram, and then mix very thoroughly.

This provides a solution of a very precise composition by mass, for example, 0.7026 gram of potassium dichromate per 202.73 grams of solution. In the equation given below the mass of the salt is M_1 and the mass of the solution is M_2. These are apparent masses; they will be corrected for the air-buoyancy effect (see Chapter 3).

Place about 60 mL of this solution in a rinsed plastic wash bottle and install the medicine dropper in a rubber stopper.

To find the approximate mass of a drop delivered from the medicine dropper, place a small beaker on the top-loader balance and deliver ten drops into it. Calculate the average drop mass. Since the titration endpoint will be determined to within 1 drop (or less), this will give you an idea of the uncertainty in the final results.

THE SAMPLE TO BE TITRATED

Weigh a small clean dry beaker (or weighing bottle) on the analytical balance and introduce 1 gram of Mohr's salt. The mass of this sample should be known to the nearest 0.1 mg to keep the uncertainty to about 1 part in 10000. Keep this sample in a secure place until it is needed in the titration procedure. Let the mass of Mohr's salt be called M_3.

PREPARATION OF THE INITIAL TITRATION SOLUTION

To a 400-mL beaker add about 150 mL of water, 50 mL of $1M$ sulfuric acid, 5 mL of 85% phosphoric acid, 10 drops of 0.2% diphenylaminesulfonate indicator, and two drops of $0.1M$ ferrous ammonium sulfate solution. The solution is now ready for a *pretitration*, which is necessary to account for (1) possible titratable impurities in the reagents and (2) the fact that the indicator requires some dichromate solution to effect its color change from colorless to purple. It is necessary to add the small amount of ferrous ion for this pretitration because the indicator does not function properly in the absence of iron.

To perform the pretitration, use the medicine dropper in the weight buret to add dichromate solution (splitting drops) until the color changes to a light but permanent purple. When this has been accomplished, add the precisely weighed sample of Mohr's salt, without loss, using a distilled-water wash bottle. The indicator will change back to its colorless form and the solution is now ready for the main titration. (Before the Mohr's salt is added, the pretitration may be repeated for practice: add another drop of the ferrous solution and titrate.)

THE MAIN TITRATION

First be sure to weigh the potassium dichromate wash bottle, with the medicine dropper installed, to the nearest 0.01 gram.

Use a *titration thief* (see TITRATIONS USING VISUAL INDICATORS) to remove about 1 mL of the solution which is to be added when the endpoint is near.

With constant stirring, add dichromate solution from the side spout of the weight buret. As the endpoint is approached, there will be purple swirls which will disappear upon stirring. Approach the endpoint (which will be a false endpoint because of the small portion of sample in the titration thief) with care, using dropwise addition from the side spout.

When the indicator changes to purple, it will be harder to see it than during the pretitration, because of the green color caused by the chromic ion formed in the titration reaction. Be alert to the first permanent change in hue of the solution from the pure green color to a purplish-green.

When the false endpoint has been reached, add the rest of the sample from the titration thief, being sure to rinse the thief into the solution. The indicator will change back to colorless, unless you have accidentally overtitrated. Finish the titration by using drops (and split drops) from the medicine dropper. In fitting the medicine dropper back into the wash-bottle neck, be careful not to cause any solution to be expelled from the side spout.

Finally, reweigh the wash bottle so that the mass of dichromate solution used may be accurately determined. Let this solution mass be called M_4.

CALCULATION OF PURITY (IN PERCENT)

Since the titration reaction involves 6 mmol of Mohr's salt (MW = 392.14) per 1 mmol of potassium dichromate (MW = 294.19), we may derive the following relationship:

$$\text{Purity, } \% = \frac{M_4 \cdot M_1 B_1 \cdot 6 \cdot 392.14 \cdot 100}{M_3 B_3 \cdot M_2 \cdot 294.19} = 799.62 \cdot \frac{M_4 \cdot M_1}{M_3 \cdot M_2},$$

where the M-values are as indicated earlier, B_1 is the air-buoyancy factor for potassium dichromate (density 2.676 g/mL, $B_1 = 1.00029$), and B_3 is the factor for Mohr's salt (density 1.864 g/mL, $B_3 = 1.00048$). The buoyancy factor for the dichromate solution does not appear because it cancels out with M_4 in the numerator and M_2 in the denominator.

POTENTIOMETRIC ENDPOINT DETECTION
(see Titrations using potentiometric measurements, p. 609)

This titration may also be carried out with a platinum-indicator electrode to follow the potential. In this case the dichromate bottle should be weighed at the point where the endpoint is close, as judged by the observed potential. Then the medicine dropper is used to add one reproducible drop at a time, recording the potential after each drop. When it is certain that the endpoint has been passed, the bottle is again weighed, and the average mass per drop is calculated. The observed potential values are plotted versus the number of drops added (see Chapter 11 for suggestions on different types of plots) in order to determine the precise endpoint. The dropwise approach should also be used for the pretitration, which will also provide an idea of what potential to expect at the endpoint.

LABORATORY REPORT 13

HIGH PRECISION ASSAY OF MOHR'S SALT BY GRAVIMETRIC TITRATION

NAME _____   LAB SECTION _____

DATE _____

Information about Mohr's salt sample:

   Brand: _____ Lot No.: _____

   Mass of sample used in the titration (M_3): _____

Information about the potassium dichromate used as titrant

   Brand: _____ Lot No.: _____

   Purity according to the label: _____ %

   Mass of $K_2Cr_2O_7$ used to prepare the titrant solution (M_1): _____

   Mass of the stock titrant solution (M_2): _____

Titration data   Trial 1   Trial 2

   Mass of wash bottle at start of titration: _____   _____

   Mass of wash bottle at endpoint of titration: _____   _____

        Mass of solution used for the titration (M_4): _____   _____

Results of assay

   Calculated purity of the Mohr's salt: _____%

LABORATORY EXPERIMENT 14

SPECTROPHOTOMETRIC DETERMINATION OF AN ACID-DISSOCIATION CONSTANT

PURPOSE

1. To apply the quantitative principles of acid–base buffers and indicators.
2. To illustrate the value of spectrophotometry as an analytical probe that does not disturb the equilibrium mixture under study.

MATERIALS NEEDED

A simple spectrophotometer, such as the Bausch & Lomb Spectronic 20

Buret and/or 10-mL graduated pipet

Three small beakers

100-mL volumetric flask

Stock solution of an acid–base indicator (for example, $10^{-4}M$ bromcresol green)

Stock solution of an acid with a pK value within one unit of the expected pK value of the indicator (for example, $0.100M$ acetic acid)

Stock solution of the conjugate base of the acid (for example, $0.100M$ sodium acetate)

GENERAL COMMENTS

The fundamental principles of spectrophotometric equilibrium are discussed in Chapter 13, which should be reviewed before attempting the experiment. The following discussion and procedure are written in fairly general terms so that they may be readily adapted to a variety of combinations of indicators and buffer systems.

Five solutions will be prepared for spectrophotometric measurement, and each of these will contain the same molar concentration of the chosen indicator. One solution will be made sufficiently acidic (low pH) so that the indicator will be virtually completely converted to its protonated form. Another solution will be sufficiently basic (high pH) to assure conversion of the indicator to its conjugate base form. By using an appropriate buffer conjugate pair, the other three solutions will be made to have pH values that are close to the pK value for the indicator. In these solutions the indicator, which will be present in very low concentration, will be forced to distribute itself between its two colored forms. The purpose of the spectrophotometric measurements is to find the precise ratio of the two forms of the indicator in the various buffer solutions.

Thus, there are two acid–base equilibrium systems operating in the solutions. The solution pH is controlled by the buffer system.

$$HB \rightleftharpoons H^+ + B,$$

conjugate   conjugate
acid   base

where ionic charges depend on the buffer pair used, and

$$[H^+]f_1 = K_{HB} \frac{[HB]}{[B]} \cdot \frac{f_{HB}}{f_B}, \tag{L14.1}$$

where activity coefficients depend on the ionic charges and on the ionic strength used.

The concentrations [HB] and [B] shown in Eq. (L14.1) are chosen by the experimenter, using the stock solutions of HB and B.

The color of the solution depends on the indicator equilibrium:

$$HInd \rightleftharpoons H^+ + Ind,$$

e.g., yellow   e.g., blue

where, again, ionic charges depend upon the substance used, and

$$K_{HInd} = \frac{[H^+]f_1[Ind]f_{Ind}}{[HInd]f_{HInd}}. \tag{L14.2}$$

In Chapter 13 it was shown that the concentration ratio [Ind]/[HInd] can be determined by spectrophotometric measurements. Recognizing that the HInd species corresponds to DX, we have from Eq. (13.8):

$$\frac{[Ind]}{[HInd]} = \frac{\epsilon_{HInd} - \epsilon}{\epsilon - \epsilon_{Ind}} = \frac{A_{HInd} - A}{A - A_{Ind}}. \tag{L14.3}$$

If the experiment is designed so that the photometric-cell path is constant and all solutions contain the same molarity of indicator, we may (Beer's Law) simply substitute the observed absorbances in place of the molar absorptivities.

Thus, the value for A_{HInd} is determined in the solution of low pH, with the indicator all in the HInd form. The value of A_{Ind} is found with the solution of high pH. For each of the buffers, we find a different value for A.

Since the goal of the experiment is to determine a reliable value for K_{HInd}, we focus attention on Eq. (L14.2) which defines this quantity. The following comments should be considered:

a) The hydrogen-ion activity $[H^+]f_1$ is to be calculated for each of the buffer solutions, using Eq. (L14.1). It is assumed that the value for K_{HB} is accurately known and that the ionic strengths of the solutions can be calculated from their known composition. The exact **buffer equation** takes into account the shift in concentrations of HB and B necessary to establish the equilibrium (see Eq. (9.20)):

$$[H^+]f_1 = K_{HB}\frac{f_{HB}}{f_B}\frac{C_{HB}-[H^+]+Q_w/[H^+]}{C_B+[H^+]-Q_w/[H^+]},\qquad\text{(L14.4)}$$

where C_{HB} and C_B are the stoichiometric concentrations of the buffer constituents calculated simply from the data on how the stock solutions were diluted in preparing the buffer. For buffers in the pH range from 4 to 10 we may assume that $[H^+]$ and $Q_w/[H^+]$ are negligible compared to C_{HB} and C_B.

b) In calculating the ratio [Ind]/[HInd] of the indicator species it is important to note that Eq. (L14.3) depends upon *differences* in absorbance. Since any absorbance value is subject to uncertainty and error, it follows that a difference between two values is subject to a higher relative error. Therefore it is important to design the experiment so that the absorbance A of a buffer solution is sufficiently different from both A_{HInd} and A_{Ind}. This means that the indicator should be shifted no more than about 75% to one form or the other in the buffer solutions.

c) We cannot make a general recommendation for the solutions needed to obtain the values for A_{HInd} and A_{Ind}. For an indicator that does not form any other conjugate species, it is convenient to use hydrochloric acid for the low-pH solution and sodium hydroxide for the high-pH solution. However, some indicators have more than one color change and caution must be taken about the pH range used for the determination. The instructor will discuss this problem for the particular indicator chosen for study.

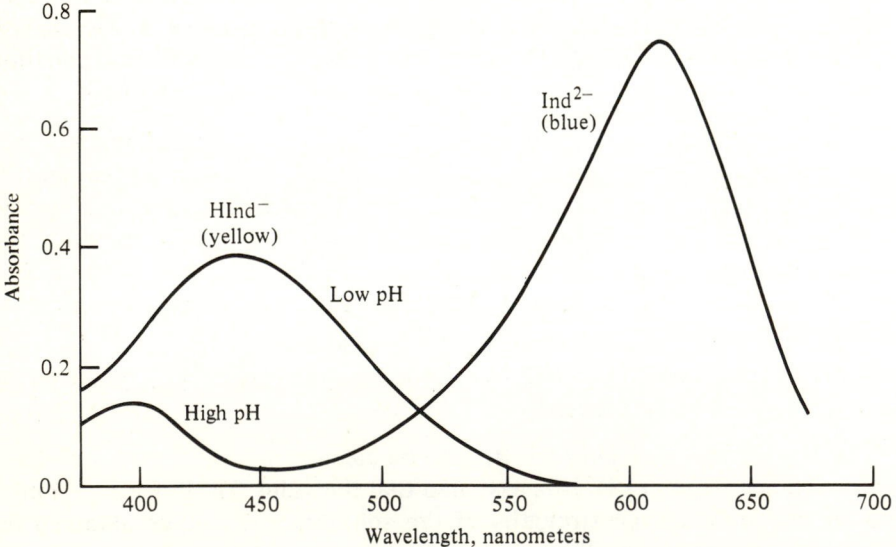

Fig. L.28 Absorption spectra for bromcresol-green indicator.

d) The absorption spectrum for the indicator is of crucial importance in this experiment, for it guides the investigator to the choice of the optimum wavelength to be used for the measurements. Referring again to the absorbance differences needed for Eq. (L14.3), we see clearly that the best wavelength is in the spectral region where the two forms of the indicator differ most in their absorptivities. It matters not whether HInd or Ind absorbs more strongly; it is the relative difference that is important. For example, consider the spectra for the indicator called bromcresol green (Fig. L.28). At a wavelength of 615 nm only the conjugate base absorbs, and this is an ideal situation. However, good results could also be obtained by working at 445 nm. At the isosbestic point corresponding to 515 nm both species have the same absorptivity.

EXPERIMENTAL PROCEDURE

1. Prepare a solution as follows: to a 100-mL volumetric flask add precisely 10 mL of the stock indicator solution and 10 mL of the stock solution of the buffer conjugate base. Dilute to the mark and mix well. This solution will contain the indicator chiefly in its conjugate-base Ind form.

2. Prepare another solution: to a 100-mL volumetric flask add precisely 10 mL of the indicator solution and 10 mL of the stock solution of the buffer conjugate acid. Dilute to the mark and mix well. This solution will contain the indicator chiefly in its conjugate acid HInd form.

At this point it is necessary to make a decision about the wavelength to be used for the absorbance measurements. Possible procedures are: (a) use solutions (1) and (2) to determine the absorption spectra, taking absorbance readings at 10-nm intervals over the visible range. Choose a wavelength according to the findings. (b) The instructor will post spectra for examination or will provide recommendations for a suitable wavelength to be used.

3. Solution (1) will be used to determine the value for A_{Ind}. Pour about 20 mL into a small beaker and measure its absorbance at the chosen wavelength. It may be that this solution is already sufficiently high in pH to be satisfactory, but as a check add a drop of $1M$ NaOH, mix well, and measure the absorbance again. If there is evidence that the indicator has shifted to a completely different form, consult the instructor. Otherwise, use the absorbance of the solution as A_{Ind}.

4. Similarly, solution (2) is used for the determination of A_{HIn}. Measure its absorbance at the chosen wavelength and then see if a drop of $1M$ HCl has an appreciable effect.

5. Mix equal (10-mL) portions of solutions (1) and (2) to form a buffer solution, stir well, and measure the absorbance. From the value obtained for A decide whether the additional buffer solutions should have higher or lower pH-values, so that the absorbance data will be optimal for Eq. (L14.3). Carry out the preparation and measurements, finally choosing three buffer solutions for use in the final calculations.

CALCULATION OF RESULTS

For each of the three buffer solutions the hydrogen-ion activity is calculated using Eq. (L14.4). The first step is to calculate the ionic strength of each buffer, given the compositions of the stock solutions provided by the instructor. If these stock solutions have the same ionic strength, then each of the buffers will have an ionic strength equal to 1/10 that of the stock solutions. If the stock solutions differ in ionic strength, each buffer will also differ.

In using (L14.2) for the calculation of K_{HInd} it is necessary to know what charge type is correct for the indicator. The instructor will provide this information if it is not available in Table 10.4 of indicators in Chapter 10.

Summarize the data and calculated results in Laboratory Report 14.

LABORATORY REPORT 14

SPECTROPHOTOMETRIC DETERMINATION OF AN ACID-DISSOCIATION CONSTANT

NAME _____   LAB SECTION _____

DATE _____

Indicator chosen for study: _____   Charge type: _____

Buffer system for pH control: _____ pK_{HB}: _____

Molarity of HB stock: _____ of B stock: _____

Absorbance of basic solution A_{Ind}: _____ of acidic solution A_{HInd}: _____

Data on buffer solutions

	Volume		Observed color	A	I	$[H^+]f_1$	K_{HInd}
	solution of B	solution of HB					
1							
2							
3							

Average value for K_{HInd}: _____ pK_{HInd}: _____

Average deviation: _____%

LABORATORY EXPERIMENT 15

PHOTOMETRIC STUDY OF THE 1:1 IRON(III)–THIOCYANATE COMPLEX

PURPOSE

To illustrate the use of spectrophotometry for determination of metal–ligand complex-formation quotients.

MATERIALS NEEDED

Spectrophotometer and cells

$1.2 \cdot 10^{-4} M$ potassium thiocyanate in $0.50 M$ perchloric acid

$0.100 M$ ferric perchlorate in $0.5 M$ perchloric acid

FORMATION OF MONOTHIOCYANATOIRON(III)

Thiocyanate ion can react with ferric ion in acid solution to form a series of thiocyanato complexes:

$$Fe^{3+} + SCN^- \rightleftarrows FeSCN^{2+}, Fe(SCN)_2^+, \ldots, Fe(SCN)_6^{3-}.$$

In the presence of high concentrations of thiocyanate the higher complexes predominate, but if the molarity of thiocyanate is very low only the monothiocyanate species is formed in appreciable amount, with the equilibrium formation quotient:

$$Q_1 = \frac{[FeSCN^{2+}]}{[Fe^{3+}][SCN^-]}.$$

In this experiment the concentration of thiocyanate ion will be about $0.0001 M$, and since the equilibrium quotient for the formation of the *di*thiocyanato species is approximately equal to 13, then the *maximum* ratio is

$$\frac{[Fe(SCN)_2^+]}{[FeSCN^{2+}]} \approx 13[SCN^-] \approx 0.001.$$

Therefore, the higher complex will never comprise more than about 0.1% of the total.

Another possible complication to be avoided is the formation of hydroxyiron(III) species, such as $FeOH^{2+}$. The aquo-ferric ion $Fe(H_2O)_6^{3+}$ is a moderately strong Brönsted acid with a dissociation quotient of about $6 \cdot 10^{-3}$ (at ionic strength 0.5). Therefore, if the concentration of monohydroxy-iron(III) ion is to be reduced to negligible amounts (0.1% or so), the hydronium-ion concentration should be of the order of $0.5 M$. This will be achieved in the experiment by using perchloric acid.

At this acidity the uncomplexed iron(III) ions do not absorb visible light appreciably, and so the thiocyanate complex is the only colored species. The wavelength of the absorption maximum is about 445 nm. It is also assumed that the thiocyanate ion does not take on protons, i.e., that thiocyanic acid is strong.

The experimental procedure will be to add small successive portions of a solution containing $0.1M$ ferric perchlorate, $0.5M$ perchloric acid to a known volume of $0.00012M$ potassium thiocyanate, $0.5M$ perchloric acid. Thus, both the ionic strength and the acidity will remain effectively constant during the experiment.

With each addition of the iron(III) solution, the concentration of the red complex will increase, and the changing absorbance can be measured. Since the complex is the only absorbing species,

$$A_{mix} = \epsilon b [FeSCN^{2+}],$$

where ϵ is the molar absorptivity, b is the cell width, and A_{mix} is the absorbance.

The general photometric expression for complexation equilibria $D + X \rightleftharpoons DX$ is discussed in Chapter 13:

$$Q = \frac{1}{[D]} \frac{A_X - A_{mix}}{A_{mix} - A_{DX}}, \qquad (L15.1)$$

where A_X is the absorbance when all X is uncomplexed, A_{DX} is the absorbance when all X is complexed with D, and A_{mix} is the absorbance at intermediate values (i.e., partly complexed). For $FeSCN^{2+}$, D corresponds to Fe^{3+} and X corresponds to SCN^-. It is impossible to determine A_{DX} directly on account of the relatively small size of Q_1 (≈ 140). That is, not enough excess Fe^{3+} could be added to the SCN^- to cause quantitative conversion to $FeSCN^{2+}$. By rearranging (L15.1) and substituting, we get

$$A_{mix} = \frac{-1}{Q_1} \cdot \frac{A_{mix}}{[Fe^{3+}]} + A_{DX}.$$

When A_{mix} is plotted versus $A_{mix}/[Fe^{3+}]$, a straight line should be obtained with $-1/Q_1$ as the slope and A_{DX} as the intercept. As a first approximation, use the total analytical concentration C_{Fe} of Fe(III) in place of $[Fe^{3+}]$ as the equilibrium concentration for your first graph. Once Q_1 has been estimated, refined values of $[Fe^{3+}]$, which allow for the amount of iron complexed, may be determined and a second plot is made of A_{mix} versus $A_{mix}/[Fe^{3+}]$. A more accurate value of Q_1 is determined from the slope. Successive approximations may be carried on until the desired consistency in Q_1 is obtained; this is an obvious place for computer iteration.

PHOTOMETRIC PROCEDURE

Pipet 50 mL of the KSCN, $HClO_4$ solution into a clean dry 250-mL beaker. Use a pipet to add successive 1-mL portions of the $0.1M$ Fe(III) (see bottle for exact concentration). After each addition of the Fe(III), *stir thoroughly*, pour into a photometer cell, measure the absorbance (or % transmittance) at

445 nm, and *return* the cell contents to the beaker, being careful not to lose any solution. Take 10 measurements in all, using 1-mL additions. The cell should be rinsed back and forth into the beaker each time.

CALCULATION AND REPORTING OF RESULTS

Calculate the initial concentration of KSCN used in the beaker. Calculate the total concentration of iron(III) present in the solution after each 1-mL increment. The absorbance measured is also affected by the dilution, therefore a dilution correction must also be made:

$$A_{cor} = A_{meas} \frac{50 + \text{Volume added}}{50}.$$

Plot A_{cor} versus A_{mix}/C_{Fe} and determine Q_1 from the slope and A_{DX} from the intercept. Successive approximations on Q_1 are optional.

LABORATORY REPORT 15

PHOTOMETRIC STUDY OF MONOTHIOCYANATOIRON(III)

NAME _____   LAB SECTION _____

DATE _____

Molarity of KSCN used:

Stock concentration of iron(III): _____

Increment	Observed absorbance	Corrected absorbance, A_{mix}	C'_{Fe}
0			
1			
2			
3			
4			
5			
6			
7			
8			
9			
10			

From the plot of A_{mix} versus A_{mix}/C_{Fe} find:

$$\frac{-1}{\text{Slope}} = Q_1 = \frac{[FeSCN^{2+}]}{[Fe^{3+}][SCN^-]} = \text{_____}$$

$$\text{Intercept} = A_{FeSCN^{2+}} = \text{_____}$$

Attach graph

LABORATORY EXPERIMENT 16

THE MERCURY ELECTRODE: CHELON EQUILIBRIA

PURPOSE

A mercury-metal electrode serves well as a probe for the concentration of mercurous (or mercuric) ions at its surface. In the first part of this experiment the standard potential for the Hg_2^{2+}/Hg couple is determined. Then the mercury electrode is used to characterize a potential-pH diagram, as discussed in Chapter 15. The data will allow calculation of (a) E^0 for the Hg_2^{2+}/Hg couple, (b) the value of K_{sp} for HgO, (c) the formation quotient for the mercury chelate, and (d) the formation quotient for the copper chelate.

SOLUTIONS NEEDED

Mercuric perchlorate ($0.01M$), perchloric acid ($0.1M$), $NaClO_4$ ($0.9M$). To prepare one liter: Weigh 2.166 g pure mercuric oxide HgO into a 250-mL beaker; add 100 mL of water and 87 mL of 70% perchloric acid; stir until dissolved and transfer quantitatively to a 1-L volumetric flask; add 45 mL of 50% NaOH, with constant stirring; dilute to the mark and mix well.

Sodium hydroxide ($0.10M$). Dilute 5.2 mL of 50% NaOH to one liter.

Perchloric acid ($0.10M$). Dilute 8.6 mL of 70% $HClO_4$ to one liter.

Copper solution ($0.0200M$). Prepare by using the sulfate, nitrate, or perchlorate (avoid chloride).

Triethylenetetramine (trien) ($0.0200M$). Alternative chelon: EGTA.

EQUIPMENT NEEDED

Two pH-meters, one of which will be used in millivolt mode. The experiment may be done with one meter, provided it is switched back and forth for measurements.

Glass electrode for pH measurement (pH = 7 buffer for standardization).

Reference electrode, such as a saturated calomel electrode.

Mercury electrode. This may be of the *cup type*, the *J-type*, or a platinum wire plated with mercury. The latter is easy to prepare and manage: mercury is plated by making Pt the cathode in a solution of a mercuric salt, with another Pt wire as anode.

Magnetic stirrer.

Two burets.

Potassium nitrate agar, in U-tube for use as a bridge. Heat 3 g granular agar in 100 mL of $2M$ KNO_3 until dispersed. Fill U-tube and let cool. Keep ends wet.

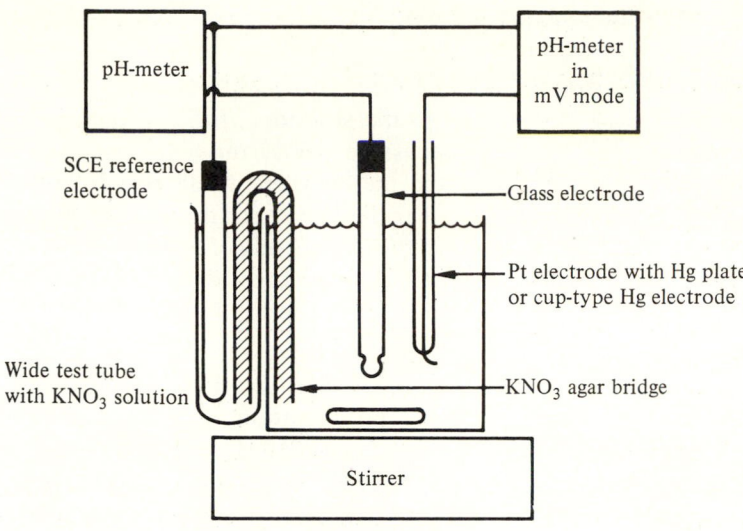

Figure L.29

EXPERIMENTAL SETUP

The essential arrangement is shown in Fig. L.29.

Note that the reference electrode can serve in two capacities, both for the pH measurement and for the mercury-electrode potential measurement. It is well to isolate it from the main solution compartment to avoid the introduction of chloride ion, which would interfere by forming calomel at the Hg electrode surface. The test tube is conveniently fastened to the beaker with tape.

PROCEDURE

1. *Determination of E^0 for the Hg_2^{2+}/Hg couple*   Pipet 25 mL of 0.01M $Hg(ClO_4)_2$ into a 250-mL volumetric flask, dilute to the mark with distilled water, and mix well. Pipet 100 mL of this solution into the 250-mL beaker to be used as the cell. This solution will have a mercuric-ion molarity of precisely 0.00100, and a pH of 2.0. The ionic strength is 0.10.

The pH-meter (with glass and SCE electrodes) should be standardized by using a reference buffer of pH ≈ 7 before assembling the apparatus.

Insert a mercury electrode and measure the potential of the cell:

$$SCE \,|\, Hg^{2+}(0.00100M) \,|\, Hg(liquid)$$

Because of the very rapid reaction of mercuric ion with mercury metal

$$Hg^{2+} + Hg = Hg_2^{2+},$$

which has an equilibrium constant of about 87, the molarity of mercurous ion at the Hg-electrode surface immediately becomes virtually 0.00100. The value for

E^0 may be calculated on this assumption, even though nearly all of the mercury in the solution remains in the mercuric-ion form.

2. *Determination of the solubility product for mercuric oxide* Use a buret to add increments of $0.10M$ NaOH to the solution, with continuous stirring. As the pH changes (try to keep successive pH changes between 0.5 and 1 pH unit), measure the potential of the mercury electrode relative to SCE. The potential will decrease in the manner described for curve II in the potential—pH diagram of Fig. 15.9 discussed in the Case Study of Section 15.8. By using Eq. (15.9), calculate the value of Q_{sp} for HgO from these data. Continue the addition of NaOH until the pH is about 12. Record the total volume added.

3. *Determination of the formation quotient for the mercury chelate* To the solution of pH $\approx$ 12, pipet 10 mL of the trien solution. This will result in the formation of HgY, and an equal concentration of excess chelon will be present. Use a buret to add increments of $0.10M$ perchloric acid, taking pH and potential measurements until the pH has dropped to between 2 and 3. Record the total volume added. It is important to obtain data at high pH, so that the mercury-electrode potential is known in the pH region where the chelon is fully deprotonated (between pH $= 10$ and 12). This part of the experiment determines curve I of Fig. 15.9, as described in the Case Study. The value for Q_{HgY} may be calculated by using Eq. (15.4) on the assumption that there is no hydrogen-ion competition for the chelon at high pH.

4. *Determination of the formation quotient for a metal ion* Pipet 10 mL of the copper-salt solution ($0.020M$) into the cell solution that has resulted from part (3). This produces an equilibrium mixture with an excess of the metal ion. Use a buret to add $0.1M$ NaOH in increments, recording pH and potential data, until the level portion of the metal-ion curve has been characterized. The value for Q_{MY} may be calculated by using Eq. (15.6) but note that it is necessary to know the molarities of the three species. This is why it was important to keep track of the total volume of the solution.

In this experiment it is fair to assume that the ionic strength remained close to 0.10, so that activity coefficients may be considered constant.

LABORATORY REPORT 16

THE MERCURY ELECTRODE: CHELON EQUILIBRIA

NAME _____   LAB SECTION _____

DATE _____

Name and formula of chelon:

Data and calculations relating to E^0 for the mercurous ion/mercury couple:

Data and calculations relating to the solubility product for mercuric oxide:

Data and calculations relating to the formation quotient for copper complex:

(Attach a graph showing the potential–pH data)

LABORATORY EXPERIMENT 17

THE SOLUBILITY PRODUCT FOR COPPER IODATE

PURPOSE

1. To test the validity of the strong-electrolyte equilibrium model for the slightly soluble salt copper iodate.

2. To illustrate the determination of an oxidant by liberation of triiodide ion, followed by thiosulfate titration. For variation, this experiment may be performed with calcium iodate.

MATERIALS NEEDED

Potassium iodate standard solution, about $0.01M$ (provided by instructor)

Saturated solutions of copper iodate (provided by instructor)

Solid sodium thiosulfate, potassium iodide

Solutions: $0.1M$ EDTA, $1M$ acetic acid

Buret, 10-mL and 5-mL pipets

DISCUSSION OF THE STRONG ELECTROLYTE MODEL

If we assume that copper iodate $Cu(IO_3)_2 \cdot H_2O$ dissociates completely in aqueous solution, then the solubility product is the only equilibrium constant needed to describe the quantitative behavior:

$$Cu(IO_3)_2 \cdot H_2O(s) = Cu^{2+} + 2IO_3^-,$$

$$K_{sp} = [Cu^{2+}][IO_3^-]^2 f_2 f_1^2.$$

To determine the numerical value for K_{sp} it is necessary to analyze the saturated aqueous solution to find the copper and iodate ion concentrations, and it is also necessary to use the Davies equation to estimate the activity coefficients. If a saturated solution is prepared in the absence of any other sources of copper or iodate ions and in the absence of any side reactions between solution species such as complexing ligands or hydrogen ions, then by stoichiometry it follows that the molar solubility S is

$$S = [Cu^{2+}] = \tfrac{1}{2}[IO_3^-],$$

and the value for K_{sp} may be written as

$$K_{sp} = S(2S)^2 f_2 f_1^2 = 4S^3 f_2 f_1^2. \tag{L17.1}$$

However, as a test of the model it is important to observe the solubility of the salt in the presence of a common ion. For example, if the solution contains a certain molarity C of copper nitrate, then this source of copper ions will repress the solubility of copper iodate. The molar solubility S will no longer be equal to $[Cu^{2+}]$, but it is still true that $S = \frac{1}{2}[IO_3^-]$. The total concentration of copper ions will be given by:

$$[Cu^{2+}] = S + C,$$

and the value for K_{sp} may be written as

$$K_{sp} = (S + C)(2S)^2 f_2 f_1^2. \tag{L17.2}$$

PRINCIPLES OF ANALYTICAL PROCEDURE

The samples to be analyzed will be portions of saturated solutions of copper iodate, with and without certain concentrations of copper nitrate present. It is desired to find the molar solubility S, so that the values for K_{sp} may be calculated by using Eqs. (L17.1) and (L17.2). Since in each case

$$S = \frac{1}{2}[IO_3^-], \tag{L17.3}$$

the analytical problem is the determination of the amount of iodate present in a sample.

Treatment with an excess of potassium iodide in the presence of an acid results in the formation of 3 mmol of triiodide ion for each millimole of iodate ion present:

$$IO_3^- + 8I^- + 6H^+ = 3I_3^- + 3H_2O.$$

The triiodide ion is then easily titrated with a standard solution of sodium thiosulfate:

$$I_3^- + 2S_2O_3^{2-} = 3I^- + S_4O_6^{2-}.$$

Thus, 6 mmol of thiosulfate are required for each mmol of iodate in the sample.

However, copper ion is also capable of oxidizing iodide ion, with the formation of a precipitate of cuprous iodide:

$$2Cu^{2+} + 5I^- = 2CuI(s) + I_3^-.$$

It would be possible to take this reaction into account, since the amount of added copper nitrate will be known. However, the thiosulfate endpoint in the presence of cuprous iodide is less accurate and it is preferable to avoid the copper–iodide reaction through masking the copper with the strongly complexing ligand, EDTA:

$$Cu^{2+} + H_2Y^{2-} = CuY^{2-} + 2H^+.$$

This stabilizes the +2 oxidation state of the copper and eliminates the reaction with iodide ion. In the presence of EDTA, only the iodate ion will be determined by the iodide–thiosulfate procedure.

Note: If calcium iodate is used, the EDTA masking becomes unnecessary.

PREPARATION OF THE EQUILIBRIUM MIXTURES

Several days before the experiment is to be performed, the instructor will prepare three solutions of the following composition:

Perchloric acid, $1.0 \cdot 10^{-4} M$

Sodium perchlorate, $0.0300 M$ (to test the ionic-strength effect)

Copper nitrate, $0.00600 M$ (to test the common-ion effect)

Large portions of these solutions will be stirred mechanically with several grams of copper iodate prepared as follows: 1 L of $0.4 M$ sodium iodate is added slowly, with stirring, to 1 L of hot $0.1 M$ nitric acid, $0.2 M$ copper nitrate. The mixture is stirred while fairly hot for a few hours. Then the blue-green solid is collected on a filter, washed with water, resuspended in a liter of water, and stirred vigorously, collected and rinsed again, and then air-dried.

After the equilibrium solutions have been stirred with the solid for 2–3 days, they will be saturated with copper iodate. The instructor will filter them and place them in labeled bottles, ready for student analysis. The temperature at the time of filtration will be noted.

EXPERIMENTAL PROCEDURE AND COMMENTS

1. Prepare one liter of $0.01 M$ sodium thiosulfate by dissolving 2.5 grams of $Na_2S_2O_3 \cdot 5H_2O$ (GFW = 248.18).

2. Standardize the thiosulfate solution by titration of a standard potassium iodate solution provided by the instructor. The iodate molarity will be about $0.01 M$ and the precise value will be posted. Pipet 5 mL of the KIO_3 solution into a flask, add about 25 mL of water and 10 mL of $1 M$ acetic acid. Add about 1 gram of solid KI, swirl to dissolve, and titrate the liberated I_3^- with the thiosulfate solution until the yellow color just disappears. The color change will be easier to see if a few milliliters of 0.2% starch indicator are added, due to the blue starch–triiodide complex. The titration should be repeated and the results should check to within about 0.2 mL. Calculate the precise molarity of the thiosulfate.

3. The copper-iodate solutions are analyzed in a similar fashion: pipet a 10-mL sample into a flask, add about 25 mL of water, 10 mL of $1 M$ acetic acid, and 10 mL of $0.1 M$ EDTA. At this point proceed with only one flask at a time, so that air oxidation of the iodide is minimal. Add about 1 gram of solid KI, swirl to dissolve, and titrate with the standard thiosulfate solution until the yellow color just vanishes. The endpoint appearance will differ from that in the standardization titration because of the presence of the blue Cu–EDTA complex. As the endpoint is neared, the solution will go through shades of green (mixture of the yellow triiodide ion and the blue copper complex). The precise endpoint is more difficult to identify, but it will be clear blue with no greenish tint. Upon standing, there may be a slow return of the green color due to air oxidation.

Perform duplicate titrations on each of the copper iodate solutions and use the average result in the calculations.

CALCULATION OF RESULTS

For each copper iodate solution, calculate the concentration of iodate present, using the relationship that 6 mmol of thiosulfate are required for each one of iodate. Calculate the solubility of copper iodate, using Eq. (L17.3).

The ionic strength of each solution is calculated by the relationship:

$$I = C_{\text{NaClO}_4} + 3C_{\text{Cu(NO}_3)_2} + 3S.$$

Find the appropriate values for activity coefficients using the Davies equation and calculate the several values for K_{sp}. Discuss the validity of the strong-electrolyte model for this chemical system.

LABORATORY REPORT 17

THE SOLUBILITY PRODUCT FOR COPPER IODATE

NAME _____   LAB SECTION _____

DATE _____

Standardization of sodium thiosulfate

Molarity of the KIO_3 standard solution: _____

Volume of thiosulfate used in titration:

(1)_____ (2)_____ Average: _____

Calculated molarity of thiosulfate: _____

Results on the copper iodate solutions

Initial solution composition	$V_{S_2O_3}$	S	I	K_{sp}

Temperature when filtered: _____

Discussion

LABORATORY EXPERIMENT 18

**DISTRIBUTION OF BROMINE BETWEEN WATER AND CARBON TETRACHLORIDE
AND DETERMINATION OF THE FORMATION CONSTANT FOR TRIBROMIDE ION**

PURPOSE

This experiment illustrates the behavior of a two-phase immiscible-solvent equilibrium system. First the fundamental behavior of molecular bromine is studied. Then, by finding the effect of potassium bromide on the distribution of the bromine between the aqueous and organic phases, it becomes possible to calculate the formation constant for the reaction $Br_2(aq) + Br^- \rightleftharpoons Br_3^-$.

MATERIALS NEEDED

The instructor will provide the following stock solutions:

$0.3M$ bromine in carbon tetrachloride. Add 1.5 mL of liquid bromine to 100 mL of CCl_4 and mix well. Place in 50-mL burets in the hood.

$0.100M$ potassium bromide. Dissolve 11.90 grams KBr and dilute to 1 L in a volumetric flask. Set up 50-mL burets to dispense this solution.

$0.100M$ nitric acid. Dilute 6.3 mL of fresh (nonyellow) concentrated HNO_3 to 1 L in a volumetric flask. Set up 50-mL burets to dispense.

$0.100M$ sodium thiosulfate. Dissolve 24.82 grams of $Na_2S_2O_3 \cdot 5H_2O$ and dilute to 1 L in a volumetric flask. Place in 250-mL bottles fitted with 25-mL pipets in cork or rubber stoppers. Students prepare their titrant solutions by diluting 25-mL portions to 250 mL in a volumetric flask.

$0.1M$ sodium sulfite. Dissolve 13 grams of Na_2SO_3 in 1 L of water. Place in clearly marked wash bottle for use in case of a bromine-solution spill or skin contact.

Each student should have two stoppered test tubes. These may be regular test tubes fitted with corks. Preferably, use screw-cap (Teflon lined) tubes such as Sargent S-79533, size F, with a capacity of about 30 mL.

0.2% starch indicator

GENERAL COMMENTS

The equilibrium principles of this experiment are quite analogous to those discussed in Chapter 19 for the iodine–triiodide ion system. An overall view of the equilibrium system is shown in the following scheme:

$$\text{Aqueous layer} \qquad Br_2(aq) + Br^- \overset{K_f}{\rightleftharpoons} Br_3^-$$

$$\updownarrow K_D$$

$$\text{CCl}_4 \text{ layer} \qquad Br_2(org)$$

Thus, there are two simultaneous equilibria to be considered with the equilibrium-constant expressions:

$$\text{Formation constant:} \quad K_f = \frac{[Br_3^-]}{[Br_2(aq)][Br^-]}$$

$$\text{Distribution constant:} \quad K_D = \frac{[Br_2(org)]}{[Br_2(aq)]}$$

The analysis of the aqueous phase for total bromine yields the sum:

$$C_w = [Br_2(aq)] + [Br_3^-],$$

whereas the analysis of the organic phase gives the concentration of molecular bromine in that phase:

$$C_o = [Br_2(org)].$$

In analogy to Eq. (19.8) we may derive the relationship between the distribution ratio D and the fundamental equilibrium constants:

$$D = \frac{C_o}{C_w} = \frac{K_D}{1 + K_f[Br^-]}.$$

From this equation it is clear that when the aqueous phase contains no bromide ion, the distribution ratio is simply the value for K_D. This is the goal of the first part of the experiment. In the second part, the aqueous phase will contain a known amount of potassium bromide, and the value for D will be smaller, in accordance with the existing value for K_f. This will be discussed in more detail later. We must realize that in the analytical procedure molecular bromine escapes very quickly from its solutions. The technique will be to deliver a pipetted portion of the bromine solution directly into a solution containing an excess of potassium iodide. The bromine will react nearly instantaneously to form an equivalent amount of triiodide ion:

$$Br_2(\text{in sample}) + 3I^-(\text{in receiving flask}) \rightarrow I_3^- + 2Br^-.$$

Since the iodine is complexed strongly by the excess iodide ion, its solution is relatively stable and may be titrated with a solution of sodium thiosulfate:

$$2S_2O_3^{2-} + I_3^- \rightleftarrows S_4O_6^{2-} + 3I^-.$$

Thus 2 mmol of thiosulfate are required for each millimole of bromine originally present in the sample.

DETERMINATION OF THE VALUE FOR THE DISTRIBUTION CONSTANT K_D

To a clean test tube add about 12 mL of $0.100M$ nitric acid and about 3 mL of $0.3M$ bromine in CCl_4. The precise volumes are not critical. (Why?) Stopper or cap the tube, grasp it with a towel to absorb any possible leakage, and *shake* the two-phase mixture *vigorously* for about five minutes. If the instructor provides a constant-temperature water bath, place the tube in it after four minutes of shaking and keep it there until it reaches the bath temperature; then shake it for an additional minute. Let the layers separate, swirling the tube to help the CCl_4 to collect and settle. Place the equilibrated tube in a beaker.

Have ready a titration flask containing 25 mL of water and 1 gram of potassium iodide. Pipet 5 mL of the aqueous phase directly into this solution, holding the pipet tip just in contact with the surface of the solution to eliminate any loss of bromine.

Have ready another titration flask with the water and KI. Use a 1-mL graduated pipet to withdraw a sample of the CCl_4 layer and deliver 0.500 mL of it into the KI solution, again keeping the pipet tip just in contact with the solution.

Restopper the test tube and shake it for another two minutes. Let it stand while you titrate the samples.

The titrant solution is $0.01M$ (be sure to note the precise molarity) sodium thiosulfate, prepared by dilution of the stock solution provided by the instructor. In titrating the aqueous-layer sample, add the titrant until the color of the triiodide ion has faded to a pale yellow. Then add 2–3 mL of starch-indicator solution and complete the titration, adding titrant with split drops until the blue color just disappears.

In titrating the CCl_4 sample solution there is a complication in that the CCl_4 forms an insoluble phase that extracts some of the iodine formed by the reaction with bromine. Add titrant until the yellow color of the triiodide ion in the aqueous phase has just disappeared. The CCl_4 layer will be purple, and it will require persistent swirling to encourage the iodine to leave the organic phase. Add starch indicator and titrate until the blue color just disappears.

Repeat these titrations with new samples of the two-phase equilibrium system. Using the average of the two sets of results, calculate C_o and C_w and the value of the distribution constant K_D.

DETERMINATION OF THE VALUE FOR THE FORMATION CONSTANT K_f

To a clean test tube add V mL of $0.100M$ KBr and $(12 - V)$ mL of $0.100M$ nitric acid. The value for V should be anywhere from 12 to 2 mL and it would be interesting to have the members of the class cover the range, so that comparisons of results could be made. Also add 3 mL of $0.3M$ Br_2 in CCl_4. By shaking perform the equilibration and analyses of the phases in the way described above for the determination of K_D.

Calculate the value for D and then find K_f by the following procedure:

a) Once the molarity of Br_2 in the CCl_4 phase has been determined, the molarity of *molecular* Br_2 in the aqueous phase may be calculated by using the value for K_D found earlier:

$$[Br_2(aq)] = C_o/K_D.$$

b) Since the analysis of the aqueous phase has provided a molarity sum, the value for the tribromide ion may now be calculated:

$$[Br_3^-] = C_w - [Br_2(aq)].$$

c) The aqueous phase originally contained a known concentration of bromide ion from the added KBr solution. The equilibrium molarity of bromide ion may be calculated by difference:

$$[Br^-] = C_{KBr, init} - [Br_3^-].$$

d) Thus, all the quantities needed for the calculation of K_f are at hand.

CLEANUP

Empty the test tubes into the special waste bottle provided in the hood rather than dumping them into the sinks. Remember that bromine gets its name from the Greek word *bromos*, stench.

COMMENT

The use of $0.100M$ nitric acid had two purposes. First of all, it seems desirable to maintain a constant ionic strength in the aqueous phases, even though the activity coefficients of the ions in the K_f expression are presumed to cancel. Second, aqueous bromine has a slight tendency to react with water:

$$Br_2(aq) + H_2O \rightleftharpoons H^+ + Br^- + HOBr.$$

The presence of acid and bromide ions helps to repress this unwanted reaction.

LABORATORY REPORT 18

DISTRIBUTION OF BROMINE

NAME _____   LAB SECTION _____

DATE _____

Molarity of the diluted ($0.01M$) sodium thiosulfate used as titrant: _____

Determination of K_D

Volume of titrant (average) for aqueous phase: _____

for CCl_4 phase: _____

Molarity of Br_2(aq): _____ Molarity of Br_2(org): _____

Value of K_D: _____

Determination of K_f

Volume of KBr solution used to prepare aqueous phase: _____

of HNO_3: _____

Volume of titrant for aqueous phase: _____ for CCl_4 phase: _____

Value for C_w: _____ for C_o: _____ for D: _____

Calculated molarities at equilibrium

Initial molarity of KBr: _____

[Br_2(aq)]: _____

[Br_3^-]: _____

[Br^-]: _____

Value of K_f: _____

LABORATORY EXPERIMENT 19

TITRATION OF CALCIUM WITH EDTA, AND
THE 1:1 CALCIUM–ACETATE COMPLEX

PURPOSE

This experiment illustrates the visual titration of calcium by using cal-magite indicator, the more precise potentiometric titration by using a mercury electrode, and the application of titration to the study of the solubility of calcium sulfate and the effect of acetate ion on solubility.

MATERIALS NEEDED

Calcium sulfate 2-hydrate, analytical-reagent grade

$0.100M$ sodium perchlorate or potassium nitrate (to serve as an inert salt)

$0.100M$ sodium acetate (13.61 g of the 3-hydrate per liter of water)

$0.01000M$ EDTA (3.723 g of $Na_2H_2EDTA \cdot 2H_2O$ per liter of water)

Buffer, pH = 9. Dissolve 80 grams of ammonium nitrate and 40 mL of concentrated ammonia, dilute to 1 L. Ammonium chloride should not be used because it interferes with the mercury electrode.

$0.10M$ mercury-EDTA solution. To 100 mL of water add 2.166 g of finely ground mercuric oxide and 3.722 g of disodium EDTA 2-hydrate. Or pipet 100 mL of $0.1000M$ disodium EDTA onto the weighed HgO. Use a magnetic stirrer or hot plate to effect solution, and filter if there is any undissolved HgO.

$0.05M$ magnesium-EDTA solution. To 50 mL of water add 1.861 g of disodium EDTA 2-hydrate and 2.50 mL of 50% NaOH. After stirring, add 1.282 g of magnesium nitrate 6-hydrate, dilute to 100 mL, and stir until dissolved, heating if necessary.

Calmagite indicator. Dissolve 40 mg of indicator in 100 mL of water.

Mercury-indicator electrode (e.g., a Pt wire plated with Hg) and a reference electrode.

pH-meter in millivolt mode.

For convenience in the attainment of solubility equilibrium, capped test tubes are recommended (e.g., Sargent S-79533, size F). For best accuracy it is good to have a water bath controlled at some definite temperature.

THE EQUILIBRIUM PRINCIPLE

When calcium sulfate dissolves in water or in a solution of an inert salt such as potassium nitrate, there are two chemical steps:

$$CaSO_4 \cdot 2H_2O(s) \rightleftarrows CaSO_4(aq) \rightleftarrows Ca^{2+} + SO_4^{2-}.$$

The first equilibrium is the intrinsic solubility:

$$Q_0 = [CaSO_4(aq)].$$

Careful studies by Yeatts and Marshall* have established the relationship between Q_0 and the ionic strength of the solution:

$$Q_0 = K_0 \cdot 10^{0.173 \cdot I^{1/2}}, \quad \text{where} \quad K_0 = 3.83 \cdot 10^{-3} \text{ at } 25°.$$

We will accept this finding and apply it later in the data interpretation.

In the inert-salt solution the molarity of calcium ion is equal to that of the sulfate ion, and we may therefore express the solubility product as:

$$Q_{sp} = [Ca^{2+}][SO_4^{2-}] = (S - Q_0)^2,$$

where S is the total molar solubility of calcium sulfate. Thus, by determining the solubility S in a particular inert-salt solution we may then calculate the value for Q_{sp} appropriate for that ionic strength.

If the solid calcium sulfate is equilibrated with a solution of a ligand that can form a complex with calcium ion, the solubility will be increased over that for the solution of an inert salt of the same ionic strength. In this experiment we will examine the effect of sodium acetate. If an increase in solubility is found, we may propose that the species $CaOAc^+$ has been formed:

$$Ca^{2+} + OAc^- \rightleftharpoons CaOAc^+, \quad Q_f = \frac{[CaOAc^+]}{[Ca^{2+}][OAc^-]}.$$

Then the total solubility may be written:

$$S_L = Q_0 + [Ca^{2+}] + [CaOAc^+] = Q_0 + [SO_4^{2-}],$$

where subscript L is a reminder that we are dealing with a solution containing a complexing ligand. The reader should derive the following relationship that allows calculation of the Q_f value given the experimental values for S and S_L:

$$Q_f = \left\{ \frac{(S_L - Q_0)^2}{(S - Q_0)^2} - 1 \right\} \frac{1}{[OAc^-]},$$

where $[OAc^-]$ is the equilibrium molarity of acetate ion in the sodium acetate solution.

ANALYTICAL PRINCIPLE (see Chapter 16 for theory of chelometric titrations)

A sample of the saturated calcium-sulfate solution is buffered to pH $= 9$ and titrated with a standard solution of EDTA. For the most precise results the titration is done potentiometrically by using a mercury-indicator electrode. However, it may also be performed by using calmagite indicator, provided some magnesium is added to sharpen the indicator color change. The pro-

---

*Yeatts, Marshall, *Journal of Physical Chemistry*, 73, 81 (1969).

cedure is written to combine these two techniques, since this is an instructional laboratory experiment.

PROCEDURE (see Chapter 11 and the lab section on potentiometric endpoints)

Into each of two test tubes place 0.5–1 gram of solid calcium sulfate 2-hydrate. To one tube add about 25 mL of 0.10M inert salt and to the other add about 25 mL of 0.10M sodium acetate. Cap the tubes and shake them vigorously for about three minutes. If a water bath of constant temperature is available, place the tubes in it for 5–10 minutes and then shake them again for three minutes. At this point the solutions should be at equilibrium with the solid.

Remove the excess solid by filtering the saturated solution through filter paper. Pipet 10 mL of the filtrate into a beaker, add about 100 mL of water and 5 mL of the ammonia, ammonium nitrate buffer.

Add four drops of the magnesium-EDTA solution. Add calmagite indicator solution dropwise until the solution has a visible pink color.

If the titration is to be performed potentiometrically, add two drops of the mercury-EDTA solution and install the mercury and reference electrodes connected to a precise millivoltmeter.

Use a titration thief to withhold about 1 mL of the solution and titrate with 0.01M EDTA until the color is just slightly changed from the pink. Then return the withheld portion and continue the titration drop by drop, recording the cell potential and noting the color changes. When the endpoint has been definitely passed, clean the beaker and repeat with a second sample.

See Chapter 11 for a discussion on the ways to find the potentiometric endpoint. Once this has been found, compare it with the observations on the visual color change.

CALCULATIONS

From the titration results on the inert-salt solution, calculate the value of Q_{sp} for ionic strength 0.10. Then, using the results on the sodium acetate solution, find the value of Q_f for the calcium acetate complex. To do this it is first necessary to assume that the equilibrium acetate-ion molarity is simply 0.100, i.e., the same as at the start of the equilibration. However, since some of the acetate was consumed by forming the complex, it is important to make a correction for this loss. To do this, use the preliminary value for Q_f obtained by assuming 0.100M acetate to estimate the molarity of the CaOAc$^+$ species. Then the corrected value for acetate ion is $0.1 - [CaOAc^+]$, and a more accurate value of Q_f may be calculated.

Question: How can you determine whether the increased solubility in sodium acetate solution is due to a series of complexes CaOAc$^+$, Ca(OAc)$_2$, etc., and not merely to a single complex, as we have assumed in the discussion?

LABORATORY EXPERIMENT 19

SOLUBILITY OF CALCIUM SULFATE

NAME _____   LAB SECTION _____

     DATE _____

Molarity of EDTA titrant solution: _____

Volume of EDTA solution used to reach titration endpoint

    Inert-salt solution: _____ mL

    NaOAc solution: _____ mL

Results

    S: _____ S_L: _____

    Q_{sp}: _____ Q_f: _____

Attach plots of E versus volume of EDTA for the potentiometric titrations.

APPENDIXES

PERIODIC SYSTEM OF THE ELEMENTS

Active metals — Transition metals — Nonmetals

Periods	I	II											III	IV	V	VI	VII	VIII
1	1 H 1.008																1 H 1.008	2 He 4.003
2	3 Li 6.939	4 Be 9.012											5 B 10.81	6 C 12.01	7 N 14.01	8 O 16.00	9 F 19.00	10 Ne 20.18
3	11 Na 22.99	12 Mg 24.31											13 Al 26.98	14 Si 28.09	15 P 30.97	16 S 32.06	17 Cl 35.45	18 Ar 39.95
4	19 K 39.10	20 Ca 40.08	21 Sc 44.96	22 Ti 47.90	23 V 50.94	24 Cr 52.00	25 Mn 54.94	26 Fe 55.85	27 Co 58.93	28 Ni 58.71	29 Cu 63.54	30 Zn 65.37	31 Ga 69.72	32 Ge 72.59	33 As 74.92	34 Se 78.96	35 Br 79.91	36 Kr 83.80
5	37 Rb 85.47	38 Sr 87.62	39 Y 88.91	40 Zr 91.22	41 Nb 92.91	42 Mo 95.94	43 Tc (97)	44 Ru 101.1	45 Rh 102.9	46 Pd 106.4	47 Ag 107.9	48 Cd 112.4	49 In 114.8	50 Sn 118.7	51 Sb 121.8	52 Te 127.6	53 I 126.9	54 Xe 131.3
6	55 Cs 132.9	56 Ba 137.3	57 La 138.9	72 Hf 178.5	73 Ta 180.9	74 W 183.9	75 Re 186.2	76 Os 190.2	77 Ir 192.2	78 Pt 195.1	79 Au 197.0	80 Hg 200.6	81 Tl 204.4	82 Pb 207.2	83 Bi 209.0	84 Po (210)	85 At (210)	86 Rn (222)
7	87 Fr (223)	88 Ra (226)	89 Ac (227)	104 Rs 257	105 Ha 260													

Inner transition metals

Lanthanum series

58 Ce 140.1	59 Pr 140.9	60 Nd 144.2	61 Pm (147)	62 Sm 150.4	63 Eu 152.0	64 Gd 157.3	65 Tb 158.9	66 Dy 162.5	67 Ho 164.9	68 Er 167.3	69 Tm 168.9	70 Yb 173.0	71 Lu 175.0

Actinium series

90 Th 232.0	91 Pa (231)	92 U 238.0	93 Np (237)	94 Pu (242)	95 Am (243)	96 Cm (247)	97 Bk (247)	98 Cf (247)	99 Es (254)	100 Fm (253)	101 Md (256)	102 No (254)	103 Lw (257)

Mass numbers of the most stable or most abundant isotopes are shown in parentheses

APPENDIX 2

TABLE OF ATOMIC WEIGHTS (1973)

Scaled to the relative atomic mass $A_r(^{12}C) = 12$

The atomic weights of many elements are not invariant but depend on the origin and treatment of the material. The footnotes to this table elaborate the types of variation to be expected for individual elements. The values of $A_r(E)$ given here apply to elements as they exist naturally on earth and to certain artificial elements. When used with due regard to the footnotes they are considered reliable to ±1 in the last digit or ±3 when followed by an asterisk *. Values in parentheses are used for certain radioactive elements whose atomic weights cannot be quoted precisely without knowledge of origin; the value given is the atomic mass number of the isotope of that element of longest known half-life.

Name	Symbol	Atomic number	Atomic weight	Footnotes
Actinium	Ac	89	(227)	
Aluminum	Al	13	26.98154	a
Americium	Am	95	(243)	
Antimony	Sb	51	121.75*	
Argon	Ar	18	39.948*	b, c, d, g
Arsenic	As	33	74.9216	a
Astatine	At	85	(210)	
Barium	Ba	56	137.34*	
Berkelium	Bk	97	(247)	
Beryllium	Be	4	9.01218	a
Bismuth	Bi	83	208.9804	a
Boron	B	5	10.81	c, d, e
Bromine	Br	35	79.904	c
Cadmium	Cd	48	112.40	
Calcium	Ca	20	40.08	g
Californium	Cf	98	(251)	
Carbon	C	6	12.011	b, d
Cerium	Ce	58	140.12	
Cesium	Cs	55	132.9054	a
Chlorine	Cl	17	35.453	c
Chromium	Cr	24	51.996	c
Cobalt	Co	27	58.9332	a
Copper	Cu	29	63.546*	c, d
Curium	Cm	96	(247)	
Dysprosium	Dy	66	162.50*	
Einsteinium	Es	99	(254)	
Erbium	Er	68	167.26*	
Europium	Eu	63	151.96	
Fermium	Fm	100	(257)	
Fluorine	F	9	18.99840	a
Francium	Fr	87	(223)	
Gadolinium	Gd	64	15.25*	
Gallium	Ga	31	69.72	
Germanium	Ge	32	72.59*	

(continuation)

Name	Symbol	Atomic number	Atomic weight	Footnotes
Gold	Au	79	196.9665	a
Hafnium	Hf	72	178.49*	
Helium	He	2	4.00260	b, c
Holmium	Ho	67	164.9304	a
Hydrogen	H	1	1.0079	b, d
Indium	In	49	114.82	
Iodine	I	53	126.9045	a
Iridium	Ir	77	192.22*	
Iron	Fe	26	55.847*	
Krypton	Kr	36	83.80	e
Lanthanum	La	57	138.9055*	b
Lawrencium	Lr	103	(260)	
Lead	Pb	82	207.2	d, g
Lithium	Li	3	6.941*	c, d, e, g
Lutetium	Lu	71	174.97	
Magnesium	Mg	12	24.305	c, g
Manganese	Mn	25	54.9380	a
Mendelevium	Md	101	(258)	
Mercury	Hg	80	200.59*	
Molybdenum	Mo	42	95.94*	
Neodymium	Nd	60	144.24*	
Neon	Ne	10	20.179*	c, e
Neptunium	Np	93	237.0482	f
Nickel	Ni	28	58.70	
Niobium	Nb	41	92.9064	a
Nitrogen	N	7	14.0067	b, c
Nobelium	No	102	(255)	
Osmium	Os	76	190.2	g
Oxygen	O	8	15.9994*	b, c, d
Palladium	Pd	46	106.4	
Phosphorus	P	15	30.97376	a
Platinum	Pt	78	195.09*	
Plutonium	Pu	94	(244)	
Polonium	Po	84	(209)	
Potassium	K	19	39.098*	
Praseodymium	Pr	59	140.9077	a
Promethium	Pm	61	(145)	
Protactinium	Pa	91	231.0359	f
Radium	Ra	88	226.0254	f, g
Radon	Rn	86	(222)	
Rhenium	Re	75	186.207	
Rhodium	Rh	45	102.9055	a
Rubidium	Rb	37	85.4678*	c
Ruthenium	Ru	44	101.07*	
Samarium	Sm	62	150.4	
Scandium	Sc	21	44.9559	a
Selenium	Se	34	78.96*	
Silicon	Si	14	28.086*	d

(continuation)

Name	Symbol	Atomic number	Atomic weight	Footnotes
Silver	Ag	47	107.868	c
Sodium	Na	11	22.98977	a
Strontium	Sr	38	87.62	g
Sulfur	S	16	32.06	d
Tantalum	Ta	73	180.9479*	b
Technetium	Tc	43	(97)	
Tellurium	Te	52	127.60*	
Terbium	Tb	65	158.9254	a
Thallium	Ti	81	204.37*	
Thorium	Th	90	232.0381	f, g
Thulium	Tm	69	168.9342	a
Tin	Sn	50	118.69*	
Titanium	Ti	22	47.90*	
Tungsten (Wolfram)	W	74	183.85*	
Uranium	U	92	238.029	b, c, e, g
Vanadium	V	23	50.9414*	b, c
Xenon	Xe	54	131.30	e
Ytterbium	Yb	70	173.04*	
Yttrium	Y	39	88.9059	a
Zinc	Zn	30	65.38	
Zirconium	Zr	40	91.22	

[a] Element with only one stable nuclide.

[b] Element with one predominant isotope (about 99 to 100% abundance) variations in the isotopic composition or errors in its determination have a correspondingly small effect on the value of $A_r(E)$.

[c] Element for which the value of $A_r(E)$ derives its reliability from calibrated measurements (i.e., from comparisons with synthetic mixtures of known isotopic composition).

[d] Element for which known variations in isotopic composition in terrestrial material prevent a more precise atomic weight being given; $A_r(E)$ values should be applicable to any 'normal' material.

[e] Element for which substantial variations in A_r from the value given can occur in commercially available material because of inadvertent or undisclosed change of isotopic composition.

[f] Element for which the value of A_r is that of the most commonly available long-lived isotope.

[g] Element for which geological specimens are known in which the element has an anomalous isotopic composition.

APPENDIX 3

SOME COMPOUNDS OF ANALYTICAL IMPORTANCE

Name	Formula	Formula weight
Acetic acid	CH_3COOH	60.05
Aluminum chloride	$AlCl_3 \cdot 6H_2O$	241.43
Aluminum oxide	Al_2O_3	101.96
2-amino-2(hydroxymethyl)-1,3-propanediol ["tris", or "THAM"]	$(HOCH_2)_3CNH_2$	121.14
Ammonia	NH_3	17.03
Ammonium chloride	NH_4Cl	53.49
Ammonium nitrate	NH_4NO_3	80.04
Antimony potassium tartrate	$K(SbO)C_4H_4O_6 \cdot \frac{1}{2}H_2O$	333.82
Antimony trioxide	Sb_2O_3	291.50
Arsenic trioxide	As_2O_3	197.84
Ascorbic acid	$C_6H_8O_6$	176.13
Barium chloride	$BaCl_2 \cdot 2H_2O$	244.28
Barium sulfate	$BaSO_4$	233.40
Benzene	C_6H_6	78.11
Bismuth trioxide	Bi_2O_3	465.96
Cadmium oxide	CdO	128.40
Calcium carbonate	$CaCO_3$	100.09
Calcium chloride	$CaCl_2 \cdot 2H_2O$	147.02
Calcium oxalate	$CaC_2O_4 \cdot H_2O$	178.12
Calcium oxide	CaO	56.08
Calcium sulfate	$CaSO_4 \cdot 2H_2O$	172.17
Carbon tetrachloride	CCl_4	153.82
Ceric ammonium nitrate	$(NH_4)_2Ce(NO_3)_6$	548.23
Chloroform	$CHCl_3$	119.38
Chloroplatinic acid	$H_2PtCl_6 \cdot 6H_2O$	517.92
Chromic nitrate	$Cr(NO_3)_3 \cdot 9H_2O$	400.15
Cobalt chloride	$CoCl_2 \cdot 6H_2O$	237.93
Copper(I) chloride	$CuCl$	99.00
Copper(II) oxide	CuO	79.55
Copper(II) sulfate	$CuSO_4 \cdot 5H_2O$	249.68
Dimethylglyoxime	$CH_3C(:NOH)C(:NOH)CH_3$	116.12
EDTA (ethylenediamine-tetraacetic acid), disodium salt	$Na_2H_2Y \cdot 2H_2O$	372.24
Ethanol	C_2H_5OH	46.07
Ferric ammonium sulfate	$FeNH_4(SO_4)_2 \cdot 12H_2O$	482.19
Ferric oxide	Fe_2O_3	159.69
Ferrous ammonium sulfate	$Fe(NH_4)_2(SO_4)_2 \cdot 6H_2O$	392.14

(continuation)

Name	Formula	Formula weight
Hexamethylenetetramine	$(CH_2)_6N_4$	140.19
Hydrazine sulfate	$N_2H_4 \cdot H_2SO_4$	130.12
Hydrochloric acid	HCl	36.46
Hydrogen peroxide	H_2O_2	34.01
Hydroxylamine hydrochloride	$NH_2OH \cdot HCl$ or $NH_3OH^+Cl^-$	69.49
Hydroxylamine sulfate	$(NH_2OH)_2 \cdot H_2SO_4$	164.14
Iodine	I_2	253.81
Lanthanum oxide	La_2O_3	325.84
Lead chromate	$PbCrO_4$	323.18
Lead nitrate	$Pb(NO_3)_2$	331.20
Lead sulfate	$PbSO_4$	303.25
Lithium carbonate	Li_2CO_3	73.89
Magnesium chloride	$MgCl_2 \cdot 6H_2O$	203.30
Magnesium oxide	MgO	40.30
Manganous sulfate	$MnSO_4 \cdot H_2O$	169.01
Mercuric oxide	HgO	216.59
Mercuric chloride	$HgCl_2$	271.50
Mercurous chloride [calomel]	Hg_2Cl_2	472.08
Nickel chloride	$NiCl_2 \cdot 6H_2O$	237.71
Nitric acid	HNO_3	63.01
Oxalic acid	$H_2C_2O_4 \cdot 2H_2O$	126.07
Perchloric acid	$HClO_4$	100.46
Phosphoric acid	H_3PO_4	98.00
Potassium biphthalate	$1\text{-}KOCOC_6H_4\text{-}2\text{-}COOH$	204.23
Potassium bitartrate	$KOCO(CHOH)_2COOH$	188.18
Potassium bromate	$KBrO_3$	167.00
Potassium bromide	KBr	119.01
Potassium chloride	KCl	74.56
Potassium chromate	K_2CrO_4	194.20
Potassium dichromate	$K_2Cr_2O_7$	294.19
Potassium ferricyanide	$K_3Fe(CN)_6$	329.26
Potassium ferrocyanide	$K_4Fe(CN)_6 \cdot 3H_2O$	422.41
Potassium hydroxide	KOH	56.11
Potassium iodate	KIO_3	214.00
Potassium iodide	KI	166.01
Potassium nitrate	KNO_3	101.11
Potassium periodate	KIO_4	230.00
Potassium permanganate	$KMnO_4$	158.04
Potassium persulfate	$K_2S_2O_8$	270.32
Potassium dihydrogen phosphate	KH_2PO_4	136.09

(continuation)

Name	Formula	Formula weight
Potassium monohydrogen phosphate	K_2HPO_4	174.18
Potassium thiocyanate	KSCN	97.18
Pyridine	C_5H_5N	79.10
8-quinolinol [8-hydroxyquinoline]	$HOC_6H_3N{:}CHCH{:}CH$	145.16
Silver acetate	$AgOCOCH_3$	166.91
Silver bromate	$AgBrO_3$	235.77
Silver bromide	AgBr	187.77
Silver chloride	AgCl	143.32
Silver iodate	$AgIO_3$	282.77
Silver iodide	AgI	234.77
Silver nitrate	$AgNO_3$	169.87
Silver oxide	Ag_2O	231.74
Sodium acetate	$NaOCOCH_3 \cdot 3H_2O$	136.08
Sodium arsenate	$Na_2HAsO_4 \cdot 7H_2O$	312.01
Sodium bicarbonate	$NaHCO_3$	84.01
Sodium bisulfate	$NaHSO_4 \cdot H_2O$	138.07
Sodium bisulfite	$NaHSO_3$	104.06
Sodium borate [borax]	$Na_2B_4O_7 \cdot 10H_2O$	381.37
Sodium bromide	NaBr	102.89
Sodium carbonate	Na_2CO_3	105.99
Sodium chloride	NaCl	58.44
Sodium dihydrogen phosphate	$NaH_2PO_4 \cdot 2H_2O$	156.01
Sodium fluoride	NaF	41.99
Sodium hydroxide	NaOH	40.00
Sodium oxalate	$Na_2C_2O_4$	134.00
Sodium perchlorate	$NaClO_4$	122.46
Sodium sulfate	Na_2SO_4	142.04
Sodium thiosulfate	$Na_2S_2O_3 \cdot 5H_2O$	248.18
Stannous chloride	$SnCl_2 \cdot 2H_2O$	225.63
Strontium carbonate	$SrCO_3$	147.63
Sulfamic acid	NH_2SO_3H	97.09
Sulfuric acid	H_2SO_4	98.08
Uranyl acetate	$UO_2(CH_3COO)_2 \cdot 2H_2O$	424.15
Urea	NH_2CONH_2	60.06
Vanadium pentoxide	V_2O_5	181.88
Zinc chloride	$ZnCl_2$	136.28
Zinc nitrate	$Zn(NO_3)_2 \cdot 6H_2O$	297.47
Zinc oxide	ZnO	81.37

APPENDIX 4

ACID-DISSOCIATION CONSTANTS (shown as pK_a values)

In this Appendix a number of acids and bases are arranged in alphabetical order according to their most commonly used name. For example, there is an entry under AMMONIA rather than under AMMONIUM ION. However, all the pK values refer to the acid dissociation of the conjugate acids, with the most acidic species shown at the left of each diagram. Each entry gives the name, the formula, an abbreviated symbolic formula if appropriate, and a molecular weight for the molecular species. To illustrate the use of the diagrams we consider the entry under ADIPIC ACID: the correct interpretation is as follows:

$$K_1 = \frac{[H^+][HAd^-]f_1^2}{[H_2Ad]} = 10^{-4.418}, \qquad K_2 = \frac{[H^+][Ad^{2-}]f_2}{[HAd^-]} = 10^{-5.142}.$$

ACETIC ACID CH_3COOH HOAc   60.05

HOAc $\xrightarrow{4.756}$ OAc$^-$
     acetate ion

ADIPIC ACID $HOOC(CH_2)_4COOH$ H_2Ad 146.14

$H_2Ad \xrightarrow{4.418}$ HAd$^- \xrightarrow{5.412}$ Ad^{2-}
                                   adipate ion

ALANINE CH_3CHNH_2COOH HAla   89.09

$H_2Ala^+ \xrightarrow{2.348}$ HAla $\xrightarrow{9.83}$ Ala$^-$

AMMONIA NH_3 17.03

NH_4^+ $\xrightarrow{9.24}$ NH_3
ammonium
ion

ARGININE $NH_2C(NH)NHCH_2CH_2CH_2CHNH_2COOH$ H_2Arg 174.20

$H_3Arg^+ \xrightarrow{1.882}$ $H_2Arg \xrightarrow{8.994}$ HArg$^- \xrightarrow{12.48}$ Arg^{2-}

ARSENIC ACID H_3AsO_4

$H_3AsO_4 \xrightarrow{2.2}$ $H_2AsO_4^- \xrightarrow{7.0}$ HAsO$_4^{2-} \xrightarrow{11.5}$ AsO$_4^{3-}$
                                            arsenate ion

ARSENIOUS ACID $HAsO_2$

$$HAsO_2 \xrightarrow{\ 9.2\ } AsO_2^-$$
arsenite ion

ASPARAGINE $NH_2COCH_2CHNH_2COOH$ HAsn   132.12

$$H_2Asn^+ \xrightarrow{\ 2.02\ } HAsn \xrightarrow{\ 8.8\ } Asn^-$$

ASPARTIC ACID $HOOCCH_2CHNH_2COOH$ H_2Asp 133.10

$$H_3Asp^+ \xrightarrow{\ 2.05\ } H_2Asp \xrightarrow{\ 3.87\ } HAsp^- \xrightarrow{\ 10.00\ } Asp^{2-}$$

BENZOIC ACID C_6H_5COOH HBz   122.13

$$HBz \xrightarrow{\ 4.20\ } C_6H_5COO^-$$
benzoate ion

BORIC ACID H_3BO_3 61.83

$$H_3BO_3 \xrightarrow{\ 9.23\ } H_2BO_3^-$$
borate ion

BROMIC ACID $HBrO_3$

$$HBrO_3 \xrightarrow{\ strong\ } BrO_3^-$$
bromate ion

CARBONIC ACID $H_2CO_3 + CO_2(aq)$

$$H_2CO_3 + CO_2(aq) \xrightarrow{\ 6.4\ } HCO_3^- \xrightarrow{\ 10.33\ } CO_3^{2-}$$
bicarbonate ion   carbonate ion

CHLORIC ACID $HClO_3$

$$HClO_3 \xrightarrow{\ strong\ } ClO_3^-$$
chlorate ion

CHLOROUS ACID $HClO_2$

$$HClO_2 \xrightarrow{\ 2.0\ } ClO_2^-$$
chlorite ion

CHROMIC ACID H_2CrO_4

$$H_2CrO_4 \xrightarrow{\ -1\ } HCrO_4^- \xrightarrow{\ 6.5\ } CrO_4^{2-}$$
$\updownarrow$ chromate ion
$Cr_2O_7^{2-}$
dichromate ion

CITRIC ACID $HOC(COOH)(CH_2COOH)_2$ H_3Cit 192.13

H_3Cit $\xrightarrow{\quad 3.220 \quad}$ H_2Cit^- $\xrightarrow{\quad 4.837 \quad}$ $HCit^{2-}$ $\xrightarrow{\quad 6.393 \quad}$ Cit^{3-}
citrate ion

CYSTEINE $HSCH_2CHNH_2COOH$ H_2Cys 121.16

H_3Cys^+ $\xrightarrow{\quad 1.85 \quad}$ H_2Cys $\xrightarrow{\quad 8.36 \quad}$ $HCys^-$ $\xrightarrow{\quad 10.72 \quad}$ Cys^{2-}

EDTA (H_4Y) (see p. 507)

H_4Y $\xrightarrow{\quad 2.0 \quad}$ H_3Y^- $\xrightarrow{\quad 2.67 \quad}$ H_2Y^{2-} $\xrightarrow{\quad 6.16 \quad}$ HY^{3-} $\xrightarrow{\quad 10.26 \quad}$ Y^{4-}

EGTA (H_4Y) (see p. 508)

H_4Y $\xrightarrow{\quad 2.0 \quad}$ H_3Y^- $\xrightarrow{\quad 2.68 \quad}$ H_2Y^{2-} $\xrightarrow{\quad 8.85 \quad}$ HY^{3-} $\xrightarrow{\quad 9.43 \quad}$ Y^{4-}

ETHANOLAMINE $HOCH_2CH_2NH_2$ B   61.09

$HOCH_2CH_2NH_3^+$ $\xrightarrow{\quad 9.50 \quad}$ B
ethanolammonium
ion

ETHYLAMINE $C_2H_5NH_2$ B   45.09

$C_2H_5NH_3^+$ $\xrightarrow{\quad 10.63 \quad}$ B
ethylammonium
ion

ETHYLENEDIAMINE $NH_2CH_2CH_2NH_2$ en   60.11

H_2en^{2+} $\xrightarrow{\quad 7.13 \quad}$ Hen^+ $\xrightarrow{\quad 9.91 \quad}$ en

FORMIC ACID   HCOOH   46.03

HCOOH $\xrightarrow{\quad 3.751 \quad}$ $HCOO^-$
formate ion

FUMARIC ACID $HOOCCH{=}CHCOOH$ (trans) H_2Fum 116.07

H_2Fum $\xrightarrow{\quad 3.095 \quad}$ $HFum^-$ $\xrightarrow{\quad 4.602 \quad}$ Fum^{2-}
fumarate ion

GLUTAMIC ACID $HOOCCH_2CH_2CHNH_2COOH$ H_2Glu 147.13

H_3Glu^+ $\xrightarrow{\quad 2.1 \quad}$ H_2Glu $\xrightarrow{\quad 4.2 \quad}$ $HGlu^-$ $\xrightarrow{\quad 9.67 \quad}$ Glu^{2-}

GLUTAMINE　　$NH_2COCH_2CH_2CHNH_2COOH$　　HGln　　146.15

H_2Gln^+ —— 2.17 —— HGln —— 9.13 —— Gln^-

GLUTARIC ACID　　$HOOC(CH_2)_3COOH$　　H_2Gl　　132.13

H_2Gl —— 4.344 —— HGl^- —— 5.420 —— Gl^{2-}
　　　　　　　　　　　　　　　　glutarate ion

GLYCINE　　NH_2CH_2COOH　　HGly　　75.07

H_2Gly^+ —— 2.351 —— HGly —— 9.779 —— Gly^-

HISTIDINE　　$HC{=}C{-}CH_2CHNH_2COOH$　　HHis　　155.16

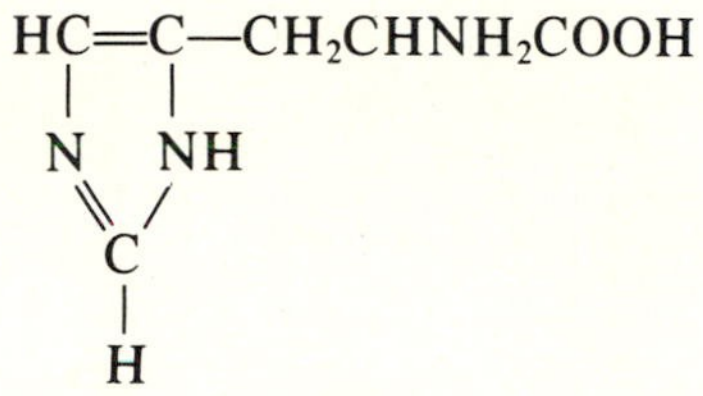

$$
\begin{array}{cc}
| & | \\
N & NH \\
\diagdown & \diagup \\
& C \\
& | \\
& H
\end{array}
$$

H_2His^+ —— 6.00 —— HHis —— 9.16 —— His^-

HYDRAZINE　　N_2H_4

$N_2H_5^+$ —— 8.0 —— N_2H_4
hydrazinium
ion

HYDROBROMIC ACID　　HBr

HBr —— strong —— Br^-
　　　　　　　　bromide ion

HYDROCHLORIC ACID　　HCl　　36.46

HCl —— strong —— Cl^-
　　　　　　　chloride ion

HYDROCYANIC ACID　　HCN

HCN —— 9.4 —— CN^-
　　　　　　　cyanide ion

HYDROFLUORIC ACID　　HF　　20.01

HF —— 3.2 —— F^-
　　　　　fluoride ion

HYDROSULFURIC ACID (HYDROGEN SULFIDE) H_2S

$$H_2S \xrightarrow{\quad 7.0 \quad} HS^- \xrightarrow{\quad 12.9 \quad} S^{2-}$$
sulfide ion

HYDROXYLAMINE NH_2OH

$$NH_3OH^+ \xrightarrow{\quad 6.0 \quad} NH_2OH$$
hydroxyl-
ammonium ion

HYPOBROMOUS ACID   HOBr

$$HOBr \xrightarrow{\quad 8.7 \quad} OBr^-$$
hypobromite ion

HYPOCHLOROUS ACID   HOCl

$$HOCl \xrightarrow{\quad 7.5 \quad} OCl^-$$
hypochlorite ion

HYPOIODOUS ACID   HOI

$$HOI \xrightarrow{\quad 11.0 \quad} OI^-$$
hypoiodite ion

HYPOPHOSPHOROUS ACID H_3PO_2

$$H_3PO_2 \xrightarrow{\quad 1.1 \quad} H_2PO_2^-$$

IODIC ACID HIO_3

$$HIO_3 \xrightarrow{\quad 0.8 \quad} IO_3^-$$
iodate ion

ISOLEUCINE $CH_3CH_2CH(CH_3)CHNH_2COOH$ HIle   131.17

$$H_2Ile^+ \xrightarrow{\quad 2.318 \quad} HIle \xrightarrow{\quad 9.758 \quad} Ile^-$$

LEUCINE $(CH_3)_2CHCH_2CHNH_2COOH$ HLeu   131.17

$$H_2Leu^+ \xrightarrow{\quad 2.328 \quad} HLeu \xrightarrow{\quad 9.744 \quad} Leu^-$$

LYSINE $NH_2(CH_2)_4CHNH_2COOH$ HLys   146.19

$$H_2Lys^+ \xrightarrow{\quad 2.18 \quad} HLys \xrightarrow{\quad 8.95 \quad} Lys^-$$

MALEIC ACID   HOOCCH=CHCOOH (cis) H_2Ma 116.07

$$H_2Ma \xrightarrow{\hspace{5pt}1.910\hspace{5pt}} HMa^- \xrightarrow{\hspace{5pt}6.332\hspace{5pt}} Ma^{2-}$$
$$\text{maleate ion}$$

MALIC ACID   HOOCCH(OH)CH$_2$COOH H_2Mac 134.09

$$H_2Mac \xrightarrow{\hspace{5pt}3.458\hspace{5pt}} HMac^- \xrightarrow{\hspace{5pt}5.097\hspace{5pt}} Mac^{2-}$$
$$\text{malate ion}$$

MALONIC ACID   HOOCCH$_2$COOH H_2Mal 104.06

$$H_2Mal \xrightarrow{\hspace{5pt}2.826\hspace{5pt}} HMal^- \xrightarrow{\hspace{5pt}5.696\hspace{5pt}} Mal^{2-}$$
$$\text{malonate ion}$$

METHIONINE   CH$_3$SCH$_2$CH$_2$CHNH$_2$COOH   HMet   149.21

$$H_2Met^+ \xrightarrow{\hspace{5pt}2.12\hspace{5pt}} HMet \xrightarrow{\hspace{5pt}9.28\hspace{5pt}} Met^-$$

NITRIC ACID   HNO$_3$ 63.01

$$HNO_3 \xrightarrow{\hspace{5pt}\text{strong}\hspace{5pt}} NO_3^-$$
$$\text{nitrate ion}$$

NITROUS ACID   HNO$_2$

$$HNO_2 \xrightarrow{\hspace{5pt}3.3\hspace{5pt}} NO_2^-$$
$$\text{nitrite ion}$$

OXALIC ACID   HOOCCOOH H_2Ox 90.04

$$H_2Ox \xrightarrow{\hspace{5pt}1.271\hspace{5pt}} HOx^- \xrightarrow{\hspace{5pt}4.266\hspace{5pt}} Ox^{2-}$$
$$\text{oxalate ion}$$

PERCHLORIC ACID   HClO$_4$ 100.46

$$HClO_4 \xrightarrow{\hspace{5pt}\text{strong}\hspace{5pt}} ClO_4^-$$
$$\text{perchlorate ion}$$

PERIODIC ACID   H$_5$IO$_6$

$$H_5IO_6 \xrightarrow{\hspace{5pt}1.6\hspace{5pt}} IO_4^-$$
$$\text{periodate ion}$$

PERMANGANIC ACID   HMnO$_4$

$$HMnO_4 \xrightarrow{\hspace{5pt}\text{strong}\hspace{5pt}} MnO_4^-$$
$$\text{permanganate ion}$$

PHENOL C_6H_5OH 94.11

$C_6H_5OH \xrightarrow{\quad 9.98 \quad} C_6H_5O^-$
phenolate ion

PHENYLALANINE $C_6H_5-CH_2CHNH_2COOH$ HPhe   165.19

$H_2Phe^+ \xrightarrow{\quad 1.83 \quad} HPhe \xrightarrow{\quad 9.31 \quad} Phe^-$

PHOSPHORIC ACID H_3PO_4 98.00

$H_3PO_4 \xrightarrow{\quad 2.12 \quad} H_2PO_4^- \xrightarrow{\quad 7.21 \quad} HPO_4^{2-} \xrightarrow{\quad 12.32 \quad} PO_4^{3-}$
phosphate ion

PHOSPHOROUS ACID H_3PO_3

$H_3PO_3 \xrightarrow{\quad 2.0 \quad} H_2PO_3^- \xrightarrow{\quad 6.6 \quad} HPO_3^{2-}$

PHTHALIC ACID $C_6H_4(COOH)_2$ (ortho) H_2Ph 166.14

$H_2Ph \xrightarrow{\quad 2.950 \quad} HPh^- \xrightarrow{\quad 5.408 \quad} Ph^{2-}$
phthalate ion

PICRIC ACID $C_6H_2(NO_2)_3OH$ 229.11

$C_6H_2(NO_2)_3OH \xrightarrow{\quad 0.19 \quad} C_6H_2(NO_2)_3O^-$
picrate ion

PROLINE $(C_4H_8N)COOH$ HPro   115.13

$H_2Pro^+ \xrightarrow{\quad 1.952 \quad} HPro \xrightarrow{\quad 10.64 \quad} Pro^-$

PROPIONIC ACID C_2H_5COOH 74.07

$C_2H_5COOH \xrightarrow{\quad 4.874 \quad} C_2H_5COO^-$
propionate ion

PYRIDINE C_5H_5N 79.10

$C_5H_5NH^+ \xrightarrow{\quad 5.17 \quad} C_5H_5N$
pyridinium
ion

SERINE $HOCH_2CHNH_2COOH$ HSer   105.10

$H_2Ser^+ \xrightarrow{\quad 2.186 \quad} HSer \xrightarrow{\quad 9.208 \quad} Ser^-$

SUCCINIC ACID $HOOCCH_2CH_2COOH$ H_2Suc 118.09

$$H_2Suc \xrightarrow{\quad 4.207 \quad} HSuc^- \xrightarrow{\quad 5.635 \quad} Suc^{2-}$$
succinate ion

SULFAMIC ACID NH_2SO_3H 97.09

$$NH_2SO_3H \xrightarrow{\quad 0.99 \quad} NH_2SO_3^-$$
sulfamate ion

SULFURIC ACID H_2SO_4 98.08

$$H_2SO_4 \xrightarrow{\quad strong \quad} HSO_4^- \xrightarrow{\quad 2.0 \quad} SO_4^{2-}$$
bisulfate   sulfate
ion   ion

SULFUROUS ACID H_2SO_3

$$H_2SO_3 \xrightarrow{\quad 1.8 \quad} HSO_3^- \xrightarrow{\quad 7.2 \quad} SO_3^{2-}$$
sulfite ion

TARTARIC ACID $HOOCCH(OH)CH(OH)COOH$ H_2Tar 150.09

$$H_2Tar \xrightarrow{\quad 3.036 \quad} HTar^- \xrightarrow{\quad 4.366 \quad} Tar^{2-}$$
tartrate ion

TETREN   (Y)

$$H_5Y^{5+} \xrightarrow{\quad 2.6 \quad} H_4Y^{4+} \xrightarrow{\quad 4.1 \quad} H_3Y^{3+} \xrightarrow{\quad 8.2 \quad} H_2Y^{2+} \xrightarrow{\quad 9.2 \quad} HY^+ \xrightarrow{\quad 10.0 \quad} Y$$

THIOCYANIC ACID   HSCN

$$HSCN \xrightarrow{\quad strong \quad} SCN^-$$
thiocyanate ion

THREONINE $CH_3CH(OH)CHNH_2COOH$ HThr   119.12

$$H_2Thr^+ \xrightarrow{\quad 2.09 \quad} HThr \xrightarrow{\quad 9.10 \quad} Thr^-$$

TRIEN   (Y)   (see p. 507)

$$H_4Y^{4+} \xrightarrow{\quad 3.32 \quad} H_3Y^{3+} \xrightarrow{\quad 6.67 \quad} H_2Y^{2+} \xrightarrow{\quad 9.20 \quad} HY^+ \xrightarrow{\quad 9.92 \quad} Y$$

TRIS $(HOCH_2)_3CNH_2$ 121.14

$$(HOCH_2)_3CNH_3^+ \xrightarrow{\quad 8.069 \quad} (HOCH_2)_3CNH_2$$
trisH$^+$ tris-hydroxymethylaminomethane

TRYPTOPHAN

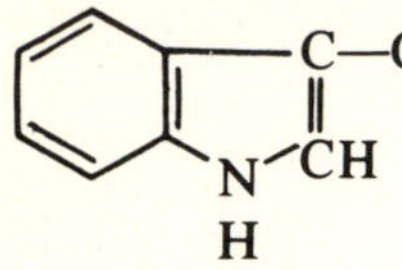

$C-CH_2CHNH_2COOH$ HTrp   204.23

$H_2Trp^+ \xrightarrow{\quad 2.46 \quad} HTrp \xrightarrow{\quad 9.41 \quad} Trp^-$

TYROSINE $HOC_6H_4CH_2CHNH_2COOH$ H_2Tyr 181.19

$H_3Tyr^+ \xrightarrow{\quad 2.2 \quad} H_2Tyr \xrightarrow{\quad 9.11 \quad} HTyr^- \xrightarrow{\quad 10.05 \quad} Tyr^{2-}$

VALINE $(CH_3)_2CHCHNH_2COOH$ HVal   117.15

$H_2Val^+ \xrightarrow{\quad 2.29 \quad} HVal \xrightarrow{\quad 9.72 \quad} Val^-$

WATER H_2O 18.02

$H_3O^+ \xrightarrow{\quad -1.7 \quad} H_2O \xrightarrow{\quad 15.7 \quad} OH^-$
hydronium   hydroxide
ion   ion

APPENDIX 5

ACID–BASE, REDOX, AND COMPLEX FORMATION EQUILIBRIA OF SELECTED ELEMENTS

For each element the values of standard-reduction potentials are shown to the right of the vertical lines connecting the two species of the redox couple. The numbers above the horizontal lines are the pK_a (acid dissociation) values for the proton-transfer couple thus connected. Ox. no. means oxidation number.

ALUMINUM (Latin, *alumen*, alum)

Comments: Aluminum metal is coated with alumina Al_2O_3 which protects it from rapid attack by oxygen and moisture in the air. The metal dissolves in hot HCl. Solutions must be kept acidic to prevent precipitation of aluminum hydroxide.

Ox. no.

+3 $Al(aq)^{3+} \xrightarrow{\ 5.0\ } AlOH^{2+} \underline{\qquad} Al(OH)_3(s),$ $\log K_{sp} = -33.5$

     aluminum   aluminum
     ion   hydroxide

     $\left|\ -1.66\ \text{V}\right.$

0   Al
     aluminum metal

Metal–ligand complexes with Al^{3+}

Ligand	Log of formation quotients				Solubility products	
	Q_1	β_2	β_3	β_4	Precipitate	$\log K_{sp}$
EDTA	16.1					
Fluoride	7.0	12.6	16.7	19.1		
Hydroxide	9.1	18.7	27.0	33.0	$Al(OH)_3$	−33.5
Oxalate	6.1	11.1	15.1			

ARSENIC (Greek, *arsenikon*, yellow orpiment, the mineral As_2S_3)

Comments: The chief analytical significance of arsenic is the reaction with iodine:

$$I_2(\text{or } I_3^-) + As(III) = 2I^- + As(V).$$

Solutions of As(III) are best prepared by dissolving pure arsenious oxide As_2O_3 in dilute NaOH followed by slight acidification. The resulting solution of $HAsO_2$ is quite stable and may be used as a titrant for iodine. Because of the acid–base characteristics summarized in the diagram, it is important to control the pH of the titrated solution. This is discussed more fully in Chapter 12.

Ox. no.

$+5 \qquad H_3AsO_4 \xrightarrow{\;2.2\;} H_2AsO_4^- \xrightarrow{\;7.0\;} HAsO_4^{2-} \xrightarrow{\;11.5\;} AsO_4^{3-}$

arsenic acid $\qquad\qquad\qquad\qquad\qquad\qquad\qquad$ arsenate ion

$\qquad\qquad\left|\; +0.56\text{ V}\right.$

$+3 \qquad HAsO_2 \xrightarrow{\;9.2\;} AsO_2^-$

arsenious acid arsenite ion

$\qquad\qquad\left|\; +0.25\text{ V}\right.$

$0 \qquad As$

arsenic

BARIUM (Greek, *barys*, heavy)

Comments: The chief analytical importance of barium ion is its ability to form a precipitate with sulfate and with carbonate.

Ox. no.

$+2$ Ba^{2+}

barium ion

$\Big|\; -2.9\ V$

0 Ba

barium metal

Metal-ligand complexes with Ba^{2+}

Ligand	Log of formation quotients	Solubility products	
	Q_1	Precipitate	$\log K_{sp}$
Carbonate		$BaCO_3$	-8.1
Chromate		$BaCrO_4$	-9.7
EDTA	7.8		
Fluoride		BaF_2	-5.8
Iodate		$Ba(IO_3)_2 \cdot 2H_2O$	-8.8
Oxalate	2.3	$BaC_2O_4 \cdot H_2O$	-7.6
Sulfate		$BaSO_4$	-10

BROMINE (Greek, *bromos*, stench)

Comments: Pure bromine is available as a liquid of high vapor pressure with a vile smell. It may be used as a water solution (solubility about 0.2*M*). Instead of using a solution of bromine as a titrant it is better to use a neutral mixture of potassium bromate and potassium bromide which, when added to an acidic medium, react quickly to form bromine *in situ*.

Hydrobromic acid is available as a 48% stock solution, and dilute solutions of HBr are quite stable.

Bromine is reddish-brown, but the other species are colorless.

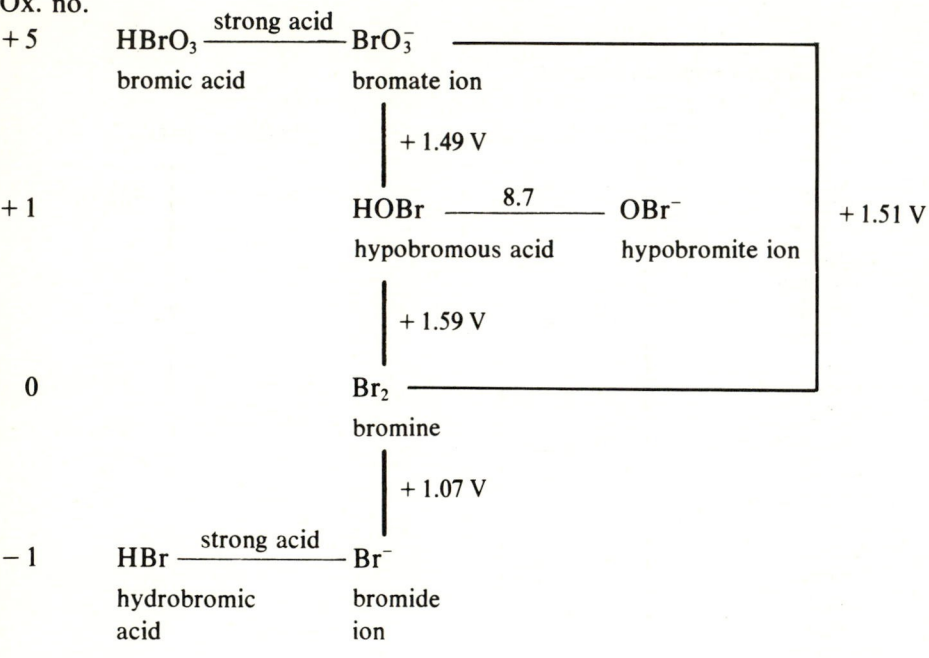

CADMIUM (Greek, *kadmeia*, ancient name for calamine)

Comments: Cadmium metal will dissolve in acid, but standard solutions may also be prepared by dissolving the oxide, CdO.

Ox. no

+2 $Cd(aq)^{2+}$ ——$\overset{11.7}{\text{———}}$—— $CdOH^+$ ————————$Cd(OH)_2(s)$, $K_{sp} = 10^{-14.4}$

$\Big|$ $-0.40\ V$

0   Cd

Metal-ligand complexes with Cd^{2+}

Ligand	Log of formation quotients				Solubility products	
	Q_1	β_2	β_3	β_4	Precipitate	$\log K_{sp}$
Acetate	1.9	3.2				
Ammonia	2.6	4.6	5.9	6.7		
Bromide	2.1	3.0	3.0	2.9		
Chloride	2.0	2.6	2.4	1.7		
Citrate	4.2					
Cyanide	6.0	11.1	15.7	17.9		
EDTA	16.5					
Ethylenediamine	5.5	10.1	12.2			
Fluoride	0.46	0.53				
Hydroxide	3.9	7.7			$Cd(OH)_2$	-14.4
Iodide	2.3	3.9	5.0	6.0		
Oxalate	3.9					
Pyridine	1.3	2.1				
Sulfate	2.3					
Sulfide					CdS	-27.8
TETREN	14.0					
Thiocyanate	1.9	2.8	2.8	2.3		

CALCIUM (Latin, *calx*, lime)

Comments: Calcium ion is a useful precipitant for oxalate ion. As a constituent of limestone and of ground water, calcium is determined typically by titration with EDTA.

Ox. no.

+2 Ca^{2+} ——————— $Ca(OH)_2(s)$, $K_{sp} = 10^{-5.2}$

calcium ion

$\Big|$ -2.9 V

0   Ca

calcium metal

Metal-ligand complexes with Ca^{2+}

Ligand	Log of formation quotients	Solubility products	
	Q_1	Precipitate	log K_{sp}
Acetate	1.2		
Carbonate		$CaCO_3$	-8.3
Citrate	3.2		
EDTA	10.7		
Fluoride	1.1	CaF_2	-10.4
Hydroxide	1.3	$Ca(OH)_2$	-5.2
Oxalate	3.0	$CaC_2O_4 \cdot H_2O$	-8.4
Phosphate		$Ca_3(PO_4)_2$	-28.7
Sulfate	2.3	$CaSO_4$	-5.9

CARBON (Latin, *carbo*, charcoal)

Comments: Of the redox couples it is the oxalic acid/carbon dioxide couple that is of most importance in analytical chemistry. Oxalic acid serves as standard for permanganate titrations.

As a component of normal air, carbon dioxide is an omnipresent contaminant of aqueous solutions. It is only slightly hydrated to form carbonic acid H_2CO_3 which then can dissociate as shown below.

The E^0 values for carbon redox couples have to be deduced from the standard Gibbs free energies of formation because the redox changes are not reversible at platinum electrodes.

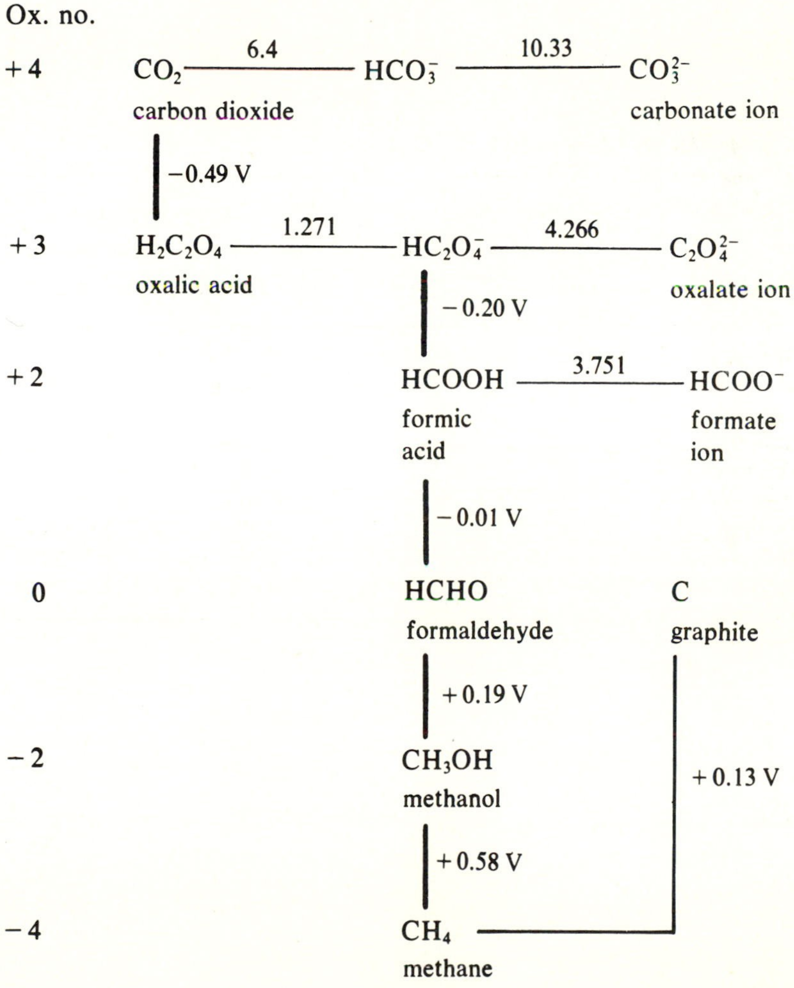

Ox. no.

+4 CO_2 —— 6.4 —— HCO_3^- —— 10.33 —— CO_3^{2-}
       carbon dioxide   carbonate ion

       -0.49 V

+3 $H_2C_2O_4$ —— 1.271 —— $HC_2O_4^-$ —— 4.266 —— $C_2O_4^{2-}$
       oxalic acid   oxalate ion

       -0.20 V

+2   HCOOH —— 3.751 —— $HCOO^-$
                             formic   formate
                             acid   ion

       -0.01 V

0   HCHO   C
                             formaldehyde   graphite

       $+0.19$ V $+0.13$ V

-2 CH_3OH
                             methanol

       $+0.58$ V

-4 CH_4
                             methane

CERIUM (named for the asteroid Ceres, which was discovered in 1801 only two years before the element)

Comments: Cerium(IV) is a useful oxidizing agent. Standard solutions are easily prepared from the salt ammonium hexanitratocerate(IV). Solutions must be acidic to prevent precipitation of ceric and/or cerous hydroxides.
    Ceric ion is yellow and cerous ion is colorless.

Ox. no.

+4 Ce^{4+} ——————— $CeOH^{3+}$ ——————— $Ce(OH)_4(s)$, K_{sp} very small
        ceric ion

        | +1.7 V

+3 Ce^{3+} ——————— $Ce(OH)_3(s)$, $K_{sp} = 10^{-19.8}$
        cerous ion   cerous hydroxide

Metal-ligand complexes with Ce^{3+}

Ligand	Log of formation quotients			Solubility products	
	Q_1	β_2	β_3	Precipitate	$\log K_{sp}$
Fluoride	3.2				
Hydroxide				$Ce(OH)_3$	−19.8
Iodate				$Ce(IO_3)_3$	− 9.5
Oxalate	6.5	10.5	11.3	$Ce_2(C_2O_4)_3 \cdot 9H_2O$	−28.5
Sulfate	1.6	2.3			

CHLORINE (Greek, *chloros*, greenish-yellow)

Comments: Perchloric acid is available as a 70.5% stock solution ($11.7M$). The concentrated acid should never be heated with organic matter or other reducing substances because a violent explosion may result. However, the acid is perfectly safe and stable in dilute solutions (for example, $2M$ or lower). Perchlorate ion is generally regarded as having virtually no tendency to form complexes with metal ions, and therefore perchlorate solutions serve as important reference points in equilibrium studies.

Hypochlorites are readily available in commercial bleach preparations.

Hydrochloric acid is perhaps the most commonly used acid in laboratory work and is available as a 37.2% stock solution ($12.1M$). At this concentration it gives off irritating HCl gas, but solutions diluted to $6M$ or less are stable. Chlorine is a yellow-green gas, but the other species are colorless.

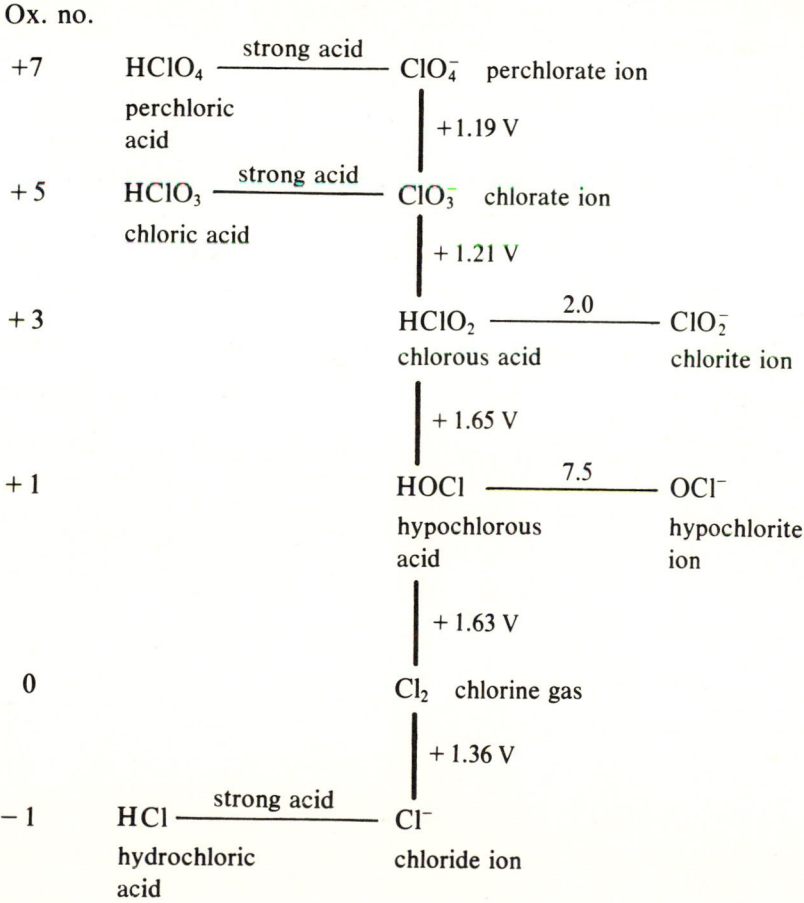

Ox. no.

+7 $HClO_4$ —— strong acid —— ClO_4^- perchlorate ion
      perchloric acid
      $+1.19\ V$

+5 $HClO_3$ —— strong acid —— ClO_3^- chlorate ion
      chloric acid
      $+1.21\ V$

+3 $HClO_2$ —— 2.0 —— ClO_2^-
      chlorous acid   chlorite ion
      $+1.65\ V$

+1 $HOCl$ —— 7.5 —— OCl^-
      hypochlorous acid   hypochlorite ion
      $+1.63\ V$

0   Cl_2 chlorine gas
      $+1.36\ V$

−1 HCl —— strong acid —— Cl^-
      hydrochloric acid   chloride ion

CHROMIUM (Greek, *chroma*, color)

Comments: Solutions of Cr(VI) are easily prepared by dissolving pure potassium dichromate. The chromate ion CrO_4^{2-} is yellow, while the other +6 species are orange. Cr(VI) solutions are very useful in the analysis as oxidants less powerful than Ce(IV) or Mn(VII).

Solutions of Cr(III) are prepared, for example, from chromic nitrate nonahydrate. The aquo Cr(III) ion is purplish. Powerful reductants, such as amalgamated zinc, will reduce Cr(III) to the chromous state, which is light blue.

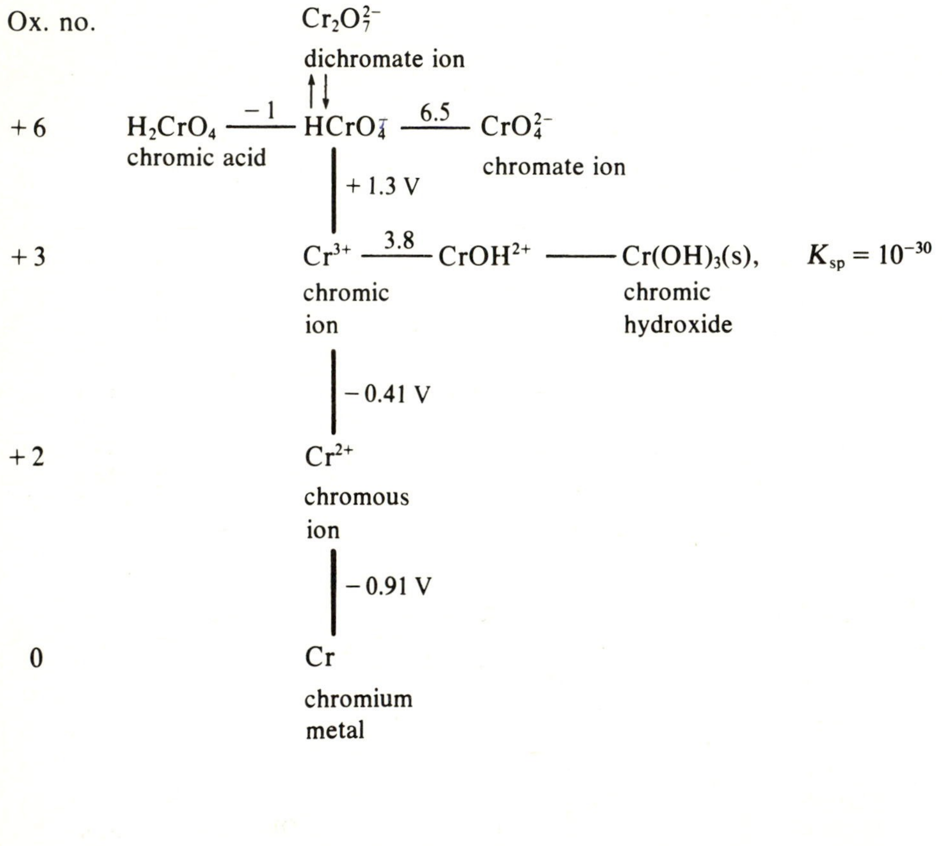

Ox. no.

$$Cr_2O_7^{2-}$$
dichromate ion

+6 $H_2CrO_4 \xrightarrow{-1} HCrO_4^- \xrightarrow{6.5} CrO_4^{2-}$
      chromic acid   chromate ion

                           + 1.3 V

+3 $Cr^{3+} \xrightarrow{3.8} CrOH^{2+} \text{———} Cr(OH)_3(s),$ $K_{sp} = 10^{-30}$
      chromic   chromic
      ion   hydroxide

                           − 0.41 V

+2 Cr^{2+}

      chromous
      ion

                           − 0.91 V

0   Cr

    chromium
    metal

COPPER (Latin, *cuprum*, from the island of Cyprus)

Comments: Solutions of cupric salts (sulfate, nitrate, perchlorate) are blue. The cuprous ion is not stable unless complexed.

Ox. no.

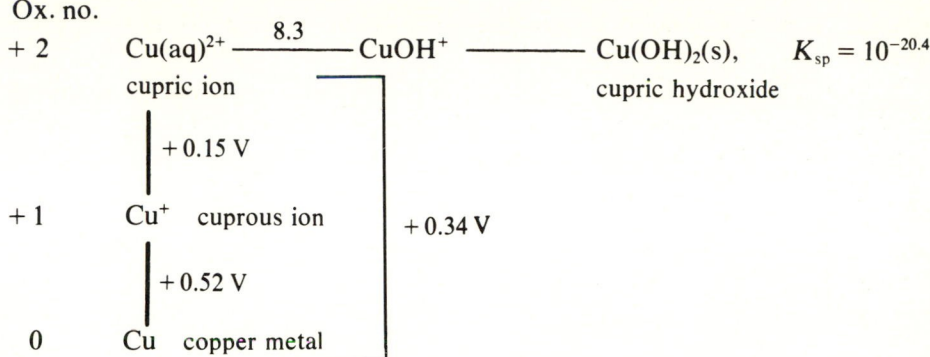

Metal–ligand complexes with Cu^+

	Log of formation quotients				Solubility products	
Ligand	Q_1	β_2	β_3	β_4	Precipitate	$\log K_{sp}$
Ammonia	5.9	10.8				
Bromide					$CuBr$	−8.3
Chloride		4.7			$CuCl$	−5.9
Cyanide		16.3	21.6	23.1	$CuCN$	−19.5
Ethylenediamine		10.8				
Iodide		8.9			CuI	−12.0
Thiocyanate					$CuSCN$	−14.3
Sulfide					Cu_2S	−47.6

Metal–ligand complexes with Cu^{2+}

	Q_1	β_2	β_3	β_4	Precipitate	$\log K_{sp}$
Acetate	2.2	3.6				
Ammonia	4.0	7.5	10.3	11.8		
Chloride	0.4					
Citrate	14.2					
EDTA	18.8					
Ethylenediamine	10.7	20.0				
Fluoride	1.2					
Hydroxide	6.3				$Cu(OH)_2$	−20.4
Iodate					$Cu(IO_3)_2 \cdot H_2O$	−7.1
Oxalate	6.2	10.3			$CuC_2O_4 \cdot \frac{1}{2}H_2O$	−7.6
Pyridine	2.4	4.3	5.4	6.0		
Sulfate	2.4					
Sulfide					CuS	−35.2
TETREN	22.9					
Thiocyanate	2.3	3.7				

FLUORINE (Latin, *fluere*, flow, or flux)

Comments: Fluorine is so highly reactive that it is never used in routine laboratory work. It even causes fingernails to burst into flame. In contrast to the other hydrogen halides, HF is a weak acid in aqueous solution. Because of the tendency of fluoride ion to form stable complexes with Al and Si, HF is useful in decomposing rock samples prior to analysis. Trace concentrations of fluoride ion are easily determined by using a lanthanum fluoride selective ion electrode.

Ox. no.

$0 \qquad F_2$
fluorine gas

$\bigg| +2.9\,V$

$-1 \qquad HF \underline{\quad 3.2 \quad} F^-$
hydrofluoric   fluoride ion
acid

HYDROGEN (Greek, *hydro*, water, and *genes*, forming)

Comments: The amphiprotic character of water is of central importance in considering acid–base reactions in aqueous solutions. The H_3O^+, H_2 couple is the primary standard for the scale of reduction potentials. The control of pH is vital in the majority of experiments in equilibrium and analysis.

Ox. no.

$+1 \qquad H_3O^+ \underline{\quad -1.7 \quad} H_2O \underline{\quad 15.7 \quad} OH^-$
hydronium ion   hydroxide ion

$\bigg| 0.000\,V$ (reference point for the scale)

$0 \qquad H_2$
hydrogen gas

IODINE (Greek, *iodes*, violet)

Comments: Iodine is my favorite element. The solubility of the solid form in pure water is only $1.3 \cdot 10^{-3}M$ at 25°, but the addition of potassium iodide greatly increases the solubility by forming I_3^- with $K_f = 720$ for $I_2(aq) + I^- = I_3^-$, and such solutions are useful as oxidizing titrants. The above species are colorless except for $I_2(aq)$ and I_3^-, which are brown. Solid I_2 is a purple-black crystalline substance. Solutions of KI slowly turn brown due to air oxidation.

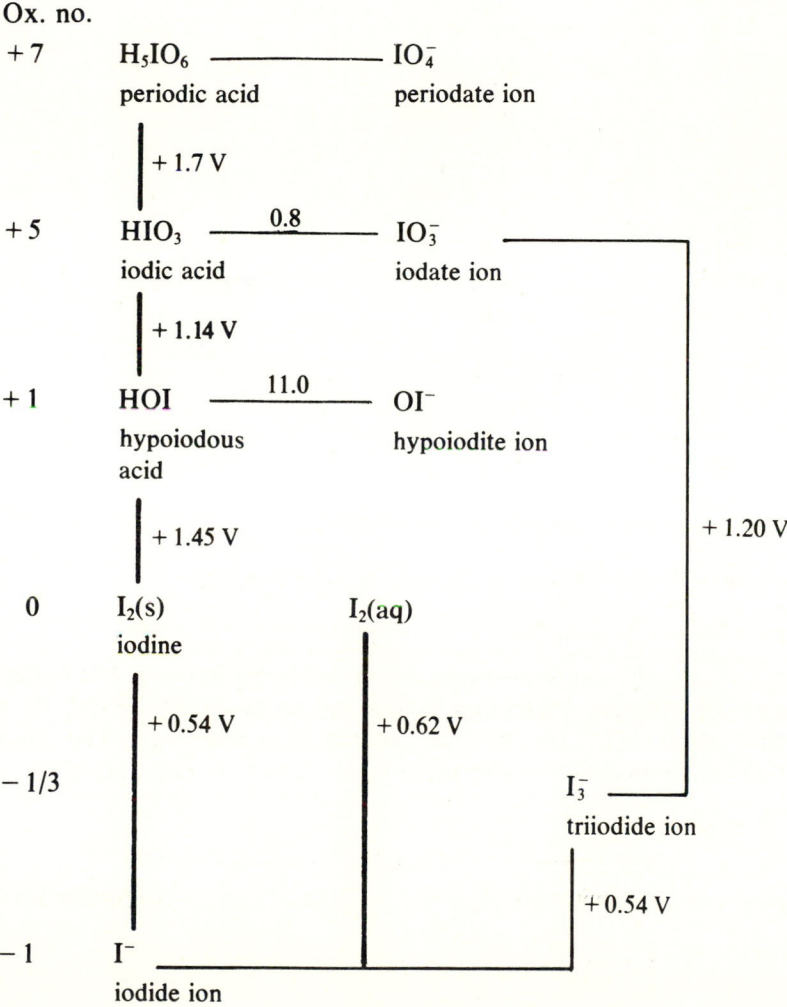

Ox. no.

+7 H_5IO_6 ——————— IO_4^-
periodic acid   periodate ion

+1.7 V

+5 HIO_3 ——0.8—— IO_3^-
iodic acid   iodate ion

+1.14 V

+1   HOI ——11.0—— OI^-
hypoiodous   hypoiodite ion
acid

+1.45 V   +1.20 V

0   $I_2(s)$ $I_2(aq)$
iodine

+0.54 V   +0.62 V

−1/3 I_3^-
triiodide ion

+0.54 V

−1 I^-
iodide ion

IRON (Latin, *ferrum*)

Comments: Ferric ion is a mild oxidant. Because of the great stability of $Fe(OH)_3$ it is necessary to use Fe(III) in fairly acidic solutions. The aquo-ferric ion is very pale purple, while ferrous ion is colorless. The ferric citrate complex is yellow, and the thiocyanate complexes are deep red.

Ox. no.

+ 3 $Fe^{3+}(aq)$ ———$\dfrac{2.8}{}$——— $FeOH^{2+}$ ——————— $Fe(OH)_3(s)$, $K_{sp} = 10^{-38.8}$

ferric ion

$\Big|$ + 0.77 V

+ 2 $Fe^{2+}(aq)$ ———$\dfrac{8.3}{}$——— $FeOH^+$

ferrous ion

$\Big|$ − 0.44 V

0   Fe   iron metal

Metal–ligand complexes with Fe^{2+}

Ligand	Log of formation quotients				Solubility products	
	Q_1	β_2	β_3	β_6	Precipitate	$\log K_{sp}$
Citrate	3.1					
Cyanide				35.4		
EDTA	14.4					
Hydroxide	4.5				$Fe(OH)_2$	−15.1
Oxalate	3.0	5.2			$FeC_2O_4 \cdot 2H_2O$	− 6.5
Sulfide					FeS	−17.2
Thiocyanate	1.3					

Metal–ligand complexes with Fe^{3+}

Ligand	Q_1	β_2	β_3	β_6	Precipitate	$\log K_{sp}$
Bromide	0.6					
Chloride	1.5	2.1				
Citrate	11.8					
Cyanide				43.6		
EDTA	25.1					
Fluoride	5.2	9.1	12.1			
Hydroxide	11.8	22.3			$Fe(OH)_3$	−38.8
Oxalate	7.5	13.6	18.5			
Sulfate	3.0	4.0				
Thiocyanate	3.0	4.6				

LEAD (Latin, *plumbum*)

Comments: Standard solutions of lead(II) are readily prepared by dissolving pure lead nitrate in slightly acidified water (pH below 4). The chromate and sulfate precipitates are both useful in analysis. Lead ion is colorless.

Ox. no.

+4 PbO_2
      lead dioxide

      | +1.46 V

+2 Pb^{2+} —————— $PbOH^+$ —————— $Pb(OH)_2(s)$, $K_{sp} = 10^{-15.2}$
      lead ion   lead hydroxide
      (plumbous ion)

      | −0.13 V

0   Pb
   lead metal

Metal-ligand complexes with Pb^{2+}

Ligand	Log of formation quotients				Solubility products	
	Q_1	β_2	β_3	β_4	Precipitate	$\log K_{sp}$
Acetate	2.7	4.1				
Bromide	1.8	2.6	3.0	2.3	$PbBr_2$	−4.4
Carbonate					$PbCO_3$	−13.5
Chloride	1.6	1.8	1.7	1.4	$PbCl_2$	−4.8
Chromate					$PbCrO_4$	−13.8
Citrate	5.7					
EDTA	17.9					
Fluoride	1.4	2.5			PbF_2	−7.44
Hydroxide	6.3	10.9	13.9		$Pb(OH)_2$	−15.2
Iodate					$Pb(IO_3)_2$	−12.5
Iodide	1.9	3.2	3.9	4.5	PbI_2	−8.2
Oxalate	4.9	6.8			PbC_2O_4	−9.3
Sulfate					$PbSO_4$	−7.8
Sulfide					PbS	−27.9
TETREN	10.5					

MAGNESIUM (*Magnesia*, district in Thessaly)

Comments: Magnesium is a very active metal: it burns in air with a bright white flame and dissolves quickly in acids. However, a surface oxide film protects it from attack by air at room temperature. As a constituent of ground water it is determined by EDTA titration. The double salt, magnesium ammonium phosphate, finds application in gravimetric determinations. Magnesium ion is colorless.

Ox. no.

$+2$ Mg^{2+} ─────────── $Mg(OH)_2(s)$, $K_{sp} = 10^{-11.2}$

magnesium ion   magnesium hydroxide

$-2.37\ V$

0 Mg

magnesium metal

Metal–ligand complexes with Mg^{2+}

Ligand	Log of formation quotients		Solubility products	
	Q_1	β_2	Precipitate	$\log K_{sp}$
Acetate	1.3			
Ammonia	0.2			
Citrate	3.2			
EDTA	8.7			
Fluoride	1.8		MgF_2	-8.2
Hydroxide	2.6		$Mg(OH)_2$	-11.2
Oxalate	2.8	4.2	$MgC_2O_4 \cdot 2H_2O$	-8.0
Phosphate			$MgNH_4PO_4$	-12.6
Sulfate	2.4			

MANGANESE (Latin, *magnes*, magnet, from magnetic properties of pyrolusite MnO_2)

Comments: Potassium permanganate has long been one of the most important analytical reagents, chiefly as a strong oxidizing titrant in acid solutions. The intense purple color of the permanganate ion allows it to serve as its own endpoint indicator. In acidic solutions only the MnO_4^-, Mn^{2+} couple need be considered. Bottles stained with MnO_2, which is slowly deposited from $KMnO_4$ solutions, may be readily cleaned with a little sodium bisulfite and acid (do it under the hood on account of the SO_2 gas).

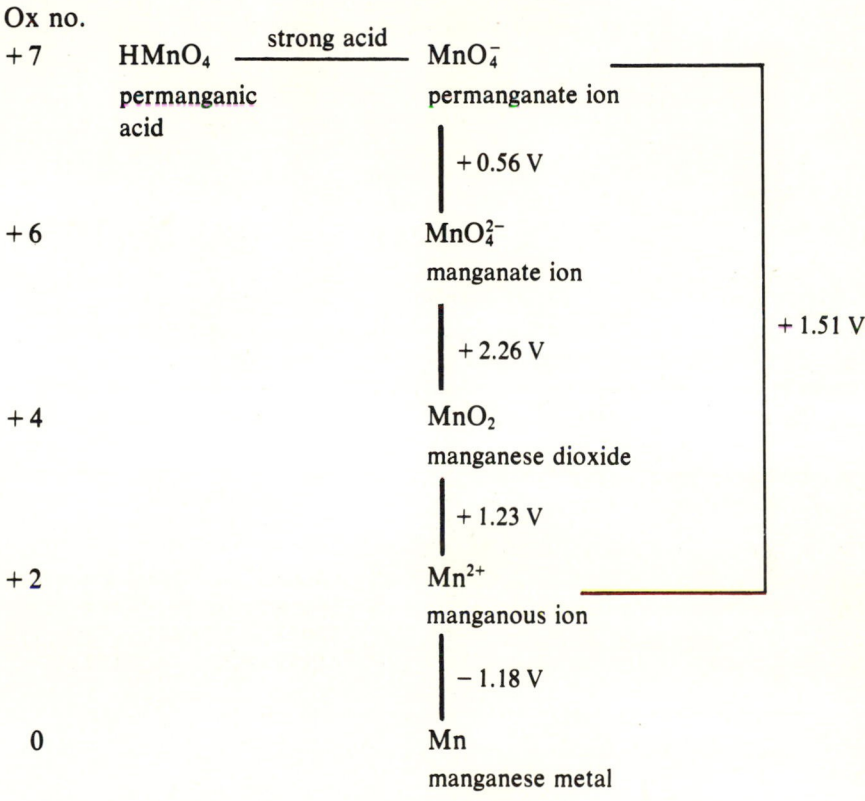

Ox no.

+7 $HMnO_4$ —— strong acid —— MnO_4^-

permanganic acid   permanganate ion

$+0.56$ V

+6 MnO_4^{2-}

manganate ion

$+2.26$ V $+1.51$ V

+4 MnO_2

manganese dioxide

$+1.23$ V

+2 Mn^{2+}

manganous ion

-1.18 V

0   Mn

manganese metal

MERCURY (planet *Mercury*, Latin,.*Hydrargyrum*, liquid silver or quicksilver)

Comments: Mercury forms very stable complexes and precipitates with many ligands. The Hg_2Cl_2, Hg couple is widely used for reference electrodes, the most important being the saturated-calomel electrode in a saturated-KCl solution. Both mercuric and mercurous ions are colorless. The dimeric nature of mercurous ion is discussed in Chapter 8.

Ox. no.

$+2$ Hg^{2+} ———— 3.7 ———— $HgOH^+$ ———————— $HgO(s)$, $K_{sp} = 10^{-25.5}$
        mercuric ion

          $+0.911$ V

$+1$ Hg_2^{2+} $Hg_2Cl_2(s)$ (calomel)
        mercurous ion

                                  $+0.266$ V (standard potential)
                                  $+0.336$ V (in $0.1M$ KCl)
          $+0.796$ V $+0.280$ V (in $1.0M$ KCl)
                                  $+0.244$ V (in saturated KCl, about $4.17M$)

0 Hg
        mercury metal

Metal–ligand complexes with Hg_2^{2+}

Ligand	Log of formation quotients				Solubility products	
	Q_1	β_2	β_3	β_4	Precipitate	$\log K_{sp}$
Acetate					$Hg_2(OAc)_2$	-10.5
Bromide					Hg_2Br_2	-22.2
Carbonate					Hg_2CO_3	-16.1
Chloride					Hg_2Cl_2	-17.9
Chromate					Hg_2CrO_4	$-\ 8.7$
Hydroxide					Hg_2O	-46
Iodate					$Hg_2(IO_3)_2$	-13.7
Iodide					Hg_2I_2	-28.4
Sulfate					Hg_2SO_4	$-\ 6.1$
Thiocyanate					$Hg_2(SCN)_2$	-19.5

Metal–ligand complexes with Hg^{2+}

Ligand	Q_1	β_2	β_3	β_4	Precipitate	$\log K_{sp}$
Acetate	5.6	9.3	13.3	17.1		
Ammonia	8.8	17.4	18.4	19.3		
Bromide	9.0	17.1	19.4	21.0	$HgBr_2$	-18.9
Chloride	6.7	13.2	14.1	15.1		
Cyanide	17.0	32.8	36.3	39.0		
EDTA	22.1					
Ethylenediamine		23.4				
Fluoride	1.6					
Hydroxide	10.6	21.8	20.9		HgO	-25.5
Iodide	12.9	23.8	27.6	29.8	HgI_2	-28.0
Oxalate	9.7				HgC_2O_4	$-\ 7.0$
Pyridine	5.1	10.0	10.4			
Sulfide					HgS	-52.4
TETREN	27.7					
Thiocyanate		17.3	20.0	21.8		
Thiosulfate		29.2	30.6			

NICKEL (German, *Nickel*, Satan, or "Old Nick")

Comments: Nickel ion is green, and its complexes with nitrogen ligands are typically violet. With dimethylglyoxime it forms a bright red precipitate used for precise gravimetric determination of nickel.

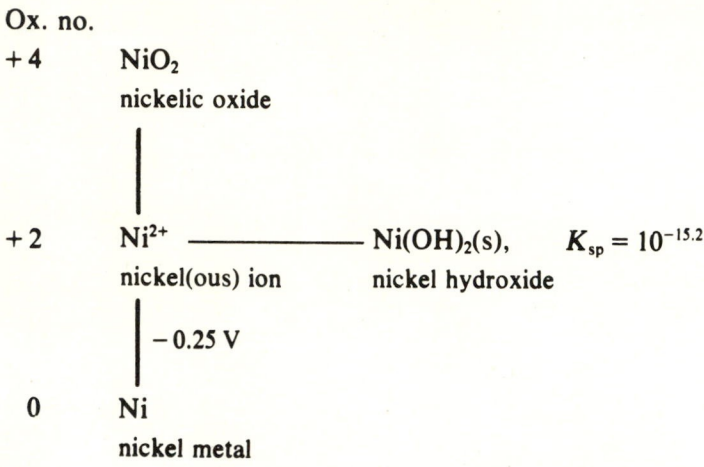

Ox. no.

+4 NiO_2
        nickelic oxide

+2 Ni^{2+} ———————— $Ni(OH)_2(s)$, $K_{sp} = 10^{-15.2}$
        nickel(ous) ion   nickel hydroxide

                -0.25 V

0   Ni
        nickel metal

Metal–ligand complexes with Ni^{2+}

Ligand	Log of formation quotients						Solubility products	
	Q_1	β_2	β_3	β_4	β_5	β_6	Precipitate	log K_{sp}
Acetate	1.4							
Ammonia	2.7	4.9	6.6	7.7	8.3	8.3		
Carbonate							$NiCO_3$	− 8.2
Cyanide				30.2				
EDTA	18.6							
Ethylenediamine	7.6	14.1	19.1					
Hydroxide	4.1	8.0					$Ni(OH)_2$	−15.2
Oxalate	5.3							
Pyridine	1.8	2.9	3.2					
Sulfate	2.4							
Sulfide							NiS	−25.7
TETREN	17.8							
Thiocyanate	1.8							

NITROGEN (Green, *nitron*, native soda; *genes*, forming)

Comments: The redox chemistry of nitrogen with its nine oxidation states is complex. Nitric acid is essentially *strong* in dilute solution. Ammonia is important in high-pH buffers and as a complexing ligand for metal ions.

Ox. no.

$+5$ HNO_3 $\xrightarrow{\quad -1.3 \quad}$ NO_3^- nitrate ion
         nitric acid

$\quad + 0.79$ V

$+4$ nitrogen dioxide $NO_2 \rightleftarrows N_2O_4$

$\quad + 1.07$ V

$+3$ nitrous acid HNO_2 $\xrightarrow{\quad 3.3 \quad}$ NO_2^- nitrite ion

$\quad + 1.0$ V

$+2$ NO nitric oxide

$\quad + 1.6$ V

$+1$ N_2O nitrous oxide

$\quad + 1.8$ V

0 N_2 nitrogen

$\quad - 1.9$ V

-1 NH_3OH^+ $\xrightarrow{\quad 6.0 \quad}$ NH_2OH
         hydroxyl-   hydroxylamine
         ammonium ion

$\quad + 1.4$ V

-2 hydrazinium ion $N_2H_5^+$ $\xrightarrow{\quad 8.0 \quad}$ N_2H_4 hydrazine

$\quad + 1.3$ V

-3 ammonium ion NH_4^+ $\xrightarrow{\quad 9.24 \quad}$ NH_3 ammonia

OXYGEN (Greek, *oxys*, sharp, acid, and *genes*, forming)

Comments: The presence of oxygen in air is a constant threat to the stability of aqueous solutions containing oxidizable species. Hydrogen peroxide is commonly available as a 30% aqueous solution which should be handled with care.

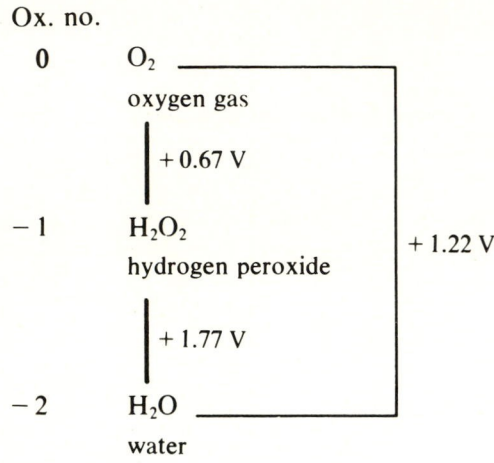

Ox. no.

0   O_2 oxygen gas

         + 0.67 V

−1 H_2O_2 hydrogen peroxide   + 1.22 V

         + 1.77 V

−2 H_2O water

PHOSPHORUS (Greek, *phos*, light; *-phorus*, bearing)

Comments: Phosphoric acid is available as a 85.5% solution (14.8 M); it is especially valuable for preparing buffers in the range near pH = 7. Various metal phosphates are insoluble. The species shown below are colorless. Elemental phosphorus ignites spontaneously in air and must be stored under water.

Ox. no.

+5 H_3PO_4 ——2.12—— $H_2PO_4^-$ ——7.21—— HPO_4^{2-} ——12.32—— PO_4^{3-}

        phosphoric   phosphate ion

        acid

        | − 0.28 V

+3 H_3PO_3 ——2.0—— $H_2PO_3^-$ ——6.6—— HPO_3^{2-}

        phosphorous

        acid

        | − 0.50 V

+1 H_3PO_2 ———— $H_2PO_2^-$

        hypophos-

        phorous acid

        | − 0.51 V

 0   P_4

        phosphorus

        | − 0.07 V

−3 PH_3

        phosphine gas

SILVER (Anglo-Saxon, *siolfur*; Latin, *argentum*)

Comments: The greatly increased price of silver has tended to discourage its frequent use in routine laboratory work. The determination of chloride, bromide, iodide and thiocyanate is best done by gravimetry methods using the silver precipitates. Silver is a catalyst for various redox reactions, probably because of the reversibility of the +2/+1 couple. Silver, silver chloride reference electrodes are often used for potentiometric studies in view of their high reproducibility and stability.

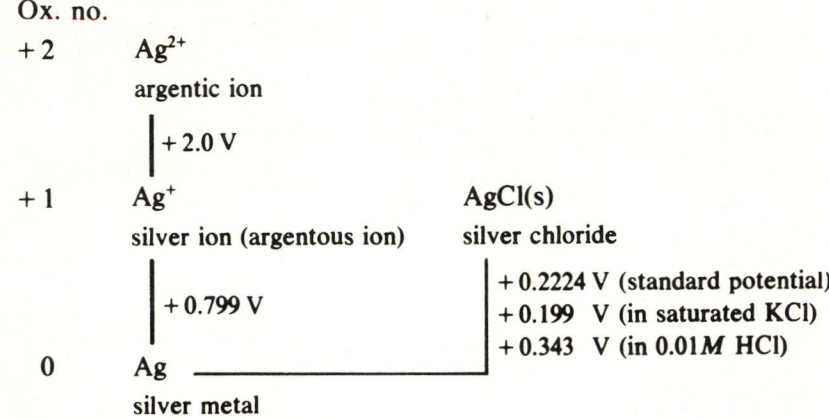

Ox. no.

+2 Ag^{2+}
      argentic ion

      | +2.0 V

+1 Ag^+ $AgCl(s)$
      silver ion (argentous ion)   silver chloride

      | +0.799 V   +0.2224 V (standard potential)
                                            +0.199 V (in saturated KCl)
                                            +0.343 V (in 0.01M HCl)

0   Ag
      silver metal

Metal–ligand complexes with Ag^+

Ligand	Log of formation quotients				Solubility products	
	Q_1	β_2	β_3	β_4	Precipitate	$\log K_{sp}$
Acetate	0.7				AgOAc	− 2.4
Ammonia	3.3	7.2				
Bromate					$AgBrO_3$	− 4.3
Bromide	4.7	7.7	8.7	9.0	AgBr	−12.3
Carbonate					Ag_2CO_3	−11.1
Chloride	3.3	5.3	6.4	6.1	AgCl	− 9.74
Chromate					Ag_2CrO_4	−12.0
Cyanide		20.5	21.4		AgCN	−15.7
Ethylenediamine	4.7	7.7				
Fluoride	0.4					
Hydroxide	2.0	4.0			AgOH	− 7.7
Iodate					$AgIO_3$	− 7.5
Iodide	6.6	11.7	13.1		AgI	−16.1
Oxalate	2.4				$Ag_2C_2O_4$	−10.5
Phosphate					Ag_3PO_4	−19.9
Pyridine	2.0	4.1				
Sulfate	0.2	0.2			Ag_2SO_4	− 4.8
Sulfide					Ag_2S	−49.2
Thiocyanate	4.8	8.2	9.5	9.7	AgSCN	−12.0
Thiosulfate	8.8	13.7	14.2			

SULFUR (Sanskrit, *sulvere*; Latin, *sulphurium*)

Comments: Sulfuric acid is the industrial chemical manufactured in the greatest annual quantity. In the analytical laboratory it is frequently used to adjust solutions to low and nearly constant pH, as in redox titrations. In the +4 state, sulfur is useful as a reducing agent. Many metal ions of insoluble sulfides and such precipitates are useful in separations. An exceptionally important application of sulfur chemistry in analysis is the use of sodium thiosulfate as a titrant for triiodide ion.

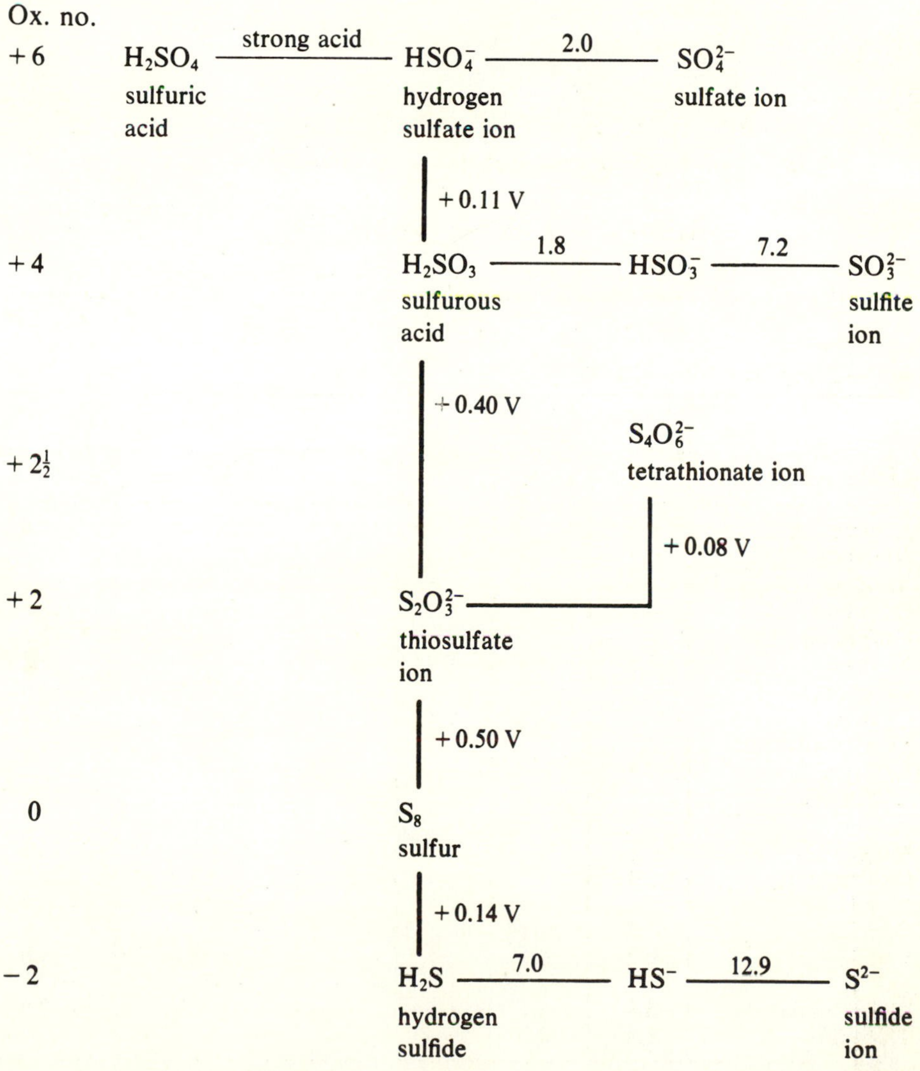

Ox. no.

+6 H_2SO_4 ——strong acid—— HSO_4^- ——2.0—— SO_4^{2-}
        sulfuric   hydrogen   sulfate ion
        acid   sulfate ion

        $+0.11$ V

+4 H_2SO_3 ——1.8—— HSO_3^- ——7.2—— SO_3^{2-}
        sulfurous   sulfite
        acid   ion

        $+0.40$ V

                                    $S_4O_6^{2-}$
+2½   tetrathionate ion

                                            $+0.08$ V

+2 $S_2O_3^{2-}$ ————————————————
        thiosulfate
        ion

        $+0.50$ V

0   S_8
        sulfur

        $+0.14$ V

−2 H_2S ——7.0—— HS^- ——12.9—— S^{2-}
        hydrogen   sulfide
        sulfide   ion

TIN (Anglo-Saxon, *tin*; Latin, *stannum*)

Comments: Tin(II) is often used as a reducing agent in strongly acidic solution. The above species are colorless. To minimize formation of oxy-species, tin solutions are usually prepared in hydrochloric acid, where chlorocomplexes are formed.

Ox. no.

$+4$ Sn^{4+} ———— 4.0 ———— $SnOH^{3+}$

stannic ion

$+0.15$ V

$+2$ Sn^{2+}

stannous ion

-0.14 V

0 Sn

tin metal

URANIUM (planet *Uranus*)

Comments: A great deal of fundamental chemistry has been learned about uranium because of its importance in nuclear power. On account of its radioactivity it does not find frequent use in routine analysis in spite of its interesting properties.

Ox. no.

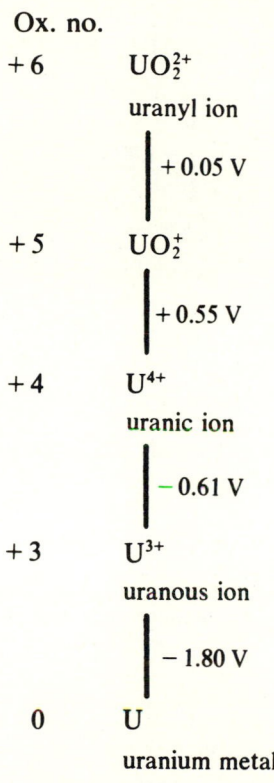

+6 UO_2^{2+}
uranyl ion

+0.05 V

+5 UO_2^+

+0.55 V

+4 U^{4+}
uranic ion

−0.61 V

+3 U^{3+}
uranous ion

−1.80 V

0   U
uranium metal

VANADIUM (Scandinavian goddess, *Vanadis*)

Comments: Amalgamated zinc is capable of reducing vanadium to the +2 state, solutions of which are useful in removing traces of oxygen from nearly pure nitrogen. Vanadium rivals chromium in its ability to form several colored species.

Ox. no.

+5 $V(OH)_4^+$ or VO_2^+
pervanadyl ion (yellow)

$$+ 1.00\ V$$

+4 VO^{2+}
vanadyl ion (blue)

$$+ 0.36\ V$$

+3 V^{3+}
vanadic ion (green)

$$- 0.26\ V$$

+2 V^{2+}
vanadous ion (violet)

$$- 1.18\ V$$

0   V
vanadium metal

ZINC (German, *Zink*, of obscure origin)

Comments: An important application of metallic zinc is in the *Jones reductor*, where a zinc amalgam (alloy with mercury) is used as a powerful reducing agent. The +2 ion is colorless, as are the complexes and precipitates of zinc.

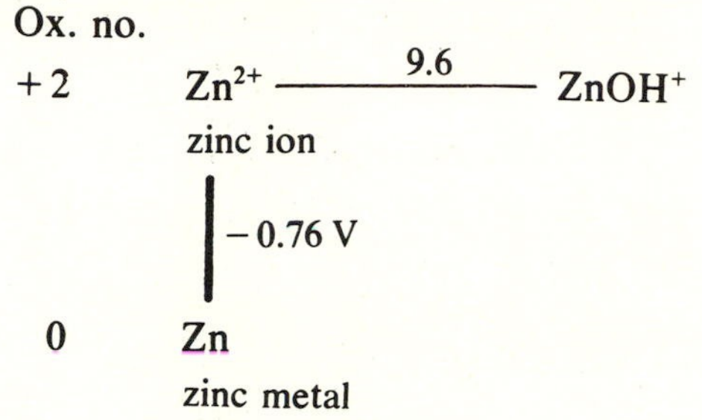

Ox. no.

$$+2 \qquad Zn^{2+} \underline{\qquad 9.6 \qquad} ZnOH^+$$

zinc ion

$$\left| \; -0.76 \; V \right.$$

$$0 \qquad Zn$$

zinc metal

Metal–ligand complexes with Zn^{2+}

	Log of formation quotients				Solubility products	
Ligand	Q_1	β_2	β_3	β_4	Precipitate	log K_{sp}
Acetate	1.6					
Ammonia	2.2	4.5	6.9	8.9		
Carbonate					$ZnCO_3$	−10.8
Cyanide		11.1	16.1	19.6	$Zn(CN)_2$	−15.5
EDTA	16.5					
Ethylenediamine	6.0	10.8	13.0			
Fluoride	1.2					
Hydroxide	5.0	11	13.6	15	$Zn(OH)_2$	−15.5
Oxalate	4.9	7.7			$ZnC_2O_4 \cdot 2H_2O$	− 7.6
Pyridine	1.0	1.5				
Sulfate	2.3					
Sulfide					ZnS	−23.8
TETREN	15.4					
Thiocyanate	1.3	1.9	2.0	1.6		

ANSWERS TO SELECTED PROBLEMS

Chapter 1 (page 16)

1. 78% **2.** Specs are met except for iron. **3.** 99.7%
6. a) $1.74 \cdot 10^{-9}$ b) -0.617 c) $2.01 \cdot 10^{-3}$ d) $1.80 \cdot 10^{-3}$
8. a) $2.92 \cdot 10^{-2}$ b) $-4.12 \cdot 10^{-2}$ c) 0.02, 0.01 e) $b^2 > 4ac$

Chapter 2 (page 43)

Aqueous solutions

1. a) 6.3 mg b) 1.76 g c) 76 μg d) 0.76 g
2. a) 21.4 b) 6.9 c) 6.6 d) 8.6
3. a) 4.97 b) 9.94 c) 5.32 d) 1.40
4. a) 0.025 b) 0.370 c) 1.71 d) 0.29
5. a) 11.5 b) 133 c) 30 d) 800
6. a) 55.3 b) 55.5 c) 1.000 d) 1000
7. a) $1.1 \cdot 10^{-3}M \; Cd^{2+}$ b) $0.070M \; NH_4^+$ c) $3.3 \cdot 10^{-5}M \; IO_3^-$ d) $0.024M \; Na^+$

Equation balancing

9. $Fe^{3+} + Cl^- + Ag = Fe^{2+} + AgCl$ **11.** $BrO_3^- + 6I^- + 6H^+ = Br^- + 3I_2 + 3H_2O$
13. $2MnO_4^- + 5H_2C_2O_4 + 6H^+ = 2Mn^{2+} + 10CO_2 + 8H_2O$

Titration stoichiometry

24. 17.39 **25.** 182 **26.** 97.0% **27.** $0.0611M$
28. 50 mL **29.** 34.4 mL **30.** $0.0400M$

Redox stoichiometry

31. NO **32.** Br^- **33.** IO_3^- **34.** N_2 **35.** NO_3^- **36.** Bi **37.** VO^{2+}

Chapter 3 (page 79)

Accuracy and precision

1. a) 141.939 ± 0.001 b) 254.2 ± 0.1 c) 64.10 ± 0.01 d) 279.55 ± 0.04
2. 0.821 ± 0.002 **3.** 0.24 ± 0.03
4. Reject 0.9138, $\bar{x} = 0.9131$, 0.08 ppt **5.** -3.8 ppt
6. Method A is more precise but has positive bias.
7. 500.6, 100.13; both outside of specs **8.** $0.11521M$ **9.** 25.0194 g

Gaussian statistics

11. 136, 4   **12.** 333, 1   **13.** $5 \cdot 10^{-5}$, $9 \cdot 10^{-5}$ **15.** ± 0.003 mL
16. "Arrrghh!"
17. For one, the number of grains of salt spilled during a lunch period.
18. b) 111 ng   **19.** a) 1.25 μg   b) 0.22 μg
20. b) 0.115, 0.0470   c) 0.005, 0.014   d) 0.35 $\pm$ 0.07
21. 5.010 mL   **22.** Both give 49.96 mL; favor direct weighing for simplicity.

Chapter 4 (page 125)

Calculation of ionic strength

3. a) 0.075   b) 0.51   c) 0.13   d) 0.03   e) 0.0225   f) 0.205   g) yes

Calculation of ionic-activity coefficients

4. 0.791, 0.807, 0.800 and 0.095, 0.306, 0.298
7. For 1:1 electrolyte, 0.78; for 2:1, 0.61; for 2:2, 0.37; for 3:1, 0.48; for 2:3, 0.23

Calculation of molecular-activity coefficients

8. Slopes are $+0.106$ and -0.042.

Conversion of K-values to Q-values

9. a) $2.5 \cdot 10^{-5}$ b) $3.2 \cdot 10^{-5}$ c) $1.65 \cdot 10^{-10}$ d) 0.037   e) $1.5 \cdot 10^{-12}$

Redox stoichiometry

10. 56.55 mL

Reactions in nonequilibrium mixtures and equilibrium calculations

11. For I_2, $2.5 \cdot 10^{-4}$; for I^-, $4.25 \cdot 10^{-3}$; for I_3^-, $7.5 \cdot 10^{-4}$.
12. For HSO_4^-, $2.34 \cdot 10^{-3}$; for H^+, $9.47 \cdot 10^{-2}$; for SO_4^{2-}, $6.64 \cdot 10^{-4}$.
13. a) For HA, 0.05; for H^+ and A^-, 0.01   b) $1.2 \cdot 10^{-3}$
    c) For H^+, 0.016; for HA, $8.9 \cdot 10^{-3}$
14. a) 0.01   b) $2.4 \cdot 10^{-3}$ c) 0.0113 g

Including ionic-activity coefficients in the equilibrium calculation

16. a) 0.05, 0.07, 0.08   b) $4.3 \cdot 10^{-5}$ c) $1.2 \cdot 10^{-4}$
17. $1.5 \cdot 10^{-4}M$ dichromate ion, or 15%   **18.** a) $1.7 \cdot 10^{-4}$ b) 30%

Determination of equilibrium constants

19. a) $8.77 \cdot 10^{-3}$ c) $8.73 \cdot 10^{-3}$ d) 0.0337   e) $5.45 \cdot 10^{-5}$

Chapter 5 (page 162)

Spectrophotometric determinations

1. b) 25.3 mL/mg · cm   c) 3500 L/mol · cm   **2.** b) $1.49 \cdot 10^{-4}$
3. b) 0.134%   **4.** MW = 327   **5.** $Q = 1.10 \cdot 10^{-7}$

Photometric titration

6. $C = 0.087$, $\epsilon_X = 1.52$, $\epsilon_T = 0$ **7.** $C_T = 0.125$

Chapter 6 (page 198)

Gravimetric determinations

1. $x = 1.98$ 2. 99.6%   3. a) $4.3 \cdot 10^{-12}$ b) $8.9 \cdot 10^{-6}$

4. $\%Hg = \dfrac{P \cdot 2 \cdot 200.59 \cdot 25 \cdot 100}{S \cdot 472.09 \cdot 5}$

5. Both Mg^{2+} and NH_4^+ are absorbed; $\%P = 1.42$

6. *Gravimetric stoichiometry*

a) 23.90	b) 19.05	c) 27.75	d) 26.63	e) 26.24	f) 15.99
g) 21.53	h) 8.30	i) 23.75	j) 5.62	k) 2.17	l) 20.02
m) 24.19	n) 18.34	o) 4.97	p) 9.55	q) 0.585	r) 4.96
s) 7.70	t) 19.54	u) 3.42	v) 25.26	w) 20.98	x) 21.65

Chapter 7 (page 212)

Using the reaction isotherm

1. a) 9.2 k cal, reaction goes left   b) -4.0 k cal, reaction goes right
2. a) $+13302$ cal   b) $8.5 \cdot 10^{13}$ c) -5708 joules   d) 640

Using standard Gibbs free energies of formation

3. $2.5 \cdot 10^{-14} M$ 4. a) $+1934$ cal   b) 0.038   c) $0.067 M$

Chapter 8 (page 250)

Galvanic cells

1. e) -0.34 h) $7 \cdot 10^{-13}$ i) Cd   2. e) $+0.76$ h) $3.6 \cdot 10^{13}$ i) Ag
3. b) $Pt|Br^-, Br_2(aq)\|Cl_2(P \text{ atm}), Cl^-|Pt$ 4. b) $Pb|Pb^{2+}\|Fe^{3+}, Fe^{2+}|Pt$
5. g) 0.010   6. e) $+0.34$ 7. g) 0.07   8. e) $+0.08$ h) $1.1 \cdot 10^4$ i) Pt
9. e) $+0.15$ i) Ag|AgCl   10. e) $+1.12$ h) 10^{38} i) V

The additivity rules

11. -0.29 12. $+0.37$ 13. -0.04 14. $K = 4.6 \cdot 10^{+6}$ 15. $E^0 = -0.30$
16. Implies $\log K_{sp} = -11.9$ instead of -12.4.   17. $E^0 = -0.15$
18. $K = 4 \cdot 10^{18}$ for $H_2O_2 = H_2O + \frac{1}{2}O_2$
19. $K = 10^{17}$, $[V^{3+}] = 2 \cdot 10^{-12}$ 20. No, K is too small
21. a) $+5830$ cal, $E^0 = +0.13$ b) $8 \cdot 10^{-9}$ c) $E^0 = -0.37$ d) $K = 5 \cdot 10^{-6}$
22. a) -0.74 b) $3 \cdot 10^{-16}$ c) no   23. a) $E^0 = -0.06$ b) $E^0 = -0.51$

The use of potentiometric data

24. a) -0.468 b) $1.4 \cdot 10^{-5}$ c) $1.5 \cdot 10^{-8}$ 25. $4.3 \cdot 10^{-3}$ 26. $[Br^-] = 0.0013$
27. a) 0.468 b) $6.3 \cdot 10^{-7}$ 28. $C = 1.46 \cdot 10^{-4}$ 29. $Pt|KI, KI_3\|HCl, O_2|Pt$ 30. 8%

Chapter 9 (page 313)

1. *Practice in conversions*

	I	f_1	pQ_w	$[H^+]$	pcH	aH	pH	$[OH^-]$	pOH
a)	0.01	0.90	13.91	$1 \cdot 10^{-6}$	6.00	$9.0 \cdot 10^{-7}$	6.05	$1.23 \cdot 10^{-8}$	7.96
b)	0.013	0.89	13.90	$5 \cdot 10^{-6}$	5.3	$4.45 \cdot 10^{-6}$	5.35	$2.52 \cdot 10^{-9}$	8.65
c)	0.03	0.85	13.86	$3.55 \cdot 10^{-11}$	10.45	$3.0 \cdot 10^{-11}$	10.52	$3.89 \cdot 10^{-4}$	3.48
d)	0.07	0.80	13.81	$1.63 \cdot 10^{-7}$	6.79	$1.3 \cdot 10^{-7}$	6.89	$9.7 \cdot 10^{-8}$	7.11
e)	0.09	0.79	13.80	$1.26 \cdot 10^{-7}$	6.90	$1.0 \cdot 10^{-4}$	4.00	$1.27 \cdot 10^{-10}$	10.00
f)	0.10	0.78	13.78	$2.28 \cdot 10^{-12}$	11.64	$1.78 \cdot 10^{-12}$	11.75	$7.2 \cdot 10^{-3}$	2.25
g)	0.13	0.89	13.90	$2.52 \cdot 10^{-12}$	11.60	$2.24 \cdot 10^{-12}$	11.65	$5.0 \cdot 10^{-3}$	2.35
h)	0.20	0.75	13.75	$2 \cdot 10^{-13}$	12.70	$1.5 \cdot 10^{-13}$	12.82	$8.8 \cdot 10^{-2}$	1.18

Fundamental pH calculations

4. a) 1.11, 2.88   b) 0.017   c) 7.00, 8.64   d) 0.18
5. a) 11.12, 5.23   b) 9.58   c) 5.24, 5.23
6. a) 8.15, 11.4   b) 9.52   **8.** 8.86, 0.96   **13.** 5.7

Buffer solutions

18. 31.45 mmol of acid, 30 mmol of base   **19.** 4.87   **20.** 780
21. 26.7 mL NaCl, 87.3 mL phenol, 250 mL NaOH   **22.** 13.4 mmol

Chapter 10 (page 352)

1. 0.0867   **2.** 3.5 g   **3.** HNO_3, 0.0582; NH_3, 0.0493   **4.** 86.2%
5. 0.0500M **6.** 48.13 mL   **7.** 87.3   **11.** 99.7%
12. a) 2.1%   b) pH = 5.5   **13.** 99.6%

Chapter 11 (page 384)

1. a) +0.280 V   b) pH = 8.55   **2.** 6.444, extrapolate to 6.442
3. a) pH = 7.80   b) pK = 7.49
4. a) 0.2389 V   b) Q = 0.0924   c) pmH = 8.197 = pQ
5. a) 20.45 mL, 0.0412M **6.** 24.40 mL, 0.0488M
7. For HA, 0.0200; for H_2B, 0.012   **9.** a) 96%   b) pK_b = 9.9
10. pK-values about 2.1, 8.3, 10.5; MW = 121.2
11. a) Endpoints at 1.9, 3.4, 5.8 mL   c) pK = 7.9   e) pH = 3.98
12. a) 3.59   b) 4.6, 9.0   d) A = 0.643 at start, 0.915 at 1st endpoint, 0.345 at 2nd
14. 2.76, 3.90   **15.** 4.68, 5.80   **16.** a) 190.7   b) pK-values 6.1, 9.1

Chapter 12 (page 421)

1. 0.02011M, 2 g, 99.3%   **2.** 0.1017M, 0.1018M **3.** 0.02222M
4. 81 times as costly   **5.** 0.3 g   **6.** 0.0490M **7.** 0.0648M **8.** 9.90%
9. WO_2^{2+} **10.** % Na = 6.61 $\cdot V \cdot C/P$ **11.** 3.88 mg   **12.** 0.075%
13. b) 121.9 mg   d) +0.23 V   **14.** 0.0568M **15.** 27.5%   **16.** 15.5%

Chapter 13 (page 439)

1. Q = 270   **2.** a) Q = $1.7 \cdot 10^{-5}$, color-change interval 3.8–5.8   b) 4.58
3. a) 0.5   b) 350 nm   c) 0.14   d) 720

Chapter 14 (page 460)

6. a) $[ML]/[M] = 160$ c) $[M]:[ML]:[ML_2] = 1:10:400$ 7. e) $F_0 = 5.2$
8. The complex with the first ligand dominates in (a), (c), and (f).
9. The complex with the first ligand dominates in (a) and (b).
11. $0.036M$ 12. Yes, the $(F_0 - 1)$ values have a ratio greater than 1000.
13. Yes, K_{sp} is exceeded. 14. About 80% is complexed.
16. $4.1 \cdot 10^{-6}$ 17. $[Ag^+]$ is about 10^{-6}; too low to exceed K_{sp}.

Chapter 15 (page 488)

1. $3.5 \cdot 10^{-16}$, $3.2 \cdot 10^{+17}$ 2. $E = -0.138$ 3. $Q_1 = 100$, $Q_2 = 30$
4. $5.3 \cdot 10^7$ 5. a) $1.4 \cdot 10^{27}$ c) for copper, $10^{22.8}$

Chapter 16 (page 522)

3. b) $1.233 \cdot 10^{-2}$ c) $9.67 \cdot 10^{-3}$ 4. a) $8.56 \cdot 10^{-3}$ mmol Ni, 0.1124 mmol Pb
4. b) $pPb = -(0.374 - E)/0.0296$ d) $F_0 = 210$, $\alpha = 6.6 \cdot 10^{-6}$

Chapter 17 (page 546)

Strong-electrolyte model

2. $3.3 \cdot 10^{-8}$ 3. $S = 0.0256$, but prediction from K_{sp} is $7 \cdot 10^{-4}$.
4. $S = 0.005$ 5. In water, $S = 8.1 \cdot 10^{-3}$; in sodium bromate, $S = 1.6 \cdot 10^{-3}$.
6. Iodate is more effective since its concentration is squared: in iodate, $S = 3.8 \cdot 10^{-5}$; in copper, $S = 1.2 \cdot 10^{-3}$.

Model with significant formation of complexes

7. $7.5 \cdot 10^{-5}$ 8. Without citrate, $S = 4 \cdot 10^{-13}$; with citrate, hypothetical $S = 6$.
9. $S = 0.0035$ 10. Q_{sp} is just equalled by ion product 12. b) $2 \cdot 10^{-7} + Q_0$.

Chapter 18 (page 566)

Effect of pH on solubility

1. $Q_{sp} = 7.4 \cdot 10^{-7}$, $Q_a = 0.46$ 2. b) $Q_0 = 1.5 \cdot 10^{-3}$, $Q_a = 1.00 \cdot 10^{-3}$
3. $Q_0 = 6.0 \cdot 10^{-6}$ and $6.5 \cdot 10^{-6}$, $Q_1 = 1.56 \cdot 10^{-3}$, $Q_2 = 1.31 \cdot 10^{-5}$
4. a) 23.36 mL NaOH, 31.32 mL HOAc to 100 mL b) plot S versus $1/[H^+]$
   c) $Q_a = 3.6 \cdot 10^{-5}$, $Q_0 = 1.3 \cdot 10^{-5}$

Potentiometric titration

5. a) Endpoint at 35.09 mL; %Cl = 48.94
   b) $Q_{sp} = 1.7 \cdot 10^{-10}$

Weak-electrolyte model with appreciable intrinsic solubility

6. $Q_0 = 4 \cdot 10^{-7}$, $Q_{sp} = 2 \cdot 10^{-10}$, $Q_d = 5 \cdot 10^{-4}$
7. $3.3 \cdot 10^7$

Chapter 19 (page 585)

1. a) 2170 b) 0.0142 d) $Q_0 = 6.0 \cdot 10^{-5}$, $K_a = 3.5 \cdot 10^{-8}$ e) $1.9 \cdot 10^3$ f) 0.33
2. b) 59 ± 1 c) 1040 d) F_1 plot is linear; $Q_1 = 1.0 \cdot 10^7$; $Q_2 = 1.14 \cdot 10^5$

INDEX